Google Earth and Virtual Visualizations in Geoscience Education and Research

edited by

Steven J. Whitmeyer
Department of Geology & Environmental Science
James Madison University
Memorial Hall, MSC 6903
Harrisonburg, Virginia 22807
USA

John E. Bailey
Scenarios Network for Alaska & Arctic Planning
University of Alaska Fairbanks
Fairbanks, Alaska 99709
USA

Declan G. De Paor
Department of Physics
Old Dominion University
Norfolk, Virginia 23529
USA

Tina Ornduff
Google Inc.
1600 Amphitheatre Pkwy
Mountain View, California 94043
USA

THE
GEOLOGICAL
SOCIETY
OF AMERICA®

Special Paper 492

3300 Penrose Place, P.O. Box 9140 ▪ Boulder, Colorado 80301-9140, USA

2012

Published by The Geological Society of America, Inc.
3300 Penrose Place, P.O. Box 9140, Boulder, Colorado 80301-9140, USA
www.geosociety.org

Printed in U.S.A.

GSA Books Science Editors: Kent Condie and F. Edwin Harvey

Library of Congress Cataloging-in-Publication Data

Google Earth and virtual visualizations in geoscience education and research / edited by Steven J. Whitmeyer...[et. al.].
    p. cm. -- (Special paper 492)
 Includes bibliographical references.
 ISBN 978-0-8137-2492-8 (pbk.)
 1. Geographic information systems. 2. Google Earth. 3. Imaging systems in geology. 4. Artificial satellites in earth sciences. I. Whitmeyer, Steven J.

 G70.212.G69 2012
 910.285--dc23

2012024960

Cover: Earth image courtesy of NASA. Image credits, clockwise from top left: (1) Photomicrograph of grain boundary migration recrystallization in quartz; photo by Carol Simpson. (2) Overlay of geologic map of Virginia, USA, and vertical cross section on Google Earth terrain image. Data courtesy of SIO, NOAA, U.S. Navy, NGA, GEBCO; image © 2012 Gnes/Spot Image; © 2012 TerraMetrics. See Chapter 11, "The new frontier of interactive, digital geologic maps: Google Earth–based multi-level maps of Virginia geology," by O.P. Shufeldt, S.J. Whitmeyer, and C.M. Bailey, doi:10.1130/2012.2492(11). (3) Google Earth terrain map of Union Island, Grenadines, showing data from layers available in Google Earth. Image © 2012 DigitalGlobe, © 2012 Google, © 2012 GeoEye. Data courtesy of SIO, NOAA, U.S. Navy, NGA, GEBCO. See Chapter 5, "Workshops, community outreach, and KML for visualization of marine resources in the Grenadine Islands," by M.E. Stewart and K. Baldwin, 10.1130/2012.2492(05).

10 9 8 7 6 5 4 3 2 1

Contents

IV. Educational Models, Learning Methods, and Assessment

The Geological Society of America
Special Paper 492
2012

Introduction: The application of Google Geo Tools to geoscience education and research

J.E. Bailey

Scenarios Network for Alaska & Arctic Planning, University of Alaska Fairbanks, Fairbanks 99709, USA

S.J. Whitmeyer

Department of Geology and Environmental Science, James Madison University, Harrisonburg, Virginia 22807, USA

D.G. De Paor

Department of Physics, Old Dominion University, Norfolk, Virginia 23529, USA

INTRODUCTION

Until relatively recently, only cosmonauts and astronauts had ever viewed the Earth as a planet suspended in space. These pioneers brought back stunning photographs of the "Blue Marble" and "Earth Rise." It wasn't quite like being out there in orbit, but nevertheless such images profoundly influenced the global human psyche and promoted awareness of the finite nature of our home in the Cosmos. Now, with the development of virtual globe technology such as Google Earth (GE), everyone who has access to the Internet can visualize the Earth as if we were all astronauts.

GE has emerged as one of the most powerful and easy-to-use tools for viewing, tracking, and analyzing geological features, surface processes, and events. Since the application's release in 2005, GE's use in the geosciences has evolved from simple fly-bys of landforms to dynamic models displaying geologic processes. The diversity of applications of GE in geoscience education and research has been highlighted at the annual meetings of the American Geophysical Union (AGU) and Geological Society of America (GSA) in dedicated and popular sessions (Bailey, 2009).

Discussions at those meetings indicated the need for a specialized forum where development of virtual globe-based educational resources and visualizations could be coordinated among the greater geoscience community. The result was a GSA Penrose Conference, which brought together educators, researchers, publishers and software developers (from both and academia and Google Inc.) to discuss recent advances in the development of educational modules and research visualizations that use the Google mapping services and related tools.

The conference was held onsite at the Google Inc. headquarters in Mountain View, California (https://sites.google.com/site/gepenrose/) in January 2011. The primary goal was to pool ideas and resources from the broader community, with the hope of stimulating new initiatives and directions in the use of GE in the geosciences, as well as encouraging the active participation of Google Inc. in the future development of geoscience education tools. The papers in this special volume highlight cutting-edge educational and research uses that were demonstrated in Mountain View, along with examples of projects that developed from collaborations established at this meeting.

The volume is organized into four sections: (*i*) Data Visualization; (*ii*) Digital Mapping; (*iii*) Virtual Field Experiences; and (*iv*) Educational Models, Learning Methods, and Assessment. The foci of each section and their importance to geoscience education and research are described in the rest of this paper.

DATA VISUALIZATIONS

Google Earth is a computer program that integrates a global digital elevation model (DEM) with base surface imagery to create a 3D, mirror-world representation of the Earth (Bailey, 2010). Technically speaking, GE is only 2.5D as the model is projected onto a 2D computer screen with the appearance of

Bailey J.E., Whitmeyer, S.J., and De Paor D.G., 2012, Introduction: The application of Google Geo Tools to geoscience education and research, *in* Whitmeyer, S.J., Bailey, J.E., De Paor, D.G., and Ornduff, T., eds., Google Earth and Virtual Visualizations in Geoscience Education and Research: Geological Society of America Special Paper 492, p. vii–xix, doi:10.1130/2012.2492(00). For permission to copy, contact editing@geosociety.org. © 2012 The Geological Society of America. All rights reserved.

being 3D. The combination of terrain and land coverage data, literally, offers a whole world for students, teachers, researchers, or anyone to explore.

Google's large investment in acquiring global high-resolution data, and efforts to make important imagery available in a timely manner (Google Inc., 2010; Bradley et al., 2011; Hennessey-Fiske, 2011), has created an archive of imagery that individual researchers would not have the resources to compile. For geoscientists, this has provided imagery good enough to perform surveys that were traditionally only achievable through field-based methods. The utility of this archive is emphasized in Tewksbury et al. (this volume, chapter 2) and Fisher et al. (this volume, chapter 1). The authors in both papers use GE imagery to map surface morphologies related to regional geology. Tewksbury et al. (chapter 2) demonstrate this for the mapping of folds, faults, and lithological units in Egypt's Western Desert. Fisher et al. (chapter 1) have used GE to define channels widths in the Himalayan Mountains, landslide properties in Haiti, and fault characteristics in California.

For those who have their own data, GE also offers a framework through which to easily view and share that data. For example, Crosby (this volume, chapter 3) describes how a growing archive of high-resolution Lidar (light and detection ranging) derived data has been made easily accessible through GE. Williams et al. (this volume, chapter 4) use GE to view other imagery, in their case derived from ground penetrating radar (GPR) and magnetic gradiometry, and describe how GE can be used to annotate these data.

GE provides a canvas to which users can add their own geospatial data to create dynamic visualizations using Keyhole Markup Language (KML). This code is a type of eXtensible Markup Language (XML) and thus is similar to Hypertext Markup Language (HTML) in form and structure (Wernecke, 2009). Like HTML, the function of many KML elements is self-evident. For example, <color>*value*</color> defines the color of an object, making KML a user-friendly and easy-to-learn coding language.

The idea of directly editing code is not a natural concept for many educators or even scientific researchers, as it lies outside their comfort zone or might involve learning time they do not have available. Fortunately, GE offers a way for non-developers to create visualizations as many (though not all) KML elements can be created directly in GE through the built-in graphical user interfaces (GUIs). Users do not need to understand the underlying code as features such as color are manipulated directly through on screen widgets (Fig. 1).

The combination of GE and KML has opened up possibilities that previously only existed for those with knowledge of, and access to, proprietary applications (e.g., ArcGIS). Stewart and Baldwin (this volume, chapter 5) describe a case where geolocated photographs and videos were shared using KML, rather than as GIS shapefiles, to address the sustainability of marine resources.

For those who are comfortable authoring computer code there are even more possibilities offered by Google's Geo applications. Specifically, a GE application programming interface (API) allows users to embed customized versions of GE directly into web pages. The JavaScript API works with a GE plug-in that needs to be installed once on a given computer, and then functions across multiple web browsers. Using JavaScript, developers can create their own interface controls, link to functional scripts, or even create KML objects with greater functionality than is seen in the GE desktop application. The utility of JavaScript is discussed in De Paor et al. (this volume, chapter 6), and an example of how it can be used to create interactive screen overlays and customized control sliders is described by Dordevic (this volume, chapter 7).

One of the much-debated questions about GE (and Google Maps) is whether they can or should be considered a type of GIS (Turner, 2008). Arguments against it being identified as such generally emphasize the lack of built-in analytical capabilities similar to those found in ArcGIS or other image analysis programs. In GE the application is only concerned with the spatial location, size and other <Style> properties of the geometry (vector data) and imagery (raster data). It does not consider the inter-relation between different KML objects or relative "values" of pixels in an image, which are core functions in ArcGIS. However, by linking GE to other Google services some of these limitations can be overcome. For vector data, Google Fusion Tables (Gonzalez et al., 2010) can provide the functionality of geospatial relational database in a user-friendly interface. For imagery, Google Earth Engine (Google Inc., 2011), while still early in its development, has exciting potential for manipulating raster images directly within the GE API.

De Paor et al. (this volume, chapter 6) briefly describe how to extend the functionality of GE by linking to Fusion Tables. Zular et al. (this volume, chapter 8) have implemented an example that combines geomorphologic observations made in GE with laboratory analyses data collated in a Fusion Table. Nunn and Bentley (this volume, chapter 9) employ further capabilities of Fusion Tables to create charts to display spatial and temporal trends in Louisiana's water usage. All of these implementations of Fusion Tables contain data with location components, and so have the option to be displayed on a map. The Google Maps API is integrated into Fusion Tables, but Fusion Tables also generate KML links and <iframe> code (the latter allows that instance of Google Maps to be embedded on a web page). The KML can be downloaded as a "current view" static file or as a network link. The advantage of the latter is that any changes made to the Fusion Table, and hence the KML, are pushed to the user when the link is refreshed, without the user having to download a new file.

Online links also allow interaction between GE and other "cloud" services, such as Google Docs and Really Simple Syndication (RSS) blog feeds. Innovative research has taken advantage of these links to generate dynamic KML that can track events in near-real time. For example, Potapov and Hronusov (this volume, chapter 10) use these services to create KML that maps the movements of radio-tagged deer and birds.

<u>Can Create</u>　　　　　　　　　<u>Cannot Create</u>

Geometry

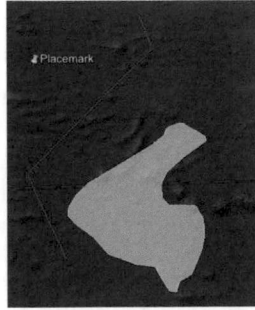

Geometry is created in KML using <Placemark>. Points, paths and polygons can all be created in Google Earth.

Models and Multi-Geometry are also <Placemark> code. The scale and location of models can be edited in GE, but must first be created in SketchUp or a similar application.

Imagery

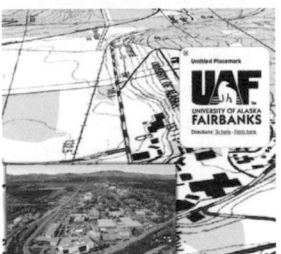

Ground Overlays, images in description balloons Photo Overlays all have authoring GUIs in Google Earth.

ScreenOverlays need to be created by directly authoring KML code or by using an external widget.

Styles

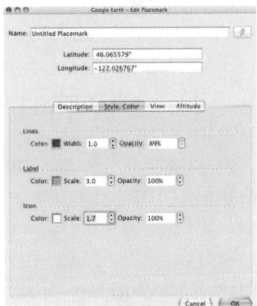

The colors, scale and transparency of icons (including icon image), lines and areas can all be directly edited.

The <BalloonStyle> cannot be edited directly in GE. This KML is used to remove the "Directions to here -from here" text.

Note: The central area of description balloons can be styled by HTML authored in the GUI.

Structure

Folder and Network Links can both be created in Google Earth.

The <Document> container cannot be created in GE, but is automatically added to any saved KML. In application Regionation is only possible using GE Pro.

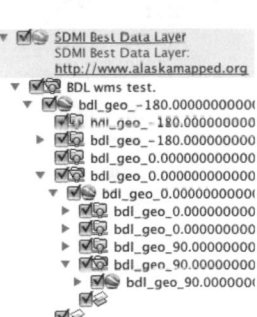

Dynamic Functions

Google Earth Tours can be created in the free and subscription versions of GE.

Note: Tours are saved as KML scripts; conversion to .avi or .mov files requires GE Pro or a screen capture software.

The Time Primitive, temporal coordinates that allow data to be animated, must be authored directly into the code

Figure 1. Summary of KML which can be created directly in the Google Earth application versus KML that must be generated using external applications or by the user directly authoring code.

DIGITAL GEOLOGIC MAPPING

While KML allows users to augment the GE landscape, the base globe itself has also developed into an important tool for education and research. GE provides a universal interface to imagery and derived products. It enables users to easily view spatial and temporal changes across images and maps, and to then mark or annotate features using KML. These functions are also possible in traditional GIS technologies, but the methods can be cumbersome and for many purposes the advantages of GE (accessible, free, simple user interface) outweigh the advantages of traditional GIS (more exact mapping, control of map projections, sophisticated analytical tools).

Shufeldt et al. (this volume, chapter 11) highlight benefits of using GE rather than printed materials for displaying geological maps. In particular, they emphasize the ability of GE to drill down into different scales of data by displaying higher resolution maps as the user's viewpoint gets closer to the ground. This tile pyramid method, which is also used by the base imagery, is the core of what makes GE special. Although this was previously possibly in desktop GIS applications, what stands GE apart is how tile pyramids (Fig. 2) have been optimized to load rapidly, transition smoothly, and work on all sizes of computing devices (smartphones, pads, laptops). This accessibility has encouraged many researchers to migrate data from GIS environments (Guth, this volume, chapter 12) and take the time to create visualizations of pre-digital maps. Simpson et al. (this volume, chapter 13) describe a case study of the latter for mapping of Vredefort Dome, South Africa.

For most geologists, remote mapping will never totally eliminate the need for fieldwork (Hasbargen, this volume, chapter 14). However, the scaling ability of GE can be combined with field-based knowledge to create maps across spatial scales that would be logistically impractical or complex on the ground (Lageson et al., this volume, chapter 15). As the archive of high-resolution imagery increases, there is also meaningful data to be found across temporal scales, e.g., landscapes shaped by active tectonic processes or surface processes. At a minimum, GE enables the identification of sites on which to focus ground-based research (Tewksbury et al., this volume, chapter 2), eliminating a need a for reconnaissance field trips.

Another advantage of using GE for remote geologic mapping is the 3D perspective and the ability to view 3D models of geologic structures. KML uses <Model> to display COLLAborative Design Activity (COLLADA) models read from Digital Asset Exchange (DAE) format files. These are easily created using SketchUp, a 3D modeling computer program that combines a tool-set with an intelligent drawing system (Fig. 3). SketchUp models can be imported into GE to show both above ground outcrops (De Donatis et al., this volume, chapter 16) and block diagrams of geological cross-sections (Karabinos, this volume, chapter17). Hill and Harrison (this volume, chapter 18) demonstrate that it is also possible to use SketchUp to fix limitations in GE's topographic model by adding terrain in areas where the resolution the data used is not high enough to show the real-life shape of the landscape.

Innovative technology is also augmenting the modern field geologist's view with a 3D perspective. Wang et al. (this volume, chapter 19) describe a systematic workflow to use an Apple iPad to manually trace geological features on 2D photos and then use them to construct 3D models of an outcrop. Tools such as these and applications such as SketchUp are changing geologists' mindsets from both an educational and a field-based research perspective.

VIRTUAL FIELD EXPERIENCES

The impact of GE on geologic mapping techniques has been described, but its use in educational settings also has helped modernize both the methods and mindsets of students toward geology. Geology is an exciting and dynamic science but it can also be limited in the classroom by the fact that traditional information sources (lectures, laboratory exercises, textbooks, and even multimedia resources) cannot always convey the scope and setting of landscapes and geologic features. The use of virtual field experiences (VFEs), especially those based in GE, can remove that limitation and provide a proxy for field trips that are not logistically feasible.

There are problems with VFEs, especially when considering their practical and pedagogical design, which are discussed by Granshaw and Duggan-Haas (this volume, chapter 20), but there are also major advantages aside from making "trips" logistically possible. It has been suggested that the ability of GE to scale from a whole-earth global view to a detailed outcrop image gives geologists a perspective and insight that was previously hard to gain (Whitmeyer et al., 2008). The same is true for GigaPans, high-resolution photographic panoramas that employ the same method of image tile pyramids that make GE so efficient (Fig. 4). Although one GigaPan is a single photomosaic, it is possible to build a series of GigaPans at different scales for the same target, e.g., outcrop to thin section. Using a common topic or theme, Piatek et al. (this volume, chapter 21) demonstrate how these GigaPan series act as VFEs that are useful for explaining geologic concepts.

A limitation of VFEs is the lack of student-to-student interactions within the virtual environment—something that is an important component during real field trips. Many students will ask their peers questions before seeking help from the instructor, especially when new concepts are being encountered. To overcome this limitation Dordevic and Wild (this volume, chapter 22) have developed a way for avatars—virtual representation of users—to communicate and cooperate while exploring field locations with the GE API. The movements and actions of these avatars can be logged allowing VFEs to be refined and to scaffold student learning.

GE offers an ideal virtual field environment as "trips" across the real-world mirror landscape can be augmented with KML to provide extra information. This information compensates for

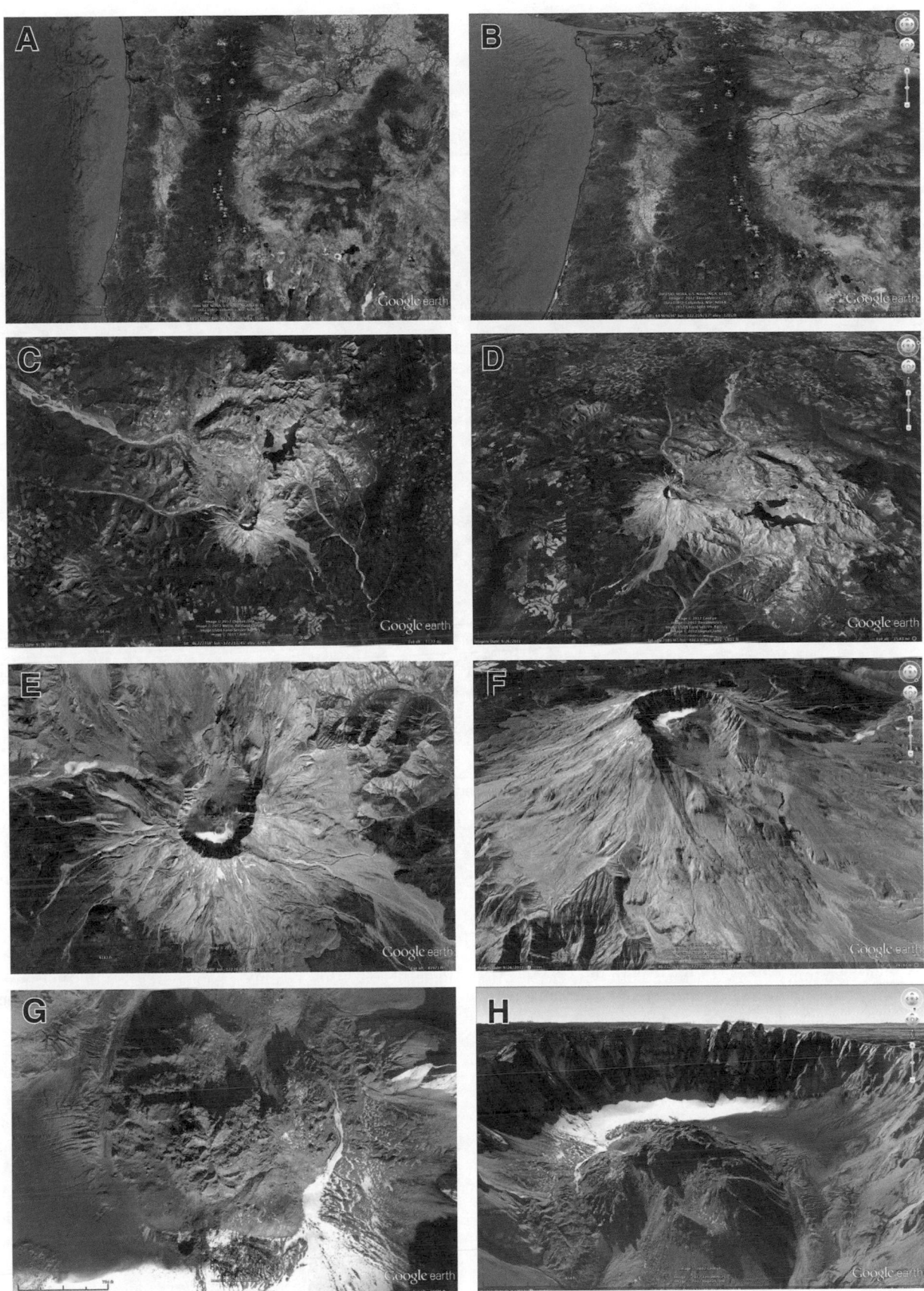

Figure 2. Visions of scale in Google Earth. (A, B) Satellite view of Cascade Volcanoes in nadir and perspective. (C, D) Regional view of Mount St. Helens in nadir and perspective. (E, F) Mount St. Helens volcanic edific in nadir and perspective. (G, H) Mount St. Helens' lava dome in nadir and perspective.

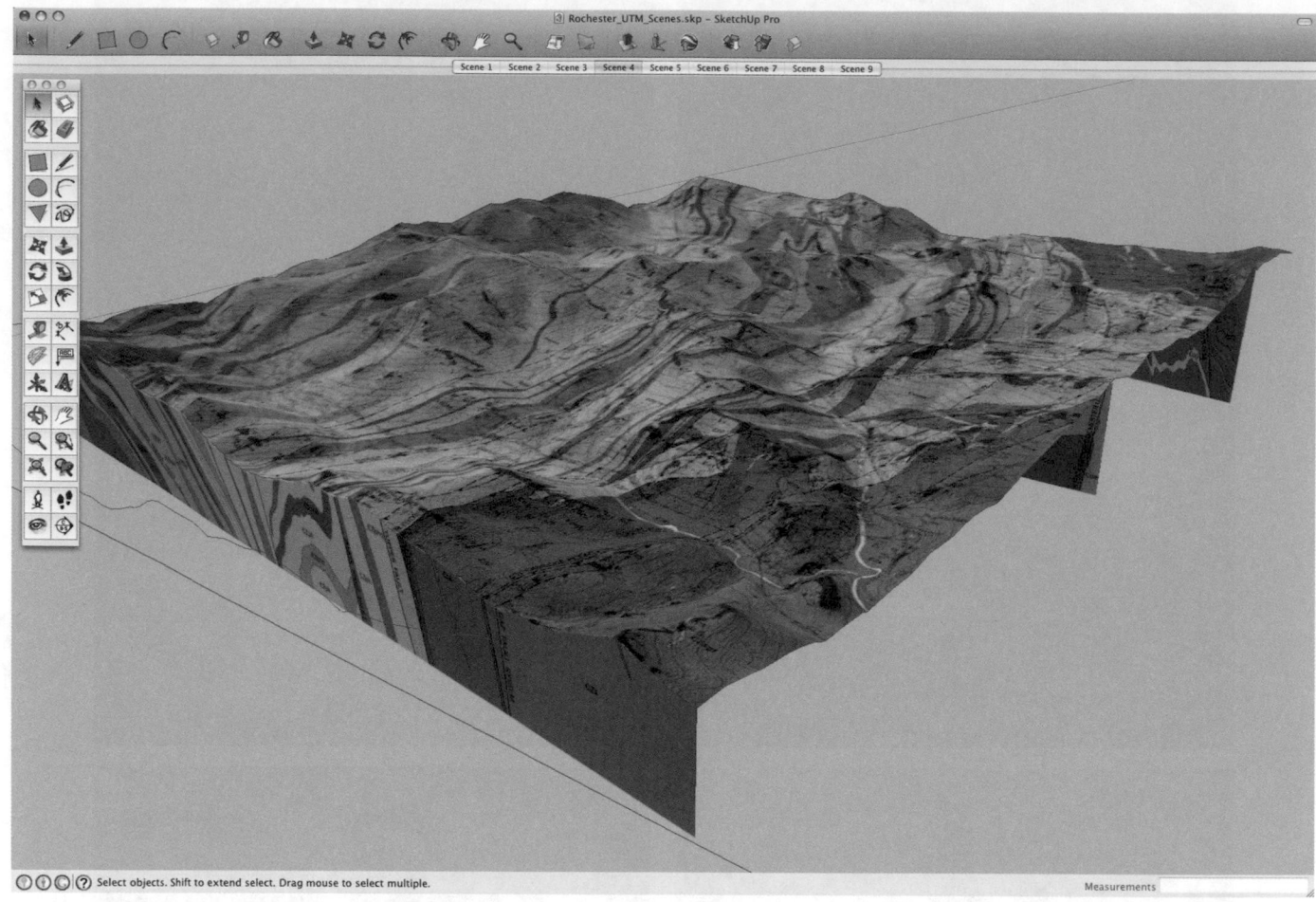

Figure 3. A 3D geological cross-section under construction in SketchUp (courtesy of Paul Karabinos, Williams College).

the lack of a field trip instructor. Most commonly, a VFE in GE displays information using different types of the <Placemark> geometry (points, lines or polygons) and associated description balloons containing text and images. The user navigates to each placemark and clicks on the object to display the balloon. For example, Lang et al. (this volume, chapter 23) describe a numbered sequence of point-placemarks, which represent "stops" on a trip around the volcanoes of Tenerife, Spain. This work was based on a real-life field trip that followed the same route.

Using a similar approach, Muller (this volume, chapter 24) created placemarks and description balloons to archive 55 years of road logs from field trips offered by the New York State Geological Association (NYSGA). Muller makes use of Fusion Tables to store, and in some cases merge, the KML created. Muller's work takes advantage of the detailed, though not always correct, geologic and route descriptions contained in the NYGSA guidebooks. By comparison, Rueger and Beck (this volume, chapter 25) created KML based on information gleaned by them and others (Clark, 2003) from historical documents and journals on an eighteenth century military march into Canada. Rueger and Beck's goal was not just to map the march's route,

but to use GE to interpret how the landscape influenced the path taken. This contrasts with Muller's NYSGA guidebook KML, which describes the geology along a specific pre-planned route. The concept that VFEs can add to the understanding of observations made in the real-world was tested by Eusden et al. (this volume, chapter 26) in an introductory geology class. They used GPS devices and cameras to collect multimedia data during a field trip and displayed that data in GE using KML created by the students. Assessment of the results was qualitative in nature, but suggested that this approach is beneficial to learning.

In the GE environment, the virtual field experience is not just confined to the terrestrial tier. Through one simple menu option in the desktop application, or one line of JavaScript in the API, the user can change the base imagery, terrain and content from Earth to Mars, the Moon or even to view the stars (Fig. 5). Members of the planetary community have appreciated these additions but wish there were more components to the "Google Universe." To aid investigations of Venus, De Paor et al. (this volume, chapter 27) have created "Google Venus," which uses global image overlays, models of geological sections (cf. Karabinos, chapter 17) and placemarks with descriptions to compile a KML collection

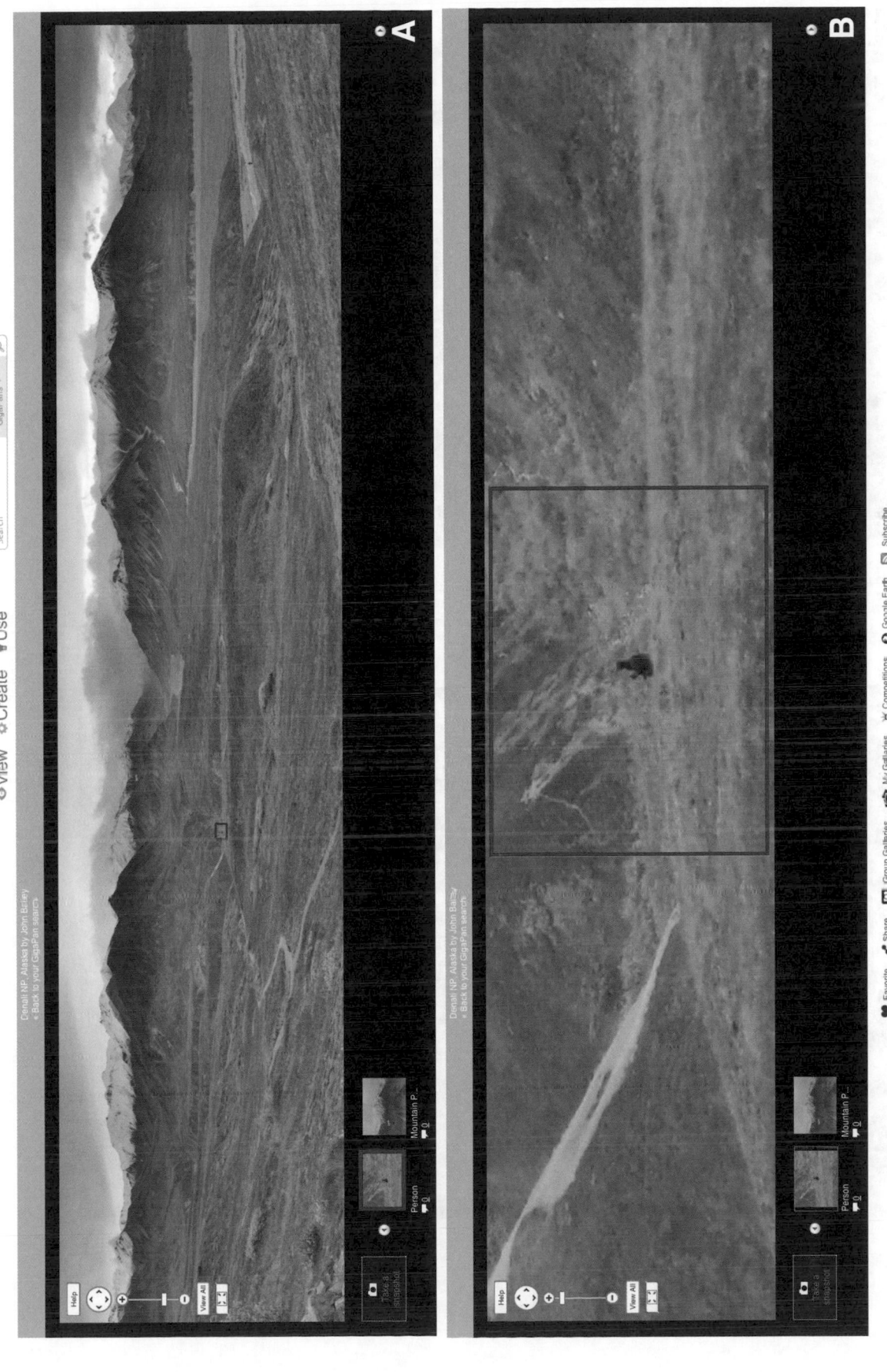

Figure 4. (A) GigaPan of Denali National Park. The red boxes indicate the area focused on in lower image (B) GigaPans use an image tile pyramid to show higher resolution detail when the user zooms in.

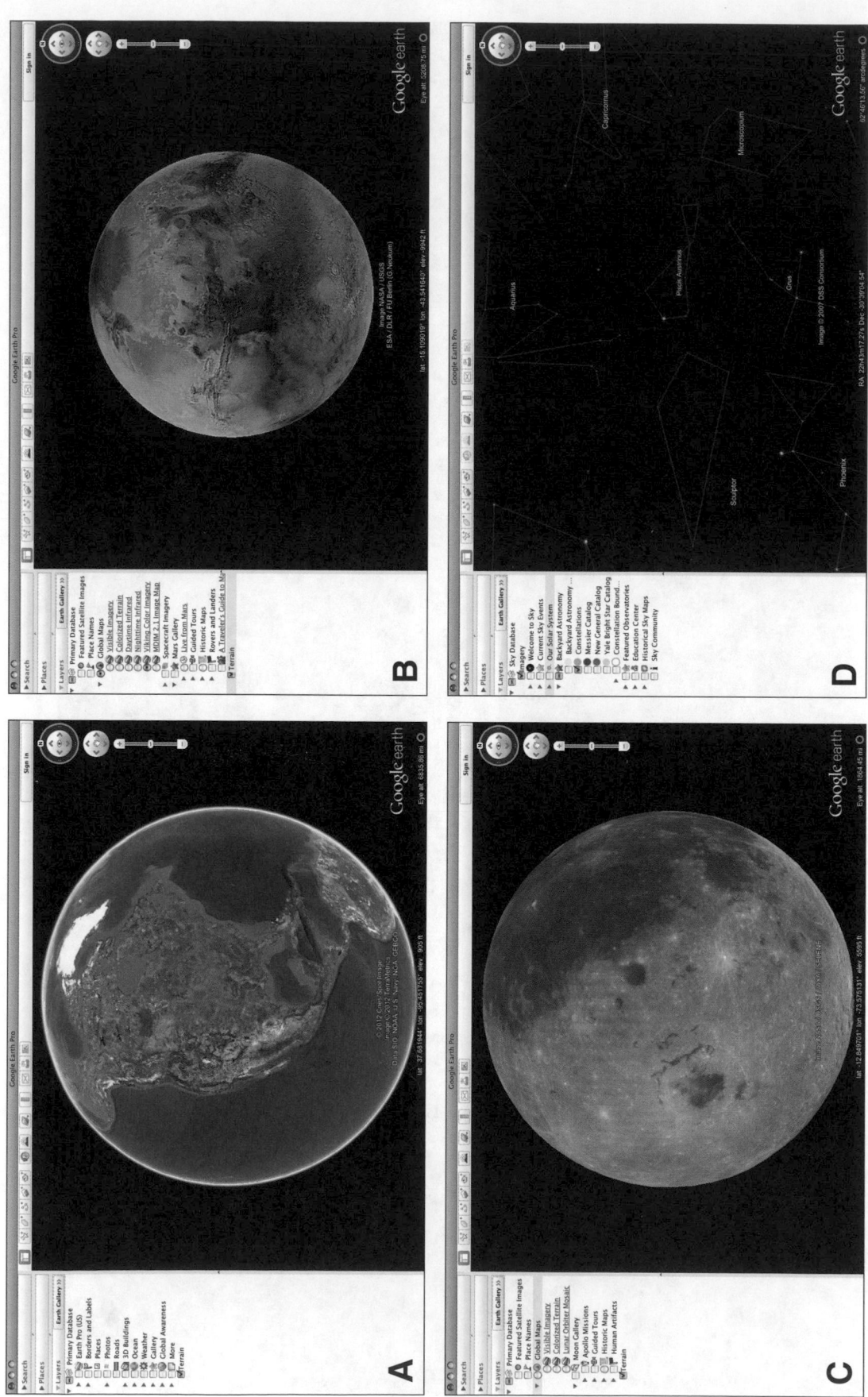

Figure 5. Google Universe. (A) The Google Earth browser. (B) Google Mars. (C) Google Moon. (D) Google Sky. The non-terrestrial globes use the Google Earth application but replace the default imagery, terrain, and layers with new content.

similar to that included by default in GE for the Moon and Mars (Fig. 5).

EDUCATIONAL MODELS, LEARNING METHODS, AND ASSESSMENT

A potential flaw of VFEs is that sometimes there is too much freedom for users, especially if there is a large environment to explore and the exploration is left open ended. While GE can contribute to this flaw, it also includes functionality that can be used to guide users though the virtual environment; Google Earth Tours. These tours are KML scripts that record movement around the 3D globe and interaction with other KML objects. A tour can be replayed to replicate that same movement and interaction. Tours can be automatically generated from a folder of placemarks (Fig. 6) or a path, or they can be freeform and used to record the user's independent navigation. Since GE Tours are composed of KML code, they can be shared as easily as any KML files, so once created they offer a distributable, "guided" VFE.

GE Tours are easy to create, but the author of a tour also needs to consider the user experience, the subject matter being illustrated, and the learning objectives. Sometimes it might not be appropriate to use a GE Tour, but in many cases they are a very useful approach if designed appropriately. Treves and Bailey (this volume, chapter 28) describe a series of design-based best practices that are recommended for developers creating GE Tours for educational purposes.

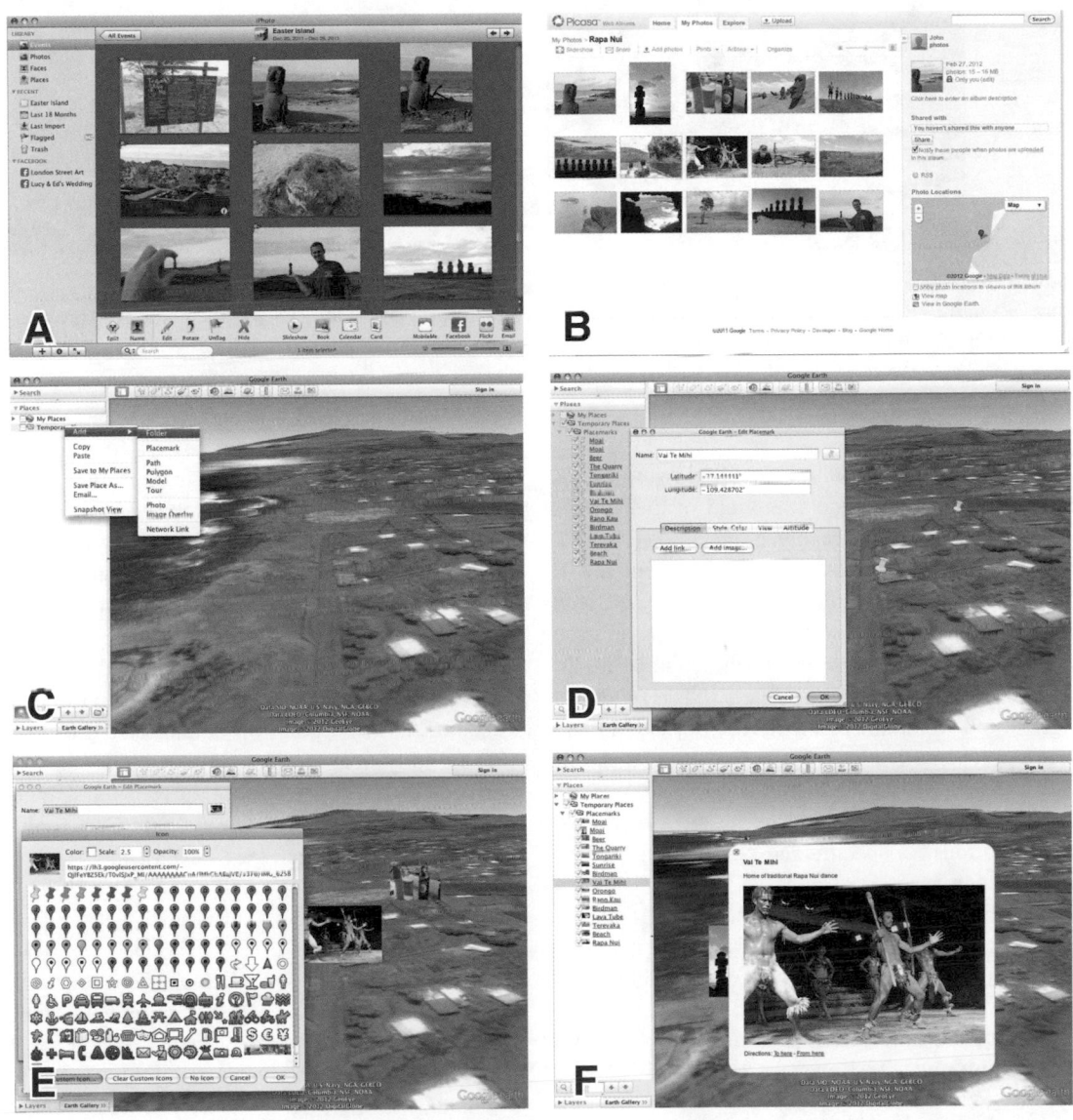

Figure 6. Brief Guide to creating an automated Point-Placemarks Tour. (A) Collect and download photos. (B) Upload photos to web. (C) Create a Folder. (D) Create a Point-Placemarks that the mark locations of the photos in Folder. (E) Customize the Placemark icons. (F) Add text and/or images to the description balloons. (*Continued.*)

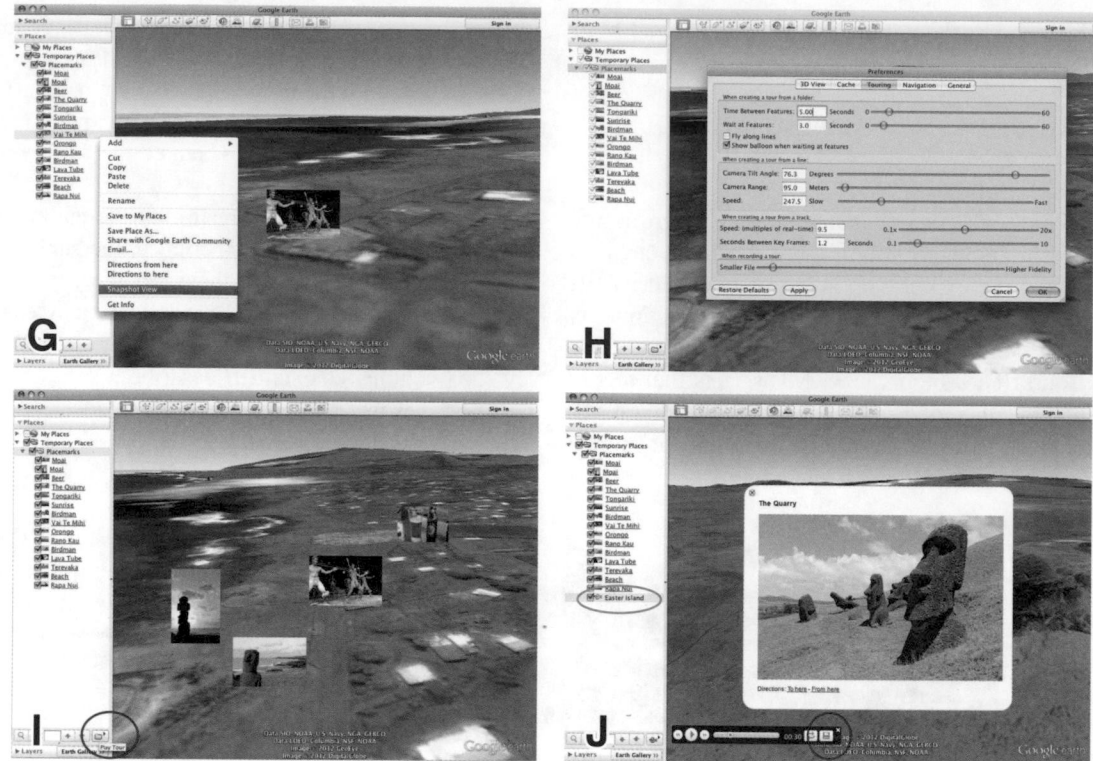

Figure 6 (*continued*). (G) Apply unique, 3D perspective views to each Placemark. (H) Change the "Touring" settings using the Google Earth > Preferences (Mac) or Tools > Options (PC) menu. (I) Select the Folder and press the play button that appears at the bottom of the Place sidebar panel (see red oval). (J) Save the tour that is generated (see red circle); it will appear as a KML object in the sidebar (see blue oval).

Design is an important consideration for all classroom uses of GE. An educator needs to consider the scale of their goals, whether the goal is to broadly encourage geo-education (Lee and Guertin, this volume, chapter 29) or to specifically integrate GE into a curriculum (Almquist, this volume, chapter 30). GE and KML also provide a common medium for collaborations between students around the world. An example of this is the Northern Environmental Education Development project (NEED), an initiative involving schools in Ireland, Norway, Iceland, and Finland that seeks to share geo-visualizations (Hennessy et al., this volume, chapter 31).

Currently, GE's most common role in the classroom is to provide additional illustrations for specific geoscience topics. There exist many "good" sources of geoscience KML. Well-known examples include: The Smithsonian's Volcanoes of the World (Venzke et al., 2006); National Snow and Ice Data Center's ice and glacier mapping (Ballagh et al., 2011); the U.S. Geological Survey's real-time earthquake locations (Blair and Ticci, 2006); the National Oceanic and Atmospheric Administration's severe weather tracking (Smith and Lakshmanan, 2011); and NASA's atmosphere profiling (Chen et al., 2009). Using these resources, or by developing new ones, educators are finding a role for GE and other Google Geo Tools within their subject areas. Examples include geomorphology (Dolliver, this volume, chapter 32),

hydrology (Habib et al., this volume, chapter 33) and oceanography (Hochstaedter and Sullivan, this volume, chapter 34).

The question remains though, "How is Google Earth influencing learning?" Answering this question is key, as it will determine the direction towards which educational developers need to head (Gobert et al., this volume, chapter 35). Some studies, especially in the area of GE-based virtual field experiences and tours, have included assessments of the impact on students (Treves and Engelbrecht, 2011; Johnson et al., 2011; Eusden et al., chapter 26), and early signs are encouraging. However, these assessments were qualitative or semi-quantitative. More studies targeting in-depth testing of the impact of Google Geo Tools are needed as new educational strategies are developed.

CONCLUSIONS

Google Earth is now a much-used and relied upon application for geoscience educators and researchers. It is in their interests to see GE improve and better fit their needs, as well as encouraging the development of associated applications (e.g., SketchUp, GigaPan), which add to the utility of GE. Continued communication among all parties—educators, researchers, developers, publishers, and others with an interest in that process—are important. The 2011 GSA Penrose Conference provided these parties with an opportunity to gather in one location and develop new collaborations. This special volume highlights collaborations and projects founded or shared at that gathering, and is a documentation of the importance of GE and related tools to geoscience education and research.

Engaging scientific visualizations are important. They have the potential to bring data to life, increase understanding of natural processes, and promote awareness of global change. We believe that GE-based visualizations are already transforming geoscience education and research and that this is only the beginning.

ACKNOWLEDGMENTS

The editors would like to express our appreciation to Google Inc. for hosting the 2011 GSA Penrose Conference that led to the publication of this special volume. We sincerely thank both the conference attendees and authors of papers in this special volume for their participation and enthusiasm. We would also especially like to acknowledge the reviewers whose time and efforts made the publication of this special volume possible:

Carlos Aiken, Steve Allard, Heather Almquist, Chuck Bailey, Greg Baker, Alan Benimoff, Callan Bentley, Andy Bobyarchick, Katherine Boggs, Andre Breton, Anna Courtier, Rónadh Cox, Chris Crosby, Holly Dolliver, Mauro De Donatis, Natalia Deligne, Don Duggan-Haas, Tyler Erickson, Dyk Eusden, Martin Feely, Ioannis Georgiou, Tod Greene, Laura Guertin, Peter Guth, Michael Harrison, Les Hasbargen, Matt Heavner, Ronán Hennessy, Otto Hermelin, Jesse Hill, Victoria Hill, Fred Hochstaedter, Eric Horsman, Micah Jessup, Paul Karabinos, Eric de Kemp, Tom Kurkowski, Dave Lageson, Wes Lauer, Nick Lang, Jack Loveless, Karen McNeal, Riley Milner, Alexandra Moore, Otto Muller, Dan Murray, Rick Murray, Jeff Nunn, Carol Ormond, Terry Pavlis, Lyman Persico, Jen Piatck, Arancha Pinan-Llamas, Phillip Prince, Eric Pyle, Bruce Rueger, Uwe Schindler, Peter Selkin, Colin Shaw, Owen Shufeldt, Jill Singer, Gary Solar, Meg Stewart, Barb Tewksbury, Dave Tewksbury, Ryan Thigpen, Peter Thompson, Sarah Titus, Rich Treves, Rich Whittecar, Crystal Wilson, Mike Winiski, Christine Witkowski, Michael Wizevich, and Andre Zular.

Special thanks are due to Chris Condit, who served as special editor in cases where the regular editors had a conflict of interests.

The Google Penrose Conference at Mountain View and this GSA Special Paper were funded in part by NSF TUES 1022755. Any opinions, findings, and conclusions or recommendations expressed in this volume are those of the authors and do not necessarily reflect the views of the National Science Foundation or Google Inc.

REFERENCES CITED

Almquist, H., Blank, L. and Estrada, J., 2012, this volume, Developing a scope and sequence for using Google Earth in the middle school earth science classroom, *in* Whitmeyer, S.J., Bailey, J.E., De Paor, D.G., and Ornduff, T., eds., Google Earth and Virtual Visualizations in Geoscience Educa-tion and Research: Geological Society of America Special Paper 492, doi:10.1130/2012.2492(30).

Bailey, J.E., 2009, Virtual Globes at AGU, http://www.snap.uaf.edu/earth/agu/ (accessed April 2012).

Bailey, J.E., 2010, Entry for "Virtual Globe," *in* Warf, B., ed., Encyclopedia of Geography: Sage Publications, 3528 p.

Ballagh, L.M., Raup, B.H., Duerr, R.E., Khalsa, S.J.S., Helm, C., Fowler, D., and Gupte, A., 2011, Representing scientific data sets in KML: Methods and challenges: Computers & Geosciences, v. 37, p. 57–64, doi:10.1016/j.cageo.2010.05.004.

Blair, J.L., and Ticci, M., 2006, Serving Bay Area geologic hazard information in Google Earth KML; a network-link approach: Eos (Transactions, American Geophysical Union), v. 87, no. 52, Fall Meeting Supplement, abstract IN43A-0896, 11–15 December, San Francisco, California, USA.

Bradley, E.S., Roberts, D.A., Dennison, P.E., Green, R.O., Eastwood, E., Lundeen, S.R., McCubbin, I.B., and Leifer, I., 2011, Google Earth and Fusion Tables in support of time-critical collaboration; Mapping the deepwater horizon oil spill with the AVIRIS airborne spectrometer: Earth Science Informatics, v. 4, no. 4, p. 169–179, doi:10.1007/s12145-011-0085-4.

Chen, A., Leptoukh, G.Z., Kempler, S., Lynnes, C., Savtchenko, A., Nadeau, D., and Farley, J., 2009, Visualization of A-Train vertical profiles using Google Earth: Computers & Geosciences, v. 35, p. 419–427, doi:10.1016/j.cageo.2008.08.006.

Clark, S., 2003, Following their footsteps: A travel guide and history of the 1775 secret expedition to capture Quebec: Shapleigh, Maine, Clark Books, 123 p.

Crosby, C.J., 2012, this volume, Lidar and Google Earth: Simplifying access to high-resolution topography data, *in* Whitmeyer, S.J., Bailey, J.E., De Paor, D.G., and Ornduff, T., eds., Google Earth and Virtual Visualizations in Geoscience Education and Research: Geological Society of America Special Paper 492, doi:10.1130/2012.2492(03).

De Donatis, M., Susini, S., and Foi, M., 2012, this volume, Geology from real field to 3D modeling and Google Earth virtual environments: Methods and goals from the Apennines (Furlo Gorge, Italy), *in* Whitmeyer, S.J., Bailey, J.E., De Paor, D.G., and Ornduff, T., eds., Google Earth and Virtual Visualizations in Geoscience Education and Research: Geological Society of America Special Paper 492, doi:10.1130/2012.2492(16).

De Paor, D.G., Whitmeyer, S.J., Marks, M., and Bailey, J.E., 2012, this volume, chapter 6, Geoscience applications of client/server scripts, Google Fusion Tables, and dynamic KML, *in* Whitmeyer, S.J., Bailey, J.E., De Paor, D.G., and Ornduff, T., eds., Google Earth and Virtual Visualizations in Geoscience Education and Research: Geological Society of America Special Paper 492, doi:10.1130/2012.2492(06).

De Paor, D.G., Hansen, V.L., and Dordevic, M.M., 2012, this volume, chapter 27, Google Venus, *in* Whitmeyer, S.J., Bailey, J.E., De Paor, D.G., and Ornduff, T., eds., Google Earth and Virtual Visualizations in Geoscience Education and Research: Geological Society of America Special Paper 492, doi:10.1130/2012.2492(27).

Dolliver, H.A.S., 2012, this volume, Using Google Earth to teach geomorphology, *in* Whitmeyer, S.J., Bailey, J.E., De Paor, D.G., and Ornduff, T., eds., Google Earth and Virtual Visualizations in Geoscience Education and Research: Geological Society of America Special Paper 492, doi:10.1130/2012.2492(32).

Dordevic, M.M., 2012, this volume, Designing interactive screen overlays to enhance effectiveness of Google Earth geoscience resources, *in* Whitmeyer, S.J., Bailey, J.E., De Paor, D.G., and Ornduff, T., eds., Google Earth and Virtual Visualizations in Geoscience Education and Research: Geological Society of America Special Paper 492, doi:10.1130/2012.2492(07).

Dordevic, M.M., and Wild, S.C., 2012, this volume, Avatars and multistudent interactions in Google Earth–based virtual field experiences, *in* Whitmeyer, S.J., Bailey, J.E., De Paor, D.G., and Ornduff, T., eds., Google Earth and Virtual Visualizations in Geoscience Education and Research: Geological Society of America Special Paper 492, doi:10.1130/2012.2492(22).

Eusden, J.D., Jr., Duvall, M., and Bryant, M., 2012, this volume, Google Earth mashup of the geology in the Presidential Range, New Hampshire: Linking real and virtual field trips for an introductory geology class, *in* Whitmeyer, S.J., Bailey, J.E., De Paor, D.G., and Ornduff, T., eds., Google Earth and Virtual Visualizations in Geoscience Education and Research: Geological Society of America Special Paper 492, doi:10.1130/2012.2492(26).

Fisher, G.B., Amos, C.B., Bookhagen, B., Burbank, D.W., and Godard, V., 2012, this volume, Channel widths, landslides, faults, and beyond: The new world order of high-spatial resolution Google Earth imagery in the study of earth surface processes, *in* Whitmeyer, S.J., Bailey, J.E., De Paor, D.G., and Ornduff, T., eds., Google Earth and Virtual Visualizations in Geoscience Education and Research: Geological Society of America Special Paper 492, doi:10.1130/2012.2492(01).

Gobert, J., Wild, S.C., and Rossi, L., 2012, this volume, Testing the effects of prior coursework and gender on geoscience learning with Google Earth, *in* Whitmeyer, S.J., Bailey, J.E., De Paor, D.G., and Ornduff, T., eds., Google Earth and Virtual Visualizations in Geoscience Education and Research: Geological Society of America Special Paper 492, doi:10.1130/2012.2492(35).

Gonzalez, H., Halevy, A., Jensen, C.S., Langen, A., Madhavan, J., Shapley, R., and Shen, W., 2010, Google Fusion Tables: Data management, integration and collaboration in the cloud, SoCC'10, 10–11 June 2010, Indianapolis, Indiana, USA, ACM 978-1-4503-0036-0/10/06.

Google Inc., 2010, Google Crisis Response—Haiti earthquake, http://www.google.com/relief/haitiearthquake/ (accessed April 2012).

Google Inc., 2011, Google Earth Engine, http://earthengine.google.org/ (accessed April 2012).

Granshaw, F.D., and Duggan-Haas, D., 2012, this volume, Virtual field-work in geoscience teacher education: Issues, techniques, and models, *in* Whitmeyer, S.J., Bailey, J.E., De Paor, D.G., and Ornduff, T., eds., Google Earth and Virtual Visualizations in Geoscience Education and Research: Geological Society of America Special Paper 492, doi:10.1130/2012.2492(20).

Guth, P.L., 2012, this volume, Automated export of GIS maps to Google Earth: Tool for research and teaching, *in* Whitmeyer, S.J., Bailey, J.E., De Paor, D.G., and Ornduff, T., eds., Google Earth and Virtual Visualizations in Geoscience Education and Research: Geological Society of America Special Paper 492, doi:10.1130/2012.2492(12).

Hasbargen, L.E., 2012, this volume, A test of the three-point vector method to determine strike and dip utilizing digital aerial imagery and topography, *in* Whitmeyer, S.J., Bailey, J.E., De Paor, D.G., and Ornduff, T., eds., Google Earth and Virtual Visualizations in Geoscience Education and Research: Geological Society of America Special Paper 492, doi:10.1130/2012.2492(14).

Habib, E., Ma, Y., and Williams, D., 2012, this volume, Development of a web-based hydrologic education tool using Google Earth resources, *in* Whitmeyer, S.J., Bailey, J.E., De Paor, D.G., and Ornduff, T., eds., Google Earth and Virtual Visualizations in Geoscience Education and Research: Geological Society of America Special Paper 492, doi:10.1130/2012.2492(33).

Hennessy, R., Arnason, T., Ratinen, I., and Rubensdotter, L., 2012, this volume, Google Earth geo-education resources: A transnational approach from Ireland, Iceland, Finland, and Norway, *in* Whitmeyer, S.J., Bailey, J.E., De Paor, D.G., and Ornduff, T., eds., Google Earth and Virtual Visualizations in Geoscience Education and Research: Geological Society of America Special Paper 492, doi:10.1130/2012.2492(31).

Hennessey-Fiske, M., 2011, Google releases first satellite images of Japan after quake: Los Angeles Times, 12 March 2011.

Hill, J.S., and Harrison, M.J., 2012, this volume, Terrain modification in Google Earth using SketchUp: An example from the Western Blue Ridge of Tennessee, *in* Whitmeyer, S.J., Bailey, J.E., De Paor, D.G., and Ornduff, T., eds., Google Earth and Virtual Visualizations in Geoscience Education and Research: Geological Society of America Special Paper 492, doi:10.1130/2012.2492(18).

Hochstaedter, A., and Sullivan, D., 2012, this volume, Oceanography and Google Earth: Observing ocean processes with time animations and student-built ocean drifters, *in* Whitmeyer, S.J., Bailey, J.E., De Paor, D.G., and Ornduff, T., eds., Google Earth and Virtual Visualizations in Geoscience Education and Research: Geological Society of America Special Paper 492, doi:10.1130/2012.2492(34).

Johnson, N.D., Lang, N.P., and Zophy, K.T., 2011, Overcoming assessment problems in Google Earth–based assignments: Journal of Geoscience Education, v. 59, p. 99–105, doi:10.5408/1.3604822.

Karabinos, P., 2012, this volume, Creating interactive 3-D block diagrams from geologic maps and cross-sections, *in* Whitmeyer, S.J., Bailey, J.E., De Paor, D.G., and Ornduff, T., eds., Google Earth and Virtual Visualizations in Geoscience Education and Research: Geological Society of America Special Paper 492, doi:10.1130/2012.2492(17).

Lageson, D.R., Larsen, M.C., Lynn, H.B., and Treadway, W.A., 2012, this volume, Applications of Google Earth Pro to fracture and fault studies

of Laramide anticlines in the Rocky Mountain foreland, *in* Whitmeyer, S.J., Bailey, J.E., De Paor, D.G., and Ornduff, T., eds., Google Earth and Virtual Visualizations in Geoscience Education and Research: Geological Society of America Special Paper 492, doi:10.1130/2012.2492(15).

Lang, N.P., Lang, K.T., and Camodeca, B.M., 2012, this volume, A geology-focused virtual field trip to Tenerife, Spain, *in* Whitmeyer, S.J., Bailey, J.E., De Paor, D.G., and Ornduff, T., eds., Google Earth and Virtual Visualizations in Geoscience Education and Research: Geological Society of America Special Paper 492, doi:10.1130/2012.2492(23).

Lee, Tsan-Kuang, and Guertin, L., 2012, this volume, Building an education game with the Google Earth application programming interface to enhance geographic literacy, *in* Whitmeyer, S.J., Bailey, J.E., De Paor, D.G., and Ornduff, T., eds., Google Earth and Virtual Visualizations in Geoscience Education and Research: Geological Society of America Special Paper 492, doi:10.1130/2012.2492(29).

Muller, O.H., 2012, this volume, Moving New York State Geological Association guidebooks into Google Earth, *in* Whitmeyer, S.J., Bailey, J.E., De Paor, D.G., and Ornduff, T., eds., Google Earth and Virtual Visualizations in Geoscience Education and Research: Geological Society of America Special Paper 492, doi:10.1130/2012.2492(24).

Nunn, J.A., and Bentley, L., 2012, this volume, Visualization of spatial and temporal trends in Louisiana water usage using Google Fusion Tables, *in* Whitmeyer, S.J., Bailey, J.E., De Paor, D.G., and Ornduff, T., eds., Google Earth and Virtual Visualizations in Geoscience Education and Research: Geological Society of America Special Paper 492, doi:10.1130/2012.2492(09).

Piatek, J.L., Kairies Beatty, C.L., Beatty, W.L., Wizevich, M.C., and Steullet, A., 2012, this volume, Developing virtual field experiences for undergraduates with high-resolution panoramas (GigaPans) at multiple scales, *in* Whitmeyer, S.J., Bailey, J.E., De Paor, D.G., and Ornduff, T., eds., Google Earth and Virtual Visualizations in Geoscience Education and Research: Geological Society of America Special Paper 492, doi:10.1130/2012.2492(21).

Potapov, E., and Hronusov, V., 2012, this volume, Extreme dynamic mapping: Animals map themselves on the "Cloud," *in* Whitmeyer, S.J., Bailey, J.E., De Paor, D.G., and Ornduff, T., eds., Google Earth and Virtual Visualizations in Geoscience Education and Research: Geological Society of America Special Paper 492, doi:10.1130/2012.2492(10).

Rueger, B.F., and Beck, E.N, 2012, this volume, Benedict Arnold's march to Quebec in 1775: An historical characterization using Google Earth, *in* Whitmeyer, S.J., Bailey, J.E., De Paor, D.G., and Ornduff, T., eds., Google Earth and Virtual Visualizations in Geoscience Education and Research: Geological Society of America Special Paper 492, doi:10.1130/2012.2492(25).

Shufeldt, O.P., Whitmeyer, S.J., and Bailey, C.M., 2012, this volume, The new frontier of interactive, digital geologic maps: Google Earth–based multilevel maps of Virginia geology, *in* Whitmeyer, S.J., Bailey, J.E., De Paor, D.G., and Ornduff, T., eds., Google Earth and Virtual Visualizations in Geoscience Education and Research: Geological Society of America Special Paper 492, doi:10.1130/2012.2492(11).

Simpson, C., De Paor, D.G., Beebe, M.R., and Strand, J.M., 2012, this volume, Transferring maps and data from pre-digital era theses to Google Earth: A case study from the Vredefort Dome, South Africa, *in* Whitmeyer, S.J., Bailey, J.E., De Paor, D.G., and Ornduff, T., eds., Google Earth and Virtual Visualizations in Geoscience Education and Research: Geological Society of America Special Paper 492, doi:10.1130/2012.2492(13).

Smith, T.M., and Lakshmanan, V., 2011, Real-time, rapidly updating severe weather products for virtual globes: Computers & Geosciences, v. 37, p. 3–12, doi:10.1016/j.cageo.2010.03.023.

Stewart, M.E., and Baldwin, K., 2012, this volume, Workshops, community outreach, and KML for visualization of marine resources in the Grenadine Islands, *in* Whitmeyer, S.J., Bailey, J.E., De Paor, D.G., and Ornduff, T., eds., Google Earth and Virtual Visualizations in Geoscience Education and Research: Geological Society of America Special Paper 492, doi:10.1130/2012.2492(05).

Tewksbury, B.J., Dokmak, A.A.K., Tarabees, E.A., and Mansour, A.S., 2012, this volume, Google Earth and geologic research in remote regions of the developing world: An example from the Western Desert of Egypt, *in* Whitmeyer, S.J., Bailey, J.E., De Paor, D.G., and Ornduff, T., eds., Google Earth and Virtual Visualizations in Geoscience Education and Research: Geological Society of America Special Paper 492, doi:10.1130/2012.2492(02).

Treves, R., and Bailey, J.E., 2012, this volume, Best practices on how to design Google Earth tours for education, *in* Whitmeyer, S.J., Bailey, J.E., De Paor, D.G., and Ornduff, T., eds., Google Earth and Virtual Visualizations in Geoscience Education and Research: Geological Society of America Special Paper 492, doi:10.1130/2012.2492(28).

Treves, R., and Engelbrecht, P., 2011, User tests on Google Earth Tour comprehension: University of Southampton, 10 p.

Turner, A., 2008, Is Google Maps GIS?, http://highearthorbit.com/is-googlemaps-gis/ (accessed April 2012).

Venzke, E., Siebert, L., and Luhr, J.F., 2006, Smithsonian volcano data on Google Earth: Eos (Transactions, American Geophysical Union), v. 87, no. 52, Fall Meeting Supplement, abstract IN43A-0900, 11–15 December, San Francisco, California, USA.

Wang, M., Rodriguez-Gomez, M.I., and Aiken, C.L.V., 2012, this volume, Interacting with existing 3D photorealistic outcrop models on site and in the lab or classroom, facilitated with an iPad and a PC, *in* Whitmeyer, S.J., Bailey, J.E., De Paor, D.G., and Ornduff, T., eds., Google Earth and Virtual Visualizations in Geoscience Education and Research: Geological Society of America Special Paper 492, doi:10.1130/2012.2492(19).

Wernecke, J., 2009, The KML handbook: Geographic visualization for the web: Upper Saddle River, New Jersey, Addison-Wesley, 368 p.

Whitmeyer, S.J., De Paor, D.G., Daniels, J., Nicoletti, J., Rivera, M., and Santangelo, B., 2008, A pyramid scheme for constructing geologic maps on geobrowsers: Eos (Transactions, American Geophysical Union), v. 89, no. 53, Fall Meeting Supplement, abstract IN41B-1140, 15–19 December, San Francisco, California, USA.

Williams, C.M., Baker, G.S., and Ault, B.A., 2012, this volume, Enhancing usability of near-surface geophysical data in archaeological surveys via Google Earth, *in* Whitmeyer, S.J., Bailey, J.E., De Paor, D.G., and Ornduff, T., eds., Google Earth and Virtual Visualizations in Geoscience Education and Research: Geological Society of America Special Paper 492, doi:10.1130/2012.2492(04).

Zular, A., Guedes, C.C.F., Mendes, V.R., Sawakuchi, A.O., Giannini, P.C.F., Tanaka, A.P.B., Fornari, M., and Nascimento, D.R., Jr., 2012, this volume, Geomorphological analysis of coastal depositional systems in SE Brazil aided by Google Earth coupled with the integration of chronological and sedimentological data by means of a Google Fusion Table, *in* Whitmeyer, S.J., Bailey, J.E., De Paor, D.G., and Ornduff, T., eds., Google Earth and Virtual Visualizations in Geoscience Education and Research: Geological Society of America Special Paper 492, doi:10.1130/2012.2492(08).

Manuscript Accepted by the Society 16 April 2012

The Geological Society of America
Special Paper 492
2012

Channel widths, landslides, faults, and beyond: The new world order of high-spatial resolution Google Earth imagery in the study of earth surface processes

G. Burch Fisher*
Department of Earth Science, University of California, Santa Barbara, California, USA

Colin B. Amos
Department of Earth and Planetary Science, University of California, Berkeley, California, USA

Bodo Bookhagen
Department of Geography, University of California, Santa Barbara, California, USA

Douglas W. Burbank
Department of Earth Science, University of California, Santa Barbara, California, USA

Vincent Godard
CEREGE, CNRS-UMR6635, Aix-Marseille Université, Aix-en-Provence, France

ABSTRACT

The past decade has seen a rapid increase in the application of high-resolution imagery and geographic-based information systems across every segment of society from security intelligence to product marketing to scientific research. Google Earth has positioned itself at the forefront of this spatial information wave by providing free access to high-resolution imagery through a simple, user-friendly interface. Whereas Google Earth imagery has been widely exploited across the earth sciences for spatial visualization, education, and place-based searches, few studies have utilized the high-resolution imagery to yield quantitative insights about the processes and mechanisms acting at the earth's surface. In this paper, we detail the benefits of the underlying high-resolution imagery available within Google Earth, review the limited published research to date, and utilize this imagery to quantitatively illuminate previously difficult and unresolved questions within the discipline of geomorphology involving: (1) channel-width variability and scaling relations in the tectonically active Himalaya; (2) landslide characteristics related to large magnitude climatic and tectonic events in Haiti; and (3) identification and quantification of laterally offset geomorphic features within eastern California. In each example, we compare analyses using

*burch@eri.ucsb.edu

Fisher, G.B., Amos, C.B., Bookhagen, B., Burbank, D.W., and Godard, V., 2012, Channel widths, landslides, faults, and beyond: The new world order of high-spatial resolution Google Earth imagery in the study of earth surface processes, *in* Whitmeyer, S.J., Bailey, J.E., De Paor, D.G., and Ornduff, T., eds., Google Earth and Virtual Visualizations in Geoscience Education and Research: Geological Society of America Special Paper 492, p. 1–22, doi:10.1130/2012.2492(01). For permission to copy, contact editing@geosociety.org. © 2012 The Geological Society of America. All rights reserved.

freely available Google Earth imagery with standard imagery and techniques (e.g., Landsat, ASTER, lidar) to demonstrate the potential benefits of using high-spatial resolution Google Earth imagery over established methodologies. In addition, we discuss the potential limitations and problems with using the imagery currently available in Google Earth and propose favorable future applications, namely studies in remote terrains and those requiring high-resolution imagery across a large spatial extent, where purchasing such imagery in an academic environment would be cost-prohibitive. Whether as a supplement, for reconnaissance, or as the primary data set, high-resolution Google Earth imagery, when properly applied, holds great promise for quantitatively tackling previously unresolved problems in the study of earth surface processes.

1. INTRODUCTION

Since the launch of Google Earth in 2005, millions of people have gained the ability to access and visualize spatial data in a historically unprecedented way. The ability to see and query the world with speed and simplicity across a broad range of spatial (and now temporal) scales was only a dream when proto–Google Earth creators Silicon Graphics (SGI) launched their "Space-to-Your-Face" demo in 1996 (M. Aubin, "Google Earth: From Space to Your Face...and Beyond," http://mattiehead.wordpress.com/tag/google-earth/, accessed May 2012). A decade and a half later, Google Earth has brought geographical information systems to the masses with hundreds of millions of recreational users exploring both human and natural ecosystems across not just the earth, but the solar system. Despite the obvious utility of Google Earth's extensive high-resolution imagery database, it has to date been largely regarded by the research community as purely for education and/or visualization purposes. The focus of this article is to demonstrate the utility of Google Earth and its high-resolution imagery as a powerful research tool to explore and quantify earth surface processes.

In this article, we begin by describing the background and underlying imagery that makes Google Earth such a powerful platform. We then discuss the limited published research to date and highlight three research examples that underscore the value of Google Earth imagery in solving problems whose solutions were previously inhibited by traditional techniques, financial constraints, and/or inaccessible terrain. Lastly, we assess some of the current limitations with using Google Earth imagery for research purposes and highlight suitable future applications of high-resolution Google Earth imagery to quantitatively study the processes shaping both earth and planetary surfaces.

2. THE BEAUTY OF GOOGLE EARTH IMAGERY

Early on, Google Earth utilized freely available Landsat imagery (30-m resolution) (landsat.gsfc.nasa.gov/) with Shuttle Radar Topography Mission (SRTM) digital elevation models (30 m or 90 m resolution) (Farr et al., 2007). With the advent of high-resolution passive optical sensors and a commercial market for that imagery, however, Google Earth has been well placed to exploit such spatial information. Presently, Google Earth contains a large range of true-color visible spectrum (400–700 nm wavelength) imagery derived from a mix of freely available public domain Landsat imagery, government orthophotos, and high-resolution commercial data sets available from DigitalGlobe™ (www.digitalglobe.com), GeoEye™ (www.geoeye.com), and SPOT™ (www.spot.com), with considerable investment in Geo-Eye™ coming directly from Google (Jones, 2008) (Fig. 1 and Table 1). The investment in high-resolution imagery has paid great dividends for Google with Google Earth now being used by millions of people around the world. Due to this popularity, earth scientists and especially geomorphologists now have access to imagery that in places reaches sub-meter resolution and costs nothing. This free access stands in stark contrast to the hundreds of dollars one would normally pay per scene (~100 km^2) for high-resolution imagery, permitting large-scale, high-resolution studies that were previously difficult or impossible to achieve. Furthermore, Google has shown great commitment to constantly improving and expanding the imagery available in Google Earth, not only in terms of the spatial resolution of the imagery, but temporally as well. With the update to Google Earth 5 in early 2009, historical imagery became a key component of the platform, and now allows users to browse through past airphotos and archival satellite images of a given area to easily detect changes through time. Historical imagery along with an ever-expanding archive of high-resolution base imagery makes Google Earth a wonderful resource for the layperson, but also a powerful database and tool for researchers in the earth sciences.

In the following section we review the minimal literature to date that has utilized Google Earth as a primary data source and show three ways in which we have leveraged Google Earth imagery to: (1) improve on current understandings of how widths of river channels adjust and scale in tectonically active orogens; (2) document landslide triggering mechanisms and characteristics in Haiti; and (3) characterize and quantify previously unidentified fault displacements in eastern California.

3. GOOGLE EARTH AS A PRIMARY DATA SOURCE

3.1. Past Work

Google Earth provides a palatable and expansive medium for visualizing, disseminating, and interacting with spatial data

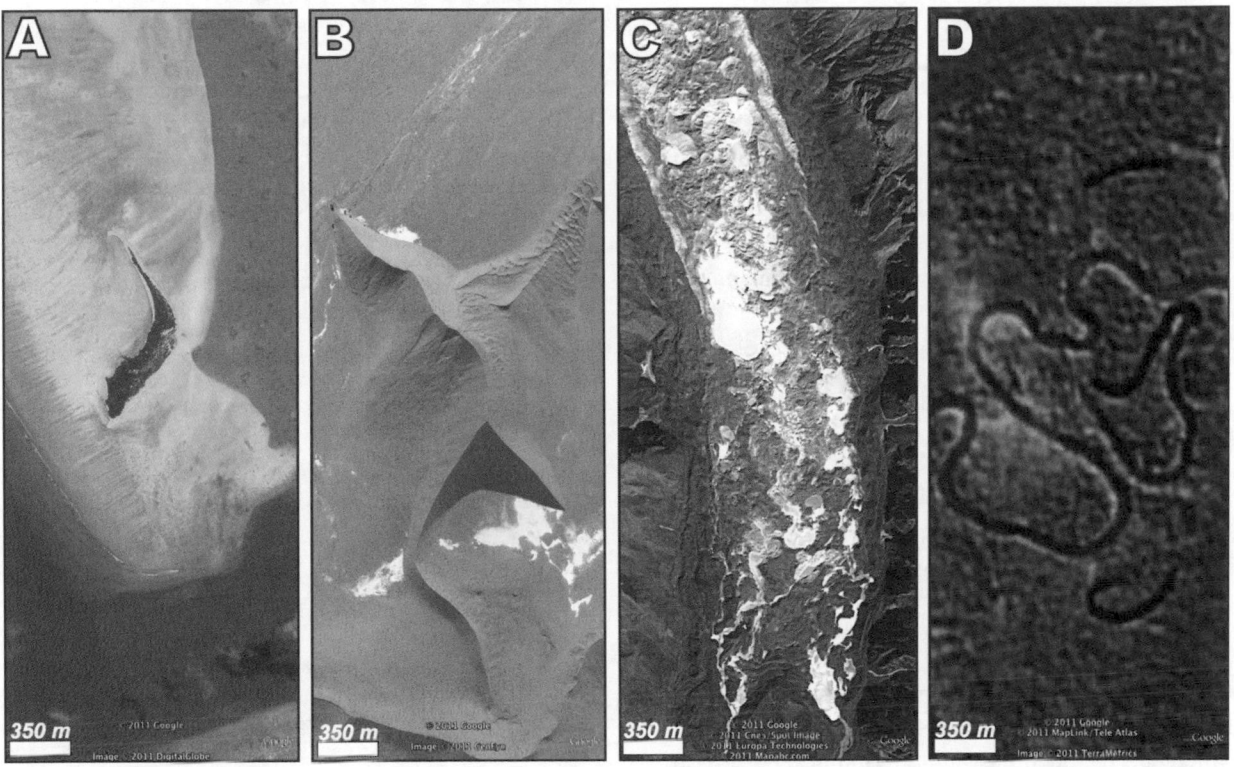

Figure 1. Examples of four predominant types of imagery available in Google Earth from a viewing height of 1500 m. (A) High-resolution (~1 m) GeoEye™ imagery of a small island in the Maldives, Indian Ocean. (B) High-resolution (~1 m) Worldview™ imagery of a star dune from the Namib Desert. (C) High-resolution (~1–5 m) CNES/Spot™ imagery of a debris covered glacial terminus in the Bhutan Himalaya. (D) Lower-resolution (pan-sharpened to 15 m) Landsat 7 ETM+ image of an Amazonian tributary near Puerto Limón, Peru. North is to the top in all images.

TABLE 1. SOME OF THE MOST COMMONLY USED IMAGERY IN GOOGLE EARTH

Type of imagery	Resolution	Major Imagery provider	Additional information
Landsat 7 ETM+	30 m or 15 m pan-sharpened	Terra Metrics, Inc. NASA	www.truearth.com landsat.gsfc.nasa.gov
SPOT, FORMOSAT-2, KOMPSAT-2, Pleiades	0.5–8 m	Spot Image S.A.	www.spot.com
Worldview-1, Worldview-2, Quickbird	0.5–2.5 m	DigitalGlobe, Inc.	www.digitalglobe.com
GeoEye-1, IKONOS	0.5–3.2 m	GeoEye, Inc.	www.geoeye.com
Aerial Imagery (USA)	0.5–2 m	U.S. Department of Agriculture; U.S. Geological Survey; Bluesky; Aerodata International Surveys; etc.	eros.usgs.gov www.fsa.usda.gov www.bluesky-world.com www.aerodata-surveys.com
Ocean and lake bathymetry	>100 m	NOAA, SIO, U.S. Navy, NGA, GEBCO	earth.google.com/ocean

Note: Google does not make public the specific type of imagery used in Google Earth, only the providers. Due to this, and the constant evolution of the platform, some of the imagery presented above may not actually be utilized in Google Earth, while other unlisted sources undoubtedly are. Additionally, the U.S. government limits the resolution of imagery made available to the public to 0.5 m despite the fact that GeoEye-1 has a maximum resolution of 0.41 m. NOAA—National Oceanic and Atmospheric Administration; SIO—Scripps Institute of Oceanography; NGA—National Geospatial Intelligence Agency; GEBCO—General Bathymetric Chart of the Oceans.

sets (Parker, 2011; Butler, 2006). Such ubiquitous use is seen even within the geomorphology community as copious articles have utilized Google Earth imagery to present spatial data (e.g., landform analysis, natural hazards, landscape structure) and/or general overviews of study areas (e.g., Shroder and Weihs, 2010; Zech et al., 2009). Yet, few of the publications to date have actually utilized the imagery available in Google Earth to yield quantitative insights about landscape processes, form, and scaling relationships. Constantine and Dunne (2008) produced one of the first data sets, to our knowledge, originating from high-resolution Google Earth imagery. Using a global data set of channel width and sinuosity measurements derived from Google Earth, they were able to successfully predict the size-frequency distributions of oxbow lakes along a given reach of the Sacramento River, as well as estimate the temporal rate of meander cutoff. To do this analysis, they measured 911 oxbow lakes and 30 channel reaches along meandering streams across the globe: a feat that if done in the field would have required considerable time and resources and if done without high-resolution imagery would have been futile. Likewise, constraints on channel widths along ~45 km of the remote Yarlung Tsangpo River in the Namche-Barwa region of the Himalaya were calculated from coarse Landsat imagery available early on in Google Earth (Finnegan et al., 2008; Finnegan et al., 2005). Due to the coarseness of the imagery, however, considerable errors in each width measurement were inescapable and precluded measurements along narrower tributaries. Most recently, Sato and Harp (2009) utilized Google Earth imagery (8-m Formosat) to rapidly assess the extent and characteristics of landslides associated with the 2008 M7.9 Wenchuan earthquake in China. To our knowledge, these studies represent the only quantitative use of Google Earth imagery in geomorphology, although cross-discipline analyses have included landcover change detection (Asner et al., 2009; Schumacher et al., 2009), digital elevation model improvement (Hoffmann and Winde, 2010), identification of archaeological looting (Contreras, 2010), complex image processing (Guo et al., 2010; Mering et al., 2010), magnetic alignment of resting mammals (Begall et al., 2008), and enhanced geologic mapping (Whitmeyer et al., 2010) to name a few. The underutilization of high-spatial resolution Google Earth imagery within the geomorphology community is the driving force of this paper, with the following three examples highlighting the potential of Google Earth imagery to illuminate previously difficult and unresolved issues within the study of earth surface processes.

3.2. Channel Width Scaling along the Goriganga River, Western Himalaya

3.2.1. Introduction to the Problem

Considerable effort to date has focused on quantitatively deriving relationships between channel geometry and other physical parameters (e.g., discharge, slope, roughness, sediment supply, uplift rate) in tectonically active bedrock rivers (cf. Yanites and Tucker, 2010). Such efforts are important because

they form the basis for how we model, interpret, and emulate river form and erosive potential across broad spatial and temporal scales in mountainous landscapes. With increased accessibility to digital elevation models (DEM) for drainage networks, considerable advances have been made in using channel slopes derived from DEMs to characterize tectonic rates and other metrics of fluvial dynamics (Godard et al., 2010; Wobus et al., 2006b; Kirby and Whipple, 2001; Seeber and Gornitz, 1983). Similarly, with an assumption that rainfall was uniform across a landscape, discharge could be estimated from a DEM as a function of the upstream area. Despite these advances, robust relationships involving channel widths and depths have remained illusive. This ambiguity is mostly due to the difficulty in accessing large spans of river in these regions and the lack of imagery available that can accurately delineate channel margins in narrow gorges and lower order drainage systems. Although collecting robust channel-depth measurements remains extremely difficult in orogenic systems, high-resolution Google Earth imagery now makes accurate channel-width delineation across large spans of river relatively simple.

Early work on channel-width scalings defined simple power-law relationships between width and discharge, whereby channel widths scale with discharge (Q) to some exponent (b) times a constant (a) that is unique to the hydrometeorology of a given region (Leopold and Maddock, 1953).

$$\text{Width} = a*Q^b \qquad (1)$$

In the case of tectonically active bedrock systems, empirical work has described *b* values ranging from ~0.3–0.6 (Kirby and Ouimet, 2011; Yanites et al., 2010; Godard et al., 2010; Whipple, 2004), with 0.5 the most commonly used empirical value (Wohl and David, 2008; Knighton, 1998). Recent work on small catchments (0.6–12.4 km^2) along the Marsyandi River in Nepal argues for the validity of such power-law scalings even in regions that are characterized by large gradients in precipitation (Craddock et al., 2007). Craddock et al.'s study produced a best-fit channel width scaling equation whereby,

$$\text{Width} = 6.2*Q^{0.38} \qquad (2)$$

where Q is the mean monsoonal discharge (m^3 s^{-1}). Whereas these simplistic mathematical descriptions of how the width of a channel may scale provide insights, they neglect many complex interactions that can vastly alter channel widths for any given discharge (i.e., slope, rock strength, roughness, width-to-depth ratio, sediment supply, rock-uplift and erosion rates, etc.).

Acknowledging the pitfalls of the simplistic power-law equations, many field, laboratory, and numerical studies over the last half-decade have attempted to distill such complexities into an all-encompassing equation in the hopes of gaining improved insight about process as well as prediction in natural systems (Yanites and Tucker, 2010; Turowski et al., 2009; Finnegan et al., 2007). The most frequently utilized attempt at predicting channel widths following these tenets uses the Manning equation and principles of mass conservation to produce the following equation:

$$\text{Width} = [\alpha(\alpha+2)^{2/3}]^{3/8}Q^{3/8}S^{-3/16}n^{3/8} \qquad (3)$$

where α is the width-to-depth ratio, Q is discharge (m^3 s^{-1}), S is channel slope (m/m), and n is roughness as defined by Manning's equation (Finnegan et al., 2005; Manning, 1891). Although Equation 3 takes a more thorough approach to predicting channel widths, it is not without its pitfalls. For example, it is difficult to estimate, much less measure, width-to-depth ratios with much reliability or accuracy over large spans of rivers in tectonically active areas. Whereas Finnegan et al. (2005) assumed a constant width to depth ratio along the ~45-km stretch of the Tsangpo River where the equation was validated, many field observations in these environments attest to the wide ranging geometries observed in tectonically active orogens (Fisher et al., 2011; Whittaker et al., 2007; Duvall et al., 2004; Lavé and Avouac, 2001). Likewise, Manning's roughness coefficient is not easily calculated where cyclical landsliding and damming can greatly alter channel bed properties over relatively short length and time scales (Korup and Montgomery, 2008; Korup, 2006; Bookhagen et al., 2005a). In addition, widths derived from the equations above are commonly key inputs for simple physics-based estimates of the geomorphic work performed on the bed of a channel by a given flow (e.g., shear stress, specific stream power). Such proxies are then commonly used to assess a host of topics ranging from tectonic rates and structural boundaries to particle and wood stabilities along fluvial profiles (Attal et al., 2011; Fisher et al., 2010; Thiede et al., 2009; Finnegan et al., 2008; Stock and Montgomery, 1999; Whipple and Tucker, 1999; Bookhagen and Strecker, 2012). Although channel width is only one term in the equation, considerable deviation in the width values can greatly alter observations of spatial variations in erosion and subsequent interpretations.

Whereas power-law and more complex equations (Equations 1–3) can provide first-order estimates of channel widths based on digital topography and a set of assumptions, the dearth of high-resolution width data to date has precluded a proper comparison between these scalings and real world data in tectonically active orogens. Copious work employing coarse satellite imagery to define channel and floodplain widths has been well utilized in large-scale systems, such as Arctic rivers and even along higher order Himalayan streams (Korup and Montgomery, 2008; Smith and Pavelsky, 2008; Lavé and Avouac, 2001). Little is known, however, as to whether these approaches can provide accurate measurements in lower order, bedrock-dominated channel systems as are typical of tectonically active orogens.

In the following analysis, we seek to illuminate both uncertainties in channel-width scalings and the utility of satellite imagery in tectonically active systems by comparing high-resolution Google Earth–derived channel widths along the Goriganga River in northwest India with those derived by freely available Landsat imagery and the previously described scaling equations (Equations 1–3) (Fig. 2). The ultimate goal of this analysis is to showcase the greatly enhanced accuracy and simplicity of utilizing high-resolution Google Earth imagery as compared to power-law scaling parameters or coarser resolution freely available satellite imagery (e.g., Landsat, ASTER).

3.2.2. Methodology

The Goriganga River was chosen as the study river due to its orthogonal orientation to major structural boundaries and lithologies in the Himalaya, as well as the availability of ubiquitous high-resolution imagery in Google Earth. In total, 88 km of the main-stem Goriganga River were digitized (with "terrain mode" off) to the highest water marks observable on both banks using both Google Earth (SPOT™, GeoEye™, and Worldview™) and Landsat imagery (Fig. 2). Channel width polygons were digitized so that polygon boundaries overlaid directly on high water lines, meaning point density was dictated by the characteristics of the channel and not held constant in order to achieve the same width accuracy along the entire study reach. While we are uncertain as to the exact recurrence interval of the floods responsible for these boundaries, we estimate that the digitized banks represent flood events on the order of 2–10 years in this monsoon-dominated catchment based on combined field and remote sensing observations. Due to the coarseness of the Landsat imagery (30 × 30 m; we did not pan-sharpen the images) only pixels exhibiting the spectral characteristics of water or exposed bars in the false-color image (bands 5,4,3) were included in the channel width mask (10/20/2001 image date). Landsat channel widths were hand-digitized in the same manner as the Google Earth imagery using ENVI software because of the inability of supervised and non-supervised classification techniques to yield a continuous, reliable data set for use with the channel-width extraction algorithm. Channel polygons were then exported, rasterized to 5 × 5 m resolutions, and manipulated using proprietary image-processing algorithms to yield channel half-widths, which were then doubled to produce a channel width at each centerline pixel (Fig. 3). Channel widths were then merged with 30 m apparent resolution Advanced Spaceborne Thermal and Reflectance Radiometer Global Digital Elevation Model (ASTER GDEM v1—property of METI and NASA) topographic data (elevation, slope, etc.), as well as mean annual discharges derived from a decade long rainfall-snowmelt model for the area (Bookhagen and Burbank, 2010). In order to assess the efficacy of the channel-width extraction algorithm, as well as to compare with the Landsat-derived channel widths, 35 hand-measured widths were taken in Google Earth along the study reach spanning the range of widths observed (Fig. 4).

3.2.3. Results

Comparison between the algorithm-based Google Earth widths and the hand-measured widths in Google Earth appear to be highly congruent ($r^2 = 0.98$) giving us confidence that the algorithm used for extracting channel widths is highly accurate (Figs.

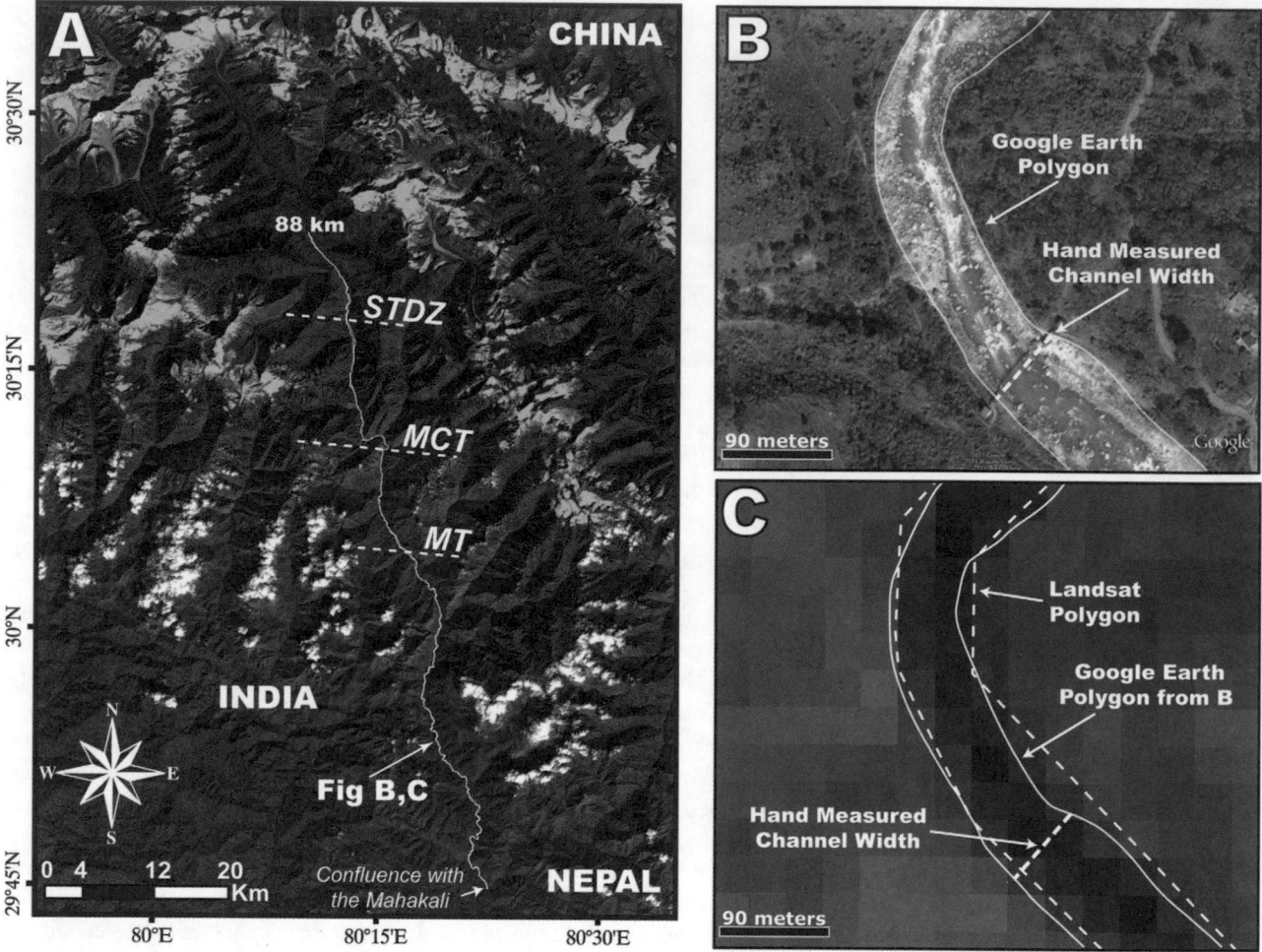

Figure 2. (A) False color (bands 5,4,3) Landsat 7 ETM+ image of the Goriganga River catchment with major structural boundaries (dashed lines). Channel widths were calculated from the confluence with the Mahakali River to 88 km upstream (white line). (B) Geo-Eye™ image (~0.5 × 0.5 m resolution) of the Goriganga River taken from a height of 388 m showing easily distinguishable channel margins and textural variability down to individual boulders within the channel bed. (C) Landsat 7 ETM + image (30 × 30 m resolution) of the same area as B, but ~3600 times lower spatial resolution. Note the difficulty in distinguishing channel margins due to low spatial resolution data and spectral mixing within individual pixels. STDZ—South Tibetan Detachment Zone; MCT—Main Central Thrust; MT—Munsyari Thrust.

3 and 4B). Whereas we don't have field measurements against which to compare the hand measured widths, both previous field research and other studies assessing the positional accuracy of Google Earth imagery argues for strong correlation between real world distances and those measured on Google Earth imagery (Constantine and Dunne, 2008; Potere, 2008).

A comparison between Landsat (30 × 30 m) and Google Earth–derived channel widths shows a continuous offset along the Goriganga River, whereby Landsat widths are consistently over-estimated along the study reach (mean and 1 standard deviation of 18.2 ± 7.4 m) (Fig. 4). When compared to the hand-measured widths from Google Earth, the importance of imagery resolution becomes readily apparent, with channel widths less than 1-pixel length nearly impossible to detect accurately (Fig. 4). With the high-resolution imagery in Google Earth, channel-width delinea-

tion appears to be dictated completely by the raster resolution (which in this case is 5 × 5 m) in these tectonically active systems where riparian vegetation is minimal. Theoretically, using a 1 × 1 m matrix with the ~1 m spatial-resolution GeoEye™ imagery could further improve the channel-width accuracy. For the sake of computational efficiency and to encompass some of the error introduced by digitizing the channel widths, however, a conservative 5 m resolution was used. Nonetheless, the improvement over Landsat-based techniques is both striking and necessary to accurately assess channel geometries in tectonically active and/or lower order channel systems (<60 m wide or drainage areas less than ~3000 km^2) where narrow, steep gorges are the norm, rather than the exception.

A closer look at the Google Earth channel-width data along the nearly 90-km-long study reach shows great variability in

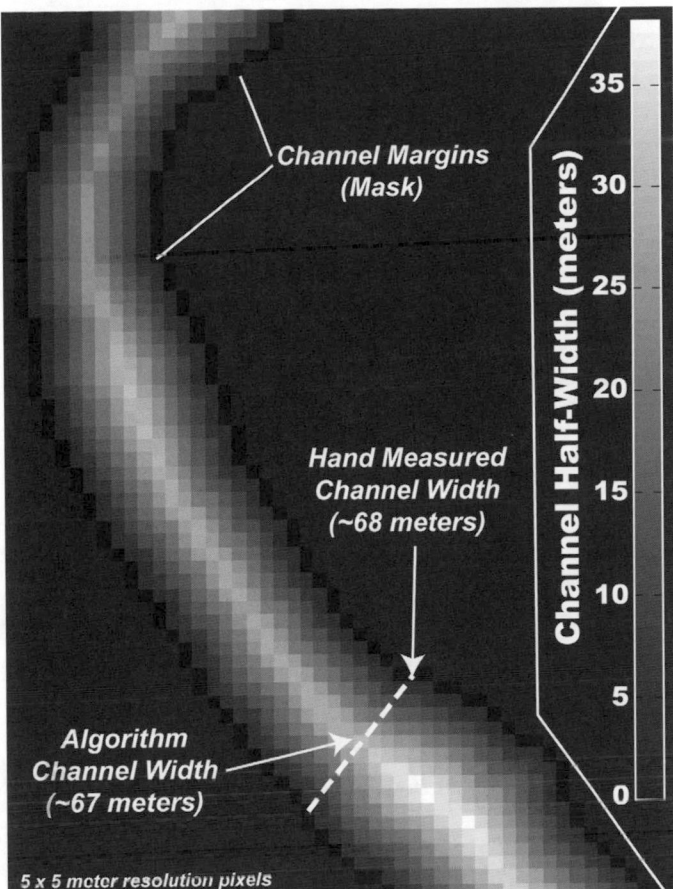

Figure 3. Example of the spatial output from the channel-width extraction algorithm used in the analysis, consistent with the river bend shown in Figure 2B and 2C. Widths are derived from a channel mask digitized in Google Earth and then rasterized to a pixel resolution of 5 × 5 m. This resolution is chosen because it is computationally efficient, yet also precise as is shown by the good agreement between hand-measured widths and those from the algorithm (see Figure 4B for all of the data).

widths that is coincident with major Himalayan tectonic units (Fig. 4 and 5). This variability makes it impossible to fit a simple power-law scaling (e.g., Craddock et al., 2007) to the entire data set that even broadly mimics the large-scale variability in channel widths, much less any small-scale perturbations (Fig. 5). Although more complex scaling relationships (Finnegan et al., 2005) may do a slightly better job, these equations include many more unknowns that must be accounted for, which in remote locations are nearly impossible to accurately assess and commonly vary greatly between sub-reaches of the river (e.g., width-to-depth ratios, roughness). The departure of these scaling laws from the actual channel widths derived using Google Earth imagery can be striking (Fig. 5). Both established power-law (Craddock et al., 2007) and mass-conservation equations (Finnegan et al., 2005) systematically under-predict the actual channel widths along the study reach, especially where the mass-conservation equation of Finnegan et al. (2005) is defined as a bedrock chan-

nel with a width-to-depth ratio of 5 (Fig. 5B). The Finnegan et al. model improves when the width-to-depth ratios are arbitrarily adjusted higher ($\alpha = 21$), consistent with a decrease in grain size (e.g., bedrock to cobble) (Finnegan et al., 2005); however, field evidence and high-resolution imagery indicate such an adjustment is unwarranted and grossly underestimates the dominant caliber of material in the channel. Furthermore, in our study, channel roughness is kept constant for simplicity ($n = 0.20$) but any reduction in grain size should coincide with a decrease in roughness, yielding even lower values and poorer agreement than is shown in this study for the width-to-depth value of 21 (blue line: Fig. 5B). These results indicate that, in many active orogenic rivers, channel-width data are greatly oversimplified by using previously established scaling laws, potentially losing invaluable information from the fluvial network about both tectonic rates and geomorphic processes affecting these landscapes.

3.2.4. Implications and Conclusions

The Google Earth–derived channel width data set presented here provides one of the most comprehensive channel-width data sets from a tectonically active orogen to date. Owing to the high-resolution data made possible by Google Earth imagery, comparisons of established scaling equations (Equations 1–3) and coarser freely available satellite imagery (Landsat) fail to adequately represent trends in channel width along the Goriganga River in northern India. Distinct channel-width domains exist within long-established Himalayan tectonic units, yet widths can vary by several folds within individual units, thereby indicating distinct channel-width responses to tectonic and geomorphic processes or characteristics (glaciation, rock types, uplift rate, etc.) within active orogenic environments (Figs. 4 and 5). These pronounced width variations serve geomorphologists as a roadmap by identifying anomalous river reaches (such as in the high Himalayan crystalline unit) where some fundamental controls are strongly changing within a restricted spatial domain. Whereas this data set leaves many questions unanswered, the availability of high-resolution Google Earth imagery and the extraction methodology presented here is sure to yield future insights about: (1) how channel geometries respond to a host of geomorphic, climatic, and tectonic forcings (Fisher et al., 2011; Yanites and Tucker, 2010; Stark et al., 2009; Wobus et al., 2006a); (2) what channel width thresholds exist and why (Fisher et al., 2011; Yanites et al., 2010; Amos and Burbank, 2007); and (3) how estimates of fluvial power can be refined using actual channel widths to better illuminate foci of erosion, transport, and incision across a host of fluvially dominated landscapes (Fisher et al., 2010; Finnegan et al., 2005; Stock and Montgomery, 1999; Magilligan, 1992).

3.3. Hurricanes, Earthquakes, and Landslides: Hillslope Response to High-Magnitude Events in Haiti

3.3.1. Introduction to the Problem

Whereas fluvial systems act as the veins of the continents transporting sediment, nutrients, and pollutants from land to

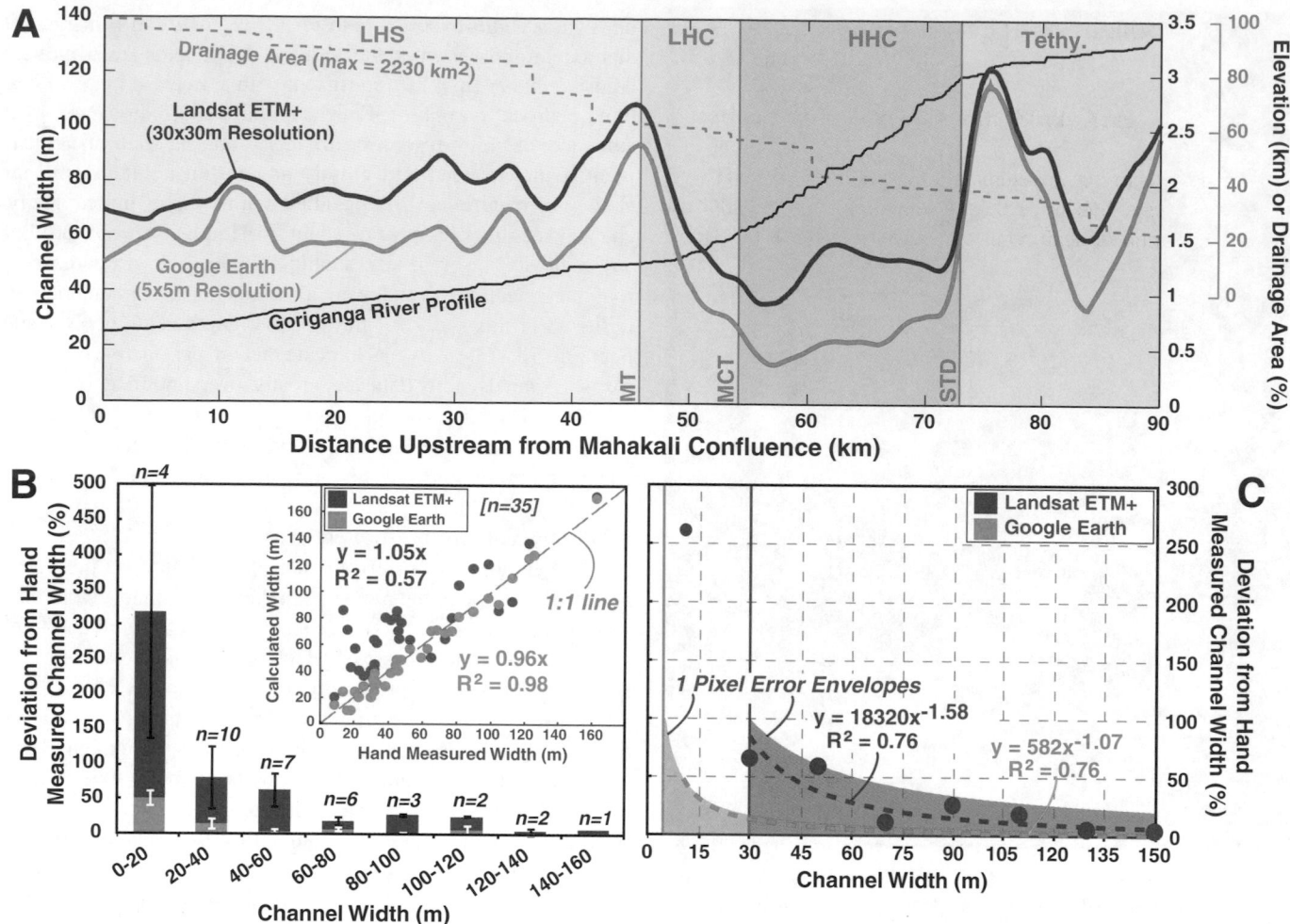

Figure 4. (A) Plot of channel width (after 5-km smoothing window) versus distance from the confluence with the Mahakali River showing the general overestimation of channel widths derived from Landsat imagery in steep, narrow gorges within tectonically active regions. Longitudinal profile and drainage area are plotted on alternate Y-axes for reference. (B) The percent deviation from hand-measured channel widths (see Fig. 2B, 2C, and Fig. 3) and fit for both data sets (see inset). (C) Fits for the average values from B (dashed lines and red points for Landsat) and the 1-pixel error envelopes (shaded regions). Note the low error associated with Google Earth (Spot™, GeoEye™, and Worldview™ imagery) widths versus Landsat ETM + (which provides little information below 1 pixel and still has ~50% error at 45 m channel widths) as shown by low percent deviation as well as high correlation in the inset plot. STD—South Tibetan Detachment Zone; MCT—Main Central Thrust; MT— Munsyari Thrust; LHS—Lesser Himalayan Sedimentary; LHC—Lesser Himalayan Crystalline; HHC—High Himalayan Crystalline; Tethy— Tethyan Series.

sea (Milliman and Syvitski, 1992), mass wasting is the dominant mechanism by which particles enter the fluvial network, especially in tectonically active regions. In moderate- to high-relief terrains, landsliding has been shown to play a critical role in landscape evolution through mass transfer related to variable tectonic rates (Clarke and Burbank, 2010, 2011; Densmore et al., 1997; Hovius et al., 1997; Burbank et al., 1996), climate perturbations (Galewsky et al., 2006; Bookhagen et al., 2005b; Gabet et al., 2004), anthropogenic effects (Lavé and Burbank, 2004), fire regimes (Roering and Gerber, 2005; Pierce et al., 2004), and increased seismicity (Parker et al., 2011; Hovius et al., 2011; Dadson et al., 2004; Hovius et al., 2000). Alternatively, landsliding has been argued to retard regressive erosion by large rivers

along the Tibetan Plateau margin (Korup et al., 2010; Korup and Montgomery, 2008) as well as shield certain glaciers from enhanced melting related to climate change by providing a protective surficial debris cover (Scherler et al., 2011a, 2011b; Santamaria Tovar et al., 2008). Because landsliding is such a diverse and integral process in landscape evolution, it is imperative to be able to identify, delineate, and develop mechanistic explanations for landsliding across broad spatial and temporal scales. Most landslide analyses to date have suffered from having spatially limited, yet high-resolution aerial photo sets (Clarke and Burbank, 2010) or spatially extensive, yet low-resolution remotely sensed optical imagery (Korup et al., 2010). Furthermore, many of these studies have been further hindered by the lack of

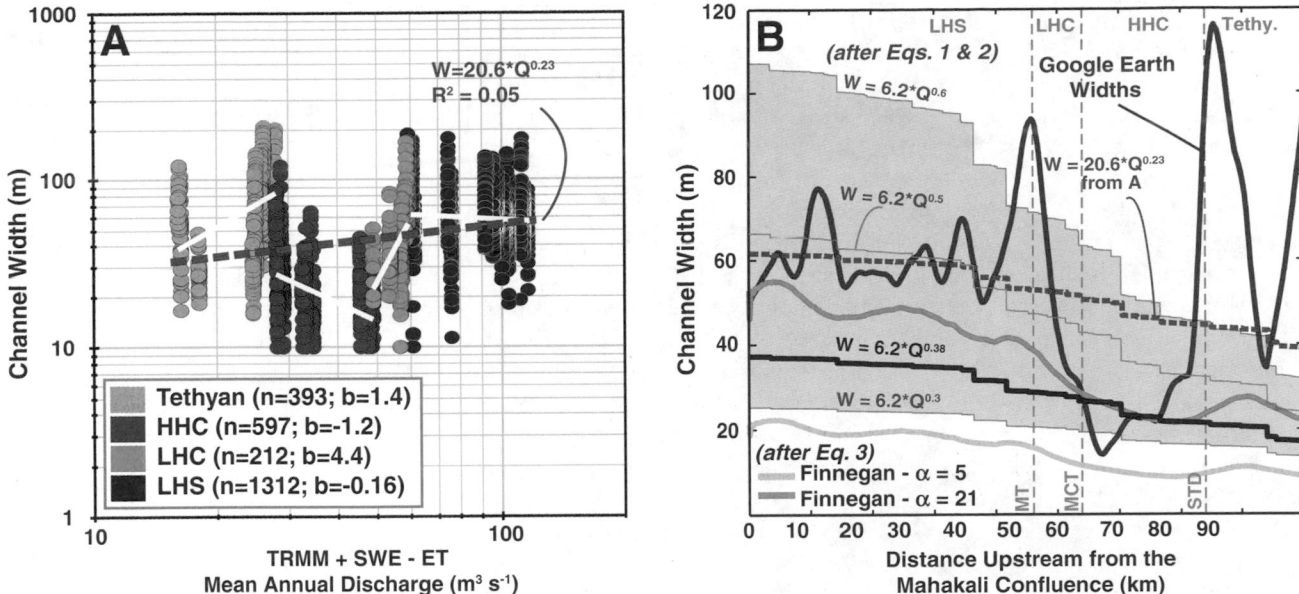

Figure 5. (A) Channel width versus mean annual discharge (for all data) for the Goriganga River with distinct tectonic units colored consistent with Figure 4A. Power-law fits are generally poor even when divided into different tectonic regimes with overall fit shown by the dashed gray line. NOTE: Log binning the data made little difference in the overall correlation and exponent of the regression. (B) Channel widths from our analysis compared to simple power-law scalings with discharge (W ~aQ^b) using calibrated width-discharge relationship from the Marsyandi region of Nepal (W = 6.2Q$^{0.38}$) (Craddock et al. 2007) as well as b values of 0.3, 0.5, and 0.6, which define the typical range found in the literature. Additionally, we plot the more complex model of Finnegan et al. (2005) (Equation 3) with two different width to depth (α) ratios and a constant Manning's n of 0.20. In general, scaling relationships only qualitatively capture the actual width distribution obtained from high-resolution Google Earth imagery, with scaling relationships further complicated by the lack of information about specific scaling components (a, b, α, Manning's n, discharge, etc.) in many tectonically active regions. The lack of parameter constraints forces researchers to arbitrarily assign values with little physical basis in the geomorphology and hydrometeorology of the area and, as shown by the Google Earth–derived width values, fails to adequately represent actual channel width trends. Weighted mean annual discharge (Q)(m^3 s^{-1}) was calculated using the rainfall-snowmelt model from Bookhagen and Burbank (2010). LHS—Lesser Himalayan Sedimentary; LHC—Lesser Himalayan Crystalline; HHC—High Himalayan Crystalline; Tethy—Tethyan Series.

temporal variability in imagery, constraining analyses and interpretations alike. Google Earth has, however, created a platform where resolution, spatial extent, and even temporal issues with imagery are greatly diminished, thereby allowing researchers to more thoroughly explore the mechanisms, characteristics, and linkages associated with landsliding across a range of environments and triggering mechanisms.

In the following analysis we utilize high-resolution Google Earth imagery portraying the area around Port-au-Prince, Haiti, to illuminate landslide-triggering mechanisms and characteristics. Haiti provides a prime opportunity to present the utility of not only recent high-resolution Google Earth imagery, but also of the rich historical image archive available for the region. The 2008 hurricane season battered Haiti with four major storms producing nearly a meter of rain over less than a month-long period. These storms caused nearly 800 deaths and billions of U.S. dollars in damages (Carroll, 2001) (Fig. 6). On 12 January 2010 Haiti experienced yet another major blow with a magnitude 7.0 earthquake located ~15 km west of the capital city of Port-au-Prince (Hayes et al., 2010) (Fig. 7). This event represented the largest earthquake to strike the area in more than 200 years and led to deaths estimated in the tens of thousands

(Associated Press, "Report challenges Haiti earthquake death toll," BBC News, retrieved 25 June 2011, http://www.bbc.co.uk/news/world-us-canada-13606720). Whereas these events are tragic, they present an unparalleled opportunity to compare and contrast the hillslope mass-wasting response to both high-magnitude climatic and seismic events. The goal of this analysis is, therefore, to gain improved insight about landslide triggers and characteristics, as well as to show the utility of Google Earth in such research ventures.

3.3.2. Methodology

The study area was chosen due to its proximity to the epicenter of the 12 January 2010 earthquake and the availability of clear, high-resolution imagery spanning both the pre- and post-2008 hurricane season, as well as imagery taken the day after the earthquake event clearly showing a large number of landslides (Fig. 7). Hurricane path and cumulative event rainfall data were obtained from the National Weather Service–National Hurricane Center (www.nhc.noaa.gov) and compared with the Tropical Rainfall Monitoring Mission (TRMM) 3B42 V6 data set for the area over the same time period (Fig. 6)(cf. Bookhagen, 2010; Huffman et al., 2007; Hong et al., 2006). Landslides were delin-

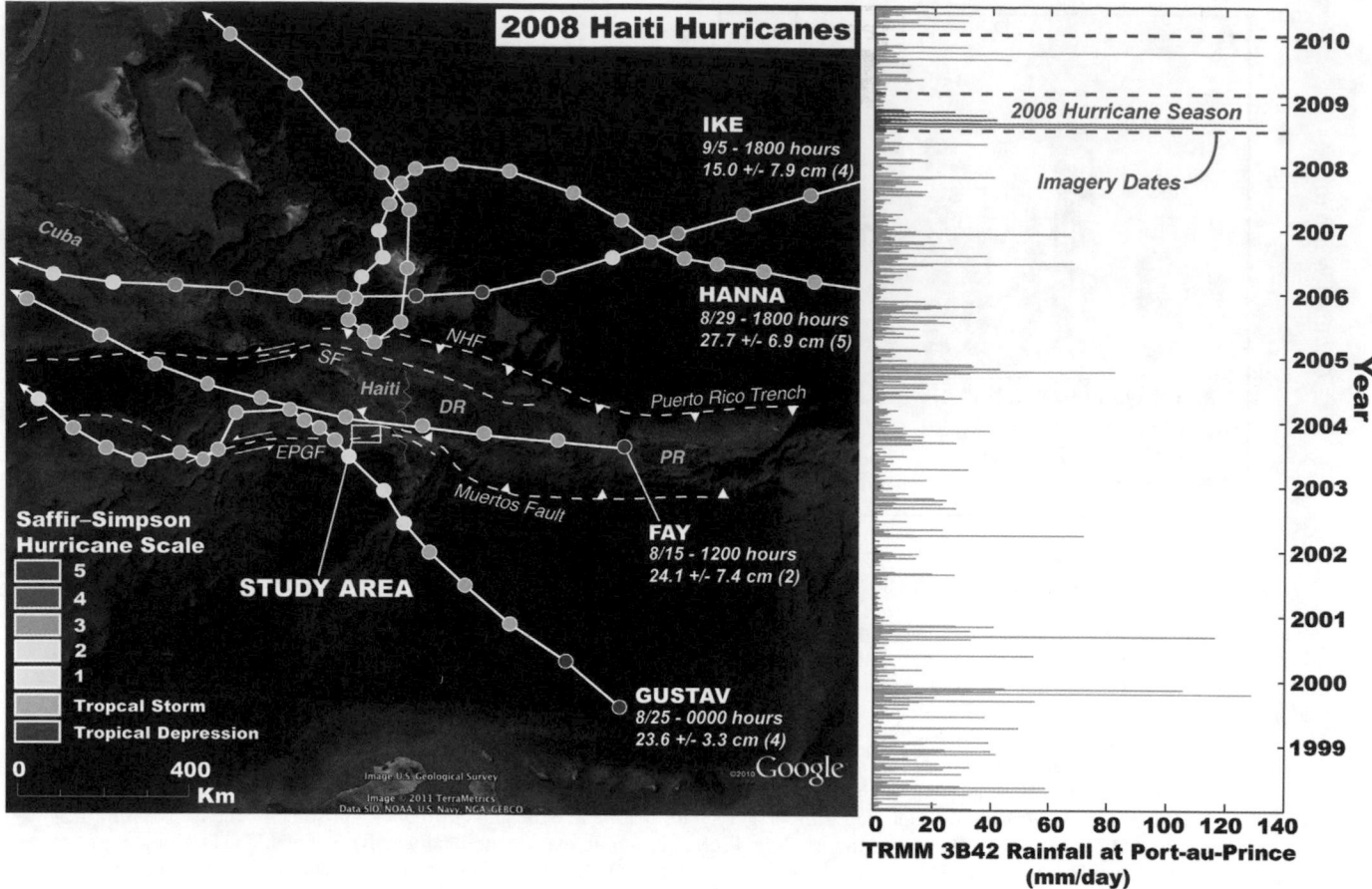

Figure 6. The 2008 hurricanes that heavily impacted Haiti and the study area. Tracks and intensities along with a tectonic overview of the Caribbean region are overlain on NASA Blue Marble: Next Generation imagery (http://neo.sci.gsfc.nasa.gov/). Total rainfall was derived from gauge stations within 150-km radius of the study area with the number of stations in parentheses. This data is biased toward the Dominican Republic side, as only one gauge in Haiti reported data for these events. Date and time are of the initiation of the storm, or in the case of Ike and Hanna, the first circle on the figure. Circles are shown for every 6 h. The plot adjacent shows the daily rainfall for Port-au-Prince, Haiti (15 km from the study area) derived from TRMM 3B42 data set collected from 1998 to 2010, with the 2008 hurricane season (pink) the biggest in ~10 years in terms of daily rainfall (Bookhagen, 2010). Hurricane data provided by the National Weather Service National Hurricane Center (www.nhc.noaa.gov). NHF—North Hispanola Fault; SF—Septentrional Fault; EPGF—Enriquillo–Plantain Garden Fault; PR—Puerto Rico; DR—Dominican Republic.

eated and digitized (with "terrain mode" off) on three images in Google Earth (Fig. 7), then exported, and merged with the 30 m apparent resolution ASTER GDEM topographic data set to derive slope angles for the study area. Proper positional accuracy between the study area and the ASTER GDEM was ensured using landscape features (e.g., hydrologic features, terrace edges), as large discrepancies will undoubtedly confound results. One potential reason for a close match in our region between the two is the availability of airborne lidar data postdating the earthquake which Google may use to orthorectify the imagery; however, this is mere speculation (cf. www.opentopography.org).

3.3.3. Results

Despite ideal conditions (namely heavy deforestation and steep slopes) for landsliding in the study area leading up to the 2008 hurricane season, only 7 slides were delineated (Fig. 7). In contrast, the 2010 magnitude 7 earthquake produced 325 land-

slides in our study area ranging in area from 10 m^2 to 27,000 m^2 (Figs. 7A, 7C, and 8). Hillslopes are generally planar to slightly convex in the study area, with landslides predominately limited to the lower, steeper toe slopes ($\geq$20 degrees) (Figs. 7 and 8). Analysis of magnitude-frequency plots for the earthquake-derived landslide data set produces a characteristic double-pareto distribution with a similar beta slope value (cf. Stark and Hovius, 2001) to previous studies, despite integrating only one event and having roughly an order of magnitude fewer mapped landslides (325 versus 1000s by Clarke and Burbank, 2010; Hovius et al., 1997)(Fig. 8B). The beta value is derived from the simple power-law function,

$$N_L = kA^{-\beta} \tag{4}$$

where N_L is the number of landslides for a given landslide area A, k is a scaling coefficient, and β defines the slope of the magnitude-frequency relationship for the log-log linear segment

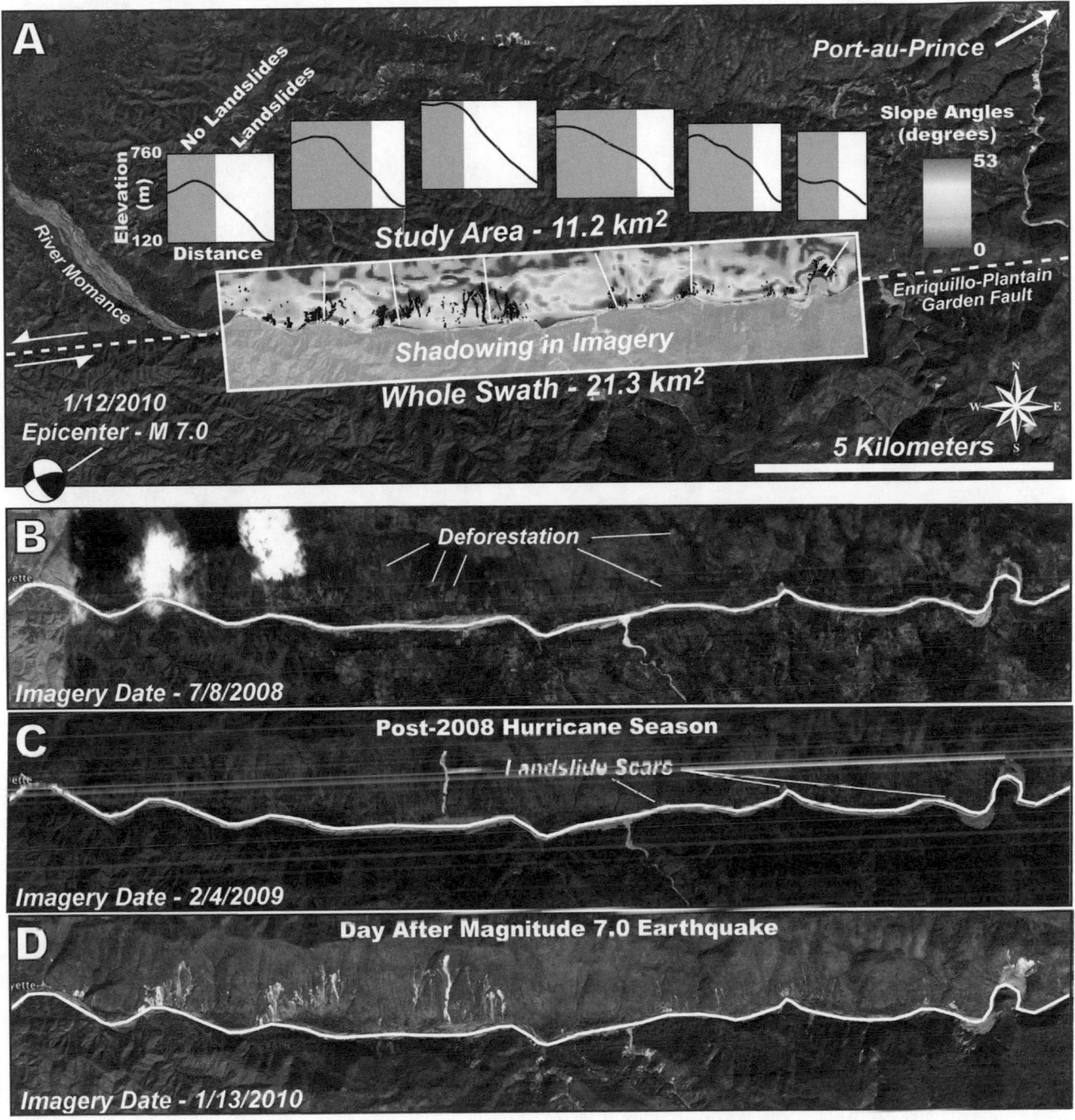

Figure 7. (A) Study area located ~15 km WSW from Port-au-Prince, Haiti. Epicenter, centroid moment tensor solution (http://earthquake.usgs.gov/), and main strike-slip system shown. Study area was reduced to north of the River Momance due to shadowing that made it difficult to accurately delineate landslides in the lower half of the swath. Delineated hurricane- and earthquake-induced landslides (black) are shown on top of a slope angle map for the study area with 6 representative elevation profiles (white lines) shown above. (B) Worldview™ image predating the 2008 hurricane season and showing extensive patchiness along the hillside from deforestation. (C) Worldview™ image taken post-dating the 2008 hurricane season and showing very limited hillslope response (*n* = 7) with only one sizeable failure in the middle of the image. (D) GeoEye™ image taken the day after the 2010 magnitude 7 earthquake showing extensive hillslope failures in the study area. The study area experienced severe shaking with peak accelerations between 36% and 45% gravity (http://earthquake.usgs.gov/).

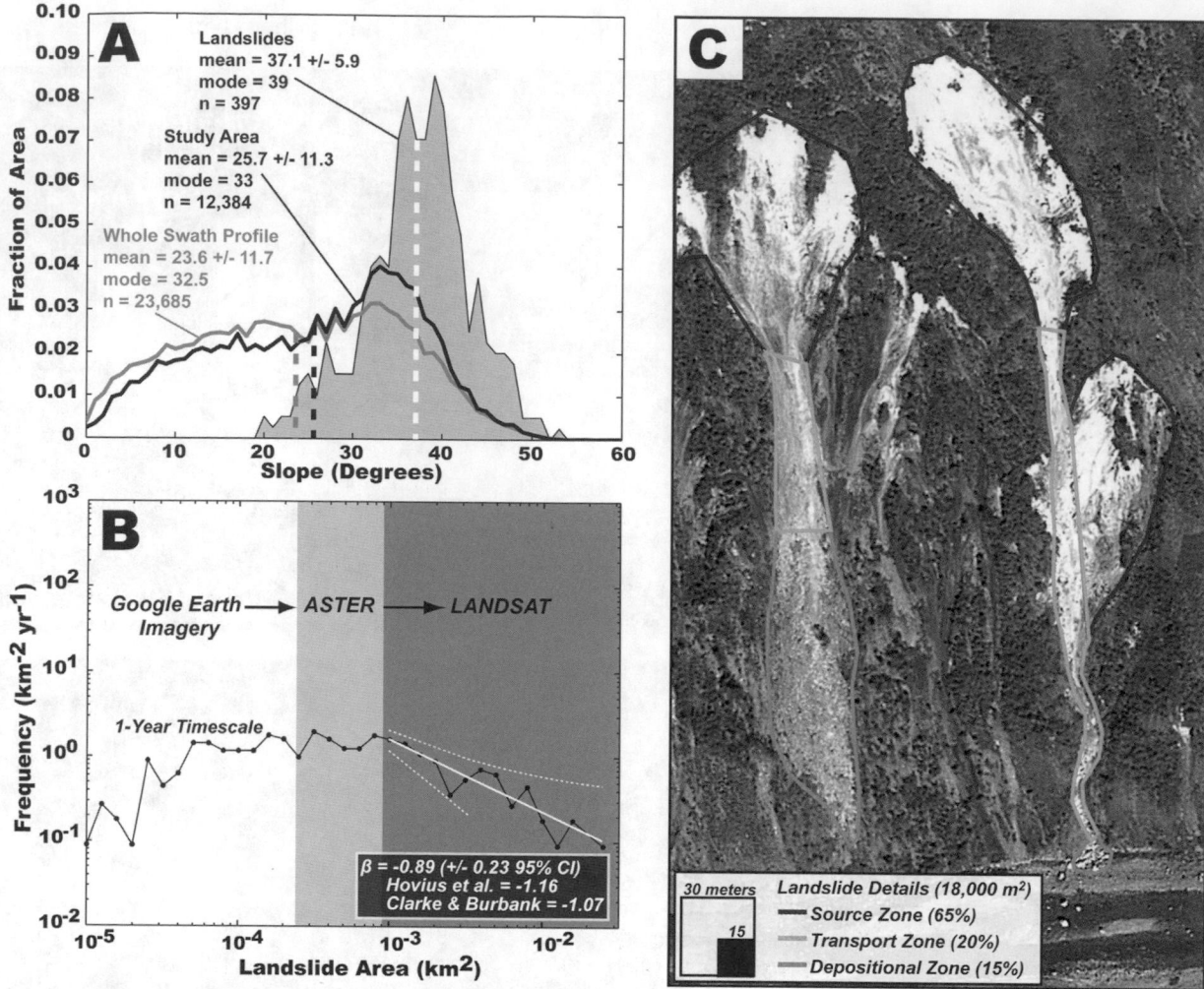

Figure 8. (A) Slope analysis (using ASTER GDEM) of those studied areas in Haiti that experienced landsliding following the 2010 magnitude 7 earthquake. Data is in 1-degree bins. All landslides are confined to areas with 20 degree slopes or greater. Dashed lines indicate mean slope values for each distribution. (B) Log-binned landslide magnitude-frequency plots showing the improved delineation of small landslides using high-resolution Google Earth imagery over coarser satellite data. Gray shading represents 1-pixel for both ASTER (225 m²—light gray) and Landsat imagery (900 m²—dark gray), the minimum threshold necessary to accurately record a reflectivity change related to landsliding. Note the beta value (slope of the regression) is (Equation 4) in general agreement with previous studies despite mapping only one event and considerably fewer landslides (325 in this study, 2211 in Clarke and Burbank, 2010, and 7691 in Hovius et al., 1997). (C) Image of two earthquake-induced landslides in the study area showing the detailed morphology of the landslides and comparison with one pixel resolution for Landsat (30 × 30 m) and ASTER (15 × 15 m) imagery.

of the landslide distribution (right side: Fig. 8B) (Hovius et al., 1997). The 7 landslides from the 2008 hurricane season were not included in Figure 8 because all but one of the landslides associated with that event were located along the banks of the river, indicating that the failures were likely caused by main-stem undercutting from flooding, and not excess pore water pressure. Due to the removal of storm-driven landslides in our magnitude-frequency analysis, the time component (generally set by the temporal range of imagery) for the earthquake-derived landslides could be considered instantaneous because it is based on only one image. For simplification we integrate over one year, which approximately accounts for the time between the last two images

in our time series (Fig. 8). When compared with coarser satellite imagery, nearly half the data falls below the resolution of both Landsat and ASTER pixel resolutions, signifying the inability of these imagery resolutions to properly identify landslides (without complex spectral mixing models) with areal extents less than 900 m² and 225 m², respectively. In addition, such coarse resolution imagery may also preclude proper delineation of the frequency apex and distribution, yielding an inaccurate power law fit (Equation 4) as well as distorting any derivations from the distribution, such as estimates of erosion rate (Stark and Hovius, 2001). Lastly, higher-resolution imagery allows the proper delineation of landslide morphologies, which oftentimes cannot be distinguished using coarser imagery and may considerably overestimate landslide area and sediment volumes in many steep terrains (Fig. 8C).

3.3.4. Implications and Conclusions

Despite hydrological evidence for intense rainfall during the 2008 hurricane season and heavy deforestation only one major failure occurred in the study area. Possible explanations for the lack of failures may include considerable overland flow due to steep slopes, lithologic and soil properties (i.e., clay rich), lack of a critical accumulation zone, localized orographic effects, and/or the low relative magnitude of the event (~10-year recurrence interval). The story is quite different following the 12 January 2010 earthquake where 325 landslides were observed. Previous work has proposed a topographic signature exists within tectonically active landscapes to distinguish between storm- and earthquake-derived landslide-dominated hillslopes (Meunier et al., 2008; Densmore and Hovius, 2000). In the Haiti study area, such simple delineations are muddled. We know that the landslides were generated coseismically, yet many of them were triggered along the lower, steeper toe slopes, consistent with a storm-induced, hydrologic accumulation model (Fig. 7A). At the same time, hillslopes are generally planar and inner gorges absent as would be predicted by a landscape dominated by earthquake-induced landsliding (Densmore and Hovius, 2000). Whereas both earthquake and storm processes are clearly active in the study area, the question becomes at what time frame and with what frequency each is dominant (e.g., Wolman and Miller, 1960). The observed earthquake event is the largest to strike the area in over two centuries, but during this same time interval, there have been numerous hurricane and storm events of greater magnitude than the 2008 hurricane season. We would expect those large storms to have caused considerable landsliding. One explanation for the lack of a landslide response to the repeated storms is the mismatch of magnitudes between the climatic and seismic events observed, whereby the heavily deforested hillslopes may already be adjusted to such climatic recurrence intervals. In all likelihood any climatic event of a similar magnitude post-dating the earthquake would exploit current earthquake-induced landslide scars, exacerbating the total volume of sediment removed from the hillslopes and providing the necessary transport mechanism to move the hillslope material into the fluvial network (Hovius et al., 2000, 2011). Another possible explanation as to why there is such a discrepancy between the hillslope response to the 2010 earthquake and 2008 hurricanes in the study area is that of the rupture style (Meunier et al., 2008). Both remote sensing observations following the 2010 earthquake and our own reconnaissance in Google Earth noted fairly localized hillslope failures concentrated in the study area (van Westen and Gorum, 2010). The first fault plane solutions to come out following the earthquake placed it as an oblique slip failure consistent with the preexisting Enriquillo–Plantain Garden Fault (EPGF). However, subsequent InSAR (Interferometric Synthetic Aperture Radar) analyses have hypothesized that the rupture may have occurred along a previously unmapped blind

thrust fault (hanging wall moved due south) that would daylight adjacent to and strike-parallel to the study area ridge (Hashimoto et al., 2011; Calais et al., 2010; Hayes et al., 2010). This geometry would then explain the relatively localized failures, where much of the energy was directly focused into the study area causing significant, localized failures on the south-facing hillslopes.

This study again underscores the complexities involved in deciphering processes and mechanisms that shape the surface of our earth. In the case of Haiti, Google Earth gives us an additional tool with which to glimpse how the landscape responds to both climatic and tectonic perturbations of varying magnitudes. Obviously, detailed field research would be needed to properly tackle the complex interactions actively shaping the landscape in this subtropical environment. Nonetheless, Google Earth imagery can provide an important starting point by making high-resolution imagery accessible, at both the temporal and spatial scales necessary to adequately undertake such questions. In this study, we have documented a time series of landsliding in Haiti using Google Earth high-resolution imagery that spans both high-magnitude climatic and tectonic events, providing valuable insights about the characteristics of landslides associated with these events, their potential mechanisms, and how high-resolution Google Earth imagery may benefit similar studies in the future.

3.4. Fault Characterization in the Eastern California Shear Zone

3.4.1. Introduction to the Problem

The eastern California shear zone and Walker Lane comprise a distributed network of dextral strike-slip and normal faults east of the Sierra Nevada in California (Fig. 9). Together, this zone of active deformation accounts for roughly one quarter of the total Pacific–North American plate translation (Dokka and Travis, 1990). An ever-expanding inventory of information on active faults within the eastern California shear zone and Walker Lane elucidates spatial and temporal patterns of strain along these structures. Comparison of these patterns with ongoing deformation measured from space geodesy (e.g., Dixon et al., 2000) and seismicity (e.g., Unruh et al., 2003) provides a unique opportunity to investigate the dynamics and evolution of this intracontinental plate-boundary fault system.

Fault studies in the eastern California shear zone typically rely on measurements of displaced Quaternary landforms and geomorphic features such as river channels, terraces, or alluvial fans to define patterns of fault displacement at a variety of spatial and temporal scales (e.g., Frankel et al., 2011). An intriguing outcome from this body of work is the apparent, twofold mismatch between the summed rates of dextral fault slip across the Mojave section of the eastern California shear zone (~6 ± 2 mm/yr, Oskin et al., 2008) and interseismic strain measured from GPS (~12 ± 2 mm/yr, Sauber et al., 1994). Individual structures within this zone, such as the Blackwater fault, also show pronounced rate discrepancies. There, dextral shear measured at decadal time scales from radar interferometry (up to ~7 mm/yr, Peltzer et al.,

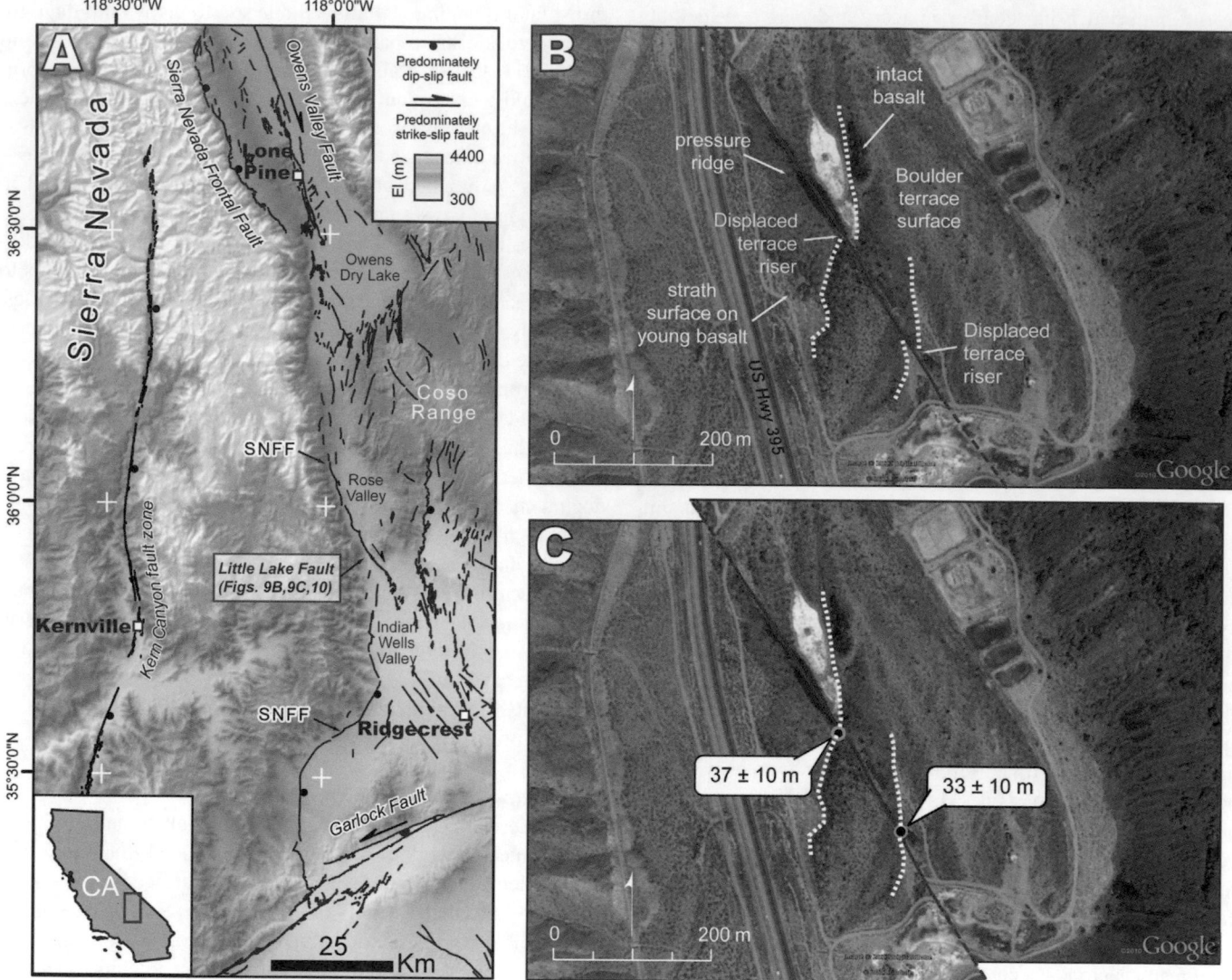

Figure 9. (A) Regional tectonic map of the southeastern Sierra Nevada of California. Active fault traces are taken from Jennings (1994), with the exception of the Kern Canyon fault (Brossy et al., 2012). SNFF—Sierra Nevada frontal fault. (B) Google Earth image of Little Lake wash (shown by arrow in A) highlighting neotectonic and geomorphic features associated with the Little Lake fault. (C) Reconstructed offset along the Little Lake fault suggests between ~33 and 37 m of dextral displacement of terrace risers cut by the fault zone.

2001) radically outpaces Quaternary-averaged rates of fault slip (~0.5 mm/yr, Oskin and Iriondo, 2004). Based on a lack of identified and described Quaternary offset landforms, comparatively little is known about how this dextral shear in the eastern California shear zone is transferred and partitioned onto structures north of the Garlock fault in the Indian Wells and Rose Valley areas (Fig. 9).

Here, we present new measurements of displaced Quaternary landforms cut by the Little Lake fault and identified using high-resolution Google Earth imagery. These images reveal a series of previously undescribed river terraces formed during the drainage of pluvial Owens Lake southward from Rose Valley into the Indian Wells Valley and China Lake basin (Fig. 9). Measured

lateral offsets of these features using Google Earth imagery compares favorably with measurements made from ground-based terrestrial laser scanning (TLS) along the Little Lake fault as part of this study. This work demonstrates the utility of Google Earth for rapidly identifying and quantifying patterns of lateral fault displacement in actively deforming landscapes.

3.4.2. Methodology

Numerous studies utilize remotely sensed imagery in arid regions for cataloging lateral offsets along strike-slip faults (e.g., Klinger et al., 2011) and for documenting the sense and style of displacement for under-characterized geologic structures (e.g., Phillips and Majkowski, 2011). Our study focuses on the western margin of the eastern California shear zone between the Sierra Nevada and the Coso Range (Fig. 9), along the northwestern continuation of interseismic dextral shearing observed from radar interferometry (Peltzer et al., 2001). Here, abundant Quaternary volcanic and geomorphic features provide markers for quantifying fault offset.

Along the Little Lake fault (Fig. 9), we used Google Earth imagery to identify a series of prominent river terraces cut by the fault immediately east of U.S. Highway 395 (Fig. 9B). These landforms are undescribed in the literature despite a well-known relative sequence of late-Pleistocene volcanism and fluvial downcutting driven by drainage of Owens Lake through the Little Lake area (Duffield and Smith, 1978). Images of the Little Lake wash collected from Google Earth were used to create a reconnaissance-level neotectonic map (Fig. 9B) and also to reconstruct the total dextral offset of this landform across the fault (Fig. 9C). These reconstructions are based on retro deformation of the mapped image, sliced along the fault trace, in order to create a "best fit" for the base of the terrace riser on either side of the fault. The riser base was chosen for these reconstructions because it provides the best visual contrast in the imagery. In order to minimize errors in our reconstruction associated with image distortion, screen captures were collected with Google Earth's "terrain" feature disengaged.

Based on this reconnaissance, we conducted field checks along the Little Lake fault to verify our geologic interpretations and to conduct a TLS survey for comparison with displacement measurements from the Google Earth imagery. TLS surveying utilized a Riegl LMS-Z420i ground-based lidar system and resulted in collection of ~22 million individual laser returns at an average density of 20 points/m². Non-ground returns from sparse brush covering the area were filtered using the Terrasolid software package, and the remaining returns were triangulated to create an equally spaced digital elevation model (DEM) at a nominal resolution of 50 cm (Fig. 10A) (cf. Perroy et al., 2010). Elevation profiles along terrace-riser margins (Fig. 10B) were extracted from the resulting DEM at a 2-m spacing and projected along an azimuth perpendicular to the average local orientation of the terrace riser (015°) on either side of the fault. This group of projected profiles were then rotated 45° clockwise onto a plane normal to the locally averaged fault strike (330°) in order to reconstruct the total dextral displacement of this feature. This reconstruction utilized the distance range in topographic midpoints from the scarp profiles (Fig. 10B), based on the assumption that the scarp midpoint undergoes the least amount of vertical change during scarp diffusion (Pelletier et al., 2006).

3.4.3. Results

Image reconstruction of offset terrace risers along the Little Lake fault suggest between 33 and 37 m of total dextral displacement of the eastern and western margins of this landform (Fig. 9C). Using the line measure tool in Google Earth for each offset riser yields similar estimates and supports the validity of these reconstructions. A nominal uncertainty of ±10 m in each mea-

surement was assigned based on the measured width of the terrace riser, encompassing the range of permissible reconstructions for each margin of the terrace surface.

Measured lateral displacement of the western margin of this terrace ranges between 30 and 42 m for the TLS profiles (Fig. 10B), in good agreement with estimations based on Google Earth imagery. This measurement incorporates the maximum and minimum distance range of riser midpoints grouped on either side of the Little Lake fault. Given the local linearity and continuity of each riser segment, as well as the fault trace, uncertainty associated with projection of these features is smaller than the observed

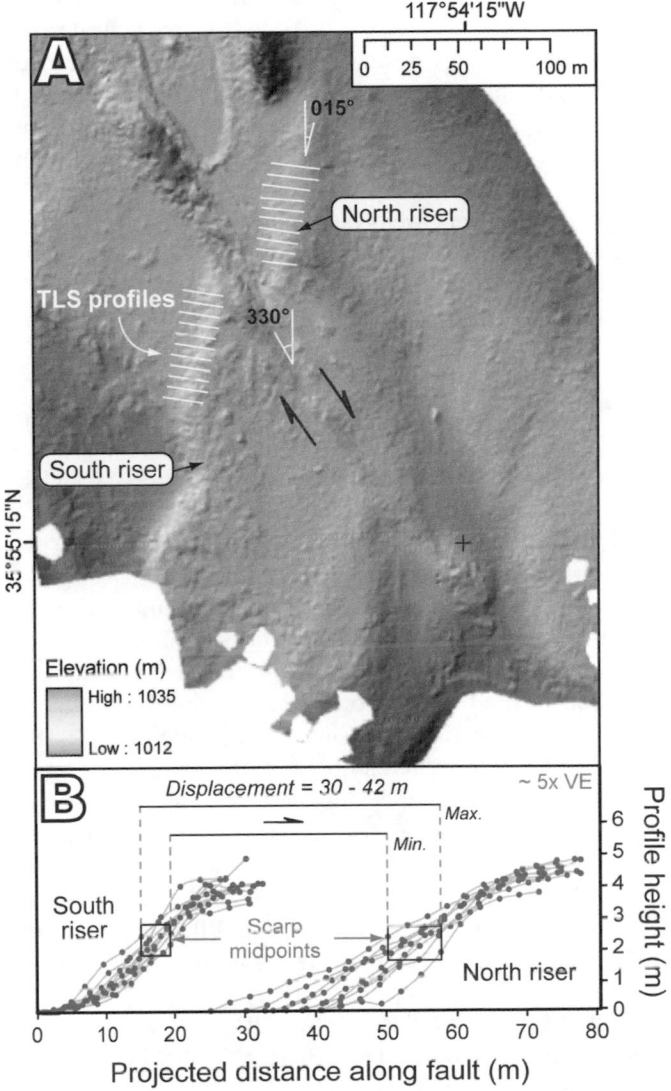

Figure 10. (A) 50-cm digital elevation model derived from terrestrial laser scanning (TLS) and overlain on a hillshade image of the displaced terrace surface in Little Lake wash. (B) Topographic profiles along the western terrace riser used to reconstruct the total dextral offset of this feature. Individual profiles were extracted perpendicular to the average riser orientation (015°), and were then rotated as a group 45° clockwise onto a plane parallel to the local fault strike (330°). Locations for each profile are presented in A.

range of riser midpoints. We do note, however, a larger variation in profile shape and midpoint location for the riser segment north of the fault, possibly due to either locally increased diffusion or a slight azimuth change to the north in this direction. To our knowledge, this study presents the first systematic comparison of fault offsets measured using Google Earth imagery and ground-based survey methods.

3.4.4. Implications and Conclusions

New measurements of fault displacement for the Little Lake fault add to our growing inventory of deformation patterns in the eastern California shear zone. Although the exact age of the offset terrace is yet unknown, several lines of evidence suggest a latest Pleistocene age for this feature. Specifically, this faulted terrace rests unconformably on a strath surface cut into the basalt of Red Hill (Fig. 9B), the youngest of several flows to spill through the Little Lake drainage (Duffield and Smith, 1978). ^{3}He exposure ages measured using olivine phenocrysts from this flow (Cerling, 1990) suggest the timing of eruption and subsequent scouring during overflow from Owens Lake between ca. 57 and ca. 15 ka, respectively. This range in ages implies a dextral slip rate of at least ~0.5 mm/yr over this interval. We emphasize the preliminary nature of this measurement and note that we are currently processing new ^{10}Be exposure age measurements to bolster our slip-rate estimate. In any case, our appraisal is within the range of dextral slip-rates described along the Owens Valley fault to the north (Bacon and Pezzopane, 2007), which transfers strain across the Coso Range in a releasing step-over onto strike-slip faults in the Indian Wells Valley (Unruh et al., 2002) (Fig. 9A).

Our measurements of dextrally offset landforms in the eastern California shear zone highlight the utility of Google Earth as a primary tool for quantifying the style and magnitude of displacement in actively deforming regions. Agreement between offset measurements from Google Earth imagery and measurements from ground-based, high-resolution survey methods indicates the quality of these displacement estimates and firmly extends the utility of this software and high-resolution imagery beyond that of a reconnaissance tool. Although these measurements are currently limited to lateral displacements, future integration of high-quality lidar terrane data (currently available for limited areas in Google Maps) will no doubt extend the range of possible measurements to dip-slip faults, providing a free and simple tool for quantifying and characterizing fault systems across the globe.

4. LIMITATIONS, PROBLEMS, AND POTENTIAL RESEARCH APPLICATIONS FOR GOOGLE EARTH IMAGERY

The previous three research examples have highlighted the utility of high-resolution Google Earth imagery as a potential research tool in geomorphology, as well as a few of the diverse research questions that can now be undertaken without regard for imagery cost or location. As with all good tools, however, a proper understanding of the limitations and pitfalls must be acknowledged in order to properly apply it. Google Earth is no exception, and presently contains key limitations and potentially problematic situations that should guide suitable applications in the future.

4.1. Standard Sensor Issues

As with all remotely sensed data sets, inherent flaws in imagery occur and are often the necessary byproduct of having an efficient, cost-effective single-satellite system. In the case of Google Earth imagery, extensive shadowing can often occur (especially in high-relief terrains) due to sensor view and/or solar incidence angles, depending on location and time of the day and year (Fig. 11). In the case of steep terrains, shadowing may completely obscure surface features (as was the case with the Haiti data set), rendering the high-resolution true-color imagery useless in those locations. Likewise, heavily vegetated regions will yield little insight about geomorphic processes and landforms (e.g., fault traces, tropical rivers, etc.), especially when compared to tools such as lidar (light detection and ranging) with which vegetation can be removed during data processing (e.g., Mackey and Roering, 2011). In addition, the lack of access to longer wavelength bands (e.g., near, short, and longwave infrared, radar) from the sensors used in Google Earth imagery prevents compensation for shadowing using the spectral range of bands normally available in multi-spectral data sets.

4.2. The Google Black Box Problem

Despite the many benefits of the Google Earth imagery, it remains a black box. For example, very little is known about the algorithms being used to process, orthorectify, overlay, interpolate, degrade, and display scenes, as these are proprietary to Google and no public documentation exists. In addition, little is known about the underlying digital elevation models used with respect to data sources, interpolation algorithms, and orthorectification basemaps. It can safely be assumed that SRTM data make up a large part of the data set where it is available, but to what extent the new ASTER GDEM (Slater et al., 2011), national mapping agency, and regional lidar (cf. Perroy et al., 2010) data sets are incorporated is unknown. Furthermore, considerable uncertainty exists about the extent to which Google reprocesses or collects new elevation data, especially given the suggestion that elevation data sets available in Google Earth are superior to publicly available data sets in certain locations (Hoffmann and Winde, 2010). There does appear to be evidence for the incorporation of lidar data in certain locations (namely urban areas); however, it is difficult to verify such assertions.

Whereas little can be done with respect to such fundamental unknowns, many problems with the imagery are blatantly visible and may preclude use of certain imagery in Google Earth. One widespread problem involves the stitching, or merging, of imagery taken at different times (Fig. 11). Telltale signs of such imagery amalgamation problems include offset or warped

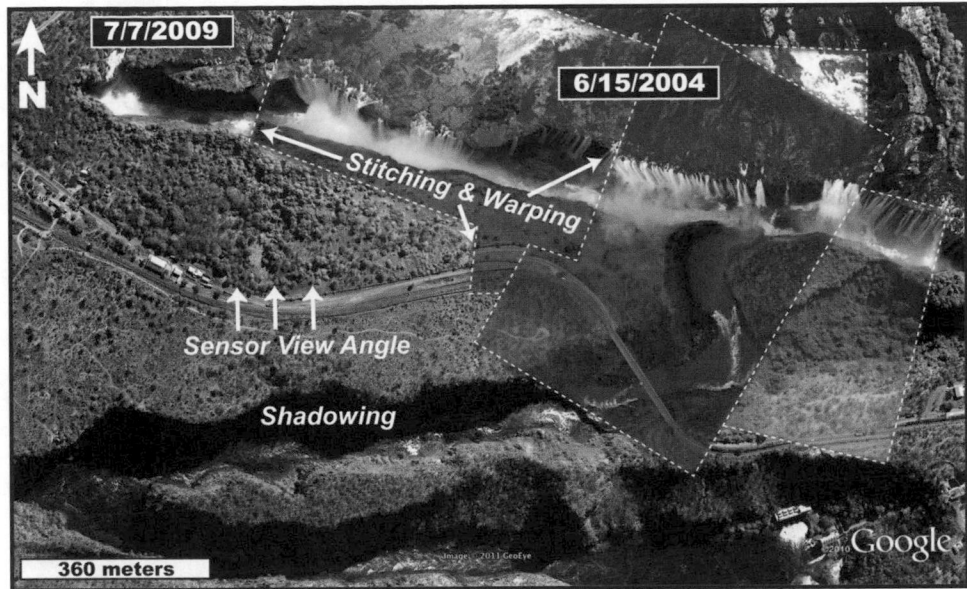

Figure 11. Image of Victoria Falls on the border of Zimbabwe and Zambia illustrating some of the potential problems with using Google Earth imagery (e.g., improper warping and stitching of imagery, shadowing due to sensor view angle and solar incidence, discrepancies in imagery dates, etc.). Arrows point to tell-tale signs of problems due to imagery amalgamation, such as diverging waterfalls, roads to nowhere, and abrupt cliff faces in the middle of the river. It should be noted that the imagery for this area was updated shortly after this figure was made, correcting all of these problems.

features along the seam of the image (broken roads, diverging waterfalls, etc.). A second major problem (especially in steep terrain) involves the unrealistic warping of imagery when using "terrain mode" in Google Earth. The spatially high-resolution optical images require nearly equally high-resolution digital elevation models, which are not readily available in most areas. The current image correction and orthorectification procedures therefore rely on lower resolution digital elevation models that often result in artifacts in high relief areas (e.g., rivers that bank up onto mountainsides and roads that go off cliffs). Because of these artifacts, any digitizing should be done with the terrain mode off, using just the optical imagery. Other current problems include: (1) the lack of spatially and temporally ubiquitous high-resolution imagery, which may dictate the area and/or scope of a research project; (2) the arbitrary availability of certain imagery, whereby imagery may disappear (Fig. 11) due to updates, causing a loss of access to imagery that may have been the foundation for previous measurements; and (3) the Google Earth user interface can be quite awkward for collecting and manipulating large amounts of data (i.e., digitizing, naming, displaying, organizing).

Lastly, the horizontal accuracy of the imagery may cause problems, especially when the digitized imagery and digital elevation models used for analyses are highly mismatched (Sato and Harp, 2009). While it is nearly impossible to assess the horizontal positional accuracy (georegistration) of the entire Google Earth imagery collection, previous work by Potere (2008) utilizing a global database of 436 control points spread across 109 cities found positional accuracy to be less than 50 m root-mean-squared error (RMSE) on the whole when compared with Landsat Geo-Cover. Satellite imagery accuracy (22.8 m RMSE) was found to have half the error of aerial imagery (41.3 m RMSE) in Google Earth and the same discrepancy was found between developing (44.4 m RMSE) and developed (24.1 m RMSE) countries.

While Potere's study (2008) is encouraging, there is no doubt that mountainous and remote regions will have greater complications with horizontal accuracy than urban areas with low relief, and in some cases the mismatch may require considerable work to properly align (Sato and Harp, 2009).

While these problems are not trivial, Google has been proactive in updating and improving the platform's utility and accuracy, as well as the imagery within it. For example, historical imagery now archives most of the past imagery for a given location. Thus, imagery used in the past can usually be accessed even when new imagery comes online. Additionally, images displaying merging issues or warping commonly are fixed with subsequent image releases (Fig. 11). Whereas proprietary algorithms and data management workflows will likely never be disclosed to the user community, many other limitations discussed above will surely continue to be improved as Google Earth expands both the user experience, content, and processing algorithms in future releases.

4.3. Suitable Research Applications

Google Earth imagery will never replace fieldwork and/or independent processing of remote sensing data by the individual, especially without access to the algorithms employed by Google. However, for many researchers, access to high-resolution optical imagery is only feasible at small spatial scales due to cost, computational ability, and/or user knowledge. Additionally, many parts of the earth preclude fieldwork due to remoteness and/or political conflicts. In these cases, Google Earth imagery provides a powerful data set that until recently was unavailable to the scientific community. Whereas Google Earth imagery has been widely underutilized to date, several key features, as utilized in section 3, may help to guide future quantitative applications in geomorphology and earth sciences alike.

The first of these features is the availability of historical imagery, which allows temporal as well as spatial change to be assessed and can supplement current research looking at land-use change, geomorphic change, event-scale impacts, etc. (Fig. 12). Second, Google has shown great commitment to providing rapid response imagery and support following large-scale natural disasters. This commitment was most recently displayed with the 2011 Japan earthquake and tsunami, but has been a major campaign of Google's since the inception of Google Earth (e.g., 2010 Haiti earthquake, 2004 Indian Ocean tsunami). Given this response, researchers now have access to both pre- and post-event imagery (one day to a few weeks) for rapid assessment and analysis that spans the gamut of processes, landforms, and events. Third, the ease of use of the imagery, specifically, the complexities of image processing and orthorectification are out of the control of the researcher. This absence can be both a burden and a benefit, depending on the project and the user. One great example of the benefit is shown with landslide imagery from New Zealand. In order to properly analyze aerial photos for landslides in Fjordland, New Zealand, Clarke and Burbank (2010) purchased and orthorectified close to 60 aerial photos, an expensive and time-consuming endeavor. This imagery recently became freely available on Google Earth, saving both processing time and money, as well as increasing the spatial

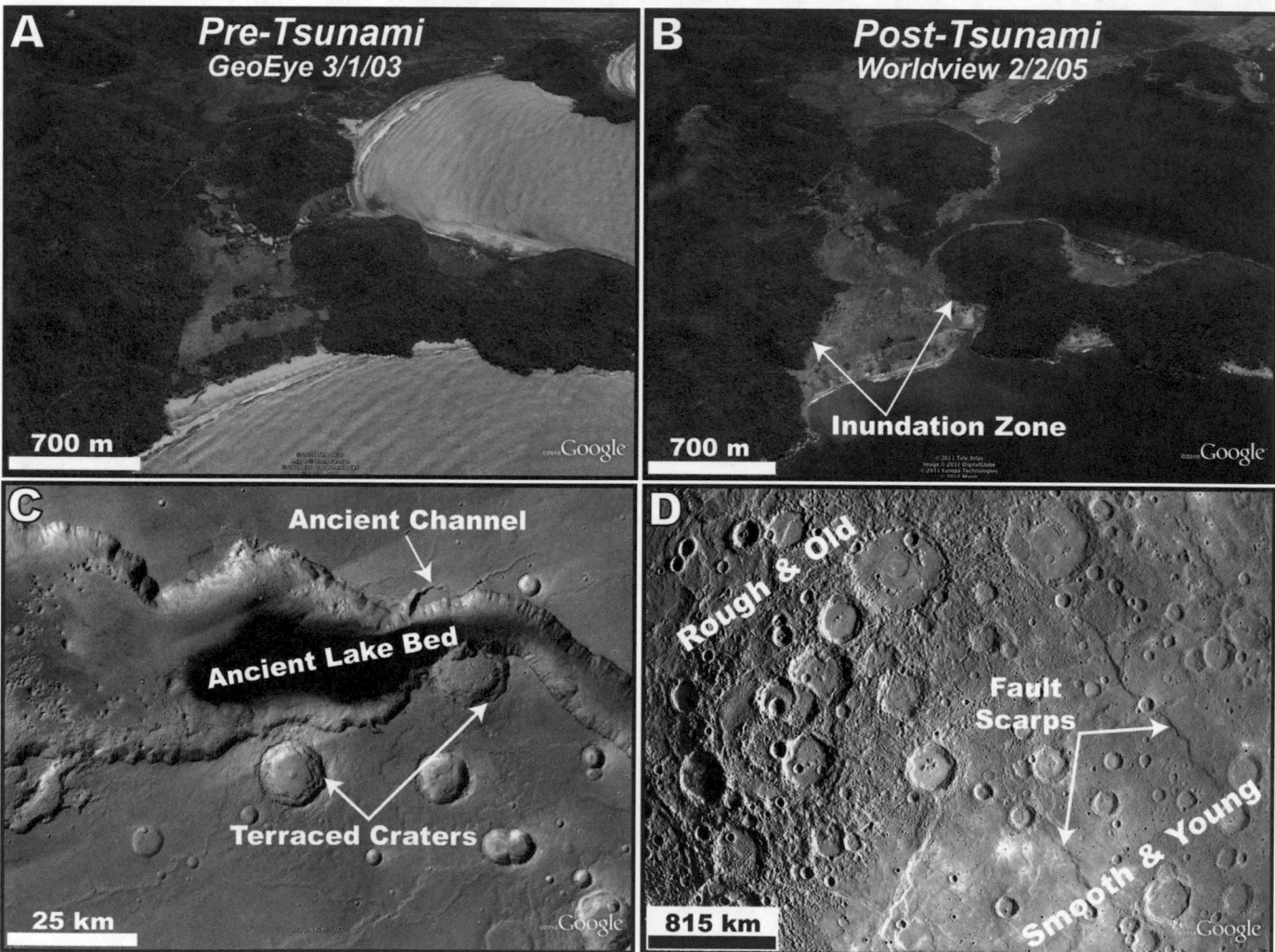

Figure 12. Images (A) (GeoEye™ image taken 1 March 2003) and (B) (Worldview™ image taken on 2 February 2005) showing the power of Google Earth's historical imagery for detecting change along a coastal area south of Banda Aceh, Sumatra, that was heavily impacted during the December 2004 Indian Ocean tsunami. (C) Image from Google Earth Mars of Shalbatana Vallis, a site argued to have direct evidence of lake strand-lines and deltaic features implying extensive water on the surface of Mars in the past (cf. Di Achille et al., 2009; Di Achille and Hynek, 2010; Head et al., 1999). Also note the variable morphology and terracing within many of the craters on the rim. (D) Image of Mercury from the MESSENGER mission showing the variation in surface texture and abundant tensional fault scarps due to crustal cooling. The fault scarp in the upper right hand corner is calculated by the Mercury Laser Altimeter to have a topographic relief of ~1.4 km (http://messenger.jhuapl.edu) (cf. Solomon, 2011).

area available for analysis due to lack of financial constraints. Although some accuracy would likely be lost by not controlling the orthorectification process, preliminary comparison of several landslides in the area indicates minimal discrepancies between the two processes.

Google has not stopped at earth, but rather expanded the digital globe to other bodies in our solar system opening up new terrains for exploration and research (Fig. 12). Currently, digital globes for the moon and Mars are standard in Google Earth, with second-party digital globe imagery now available for Mercury (Appendix). Moreover, as imagery and topographic data are acquired from across our solar system and beyond, the spatial information available in Google Earth will certainly expand along with the research possibilities. Although Google Earth imagery has many limitations, it provides one of the only feasible high-resolution spatial data sets currently available that allows user interaction for many remote areas and projects necessitating high-resolution imagery across large spatial extents (see also Bing Maps in ESRI ArcGIS). In cases where field access is plausible, Google Earth imagery may act as supplemental tool to compare data sets through time and space. The most beneficial aspect of high-resolution Google Earth imagery, however, may be the serendipitous discoveries and ideas generated from simply exploring the prolific imagery. The combination of curiosity-driven hypothesis creation and the ability to quantitatively test those hypotheses makes Google Earth imagery beneficial, when properly applied, as a supplement, for reconnaissance, or as the primary data set across a wide range of studies in earth surface processes.

5. CONCLUSIONS

The utility of high-resolution imagery has permeated nearly every segment of society over the past decade, with Google Earth well-positioned at the forefront by providing freely available high-resolution imagery within a simple, user-friendly interface. Whereas Google Earth imagery has been widely exploited as a visualization and education tool across the earth sciences, few quantitative studies have utilized the high-resolution imagery freely available within Google Earth to yield insights about the processes and mechanisms acting upon the surface of the earth. In this paper, we have discussed the underlying high-resolution imagery within Google Earth and provided three research examples using this imagery as the primary datasource in order to demonstrate the benefits of the imagery over established methods, and to illuminate previously unresolved questions involving: (1) channel-width variability in the tectonically active Himalaya; (2) landslide characteristics related to large magnitude climatic and tectonic events in Haiti; and (3) distributed fault characteristics within the eastern California shear zone. Lastly, we have highlighted the potential limitations and problems with using the imagery currently available in Google Earth and proposed favorable future applications with respect to the study of earth surface processes.

APPENDIX. HELPFUL LINKS AND INFORMATION RELATED TO GOOGLE EARTH

Google Earth: http://www.google.com/earth/
Google Earth Engine: http://www.google.org/earthengine/
Google Fusion Tables: http://www.google.org/fusiontables/
Google Earth Help Forum: http://www.google.com/support/forum/p/earth
Google Earth Development Team Blog: http://google-latlong.blogspot.com/
NASA Blue Marble: Next Generation KML file: http://ti.arc.nasa.gov/tech/asr/intelligent-robotics/planetary/earth/
Mercury Messenger Virtual Globe: http://messenger.jhuapl.edu/the_mission/google.html

ACKNOWLEDGMENTS

This work was generously supported by a NASA Earth and Space Science Fellowship (NNX09AO24H) to G.B. Fisher and B. Bookhagen, NASA (NNX08AG05G) and NSF grants (EAR0819874) to D.W. Burbank and B. Bookhagen, an NSF Postdoctoral Research Fellowship (EAR0847990) to C.B. Amos, and a Marie Curie Fellowship to V. Godard (Geocycl-219662). Additional support from Google and Oxford Press to reduce the cost of attendance to the Google Earth Penrose Conference for G.B. Fisher is greatly appreciated and led to the formulation of this paper. Helpful reviews by Phillip Prince and Wes Lauer greatly improved an earlier version of this manuscript.

REFERENCES CITED

Amos, C.B., and Burbank, D.W., 2007, Channel width response to differential uplift: Journal of Geophysical Research, v. 112, no. F02010, 11 p., doi:10.1029/2006JF000672.
Asner, G.P., Rudel, T.K., Aide, T.M., Defries, R., and Emerson, R.A., 2009, Contemporary assessment of change in humid tropical forests: Conservation Biology, v. 23, p. 1386–1395, doi:10.1111/j.1523-1739.2009.01333.x.
Attal, M., Cowie, P.A., Whittaker, A.C., Hobley, D., Tucker, G.E., and Roberts, G.P., 2011, Testing fluvial erosion models using the transient response of bedrock rivers to tectonic forcing in the Apennines, Italy: Journal of Geophysical Research, v. 116, no. F02005, 17 p., doi:10.1029/2010JF001875.
Bacon, S.N., and Pezzopane, S.K., 2007, A 25,000-year record of earthquakes on the Owens Valley fault near Lone Pine, California: Implications for recurrence intervals, slip rates, and segmentation models: Geological Society of America Bulletin, v. 119, p. 823–847, doi:10.1130/B25879.1.
Begall, S., Cerveny, J., Neef, J., Vojtech, O., and Burda, H., 2008, Magnetic alignment in grazing and resting cattle and deer: Proceedings of the National Academy of Sciences of the United States of America, v. 105, p. 13451–13455, doi:10.1073/pnas.0803650105.
Bookhagen, B., 2010, Appearance of extreme monsoonal rainfall events and their impact on erosion in the Himalaya: Geomatics, Natural Hazards, and Risk, v. 1, p. 37–50, doi:10.1080/19475701003625737.
Bookhagen, B., and Burbank, D.W., 2010, Toward a complete Himalayan hydrological budget: Spatiotemporal distribution of snowmelt and rainfall and their impact on river discharge: Journal of Geophysical Research, v. 115, no. F03019, 25 p., doi:10.1029/2009JF001426.
Bookhagen, B., and Strecker, M.R., 2012, Spatiotemporal trends in erosion rates across a pronounced rainfall gradient: Examples from the south-central Andes: Earth and Planetary Science Letters, v. 327-328, p. 97–110, doi:10.1016/j.epsl.2012.02.005.
Bookhagen, B., Thiede, R.C., and Strecker, M.R., 2005a, Late Quaternary intensified monsoon phases control landscape evolution in the northwest Himalaya: Geology, v. 33, p. 149, doi:10.1130/G20982.1.

Bookhagen, B., Thiede, R.C., and Strecker, M.R., 2005b, Abnormal monsoon years and their control on erosion and sediment flux in the high, arid northwest Himalaya: Earth and Planetary Science Letters, v. 231, p. 131–146, doi:10.1016/j.epsl.2004.11.014.

Brossy, C.C., Kelson, K.I., Amos, C.B., Baldwin, J.N., Kozlowicz, B., Simpson, D., Ticci, M.G., Lutz, A.T., Kozaci, O., Streig, A., Turner, R., and Rose, R., 2012, Map of the late Quaternary active Kern Canyon and Breckenridge faults, Southern Sierra Nevada, California: Geosphere, doi:10.1130/GES00663.1.

Burbank, D., Leland, J., Fielding, E., Anderson, R., Brozovic, N., Reid, M., and Duncan, C., 1996, Bedrock incision, rock uplift and threshold hillslopes in the northwestern Himalayas: Nature, v. 379, p. 505–510, doi:10.1038/379505a0.

Butler, D., 2006, Virtual globes: The web-wide world: Nature, v. 439, p. 776–778, doi:10.1038/439776a.

Calais, E., Freed, A., Mattioli, G., Amelung, F., Jónsson, S.F.J., Jansma, P., Hong, S.-H., Dixon, T., Prépetit, C.P., and Momplaisir, R., 2010, Transpressional rupture of an unmapped fault during the 2010 Haiti earthquake: Nature Geoscience, v. 3, p. 794–799, doi:10.1038/ngeo992.

Carroll, R., 2008, We are going to disappear one day: The Guardian, 8 November 2008 (retrieved 25 June 2011).

Cerling, T.E., 1990, Dating geomorphologic surfaces using cosmogenic He-3: Quaternary Research, v. 33, p. 148–156, doi:10.1016/0033-5894(90)90015-D.

Clarke, B.A., and Burbank, D.W., 2010, Bedrock fracturing, threshold hillslopes, and limits to the magnitude of bedrock landslides: Earth and Planetary Science Letters, v. 297, p. 577–586, doi:10.1016/j.epsl.2010.07.011.

Clarke, B.A., and Burbank, D.W., 2011, Quantifying bedrock-fracture patterns within the shallow subsurface: Implications for rock-mass strength, bedrock landslides, and erodibility: Journal of Geophysical Research–Earth Surface, v. 116, F04009, doi:10.1029/2011JF001987.

Constantine, J.A., and Dunne, T., 2008, Meander cutoff and the controls on the production of oxbow lakes: Geology, v. 36, p. 23–26, doi:10.1130/G24130A.1.

Contreras, D.A., 2010, Huaqueros and remote sensing imagery: assessing looting damage in the Viru Valley, Peru: Antiquity, v. 84, p. 544–555.

Craddock, W.H., Burbank, D.W., Bookhagen, B., and Gabet, E.J., 2007, Bedrock channel geometry along an orographic rainfall gradient in the upper Marsyandi River valley in central Nepal: Journal of Geophysical Research, v. 112, no. F03007, 17 p., doi:10.1029/2006JF000589.

Dadson, S.J., Hovius, N., Chen, H., Dade, W.B., Lin, J.-C., Hsu, M.-L., Lin, C.-W., Horng, M.-J., Chen, T.-C., Milliman, J., and Stark, C.P., 2004, Earthquake-triggered increase in sediment delivery from an active mountain belt: Geology, v. 32, p. 733, doi:10.1130/G20639.1.

Densmore, A., and Hovius, N., 2000, Topographic fingerprints of bedrock landslides: Geology, v. 28, p. 371–374, doi:10.1130/0091-7613(2000)28<371:TFOBL>2.0.CO;2.

Densmore, A.L., Anderson, R.S., McAdoo, B.G., and Ellis, M.A., 1997, Hillslope evolution by bedrock landslides: Science, v. 275, p. 369–372, doi:10.1126/science.275.5298.369.

Di Achille, G., and Hynek, B.M., 2010, Ancient ocean on Mars supported by global distribution of deltas and valleys: Nature Geoscience, v. 3, p. 459–463, doi:10.1038/ngeo891.

Di Achille, G., Hynek, B.M., and Searls, M.L., 2009, Positive identification of lake strandlines in Shalbatana Vallis, Mars: Geophysical Research Letters, v. 36, no. L14201, 5 p., doi:10.1029/2009GL038854.

Dixon, T.H., Miller, M., Farina, F., Wang, H., and Johnson, D., 2000, Present-day motion of the Sierra Nevada block and some tectonic implications for the Basin and Range province, North American Cordillera: Tectonics, v. 19, p. 1–24, doi:10.1029/1998TC001088.

Dokka, R.K., and Travis, C.J., 1990, Role of the Eastern California shear zone in accommodating Pacific-North-American plate motion: Geophysical Research Letters, v. 17, no. 9, p. 1323–1326, doi:10.1029/GL017i009p01323.

Duffield, W.A., and Smith, G.I., 1978, Pleistocene history of volcanism and Owens River near Little-Lake, California: Journal of Research of the U.S. Geological Survey, v. 6, no. 3, p. 395–408.

Duvall, A., Kirby, E., and Burbank, D., 2004, Tectonic and lithologic controls on bedrock channel profiles and processes in coastal California: Journal of Geophysical Research, v. 109, no. F03002, 18 p., doi:10.1029/2003JF000086.

Farr, T.G., Rosen, P.A., Caro, E., Crippen, R., Duren, R., Hensley, S., Kobrick, M., Paller, M., Rodriguez, E., Roth, L., Seal, D., Shaffer, S., Shimada, J., Umland, J., Werner, M., Oskin, M., Burbank, D., and Alsdorf, D., 2007, The shuttle radar topography mission: Reviews of Geophysics, v. 45, Rg2004, doi:10.1029/2005rg000183.

Finnegan, N.J., Roe, G.H., Montgomery, D.R., and Hallet, B., 2005, Controls on the channel width of rivers: Implications for modeling fluvial incision of bedrock: Geology, v. 33, p. 229–232, doi:10.1130/G21171.1.

Finnegan, N.J., Sklar, L.S., and Fuller, T.K., 2007, Interplay of sediment supply, river incision, and channel morphology revealed by the transient evolution of an experimental bedrock channel: Journal of Geophysical Research, v. 112, no. F03S11, 17 p., doi:10.1029/2006JF000569.

Finnegan, N.J., Hallet, B., Montgomery, D.R., Zeitler, P.K., Stone, J.O., Anders, A.M., and Yuping, L., 2008, Coupling of rock uplift and river incision in the Namche Barwa-Gyala Peri massif, Tibet: Geological Society of America Bulletin, v. 120, p. 142–155, doi:10.1130/B26224.1.

Fisher, G.B., Magilligan, F.J., Kaste, J.M., and Nislow, K.H., 2010, Constraining the timescales of sediment sequestration associated with large woody debris using cosmogenic ^{7}Be: Journal of Geophysical Research, v. 115, no. F01013, 19 p., doi:10.1029/2009JF001352.

Fisher, G.B., Bookhagen, B., and Burbank, D.W., 2011, High-resolution channel widths and erosion along the entire Indus River: Eos (Transactions, American Geophysical Union), Fall Meeting Supplement, abstract EP23C-0762.

Frankel, K.L., Dolan, J.F., Owen, L.A., Ganev, P., and Finkel, R.C., 2011, Spatial and temporal constancy of seismic strain release along an evolving segment of the Pacific-North America plate boundary: Earth and Planetary Science Letters, v. 304, p. 565–576, doi:10.1016/j.epsl.2011.02.034.

Gabet, E., Burbank, D., Putkonen, J., Pratt-Sitaula, B., and Ojha, T.P., 2004, Rainfall thresholds for landsliding in the Himalayas of Nepal: Geomorphology, v. 63, p. 131–143, doi:10.1016/j.geomorph.2004.03.011.

Galewsky, J., Stark, C.P., Dadson, S., Wu, C.-C., Sobel, A.H., and Horng, M.-J., 2006, Tropical cyclone triggering of sediment discharge in Taiwan: Journal of Geophysical Research, v. 111, p. F03014, doi:10.1029/2005JF000428.

Godard, V., Lavé, J., Carcaillet, J., Cattin, R., Bourlès, D., and Zhu, J., 2010, Spatial distribution of denudation in Eastern Tibet and regressive erosion of plateau margins: Tectonophysics, v. 491, p. 253–274, doi:10.1016/j.tecto.2009.10.026.

Guo, J., Liang, L., and Gong, P., 2010, Removing shadows from Google Earth images: International Journal of Remote Sensing, v. 31, p. 1379–1389, doi:10.1080/01431160903475316.

Hashimoto, M., Fukushima, Y., and Fukahata, Y., 2011, Fan-delta uplift and mountain subsidence during the Haiti 2010 earthquake: Nature Geoscience, v. 4, p. 255–259, doi:10.1038/ngeo1115.

Hayes, G.P., Briggs, R.W., Sladen, A., Fielding, E.J., Prentice, C., Hudnut, K., Mann, P., Taylor, F.W., Crone, A.J., Gold, R., Ito, T., and Simons, M., 2010, Complex rupture during the 12 January 2010 Haiti earthquake: Nature Geoscience, v. 3, p. 800–805, doi:10.1038/ngeo977.

Head, J., Hiesinger, H., Ivanov, M., Kreslavsky, M., Pratt, S., and Thomson, B., 1999, Possible ancient oceans on Mars: Evidence from Mars Orbiter Laser Altimeter data: Science, v. 286, p. 2134–2137, doi:10.1126/science.286.5447.2134.

Hoffmann, E., and Winde, F., 2010, Generating high-resolution digital elevation models for wetland research using Google Earth™ imagery—an example from South Africa: Water S.A., v. 36, p. 53–68.

Hong, Y., Adler, R., and Huffman, G., 2006, Evaluation of the potential of NASA multi-satellite precipitation analysis in global landslide hazard assessment: Geophysical Research Letters, v. 33, L22402, 5 p., doi:10.1029/2006GL028010.

Hovius, N., Stark, C., and Allen, P., 1997, Sediment flux from a mountain belt derived by landslide mapping: Geology, v. 25, p. 231–234, doi:10.1130/0091-7613(1997)025<0231:SFFAMB>2.3.CO;2.

Hovius, N., Stark, C., Chu, H., and Lin, J., 2000, Supply and removal of sediment in a landslide-dominated mountain belt: Central Range, Taiwan: The Journal of Geology, v. 108, p. 73–89, doi:10.1086/314387.

Hovius, N., Meunier, P., Ching-Weei, L., Hongey, C., Yue-Gau, C., Dadson, S., Ming-Jame, H., and Lines, M., 2011, Prolonged seismically induced erosion and the mass balance of a large earthquake: Earth and Planetary Science Letters, v. 304, p. 347–355, doi:10.1016/j.epsl.2011.02.005.

Huffman, G.J., Bolvin, D.T., Nelkin, E.J., Wolff, D.B., Adler, R.F., Gu, G., Hong, Y., Bowman, K.P., and Stocker, E.F., 2007, The TRMM Multisatellite Precipitation Analysis (TMPA): Quasi-Global, Multiyear, Combined-Sensor Precipitation Estimates at Fine Scales: Journal of Hydrometeorology, v. 8, p. 38–55, doi:10.1175/JHM560.1.

Jennings, C.W., 1994, Fault activity map of California and adjacent areas, Geologic Data Map 6: California Division of Mines and Geology, Sacramento.

Jones, K.C., 2008, Google Launches Mapping Satellite: InformationWeek.com, 8 September 2008 (retrieved 15 February 2011).

Kirby, E., and Ouimet, W., 2011, Tectonic geomorphology along the eastern margin of Tibet: insights into the pattern and processes of active deformation adjacent to the Sichuan Basin, *in* Gloaguen, R., and Ratschbacher, L., eds., Growth and Collapse of the Tibetan Plateau: Geological Society of London Special Publication 353, p. 165–188, doi:10.1144/SP353.9.

Kirby, E., and Whipple, K., 2001, Quantifying differential rock-uplift rates via stream profile analysis: Geology, v. 29, p. 415–418, doi:10.1130/0091-7613(2001)029<0415:QDRURV>2.0.CO;2.

Klinger, Y., Etchebes, M., Tapponnier, P., and Narteau, C., 2011, Characteristic slip for five great earthquakes along the Fuyun fault in China: Nature Geoscience, v. 4, p. 389–392, doi:10.1038/ngeo1158.

Knighton, D.A., 1998, Fluvial forms and processes: Edward Arnold, London, 377 p.

Korup, O., 2006, Rock-slope failure and the river long profile: Geology, v. 34, no. 1, p. 45–48, doi:10.1130/G21959.1.

Korup, O., Densmore, A.L., and Schlunegger, F., 2010, The role of landslides in mountain range evolution: Geomorphology, v. 120, p. 77–90, doi:10.1016/j.geomorph.2009.09.017.

Korup, O., and Montgomery, D.R., 2008, Tibetan plateau river incision inhibited by glacial stabilization of the Tsangpo gorge: Nature, v. 455, p. 786–789, doi:10.1038/nature07322.

Lavé, J., and Avouac, J., 2001, Fluvial incision and tectonic uplift across the Himalayas of central Nepal: Journal of Geophysical Research—Solid Earth, v. 106, p. 26561–26591, doi:10.1029/2001JB000359.

Lavé, J., and Burbank, D.W., 2004, Denudation processes and rates in the Transverse Ranges, southern California: Erosional response of a transitional landscape to external and anthropogenic forcing: Journal of Geophysical Research, v. 109, no. F01006, 31 p., doi:10.1029/2003JF000023.

Leopold, L., and T. Maddock, 1953, The hydraulic geometry of stream channels and some physiographic implications: U.S. Geological Survey Professional Paper 252.

Mackey, B.H., and Roering, J.J., 2011, Sediment yield, spatial characteristics, and the long-term evolution of active earthflows determined from airborne LiDAR and historical aerial photographs, Eel River, California: Geological Society of America Bulletin, v. 123, p. 1560–1576, doi:10.1130/B30306.1.

Magilligan, F.J., 1992, Thresholds and the spatial variability of flood power during extreme floods: Geomorphology, v. 5, p. 373–390, doi:10.1016/0169-555X(92)90014-F.

Manning, R., 1891, On the flow of water in open channels and pipes: Institute of Civil Engineers of Ireland Transactions, v. 20, p. 161–207.

Mering, C., Baro, J., and Upegui, E., 2010, Retrieving urban areas on Google Earth images: application to towns of West Africa: International Journal of Remote Sensing, v. 31, p. 5867–5877, doi:10.1080/01431161.2010.512311.

Meunier, P., Hovius, N., and Haines, J.A., 2008, Topographic site effects and the location of earthquake induced landslides: Earth and Planetary Science Letters, v. 275, p. 221–232, doi:10.1016/j.epsl.2008.07.020.

Milliman, J., and Syvitski, J., 1992, Geomorphic tectonic control of sediment discharge to the ocean—The importance of small mountainous rivers: The Journal of Geology, v. 100, p. 525–544, doi:10.1086/629606.

Oskin, M., and Iriondo, A., 2004, Large-magnitude transient strain accumulation on the Blackwater fault, Eastern California shear zone: Geology, v. 32, p. 313–316, doi:10.1130/G20223.1.

Oskin, M., Perg, L., Shelef, E., Strane, M., Gurney, E., Singer, B., and Zhang, X., 2008, Elevated shear zone loading rate during an earthquake cluster in eastern California: Geology, v. 36, p. 507–510, doi:10.1130/G24814A.1.

Parker, J.D., 2011, Using Google Earth to teach the magnitude of deep time: Journal of College Science Teaching, v. 40, p. 23–27.

Parker, R.N., Densmore, A.L., Rosser, N.J., de Michele, M., Li, Y., Huang, R., Whadcoat, S., and Petley, D.N., 2011, Mass wasting triggered by the 2008 Wenchuan earthquake is greater than orogenic growth: Nature Geoscience, v. 4, p. 449–452, doi:10.1038/ngeo1154.

Pelletier, J.D., DeLong, S.B., Al-Suwaidi, A.H., Cline, M., Lewis, Y., Psillas, J.L., and Yanites, B., 2006, Evolution of the Bonneville shoreline scarp in west-central Utah: Comparison of scarp-analysis methods and implications for the diffusion model of hillslope evolution: Geomorphology, v. 74, p. 257–270, doi:10.1016/j.geomorph.2005.08.008.

Peltzer, G., Crampe, F., Hensley, S., and Rosen, P., 2001, Transient strain accumulation and fault interaction in the Eastern California shear zone: Geology, v. 29, p. 975–978, doi:10.1130/0091-7613(2001)029<0975:TSAAFI>2.0.CO;2.

Perroy, R.L., Bookhagen, B., Asner, G.P., and Chadwick, O.A., 2010, Comparison of gully erosion estimates using airborne and ground-based LiDAR on Santa Cruz Island, California: Geomorphology, v. 118, p. 288–300, doi:10.1016/j.geomorph.2010.01.009.

Phillips, F.M., and Majkowski, L., 2011, The role of low-angle normal faulting in active tectonics of the northern Owens Valley, California: Lithosphere, v. 3, p. 22–36, doi:10.1130/L73.1.

Pierce, J.L., Meyer, G.A., and Jull, A.J.T., 2004, Fire-induced erosion and millennial-scale climate change in northern ponderosa pine forests: Nature, v. 432, p. 87–90, doi:10.1038/nature03058.

Potere, D., 2008, Horizontal Positional Accuracy of Google Earth's High-Resolution Imagery Archive: Sensors (Basel, Switzerland), v. 8, p. 7973–7981, doi:10.3390/s8127973.

Roering, J.J., and Gerber, M., 2005, Fire and the evolution of steep, soil-mantled landscapes: Geology, v. 33, p. 349, doi:10.1130/G21260.1.

Santamaria Tovar, D., Shulmeister, J., and Davies, T.R., 2008, Evidence for a landslide origin of New Zealand's Waiho Loop moraine: Nature Geoscience, v. 1, p. 524–526, doi:10.1038/ngeo249.

Sato, H.P., and Harp, E.L., 2009, Interpretation of earthquake-induced landslides triggered by the 12 May 2008, M7.9 Wenchuan earthquake in the Beichuan area, Sichuan Province, China using satellite imagery and Google Earth: Landslides, v. 6, p. 153–159, doi:10.1007/s10346-009-0147-6.

Sauber, J., Thatcher, W., Solomon, S.C., and Lisowski, M., 1994, Geodetic slip rate for the eastern California shear zone and the recurrence time of Mojave Desert earthquakes: Nature, v. 367, p. 264–266, doi:10.1038/367264a0.

Schumacher, J., Luedeling, E., Gebauer, J., Saied, A., El-Siddig, K., and Buerkert, A., 2009, Spatial expansion and water requirements of urban agriculture in Khartoum, Sudan: Journal of Arid Environments, v. 73, p. 399–406, doi:10.1016/j.jaridenv.2008.12.005.

Scherler, D., Bookhagen, B., and Strecker, M.R., 2011a, Hillslope-glacier coupling: The interplay of topography and glacial dynamics in High Asia: Journal of Geophysical Research, v. 116, no. F02019, 21 p., doi:10.1029/2010JF001751.

Scherler, D., Bookhagen, B., and Strecker, M.R., 2011b, Spatially variable response of Himalayan glaciers to climate change affected by debris cover: Nature Geoscience, v. 4, p. 156–159, doi:10.1038/ngeo1068.

Seeber, L., and Gornitz, V., 1983, River profiles along the Himalayan Arc as indicators of active tectonics: Tectonophysics, v. 92, p. 335–367, doi:10.1016/0040-1951(83)90201-9

Shroder, J.F.J., and Weihs, B.J., 2010, Geomorphology of the Lake Shewa Landslide Dam, Badakhshan, Afghanistan, Using Remote Sensing Data: Geografiska Annaler, v. 92 A, p. 469–483.

Slater, J.A., Heady, B., Kroenung, G., Curtis, W., Haase, J., Hoegemann, D., Shockley, C., and Tracy, K., 2011, Global assessment of the new ASTER Global Digital Elevation Model: Photogrammetric Engineering and Remote Sensing, v. 77, p. 335–349

Smith, L.C., and Pavelsky, T.M., 2008, Estimation of river discharge, propagation speed, and hydraulic geometry from space: Lena River, Siberia: Water Resources Research, v. 44, doi:10.1029/2007WR006133.

Solomon, S., 2011, A new look at the planet Mercury: Physics Today, v. 64, p. 50–55, doi:10.1063/1.3541945.

Stark, C.P., and Hovius, N., 2001, The characterization of landslide size distributions: Geophysical Research Letters, v. 28, p. 1091–1094, doi:10.1029/2000GL008527.

Stark, C.P., Foufoula-Georgiou, E., and Ganti, V., 2009, A nonlocal theory of sediment buffering and bedrock channel evolution: Journal of Geophysical Research, v. 114, p. F01029, doi:10.1029/2008JF000981.

Stock, J.D., and Montgomery, D.R., 1999, Geologic constraints on bedrock river incision using the stream power law: Journal of Geophysical Research. Solid Earth, v. 104, p. 4983–4993, doi:10.1029/98JB02139.

Thiede, R.C., Ehlers, T.A., Bookhagen, B., and Strecker, M.R., 2009, Erosional variability along the northwest Himalaya: Journal of Geophysical Research–Earth Surface, v. 114, F01015, 19 p., doi:10.1029/2008jf001010.

Turowski, J.M., Lague, D., and Hovius, N., 2009, Response of bedrock channel width to tectonic forcing: Insights from a numerical model, theoretical considerations, and comparison with field data: Journal of Geophysical Research, v. 114, no. F03016, 16 p., doi:10.1029/2008JF001133.

Unruh, J., Humphrey, J., and Barron, A., 2003, Transtensional model for the Sierra Nevada frontal fault system, eastern California: Geology, v. 31, p. 327–330, doi:10.1130/0091-7613(2003)031<0327:TMFTSN>2.0.CO;2.

Unruh, J.R., Hauksson, E., Monastero, F.C., Twiss, R.J., and Lewis, J.C., 2002. Seismotectonics of the Coso Range-Indian Wells Valley region, California: Transtensional deformation along the southeastern margin of the Sierran microplate, *in* Glazner, A.F., Walker, J.D., and Bartley, J.M., eds., Geologic Evolution of the Mojave Desert and Southwestern Basin and Range: Geological Society of America Memoir 195, p. 277–294.

van Westen, C., and Gorum, T., 2010, Preliminary results on earthquake triggered landslides for the Haiti earthquake (January 2010): Geophysical Research Abstracts, v. 12, p. EGU2010–EGU1153.

Whipple, K.X., 2004, Bedrock rivers and the geomorphology of active orogens: Annual Review of Earth and Planetary Sciences, v. 32, no. 1, p. 151–185, doi:10.1146/annurev.earth.32.101802.120356.

Whipple, K.X., and Tucker, G.E., 1999, Dynamics of the stream-power river incision model: Implications for height limits of mountain ranges, landscape response timescales, and research needs: Journal of Geophysical Research, Solid Earth, v. 104, p. 17661–17674, doi:10.1029/1999JB900120.

Whitmeyer, S., Nicoletti, J., and De Paor, D., 2010, The digital revolution in geologic mapping: GSA Today, v. 20, p. 4–10, doi:10.1130/GSATG70A.1.

Whittaker, A.C., Cowie, P.A., Attal, M., Tucker, G.E., and Roberts, G.P., 2007, Bedrock channel adjustment to tectonic forcing: Implications for predicting river incision rates: Geology, v. 35, no. 2, p. 103, doi:10.1130/G23106A.1.

Wobus, C.W., Tucker, G.E., and Anderson, R.S., 2006a, Self- formed bedrock channels: Geophysical Research Letters, v. 33, no. L18408, 6 p., doi:10.1029/2006GL027182.

Wobus, C., Whipple, K., Kirby, E., Snyder, N., Johnson, J., Spyropolou, K., Crosby, B., and Sheehan, D., 2006b, Tectonics from topography: Procedures, promise and pitfalls, *in* Willett, S., Hovius, N., Brandon, M., and Fisher, D., eds., Tectonics, Climate, and Landscape Evolution: Geological Society of America Special Paper 398, p. 55–74, doi:10.1130/2006.2398(04).

Wohl, E., and David, G.C.L., 2008, Consistency of scaling relations among bedrock and alluvial channels: Journal of Geophysical Research, v. 113, no. F04013, 16 p., doi:10.1029/2008JF000989.

Wolman, M.G., and Miller, J.P., 1960, Magnitude and frequency of forces in geomorphic processes: The Journal of Geology, v. 68, no. 1, p. 54–74.

Yanites, B.J., and Tucker, G.E., 2010, Controls and limits on bedrock channel geometry: Journal of Geophysical Research, v. 115, no. F04019, 17 p., doi:10.1029/2009JF001601.

Yanites, B.J., Tucker, G.E., Mueller, K.J., Chen, Y.G., Wilcox, T., Huang, S.Y., and Shi, K.W., 2010, Incision and channel morphology across active structures along the Peikang River, central Taiwan: Implications for the importance of channel width: Geological Society of America Bulletin, v. 122, no. 7–8, p. 1192–1208, doi:10.1130/B30035.1.

Zech, R., Zech, M., Kubik, P.W., Kharki, K., and Zech, W., 2009, Deglaciation and landscape history around Annapurna, Nepal, based on [10]Be surface exposure dating: Quaternary Science Reviews, v. 28, no. 11–12, p. 1106–1118, doi:10.1016/j.quascirev.2008.11.013.

Manuscript Accepted by the Society 16 April 2012

The Geological Society of America
Special Paper 492
2012

Google Earth and geologic research in remote regions of the developing world: An example from the Western Desert of Egypt

Barbara J. Tewksbury*
Geosciences Department, Hamilton College, 198 College Hill Road, Clinton, New York 13323, USA

Asmaa A.K. Dokmak
Geology Department, Faculty of Science, Alexandria University, Alexandria, Egypt

Elhamy A. Tarabees
Geology Department, Faculty of Science, Damanhour University, Damanhour, Egypt

Ahmed S. Mansour
Geology Department, Faculty of Science, Alexandria University, Alexandria, Egypt

ABSTRACT

Remote sensing is an important option for finding interesting research problems in remote regions of the world, but existing freely available imagery, such as Landsat imagery, has limitations in terms of resolution. In some remote areas, recently available high-resolution imagery in Google Earth has the potential to revolutionize the kind of research that can be initiated and carried out. This paper details an example from a remote region of Egypt's Western Desert.

Work by others on Eocene carbonates of the Drunka and El Rufuf Formations has focused on lithologic and paleontologic aspects, and previous mapping of the contact between the two formations in the Western Desert using early Landsat imagery (69 m/pixel) shows a simple contact. High-resolution imagery in Google Earth (~1 m/pixel) shows, however, that the contact is both folded and faulted. We used high-resolution images in Google Earth to define mappable subunits and to do detailed mapping of folds and faults in a 400 km² study area. Subsequent field work confirmed the accuracy of lithologic and structural mapping in Google Earth, targeted critical areas for field data collection, and provided ground truth for extending mapping into remote areas.

Freely available, high-resolution satellite imagery in Google Earth not only allows identification of research questions but is also critical in pre–field work mapping, targeting sites for field work, and disseminating research results in areas of the world where field work is difficult, funding is poor, and access to dissemination of research results outside the region is limited.

*btewksbu@hamilton.edu

Tewksbury, B.J., Dokmak, A.A.K., Tarabees, E.A., and Mansour, A.S., 2012, Google Earth and geologic research in remote regions of the developing world: An example from the Western Desert of Egypt, *in* Whitmeyer, S.J., Bailey, J.E., De Paor, D.G., and Ornduff, T., eds., Google Earth and Virtual Visualizations in Geoscience Education and Research: Geological Society of America Special Paper 492, p. 23–36, doi:10.1130/2012.2492(02). For permission to copy, contact editing@geosociety.org. © 2012 The Geological Society of America. All rights reserved.

INTRODUCTION

Remote regions of the world have literally millions of square kilometers of inaccessible terrain. Remote sensing is one of the only options for finding interesting research problems in such areas and for carrying out research both at reconnaissance and detailed levels. Desert areas in particular are ideal for remote sensing analysis because they have essentially no obscuring vegetation, water, ice, or cultural features. In many of these remote areas, satellite images are, in fact, the only imagery available for use in remote sensing analyses, as aerial photography has never been flown or is unavailable. Research in such areas bears striking similarities to research on other planets, both in terms of types of data and extremely limited access for ground truthing.

Egypt is a country with vast tracts of inaccessible desert that have comparatively little sand cover to obscure interesting bedrock features, and satellite imagery and data from orbit have been used successfully for studying geology in Egypt for over 30 years. Previous workers have used Landsat, ASTER, SIR-C, MODIS, and SRTM data to investigate topics ranging from bedrock structure (e.g., Abdelsalam et al., 2000, 2008; Alfarhan et al., 2006; El-Baz and El-Etr, 1979; El-Etr and El-Baz, 1979; El-Etr and Moustafa, 1981, 1980; McCauley et al., 1997; Stern and Abdel Salam, 1996) to surface processes (e.g., Luo et al., 1997), Nile evolution (e.g., Gani et al., 2009), and water resources (e.g., Masoud and Koike, 2006). In addition, the current 1:500,000 geologic map sheets of Egypt relied heavily on Landsat MSS images from the 1970s (Klitzsch et al., 1987).

Resolution of satellite imagery used in previous studies ranges from ~15 m/pixel (Landsat ETM panchromatic band) to ~250 m/pixel (MODIS). Despite the progress that has been made in using satellite imagery to study bedrock geology in Egypt, the comparatively low resolution of the imagery has limited research to features that can be successfully studied in images with a maximum resolution of ~15 m/pixel. Until recently (late 2008), Google Earth displayed images at these lower resolutions. As Google has replaced older Landsat imagery in Egypt with high-resolution commercial satellite imagery, it has become possible to study a whole host of features that were too small to be resolved at Landsat resolutions. High-resolution satellite imagery is now available in Google Earth for all of Egypt.

Google Earth offers a unique combination of freely available, high-resolution imagery. This paper focuses on an example of research on previously unrecognized bedrock structures in the Western Desert of Egypt that escaped previous study because they are too small for study in the Landsat, ASTER, and other imagery that has been used previously to study remote areas of the Western Desert.

THE NEED FOR HIGH-RESOLUTION IMAGERY

The structures of interest in this study range in scale from under 100 m to ~1 km in size, and features that are critical for analyzing these structures are typically a few meters to tens of meters in size. Figure 1 illustrates the crucial difference that high-resolution images make in study of these structures. Figure 1A shows a portion of our study area at the best available Landsat resolution (15 m/pixel, Landsat ETM panchromatic band). The patterns suggest that something interesting might be there, but the resolution is inadequate for analysis. Figure 1B shows the same area in Google Earth with Digital Globe imagery at ~1 m/pixel. Not only are the patterns recognizable as bedrock structures, but the structures can be analyzed in the satellite imagery. Features that were unrecognizable when represented by 100 pixels in a 10×10 pixel square are clear when represented in Google Earth by 10,000 pixels in a 100×100 pixel square.

Not all high-resolution images of Egypt in Google Earth are currently at a resolution of 1 m/pixel. SPOT imagery at ~3 m/pixel is the best that is available in some places. Although this is a significant improvement over Landsat resolutions, the contrast between what can be seen in Google Earth's highest resolution imagery from Digital Globe versus what can be seen in the slightly lower resolution SPOT images (Fig. 1C) is striking and underscores why structures such as these have not been studied previously—they have simply been "invisible" at the resolution of previous imagery.

THE ROLE OF GOOGLE EARTH

Studies such as the one described in this paper require resolutions of a few meters per pixel or less, which is an order of magnitude or more higher resolution than either Landsat or ASTER. The problem arises from the fact that, while researchers can obtain free access to NASA's Landsat and ASTER imagery, those who wish to acquire higher resolution images must purchase them from commercial vendors such as SPOT, GeoEye, or Digital Globe at prices as high as US$25/km^2, even with an educational discount.

The cost of commercial satellite imagery is affordable for detailed work on a known problem of limited extent. Purchasing high-resolution imagery for huge areas for the purpose of searching for interesting research problems, on the other hand, is outside the realm of possibility for most researchers. The problem, then, is that resolution of existing free NASA imagery is inadequate for finding interesting research problems involving small-scale features but the cost is prohibitive for purchasing large areas of high-resolution commercial imagery to find and carry out research on such structures.

Enter Google Earth. Freely available high-resolution imagery in Google Earth makes it possible to browse large areas in remote regions to find interesting research problems. In the course of our work in Egypt, one of us (Tewksbury) has combed more than one hundred thousand square kilometers of Google Earth imagery in Egypt that would have cost quite literally millions of dollars on the commercial market.

Google Earth can also be used to carry out the research, to choose critical areas for targeted "boots on the ground" field work, and to make selective purchases of commercial imagery

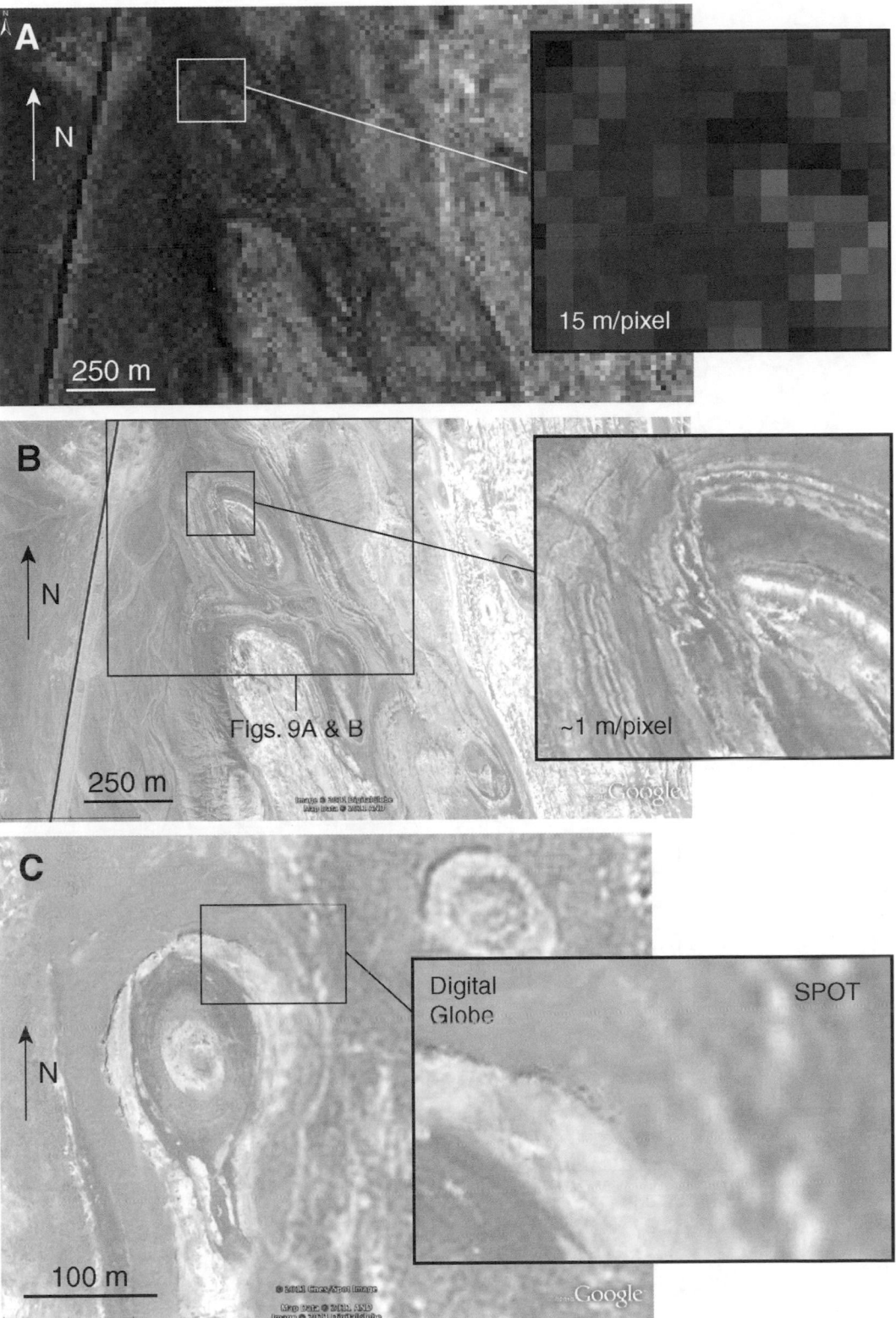

Figure 1. (A) A portion of our study area at highest Landsat resolution, with an inset showing the 15 m pixels of the Landsat ETM panchromatic band. The dark diagonal line at far left is the Asyut-El Kharga Road. (B) The same area as (A) but in Google Earth with Digital Globe imagery. The inset shows the same area as the inset in (A). Features that are enigmatic at best at Landsat resolution are mappable in Google Earth where imagery has resolution of ~1m/pixel. The red outline shows the area in Figures 9A and 9B. (C) Digital Globe imagery in Google Earth at ~1 m/pixel (left side of image capture) allows analysis of small-scale features that remain enigmatic even in SPOT imagery at ~3 m/pixel (right side of image capture). Image centers: N26.264758, E30.718018 (A and B); N26.279837, E30.801195 (C).

for stereo or multi-spectral coverage and image analysis. Furthermore, Google Earth offers many ways of disseminating research results, maps, and photographs in the developing world.

The remainder of this paper will detail an example from the Western Desert of Egypt. Although we will provide some research results to illustrate the validity of this research strategy, the paper will focus on the methodology, which is transportable to other remote regions.

EXAMPLE FROM THE WESTERN DESERT OF EGYPT

Location and Remoteness of the Study Area

The Western Desert encompasses all of Egypt west of the Nile River floodplain (Fig. 2A, shaded area), an area of ~650,000 km². Although the area has a number of oasis

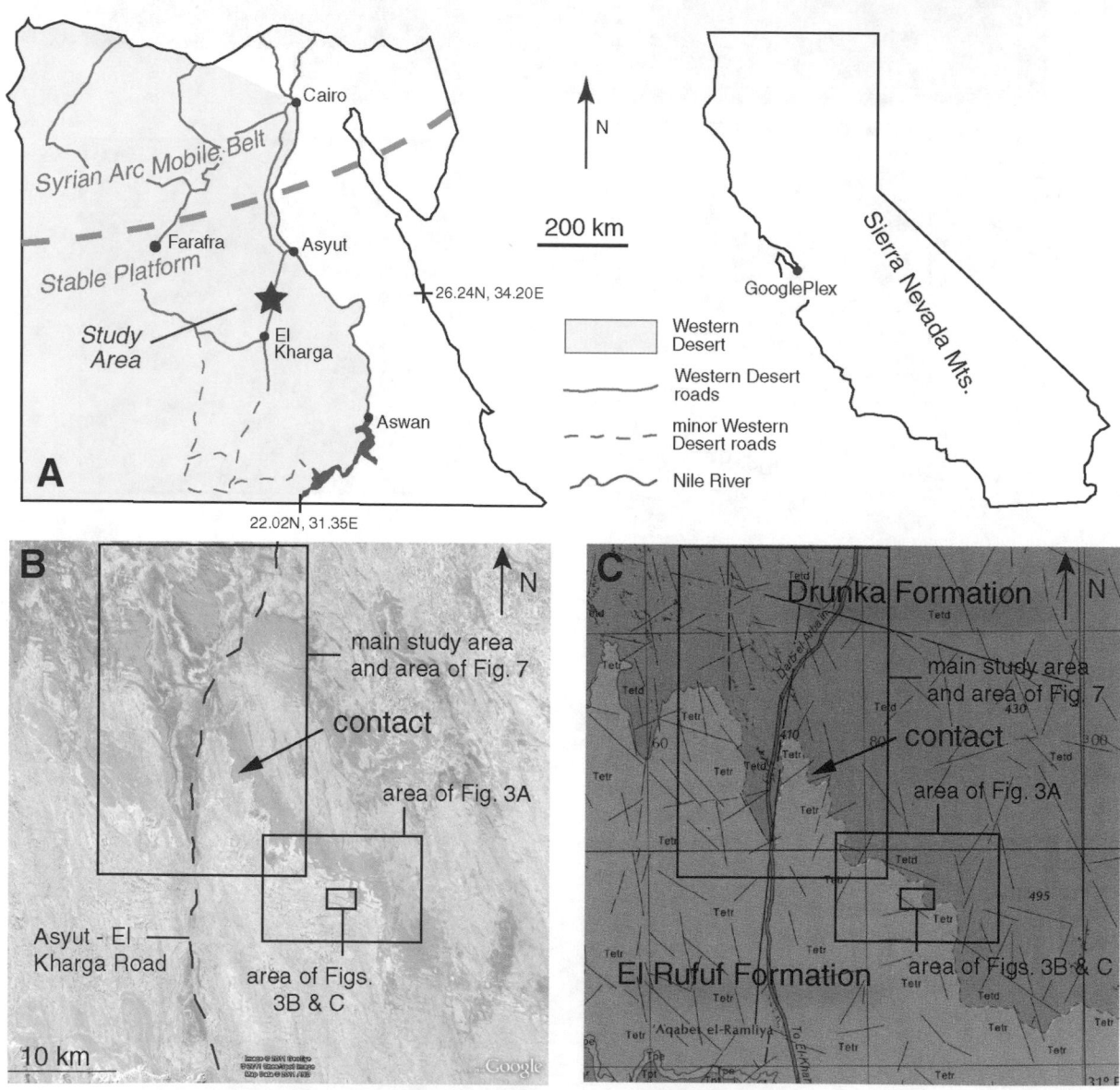

Figure 2. (A) Outline maps of Egypt and California at the same scale. The Egypt map shows our study area (red star) in the Western Desert. The general boundary between the Stable Platform and the Syrian Arc Mobile Belt is based on Bosworth et al. (1999; also referred to as the Stable versus Unstable Shelf, e.g., Meshref, 1990; Youssef, 2003). (B) Google Earth image of our main study area and surrounding region, which is bisected by the Asyut-El Kharga Road. Surface materials in the Google Earth satellite imagery range in color from white to pale brown. (C) A portion of the 1:500,000 Asyut Sheet of the Geological Map of Egypt (Klitzsch et al., 1987) covering the same area as the Google Earth image in (B). The contact between the Eocene El Rufuf Formation and the younger Drunka Formation is clearly visible in the Google Earth imagery in (B) as a change in color from white to pale brown. Solid and dashed black lines on the geologic map are lineaments based on Landsat MSS imagery. Image centers: N26.229491, E30.810484 (B and C).

settlements linked by three major roads, the vast majority of the area is roadless and untracked. Figures 2A puts into perspective just how inaccessible much of the Western Desert actually is. The Cairo-Farafra Road and the Asyut–El Kharga Road, which are spaced ~300 km apart, are the *only* roads in the central part of the Western Desert—Lake Ontario would fit length-wise between them with room to spare. Figure 2A also shows, at the same scale, the state of California, where roads near Google Headquarters in Mountain View are spaced less than 0.25 km apart. A road spacing in California comparable to the road spacing in the Western Desert would place one road along the California Coast and one road east of the Sierra Nevada Mountains!

Our study area (red star in Fig. 2A and larger red outline in Fig. 2B) lies ~125 km SW of Asyut along the Asyut–El Kharga Road. The area that we mapped on Google Earth images for this project covers an area of ~400 km^2 and extends ~10 km both east and west of the Asyut–El Kharga Road. Ground truthing in the field was limited by difficulty of access to locations within ~1 km of the road.

Previous Work and Project Evolution

The Stable Platform of central Egypt (Fig. 2A) consists of a sequence of Late Mesozoic through Miocene sediments lying unconformably on Precambrian basement. The sequence generally dips very gently north, exposing the oldest rocks in the south. The latest Cretaceous through Miocene section consists almost exclusively of shallow marine carbonates (Said, 1990).

The study area lies in Eocene carbonates of the Drunka and El Rufuf Formations. Previous work by others has targeted general sedimentology, stratigraphy, and paleontology (e.g., Hassaan et al., 1993; Keheila and El-Ayyat, 1990; Rashed and Sediek, 1997) and includes focused studies on the origin of chert in the carbonates (Elshishtawy et al., 1997) and unusual large concretions in the Drunka Formation (McBride et al., 1999). Measured sections are limited to one drill hole (Barakat and Asaad, 1965) and to rare areas with significant vertical exposure along the Nile escarpment (e.g., Kenawy et al., 1988), in small quarries along the Asyut–El Kharga Road, and in the escarpment bordering the El Kharga Valley (Hermina, 1967). Very limited work has been published on yardangs in the area of the Western Desert (Grolier et al., 1980). The contact between the Drunka and El Rufuf Formations has been mapped on the Asyut Sheet of the 1:500,000 Geological Map of Egypt (Klitzsch et al., 1987) (Fig. 2C). Essentially the only structural work that has been done in the study area is the mapping of lineaments for the Asyut regional geology sheet using Landsat MSS imagery.

While exploring new high-resolution imagery in Google Earth in 2009 along the Drunka–El Rufuf contact, Tewksbury was intrigued by the complexity of patterns (Fig. 3A) that suggest that the contact is not as simple as portrayed on the Asyut geologic map sheet. Both vertical and oblique views of SPOT images in Google Earth (Figs. 3C and 3D) suggest that the pattern results from the presence of open folds, something that is not at all clear at Landsat resolutions (Fig. 3B). Reconnaissance mapping by Tewksbury and Hamilton undergraduate Devin Farkas using SPOT imagery in Google Earth confirmed the presence of many open, shallowly plunging narrow synclines trending both WNW and NS and separated by broad anticlinal or domical structures (Tewksbury et al., 2009).

The study area for the work in Google Earth described in the previous paragraph lies ~20 km east of the Asyut–El Kharga Road in an area not easily accessible in the field. When the opportunity arose for Alexandria University Masters student Asmaa Dokmak to do both work in Google Earth and field ground truthing, we chose an area along the same contact but farther west where the study area is accessible from the Asyut–El Kharga Road and where Digital Globe imagery offers higher resolution than the SPOT images shown in Figure 3. The purpose of this study has been to conduct pre–field work mapping and structural analysis over a large area using imagery in Google Earth, to field check critical areas where they are accessible along the road, and to generalize those field results to validate and improve our mapping in the inaccessible portions of the area.

Methods for Mapping Bedrock Structures in Google Earth

Conducting Virtual Field Work

We carried out our reconnaissance mapping using a combination of on-screen Google Earth and large format prints of our study area. Tewksbury and Dokmak were in the United States and Egypt, respectively, and each needed to be able to easily understand what the other was doing. We took all of our "field" notes in Google Earth placemarks and saved them as KMZ files (Keyhole Markup Language [KML] files that have been zipped). Not only is it easy to email KMZ files back and forth, but each placemark and its data are linked directly to a spot in Google Earth, making it easy to see what colleagues are thinking and to have discussions using Skype. Furthermore, data and observations in individual placemarks can easily be added to and updated as work proceeds.

We use a standard numbering scheme for Google Earth placemarks, with unique project and worker identifiers, as one would do with standard field notes. In the past two years, Tewksbury has used this same note-taking strategy successfully for mapping in Google Earth on four other projects in Egypt with nine collaborators spread over seven institutions in the United States and Egypt (Tewksbury et al., 2010, 2011a, 2011b). As with any electronic note-taking scheme, saving often and making electronic backups with clear date designations or version numbers is essential. To make a printable file of a KML file with placemarks, it is a simple matter to import the KML file into Google Fusion Tables. Each placemark becomes a row in the Fusion Table, and placemark name, description, and geometry appear in separate columns.

We divided our reconnaissance mapping area into sections and used the capability of Google Earth Pro to save a high-resolution image of each section. (Note: for adjacent images to match perfectly, terrain must be turned *off* when saving the image.) We

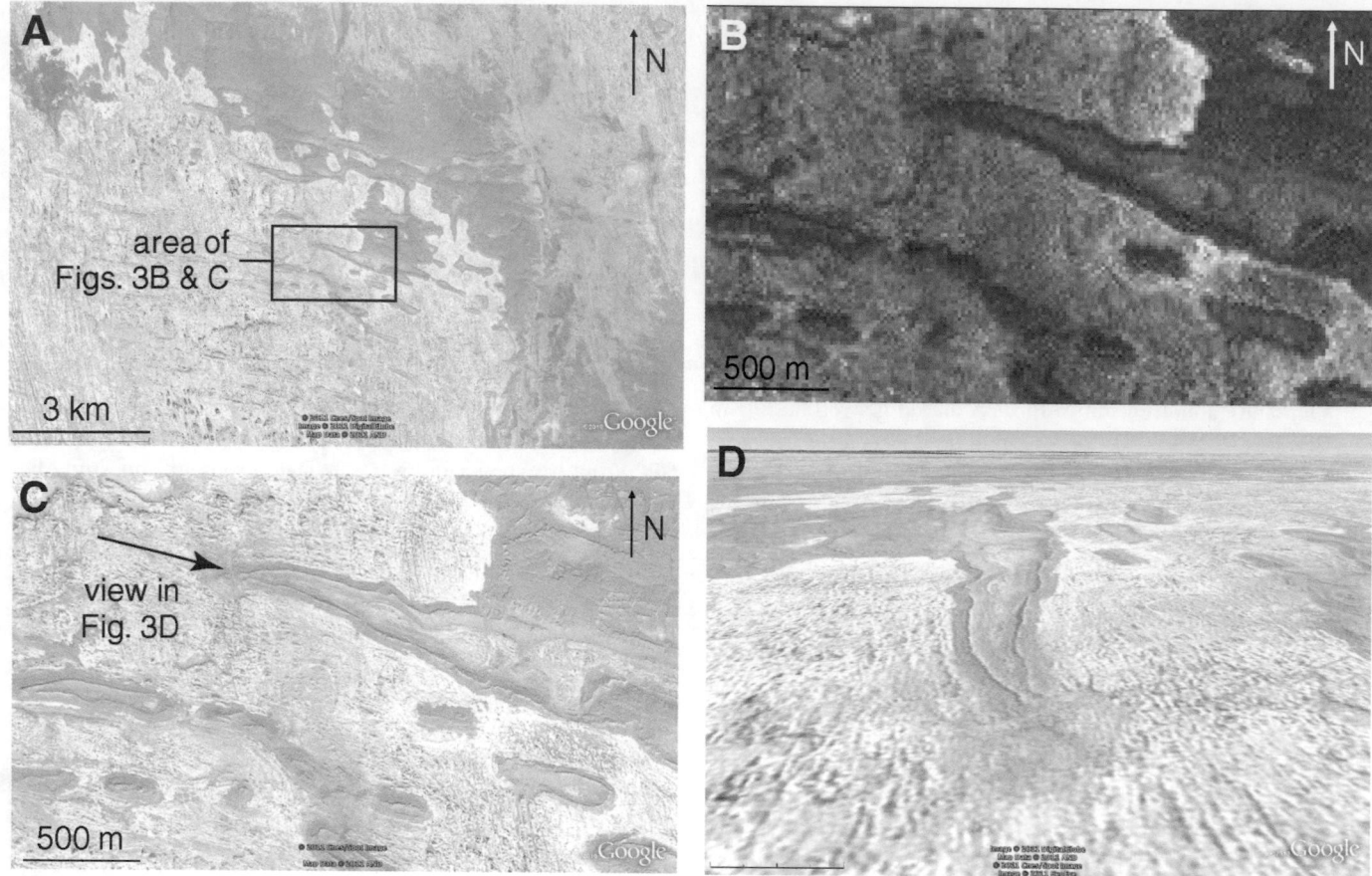

Figure 3. (A) In high-resolution imagery in Google Earth, the outcrop trace of the contact between the El Rufuf Formation (white) and the Drunka Formation (pale brown) displays a complex pattern. The location box for this figure is shown in Figures 2B and 2C. (B) Landsat ETM panchromatic imagery (15 m/pixel) does not have high enough resolution to resolve the nature of the complexity of the outcrop trace. (C) Imagery in Google Earth has high enough resolution to show that the elongate pale-brown patches of the Drunka Formation lie in the keels of narrow, open, WNW-ESE–trending synclines (Tewksbury et al., 2009). (D) View in Google Earth looking ESE along the main synclinal keel in (C). Image centers: N26.169381, E30.850278 (A); N26.163600, E30.848134 (B and C); N26.163840, E30.853516 (D).

printed each image out at 55 cm × 72 cm using a large-format color printer and did our mapping on high-transparency Mylar overlays. We found that having Google Earth up on the computer screen for adding notes in placemarks, coupled with mapping by hand on an overlay, was a good combination for this complex area. This strategy allowed us to zoom in closely on the computer screen but to do the mapping in a larger context on the paper printout. Mapping by hand in pencil also allowed us to make tentative mapping decisions in uncertain areas and to make changes and additions easily, both of which are more difficult when mapping digitally.

Data Source and Quality

In the study area, current Google Earth imagery comes from Digital Globe and has a resolution of ~1 m/pixel or less. Digital Globe's QuickBird and WorldView 2 platforms are both Sun-synchronous, and both image this part of Egypt at ~10:30 a.m. local time. This means that Sun illumination is consistently from the SE quadrant. Elevation data are available for the study area only from the Shuttle Radar Topography Mission (SRTM;

90 m/pixel horizontal resolution and 16 m vertical resolution) and from the ASTER G-DEM (30 m/pixel horizontal resolution and 20 m vertical resolution).

Determining Dip Direction

Mapping fold structures in the study area on Google Earth images requires reliable determination of dip direction. In the absence of high-resolution elevation data, dip direction must be inferred from the pattern of outcrop traces of contacts where they intersect the topographic surface. These interpretive techniques are typically used in areas with moderate to steep dips where tens to hundreds of meters of topographic relief produce significant valleys, hogbacks, and flatirons (Fig. 4A).

In our study area, however, the local topographic relief is generally very low, commonly only a few meters, and the area is not dissected by integrated drainage networks. In such a terrain, one might think that it would be ineffective to use outcrop patterns in imagery to infer dip direction. Surprisingly, slight differences in erosional resistance in the interlayered carbonates produce low ridges that are cut by tiny wadis, producing mini-hogbacks, mini-flatirons, and mini-scalloped ridges (Fig. 4B). Despite the fact that these flatirons are one to two orders of magnitude smaller than those typically seen in mountainous regions, they serve as good dip indicators in the Western Desert. Our study area has a fortuitous combination of shallow dip and very low topographic relief. If dips were steeper, the very low local relief would not be enough to create interpretable outcrop patterns. If the local relief were higher and the area more thoroughly dissected by integrated drainage networks, the subtle outcrop patterns would be masked.

We employed a variety of strategies to interpret dip direction in the Google Earth images. Figures 5A and 5B show the typical pattern of Vs that develops in the outcrop traces of slightly

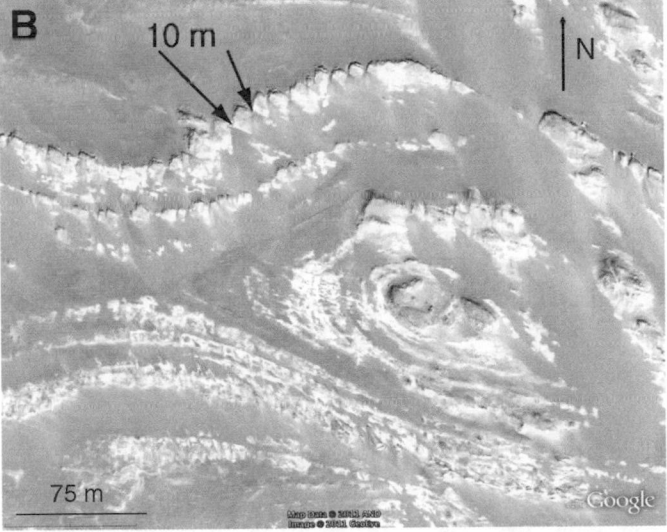

Figure 4. (A) The flatirons on the flank of this whaleback anticline in the Zagros Mountains are typical of those in areas of moderate to steep dips and significant topographic relief. (B) Mini-flatirons occur in areas of shallowly dipping sedimentary layers and low topographic relief in the Western Desert. Despite the fact that these flatirons are one to two orders of magnitude smaller than those typically seen in mountainous regions, they serve as good dip indicators in the Western Desert. Image centers: N27.455276, E55.562232 (A); N26.238550, E31.004897 (B).

more resistant units where they are eroded by small wadis. The Google Earth image in Figure 5C shows a scalloped ridge of mini-hogbacks at "a" on the north side of an eye-shaped structure. The locations of wadis with respect to scallops and flatirons indicate a southward dip. With very few exceptions in our study area, the outcrop patterns follow the "Rule of Vs," i.e., the Vs in the valleys point down dip. Local relief is not high enough relative to dip amounts to create the exception to the rule.

Figure 5 also illustrates how the brightness of scarps and dip slopes in eroded dipping layers of different resistance depends on the Sun illumination direction relative to the dip direction. Where layers dip toward the Sun, (Fig. 5A), the Sun-facing dip slope is brightly but broadly illuminated, and a very sharp narrow shadow occurs where the scarp that cuts across the unit faces away from the Sun. Where resistant layers dip away from the Sun (Fig. 5B), the scarp faces the Sun and is narrow and brightly lit, whereas the dip slope, which faces away from the Sun, is broad and more diffusely lit (and commonly somewhat shadowed). Figure 5C shows a dark scarp and bright dip slope at "A," and a bright scarp and partly shadowed dip slope at "B," indicating dips to the SSW and NW respectively. Both are consistent with the dip direction determined from the Vs in the small wadis, and all of the dip indicators taken together suggest that the structure in Figure 5C is a very small, elongate, and somewhat cuspate basin.

Figure 5D also illustrates another technique that we used where scarps and flatirons are not present. The Google Earth image shows the typical straight line outcrop trace of the upper contact of a shallowly dipping unit where it meets a less resistant overlying unit. By contrast, the outcrop trace of the lower contact is typically ragged and irregular or scalloped. This pattern is particularly useful where layers are thin and scarps are too low to cast significant shadows.

The critical requirement for determining dip direction using outcrop patterns in this area is high-resolution imagery. Figure 1C shows an area on the eastern margin of our study area where high-resolution Digital Globe imagery in Google Earth meets somewhat lower resolution SPOT imagery (1 m/pixel as opposed to 3 m/pixel). Dip slopes, small scarps, and even individual concretions are visible in the higher resolution imagery, and features suggest shallow inward dips in a slightly elongate basin. The resolution in the SPOT imagery to the east would be inadequate to make that determination.

In our study area, we are able to determine dip direction in Google Earth images but not dip amount. Google Earth elevation data in this part of the world come from the SRTM data set and have a horizontal resolution of 90 m/pixel. Given the small horizontal scale of flatirons and Vs in our study area, coupled with less than a few meters of vertical relief, we would need elevation data at typical LiDAR (light detection and ranging) resolutions in order to calculate dip amounts.

Determining Mappable Sub-units

Mapping structures in our study area in Google Earth required developing a stratigraphy of sub-units so that we could

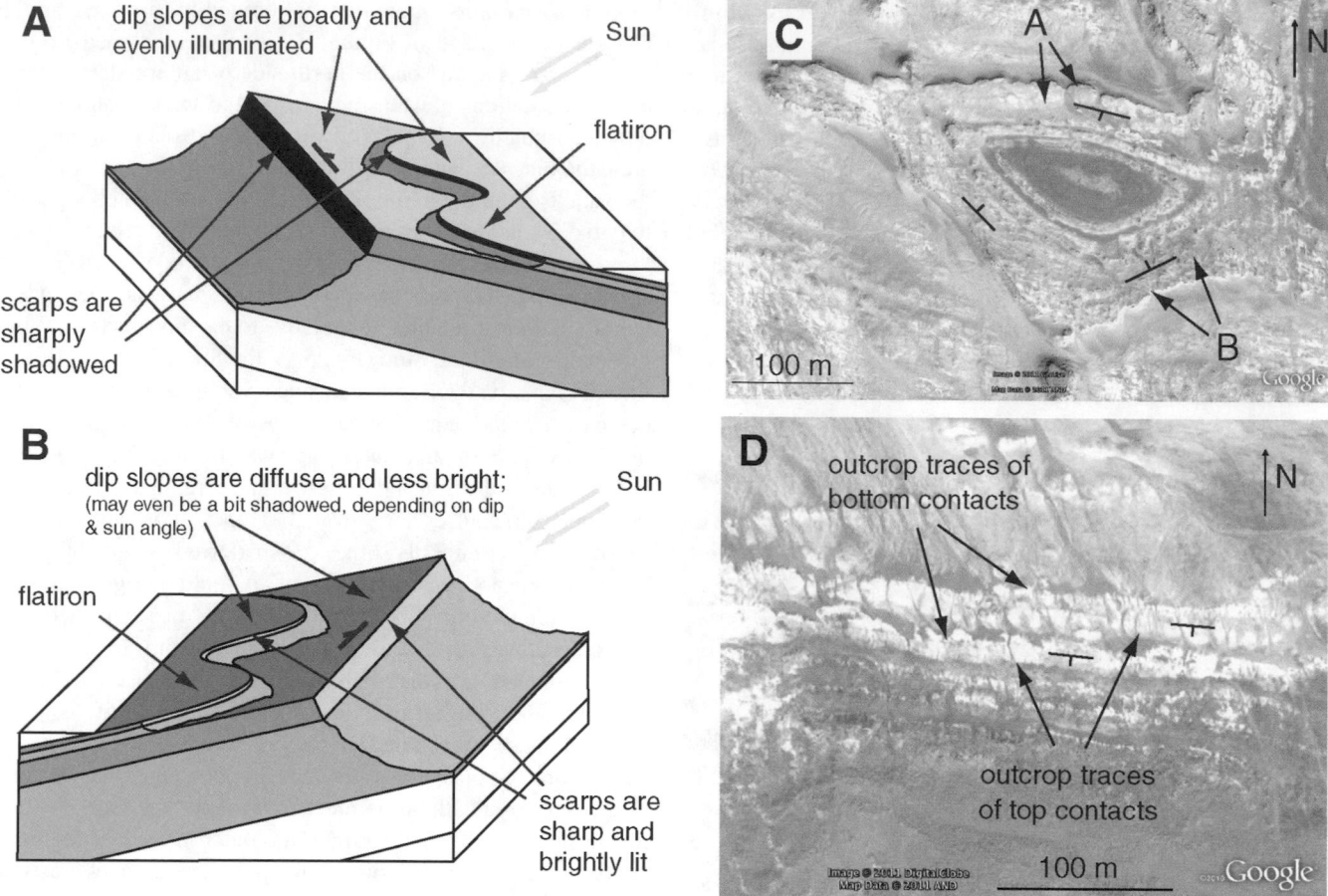

Figure 5. (A and B) Subtle differences in erosional resistance allow determination of dip direction in our bedrock layers despite the fact that both topographic relief and dip are low. Features include mini-flatirons and scalloped ridges with down-dip Vs in wadis and differences in brightness of scarps and dip slopes depending on dip direction. (C) This small eye-shaped structure has mini-flatirons, a scarp shadow, and bright dip slopes on the north side, and bright scarps and somewhat shadowed dip slopes on the south side, indicating inward dips in a very small, elongate, and slightly cuspate basin. (D) South-dipping layers showing ragged and irregular traces of lower contacts and straight traces of upper contacts. Image centers: N26.276515, E31.032348 (C); N26.293425, E30.688310 (D).

map not only the Drunka–El Rufuf contact but also structures within both the upper El Rufuf and the lower Drunka Formations on Google Earth images. Both formations in our study area have been described in the literature as lagoonal and platform limestones with concretion horizons and local chert bands, particularly in the Drunka Formation. Google Earth images of our study area show clearly that, within these limestones, sub-units of slightly different character are recognizable and can be traced over long distances.

Existing published descriptions of these two formations are either too general or are written for sections too far removed from our study area to allow us to correlate sub-units that we could see in Google Earth imagery with descriptions based on field data collected by previous workers. Consequently, we defined our own local stratigraphy and set of sub-units for mapping based on characteristics that we could see in Google Earth images.

The Google Earth images show three different types of units within our study area (Fig. 6): (1) a white unit (Fig. 6B), commonly

characterized by prominent jointing and, where thick enough, prominent yardangs; (2) a thin, pale-brown unit that holds up mini-hogbacks with small scarps on one side and, on the other side, prominent dip slopes decorated with a distinctive speckled texture (Figs. 6B and 6C); and (3) a gray-brown unit with no scarps and hogbacks and no yardangs (Fig. 6D, units 5, 7, and 9).

Without elevation data, we were unable to determine sub-unit thicknesses from the image data, but we inferred from the outcrop patterns that most of our sub-units are probably meters thick, rather than many 10s to 100s of meters thick. Outcrop widths of our units vary across the study area from a few meters to several kilometers. We found nothing in the literature to suggest that actual layer thicknesses varies significantly, and we reasoned that the narrowest outcrop widths likely occur where dips are steeper and that extreme outcrop widths occur where bedding surfaces of a more resistant layer have been stripped of overlying softer layers and exposed as a horizontal or slightly undulating surface over hundreds of meters to many kilometers.

The Google Earth imagery also shows two different surficial deposits, a tan unit and a dark-gray unit, that locally blanket the bedrock (Fig. 6B, arrows). We interpreted the pale-tan, smooth-textured material as aeolian sand. The dark-gray deposit also appears to reflect prevailing wind direction and is commonly associated with the aeolian sand, but we were unable to determine its origin from the Google Earth images. Because we are interested in bedrock structure, we did not map either deposit in our reconnaissance work.

Results of Pre–Field Work Mapping

By combining dip determinations in Google Earth with mappable units, we rapidly realized that our stratigraphy consists of nine sub-units (Fig. 6A): four different white rock units (1, 3, 6,

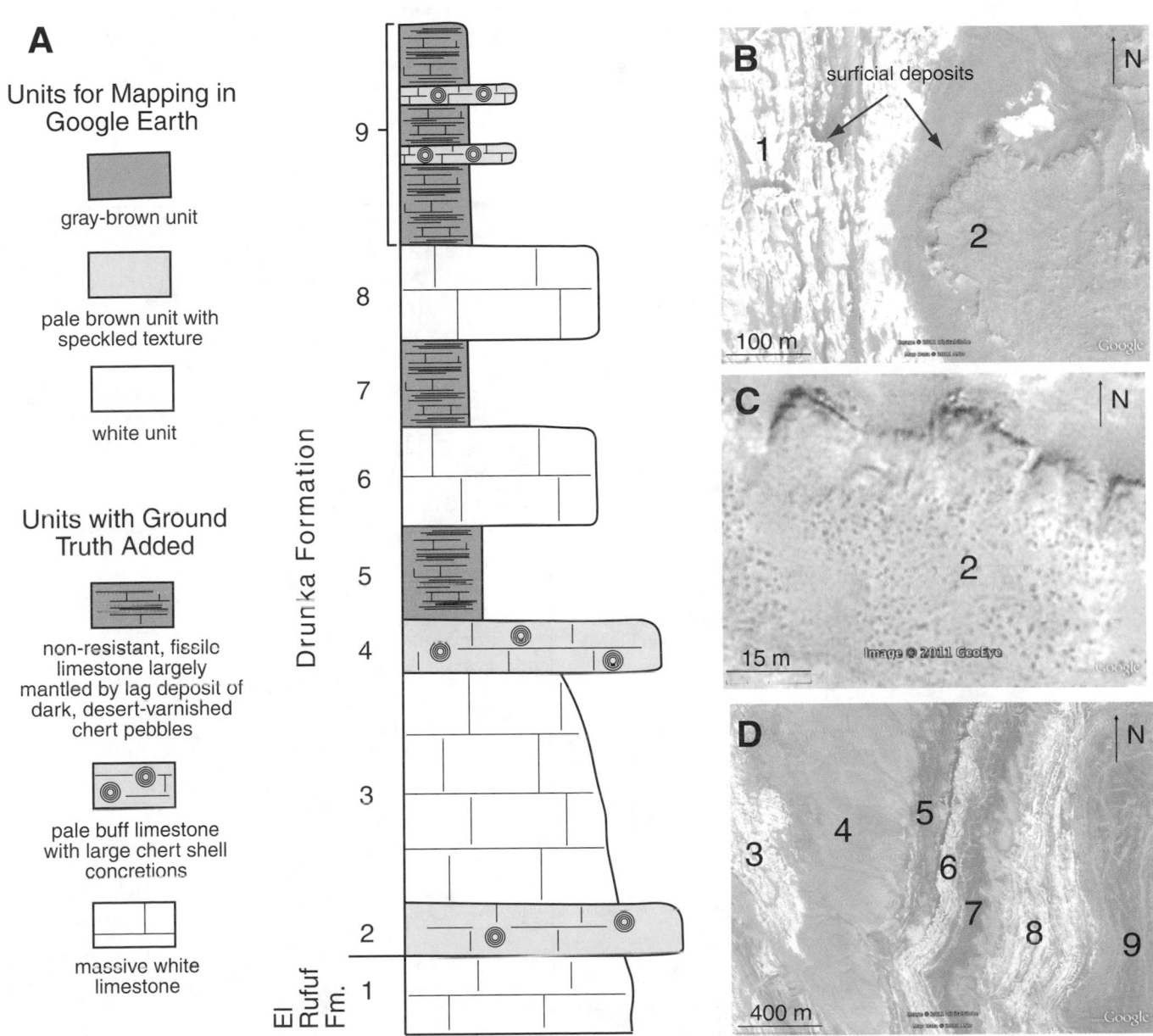

Figure 6. (A) For mapping in Google Earth, we established a stratigraphy consisting of one subunit in the El-Rufuf Formation and eight subunits in the Drunka Formation. The top legend and images B–D show the characteristics of the three different surface types that define these subunits, and the stratigraphic column shows the order of the subunits. The bottom legend shows ground truth information on each of the unit types, which are pictured in the field in Figure 8. (B) Google Earth image showing Unit 1 (one of the white units with prominent yardangs), and Unit 2 (one of the thin, pale-brown resistant units). Arrows point to tan and dark-gray surficial deposits (which we did not map) that reflect prevailing wind direction. (C) Google Earth close-up showing the speckled texture of units 2 (also characteristic of unit 4). (D) Google Earth image showing Units 3, 6, and 8 (white units), Unit 4 (a pale-brown unit), and Units 5, 7, and 9 (gray-brown units). Image centers: N26.283681, E30.744191 (B); N26.311928, E31.101309 (C); N26.293425, E30.688310 (D).

and 8), two different pale-tan, speckled units (2 and 4), and three different gray-brown units (5, 7, and 9). All but Unit 1 are subunits within the Drunka Formation. Combining dip direction and stratigraphy allowed us not only to map small-scale structures within the lower part of the Drunka Formation but also larger-scale structures, including faults, across the full study area.

Prior to going into the field in December of 2010, we had mapped the entire study area on the Google Earth images. On a casual look in Google Earth, this area looks splotchy and irregularly patterned (Fig. 7A). Our mapping revealed, however, that the color variations reflect the stratigraphy and that the splotchiness is fundamentally structural. The geologic map in Figure 7B shows a generalized version of our mapping results, emphasizing

large-scale features and stratigraphy. Unit 4 cores two very broad north-plunging anticlines and one broad, central, low-amplitude dome. The central dome is separated from the two broad anticlines by open, narrower, synclinal structures, one trending NNW and one trending NNE, that merge south into one syncline. The southern syncline lies parallel to a major NS-striking fault that continues south to bound the eastern side of the El Kharga Valley. Furthermore, the pervasive "stripiness" of much of the western and northern portions of the study area (and, indeed, much of the Western Desert between here and the Nile) is due to the presence of narrow, open, WNW-ESE–trending, synclinal structures that are typically several 100 m wide separated by broader, low amplitude, anticlinal structures (Fig. 7C).

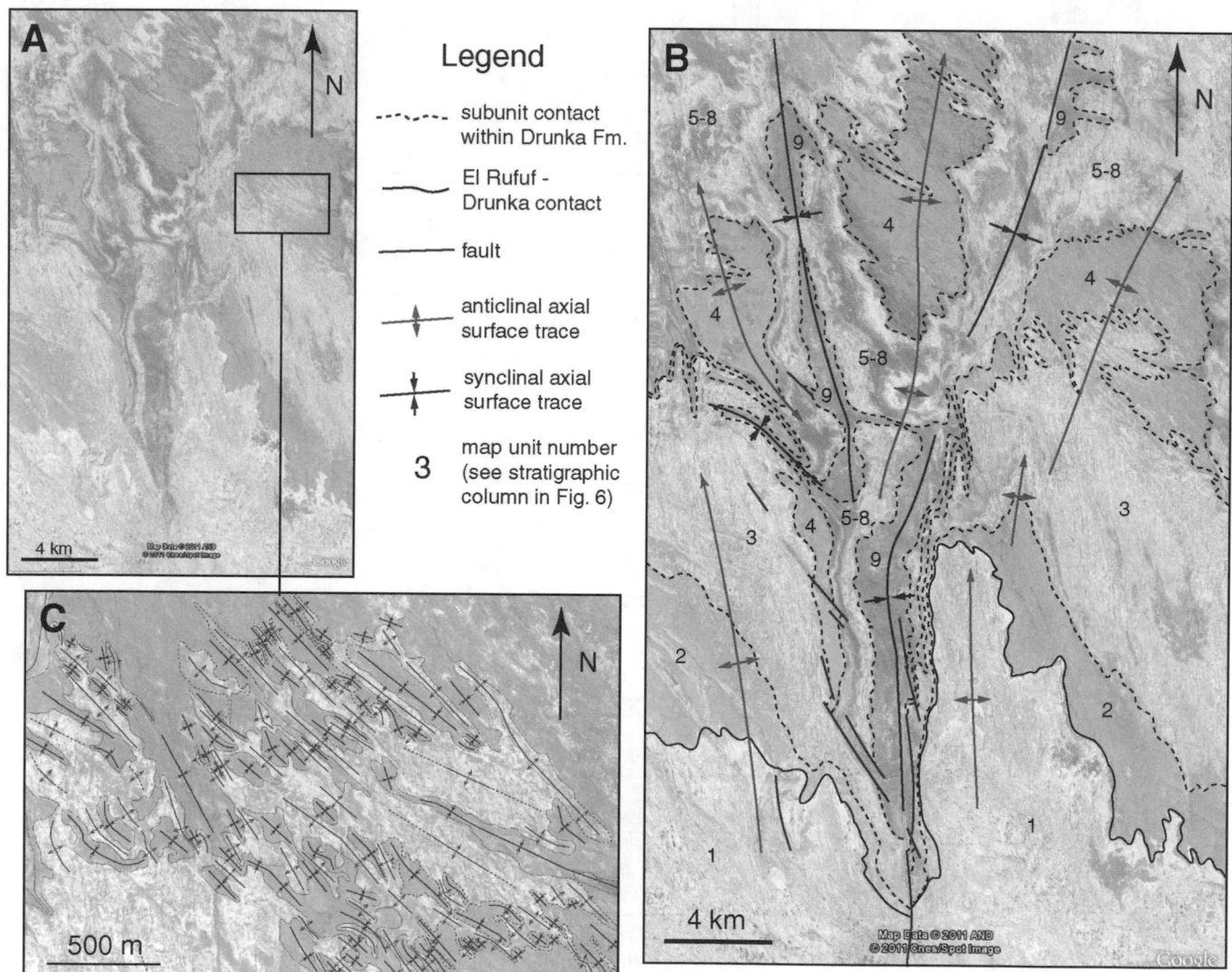

Figure 7. (A) The Google Earth image of our map area shows prominent broad NS trends in color patterning and a finer-scale WNW-ESE "stripiness." (B) A generalized geologic map, which shows only main structures and contacts, illustrates that the broad NS patterning results primarily from broad, open anticlinal and synclinal structures that repeat the stratigraphy, with faults becoming more prominent to the south. See Figure 6 for the stratigraphic column. (C) A more detailed geologic map of a portion of the area reveals that the WNW-ESE patterning is due to dozens of narrow, open synclines and broader anticlines. Image centers: N26.325800, E30.721515 (A and B); N26.357718, E30.778587 (C).

The high resolution of imagery in Google Earth also clearly illuminates the nature of the lineaments mapped on the 1987 1:500,000 geologic map sheets (Klitzsch et al., 1987) (see, for example, the lineaments in Fig. 2C). Because the map authors had access only to low-resolution Landsat MSS images (at 79 m/pixel), they mapped lineaments wherever they saw linear trends in the satellite imagery. Our study has revealed that the vast majority of these lineaments are, in fact, the narrow synclinal structures described above, rather than faults.

Google Earth and Field Work

After having mapped the study area using the imagery in Google Earth, we were fortunate to have an opportunity to field check portions of our mapping in order to determine whether our basic strategy for dip determination in the Google Earth images produced accurate results, to verify that our maps units are indeed stratigraphic units, and to test our interpretation of both fold and fault structures with data collected in the field. Because field access to this area was very limited both in time and in distance that we could travel from the Asyut–El Kharga Road, our strategy was to choose a small number of carefully targeted areas to validate critical interpretations so that we could extend these, in principle, across the inaccessible portions of the study area where similar features occur.

Google Earth played a crucial role in our field work. Each night prior to going into the field, we determined the critical target for the following day based on our pre–field work mapping (Fig. 9A), and we used Google Earth to determine coordinates for the locations. Because we did not have reliable Internet access in the field itself, we wrote the coordinates directly on the back of our high-resolution Google Earth printouts and used a hand-held GPS device to locate our targets in the field.

Most critically, our field work showed clearly that our interpretations of dip directions and fold structures based on high-resolution imagery in Google Earth were spot on. Structures that we had mapped as folds are, in fact, folds. Dips are shallow in most places, as we had inferred, and units are on the order of meters, rather than 10s or 100s of meters thick. Features that we had inferred as faults on the Google Earth images have fault characteristics in the field.

Our map units, on the other hand, held a few surprises. The white rock of map units 1, 3, 6, and 8 is, indeed, a dense, white, crystalline limestone that is spectacularly wind fluted in many places (Fig. 8A). The thin, pale-brown resistant rock of map units 2 and 4 is a pale-buff, slightly purplish, siliceous limestone with large concretions up to 1 m or more in diameter scattered across the dip slopes (Fig. 8B). These concretions are what give rise to the distinctive and widespread speckled texture in the high-resolution Google Earth images.

Gray-brown map units 5, 7, and 9, however, turned out to be surficial lag deposits, rather than bedrock. The dark color is due to abundant chert fragments coated with dark desert varnish (Fig. 8C). In a few rare places, we encountered small outcrops

Figure 8. (A) The white rock unit in the field is a dense, white, crystalline limestone occurring as both low, wind-fluted outcrops (inset, foreground) and as yardangs (inset, background). The main image shows a thin, shallowly dipping layer of the Unit 3 white limestone and a line of large concretions marking the outcrop of dipping Unit 4 in the middle distance. (B) The speckled pale-brown unit in the field is a pale buff, slightly purplish, siliceous limestone with large concretions up to 1 m or more in diameter. In the main image, the people are walking on a dip slope in Unit 2. The middle distance shows yardangs in the white rock of the underlying Unit 1, and the far distance shows resistant Unit 2 holding up the near rim of an elongate, open basin trending parallel to the plane of the photograph. The inset photo shows a close-up of the large concretions weathered out of Unit 2 (on the skyline) and accumulating at the base of the slope on top of the white limestone of Unit 1 (foreground). (C) The gray-brown unit turned out in the field to be a surficial deposit consisting of abundant chert fragments coated with dark desert varnish. Rare outcrops of pale-buff, fissile limestone suggest that the chert lag deposit blankets a largely non-resistant sequence in the stratigraphy. White limestone of Unit 6 appears at the left in the main image.

of pale buff, fissile, slightly marly limestone poking up through the surface lag, from which we infer that the chert lag deposit blankets a largely non-resistant sequence in the stratigraphy (Fig. 6A). Where this dark lag deposit is confined between the outcrop areas of two adjacent bedrock units, it is a good proxy for the non-resistant bedrock unit that immediately underlies it (Fig. 9) and allows accurate mapping of bedrock units in Google Earth

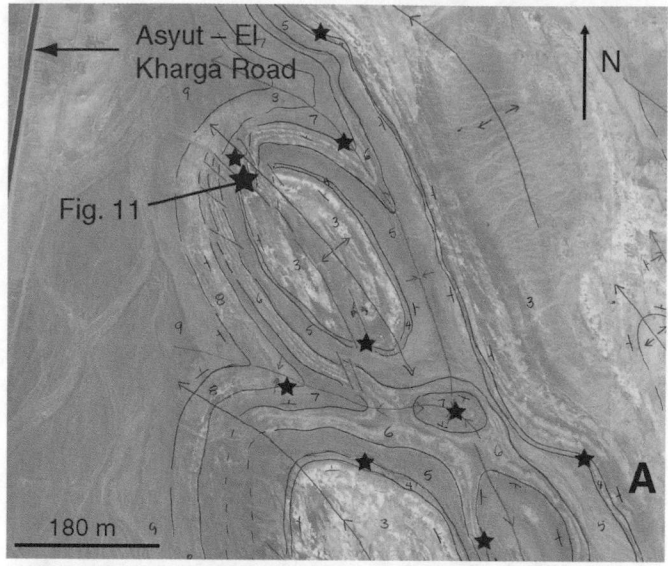

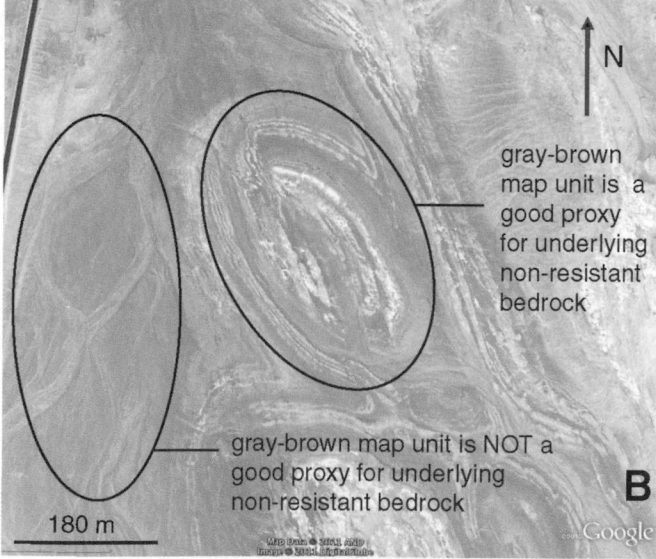

Figure 9. (A) Preliminary mapping in Google Earth defined stratigraphy and structure and allowed us to choose target areas (brown stars) to ground truth both the mapping units and the structures. (B) Despite the fact that our gray-brown mapping unit turned out to be a surficial deposit, it still proved useful in mapping where it was confined between the outcrop areas of two adjacent bedrock units, as in the area circled in the center of (B). Where the dark lag deposit has been redistributed by past sheetwash and fluvial activity, it is not a good proxy for an underlying non-resistant bedrock layer and helps explain why both structure and stratigraphy are impossible to map in areas such as that circled at the left of (B). Image center: N26.265632, E30.716359.

images. In some of the wider low areas in the study area, however, this dark lag deposit has been redistributed by past sheetwash and fluvial activity and, in such places, is not a good proxy for an underlying non-resistant bedrock layer (Fig. 9). This kind of field data is crucial to feed back into the mapping process for remote areas in Google Earth and helps explain why structures are commonly difficult to sort out in the gray-brown areas.

WHAT DIFFERENCE DID GOOGLE EARTH MAKE?

The fact that these structures have been missed previously by geologists in the field is a function, in part, of the remoteness of the area and the incredibly difficult access to virtually the entire area. Geologists have simply not combed this area over the course of decades as they have other areas with better accessibility. But even where roads cross these structures, the nature of both the terrain and the bedrock geology conspire against a geologist noticing these structures from the ground. The Asyut–El Kharga Road traverses a region with three types of terrain: (1) areas dominated by wind-sculpted yardangs (Fig. 10A); (2) vast, flat expanses of desert pavement with irregular, very low exposures of white limestone (Fig. 10B); and (3) surfaces with scattered large concretions (Fig. 10C). Bedding in the limestones is commonly massive and difficult to see at a distance, and many outcrops are dominated by wind-scoured surfaces that cut across dip. These features catch the eye from road level, not the structures that show up so clearly in Google Earth.

When these terrain features are combined with the fact that the structures have low dips and are fairly large (typically 100–300 m across), it is easy to see why they were missed from the ground by other geologists. For example, the faulted dome in Figure 9 lies less than 250 m from the Asyut–El Kharga Road but is difficult to see even when one is standing on top of the structure (Fig. 11), let alone from the road, because the wind-scoured bedrock exposures are only a few centimeters high and the structure is large enough that it is impossible to see in its entirety from one vantage point.

Google Earth was also crucial for mapping these structures efficiently. Even in the accessible portions along the Asyut–El Kharga Road, it would have taken weeks to months of careful field work to trace contacts and structures. By contrast, Google Earth allowed us to map structures ahead of time and to plan specific points for sampling and data collection both to field check and to augment our mapping in Google Earth (Fig. 9B).

CONCLUSIONS: THE POTENTIAL OF GOOGLE EARTH

Google Earth proved instrumental in many ways in this research project. Perhaps most important is the role that Google Earth played in finding this research problem in the first place. As far as we can determine, these structures have never been recognized before, even in satellite images. As mentioned at the start of this paper, these structures are just small enough to have been enigmatic at best, if not essentially invisible, in older freely available satellite imagery. Although high-resolution commercial satellite imagery can be purchased at significant cost, the paradox remains that one would be unlikely to purchase costly high-resolution satellite imagery for large areas to look for structures that one didn't know were there in the first place. High-resolution imagery in Google Earth has changed the landscape of finding structures such as these, and for studying them as well.

Finding interesting research problems in high-resolution Google Earth imagery is not limited to bedrock structures or to Egypt. Spending any time in Google Earth in the desert regions of North Africa reveals a whole host of interesting features in both the surficial and bedrock geology that have been little studied previously, and the same can be said for other remote regions of the world.

Reconnaissance using Google Earth imagery is also an outstanding and virtually cost-free strategy for collecting enough data to make a credible proposal for research funding. Reconnaissance work that Tewksbury and several Hamilton undergraduates did on structures of the Sinn El Kaddab Plateau (Tewksbury et al., 2009) formed the basis for a successful NSF-IRES (National Science Foundation–International Research Experiences for Students) proposal to fund three years of mapping in Google Earth and field work in Egypt (Tewksbury et al., 2011b). Additional work in El Sett Tellaal in Egypt (Tewksbury et al., 2010) has provided a strong rationale and supporting data for another proposal. Neither would have been possible without the freely available, high-resolution imagery in Google Earth.

Using Google Earth to find research projects that can be carried out essentially entirely in Google Earth is particularly valuable for geologists from the developing world. Whereas support is commonly available for applied geological research (e.g., water resources, engineering geology), funding is more difficult to come by for basic geologic research. The ready availability of the Internet in countries such as Egypt makes a Google Earth–based approach particularly attractive for basic research and collaboration. Furthermore, Google Earth offers a platform for wide and low-cost dissemination of research results in the developing world, including maps, cross sections, field and sample data, and

Figure 10. The three terrain types in the study area conspire against a geologist on the ground noticing the structures that we mapped easily in Google Earth. (A) Bedding is difficult to see in wind-sculpted yardangs in the massive white limestone. (B) Open fold structures several 100 m across are almost impossible to see in the vast, flat expanses of desert pavement with irregular, very low exposures of wind-fluted white limestone. In this image, the dark patch in the middle distance cores an elongate structural basin ~200 m wide and 600 m long. Dips in the limestones in the foreground toward the center of the basin are almost impossible to see because of wind fluting that cuts across bedding. (C) Concretions up to 1 m across, such as these at the Valley of the Watermelons, occur both on bedding surfaces and on surfaces that cut subtly across bedding where weathering and erosion has released resistant concretions from overlying layers.

Figure 11. The prominent dome structure shown in Figure 9A lies in an area with essentially no topographic relief. This photograph was taken at the large red star in Figure 9A and shows that, although the dome shows up spectacularly well in Google Earth, it is essentially impossible to see from the ground, let alone from the Asyut-El Kharga Road, which lies less than 250 m away. The two small black specks (circled) on the skyline in the left hand part of the photo are two vehicles on the road.

photographs, using strategies discussed by Whitmeyer et al. (2010) and in this volume. Online resources that outline specific methods are available at http://serc.carleton.edu/NAGTWorkshops/ google_earth/index.html.

ACKNOWLEGMENTS

Our work has benefited from contributions to these strategies for mapping using Google Earth by a number of individuals, including Hamilton undergraduates Devin Farkas, Stephen Kemp, Tucker Keren, Nicholas Kernan, and Anoop Pandey, and U.S. Air Force Capt. Carolyn Tewksbury-Christle. Dave Tewksbury was instrumental in providing technical support. The authors are also deeply grateful for the support that we have received from Hamilton College, Alexandria University, and Damanhour University and from our colleagues and families.

REFERENCES CITED

Abdelsalam, M.G., Robinson, C., El-Baz, F., and Stern, R.J., 2000, Applications of orbital imaging radar for geologic studies in arid regions; the Saharan testimony: Photogrammetric Engineering and Remote Sensing, v. 66, p. 717–726.

Abdelsalam, M.G., Tsige, L., Yihunie, T., and Hussien, B., 2008, Terrane rotation during the East African orogeny; evidence from the Bulbul shear zone, south Ethiopia: Gondwana Research, v. 14, p. 497–508, doi:10.1016/j.gr.2008.05.001.

Alfarhan, M.S., Arafat, S.M., and Abdelsalam, M.G., 2006, Interplay of Cretaceous-Quaternary faulting and folding in the south desert of Egypt: Insights from remote sensing analysis: Geological Society of America Abstracts with Programs, v. 38, no. 1, paper no. 3-15, p. 9.

Barakat, M.G., and Asaad, Fakhry A., 1965, Geological results of the Assiut-Kharga well: Journal of the Geology of the U.A.R., v. 9, no. 2, p. 81–87.

Bosworth, W., Guiraud, R., and Kessler, L.G., II, 1999, Late Cretaceous (ca. 84 Ma) compressive deformation of the stable platform of northeast Africa (Egypt): far-field stress effects of the "Santonian event" and origin of the Syrian arc deformation belt: Geology, v. 27, no. 7, p. 633–636, doi:10.1130/0091-7613(1999)027<0633:LCCMCD>2.3.CO;2.

El-Baz, F., and El-Etr, H.A., 1979, Color zoning in the Western Desert of Egypt: NASA Special Publication, Issue SP-412, p. 203–218.

El-Etr, H.A., and El-Baz, F., 1979, Utilization of ASTP photographs in the study of small structures in Abu Rawash and Wadi el Natrun, Egypt: Apollo-Soyuz Test Project, Summary Science Report; Volume II, Earth Observations and Photography NASA Special Publication, p. 107–118.

El-Etr, H.A., and Moustafa, A.A.R., 1981, Lineation patterns of the central Western Desert: Annals of the Geological Survey of Egypt, v. XI, p. 51–56.

El-Etr, H.A., and Moustafa, A.R., 1980, Utilization of orbital imagery and conventional aerial photography in the delineation of the regional lineation pattern of the central Western Desert of Egypt with particular emphasis on the Bahariya Region: Symposium on the Geology of Libya, Issue 2, p., 933–953.

Elshishtawy, A.M., Abd Elhameed, A.E.T., Abd Elwahab, A.A., and Abu Shama, A.A., 1997, Length-slow chalcedony and origin of quartz geodes in lower Eocene limestones, Gebel Duwi, Egypt: Sedimentology of Egypt, v. 5, p. 21–30.

Gani, N.D., Abdelsalam, M.G., Gera, S., and Gani, M.R., 2009, Stratigraphic and structural evolution of the Blue Nile Basin, northwestern Ethiopian Plateau: Geological Journal, v. 44, p. 30–56, doi:10.1002/gj.1127.

Grolier, M.J., McCauley, J.F., Breed, C.S., and Embabi, S., 1980, Yardangs of the Western Desert: The Geographical Journal, v. 146, no. 1, p. 86–87, doi:10.2307/634077.

Hassaan, M.M., Mohamed, M.H., Hathout, M.H., and Abdel Moneim, S.M., 1993, Eocene chemo-litho-facies maps and significance to depositional pattern, Nile Valley, Egypt: Sedimentology of Egypt, v. 1, p. 67–85.

Hermina, M.H., 1967, Geology of the North-Western Approaches of Kharga: Geological Survey of Cairo Report, no. 44, 88 p.

Keheila, E.A., and El-Ayyat, A.A.M., 1990, Lower Eocene carbonate facies, environments and sedimentary cycles in upper Egypt; evidence for global sea-level changes: Palaeogeography, Palaeoclimatology, Palaeoecology, v. 81, p. 33–47, doi:10.1016/0031-0182(90)90038-9.

Kenawy, A.I., Bassiouni, M.A., Khalifa, H. and Aref, M.A.M., 1988, Stratigraphy of the Eocene outcrops between Assiut and Beni Suef, Nile Valley, Egypt: Bulletin of the Faculty of Science, Assiut University, v. 17, p. 161–193.

Klitzsch, E., List, F.K., Pohlman, G., Handley, R., Hermina, M., and Meissner, B., 1987, Geological map of Egypt, 1:500,000, Asyut sheet NG 36 NW: Conoco and the Egyptian General Petroleum Company, 1 sheet.

Luo, W., Arvidson, R.E., Sultan, M., Becker, R., Crombie, M.K., Sturchio, N.C., and El Alfy, Z., 1997, Ground-water sapping processes, Western Desert, Egypt: Geological Society of America Bulletin, v. 109, p. 43–62, doi:10.1130/0016-7606(1997)109<0043:GWSPWD>2.3.CO;2.

Masoud, A., and Koike, K., 2006, Tectonic architecture through Landsat-7 ETM+/SRTM DEM-derived lineaments and relationship to the hydrogeologic setting in Siwa region, NW Egypt: Journal of African Earth Sciences, v. 45, p. 467–477, doi:10.1016/j.jafrearsci.2006.04.005.

McBride, E.F., Abdel-Wahab, A., and El-Younsy, A.R.M., 1999, Origin of spheroidal chert nodules, Drunka Formation (Lower Eocene), Egypt: Sedimentology, v. 46, p. 733–755, doi:10.1046/j.1365-3091.1999.00253.x.

McCauley, J.F., Breed, C.S., Issawi, B., Schaber, G.G., El-Hinnawi, M., and El-Kelani, A., 1997, Spaceborne imaging radar (SIR) geologic results in Egypt, a review: 1982–1997: Proceedings of the Geological Survey of Egypt Centennial Conference, p. 489–527.

Meshref, W.M., 1990, Tectonic framework, in Said, R., ed., The geology of Egypt: Rotterdam, Netherlands, A.A. Balkema, p. 113–156.

Rashed, M.A., and Sediek, K.N., 1997, Petrography, diagenesis and geotechnical properties of the El-Rufuf Formation (Thebes Group), El-Kharga Oasis, Egypt: Journal of African Earth Sciences, v. 25, p. 407–423, doi:10.1016/S0899-5362(97)00113-9.

Said, R., ed., 1990, The geology of Egypt: Rotterdam, Netherlands, A.A. Balkema, p. 451–486.

Stern, R.J., and Abdel Salam, M.G., 1996, The origin of the great bend of the Nile from SIR-C/X-SAR imagery: Science, v. 274, p. 1696–1698, doi:10.1126/science.274.5293.1696.

Tewksbury, B.J., Abdelsalam, M.G., Tewksbury-Christle, C.M., Hogan, J.P., Pandey, A.R., and Jerris, T.J., 2009, Reconnaissance study of domes and basins in Tertiary sedimentary rocks in the Western Desert of Egypt using high resolution satellite imagery: Geological Society of America Abstracts with Programs, v. 41, no. 7, p. 458.

Tewksbury, B.J., and Hogan, J.P., Kemp, S.M., Keren, T.T., Tewksbury-Christle, C.M., Schultz, R.A., and Mehrtens, C., 2010, Deformation bands and the expression in siliciclastic cover rocks of slip on basement faults in southern Egypt: Geological Society of America Abstracts with Programs, v. 42, no. 7, p. 473.

Tewksbury, B., Dokmak, A., and Tarabees, E., Mansour, Ahmed Sadek M., and Rashed, Mohamed A., 2011a, A previously unrecognized system of folds and related faults in Stable Platform limestones of the El Rufuf and Drunka Formations, Western Desert, Egypt: Geological Society of America Abstracts with Programs, v. 43, no. 7, p. 98.

Tewksbury, B., Kattenhorn, S., Sayler, F., Tewksbury-Christle, C., and Saint-Jacques, D., 2011b, Polygonal patterns and desert eyes: reconnaissance satellite image study of fold and fault structures in Late Cretaceous and Early Tertiary limestones of the Western Desert, Egypt: Geological Society of America Abstracts with Programs, v. 43, no. 7, p. 98.

Whitmeyer, S.J., Nicoletti, J., and De Paor, D.G., 2010, The digital revolution in geologic mapping: GSA Today, v. 20, no. 4, p. 4–10, doi:10.1130/GSATG70A.1.

Youssef, M.M., 2003, Structural setting of central and South Egypt; an overview: Micropaleontology, v. 49, Suppl. 1, p. 1–13, doi:10.2113/49.Suppl_1.1.

Manuscript Accepted by the Society 16 April 2012

The Geological Society of America
Special Paper 492
2012

Lidar and Google Earth: Simplifying access to high-resolution topography data

Christopher J. Crosby*

San Diego Supercomputer Center, University of California, San Diego, La Jolla, California 92093-0505, USA

ABSTRACT

High-resolution topography data acquired with lidar (light detection and ranging) technology are revolutionizing the way we study Earth surface processes. These data permit analysis of the mechanisms that drive landscape evolution at resolutions not previously possible yet essential for their appropriate representation. Unfortunately, the volume of data produced by the technology, software requirements, and a steep learning curve are barriers to lidar utilization. To encourage access to these data we use Keyhole Markup Language (KML) and Google Earth to deliver lidar-derived visualizations of these data for research and educational purposes. Display of full-resolution images derived from lidar in the Google Earth virtual globe is a powerful way to view and explore these data. Through region-dependent network linked KML (a.k.a., super-overlay), users are able to access lidar-derived imagery stored on a remote server from within Google Earth. This method provides seamless, Internet-based access to imagery through the simple download of a small KML-format file from the OpenTopography Facility portal. Lidar-derived imagery in Google Earth is the most popular product available via OpenTopography and has greatly enhanced the usability and thus impact of these data. Users ranging from scientists to K–12 educators have downloaded KML files ~12,000 times during the first eight months of 2011. The overwhelming usage of these data products demonstrates the impact of this simple yet novel approach for delivering easy to use lidar data visualizations to Earth scientists, students, and the general public.

INTRODUCTION

High-resolution topography data collected with lidar (light detection and ranging) remote sensing technology (e.g., Carter et al., 2007) is revolutionizing Earth science research (Carter et al., 2001). Typically collected from airborne, terrestrial and mobile platforms, lidar topography data represent the Earth's surface, overlying vegetation and built environment in three dimensions, with meter-scale detail over large (10^2 to 10^5 km²) spatial extents. As a result, these data permit analysis of the mechanisms that drive landscape evolution at resolutions not previously possible yet essential for their appropriate representation. With rapid growth over the past ten years in the collection of lidar data by the National Science Foundation, as well various state and local agencies, public domain lidar data have enabled new insights into a wide range of geologic phenomena (e.g., Frankel, et al., 2007;

---

*Now at UNAVCO, Inc., Boulder, Colorado 80301, USA; crosby@unavco.org.

Crosby, C.J., 2012, Lidar and Google Earth: Simplifying access to high-resolution topography data, *in* Whitmeyer, S.J., Bailey, J.E., De Paor, D.G., and Ornduff, T., eds., Google Earth and Virtual Visualizations in Geoscience Education and Research: Geological Society of America Special Paper 492, p. 37–47, doi:10.1130/2012.2492(03).

Walter and Merritts, 2008; Perron, et al., 2009; National Research Council, 2010; Ewing and Kocurek, 2010; DeLong, et al., 2011). Given the amazing representation of natural landforms provided by lidar topography, these data also have significant potential in Earth science education.

Airborne lidar is an active remote sensing technology that combines a scanning laser altimeter operating at 10s to 100s of kilohertz with differential GPS, and a high-precision inertial measurement instrument on an aircraft or helicopter to record dense measurements of the landscape (Carter et al., 2007). The resulting data set, a heterogeneous collection of measurements in geo-referenced *x, y, z* coordinate space known as a "point cloud," provides a 3-dimensional representation of natural and anthropogenic features at fine resolution (one or more measurement per m^2). Lidar instruments record multiple returns from each outgoing laser pulse, making it possible to differentiate ground returns from vegetation returns by applying various filtering algorithms (e.g., Sithole and Vosselman, 2004). The ability to segregate the point data based upon the origin of the return significantly enhances the utility of these data; for example, by permitting the production of a "bare earth" terrain model stripped of vegetation.

For many Earth science applications, the 3-dimensional point cloud is represented as a continuous surface of elevation values (a digital elevation model, DEM) estimated through the process of gridding or interpolation on a regular *xy* grid where *z* is a single-valued function of *x* and *y* ("2.5 dimensional"). The resulting raster data set can be visualized and analyzed in geospatial software. Derivatives such as "hillshades"—produced by imposing an artificial sun illumination on the landscape— and maps of landscape slope can be generated (El-Sheimy et al., 2005).

Unfortunately, the volume of data produced by lidar technology, software requirements, and a steep learning curve can be barriers to utilization of these data for many scientists, educators, and the general public. With typical lidar point cloud data sets consisting of 100s of millions to several billion individual returns—occupying 100s of gigabytes on disk—these data sets can be very challenging to share, visualize, and analyze for users who lack expertise in geospatial data. Although there are commercial and open source software options for working with and analyzing lidar topography in point cloud and DEM form, many of the common geospatial and scientific computing software packages available to Earth scientists struggle with large volumes of lidar data. For non-experts, students, educators, and the general public, acquiring and becoming familiar with new software are still fundamental barriers to use of lidar data.

Virtual globe environments such as Google Earth offer an opportunity to simplify access to lidar topography-derivatives, thereby allowing non-experts to begin exploring these data. The Google Earth virtual globe provides a freely available, easily navigated, and familiar viewer that enables quick integration of lidar-derived visualizations with native high-resolution imagery,

geographic layers, and other geospatial data widely available in the international standard Keyhole Markup Language (KML) format (Google Inc., 2008).

In this paper I document the approach developed by the National Science Foundation–funded OpenTopography Facility to generate high-resolution imagery layers from large volumes of lidar point cloud data for display in Google Earth. The processing workflow discussed in the *Methodology* section below results in a KML file that seamlessly provides access to several layers of lidar-derived imagery stored remotely on OpenTopography servers at the San Diego Supercomputer Center at University of California, San Diego. I'll also detail statistics that show how these lidar KML files are being utilized and their impressive impact in terms of improving access to these data for non-expert users. Finally, this paper concludes with several specific examples of how the lidar KML files are being used by scientists and educators in research and the classroom.

APPROACH

The initial effort to bring lidar topography data into Google Earth was undertaken as a result of the 5600 km^2 lidar data collection performed by the National Science Foundation–funded EarthScope project over active faults in the western United States in 2007–2009 (Prentice et al., 2009). These data, collected and immediately released as a resource for the tectonics and earthquake science communities, were exceptionally popular, but it quickly became apparent that for a certain subset of the research community, the technical challenges associated with the data were a barrier to their utilization. Despite web-based access and processing tools provided by OpenTopography (Crosby et al., 2011; Krishnan et al., 2011) to streamline access to lidar point cloud and DEM products, many users simply wanted basic data visualizations that they could browse for field site selection (e.g., paleoseismic trenching locations) or upon which they could map faults and related landforms. Further, we recognized that the education and outreach potential for these data were significant if they could be made easily accessible.

Google Earth, with its familiar and intuitive user interface, 3-dimensional visualizations, and native high-resolution imagery has been widely adopted in the Earth sciences as a platform for data visualization, integration, and education and outreach (Bailey and Chen, 2011). These factors, combined with capabilities included in the KML format for serving large imagery data sets, made Google Earth the obvious candidate for delivering lidar topography data for research and teaching.

To enable access to lidar topography data in Google Earth, I use region-dependent network linked ("super-overlay") KML (Google Inc., 2011) to deliver lidar-derived imagery into the Google Earth virtual globe (Fig. 1). The super-overlay KML allows streaming of remotely stored imagery through a small (typically less than 1 MB) KML file downloaded by the user and opened in Google Earth. This KML then provides seamless streaming access to several layers of full-resolution hillshade

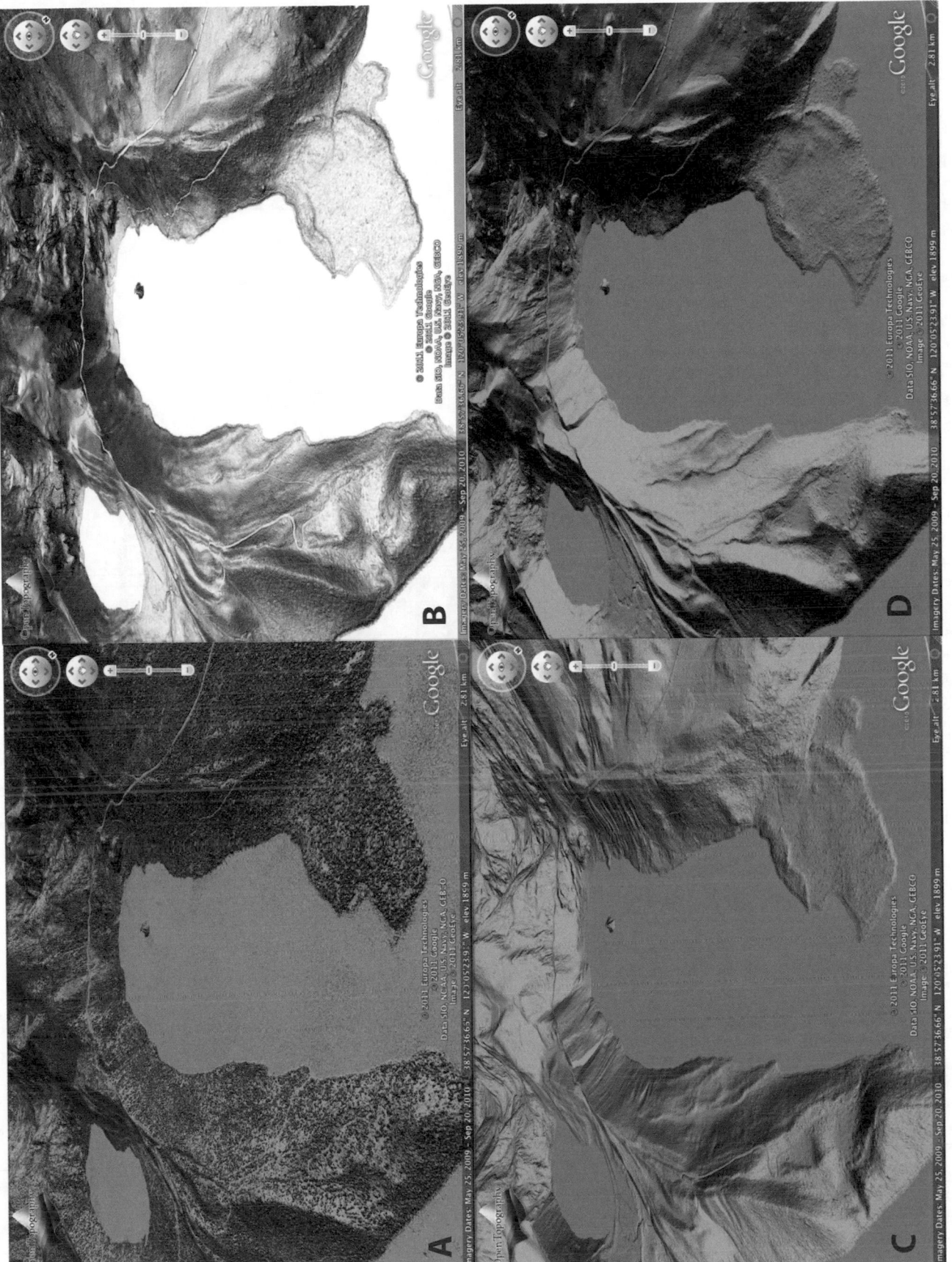

Figure 1. Google Earth image overlays for the Lake Tahoe Basin lidar data set at Emerald Bay. View is to the southwest. (A) Hillshade of the highest hit surface. (B) Bare earth slope-shade. (C) Hillshade of the bare earth surface—sun angle: 45° azimuth, 45° elevation. (D) Hillshade of the bare earth surface—sun angle: 315° azimuth, 45° elevation.

Crosby

and slopeshade images derived from lidar DEMs produced from the original lidar point cloud data set. The KML super-overlay approach subdivides a single massive imagery data set into a hierarchy of tiles of different resolutions, where each tile is associated with a given region. Google Earth then dynamically loads the appropriate imagery as a function of which portion of the data set is displayed in the Google Earth viewer and the elevation of the user's viewpoint. As the user zooms closer to the Earth's surface, higher resolution imagery is streamed across the network and displayed in Google Earth (Fig. 2).

It is important to note that the approach described here is restricted to rendering high-resolution imagery derived from lidar topography data draped on top of the Google Earth globe. Because Google Earth (free and Pro versions) does not permit a user to supplement the relatively low-resolution native topographic "mesh," we are unable to manipulate the actual topography a user sees when performing a perspective (3-dimensional) view in Google Earth. Instead, the approach I discuss here uses several common terrain visualization techniques such as hillshading and slope mapping (e.g., El-Sheimy et al., 2005), to produce imagery from the original lidar digital elevation models. Thus, the raster data delivered by the OpenTopography super-overlay KMLs present an xy grid where z is a color value (grayscale with integer value 0–255) generated as a derivative of the topography at that given location in the DEM. For example, the z value in a hillshade image is a function of the illumination and shading at a given location in the landscape based on pre-defined lighting conditions (e.g., northwestern illumination, 345°, with sun at 45° above horizon).

Although supplementation of the native topographic mesh in Google Earth would be ideal, as it would permit users to explore the high-resolution topography in three dimensions, imagery derived from the lidar topography is surprisingly effective at appearing to "improve" the resolution of the native topography when viewed in perspective (Fig. 1). Draping the hillshade and slopeshade imagery over the native topographic mesh brings out subtle details in the landscape such as fault and landslide scarps, fluvial channels, marine and fluvial terraces, and other landscape features that are indicative of geologic and geomorphic processes.

METHODOLOGY

Overview

The workflow through which the super-overlay KML files and associated lidar imagery cache is created consists of four basic steps: (1) DEM generation; (2) derived imagery production; (3) super-overlay creation; and (4) final KML construction (Fig. 3). The workflow I document here reflects the process used to produce the OpenTopography super-overlay KML files, but the process is non-unique and there are several approaches and software options that could be utilized at each step to construct a similar workflow. The workflow discussed here also represents a

Figure 2. Diagram depicting network link connection between Google Earth and the OpenTopography multi-resolution image cache located on servers at the San Diego Supercomputer Center. Users download simple KML file from the OpenTopography website (http://opentopography.org/kml) and open it in Google Earth. Imagery at the appropriate resolution are accessed and displayed seamlessly as a function of the region being viewed by the user.

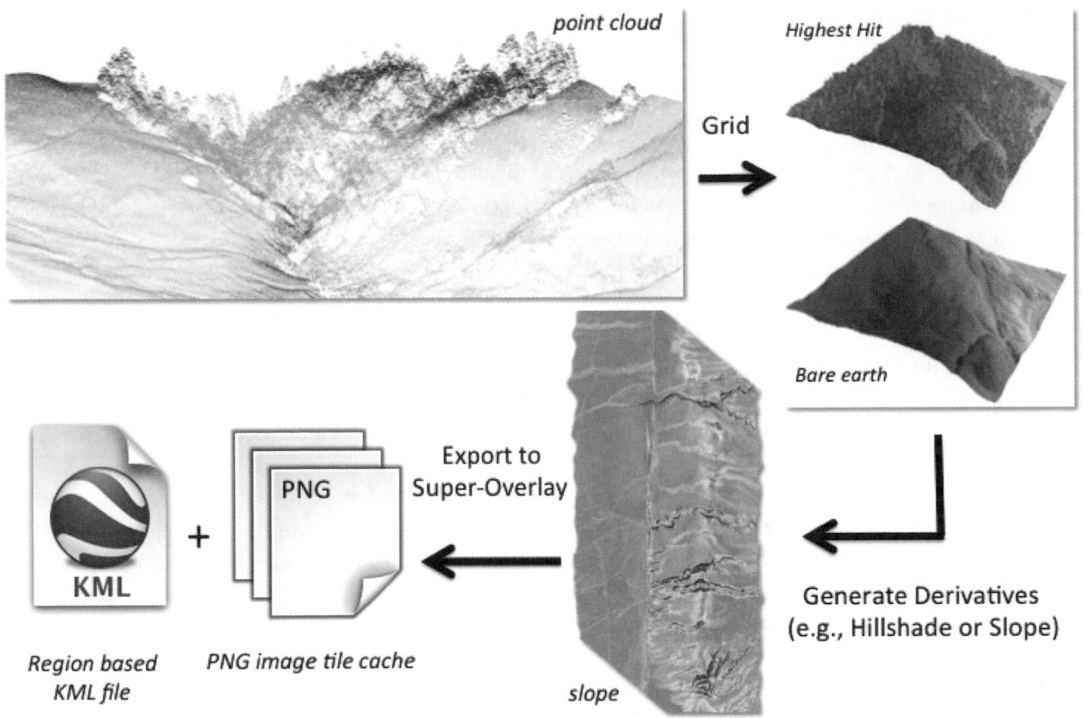

Figure 3. Overview of the lidar KML processing workflow. Point cloud data are gridded into bare earth and highest hit surface models. Derivative products are produced from these digital elevation models. The terrain visualizations are then exported to a tile cache of several levels of detail imagery and an associated KML file, which contains region tags to define when a given image is loaded.

generic model for how KML super-overlays could be generated for other large raster data sets such as remotely sensed imagery or global topography.

DEM Generation

The first step in the lidar imagery workflow is to produce a DEM from the "raw" lidar point cloud data. In many ways this is the most challenging step in the process since it may involve working with billions of lidar measurements. Fortunately, there are both commercial and open source software tools available for gridding lidar point cloud data. OpenTopography relies on our own Points2Grid algorithm (http://www.opentopography .org/index.php/resources/otforge/points2grid) and the LASTools StreamingTIN (Isenburg et al., 2006) to produce 0.5–1 m resolution DEMs from the point cloud data. A full discussion of the subtleties of gridding lidar point cloud data is beyond the scope of this paper, but additional discussion may be found in the literature (e.g., Mitasova and Hofierka, 1993; El-Sheimy et al., 2005; Isenburg et al., 2006).

In many cases the lidar data vendor will deliver pre-computed DEMs as standard data products that arrive alongside the point cloud data. These grids can be used as input into the super-overlay generation process, thereby bypassing the DEM generation step. In most cases these DEMs are tiled (e.g., as 1 km² or USGS Digital Orthophoto Quarter Quads, DOQQ) and thus will need to be mosaicked before derived imagery is created. DEM mosaicking is a common geospatial-processing task supported by most GIS software packages but may be computationally intensive for large data sets. The Global Mapper (http://www.globalmapper. com) software product discussed below as well as the Geospatial Data Abstraction Library (GDAL) open source package both permit "virtual" mosaicking of DEMs—creating a seamless large grid from many small grid files without physically creating a new DEM—a time-saving feature that can streamline the DEM handling process when working with many gigabytes of grid data.

Derived Imagery Production

Once a seamless high-resolution DEM has been created from the lidar point cloud data, derivatives of the terrain model are created by processing the DEMs using standard GIS techniques to create new representations of the topography. Any number of common terrain-derived metrics (El-Sheimy et al., 2005) could be utilized, but for the purposes of lidar display in Google Earth, OpenTopography favors hillshade (a.k.a. shaded-relief) and slopeshade derivatives. A hillshade is simply a grayscale raster where the value at each pixel reflects the illumination as a function of topographic position for a given artificial sun orientation (azimuth and elevation above the horizon). Similarly, a slopeshade is constructed by taking the slope of the topography and applying a grayscale color ramp that mimics the illumination effect of a hillshade—low slopes colored white, steep slopes shaded black. The slopeshade approach removes the illumination bias imposed by the artificial illumination angle and can be especially advantageous in steep and high topography where large shadows can be cast on more subtle landforms.

Using both the full resolution bare earth and highest hit terrain models as input, I typically generate four imagery layers (Fig. 1):

Highest Hit:

(A) 315° azimuth, 45° elevation hillshade

Bare Earth:

(B) 315° azimuth, 45° elevation hillshade

(C) 45° azimuth, 45° elevation hillshade

(D) slopeshade

These derivatives are produced at the same resolution as the input DEMs—typically 0.5–1 m resolution. The three bare earth imagery products emphasize detail in the de-vegetated terrain model that are typically of greatest interest to earth scientists. The orthogonal illumination angles—northwest and northeast—of the two hillshades plus the slopeshade are typically sufficient to reveal subtle topographic details such as fault and landslide scarps, fluvial channels, and terrace risers.

Region-Dependent Network Linked KML (Super-Overlays)

The lidar-derived raster layers, although representing a significant reduction in the total data volume compared to the original lidar point cloud data set, are still typically many gigabytes in size and not directly suited to display in a virtual globe environment such as Google Earth. However, KML conveniently offers a mechanism through which these large raster layers can be optimized for display. Further, super-overlay KML allows the imagery data to be stored remotely and accessed via simple HTTP request across the network, thus removing the need to download large volumes of data initially.

The region-dependent network linked image overlay KML (a.k.a. super-overlay; Google, 2011) approach subdivides a single large imagery layer into a hierarchy of tiles of different resolutions or *levels of detail*, where each tile is associated with a given display *region* which defines the conditions under which that image should be shown (Fig. 4). Google Earth then dynamically loads the appropriate imagery as a function of which portion of the data set is displayed in the Google Earth viewer and the elevation of the user's viewpoint. As the user zooms closer to the Earth's surface, higher resolution imagery is streamed across the network and displayed in Google Earth (Fig. 2). This means that when a user is viewing a large region and there are not enough pixels on the screen to support the display of a full high-resolution image, instead a lower-resolution image that is smaller in terms of file size (total bytes) is displayed by Google Earth (Fig. 4). In order to achieve this nested, hierarchical display of images of different levels of detail, the original high-resolution imagery must be preprocessed into tiles of a

specific resolution, and each image needs to receive the appropriate KML region tags to define when that image should be displayed in Google Earth. KML authors have extensive control over the regions and levels of detail to optimize data display. An extensive discussion of KML region concept as well as how to fully implement super-overlays can be found in Google's (2011) online documentation at: http://code.google.com/apis/kml/documentation/regions.html#superoverlays

In the specific case of the OpenTopography lidar KML files, the lidar-derived imagery is pre-processed into hierarchical tiles of 1024 × 1024 pixel PNG images (see Fig. 4). Nine different levels of detail are created from the original input imagery layer. The code snippet below shows the KML code for a single image within one of the OpenTopography lidar KML files, and nicely illustrates the mechanics of super-overlays. A simple deconstruction of this snippet shows that the contents of the *<Region>* tag defines the spatial extent of the region for which the image is image is displayed (*<LatLonAltBox>*), as well as the level of detail (*<Lod>*)—or amount of screen space that must be occupied—for the image to be displayed. If the *<Region>* display conditions are met, Google Earth calls the *<GroundOverlay>* URL to load the PNG image that applies to that area. For any given OpenTopography lidar KML file, thousands of hierarchical images are described in the same manner as the snippet below and thus are seamlessly delivered to the user. Additional discussion of the KML elements below can be found in Google's (2011) online documentation at: http://code.google.com/apis/kml/documentation/regions.html#superoverlays.

```
<Folder>
  <name>kml_image_L9_2_1</name>
  <Region>
    <LatLonAltBox>
      <north>39.32675572725105</north>
      <south>39.32201922239017</south>
      <east>-120.2478598581477</east>
      <west>-120.2539744805725</west>
      <minAltitude>0</minAltitude>
      <maxAltitude>0</maxAltitude>
    </LatLonAltBox>
    <Lod>
      <minLodPixels>512</minLodPixels>
      <maxLodPixels>-1</maxLodPixels>
      <minFadeExtent>0</minFadeExtent>
      <maxFadeExtent>0</maxFadeExtent>
    </Lod>
  </Region>
  <GroundOverlay>
    <drawOrder>9</drawOrder>
    <Icon>
    <href>http://opentopo.sdsc.edu/lidarfiles/
    tahoe/tahoe_be_he_315/kml_image_L9_2_1.png</href>
    </Icon>
    <LatLonBox>
      <north>39.32675572725105</north>
      <south>39.32201922239017</south>
      <east>-120.2478598581477</east>
      <west>-120.2539744805725</west>
    </LatLonBox>
  </GroundOverlay>
</Folder>
```

For efficiency, producing several layers of imagery at a range of levels of detail and the associated KML markup should be automated. Several software products are available to take an input image at full resolution and produce super-overlay KML. I use Global Mapper, a commercial GIS software product, which supports export to KML through the *Export Web Formats* (*Google Maps*, *VE*, *WW*, etc.) option (Global Mapper Software, LLC, 2011). Other products that offer similar capabilities include the commercial *Superoverlay* (http://superoverlay .geoblogspot.com/), and the free and open source command line-based GDAL2Tiles (http://www.klokan.cz/projects/gdal2tiles/) and its graphical user interface sibling MapTiler (http://www. maptiler.org/). The advantage to Global Mapper is that it can also be used to virtually tile a large number of DEMs and produce a seamless hillshade or slopeshade image. Thus, once I have lidar-derived DEMs for a given data set, Global Mapper provides a quick workflow for ingesting those grids, mosaicking them, producing derivatives, and exporting to super-overlay KML. The image tile size (e.g., 1024 × 1024 pixels) is user definable at export to optimize super-overlay performance.

At export, Global Mapper by default embeds the hierarchical tiled PNG imagery along with the KML file inside a compressed format known as a KMZ. Thus, the final step in constructing a network-linked overlay file is to divorce the imagery from its embedded location adjacent to the KML, and to relocate it to an OpenTopography web server. By uncompressing the KMZ file, the directory of imagery can be extracted and uploaded to a server. Once the imagery are posted to the server, each *<href>* tag in each *<GroundOverlay>* in the KML must be updated to point to the server. A simple find-and-replace routine run on the KML file automates this process.

Building the Final KML

The final step in the creation of the lidar super-overlay KML files is to construct a single KML that includes all layers of lidar-derived imagery, data set overview language in the KML pop-up, and screen overlays with an OpenTopography logo (Fig. 5). These steps are largely manual and accomplished using a combination of the Google Earth application and a text editor. Because each layer of imagery is referenced from its own KML file, the layers must be aggregated into a single KML for distribution. This compilation can be performed within the Google Earth application, and html content describing the KML can be added (Fig. 5). A data set outline layer is also added to the KML to help users identify the bounds of the data set when viewing it in Google Earth. Additional styling of the KML such as adding a screen overlay logo can be done through direct edits to the KML code. Once the construction and styling of the final KML product is complete, the file is saved out as a KMZ to reduce its size, and posted to an OpenTopography server for users to download.

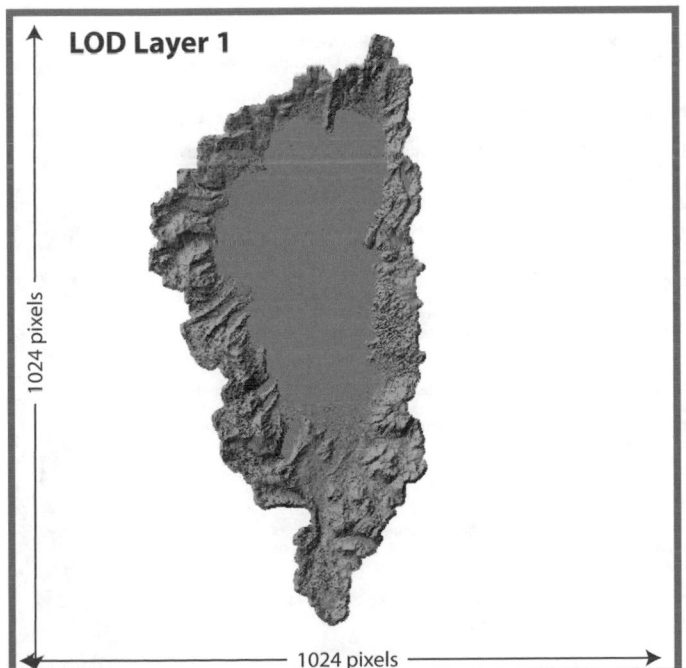
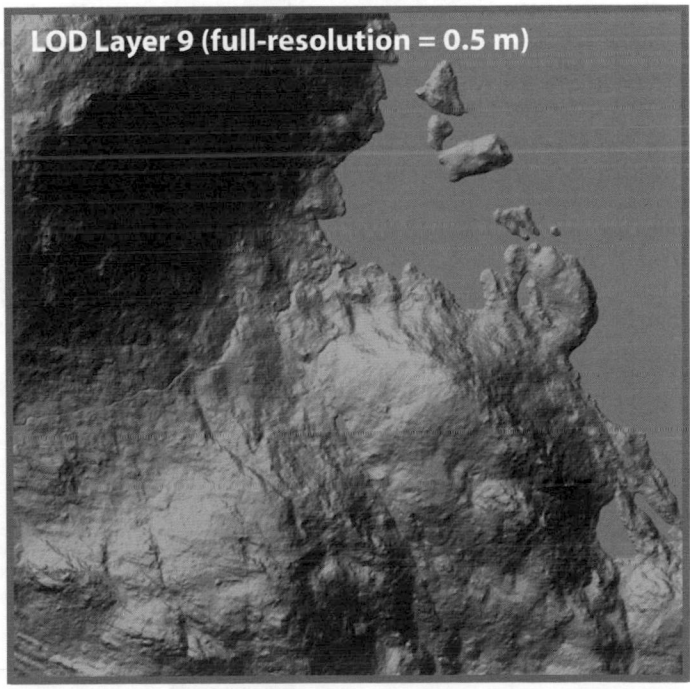

Figure 4. Comparison of range of images produced during the tiled and pyramided image export process used to produce the lidar super-overlay KML files. The left image depicts the full Lake Tahoe Basin data set in a single 1024 × 1024 pixel image and represents level of detail (LOD) layer 1. To the right, the same1024 × 1024 pixel PNG image can depict only a very small portion of the Tahoe data set when forced to represent the data set at full native resolution (LOD layer 9 = 0.5 m). Through this tiled and multi-resolution representation of a given imagery layer, the display of data is optimized in Google Earth, showing only the imagery necessary for a given region shown on the screen.

STATISTICS AND USAGE

As of August 2011, OpenTopography provides access to 39 GB of lidar-derived imagery covering 8253 km^2. Each super-overlay KML file corresponds to a specific data set and provides access to several layers of imagery, typically including bare earth and highest hit surfaces hillshaded from northwest and northeast sun angles, and a bare earth slopeshade (Fig. 1). Table 1 includes a list of data sets currently offered by OpenTopography in KML format; in general, OpenTopography generates lidar KML files for large and community-oriented data sets where large numbers of users are expected.

Lidar imagery super-overlay KML files have proven to be the most popular product available via the OpenTopography Facility. Collectively the files were downloaded over 12,000 times in the first 8 months of 2011; we've tracked similarly impressive yearly usage since OpenTopography began to release the files for the EarthScope lidar data set in 2008. Most impressive, however, is the diversity of users who are viewing the lidar KML in Google Earth. We know from both usage logging and through direct communication with users that the lidar Google Earth files are being utilized by academic scientists, government agency staff, the commercial sector, and K–12 educators. The files have proven popular for non-expert lidar users in the commercial sector; at federal, state, and local government agencies; and by graduate, undergraduate and even secondary school level students. Below I provide a brief overview of several the lidar use cases.

Research Applications

The Earth science research community has heavily utilized the lidar KML files. Although not often explicitly referenced in the methods sections of their papers, direct conversations with OpenTopography users have shown that the lidar KMLs are a common starting point for many scientists who plan to work with lidar data. The KMLs are used for synoptic browsing of

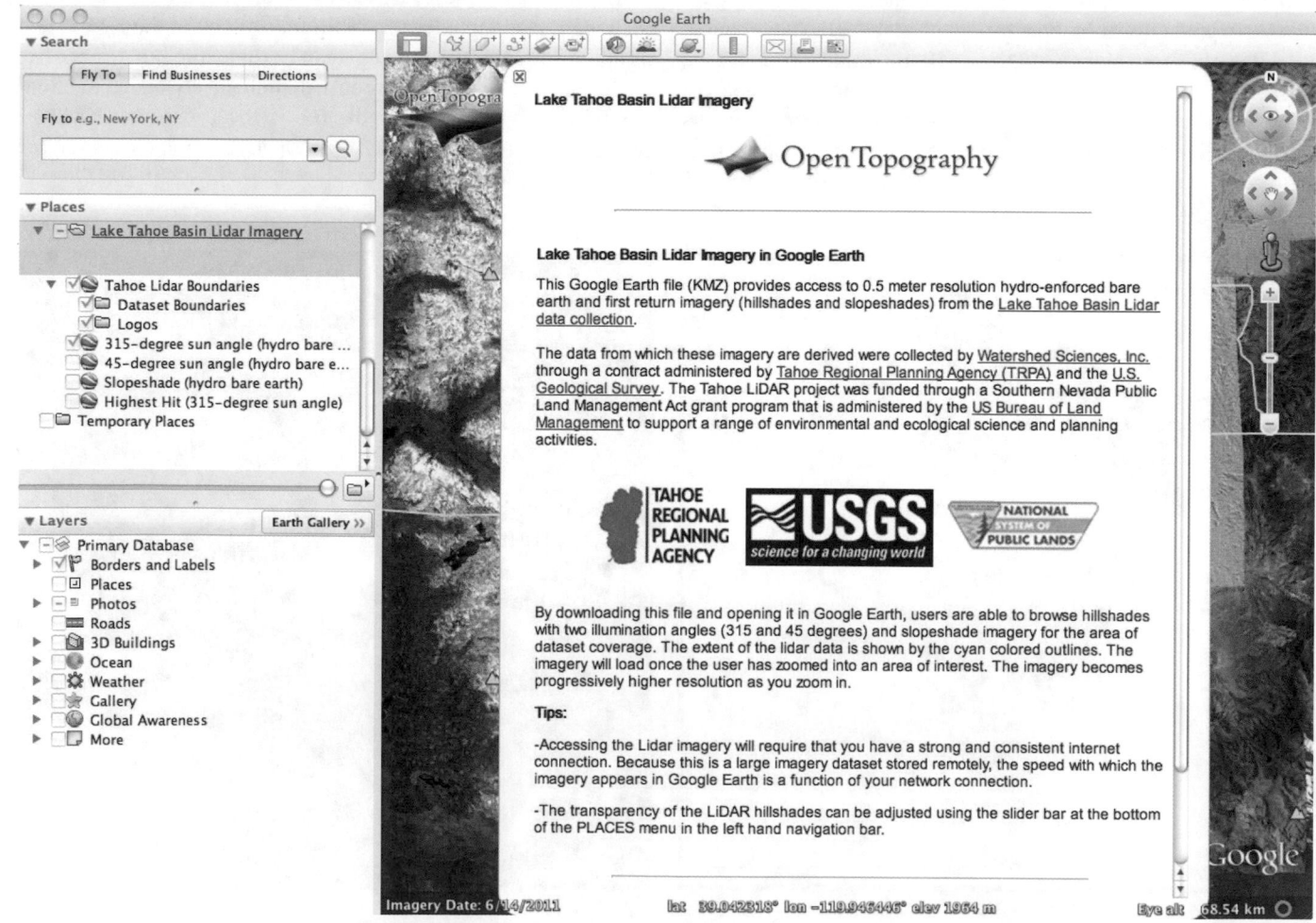

Figure 5. Screen capture of the Lake Tahoe Basin lidar KML file open in Google Earth. The *Places* menu in the upper left shows the compiled layers of imagery available through the file, including a data set boundary and four imagery layers. Also shown in the main window is the data set overview information and basic instructions included as part of the KML pop-up, viewed by clicking on the KML name in the *Places* menu. A screen overlay image in the upper left of the viewer identifies OpenTopography as the source of the KML.

TABLE 1. LIDAR IMAGERY KML FILES AVAILABLE AS OF AUGUST 2011 FROM THE
OPENTOPOGRAPHY FACILITY (http://opentopography.org/kml)

Dataset	Area (km^2)
EarthScope Northern California Lidar Project	2448
EarthScope Intermountain Seismic Belt Lidar Project	830
EarthScope Southern and Eastern California Lidar Project	1683
EarthScope Alaska Denall Totschunda Lidar Project	410
EarthScope Yakima Lidar Project	371
World Bank—ImageCat Inc.—RIT Haiti Earthquake Lidar dataset	838
El Mayor–Cucapah earthquake (4 April 2010) rupture Lidar scan	372
Lake Tahoe Basin Lidar	941
Teton Conservation District Wyoming Lidar	365
TOTAL	8253

Note: RIT—Rochester Institute of Technology.

the data set to identify research sites (e.g., paleoseismic trenching sites) and to look for landforms indicative of process (e.g., fault scarps, fluvial and marine terraces, landslides). Because the super-overlay approach enables seamless browsing of large data sets, the KML files are undoubtedly the fastest method to browse through a data set.

In some cases, non-expert lidar and GIS users have utilized the KML files for more advanced mapping. In this case the lidar imagery is displayed in Google Earth and the drawing tools are used to build fault strip maps and to perform geologic or geomorphic map compilations. Although OpenTopography typically encourages users to utilize the DEM and point cloud data for their scientific analysis, for certain applications the imagery in Google Earth is an adequate mapping platform.

Educational Applications

For the same reasons that the lidar super-overlay KML approach is appealing to researchers—seamless browsing of large data sets in familiar software and high-resolution imagery that powerfully depicts geologic and geomorphic processes—the files also have significant potential for use in the classroom. Faculty have reported using the KML files in their geomorphology, earthquake hazards, and introductory geology classes to allow students to view and explore landforms in real data.

Robinson (2011) used the OpenTopography KML files to explore the question of whether lidar-derived imagery in the classroom would "enhance a student's ability to understand geologic concepts such as plate tectonics, the earthquake cycle, strike-slip faults, and geomorphology." The lidar KML files were integrated into an undergraduate level activity on the San Andreas Fault designed to introduce students to the earthquake cycle; the activity was tested in undergraduate classrooms in order to determine its effectiveness. A full discussion of this work is beyond the scope of this manuscript, but Robinson (2011) concluded that the inclusion of the lidar imagery led to an increase in understanding of the earthquake cycle from pre-test to post-test results. In part,

she attributes this to the fact that the grayscale lidar hillshade imagery "helps reduce the visual distracters to a student when viewing the topography while also increasing their ability to see more subtle features within the topography that are important for interpreting the geologic, geomorphologic, and tectonic history." Thus, her results suggest that the use of lidar derived imagery in teaching about geologic processes is beneficial for students.

Primary and secondary students and instructors have also made use of the lidar KML imagery. In an exceptional example, "Will from Journeys School" posted to the Google Earth Community discussion forum in 2011 (http://bbs.keyhole.com/ubb756/ubbthreads.php?ubb=cfrm) with an example of his 5th Grade Capstone Project that utilized the OpenTopography lidar KML file from the EarthScope Intermountain Seismic Belt to explore earthquake hazards in the Jackson, Wyoming, region. He constructed a Google Earth flyover of the data set and writes:

First it shows the relationship in distance between our school and the Jackson Lake dam. Then the relationship between the dam and the Teton Fault. My project was looking at the potential impacts a large earthquake could have on Jackson Hole.

I then focus on a special data set which is the lidar hillshading from opentopography.org. I do a flyover which allows the viewer to get a clearer picture of where the Teton Fault really is. (http://bbs.keyhole.com/ubb/ubbthreads.php?ubb=showflat&Number=1438154&#Post1438154)

This example illustrates the power and usability of presenting high-resolution scientific data via an intuitive platform like Google Earth. The combination simplifies access to the data to a level where it is accessible to not only non-expert science users and graduate or undergraduate students, but it also makes the data accessible to children and the general public in a way that dramatically extends the impact of these massive and research-oriented data sets.

Post-Event Earthquake

The power of Google Earth visualizations of lidar data in post-event response situations has been demonstrated through the impact of OpenTopography-hosted KML files produced soon after the 2010 earthquakes in Haiti (12 January 2010) and northern Baja California (the El Mayor–Cucapah earthquake, 4 April 2010). In the case of the Haiti event, a rapid lidar data and imagery collection was organized by the Rochester Institute of Technology and Kucera International, under sub-contract to ImageCat Inc. and funded by the Global Facility for Disaster Reduction and Recovery (GFDRR) hosted at The World Bank. The lidar data were processed quickly and initially posted to a simple FTP site for distribution to users. Using the data products on the FTP, I quickly ran the lidar KML generation routine and posted the KML file to OpenTopography for community use (Crosby, 2010a). This KML file and the associated imagery were accessed extensively in the months following the earthquake and remain popular today.

Similarly, following the El Mayor–Cucapah event, colleagues at CICESE in Ensenada, Mexico, provided a lidar data set that had fortuitously been collected several years before the event. In this case, the lidar data were processed into super-overlays and again posted to OpenTopography for scientists performing initial mapping of the ground rupture and related deformation (Crosby, 2010b). Because much of the work being done immediately after the earthquake was field-based and thus researchers were not connected to a network, I also provided a set of super-overlay KML files where the imagery was embedded locally and thus did not require a network connection to access. Since this event, OpenTopography has provided a copy of the EarthScope southern and eastern California lidar KML files with imagery embedded to the U.S. Geological Survey Earthquake Hazards program office in Pasadena, California, for use immediately following a potential future southern California event.

Both of these event response examples illustrate the power of the simple, easily accessible lidar visualizations provided by the super-overlay KML files. Immediately following a natural disaster, high-resolution remotely sensed imagery products are commonly collected, and simple approaches to quickly process those data and to get them into the hands of responders and researchers are invaluable.

CONCLUSIONS

The approach presented here for serving large volumes of high-resolution lidar topography-derived imagery into Google Earth for research, education, outreach, and response harnesses standard and documented super-overlay capabilities within the KML specification generated through off-the-shelf software. Although not exceptionally novel technologically, the approach and widespread usage of the OpenTopography lidar KML files demonstrates the impact of this simple method for delivering these normally unwieldy scientific data sets to users. Utilized by communities ranging from research scientists to primary school students, the lidar super-overly KML files served by OpenTopography have greatly enhanced the usability and thus impact of these data. Although easily accessible and user friendly, the lidar imagery in Google Earth does not necessarily remove the need for scientific users to return to the original point cloud data to further explore how the lidar data can be fully optimized to represent the landscape. Although specifically focused on high-resolution lidar topography data, the approach I present here could be modified relatively easily to distribute other large remotely sensed data sets collected by research communities.

ACKNOWLEDGMENTS

The OpenTopography Facility is funded by the National Science Foundation's Earth Sciences Instrumentation and Facilities (EAR/IF) program and the Office of Cyberinfrastructure (OCI), under award numbers 0930731 and 0930643. Initial development of the Google Earth lidar files was performed under funding from NSF's Information Technology Research project, The Geosciences Network (GEON), award numbers 0225673, 0744229, and 0225543. James Luke Blair, USGS Earthquake Hazards program, provided an introduction to network-linked KML and processed an initial batch of Northern California EarthScope lidar imagery. J Ramon Arrowsmith (Arizona State University) provided valuable feedback on and been a strong advocate for the lidar KML files. The Lake Tahoe Basin lidar data set was provided by the Tahoe Regional Planning Agency. EarthScope lidar is based on services provided to the Plate Boundary Observatory by NCALM (http://www.ncalm.org). PBO is operated by UNAVCO for EarthScope (http://www.earthscope.org) and supported by the National Science Foundation (No. EAR-0350028 and EAR-0732947).

REFERENCES CITED

Bailey, J.E., and Chen, A., 2011, The role of Virtual Globes in geoscience: Computers & Geosciences, v. 37, no. 1-2, doi:10.1016/j.cageo.2010.06.001.

Carter, W.E., Shrestha, R.L., Tuell, G., Bloomquist, D., and Sartori, M., 2001, Airborne Laser Swath Mapping Shines New Light on Earth's Topography: Eos (Transactions, American Geophysical Union) v. 82, p. 549–550, 555.

Carter, W.E., Shrestha, R., and Slatton, K.C., 2007, Geodetic Laser Scanning: Physics Today, v. 60, no. 12, p. 41–47, doi:10.1063/1.2825070.

Crosby, C.J., 2010a, Haiti LiDAR Imagery in Google Earth: OpenTopography Blog post, http://www.opentopography.org/index.php/blog/detail/haiti_lidar_imagery_in_google_earth (11 February 2010).

Crosby, C.J., 2010b, LiDAR data for N. Baja, Mexico in Google Earth: pre–M 7.2 El Mayor–Cucapah earthquake: OpenTopography blog post, http://www.opentopography.org/index.php/blog/detail/lidar_data_for_n_baja_mexico_in_google_earth_pre_m_72_el_mayor_cucapah_eart (4 May 2010).

Crosby, C.J., Arrowsmith, J R., Nandigam, V., and Baru, C., 2011, A geoinformatics approach to online access and processing of LiDAR topography data, *in* Keller, R., and Baru, C., eds., Geoinformatics: London, Cambridge University Press, p. 251–265.

DeLong, S.B., Prentice, C.S., Hilley, G.E., and Ebert, Y., 2011, Multitemporal ALSM change detection, sediment delivery, and process mapping at an active earthflow: Earth Surface Processes and Landforms, v. 37, p. 262–272, doi:10.1002/esp.2234.

El-Sheimy, N., Valeo, C., and Habib, A., 2005, Digital terrain modeling: acquisition, manipulation, and applications: Boston, Massachusetts, Artech House, 257 p.

Ewing, R.C., and Kocurek, G.A., 2010, Aeolian dune interactions and dune-field pattern formation: White Sands Dune Field, New Mexico: Sedimentology, v. 57, p. 1199–1219.

Frankel, K.L., Brantley, K.S., Dolan, J.F., Finkel, R.C., Klinger, R.E., Knott, J.R., Machette, M.N., Owen, L.A., Phillips, F.M., Slate, J.L., and Wernicke, B.P., 2007, Cosmogenic (10)Be and (36)CI geochronology of offset alluvial fans along the northern Death Valley fault zone: Implications for transient strain in the eastern California shear zone: Journal of Geophysical Research–Solid Earth, v. 112, p. B06407, doi:10.1029/2006JB004350.

Global Mapper Software, LLC, 2011, Global Mapper User's Manual—Export KML/KMZ Command, http://www.globalmapper.com/helpv13/Help_MenuBarAndToolBar.html#file_me nu_export_kml_raster.

Google Inc., 2008, KML 2.2 Reference Document, http://code.google.com/apis/kml/documentation/kml_tags_beta1.html.

Google Inc., 2011, KML Documentation: Super-Overlays, http://code.google.com/apis/kml/documentation/regions.html#superoverlays (accessed 30 August 2011).

Isenburg, M., Liu, Y., Shewchuk, J., Snoeyink, J., and Thirion, T., 2006, Generating Raster DEM from Mass Points via TIN Streaming: GIScience'06 Conference Proceedings, p. 186–198.

Krishnan, S., Crosby, C.J., Nandigam, V., Phan, P., Cowart, C., Baru, C., and Arrowsmith, J R., 2011, OpenTopography: a services oriented architecture for community access to LIDAR topography, *in* Proceedings of the 2nd International Conference on Computing for Geospatial Research & Applications (COM.Geo '11), AMC, 8 p., doi:10.1145/1999320.1999327.

Mitasova, H., and Hofierka, J., 1993, Interpolation by Regularized Spline with Tension II: Application to Terrain Modeling and Surface Geometry Analysis: Mathematical Geology, v. 25, p. 657–667.

National Research Council, 2010, Landscapes on the Edge: New Horizons for Research on Earth's Surface: National Academies Press, 180 p.

Perron, J.T., Kirchner, J.W., and Dietrich, W.E., 2009, Formation of evenly spaced ridges and valleys: Nature, v. 460, p. 502–505, doi:10.1038/nature08174.

Prentice, C.S., Crosby, C.J., Whitchill, C.S., Arrowsmith, J R., Furlong, K.P., and Phillips, D.A., 2009, Illuminating Northern California's Active Faults: Eos (Transactions, American Geophysical Union) v. 90, no. 7, p. 55–56, doi:10.1029/2009EO070002.

Robinson, S.E., 2011, Integrating LiDAR Topography into the Study of Earthquakes and Faulting [M.S. thesis]: Tempe, Arizona, Arizona State University.

Sithole, G., and Vosselman, G., 2004, Experimental comparison of filter algorithms for bare-Earth extraction from airborne laser scanning point clouds: ISPRS Journal of Photogrammetry and Remote Sensing, v. 59, p. 85–101, doi:10.1016/j.isprsjprs.2004.05.004.

Walter, R.C., and Merritts, D.J., 2008, Natural streams and the legacy of water-powered mills: Science, v. 319, p. 299–304, doi:10.1126/science.1151716.

Manuscript Accepted by the Society 16 April 2012

The Geological Society of America
Special Paper 492
2012

Enhancing usability of near-surface geophysical data in archaeological surveys via Google Earth

Caitlyn M. Williams
Gregory S. Baker*
Department of Earth & Planetary Sciences, University of Tennessee, Knoxville, Tennessee 37996-1410, USA

Bradley A. Ault
Department of Classics, State University of New York at Buffalo, Buffalo, New York, 14260, USA

ABSTRACT

Conventional archaeological excavation methods are, by design, extremely invasive and result in culturally sensitive areas being irrevocably altered. For this reason, near-surface geophysical techniques have been incorporated into archaeological investigations to aid in locating buried features and developing specific excavation plans with minimal damage to the sites. The objective of our research was to conduct a geophysical surveying campaign at a test site in Knoxville, Tennessee, to develop a workflow for an improved data management methodology that would be applied to data acquired at an active archaeological site in Cyprus.

A multi-tool geophysical survey was completed as a first case study at a control site with known subsurface features on the University of Tennessee Agricultural Campus using both ground penetrating radar and magnetic gradiometry. Using real-time differential corrected GPS data, we systematically imported the images into Google Earth as accurately georeferenced overlays on existing topographic maps and air photos. We added placemarks where we interpreted subsurface anomalies based on the data, exported waypoints for the features into spreadsheet software, and correlated the results to the known locations. We next tested this methodology with data from an active archaeological site in southern Cyprus. Data were displayed in Google Earth and accurate GPS coordinates for features were exported into a spreadsheet file. We were able to share a tested, easily accessible final product that was immediately useful and accessible to the archaeologists on the team and the broader archaeological community.

*Corresponding author: gbaker@utk.edu.

Williams, C.M., Baker, G.S., and Ault, B.A., 2012, Enhancing usability of near-surface geophysical data in archaeological surveys via Google Earth, *in* Whitmeyer, S.J., Bailey, J.E., De Paor, D.G., and Ornduff, T., eds., Google Earth and Virtual Visualizations in Geoscience Education and Research: Geological Society of America Special Paper 492, p. 49–62, doi:10.1130/2012.2492(04). For permission to copy, contact editing@geosociety.org. © 2012 The Geological Society of America. All rights reserved.

INTRODUCTION

Our research is focused on improving the standard methodologies for actively integrating geophysical data into active archaeological investigations. While numerous multi-tool geophysical surveys have been executed successfully in the past (e.g., Chianese et al., 2010; Kamei et al., 2002), problems have arisen with the data acquisition, processing, and interpretation workflow. Typically, a significant amount of time (weeks to months to years) can elapse between the completion of the survey and the actual usage of the data in the field. By developing new techniques for rapid geophysical data integration, archaeologists will have near-real-time access to accurately positioned geospatial data and may be able to revise excavation plans within their current field season accordingly.

Motivation

Archaeology has integrated geophysical surveys as a means to maintain site integrity (e.g., Baker and Ambrose, 2007; Abdallatif et al., 2010). Traditional archaeological surveying methods are extremely invasive to the site area (Wynn, 1986); conventional methods utilize trowels, shovels, and occasionally heavy machinery to excavate, and these methods essentially destroy the site area for the sake of research. Thus archaeologists have begun using near-surface geophysics as a non-invasive means to detect features or artifacts in the subsurface.

Within the past decade, there has been little change in archaeogeophysics, with the exception of improving instrumentation. Current limitations to using geophysics are due to the time and money needed to complete a survey and process the data. Generally, there is only a finite amount of time to complete a survey within the given field season and it costs money to hire field personnel, rent equipment, and ship it to the site to be surveyed.

Another problem with the process is that there is often a large gap in time between data acquisition and their physical use in the field; data that can be easily used and shared enhance the flow of information. There is a distinct disconnect between archaeologists and geophysicists in the field. Geophysicists are more concerned with the survey design, data acquisition tools, and processes, whereas an archaeologist is more interested in the presence (or absence) of discrete targets in the subsurface (i.e., artifacts).

Figure 1 represents a traditional workflow for a geoarchaeological survey. The initial step is that survey boundaries are delineated with help from an archaeologist, typically based on the presence of surface remains, and GPS coordinates are taken for the survey area. Then, a geophysical survey is planned and executed with one or more techniques, and data are uploaded onto a field computer. Software specific to each instrument is then used to process the data, often involving multiple, expensive programs. Data must then be interpreted in order to identify any subsurface anomalies, and interpretations are made at the discretion of the geophysicist in a variety of software packages. Finally, the data are given to the archaeologist in some format and then he or she must decipher it with or without the help of the geophysicist. *It is this disconnect between archaeologist and geophysicist that undermines the usefulness of the data presented.*

Objective and Hypothesis

The objective of this research is to help reduce some inefficiencies in the geophysical surveying process when applied to archaeological projects. Specifically, we are attempting to streamline the data interpretation workflow that will allow for near real-time feedback between archaeologist and geophysicist. In order to work toward this aim, we first incorporate Google Earth with data from a control site (in Knoxville, Tennessee), and then apply the methodology at an active archaeological site (Akrotiri Peninsula, Cyprus) to create an improved geophysicist-archaeologist interface. By using Google Earth, we contend that archaeologists are provided with near-real-time

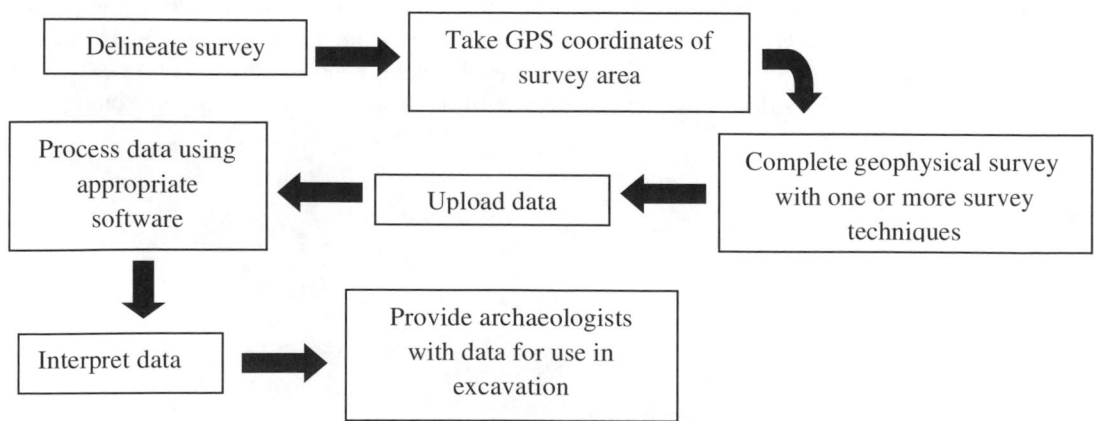

Figure 1. Diagram showing the conventional workflow for geophysical data acquisition at an archaeological site. The disconnect in geoarchaeological surveying often occurs with the last three steps of the workflow; commonly a long time lapse in completing the final two steps makes data less effective.

feedback about the locations of possible features in the subsurface it a streamlined fashion. Our hypothesis is that the effectiveness and efficiency of multi-tool, near-surface geophysical surveys for archaeological applications can be improved by displaying data with accurate GPS coordinates using a virtual globe, for example Google Earth.

CONTROL SITE

The purpose of this control study was to test the viability of Google Earth as an effective platform for displaying near-surface geophysical data and to test its accuracy in providing precise GPS coordinates to locate targets. A geophysical survey was completed during the spring of 2011 at a site on the University of Tennessee campus in order to develop a new data management workflow.

Control Site Background

The B4 Plot is located between Alcoa Highway 129 and the Tennessee River, approximately two miles south of the University of Tennessee main campus in Knoxville, Tennessee, on the Experimental Agricultural Research Station (Fig. 2). Soil conditions across the site vary from residual soils developed directly on Ordovician sedimentary bedrock (near the highway) to loamy soils developed on alluvial terraces at different elevations above the river. Relative permittivity—a unitless parameter that controls electromagnetic signals in the Earth—of loamy, dry soil lies anywhere in a range from four to six and for loamy, wet soil it ranges from 15 to 30 (Baker et al., 2007).

Silt or sandy silt dominates the top 6.1 m of soil, which overlie ~0.9–1.5 m of fine to medium sand and cemented sand. Sediments are underlain by fractured shale and limestone

Figure 2. (A) Google Earth image showing the location of the B4 Plot in relation to the University of Tennessee main campus in Knoxville, Tennessee. (B) Photo taken of the B4 Plot, looking to the south. The fenced-in hydrogeology station can be seen on the right-hand side of this photograph.

bedrock. Bedrock is Ottossee Shale, which is a Middle Ordovician member of the Chicamauga Group. As a whole, it is generally characterized by fine-grained calcareous shale with some interbedded limestone. As depth of penetration did not surpass six meters, bedrock was not considered to be a key factor in this investigation.

The Plot B4 contains 24 known targets buried in the spring of 1999. The locations of the objects are accurately known, as GPS coordinates for each target were documented upon burial and data have been recorded on the size, shape, and orientation of each object. Depths to buried objects range from 0.30 to 1.11 m. It is assumed that there has been sufficient time for the ground to settle, and any major disturbance to the subsurface and resulting signal in the data is minimized. This test site and the associated targets are used for a broad range of courses, seminars, and workshops in near-surface geophysics; thus, the specifics of the targets and their coordinates are not published here.

Control Site Methodology

Geophysical Techniques

Ground penetrating radar (GPR) and magnetic gradiometry were used together as a part of a comprehensive multi-tool survey. These techniques have been successfully used in previous geoarchaeological surveys (e.g., Chianese et al., 2010; Kamei et al., 2002) and the equipment was readily available for use in this investigation. A Pulse EKKO Pro Smart Cart (manufactured by Sensors and Software, Inc.) was used to complete the GPR survey. This particular model was chosen since the University of Tennessee Near-Surface and Environmental Geophysics Lab owns a unit, and the Smart Cart is a versatile piece of equipment that can be used to survey in a variety of terrains. The entire area was surveyed with the 100 MHz antenna in order to obtain the best possible depth of penetration and to maintain consistency with the data. For the magnetic survey, a Bartington Grad 601-2 Dual Gradiometer was used. In addition to the field crew's familiarity with its operation, the Bartington gradiometer has a reputation for being extremely portable and easy to use in the field. The B4 Plot has an area of 40 m by 50 m and the entire area was surveyed with the exception of a small fenced-in area in the southwest corner of the survey area.

Data Collection

For the magnetic survey, the site area was divided up into three 20 m by 20 m grids to ease data processing. The Bartington has strict, pre-set survey parameters and the whole area could not be surveyed at one time. Data were collected for each grid in an alternating pattern. We collected GPR data in a similar alternating pattern, but kept spacing between transects constant at 0.5 m. Data over the entire 40 m by 50 m were collected as one cohesive grid and the fenced-in area was simply omitted from the survey area by making lines 15 through 37 shorter than the others (Fig. 3).

Data Processing

GPR data were first imported into a program called GFP Edit (Sensors and Software) as Y-lines with spacing between transects set to 0.5 m. The length of profiles 1–14 and 38–99 was set to 40 m and the length of lines 15–37 was set to 26.15 m to allow for the omission of the fenced area which was not surveyed. The start direction of every other line was flipped to account for the zigzag pattern of data collection. Once we created the GFP file, we opened it using EKKO Mapper 3 (Sensors and Software). Slice resolution was set to 0.50 m and the velocity was set to 0.086 m/ns.

Magnetic gradiometer data were processed using proprietary ArcheoSurveyor software (DW Consulting). Functions of destagger and clip were applied to get the best data possible, which is accomplished by aligning known straight features in the data. Destaggering compensates for data collection errors caused by the operator starting recording of each traverse too soon or too late. It shifts each traverse forward or backward by a specific number of intervals (DW Consulting, 2010). Clipping replaces all values outside a specified minimum and maximum with those values. This process removed extreme data points (created by

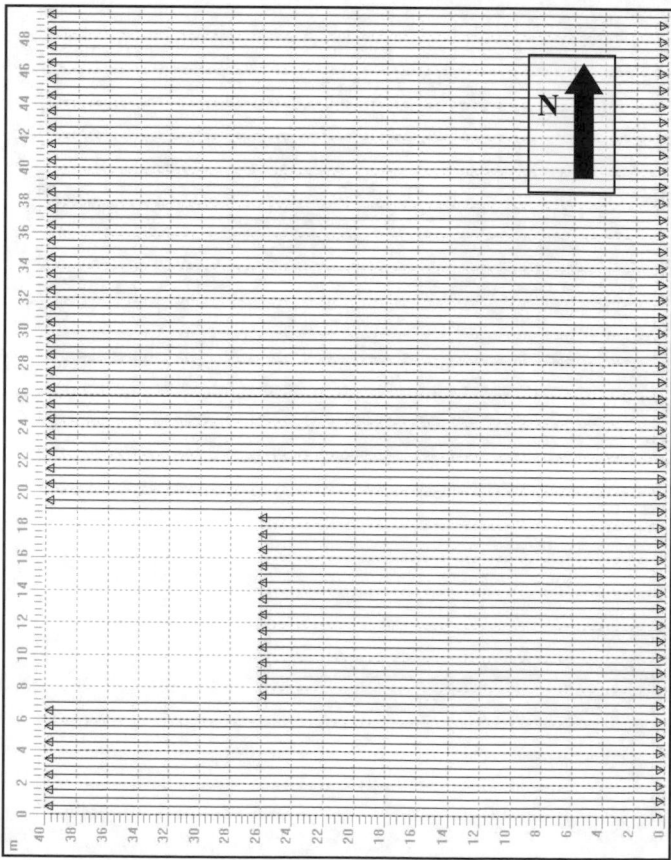

Figure 3. Map-view schematic of the survey geometry for GPR data collection at the B4 Plot with distances shown in meters. Note the "zigzag" pattern in which the data were collected and the shorter transects run around the fenced-in area.

metallic objects at the surface, like well heads) and highlighted the fine details in the data.

Geophysical Results

In looking at these data, we were primarily concerned with high-amplitude anomalies that presented themselves as hot colors in the data sets. In the GPR data there were a few discrete anomalies that were not chosen as targets. Primarily, at the B4 Plot, there are metallic well-heads at the surface, which show up very prominently in the data. As we have the GPS coordinates for the locations of these objects, and they were visible on the surface, it was not difficult to rule them out as possible subsurface features. There were also two linear features in the GPR data that were not chosen as targets. These features were a result of the transmitting antenna "misfiring," and the locations of these transects were accurately recorded in a field notebook for later use. The depth slice was selected from 1.250 to 1.500 m that shows the most features buried in the subsurface (Fig. 4). Out of the 24 buried targets at the site, a total of 15 targets were identified in the data. The image was saved as a JPEG for ease in later data manipulation

In the magnetic data, we were mainly concerned with the dipole anomalies present. These anomalies are characterized by having both a positive (red) and a negative (blue) anomaly.

Any anomaly that did not have this pairing was not interpreted to be a feature.

Differences between the two data sets can easily be explained by the presence of both metallic and non-metallic buried objects. The GPR is capable of detecting all targets at the B4 Plot, while the magnetometer is only able to detect metallic targets. Because of this fact, and because we know for certain that there are both metallic and non-metallic targets buried there, we can say that all anomalies are real. The three 20 m by 20 m grids were stitched together as one composite grid and a total of 13 targets were identified out of 24 (Fig. 5). The composite grid was then saved as a JPEG file for later ease with data manipulation.

Discussion on Incorporating Data into Google Earth for the Control Site

Ground overlays are most appropriate when dealing with geoarchaeological data as they are images draped over Google Earth's terrain, on top of the base satellite imagery, and follow the natural curvature of the earth. Images can be in a variety of formats (e.g., .jpg, .tif, .bmp etc.), which is one reason Google Earth is such a useful and versatile tool for data display (Weinecke, 2009).

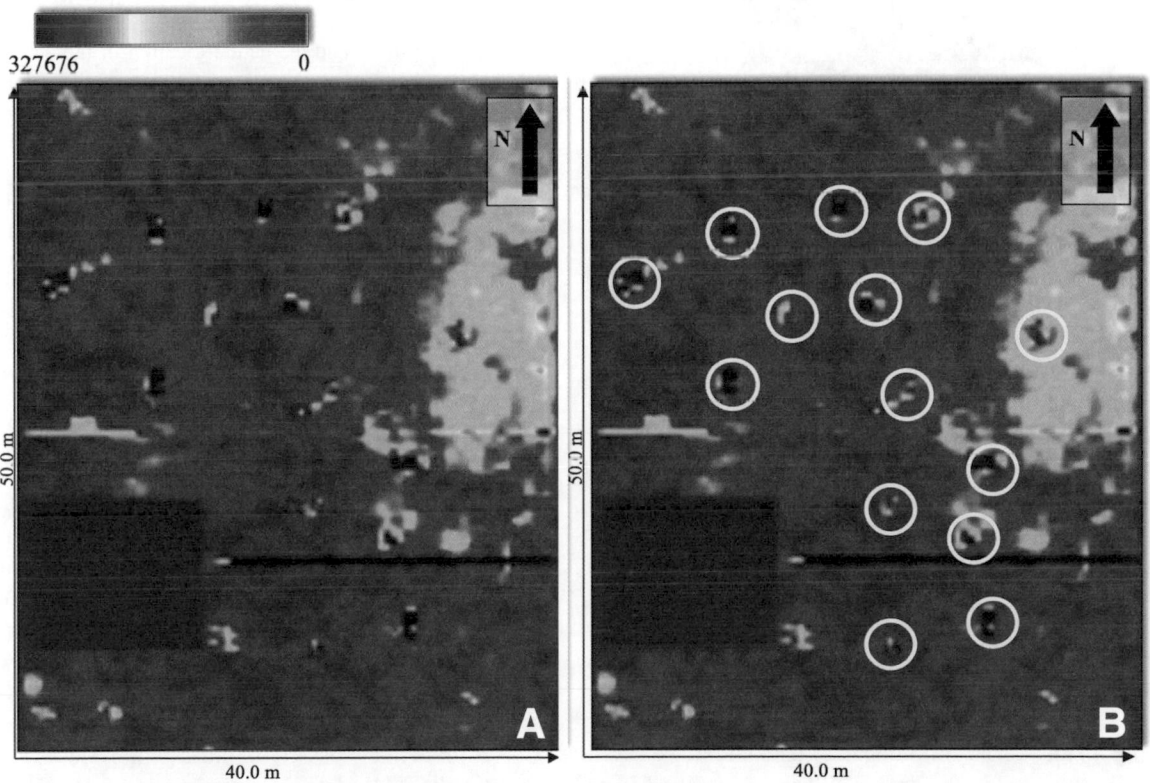

Figure 4. Uninterpreted (A) and interpreted (B) images of the GPR data collected at the B4 Plot (A). The dark-blue rectangle is representative of the fenced-in area where no data were collected.

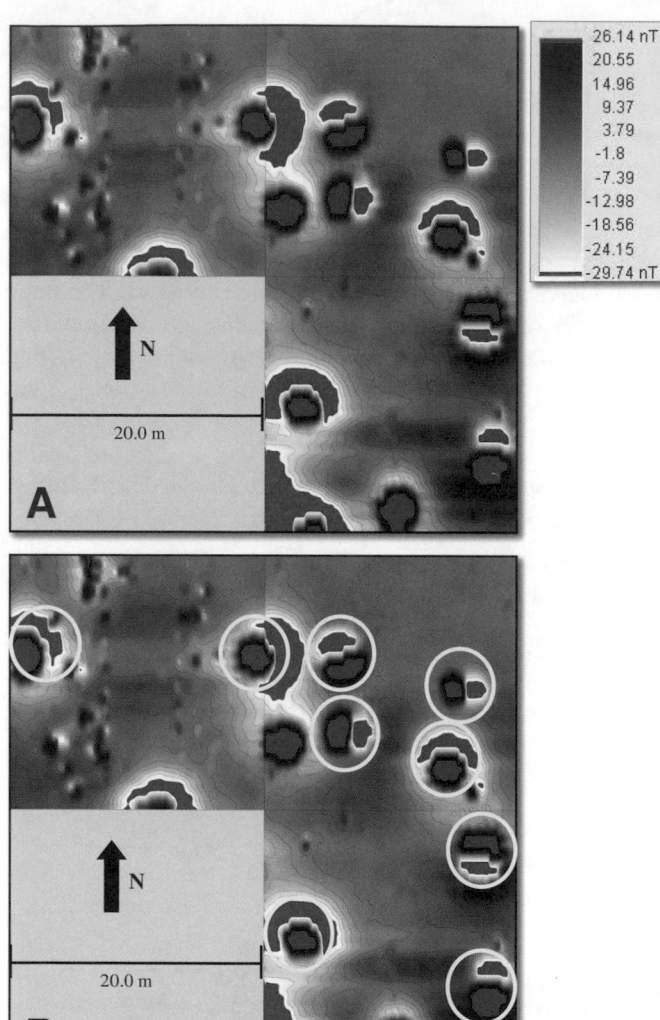

Figure 6. Georeferenced image of GPR data displayed in Google Earth. Targets are identified by white placemarks. Distance was measured between these and the actual location of targets and then error was calculated.

Figure 5. Image of the magnetic data collected at the B4 Plot. (A) The uninterpreted magnetic data. (B) Possible targets are circled in yellow in the interpreted data. The green rectangle in the lower left of the photo is the location of the fenced-in area where no data were collected.

Displaying data in Google Earth is a relatively simple process. As a default, Google Earth will display latitude and longitude coordinates as degrees, minutes, and seconds, but for our purposes the coordinate system was changed to display in decimal degrees for manipulation ease. Placemarks were created for the corner points of each grid using the GPS data uploaded from the TerraSync software (Fig. 6). Icon style and size was changed to the user's preference in order to make the icons easily visible on the map. Once all placemarks were created, they were saved in a folder under the Places menu.

Creating overlays is the next step in displaying data. For ease in setting up an overlay, all images of data should be saved as the same file type and in the same folder at a specified location on the computer. To create a map of the data, click the overlay button and browse for the location of the appropriate photo in the "Link" space. While the overlay menu is still open, green corners appear on the image and are used to re-size and rotate the images so that it is georeferenced to the GPS coordinates. The opacity is not changed and no extra description was necessary. All files were again saved in their own folder in the "Places" menu.

Final interpretations from Google Earth are saved as Keyhole Markup Language (KML) files for export. KML is an extremely versatile programming language that can be read by a variety of applications (e.g., Microsoft Virtual Earth, ESRI ArcGIS Explorer) and these files can easily be emailed between users and opened in Google Earth (Wernecke, 2009). Interpretations are made on these overlays using either the path or placemark option depending on the geometry of the features. In this example, placemarks were used, as the features were not linear. All interpretation placemarks were saved in their own folder in Google Earth. Because the geophysical images used for the interpretation on the local computer may be of use to the archaeologists (or more likely passed to future geophysicists by the archaeologists), the KML data can be saved as a .kmz file such that the geophysical images can be passed along in addition to the KML placemark/points of the interpretation.

To manipulate the final geophysical data, Microsoft Excel was used to display the coordinates of features exported from Google Earth (in recognition that archaeologists do not always have access to specialized software). To import a KML file into Excel, data are imported as XML data (XML stands for Extensible Markup Language, a coding language that encodes documents in a machine-readable form). This is accomplished simply by manually changing (or writing a small script to change, if a large number of files exist) the .kml extension to a .xml extension that will render the file readable by most spreadsheet programs (e.g., Microsoft Excel). After importing the appropriate file, data are displayed in an easy-to-read table. Some cleanup is required to condense data to the pertinent information, but that task is easily accomplished by deleting the rows and columns that are not needed.

Once data are organized into latitude and longitude columns, coordinates can be imported as waypoints into many GPS software programs. The GPS used in this investigation was a Trimble Ranger and coordinates were uploaded to the Pathfinder Office software using the ASCII import function. The GPS can then be used to physically map out the locations of features while in the field and their locations marked with PVC flags or spray paint. By mapping features on the surface archaeologists can alter excavation plans as needed in order to have the most efficient field season.

The GPR data shown in Figure 5 were displayed in Google Earth using the method described above. Placemarks were added where targets were identified in the data (Fig. 6) with the placemark being located on the brightest pixel of the anomaly. Additional placemarks were added for the actual location of the objects based on the information recorded in 1999, when the targets were buried. Having both sets of data on the same map allowed for the distance between buried and measured locations to be calculated by using the ruler function in Google Earth.

Using the distances obtained from Google Earth, error was calculated in a spreadsheet in order to specify the in-line and cross-line error and how accurate Google Earth is as a platform for displaying near-surface geophysical data and exporting waypoints. From each data set, both in-line and cross-line error were calculated. Using the manner in which the data was collected, we set north as the positive Y-direction and east as the positive X-direction (south and west were the negative X and Y directions, respectively). Error values were calculated by measuring the distance from the interpreted target to the real target, first in only the X-direction and then in the Y-direction. Then the average of those values was calculated as well as two standard deviations in order to have a 95% confidence level.

The average in-line error was calculated to be 0.004 m ± 1.1 m, and the average cross-line error was calculated to be −0.22 m ± 1.6 m. Recognize that these error reporting results indicate that although there was some scatter of the correlation between real and interpreted locations (in this case, smaller than 1.1 m in-line and 1.6 m cross-line positions with 95%

confidence) there was no significant systematic bias of the points away from the expected positions. If such a systematic shift (e.g., translation of the overlay images, etc.) occurred, then the error values would deviate strongly from zero, but instead there were 0.004 m in-line and 0.22 m in the cross-line position. In other words, the measured errors only included a random positional component and not a systematic translational component. This means that the dominant error was not in the Google Earth component of the data processing stream (where translations are the most common error) and are instead more representative of standard errors observed in geophysical acquisition and interpretation.

The magnetic gradiometry data are shown in Figure 7 with a total of 13 targets identified, and each was assigned a placemark. Using the same method as given in the above section, in-line and cross-line error were calculated for this data set. The in-line error was calculated to be 0.075 m ± 1.4 m and the cross-line error was calculated to be 0.28 m ± 1.5 m. As discussed previously, the random positional component errors were reasonable, and there appears to be no significant systematic translational shift in the interpretations relative to the known targets.

TEST STUDY

The purpose of this test study was to apply the newly developed data management workflow on archaeological data collected at an active archaeological site in Cyprus in 2010. Geophysical surveying on the Akrotiri Peninsula took place from 9 June 2010 to 13 June 2010.

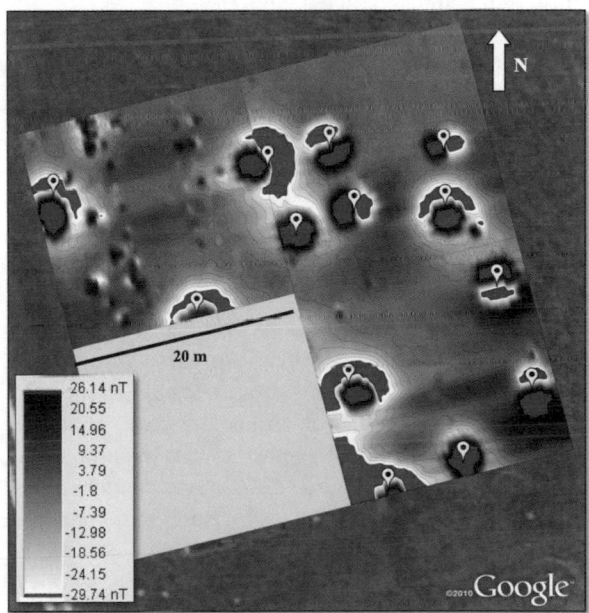

Figure 7. Georeferenced image of magnetic data collected at the B4 Plot and displayed in Google Earth. Yellow placemarks indicate the location of possible targets.

Test Site Background

Cyprus is the third-largest island in the Mediterranean Sea and lies ~386 km north of Egypt and 64 km south of Turkey. While most of the island is relatively flat, the main topographic features are the Troodos Mountains that lie in the central part of the island. The Mediterranean climate is best described as hot and dry from June to September. For the remainder of the year, the island has rainy winters (Solsten, 1993).

The test site, which includes two specific areas of interest, is located along the southern coast of the Akrotiri Peninsula in Cyprus, near the modern city of Akrotiri (Fig. 8). The peninsula is ~12 km north to south and 9 km wide and is flanked on the sides by the Akrotiri Bay to the east and the Episkopi Bay to the west. The test site lies within the confines of the Akrotiri Royal Air Force Base, which is considered British Sovereign Territory.

The peninsula itself is a relatively recent geomorphic feature, geologically speaking, with the first sediments being deposited during the Pliocene (Heywood, 1982). The coast of the southernmost portion of the peninsula is made up of Miocene sandstone and marl cliffs that rise ~64 m above the Mediterranean Sea. Moving inland, the remains of the Akrotiri Forest cover the terrain, dominated by low, shrubby plants. The Akrotiri Salt Lake lies in the low-lying flat interior of the peninsula. Approximately 4 km in diameter, it was likely once a protected bay open to the sea on the east (Heywood, 1982).

Since the region under investigation lies within the confines of a military base, it presents a unique opportunity to study an undisturbed archaeological landscape: the combination of restricted access to the site and a specific construction plan that was conscientious of the location of surface remains led to the protection of the archaeological landscape.

The location at Akrotiri appears to have served as a port and maritime trans-shipment point likely associated with the modern city of Kourion (not to be confused with Kourias, which is an ancient city with its current location unknown) that lies 13 km to the northwest. A port was likely established due to its locale, being both militarily and economically strategic as it served as a waypoint between Greece, Egypt, and Turkey.

Past studies on the Akrotiri Peninsula have been purely archaeological (with no associated geophysics) in nature dealing with surficial features and have yielded artifacts both offshore and on land. Most significant of these recent finds is a quarry, located on a cliff ~30 m above the bay. Associated with the quarry are rutted tracks of an ancient roadway utilized to cart stone to the nearby settlement (current location unknown) of Kourias (Ault and Leonard, 2009). Archaeologists believe the remains of this settlement lie within the boundaries of the modern Akrotiri RAF Base. The test site consists of two smaller regions, separated by less than 2 km, known locally as Dreamer's Bay and Saint Mark's that are anticipated to contain different archaeological structures. Geophysics and associated Google Earth images were generated for these two smaller areas within the test site.

At Dreamer's Bay, two structures with partial surficial expression had previously been identified as warehouses given their location and geometry. Sections of eroded walls are also visible at the surface here, though no statement has been made as to their purpose. The majority of the remains visible at Dreamer's Bay date to the late Roman/Early Byzantine period, approximately during the fourth to seventh centuries AD (Ault and Leonard, 2009).

St. Mark's had previously been identified as an area of archaeological significance as a potential site of an ancient church known locally as S. Mercourios. Features visible on the surface include loose building debris as well as areas of mortared rubble "floor" construction. Many wall alignments and possible paved surfaces are scattered across the areas of high terrain, but no complete structures have yet been found. Also located at this site is a 9-m length of tunnel cut into the rock and finished with plaster. The tunnel is open to the surface at either end, but is filled with debris, leaving only the uppermost 1–1.5 m visible. Archaeologists speculate that this tunnel is the remains of a crypt (Ault and Leonard, 2009).

Geophysical Results

The magnetic gradiometry data for Dreamer's Bay showed many possible features in the subsurface. An examination of the uninterpreted data (Fig. 9) yields linear high-amplitude trends of lighter gray shading that have been interpreted to be the remnants of additional warehouse-like buildings, as they appear to have similar dimensions to the warehouses already exposed at the surface. These remains were highlighted using the path tool as described above in order to spatially describe their locations in the subsurface. Waypoints from the endpoints and corners of each feature were exported to Excel and displayed in columns of latitude and longitude.

Results from the GPR at Dreamer's Bay were not as spectacular as those of the gradiometer (Fig. 10). Total depth of penetration was ~1.5 m for the entire area. Due to the presence of salt and the calcareous nature of the soils the signal attenuated quickly and resolution for the most part was fairly poor. Despite these setbacks, a few linear features can be interpreted from the data. As with the magnetic data, waypoints for the endpoints and corners of each feature were exported to Excel and displayed in an easy-to-read table.

Figure 11 shows both geophysical data sets at Dreamer's Bay displayed together, with the GPR set to 40% transparency. It is important to remember that both instruments will detect different changes in the subsurface. The magnetometer is sensitive to changes in the magnetic susceptibility of the target versus the background material, while the GPR images show contrasts in the physical properties that are likely created by disturbances in the sediment. While all features are considered real, we can hypothesize that those detected by the magnetometer are likely walls or remnants of buildings and that those detected by the GPR are trenches dug for foundations or storage.

Figure 8. Google Earth image of the island of Cyprus, with the area of interest zoomed in the white rectangle. The location of the modern cities of Akrotiri (north) and Kourion (northwest) in relation to the sites at St. Mark's and Dreamer's Bay are shown in the inset. The city of Kourion has been made famous by the spectacular archaeological remains discovered there.

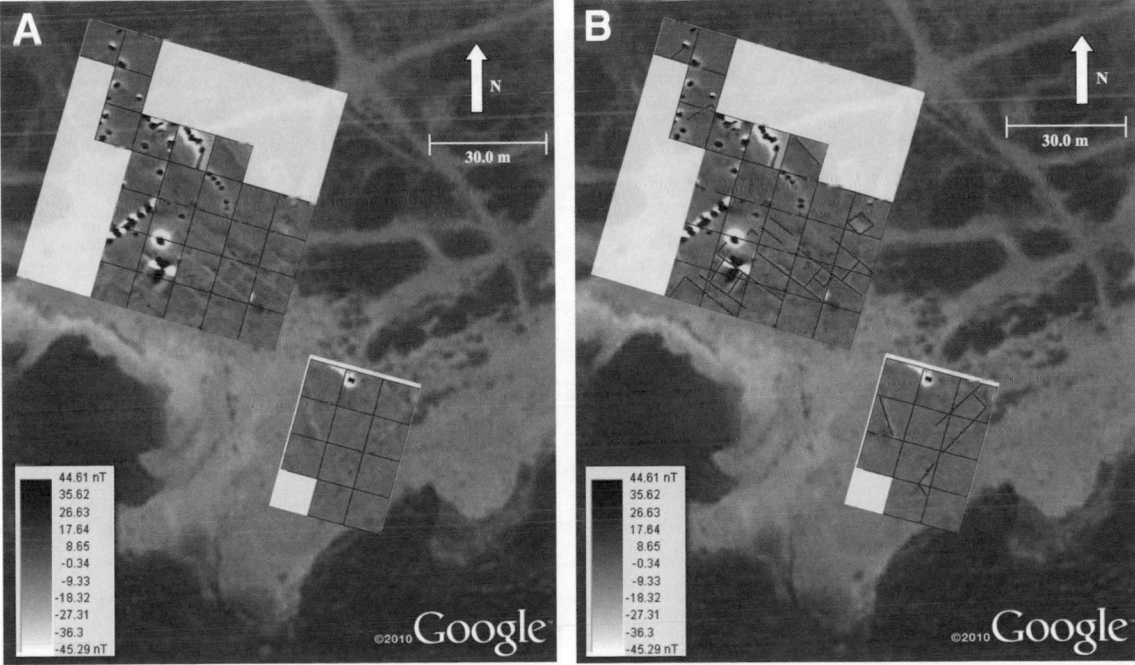

Figure 9. Uninterpreted (A) and interpreted (B) magnetic gradiometry data for the site at Dreamer's Bay. Black and white dipoles that show up in the data were interpreted to be metallic debris left on the surface from historic military operations. The features of interest are the light gray linear features seen in the southeast portions of the data. Each small square represents a minor grid and is 10 m by 10 m.

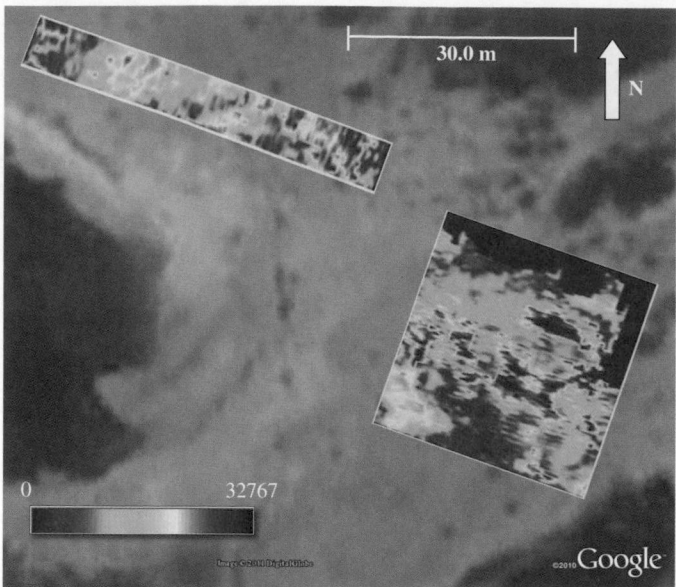

Figure 10. Uninterpreted GPR data collected at Dreamer's Bay with the 100 MHz antenna. Data were accurately georeferenced in Google Earth using GPS data collected at the site.

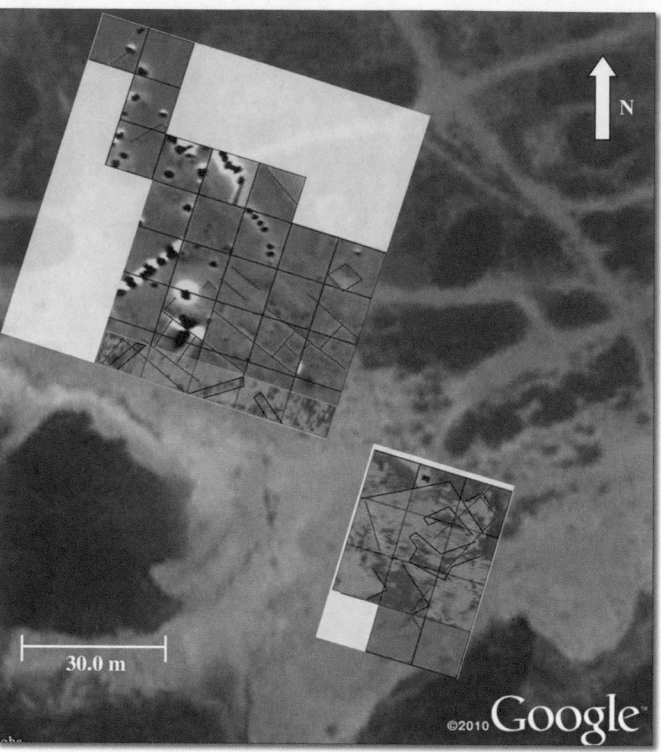

Figure 11. Google Earth image displaying the geometry of features found in both data sets. Transparency of the GPR data was set to 40% in order to be able to see both sets of data. All features seen in the data are interpreted to be real; however, we can hypothesize that features detected by the magnetometer are walls while those detected by the GPR are trenches.

For St. Mark's, uninterpreted magnetic data were accurately georeferenced in Google Earth using the GPS coordinates recorded at the site (Fig. 12). Data collected at St. Mark's show two large, L-shaped, high-amplitude features in the northern portion of the survey area. Also shown in the data are some magnetic dipoles (paired positive/negative anomalies) that appear to be in a consistent linear pattern. Coordinates for placemarks inserted at the ends and corners of each linear feature were exported as XML files into a spreadsheet program for later use in planning a future excavation.

Depth of penetration for GPR at St. Mark's was ~1.5 m. The depth slice for the northern part of the survey area is ~0.75–0.95 m and the depth slice for the southern part of the survey area is from 0.84 to 1.04 m (Fig. 13). GPR data for St. Mark's show different structures than the magnetic data. A Microsoft Excel file was created which displays the coordinates of features in the subsurface that were exported from Google Earth.

Figure 14 shows both geophysical data sets at St. Mark's displayed together, with the GPR set to 40% transparency. While all features are considered real, we can hypothesize that those detected by the magnetometer are likely walls or remnants of buildings and that those detected by the GPR are trenches dug for foundations or storage.

Discussion on Incorporating Data into Google Earth at the Test Site

All data for Dreamer's Bay and St. Mark's were imported and georeferenced in Google Earth as described previously for the control site. GPS coordinates were displayed as pink placemarks and overlays were made slightly transparent in order to let some of the satellite image show through the data image. Interpretations were made on overlays using the path option as features found were linear (e.g., building walls, foundations) as opposed to single targets. Paths are created by pointing and clicking where features are in order to highlight them. The color of the paths was set to red and the width was set to 2.0 to make the interpretations easier to see. Once the interpretations were made, the paths were saved in their own folder on the "Places" menu. Placemarks were added at key locations along the paths (i.e., beginning and end of a path or where paths intersected) to roughly map out the locations of subsurface features. All placemarks were saved in their own folder and this folder was saved as a .kmz archive on the Desktop. This file was treated in the same fashion as the data points collected at the B4 Plot and coordinates for the locations of subsurface features were displayed in Excel.

Looking at the data images, one might notice, especially at Dreamer's Bay, that some of the images appear to be skewed relative to the background image (e.g., some data appearing to be in the Mediterranean Sea). This offset is due to the error in the satellite images uploaded by Google. As this particular location is on an air force base, some intentionally introduced error was expected as security is a top priority. It is important to

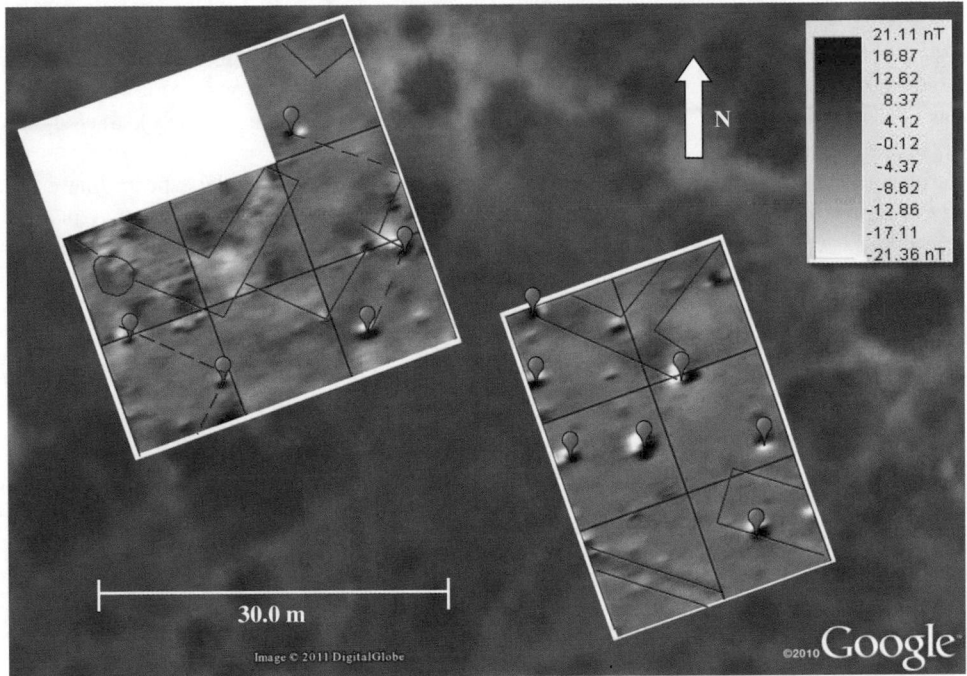

Figure 12. Interpreted magnetic data collected at St. Mark's. Interpretations were added with both the path and the placemark tool. The solid red lines indicate linear, subsurface features and the dashed red lines are more subtle features. Blue placemarks indicate discrete features that are not linear, but appear as dipole anomalies in the data. Each square represents a minor grid and is 10 m by 10 m.

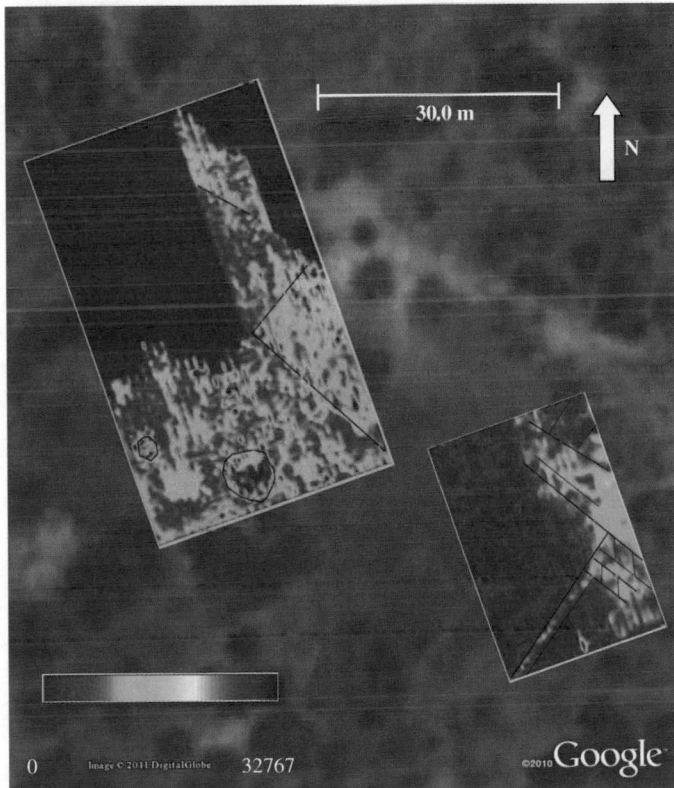

Figure 13. Interpreted GPR data for St. Mark's shows fewer subsurface features than the magnetic data. Black lines are interpreted linear features in the subsurface. Thicker lines indicate two linear features which were located very close together. Black circles encompass potential features which are not linear. Depth to features ranges from 0.75 m to 1.0 m.

Figure 14. Google Earth image displaying the geometry of features found in both data sets. Transparency of the GPR data was set to 40% in order to be able to see both sets of data. All features seen in the data are interpreted to be real, however, we can hypothesize that features detected by the magnetometer are walls while those detected by the GPR are trenches.

note that while the satellite photos may be skewed with respect to the data, the data images are accurately georeferenced to the correct GPS coordinates. Maps were created for use by the consulting archaeologist where the image was correct but the GPS coordinates were wrong, in order to prepare a report for the Cypriot Department of Antiquities illustrating the general location of features in the subsurface, however, those are not displayed here. The purpose of these maps presented is to extrapolate accurate data points for features located in the subsurface.

While georeferenced images of near-surface geophysical data may be useful to geophysicists, maps that display the locations of subsurface features on the surface (think earthquake focus versus epicenter) are more useful to archaeologists. Using these maps, archaeologists can create or alter excavation plans as needed to allow for a more productive field season. Feature maps were created of Dreamer's Bay (Fig. 15) and St. Mark's (Fig. 16) by simply turning off the data overlay layer.

Unfortunately, with the Cyprus data set, we had no ground truth information, due to permit restrictions, and thus we could not calculate the error in our interpretations as we had done with the data collected at the B4 Plot. The peak value of the magnetometer data is ~45 nT, and as the error of the instrument is ± 0.1 nT, amplitude error is not a concern for these particular data sets. When making interpretations in the GPR data, features are typically mapped between a high amplitude and low amplitude region in the data (this would denote a change in the electromagnetic properties of the subsurface, indicating either a "boundary" had been crossed or the detection of an anomaly), thus signal error is not a factor.

Using the B4 study as a proxy, we select the in-line error associated with the magnetometer is approximately equal to 0.075 m ± 1.36 m and cross-line error as ~0.28 m ± 1.47 m, while those values for the GPR data are approximately equal to 0.0038 m ± 1.08 m and −0.22 m ± 1.66 m, respectively.

CONCLUSIONS

To reiterate, the objective of this research was to identify and mitigate inefficiencies during the process of completing geophysical surveys at active archaeological sites and enhance the usability of geophysical data for archaeologists. Our hypothesis was that the effectiveness and efficiency of multi-tool, near-surface geophysical surveys for archaeological applications can be improved by displaying data with accurate GPS coordinates using a virtual globe. The success of this study was based on the error calculated at the control site, and comparing that with traditional excavation parameters.

Traditionally, geophysical surveys are time-consuming and taxing on both funds and personnel, which has a limiting effect on the lifespan of any geoarchaeological project. Normally, inefficiencies are encountered while in the field and the subsequent processing and interpretation of the data. Once the data are interpreted, they are often passed off to the archaeologist with little to no guidance from the geophysicist, and data manipulation generally requires computer programs that are expensive and difficult to learn. It is this final step of data manipulation that becomes the most draining on resources; thus, we incorporated Google Earth to display near-surface geophysical data

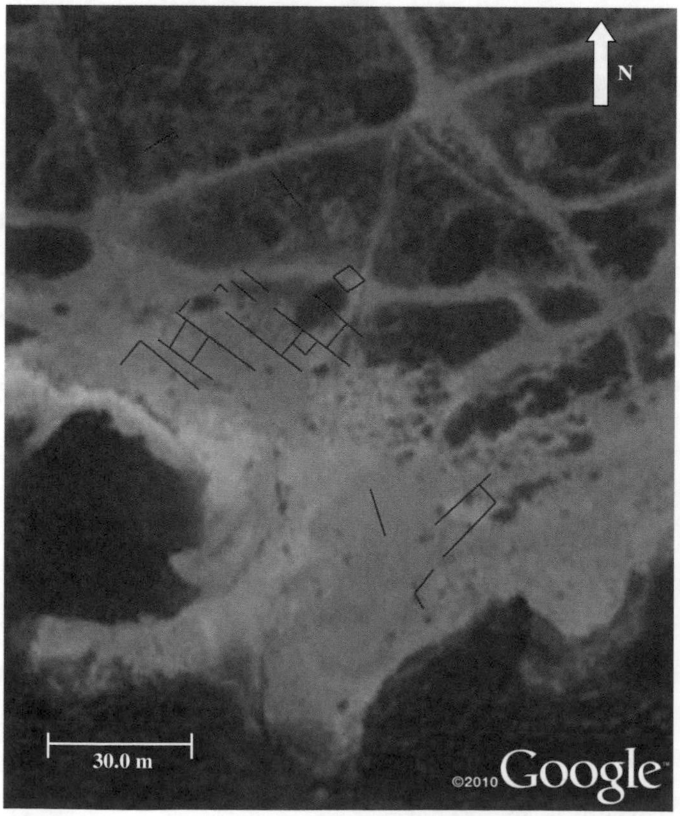

Figure 15. Map of Dreamer's Bay with subsurface features appearing on the surface. This map is useful to archaeologists in order to develop excavation plans.

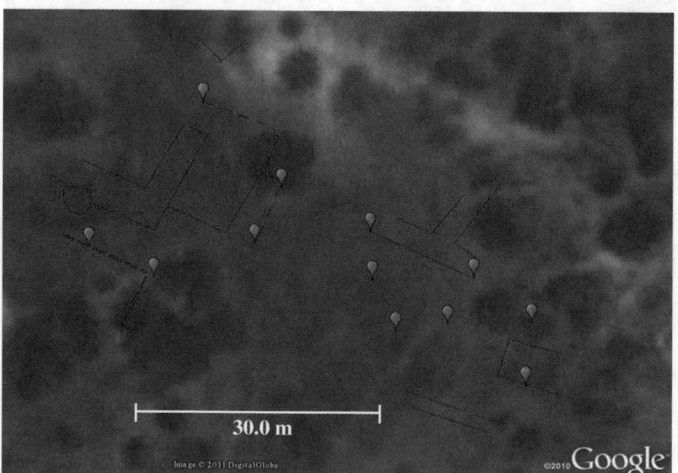

Figure 16. Map of subsurface features at St. Mark's. This map provides spatial data about features to archaeologists and aids in excavation plans.

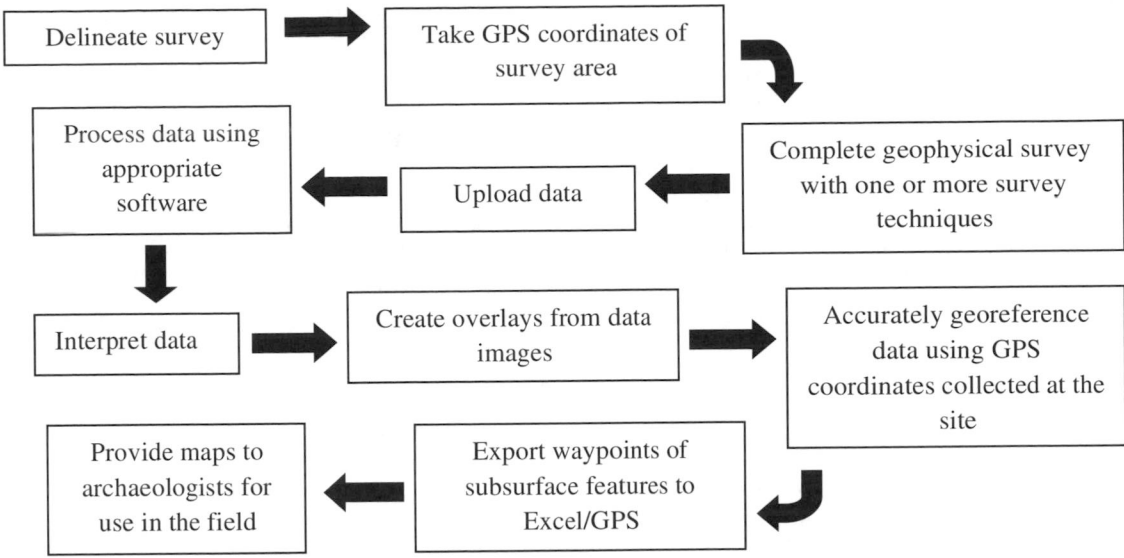

Figure 17. Diagram showing the new data management workflow developed at the B4 Plot in Tennessee. The first six steps of the workflow remain unchanged, as the focus of this research was expediting data processing and interpretation. Changes were made to the final steps of the workflow in order to provide archaeologists with a more useful final product.

and to create a user-friendly data manipulation interface. The archaeogeophysical control study conducted at the University of Tennessee Agricultural Campus was idea because targets had been buried and their locations accurately recorded. This control study was performed in order to test a new data management workflow involving Google Earth. The survey was completed using two different geophysical techniques: ground penetrating radar and magnetic gradiometry.

To create a map, GPS coordinates of the survey area were imported into Google Earth as placemarks, and data images were imported as overlays and accurately georeferenced using the GPS coordinates. Targets were identified in the data, and waypoints for these data points were exported to an Excel file for ease with later manipulation. Error was calculated between the actual data points which were recorded when the targets were buried versus the data points calculated in Google Earth. For the GPR data, the error was calculated to be 0.0038 m ± 1.08 m (in-line) and −0.22 m ± 1.66 m (cross-line) and error in the magnetic data was calculated to be 0.075 m ± 1.36 m (in-line) and 0.28 m ⊥ 1.47 m (cross-line). These values fall well within the average size of an archaeologist's excavation square that is typically five meters by five meters.

Given the success of this test study, a new workflow has been developed (Fig. 17). The main difference lies in the latter half of the workflow, and while there are more steps, they are more efficient at processing data. After data are interpreted, images are created of the subsurface data and are then accurately georeferenced in Google Earth. Waypoints can then be exported from Google Earth into Microsoft Excel and subsequently uploaded onto a GPS unit. This entire process can be done in a few hours, once data have been collected, and data are immediately acces-

sible and useful to archaeologists during the time limits of their field season.

This new methodology was tested on data collected via a multi-tool survey completed on the Akrotiri Peninsula, Cyprus, in 2010. Using the error calculated from the B4 Plot as a proxy, the geophysicists deemed this exercise a success as waypoints for features could be exported from Google Earth that were within the boundaries of the square size that an archaeologist would use for excavation.

Using Google Earth in this innovative way helps to expedite data processing, cut costs, and shorten the length of time needed for the execution of geoarchaeological surveys. Not only is the program free, but it is easy to use and can provide accurate data to archaeologists in the field. Giving accurate waypoints to archaeologists will enable them to develop a streamlined and potentially less-invasive excavation plan centered on known features in the subsurface.

ACKNOWLEDGMENTS

The authors would like to thank the Jones/Bibee Endowment for providing funding for field work in June 2010. Also thanks go to the Geological Society of America and Geometrics for funding to travel to conferences to present this research. The University of Tennessee Near-Surface and Environmental Geophysics Lab provided the equipment and lab space. Thanks also go to Greg Johnston from Sensors and Software for invaluable technical support, Rachel Storniolo and Christian Hunkus for being wonderful field assistants, and Frank Garrod and the WSBA Archaeological Society for allowing us to survey on their turf.

REFERENCES CITED

Abdallatif, T., El Emam, A.E., Suh, M., El Hemaly, I.A., Ghazala, H.H., Ibrahim, E.H., Odah, H.H., and Deebes, H.A., 2010, Discovery of the causeway and the mortuary temple of the Pyramid of Amenemhat II using near-surface magnetic investigation, Dahshour, Giza, Egypt: Geophysical Prospecting, v. 58, p. 307–320, doi:10.1111/j.1365-2478.2009.00814.x.

Ault, B.A., and Leonard, J.R., 2009, The Akrotiri-Dreamer's Bay Ancient Port Project: Ancient Kourias Found? *in* Herscher, E., ed., The Ancient Kourion Area: Penn Museum's Legacy and Recent Research in Cyprus: Philadelphia, University Museum Publications, 211 p.

Baker, G.S., and Ambrose, H.M., 2007, Ground penetrating radar imaging of a 4th Century Roman Fort, Humayma, Jordan: 4th International Workshop on Advanced Ground Penetrating Radar, no. 5, p. 4–59.

Baker, G.S., Jordan, T.E., and Talley, J., 2007, An introduction to ground penetrating radar (GPR), *in* Baker, G.S., and Jol, H.M., ed., 2007, Stratigraphic Analyses Using GPR: Geological Society of America Special Paper 432, p. 1–18, doi:10.1130/2007.2432(01).

Chianese, D., Lapenna, V., Di Salvia, S., Perrone, A., and Rizzo, E., 2010, Joint geophysical measurements to investigate the Rossano of Vaglio archaeological site (Basilicata Region, Southern Italy): Journal of Archaeological Science, v. 37, p. 2237–2244, doi:10.1016/j.jas.2010.03.021.

DW Consulting, 2010, ArcheoSurveyor User Manual: DW Consulting, 112 p.

Heywood, H.C., 1982, The Archaeological Remains of the Akrotiri Peninsula, *in* Swiny, H.W., ed., An Archaeological Guide to the Ancient Kourion Area and the Akrotiri Peninsula: Nicosia: Department of Antiquities, Cyprus, 235 p.

Kamei, H., Atya, M.A., Abdallatif, T.F., Mori, M., and Hemthavy, P., 2002, Ground-penetrating radar and magnetic survey to the west of Al-Zayyan Temple, Kharga Oasis, Al-Wadi Al- Jadeed (New Valley), Egypt: Archaeological Prospection, no. 9, p. 93–104.

Solsten, E., 1993, Cyprus: a country study: Washington D.C., Department of the Army, 383 p.

Wernecke, J., 2009, The KML Handbook: Addison-Wesley, 231 p.

Wynn, J.C., 1986, A review of geophysical methods used in archaeology: Geoarchaeology, v. 1, p. 245–257, doi:10.1002/gea.3340010302.

MANUSCRIPT ACCEPTED BY THE SOCIETY 16 APRIL 2012

The Geological Society of America
Special Paper 492
2012

Workshops, community outreach, and KML for visualization of marine resources in the Grenadine Islands

Meg E. Stewart*
Kimberly Baldwin
Centre for Resources Management and Environmental Studies, University of the West Indies, Cave Hill campus, Barbados, West Indies

ABSTRACT

The Grenadine Islands and the marine environment surrounding the islands were mapped over a five-year span. The project—Grenadines Marine Resource and Space-Use Information System (MarSIS)—involved merging local knowledge with existing scientific data into a geographic information system (GIS). Located in the Caribbean, the Grenadines share an international boundary between Grenada and St. Vincent and the Grenadines, creating numerous challenges for not only collecting data but sharing those data with the residents of the islands.

Project geospatial information was collected in a GIS, but Google Earth was used as a way to share the findings on the web and through a series of tutorials and workshops. Though project GIS shapefiles will be made available through the project website, Google Earth was used as a ready delivery tool because it is cross platform, easy to use, and free. Using aftermarket GIS extensions, shapefile layers were exported from ArcGIS into Keyhole Markup Language (KML) layers. Over 400 photographs and videos were geolocated in the project KML.

Once the Grenadines marine map was assembled as a KML project, we gave workshops on various islands. From user feedback following the first series of tutorials, we modified the KML by fixing problems, correcting mistaken information, and making the KML project file more understandable. When the project was finalized we put the KML on the MarSIS project web page and sent it as an attachment to the project email list. We traveled a second time to the Grenadine Islands to give another series of tutorials and workshops. We also created a video to help users navigate the project KML.

---

*meg.stewart@fulbrightmail.org

Stewart, M.E., and Baldwin, K., 2012, Workshops, community outreach, and KML for visualization of marine resources in the Grenadine Islands, *in* Whitmeyer, S.J., Bailey, J.E., De Paor, D.G., and Ornduff, T., eds., Google Earth and Virtual Visualizations in Geoscience Education and Research: Geological Society of America Special Paper 492, p. 63–76, doi:10.1130/2012.2492(05). For permission to copy, contact editing@geosociety.org. © 2012 The Geological Society of America. All rights reserved.

INTRODUCTION

The Grenadine Islands and the marine environment surrounding the islands were mapped over a five-year span. The project, called Grenadines Marine Resource and Space-Use Information System (MarSIS), is part of a regional effort to document local marine-based resources, such as fishing, reef environments, and aspects of beach tourism, and plan for equitable and sustainable use of those resources. Local residents were included in the decision-making process (Sustainable Grenadines Project, 2006). Located in the Caribbean, the Grenadine Islands include an international boundary between Grenada and St. Vincent and the Grenadines (Fig. 1). This geographic distribution creates numerous challenges not only for collecting data but sharing those data with the residents of the islands. The more than fifty islands, islets, and cays that make up the Grenadines are of significant ecological importance because of their extensive coral reef and related habitats, which are known to have vulnerable ecosystems with endangered wildlife (McGann and Creary, 2008; Chakalall et al., 2005; Mahon et al., 2004). The islands have a growing tourism industry that has led to unregulated development (Bresson and Logossah, 2011; Conway and Timms, 2010; Economic Commission for Latin America and the Caribbean [ECLAC], 2004). Coupled with the increasing needs of the local population and limited environmental controls, these development pressures have led to pollution, deforestation, over-

fishing, and overall marine and terrestrial ecosystem degradation (DeGeorges et al., 2010; ECLAC, 2004).

Marine-based activities, such as fishing, shipping, and tourism, are vitally important to the people of the Grenadine Islands (Baldwin et al., 2006; Mills, 2001). MarSIS is a detailed documentation of existing marine resources collected using a GIS (geographic information system). The second author worked with a range of parties including marine resource users, non-government organizations (NGOs), government agencies, and members of the community to collaboratively identify, document, and quantify sea-based resources in the Grenadines. This data collection research amalgamated marine information with the intent to foster a greater understanding of the location and distribution of key resources and what space-use patterns exist, in order to provide a comprehensive information base to assist with environmentally sustainable development across the Grenadines (Baldwin, 2006). Local knowledge of existing sea-based resources and their use patterns was used throughout the research. Detailed locally derived data were combined with scientific, publicly available data sets to build the MarSIS. Project management includes making the GIS database available to help the Grenadine Island communities and agencies make informed planning decisions.

Information provided for community use should be in an understandable format, useable, and accessible to stakeholders in order to facilitate a more equitable, transparent, and collaborative decision-making environment (McCall, 2003; Sieber, 2006; Rambaldi et al., 2006). GIS data sets will be provided to stakeholders through the project website (http://grenadinesmarsis.com/) and other means. However, there is limited capacity within the Grenadine Islands to utilize geodata, as there are very few licenses of ArcGIS software and still fewer people with the technical skills to analyze and manipulate the geodatabase and shapefiles (Opadeyi, 2007). For these reasons, Google Earth software was seen as a valuable platform to provide the community with the collected information found in the marine map of Grenadines MarSIS, and the GIS data sets were translated into Google Earth format as KML (Keyhole Markup Language) files. Google Earth is a virtual globe comprising satellite imagery, aerial photography, and 3-D terrain models (Ratliff, 2007; Patterson, 2007; Tulloch, 2007). These data are presented at increasingly high resolution as the user narrows the field of view in any region of the world. Additional user-generated content can be added to the globe, increasing its functionality. The ease of use and wide availability of Google Earth make it a functional platform for sharing spatial data with a community of interested users (Ratliff, 2007; MapAction, 2008). By using the understandable and intuitive Google Earth, exploration of the scientific data of MarSIS allows for enhanced learning of the marine environment and the resources therein (Ballagh et al., 2011).

To orient and educate the community about the availability of the MarSIS data, workshops were held over two different periods of time on various islands in the Grenadines. The agenda for workshop days included an overview of the MarSIS

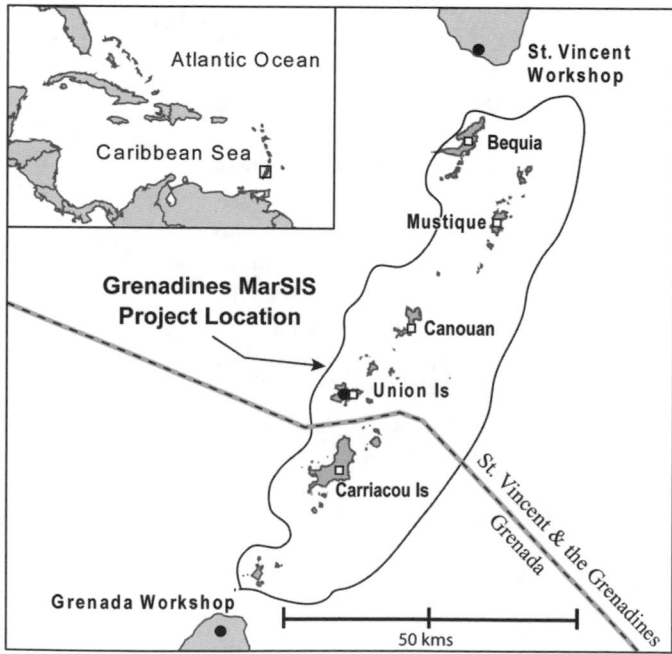

Figure 1. Location map of the Grenadines MarSIS project area in the Caribbean Sea showing the international transboundary line between Grenada and St. Vincent and the Grenadines. The MarSIS study boundary is outlined. The locations of the November 2009 workshops are shown with black dots and with squares for the June 2010 workshops.

TABLE 1. EXAMPLES OF THE TYPES OF GEOSPATIAL INFORMATION
FOUND IN THE MarSIS GEODATABASE

Classification scheme	Types of information layers
Bathymetry	Sonar-derived contour lines for Grenada Bank
Infrastructure	Jetties, seaports, and airports
Marine habitats	Reefs, sea grasses, and mangroves (shallow/deep)
Marine resources	Turtle nesting beaches and seabird roosting sites
Marine resources users	Day-tour operations, water taxis, and fishers
Space-use patterns	Dive sites, aquaculture, anchorages, and shipping lanes
Areas of threats	Illegal dump sites, desalination plants, and sand mining operations
Other	Marine protected areas, local names, and Exclusive Economic Zones

to-date, a discussion and demonstration of available GIS data sets, a preview of the same data sets using KML in Google Earth software, a tutorial in using Google Earth, and an interactive set of guided exercises using the MarSIS KML file (Appendix I). We created a video to show users of the MarSIS KML some of the available layers and other usage tips. Our paper highlights the use of Google Earth to communicate issues of interconnectedness of marine ecosystems to a wide audience, many of whom derive their livelihood from the very resources that this project addresses.

METHODS

GIS Data Collection

The MarSIS geodatabase was built using ESRI's ArcGIS 9.2 software 3D and Spatial Analysts extensions. Key spatial information is shown in Table 1. The MarSIS is useful for progressively identifying the locations of spatial importance for marine conservation and livelihoods as well as highlight emergent areas that are currently or potentially at-risk for space-use conflict. For detailed information on project planning, mapping methodology, and data structuring used to collect geospatial data for the Grenadines MarSIS research, see Baldwin (2012).

GIS to Google Earth

In order to provide the Grenadine Island community of stakeholders with the geodata compiled in the MarSIS marine map, the GIS layers were exported and reassembled as a project file in Google Earth. The final MarSIS project file is in a Google Earth KMZ (Keyhole Markup Language—zipped) format and can be found on the project website (http://grenadinesmarsis.com/Google_Earth.html). Figure 2 shows the available data layers. We exported 45 vector-based layers from the geodatabase. To keep the file size small enough for easy email attachment, raster-based layers were not included with the Google Earth project. Additionally, most of the raster data collected for the MarSIS were proprietary. The goals for the MarSIS Google Earth file were (1) ease of sharing, (2) to create a project that looked attractive but still contained all the scientifically derived data sets, and (3) to package the KMZ file just like the MarSIS GIS database, in the same order.

The methods used to export the GIS shapefiles to create Google Earth KML are found in Appendix II.

WORKSHOPS

The MarSIS data collection and research analysis was finalized by September 2009. We led on-island workshops to discuss and share with the community various aspects of the work at that point. The workshops were announced broadly in an effort to provide community members and stakeholders an update of the Grenadines MarSIS research and findings, an overview of available GIS layers within the MarSIS, a hands-on tutorial on how to use the Google Earth application, and interactive exercises using the MarSIS KMZ file (Fig. 3). The goals for the workshops were to: (1) show the community the MarSIS research as a GIS and demonstrate how to make spatial analyses of the data in the GIS software; (2) exhibit to attendees the MarSIS data as a KMZ in Google Earth and what layers were available to them; (3) teach people how to use Google Earth software; and (4) illustrate how to use the MarSIS KMZ as a planning tool through guided examples (Appendix I). We wanted attendees to feel they had ownership

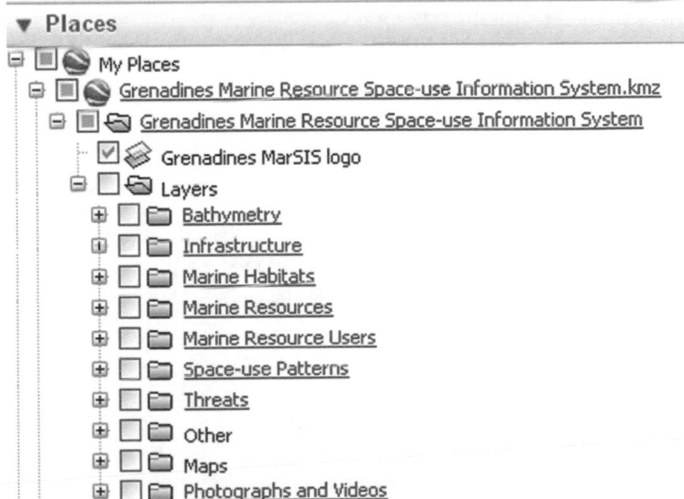

Figure 2. In Google Earth, the Grenadines Marine Resource Space-use Information System is a KMZ file with ten layers that have been exported from the MarSIS project GIS geodatabase. There are 45 sub-layers plus photographs and videos.

of the data. We believe these skills are useful for not only viewing and understanding the MarSIS KMZ but also using the KMZ as a means for decision-making and long-range planning.

Workshops in November 2009

The public discussion of the finalized MarSIS research and demonstration the KMZ file were held in November 2009. Three one-day workshops were led on the islands of St. Vincent, Union, and Grenada (Fig. 1). The workshops followed the agenda shown in Figure 3 and gave attendees a hands-on opportunity to visualize and use the Grenadines MarSIS using the Google Earth interface as well as provide the leaders with direct user feedback on the usability of the MarSIS KMZ. To help participants become acquainted with using Google Earth, basic training in the use of the software, modeled after the humanitarian mapping program of MapAction (2009), was provided. Attendees explored and evaluated the project file by using guided exercises using the MarSIS KMZ (see Appendix I—this is the most recent version of the exercises).

All three workshops were well attended by representatives from a wide range of marine-related government agencies, local NGOs and members of the Grenadine community (teachers, fishers, rangers, etc.) involved in the research. A total of 55 participants (25 Grenadian and 30 Vincentian) attended the workshops drawn from 23 different agencies or groups. Of the 55 participants, 30 persons represented government agencies, 14

were from the Grenadines community and 11 from local NGOs. Participants were asked to bring their own laptops, though we brought extra computers in case some did not (Fig. 4). Most people shared laptops and worked collaboratively.

The workshop attendees were introduced to Google Earth and shown how to search and navigate, how to create placemarks (points) and paths (lines), how to save their placemarks as a KML, how to create a map with a title and several features and save it as a JPEG (Fig. 5), and how to share their KMLs and JPEGs by email attachments (Appendix I).

Both oral and written evaluation techniques were utilized to collect feedback on a range of topics from the attendees. Using an oral focus group discussion format, stakeholders were questioned throughout the sessions on the overall design and usefulness of the MarSIS KMZ project. As this was the first public launch of the KMZ, participants found mistakes in the data set as well and made suggestions on how to improve the final product. We were given feedback on the Google Earth demonstration and the provided exercises (Appendix I), as well as any suggestions for future community training workshops. Workshop participants gave very positive feedback to the practical nature of the hands-on training and associated exercises. Many expressed surprise that after a short period of time the Google Earth application could be mastered. Participants were impressed by the ease with which one could successfully access the MarSIS information using Google Earth and create their own maps (Fig. 5). In all three workshops, attendees requested that additional training be provided for other interested persons including government agencies, schools, and communities, and that the training be in a similar interactive, hands-on workshop format.

Upon exit of the workshops, the November 2009 participants were asked to complete a survey evaluating the process of

---

Workshop Agenda

9:00	Welcome and Introductions
9:15	Review of MarSIS Research Activities (2005-2009)
9:35	Key MarSIS Findings and Implications of Research
10:15	Coffee Break
10:30	Google Earth Introduction and Training
12:00	Lunch
1:00	Participants use MarSIS for decision-making exercises in Google Earth or ArcGIS
2:30	Evaluation of MarSIS and Research Methods
3:15	Thanks and Closing

If you brought a laptop and do not have Google Earth installed, please be sure to let us know this morning. We will load the Google Earth software and MarSIS (KMZ) data during the workshop for use in the afternoon Google Earth exercises.

Figure 3. Review of the workshop agenda for the Grenadines MarSIS update meetings in 2009 and 2010. See Appendix I for details of workshop Google Earth exercises.

Figure 4. Workshops participants in St. Vincent Office of Fisheries using laptops and Google Earth to visualize the MarSIS marine map geodata.

Figure 5. Example of a workshop participant's map using Google Earth and the MarSIS KMZ file. This map shows the location of hotels and baitfish bays in Friendship Bay on the island of Bequia and was saved as a JPEG and sent to the workshop leaders.

using participatory techniques, information communication and exchange mechanisms used throughout the research. There was 78 percent survey participation. Of those who completed the survey, there was 100 percent agreement that the MarSIS will be of use for their work and allow for informed decision making in management and planning of coastal and marine resources. There was strong agreement that MarSIS data will be a good educational resource that highlights the importance of the sea to the people of the Grenadines. In terms of using Google Earth as a platform for dissemination of MarSIS information to the wider public, 84 percent believe that Google Earth is currently the most appropriate application for the local technological capacity.

All three November 2009 workshops were considered successful. Additional training in both Google Earth and ArcGIS was requested by workshop attendees and will be needed in order to allow for wider dissemination and use of the Grenadines MarSIS information by the general public, especially within the school system.

From the three November 2009 workshops we created a short video showing brief portions of meetings. (http://www.youtube.com/watch?v=GH401suGdA4) This video provides a context for the community-based MarSIS research and Google Earth workshops.

Follow-up Workshops—June 2010

In June 2010 we led a second round of training workshops, this time for teachers at the 14 primary and 5 secondary schools of the Grenadine Islands. Day-long workshops took place in Car-

riacou, Union Island, Canouan, Mustique, and Bequia (Fig. 1). In these workshops we used the final version of the KMZ for which we corrected all errors and incorporated suggested changes that we received from participants during the first workshops. The agenda for June 2010 workshops did not vary from the November 2009 workshops and is shown in Figure 3. A total of 53 teachers and 20 students attended the five workshops.

From suggestions we received in November 2009, we modified the original Google Earth tutorial exercise sheet to include more screen captures and more questions. Appendix I is our final tutorial. As with the previous workshops, the June 2010 meetings gave the educator participants practical experience using Google Earth in the first part of the session and a guided set of exercises using the MarSIS KMZ in the second part.

Each workshop ended with time for written evaluations as well as a short group discussion on how the Grenadines MarSIS KMZ and Google Earth can be used in the teaching curriculum. All participants agreed that Google Earth as well as the local Grenadines MarSIS database will be assets to their curriculum and benefit their students. Ideas generated included: integrating MarSIS and Google Earth into local and global geography lesson plans, using spatial information in environmental education, building and expanding teacher knowledge of local and global issues using Google Earth, and thoughts on how students and educators can add to the Grenadines MarSIS geodata. The teachers desired acquiring more information technology and computer skills for themselves and their students.

Unfortunately many of the June 2010 workshop participants did not fill in the exit evaluation form. A total of 29 of the

53 attendees completed the survey. Half (or 14) had never used Google Earth prior to the workshop. All participants agreed that the workshop was informative and that the exercises were easy to follow. Twenty percent of respondents felt that the workshop was too short and wished that they had a second day to practice their newly learned skills. All of the teachers felt that the Grenadines MarSIS KMZ and the use of Google Earth technology will be extremely useful in their teaching. All reported that they will be able to use exercises similar to those used in the workshop (Appendix I) in their teaching curricula. All respondents would recommend this type of training to colleagues as well as to family and friends. In particular, respondents identified the need for additional training workshops for information dissemination to students, fishers, divers, the tourism industry and marine-based government workers, as well as the public.

We note here that the technological infrastructure at the schools was in need of upgrading. Computers on-site were either out-of-date or in need of repair and thus limited the available hardware to carry out the afternoon practical exercises during the workshops.

ADDITIONAL OUTREACH

To help make sense of the Google Earth project file for MarSIS, which contains ten layers and 45 sub-layers, we created a video tutorial. The video shows users of the MarSIS KMZ how to use Google Earth and view the data found within the KMZ. The link is on YouTube (http://www.youtube.com/watch?v=ySVzDZLxAS0) and is also found on the Grenadines MarSIS project website.

UPDATE ON MARSIS KMZ COMMUNITY USE

Since the release of the Grenadine MarSIS KMZ file June 2010, we have heard of many examples of its use in planning and teaching. The KMZ was used by the NGO, Friends of the Tobago Cays, to contest a dredging and sand reclamation project in Canouan. In Grenada, People in Action NGO used the spatial data of MarSIS to rally against a port on Carriacou and added land-based geodata to strengthen their geo-tourism initiative. Governmental planning departments and consultants have also used MarSIS geodata for informed environmental decision-making. We have heard that teachers are using MarSIS KMZ in classes, but we do not have specific details of the outcomes.

Because of the volume of hits and downloads of the KMZ, we know that there is use and interest in the MarSIS KMZ file. The MarSIS project website had to be upgraded.

DISCUSSION

Community-based mapping can be used to incorporate local knowledge of marine resources and space-use patterns into planning and decision-making. Ideally, for these research efforts to be successful, information gathered must be useful to the local community (McCall, 2003; Rambaldi et al., 2006). GIS software has many benefits, but without extensive GIS skills, user-friendliness for training purposes is not one (Sieber, 2006; Patterson, 2007). Google Earth is a free and easy-to-use mapping application that can be used to share data with a large range of stakeholders with varying levels of technical skills (Ratliff, 2007; MapAction, 2008). Although initial data gathering involves building the information in a GIS, the data can be disseminated to the community as a functional Google Earth project file (KMZ) resulting in a useful and relevant end product (Tulloch, 2007). The Google Earth application was used for MarSIS to enable presentation of the complex interrelationships inherent in marine ecosystem data to an audience that includes those with limited technological skill or capacity. Google Earth has limitations. It is not a true GIS because one cannot perform spatial analyses, queries, or a nearest-neighbor function, to name a few. To use Google Earth, one needs Internet access. In addition, there is a general lack of high-resolution or cloud-free imagery for smaller or developing islands. It is extremely useful, though, as a visual aid for project-specific geodata or for computer cartography, with limited GIS functionality (Patterson, 2007; Stewart et al., 2007). Community members can make environmentally based decisions and create placemarks, send them to an interested colleague or a project leader, and the spatial information can be added back into the larger data set. The use of Google Earth in mapping projects allows for greater access to geospatial data information than by more traditional approaches such as providing GIS data layers or use of paper maps alone (MapAction, 2008; Tulloch, 2007; Rambaldi et al., 2006).

Future directions for the MarSIS may include the implementation of an interactive web map, one in which community members have a single user-interface to view MarSIS data but have the ability to add to the data set as well. Another consideration is staffing. If the project is to be truly sustainable, a person needs to be staffed to manage the data.

APPENDIX I. WORKSHOP EXERCISES

Google Earth Workshop—June 2010

We will give a short overview on how to use Google Earth to begin. The Grenadines MarSIS geodata that we are showing today are available for viewing in Google Earth. The second half of the workshop will give you time to explore the MarSIS data.

The morning Google Earth tutorial can be found at this link: www.grenadinesmarsis.com on the "Google Earth" page.

I. Topics for Morning Workshop

- Getting Started with Google Earth
- Navigation
- Searching
- Viewing Layers
- Measure Tool
- Setting Up Placemarks
- Adding a Polygon & Lines
- Making Your Own Map
- Sending Google Earth Files
- Saving and Sending Maps

II. Topics for Afternoon Workshop

Exploring the Grenadines MarSIS Database
Be sure you have Google Earth downloaded on your computer. We will provide you with the Google Earth file for the Grenadines MarSIS, but the file can also be found on the project web page under the Google Earth tab (www.grenadinesmarsis.com/Google _Earth.html). Look for the link that says "Grenadines MarSIS Data (.kmz)" and click on that; this will open up the MarSIS in Google Earth. A 6-minute video tutorial link is also available on this webpage.

The MarSIS Google Earth files are set up as folders according to topics, which you can open by clicking the + sign next to each of the folders.

Explore MarSIS
Take some time to look through the various datasets. Expand the **Layers Folder** and review each of the different datasets within each folder to get familiar with the data.

Exercises
(1) Analyze and Measure Distances
Goal: Understanding straight-line and curved-path distances.

Using the **Ruler** tool (circled above) found on the Toolbar, answer these questions.

(a) Navigate to **Petit Martinique**.

Measure the longest distance in the north-south direction on Petit Martinique. Answer: _____ kilometers.

How far across is the island in the east-west direction? _____ kilometers
(HINT: use the **Line** tab on the **Ruler** tool. "Heading" indicates direction and north-south heading is 180 degrees.)

(b) In the **Other** Folder: Turn on **Local Names**. Find **Mayreau** and zoom in close enough to see the road.

Measure the distance from Salt Whistle Bay to Saline Bay following the main road across the island. Answer: _____ miles (HINT: For this you will use the **Path** tab on the **Ruler** tool. Use mouse clicks along the road in Mayreau to measure the total path distance.)

(2) Proximity Analysis
Goal: Assessing how close (or far) features are to one another.

Zoom to **Bequia**. Turn on two layers only: Under the **Infrastructure** folder, turn on the **Hotels** layer, and under **Space-Use Patterns** folder, turn on the **Historic Sites** layer. Using the **Ruler** tool as you did in the previous exercise, find which hotel, apartment or guest house that has the greatest number of historic sites within a 1 kilometer radius. (HINT: Use the Line feature in the **Ruler** tool. Click on a hotel, pull the ruler line out to 1 km and swing the line around the hotel point in a circle to observe how many historic sites occur within a 1 km circle.

Which hotel has the most historic sites nearby? _____

What are the *types* of features found at these historic sites? _____

_____

(3) Use and Distribution of Important Resources
Goal: Delimiting regions of significance, understanding their importance and evaluating new significant regions.

Under the **Marine Habitats** folder, turn on **Shallow Water Habitats**. First, click on the blue underlined Marine Habitats folder name. A window will pop up with photos that show what the different types of marine habitats look like. To close this window, click the check box at the top. Now under the **Other** folder, turn on **Protected Areas**. These are regions either designated or proposed for marine protection. Use the legend on the left to understand the different marine habitat types.

Which protected area has the most of all three (reef, sea grass beds and mangrove) habitats needed for good reef ecosystem function?

_____

_____

Now you will find a site and propose it for marine protection. Turn off the **Protected Areas** layer only. Explore the various islands and find an area of good marine reef habitat quality. Again, these are areas that should have all three habitats (reef, sea grass beds, and mangroves) close together as these provide for good reef ecosystem function. Turn on the **Photographs** layer (under Photographs and Videos) and explore the various habitats to identify potential areas that have good quality according to the photos.

Now you will make a map of your newly proposed protected area. Under the **Other** folder, turn on the **Local Names** layer. Using the **Polygon** tool (circled above) draw a polygon around your new "potential marine protected area" site. Your mouse will look like a square with cross hairs. You can create a polygon by drawing (left clicking) around the area that you want to map. You do not need to close the loop by clicking on the first mouse click; simply double click your polygon line to end it. You can always edit your polygon by right-clicking to access the Properties dialogue box. In the Style, Color tab you can get rid of the fill color on your polygon by choosing the Area as **"Outlined."** Make the line of your newly proposed marine protected area thicker and change the color so it is noticeable on the map.

Now add a title to your map. Click the **Placemark** icon (circled above) and type in "Proposed Marine Protected Area." Click on the Style, Color tab and increase the label size by increasing the scale (this makes the font larger). One last step: **Save** your map as a picture (.JPEG file) for easy sharing. Go to **File > Save > Save Image** and give your map a name. Be sure to remember where you saved it. You can email this to yourself, to your students, or to other teachers as an attachment. Please be sure to email your map to baldwin.kimberly@gmail.com.

(4) Physical Characteristics of a Place
Goal: Understanding the shape of the land and directional orientation.

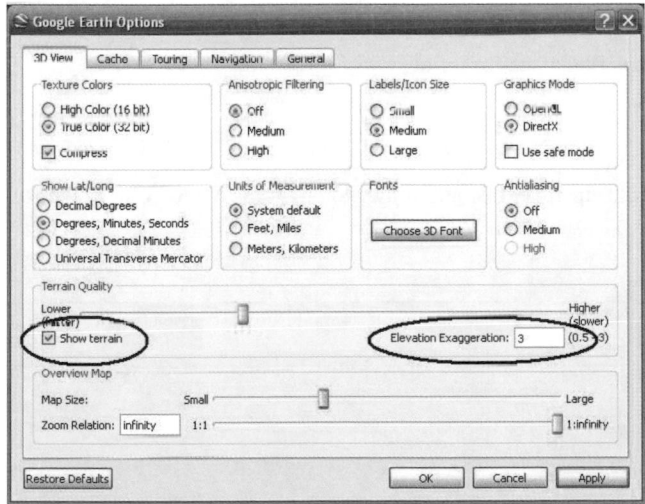

Zoom to **Carriacou**. At the top in the **Menu** bar go to **Tools > Options** (you'll see the window to the left). In the **3D View** tab, be sure Terrain is checked and change the elevation exaggeration to 3 (both circled).

Using a combination of zooming and using the upper circle in the navigation area (it has an "N" on the outer ring and an eye in the middle), zoom closer into Carriacou so that you can see the peaks of hills and the shape of the island's landscape. Notice when you hover your mouse over land, you can see the elevation for that spot shown in the lower bar of the Google Earth window (labeled "elev" next to latitude and longitude coordinates).

There are three peaks on Carriacou. What is the elevation of highest peak? _____ feet

Now you will create a .KML (the file extension of Google Earth—it stands for Keyhole Markup Language) for the three peaks. First, create a folder to store your placemarks by right-clicking on the **My Places** icon and go to **Add > Folder**. A window will pop up so you can designate the **New Folder** features. Name your folder "Peaks on Carriacou." Next find one of the peaks. Right-click on your folder and go to **Add > Placemark**. You will see a blinking yellow push pin. Move the push pin icon to rest exactly at the top of the peak. Give the placemark a name and add a description (such as the elevation in feet) and click **OK**. Now do the next two peak features in this same way. Then after you have all three placemark peaks in your folder, right-click on the folder name and go to **Save place as...** and give your KML file a name and click **Save**. Be sure to remember where you saved this KML file so you can attach it and send in an email to yourself, another teacher, or your students. Again, please email your KML file with the peaks to baldwin. kimberly@gmail.com.

Navigate as though you are standing on top of the highest peak.

Which island has the highest peak that is due north ("N" is at the top of the navigation circle)? _____

_____

What is the highest island due south ("N" is at the bottom of the navigation circle)? _____

_____

(5) How Human Activities Alter the Physical Environment
Goal: Understanding the spatial relationships between natural resources and the impacts of human uses of resources.

Navigate to **Union Island**. Turn off all layers except for, under the **Marine Habitats** Folder, **Sea Turtle Nesting Beaches,** and under the **Threats** folder, **Illegal Sand Mining** operations. Zoom around the coastal areas of Union and look for areas where illegal sand mining operations and turtle nesting beaches overlap.

How many known nesting beaches on Union Island occur where there is illegal sand mining? _____

_____

Human Health Considerations: In the **Threats** Folder, turn on **Illegal Dumping Sites,** and under the **Marine Space Use** folder, turn on the **Recreational Areas** layer.

What is the correlation between these layers? _____

How many areas are affected? _____

APPENDIX II: METHODS FOR CREATING KMZ FROM PROJECT DATA

To compile the MarSIS data KMZ from the MarSIS geodatabase we used a variety of ArcGIS extensions, plug-ins and custom code. This appendix details some of the steps taken to export GIS shapefile and add imagery to build a KMZ.

Exporting Layers from ArcGIS

Individual shapefiles or layers were exported out of ArcGIS as KML (Keyhole Markup Language) files and assembled in Google Earth as a KMZ. Two methods for exporting GIS layers were used. At the time of building the Google Earth project (September to November 2009), we were using ArcGIS 9.2 and we did not have access to the GIS import functions of Google Earth Pro. To export point, line, and polygon shape-files out of ArcGIS 9.2 we used a combination of a free ESRI ArcScript called Export to KML and a fee-based extension called XTools Pro. The methods for creating KML files out of shapefiles allow for exporting all attribute data associated with each feature. Once the individual KML file was created, some layers were re-symbolized and re-colorized. Most colors exported appropriately out of the GIS software but to keep color consistency within, for example, the shallow water habitat map and the deep water habitat map, we modified the color scheme from within Google Earth (Fig. A1). We added a legend for the polygonal habitat maps using Adobe Illustrator and created a screen overlay in Google Earth. If using ArcGIS 10 software, the export process is fully integrated into the standard version of the software. In addition, Google Earth Pro allows for direct importing of GIS shapefiles.

To ease the data collection process during the five years of the MarSIS data collection and add consistency to values that were repeated within and across data layers, subtypes and domain codes were developed and implemented for the geodatabase. Problems occurred, however, when points were exported out of ArcGIS and the attribute values would be assigned the numeric value of the domain which would show in the placemark (point) balloon and not the actual text value of the point itself. For example, a balloon in Google Earth for an island would read "Island = 1" when what is meant is "Island = Mayreau" (Fig. A2). We recreated new shapefiles substitut-ing the numeric values for the proper text values for all of the shape-files that utilized domain and subtypes.

Adding Project Photographs and Videos

As discussed above, shallow and deep water habitat maps were created. The shallow water map was derived using a "mixed-method" of conventional remote sensing and ground-truthing in the field where photographs were taken and locations were collected by GPS. The deep water habitat map was created by taking direct field measure-ments using a standardized sampling grid and remote video to interpo-late marine habitat. The 200 geolocated photographs and 190 videos were included in the MarSIS Google Earth project. We used Google Spreadsheet Mapper 2.0 to create placemarks for all of the imagery. Spreadsheet Mapper provides several templates to create consistent and attractive placemark balloons and is useful in creating a custom look for project photographs (Fig. A3). In Spreadsheet Mapper, we entered the latitude and longitude, a web link URL for the image, related field notes, and a title for each photograph. A KML network link was then generated. This KML was added to the Google Earth KMZ project (see Fig. 2—"Photographs and Videos"). There are two drawbacks to Spreadsheet Mapper: (1) only photographs can be used in the spreadsheet, and (2) a maximum of 400 photographic images can be in the spreadsheet. To get around not being able to embed video into placemarks, we used simple screen captures of each video and used that static JPEG image as a placeholder in Spreadsheet Mapper. Once the image placemark for the video was created, using an HTML editor we customized each individual placemark balloon by removing the image URL information and copying the video embed code (Fig. A4). All MarSIS project videos are on YouTube (http://www.youtube .com/user/grenadinesmarsis). The process of creating placemark bal-loons for viewing imagery in Google Earth can now be done with Google Fusion Tables and there is no limit to the number of photos and videos to geolocate through a Fusion Table.

Digital Raster Maps

Though the MarSIS geodatabase includes many digitized raster maps, mostly topographic and nautical charts, because of proprietary limitations, we made available just one raster file in the Google Earth MarSIS project (Fig. A5). The file is a declassified U.S. Department of Defense 1:250,000 scale chart of Grenada and the Lesser Antilles. The scanned and georectified file is 110 MB. To keep the MarSIS Google Earth project KMZ as small as possible, we did not include the nautical

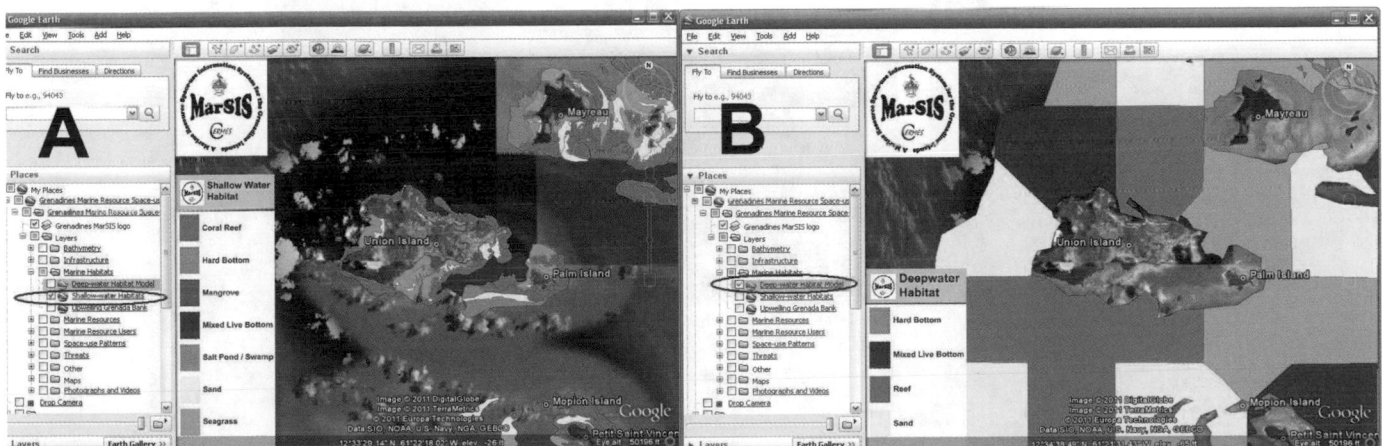

Figure A1. Symbology for the two habitat maps was configured in Google Earth. Map A shows the shallow water habitat map and Map B shows the deep water habitat map. The MarSIS project logo (upper left hand side of window) and the two habitat legends were created in Adobe Illustra-tor and the Google Earth screen overlay function was used to link them to the corners.

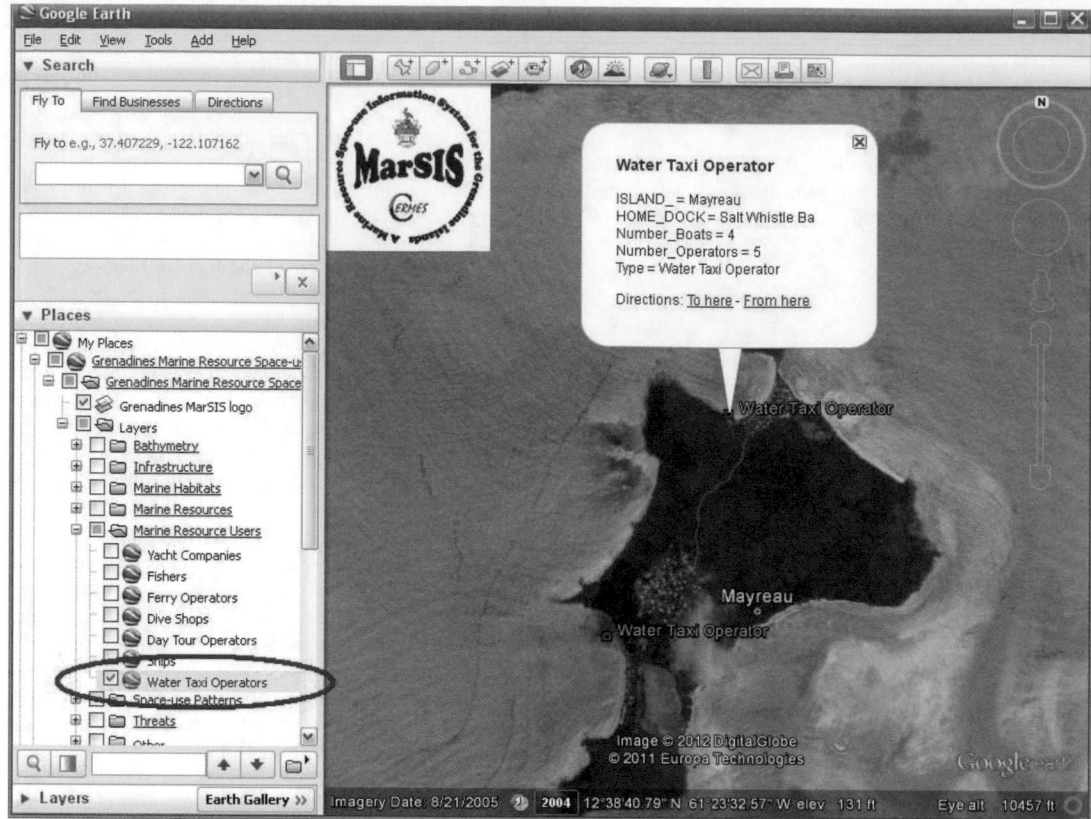

Figure A2. An example of a placemark (point) balloon from the Water Taxi Operators layer. All data fields for the point export into the KML. The MarSIS geodatabase design utilized domains and subtypes to ease the data collection process when assigning attribute information. The island Mayreau (shown above) was set as "1" and exported into Google Earth as "ISLAND = 1." To correct this and make the placemark information useful as well as correct, shapefiles using domain and subtypes were recreated using text rather than numeric values.

Figure A3. Photograph placemark balloon showing a photo taken off shore of Petite Martinique. We used a template in Google Spreadsheet Mapper 2.0 to create a consistent look to each project photograph. The spreadsheet includes the latitude and longitude for each image, a URL link to the photographs, and descriptions taken from field observations.

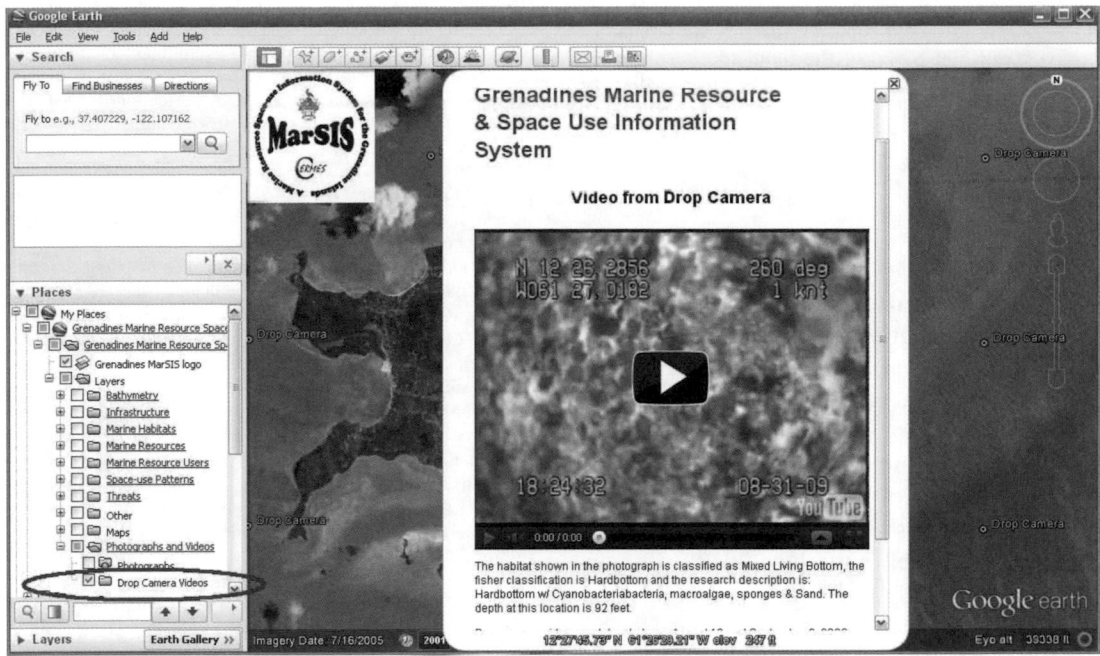

Figure A4. Videos for the MarSIS were collected for deep water habitat interpolation. All videos were uploaded to YouTube and the embed code was used for each geolocated image placemark created in Spreadsheet Mapper. Some HTML customization was required but the balloon template stayed consistent and the field description remained within the placemark balloon.

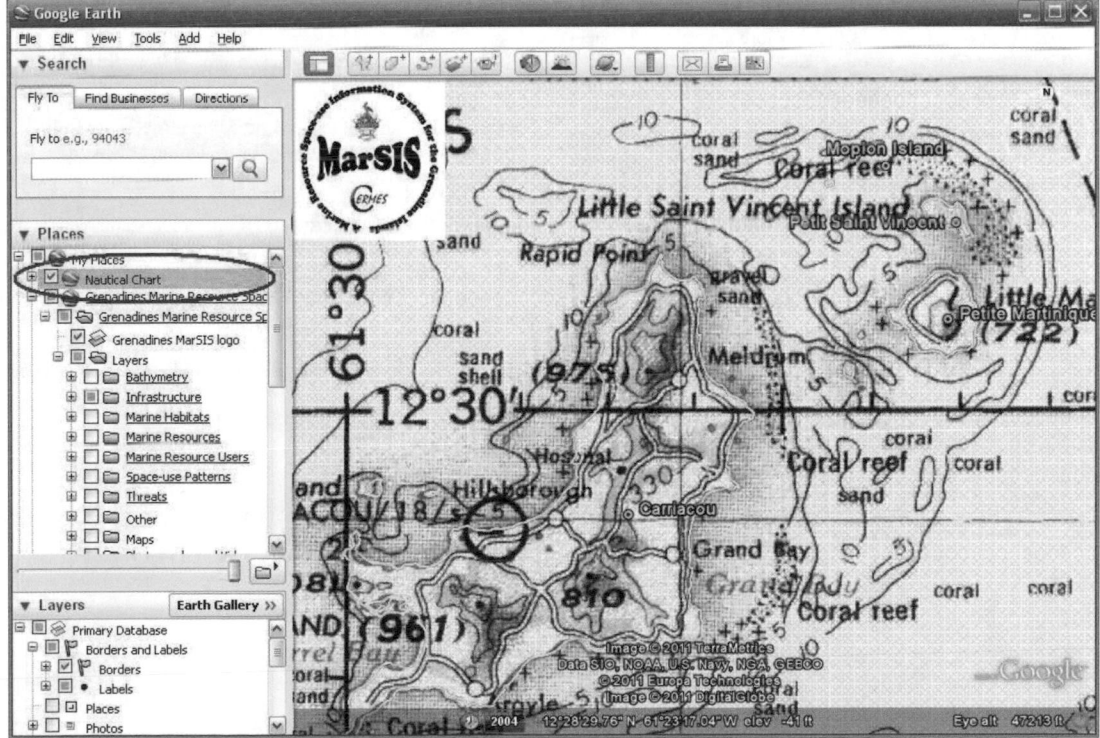

Figure A5. Nautical chart is a digital raster graphic zoomed into the islands of Carriacou, Petit St. Vincent, and Petite Martinique. The scale is 1:250,000 and is from the U.S. Department of Defense chart of Grenada and the Lesser Antilles. Collar information along the sides of the chart is included in the KMZ.

chart in the final product but link to it from within the MarSIS KMZ (under "Maps" in Fig. 2) and have a separate link on the Grenadines MarSIS web site. To create the KMZ of the nautical chart, we used Valery Haronusov's SuperOverlay program, which takes a digitized raster graphic and related world file (i.e., *.jpw or *.tfw) and tiles the image into smaller rasters which greatly reduces the KMZ file size. The KMZ of the nautical chart is 10.5 MB.

ACKNOWLEDGMENTS

We thank the Council for International Exchange of Scholars for support of this project through a Fulbright Scholarship at the Centre for Resources Management and Environmental Studies, University of the West Indies, Cave Hill campus, to M. Stewart. We also thank the University of the West Indies Graduate Studies Research Awards for providing funding (to K. Baldwin) to hold the November 2009 workshops and to the U.S. Embassy Bridgetown, Barbados–Public Affairs Section (to K. Baldwin) for helping fund travel to the June 2010 workshops. The manuscript was significantly improved by careful reading from Alfred Hochstaedter, an anonymous reviewer, and the editor, Steven Whitmeyer. We thank the reviewers.

REFERENCES CITED

Baldwin, K., 2006, Summary report for the data scoping of the Grenadines marine space use information system: A summary of preliminary observations and key informant interviews from Grenada, St. Vincent and the Grenadine Islands: Centre for Resource Management and Environmental Studies, University of the West Indies, Barbados, 13 p.

Baldwin, K., Gill, D., Cooke, A., Staskiewicz, T., Lizama, D., Mahon, R., McConney, P., and Oxenford, H.A., 2006, A socio-economic and space-use profile of Grenadine marine resource users: Centre for Resource Management and Environmental Studies, University of the West Indies, Barbados, 116 p.

Baldwin, K., 2012, A participatory marine resource and space-use information system for the Grenadine Islands: An ecosystem approach to collaborative planning and management of transboundary marine resources [Ph.D. dissertation]: Barbados: University of the West Indies, 372 p.

Ballagh, L.M., Raup, B.H., Duerr, R.E., Khalsa, S.S., Helm, C., Fowler, D., and Gupte, A., 2011, Representing scientific data sets in KML: Methods and challenges: Computers & Geosciences, v. 37, p. 57–64, doi:10.1016/j.cageo.2010.05.004.

Bresson, G., and Logossah, K., 2011, Crowding-out effects of cruise tourism on stay-over tourism within the Caribbean: Non-parametric panel data evidence: Tourism Economics, v. 17, p. 127–158, doi:10.5367/te.2011.0028.

Chakalall, Y., Mahon, R., and Oxenford, H., 2005, Activities of trading vessels and supplying fishers in the Grenadine Islands, Lesser Antilles: Proceedings of the Gulf and Caribbean Fisheries Institute, v. 47, p. 236–263.

Conway, D., and Timms, B.F., 2010, Re-branding alternative tourism in the Caribbean: The case for "Slow Tourism": Tourism and Hospitality Research, v. 10, p. 329–344, doi:10.1057/thr.2010.12.

DeGeorges, A., Goreau, T.J., and Reilly, B., 2010, Review: Land-sourced pollution with an emphasis on domestic sewage: Lessons from the Caribbean and implications for coastal development on Indian Ocean and Pacific coral reefs: Sustainability, v. 2, p. 2919–2949, doi:10.3390/su2092919.

Economic Commission for Latin America and the Caribbean (ECLAC), 2004, Yachting in the eastern Caribbean: ECLAC Technical Report LC/CAR/R.75, 286 p.

MapAction, 2008, Google Earth and its potential in the humanitarian sector: a briefing paper, April, 10 p., http://www.humanitarian.info/wp-content/uploads/2008/04/google-earth-and-its-potential-in-the-humanitarian-sector.pdf (accessed 30 December 2011).

MapAction, 2009, Field guide to humanitarian mapping, 1st edition, March, 114 p., http://tinyurl.com/MapActionFieldGuid (accessed 30 December 2011).

Mahon, R., Almerigi, S., McConney, P.R., and Whyte, B., 2004, Coastal resources and livelihoods in the Grenadine Islands: Facilitating change in self-organizing systems: Proceedings of the Gulf and Caribbean Fisheries Institute, v. 55, p. 56–67.

McCall, M., 2003, Seeking good governance in participatory-GIS: a review of processes and governance dimensions in applying GIS to participatory spatial planning: Habitat International, v. 509, p. 1–26.

McGann, J., and Creary, M., 2008, Coral reef monitoring for the Organization of Eastern Caribbean States (OECS) and Tobago: Proceedings of the 11th International Coral Reef Symposium, Fort Lauderdale, Florida, no. 18, p. 711–715.

Mills, A.P., 2001, St. Vincent and the Grenadines: Marine Pollution Bulletin, v. 42, no. 12, p. 1208–1220, doi:10.1016/S0025-326X(01)00237-5.

Opadeyi, J., 2007, Road map towards effective mainstreaming of GIS for watershed management in the Caribbean: Integrated Watershed and Coastal Areas Management, 58 p.

Patterson, T.C., 2007, Google Earth as a (not just) geography education tool: The Journal of Geography, v. 106, p. 145–152, doi:10.1080/00221340701678032.

Rambaldi, G., McCall, M., Kyem, P., and Weiner, D., 2006, Participatory spatial information management and communication in developing countries: The Electronic Journal on Information Systems in Developing Countries, v. 25, no. 1, p. 1–9, http://www.ejisdc.org/ojs2/index.php/ejisdc/article/viewFile/237/158 (accessed 30 December 2011).

Ratliff, E., 2007, Google Maps is changing the way we see the world: Wired Magazine, v. 15.07.

Sieber, R., 2006, Public participation geographic information systems: A literature review and framework: Annals of the American Association of Geography, v. 96, no. 3, p. 491–507, doi:10.1111/j.1467-8306.2006.00702.x.

Stewart, M.E., Cunningham, M.A., Menking, K., and Bolton, M., 2007, Sharing project data using Google Earth: Doing GIS without learning GIS [abs.]: Association of American Geographers National Meeting, 17–21 April 2007, San Francisco, California.

Sustainable Grenadines Project, The, 2006, Sustainable integrated development and biodiversity conservation in the Grenadine Islands – overview, Clifton, Union Island, St. Vincent and the Grenadines: Project Implementation Unit Clifton, Union Island St. Vincent and the Grenadines, 7 p.

Tulloch, D. L., 2007, Many, many maps: Empowerment and online participatory mapping: First Monday, v. 12, n. 2, http://firstmonday.org/htbin/cgiwrap/bin/ojs/index.php/fm/article/view/1620/1535 (accessed 30 December 2011).

MANUSCRIPT ACCEPTED BY THE SOCIETY 16 APRIL 2012

The Geological Society of America
Special Paper 492
2012

Geoscience applications of client/server scripts, Google Fusion Tables, and dynamic KML

Declan G. De Paor
Department of Physics, Old Dominion University, Norfolk, Virginia 23529, USA

Steven J. Whitmeyer
Department of Geology & Environmental Science, James Madison University, Memorial Hall, MSC 6903, Harrisonburg, Virginia 22807, USA

Mano Marks
Developer Advocate, Google Geo Team, Google Inc., Mountain View, California, USA

John E. Bailey
Scenarios Network for Alaska & Arctic Planning, University of Alaska Fairbanks, Fairbanks, Alaska 99709, USA

ABSTRACT

Keyhole Markup Language (KML)—a type of extensible markup language (XML)—is the key to the extensibility of Google Earth for geoscience applications. Static KML code may be saved to a file from the Google Earth desktop application, handwritten with a text editor, or generated by running a custom computer program. Many Google Earth visualizations are limited to static KML developed with the desktop application's user interface. The purpose of this paper is to highlight how much more is possible with the implementation of additional applications. Geoscience learning resources may be taken to the next level with the interactive generation and animation of graphics and models both in the desktop application and using the Google Earth web browser plug-in and its JavaScript application programing interface. Dynamic KML may be generated on-the-fly by means of client-side or server-side scripts, or with the aid of Google Fusion Tables and network links.

*ddepaor@odu.edu

De Paor, D.G., Whitmeyer, S.J., Marks, M., and Bailey, J.E., 2012, Geoscience applications of client/server scripts, Google Fusion Tables, and dynamic KML, *in* Whitmeyer, S.J., Bailey, J.E., De Paor, D.G., and Ornduff, T., eds., Google Earth and Virtual Visualizations in Geoscience Education and Research: Geological Society of America Special Paper 492, p. 77–104, doi:10.1130/2012.2492(06). For permission to copy, contact editing@geosociety.org. © 2012 The Geological Society of America. All rights reserved.

INTRODUCTION TO GOOGLE EARTH IN GEOSCIENCE EDUCATION AND RESEARCH

The geosciences are concerned with global topics such as climate change and resource conservation that pertain to the entire planet Earth and that vary on a range of length and time scales. The use of flat paper maps in an attempt to convey ideas spanning four dimensions has always been a case of fitting a round peg into a square hole. In the past, mapping was, of necessity, two dimensional, but that approach changed first with the advent of geographic information systems (GIS) in the 1970s and then with the evolution of geo-browsers (also known as virtual globes) such as Google Earth in the 2000s. In less than a decade, the impact of Google Earth on the classroom at all levels from elementary school to college has been truly revolutionary; it has set in motion a paradigm shift in the geosciences.

The "Google Earth for Educators" website (http://sitescontent .google.com/google-earth-for-educators) hosts a range of useful resources for teachers and students of the geosciences (Fig. 1) These resources were created by Google engineers and a small number of computer-savvy academic colleagues. By comparison, when the World Wide Web first became popular in the 1990s, there were a handful of content creators from a programming background who wrote hypertext markup language (HTML) by hand and a host of users who passively viewed that content with their web browsers, without ever viewing the underlying source code. Today, many, if not most, web users upload their own textual and graphical content using tools such as Facebook, Twitter, iWeb, or even MS Word that do not require advanced computer skills or HTML coding. There is a clear need for equivalent tools for easy geospatial content development (call it "Dreamweaver for KML"). Meanwhile, much can be achieved with a moderate level of code authoring.

Most authors of educational content for Google Earth use only a fraction of its full potential. The geo-browsing experience can be greatly enhanced with the aid of KML authoring and especially with the Network Link feature. Use of the Google Earth plug-in and its JavaScript application programming interface (API) allow instructors and enthusiasts to link web page content to a virtual globe in interesting ways and to change content with time. Server-side scripts written in PHP, Python, or Ruby can generate new content and push it to students' desktops, either in real time or in response to their database queries. The newly developed Google Fusion Tables add GIS-style database functionality to Google Earth, making it a true competitor to systems such as ESRI's ArcGIS, and WxAnalyst permits COLLADA (collaborative design activity) models to be queried.

In this paper, we explore some of these exciting new possibilities for creating engaging and effective four-dimensional learning experiences using Dynamic KML to drive Google Earth to new heights, and we suggest ways in which KML authoring can be made accessible to many more people. A multi-volume book would be required to cover comprehensively all the topics touched on here. Our primary goal is to help readers to get started and point them toward useful resources for further advancement.

GETTING STARTED WITH KML FOR THE GEOSCIENCES

Keyhole Markup Language (KML) is an extensible markup language (XML)–based scripting language designed specifically for creating content on geo-browser applications such as Google Earth. Because it is an open-source scripting language, files with the ".kml" suffix are human-readable and may be directly authored in a text editor such as MS Wordpad, or Apple TextEdit. Better still is a purpose-built code editor such as BBEdit, EditRocket, or jEdit. These code editors have very useful features, including XML validation; language-specific syntax checking; automatic indentation and formatting; and element expand/collapse options that help the author to scan the large-scale structure of a file.

A KML script consists of elements in which text is bounded by opening and closing tags. Tags are paired, angle-bracketed keywords with a leading slash distinguishing the closing tag from the opening one. A bare-bones example is shown in Table 1. After the XML version declaration, all other elements must be nested between the opening <kml> and closing </kml> tags. The "Placemark" element in this example puts a Google Earth map pin (yellow by default) at the location indicated by the "coordinates" element and labels it with the "name" element. The point's coordinates are comma delimited longitude and latitude—altitude

---

Google Earth for Educators

Welcome to the Google Earth for Educators Community, a site to share, connect & learn

This site is brought to you by Google and made especially for Google Earth educators and students. Come join and help us build it!

 Classroom Resources
Get lesson plans and ideas on how Google Earth can help students learn in your specific subject.

 Tutorials & Tips
View videos, tutorials and an interactive game to help teach Google Earth basics to your students.

 Talk Teacher-to-Teacher New!
Connect with other educators through our forum and mailing list to share tips and get ideas.

 Google Earth Pro Grants New!
Use our Pro Grants Wizard to see if you are qualified to apply for a Google Earth Pro Grant.

 Student Work Showcase
See some great examples of student work created in Google Earth, and submit your own students work.

 What Educators are Saying
See quotes about Google Earth in the classroom and share how you use Google Earth at your school.

Figure 1. The Google Earth for Educators web page: http://sitescontent.google .com/google-earth-for-educators/.

TABLE 1. A BARE-BONES KML SCRIPT

```xml
<?xml version="1.0"?>
<kml>
    <Document>
        <Placemark>
            <name> Santa's Hut </name>
            <Point>
                <coordinates>
                    0,90
                </coordinates>
            </Point>
        </Placemark>
    </Document>
</kml>
```

is optional. Note that element names are case sensitive. Capitalization denotes parent elements that can contain nested child elements whereas un-capitalized elements contain only data values (the root-level "kml" element is one of a few odd exceptions to this rule, however). The "Document" element is not necessary in this example but would be required if the script contained more than one placemark as there can be only one root level element.

When the script in Table 1 is saved to file as, say, "doc.kml" and then opened in Google Earth, the result is as shown in Figure 2.

In addition to the handwritten approach outlined above, KML scripts can be generated automatically using tools built into the Google Earth desktop application. Custom content such as placemarks and images can be added using a graphical user interface that is activated by menu and toolbar commands (for example, "Add placemark") and then saved to a file with the suffix ".kml" on the user's desktop using the application's "Save place as…" option. Google Earth also lets you save places in zip archives with the ".kmz" suffix. These archives may be unzipped to reveal an enclosed "doc.kml" file and a folder of associated images or models. KMZ archives are potentially useful for working with slow or no Internet access. Keeping copies of images or other resources bundled with the KML document avoids the kind of broken link that occurs when people move or rename files on the web. However, KMZ archives have their own sources of the red "X" that appears in place of an image when a link fails. For example, if a user downloads a KMZ file attached to mail, adds it to other items in their "Places" sidebar, saves a new KMZ archive,

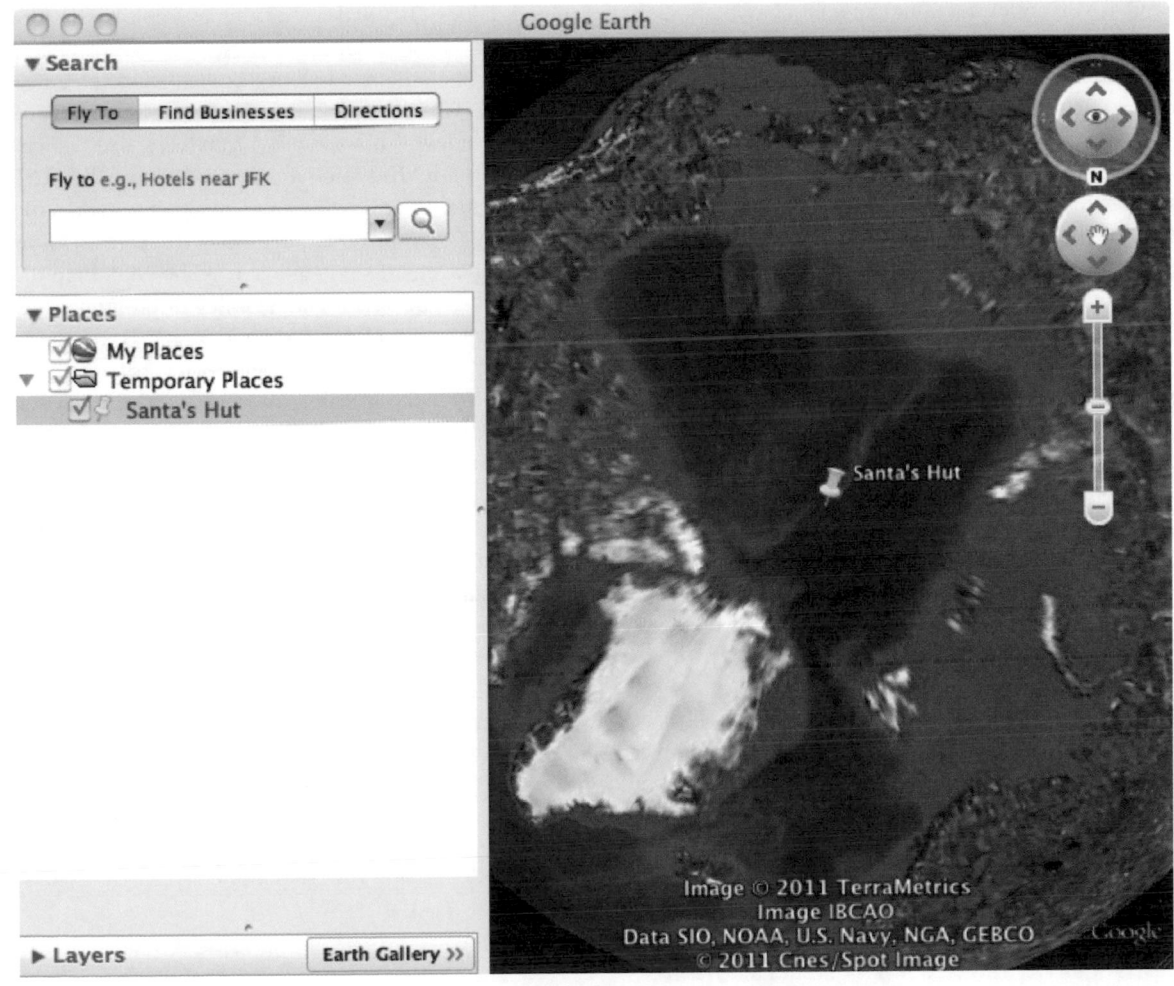

Figure 2. Script from Table 1 loaded in Google Earth.

De Paor et al.

TABLE 2. KML SCRIPT SAVED FROM GOOGLE EARTH AND OPENED WITH A TEXT EDITOR

```
<?xml version="1.0" encoding="UTF-8"?>
<kml xmlns="http://www.opengis.net/kml/2.2" xmlns:gx="http://www.google.com/kml/ext/2.2"
xmlns:kml="http://www.opengis.net/kml/2.2" xmlns:atom="http://www.w3.org/2005/Atom">
    <Placemark>
        <name>Santa&apos;s Hut</name>
        <Point>
         <coordinates>0,90,0</coordinates>
        </Point>
    </Placemark>
</kml>
```

and distributes the latter archive, end users may find that the links they receive point to the inaccessible sender's mail download folder, not to images bundled with the new archive. The best solution, whenever possible, is to carefully maintain server links that employ absolute uniform resource locators (URLs; absolute URLs begin with the "http://" protocol).

One strategy for learning KML quickly is to create and save content using the Google Earth application graphical user interface and then open the saved file with a text editor. By way of example, the code in Table 2 was saved from the Google Earth "Places" sidebar after the document in Table 1 had been loaded. The XML and KML boilerplate, and special characters such as the apostrophe in the "name" element, are not identical to Table 1 but such differences can generally be ignored. Note that the "coordinate" element in Table 2 is written in-line whereas in Table 1 it was spread out over three indented lines. These different structures of the element are interpreted identically by Google Earth—white space is inserted for readability only. The computer would parse the entire file without a problem even if it were all written in-line, but that would make it very difficult for humans to parse.

In addition to examining whole files, individual elements may be right-clicked (control-clicked with a single-button mouse) and copied in the Google Earth "Places" sidebar, and then pasted into a text file. Much can be learned this way—indeed, many hours may be whiled away. Pretty soon, a serious content creator will need a comprehensive guide such as Wernecke (2009). Lots of help is also available on the Google web pages, http://earth.google.com, including KML tutorials and a programmer's reference (http://code.google.com/apis/kml/).

KML CODE GENERATION BY STAND-ALONE DESKTOP APPLICATIONS

KML is not designed to be a computer programming language like FORTRAN, BASIC, C++, or Java. Many users would want to be able to write a KML script such as the pseudocode in Table 3 including its loop structure. Indeed, the need for a do-loop becomes acute when the KML "Timespan" element is employed repeatedly, as when the time slider is used to animate a geologic map or to elevate segments of the Earth's crust (De Paor and Williams, 2006; De Paor, 2008). Conditional "if…then…else" structures, algebraic expressions, and other such program-ming elements are not available to the KML author. We therefore need a way to generate repetitive code elements efficiently.

To overcome the above limitations of KML, one can take a brute-force approach, for example by copying and repeatedly pasting elements in a text editor and then making incremental changes to successive elements. However, if dozens or hundreds of elements are involved, it becomes worthwhile to write routines in other programming languages that generate KML output. For example, Table 4 contains a code snippet written in Adobe Action-script, the language used to generate Flash content for the Web. The variables specifying longitude and latitude are set to arbitrary initial values (10° and 35° in this example). The for-loop increments an integer counter i and creates a set of "Placemark" elements with enumerated names as well as incremental changes in longitude. Code is output to a Flash text field called "KML" (the output window is not shown). From there, text can be copied and saved into a KML file on the user's desktop (Table 5), this being more secure than allowing an application to send a file directly to the user's desktop. When this file is loaded into Google Earth, a string of placemarks is displayed, as shown in Figure 3. The same approach can be taken with a variety of programming and scripting languages. In FORTRAN and Visual Basic for example, the code snippets in Table 6 generate KML scripts.

For someone learning to program or to write scripts for the first time, we would certainly not recommend that they start with any of the above-mentioned computer languages, because modern alternatives such as Python are more powerful (see below). However, there may be readers who learned to program in FOR-TRAN at the same time that they mastered lecture presentation with acetate overheads and image processing with developer and fixer, and who have limited enthusiasm for learning new languages. It is possible that many more people would generate KML if they knew that they could use the programming language of their choice, or of their youth.

For those who do not plan to write any code whatsoever, there are commercial packages available such as EarthPoint

TABLE 3. LOOPS ARE NOT SUPPORTED DIRECTLY IN KML

```
for (i = 1; i < 100; i++){
    ...
        <!-- you can't do this in KML :( -->
    ...
}
```

TABLE 4. A SIMPLE FLASH ACTIONSCRIPT FUNCTION
TO WRITE WELL-FORMATTED KML TO A TEXT FIELD

```
function writeKML(){
        var lon = 10;
        var lat = 35;
        var tab1 = "\t";
        var tab2 = "\t \t";
        var tab3 = "\t \t \t";
        'note \n inserts a new line';
        ' and \" inserts a literal quotation mark';

        KML.text   = "<?xml version = \"1.0\"?> \n";
        KML.text += "<kml> \n";
        KML.text += "<Folder> \n";
        KML.text += tab1 + "<open> 1 </open> \n";
        for (i = 1; i <= 10; i++){
                KML.text += tab1 + "<Placemark> \n";
                KML.text += tab2 + "<name>My Place " + i + "</name> \n";
                KML.text += tab2 + "<Point> \n";
                KML.text += tab3 + "<coordinates>";
                KML.text += lon*i + "," + lat;
                KML.text += "</coordinates> \n";
                KML.text += tab2 + "</Point> \n";
                KML.text += tab1 + "</Placemark> \n";
        }
        KML.text += "</Folder> \n";
        KML.text += "</kml>";
}

writeKML();
stop();
```

TABLE 5. CONTENT OF THE KML OUTPUT TEXT FIELD
GENERATED BY THE FLASH ACTIONSCRIPT
CODE FROM TABLE 4

```
<?xml version = "1.0"?>
<kml>
<Folder>
     <open> 1 </open>
     <Placemark>
         <name>My Place 1</name>
            <Point>
                <coordinates>10,35</coordinates>
            </Point>
     </Placemark>

     <Placemark>
         <name>My Place 2</name>
         <Point>
             <coordinates>20,35</coordinates>
         </Point>
     </Placemark>
  ...
  ...

     <Placemark>
         <name>My Place 10</name>
         <Point>
             <coordinates>100,35</coordinates>
         </Point>
     </Placemark>

</Folder>
</kml>
```

Note: Placemarks 3 through 9 are omitted at the ellipses. Note the enumerated placemark names and sequential longitudes.

(http://www.earthpoint.us/ExcelToKml.aspx), which exports Excel spreadsheet data to KML and Arc2Earth/Arc2Cloud (http://www.arc2earth.com/) both of which help in the transfer of ArcGIS data (ArcGIS v. 9 and 10 have built in support for export to KML).

NETWORK LINKS AND DYNAMIC KML

One of the most important KML elements for teaching with Google Earth is the "NetworkLink." Network links are analogous to "include" statements in many computer programming languages. They set up an HTTP link to an external KML file and implement the linked code as if it were embedded in the source document at the line occupied by the network link. As an example of their usefulness, students can be given a local KML file here called "student_client.kml" (Table 7). This file can be distributed on a flash memory stick or sent to students' desktops as a mail attachment. This local file simply links to a file called "teacher_server.kml" on the teacher's server (Table 8) and checks for changes at five-second intervals. There are a number of options for controlling how the link is refreshed and the addition of a "flyToView" element allows the teacher to take control of the camera angle on the students' computer, which has the potential to keep wandering students on task. Using the "NetworkLinkControl" element, the teacher can send a message that will appear as an alert on the students' Google Earth desktop application (Fig. 4). In this example from De Paor et al. (2007),

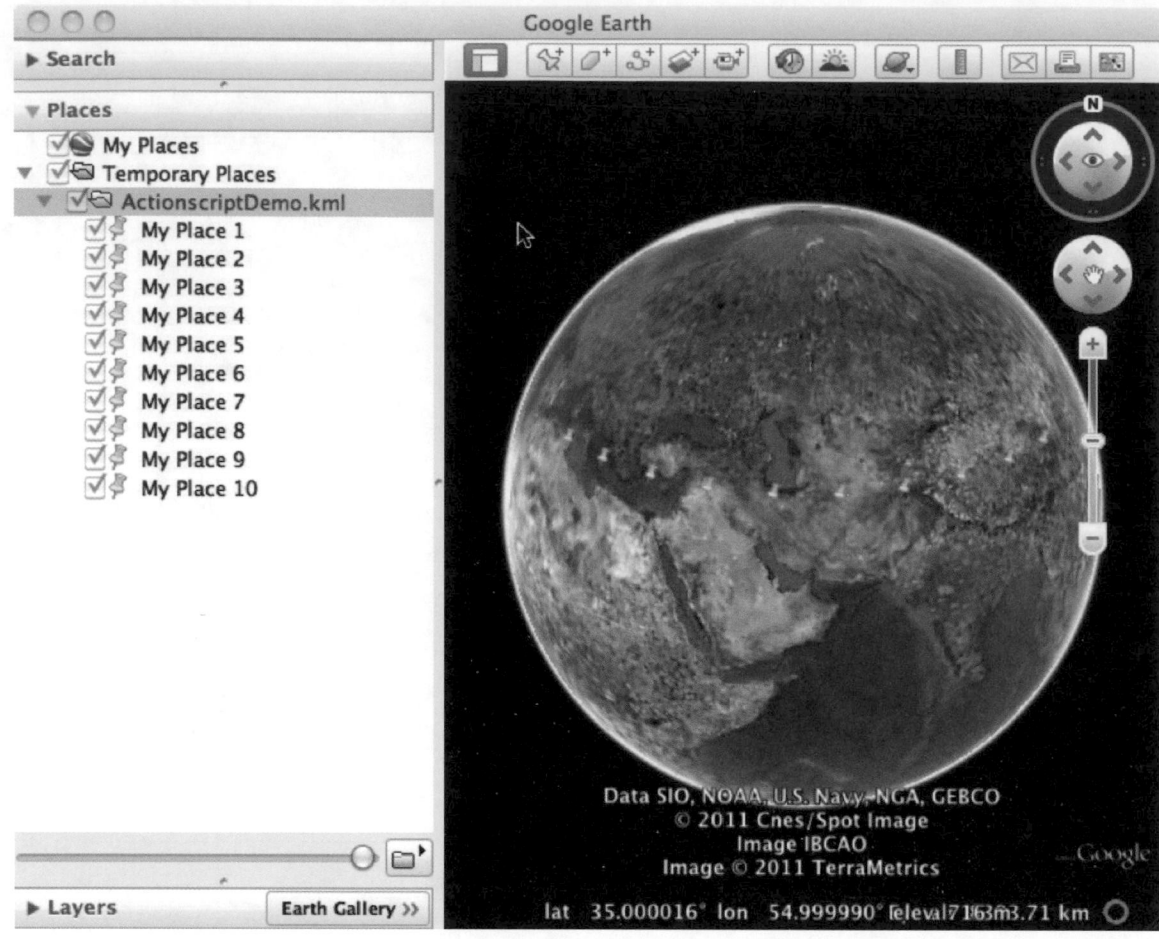

Figure 3. KML from Table 5 loaded into Google Earth generates a series of ten placemarks.

TABLE 6. CODE SNIPPETS FROM FORTRAN AND VISUAL BASIC

```
FORTRAN:

write(1,*) '<Folder>'
write(1,*) '<open>1</open>'
do i = 1,10
        write(1,*) '<Placemark>'
        ...
        ...
end do
```

```
Visual Basic:

KML.WriteStartElement("Folder")
KML.WriteString("insert data here")
KML.WriteEndElement()
```

TABLE 7. A SIMPLE KML DOCUMENT CALLED STUDENT_CLIENT.KML
LINKS TO THE TEACHER FILE IN TABLE 8

```
<?xml version="1.0" encoding="UTF-8"?>
<kml>
    <NetworkLink>
        <Link>
         <href>http://www.digitalplanet.org/gsa_sp/teacher_server.kml</href>
            <refreshMode>onInterval</refreshMode>
            <refreshInterval>5</refreshInterval>
        </Link>
    </NetworkLink>
</kml>
```

TABLE 8. THIS TEACHER_SERVER.KML FILE RESIDES ON THE TEACHER'S SERVER

```xml
<?xml version="1.0" encoding="UTF-8"?>
<kml>
    <NetworkLinkControl>
        <message>
            Watch for Images at bottom left.<br/>
            Search the globe for a match
        </message>
    </NetworkLinkControl>

    <ScreenOverlay>
        <name>Where on Earth is This?</name>
        <Icon>
            <href>files/Tasmania.tiff</href>
            <!-- <href>files/Tasmania.tiff</href> -->
            <!-- <href>files/Kuril.tiff</href> -->
            <!-- <href>files/CapeVerde.tiff</href> -->
        </Icon>

        <overlayXY x="0" y="0" xunits="fraction" yunits="fraction"/>
        <screenXY x="0" y="0.03" xunits="fraction" yunits="fraction"/>
        <rotationXY x="0" y="0" xunits="fraction" yunits="fraction"/>
        <size x=".3" y="0" xunits="fraction" yunits="fraction"/>
    </ScreenOverlay>
</kml>
```

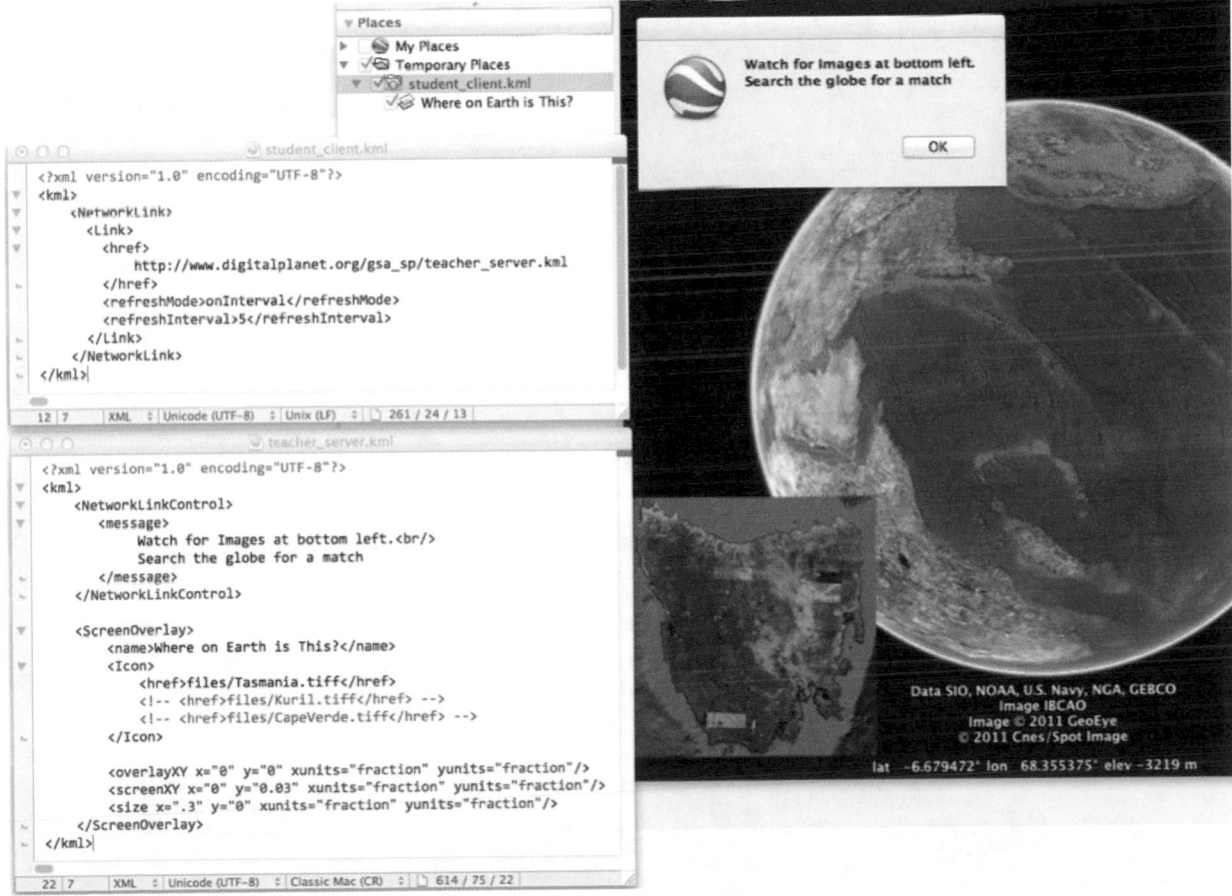

Figure 4. Teacher-student interaction with NetworkLinks in the desktop application. The two windows on the left are open in a text editor. The alert is running in the Google Earth desktop application in the background. © 2011 Google.

the teacher instructs students to search for islands displayed in a screen overlay. When the teacher_server file is changed (for example, to link to a different screen overlay image), the student _client file responds within the stated time interval. The screen overlay can also be changed by replacing the image file on the server while retaining its file name (a generic name such as "img .png" would be best in this case). Indeed, if student_client files are enumerated ("student_client_1.kml," "student_client_2.kml," etc.) and distributed to different students, then different screen overlays and alerts can be sent by a teacher to these different students in order to customize help based on student need. The implications for distance education are significant.

Small local KML files linking to larger server-side files are useful in general. They cut down on the size of file that needs to be distributed and downloaded to the user's desktop and they allow the content creator to retain control, expand features, and fix errors without the need to distribute a new file to users. A detailed tutorial on Dynamic KML is available here:

http://www.youtube.com/watch?v=QzS_shIzfcM

CLIENT-SIDE GENERATION OF KML USING JAVASCRIPT

One scripting language is more important than all others— namely JavaScript—because it permits KML code generation that can be viewed in a web browser (Yamagishi et al., 2008, 2010). JavaScript is a popular language for developing web content and

TABLE 9. KML CODE GENERATION AND FORMATTING USING CLIENT-SIDE JAVASCRIPT

```
<html>
<head>

    <script type="text/JavaScript">
        function indent(numTabs){
            var str = '';
            var tabSize = 5;
            for (var i = 0; i < numTabs; i++){
                for (var j = 0; j <= tabSize; j++){
                    str +=' ';
                }
            }
            return str;
        }
        //-------------------------------------
        function tag(slash,tagName){
            //wrap tagName with angle brackets + optional slash
            //return null if tagName is null.
            var str = '';
            if (tagName){
                str = '&lt;'+slash+tagName+'&gt;';
            }
            return (str);
        }
        //-------------------------------------
        function inlineElement(tagName,data,level){
            //tagName can be 'comment', valid kml, or null ('') for no tags
            var str = '';
            str += indent(level);
            if (tagName == 'comment'){
                str += '&lt;!--'+ data +'--&gt;';
            }else{
                str += tag('',tagName) + data + tag('/',tagName);
            }
            return (str);
        }
        //-------------------------------------
        function multilineElement(tagName,child,level){
            var str = '';
            str = indent(level++) + tag('',tagName);
            str += indent(level) + '<br/>' + child + '<br/>';
            str += indent(--level) + tag('/',tagName);
            return (str);
        }
        //-------------------------------------
```

(continued)

controlling web page user actions (e.g., Brady et al., 2002). It may be run on the server or client side. The simplest way to tell the difference is to load a JavaScript-enabled web page and then turn off wifi or unplug the Ethernet. If the script still works, it is running on the local computer, or client side. JavaScript may be embedded in either the "head" or "body" element of an HTML document or it may be accessed via a link to an external document with a ".js" suffix.

JavaScript code is embedded in an HTML document between the <script type="text/JavaScript"> and </script> tags. Table 9 shows a generic script that generates formatted KML as shown in Table 10. The trick is to write the most deeply nested elements first and sequentially include them up to the root level. Note that the element names "child 1," etc. in Tables 9 and 10 are placeholders, not valid KML; you can substitute any valid KML element name in these locations.

TABLE 9. (*continued*)

```
function writeKML(){
        //send kml to output window - users cut & paste into a kml doc.
        var HTML_headers = "<html><head><title>KML</title></head><body>";
        var XML_header = '&lt;?xml version=\"1.0\" encoding=\"UTF-8\"?&gt;<br/>';
        var xmlns= 'xmlns=\"http://earth.google.com/kml/2.2\"';
        var KML_header = tag('','kml ' + xmlns) + '<br/>'
        var KML_footer = '<br/>'+tag('/','kml');
        var HTML_footers = "</body></html>";

        var output = window.open('','name','scrollbars=yes');
        with(output){
            document.write(HTML_headers);
            document.write(XML_header);
            document.write(KML_header);

            var child1 = inlineElement('comment','this is a child element',4);
            child1 += '<br/>' + inlineElement('', 'this is inline data',4);
            child1 += '<br/>' + inlineElement('', 'this is more data',4);

            var child2 = inlineElement('comment','another child element',4);
            child2 += '<br/>' + inlineElement('', 'this is inline data',4);
            child2 += '<br/>' + inlineElement('', 'this is more data',4);

            var parent1 = multilineElement('child 1', child1,3);
            parent1 += '<br/>' + multilineElement('child 2', child2,3);

            var parent2 = multilineElement('child 1', child1,3);
            parent2 += '<br/>' + multilineElement('child 2', child2,3);

            var root = inlineElement('comment','root level',2);
            root += '<br/>' +multilineElement('Parent 1',parent1,2);
            root += '<br/>' + multilineElement('Parent 2',parent2,2);

            var doc = multilineElement('Document',root,1);
            document.write(doc);
            document.write(KML_footer);
            document.write(HTML_footers);
            document.close();
        }
    }
</script>

</head>

<body>

    <form name="generate KML">

        <input type="button" onclick="JavaScript:writeKML();return false;" value="Write
KML"/>

    </form>

</body>

</html>
```

TABLE 10. KML CODE GENERATED BY TABLE 9

```
<?xml version="1.0" encoding="UTF-8"?>
<kml xmlns="http://earth.google.com/kml/2.2">
    <Document>
            <!--root level-->
            <Parent 1>
                <child 1>
                    <!--this is a child element-->
                    this is inline data
                    this is more data
                </child 1>
                <child 2>
                    <!--another child element-->
                    this is inline data
                    this is more data
                </child 2>
            </Parent 1>
            <Parent 2>
                <child 1>
                    <!--this is a child element-->
                    this is inline data
                    this is more data
                </child 1>
                <child 2>
                    <!--another child element-->
                    this is inline data
                    this is more data
                </child 2>
            </Parent 2>
    </Document>
</kml>
```

The code in Table 11 uses the "TimeSpan" element to elevate a COLLADA model of Mt. Fuji. First a model of the surface was created in SketchUp (Fig. 5) and saved to the URL indicated by the "href" element. A prototype "Placemark" element is formatted with temporary non-KML terms "replaceBeginTime," "replaceEndTime," and "replaceAltitude." Then a for-loop runs from the minimum to the maximum desired altitude (3500–5500 m in this case) and replaces these terms with sequential numerical values. The result is seen in Table 12 and Figure 6.

Typically the time slider only displays one moment in time, so one copy of the model is shown at a given elevation, but the slide markers have been split to show a range of time, and therefore multiple instances of the model.

Repurposing the Google Earth time slider to change the altitude tag of COLLADA models of subsurface structure, as in this "Moving Mt. Fuji" example is just one possibility. Many KML elements can be varied in a similar fashion. For example, the heading, tilt, and roll elements of a model can be varied to turn a model about three geographic axes. Applications in the geosciences abound: for example in structural geology (e.g., folding and faulting, strain ellipsoids) and geophysics (e.g., magnetic vectors, Euler tectonic plate rotations). De Paor and Pinan-Llamas (2006) used the time slider to make focal mechanism solutions for recent Andean earthquakes appear sequentially based on the seismic event time (Fig. 7). Each geophysical beach ball hovers over its hypocenter and is controlled by heading, tilt, and roll elements. Figure 8 shows a second example—scenes from an animation of sea stack formation and Figure 9 from oxbow lake evolution. Both examples are available for download from http://www.digitalplanet.org.

An alternative approach to cross-section elevation uses Google Earth's tour feature rather than its time slider shown in the code in Table 13, which was generated using Steve Whitmeyer's JavaScript web form (Fig. 10):

http://www.digitalplanet.org/DigitalPlanet/Tools.html

The web form produces KML code to position a COLLADA model of a cross section in a specified location on the Google Earth terrain. The option to create a Google Earth tour to elevate the cross section is shown as a button near the bottom of the web form. Here the tour function is used to control the gradual emergence of the model by incrementally

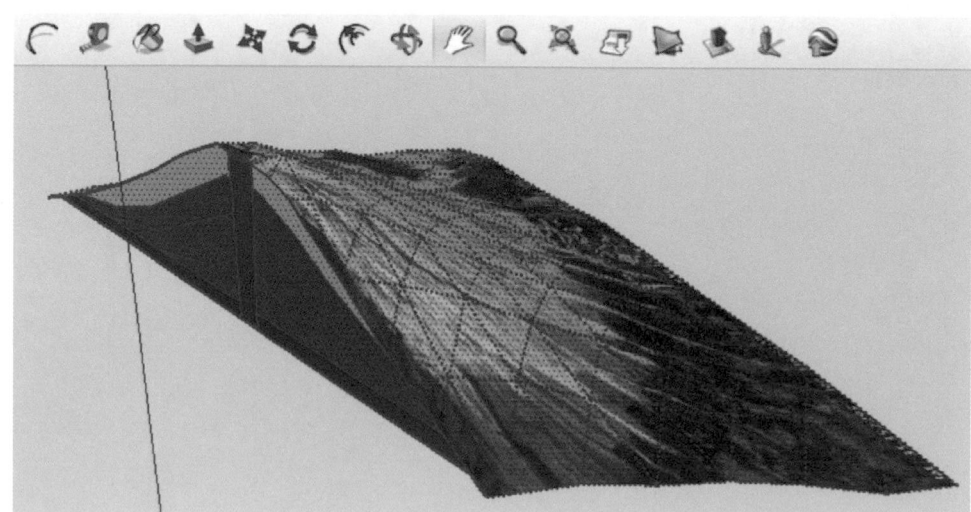

Figure 5. Creating a model of Mt. Fuji in SketchUp.

TABLE 11. EMERGENT MODEL FUNCTION; SUBSTITUTE FOR THE GENERIC WRITEKML FUNCTION IN TABLE 7

```
function makeEmergentModel(){
        //send kml to output window - users cut & paste into a kml doc.
            var HTML_headers = "<html><head><title>KML</title></head><body>";
            var XML_header = '&lt;?xml version=\"1.0\" encoding=\"UTF-8\"?&gt;<br/>';
            var xmlns= 'xmlns=\"http://earth.google.com/kml/2.2\"';
            var KML_header = tag('','kml ' + xmlns) + '<br/>'
            var KML_footer = '<br/>'+tag('/','kml');
            var HTML_footers = "</body></html>";

        var output = window.open('','name','scrollbars=yes');
        with(output){
            document.write(HTML_headers);
            document.write(XML_header);
            document.write(KML_header);
            var root = inlineElement('name','Moving Mt. Fuji',2);

            var LookAtChildren = inlineElement('longitude','138.6606047169506',3);
            LookAtChildren += '<br/>' + inlineElement('latitude','35.42853192819116',3);
            LookAtChildren += '<br/>' + inlineElement('altitude','1174',3);
            LookAtChildren += '<br/>' + inlineElement('range','5088',3);
            LookAtChildren += '<br/>' + inlineElement('tilt','76',3);
            LookAtChildren += '<br/>' + inlineElement('heading','147',3);
            root += '<br/>' +multilineElement('LookAt',LookAtChildren,2);

            var TimeSpanChildren = inlineElement('begin','replaceBeginTime',4);
            TimeSpanChildren += '<br/>' + inlineElement('end','replaceEndTime',4);
            var PlacemarkChildren = multilineElement('TimeSpan',TimeSpanChildren,3);

            var ModelChildren = inlineElement('altitudeMode','absolute',4);
            var LocationChildren = inlineElement('longitude','138.732278',5);
            LocationChildren += '<br/>' + inlineElement('latitude','35.352621',5);
            LocationChildren += '<br/>' + inlineElement('altitude','replaceAltitude',5);
            ModelChildren += '<br/>' + multilineElement('Location',LocationChildren,4);

            var OrientationChildren = inlineElement('heading','0',5);
            OrientationChildren += '<br/>' + inlineElement('tilt','0',5);
            OrientationChildren += '<br/>' + inlineElement('roll','0',5);
            ModelChildren += '<br/>' + multilineElement('Orientation',OrientationChildren,4);

            var ScaleChildren = inlineElement('x','1',5);
            ScaleChildren += '<br/>' + inlineElement('y','1',5);
            ScaleChildren += '<br/>' + inlineElement('z','1',5);
            ModelChildren += '<br/>' + multilineElement('Scale',ScaleChildren,4);

            var href = "http://www.digitalplanet.org/gsa_sp/files/MtFujiMiyaji.dae";
            var LinkChild = inlineElement('href',href,5);
            ModelChildren += '<br/>' + multilineElement('Link',LinkChild,4);
            PlacemarkChildren += '<br/>' + multilineElement('Model',ModelChildren,3);

            var nextPlacemark = '';
            for (var i = 3500; i < 5500 ; i += 100){
                nextPlacemark = PlacemarkChildren.replace('replaceBeginTime',i);
                nextPlacemark = nextPlacemark.replace('replaceEndTime',i+100);
                nextPlacemark = nextPlacemark.replace('replaceAltitude',i+100);
                root += '<br/>' +multilineElement('Placemark',nextPlacemark,2);
            }

            var doc = multilineElement('Document',root,1);
            document.write(doc);
            document.write(KML_footer);
            document.write(HTML_footers);
            document.close();
    }
}
```

De Paor et al.

TABLE 12. PLACEMARK SEQUENCE GENERATED BY EMERGENT MODEL FUNCTION

```
<Placemark>
        <TimeSpan>
                <begin>3500</begin>
                <end>3600</end>
        </TimeSpan>
        <Model>
                <altitudeMode>absolute</altitudeMode>
                <Location>
                        <longitude>138.732278</longitude>
                        <latitude>35.352621</latitude>
                        <altitude>3600</altitude>
                </Location>
                <Link>
                <Link>

<href>http://www.digitalplanet.org/gsa_sp/files/MtFujiMiyaji.dae</href>
                </Link>

                </Link>
        </Model>
</Placemark>

<Placemark>
        <TimeSpan>
                <begin>3600</begin>
                <end>3700</end>
        </TimeSpan>
        <Model>
                <altitudeMode>absolute</altitudeMode>
                <Location>
                        <longitude>138.732278</longitude>
                        <latitude>35.352621</latitude>
                        <altitude>3700</altitude>
                </Location>
                <Link>
                <Link>

<href>http://www.digitalplanet.org/gsa_sp/files/MtFujiMiyaji.dae</href>
                </Link>

                </Link>
        </Model>
</Placemark>

            ...
            ...

<Placemark>
        <TimeSpan>
                <begin>5400</begin>
                <end>5500</end>
        </TimeSpan>
        <Model>
                <altitudeMode>absolute</altitudeMode>
                <Location>
                        <longitude>138.732278</longitude>
                        <latitude>35.352621</latitude>
                        <altitude>5500</altitude>
                </Location>
                <Link>
                  <Link>

<href>http://www.digitalplanet.org/gsa_sp/files/MtFujiMiyaji.dae</href>
                </Link>

                </Link>
        </Model>
</Placemark>
```

Figure 6. Emergent model of Mt. Fuji generated from the code in Table 9. © 2011 Google, DigitalGlobe, MIRC/JHA, GeoEye, Digital Earth Technology.

Figure 7. Elevated first motion geophysical "beach balls" for recent Andean earthquakes. Colors are coded to depth; note deeper earthquake to the east. © 2012 Google, Cnes/Spot Image. Image courtesy of USGS, data courtesy of SIO, NOAA, U.S. Navy, NGA, GEBCO.

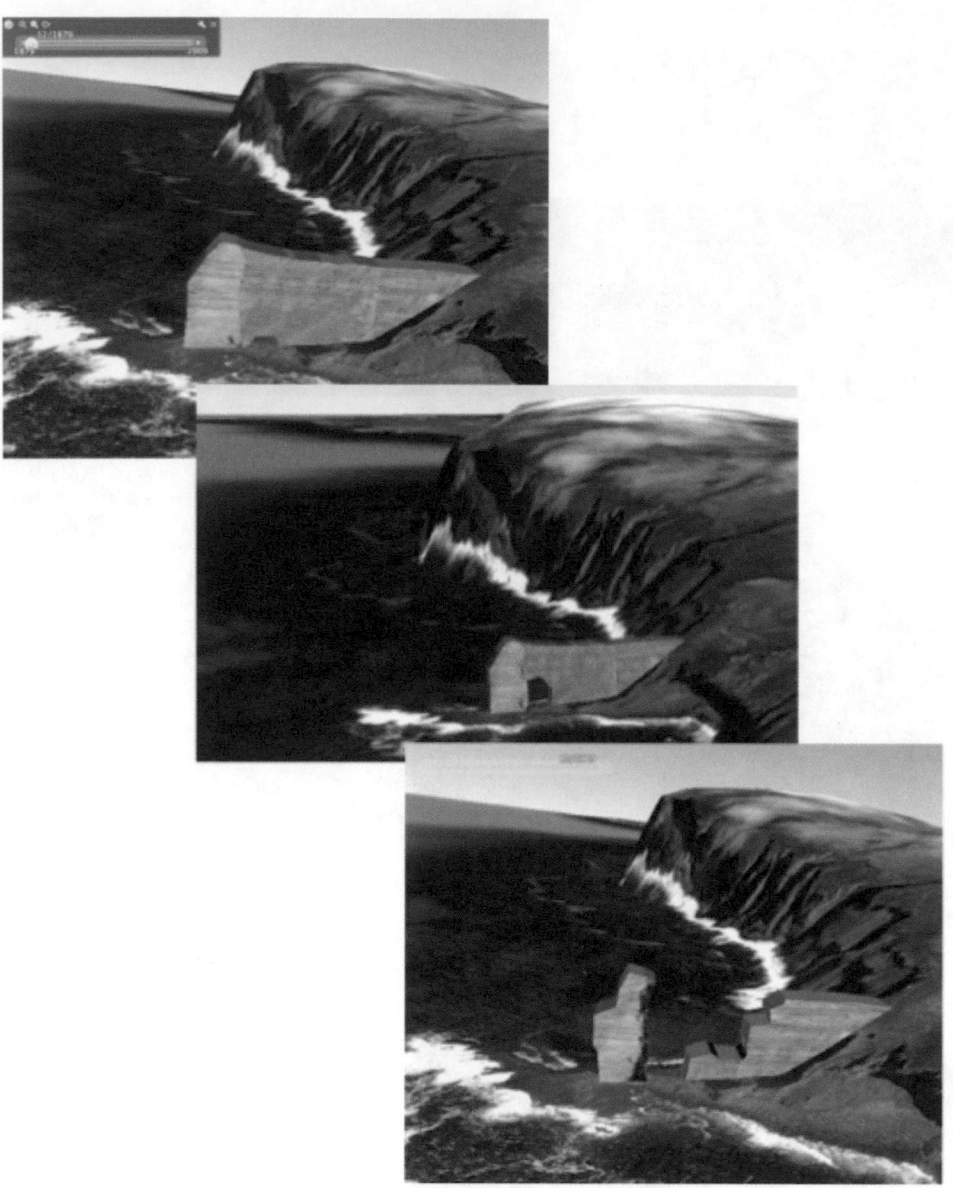

Figure 8. Animation of sea stack formation, Old Man of Hoy, Scotland. © 2012 Google, TerraMetrics, GeoEye, Getmapping plc. Data courtesy of SIO, NOAA, U.S. Navy, NGA, GEBCO.

changing its altitude datum. The "gx:" prefix in the <gx:Tour> tag indicates a Google Earth–specific KML element that may not work with other virtual globes such as NASA World Wind. Beginning with version 5.0, Google Earth began supporting KML elements that go beyond the OGC standard (http://www.opengeospatial.org) as gx extensions. Nested inside the gx:Tour and gx:Playlist elements are a series of gx:AnimatedUpdate elements that sequentially change a component of the "Placemark" element designated by the <targetId> tag component. For a cross-section elevation tour the targeted placemark contains the cross-section model, cross-section image, and location information. During the tour, the location of the cross-section model stays fixed, and the only component that changes is the altitude of the model. The

<gx:duration> tags are set to a minimum time, 0.1 seconds, for each iteration of the tour. However, large cross-section models may require a longer duration for smooth performance.

The cross-section elevation tour is shown as a separate item in the "Places" window (Fig. 10). When the tour is activated (double-clicked) the standard tour controls appear in the bottom left of the main Google Earth window, and the cross section rises to the specified altitude. The tour controls have only one slider marker, (thumb) which may be easier for users to manipulate. One potential disadvantage to this method is that Google Earth will always return the user to the "LookAt" viewpoint specified in the KML code when the tour is run. Thus, the user cannot change their viewpoint while the tour controls are active and then run the tour.

Figure 9. Animation of oxbow lake formation, Mississippi, USA. © 2012 Google, TerraMetrics, DigitalGlobe, USDA Farm Service Agency, GeoEye.

THE GOOGLE EARTH PLUG-IN AND API

The Google Earth desktop application is limited to one time slider or one tour slider and when either is used for other purposes, such as making COLLADA models emerge, there may be a conflict with genuine time-based functions such changing sunlight illumination or built-in historical imagery. In order to simultaneously manipulate multiple models, ground overlays, screen overlays, etc., it is necessary to use the Google Earth browser plug-in (http://code.google.com/apis/earth). A developer can embed one or more instances of the plug-in in an HTML web page and control its/their appearance with JavaScript commands. To use the Google Earth application programming interface (API), you simply insert the code from Table 14 in the head element of your HTML document. The result of this bare bones script is shown in Figure 11.

Table 15 lists a more detailed script that allows the viewer to create "Placemark" elements by clicking a button (Fig. 12). There are pedagogical advantages to limiting user actions, as opposed to the Google Earth desktop application where students can more easily wander off task. Other examples of API use from De Paor et al. (2011) may be accessed from:

http://www.digitalplanet.org/DigitalPlanet/API.html

Table 16 shows the organization of such complex JavaScript-controlled plug-in instance. In this case, we purchased a professional version of the Tigra slider control from http://www.softcomplex.com/products/tigra_slider_control/ in order to avoid issues with the free version (however, better custom sliders are described by Dordevic, this volume, Chapter 7). For ease of file management, functions are split among several external JavaScript files (identified by the suffix ".js").

SERVER-SIDE PHP SCRIPTS

Although JavaScript can be run on the server side, there are limitations on its ability to create or write data to server files. Therefore, we need a more powerful server-side scripting language. The three top contenders are PHP, Python, and Ruby. We will illustrate each in turn in the sections that follow. It is certainly not necessary to learn more than one server-side scripting solution, and the authors of this paper differ in their personal preferences. One prefers PHP because he learned to use it before the others were invented, whereas another finds Python more efficient—it achieves more with less coding effort.

Tables 17 and 18 illustrate the process of uploading a PNG image to a server using a PHP script (see, for example, Castagnetto et al., 1999). Once on the server, this image can be embedded in

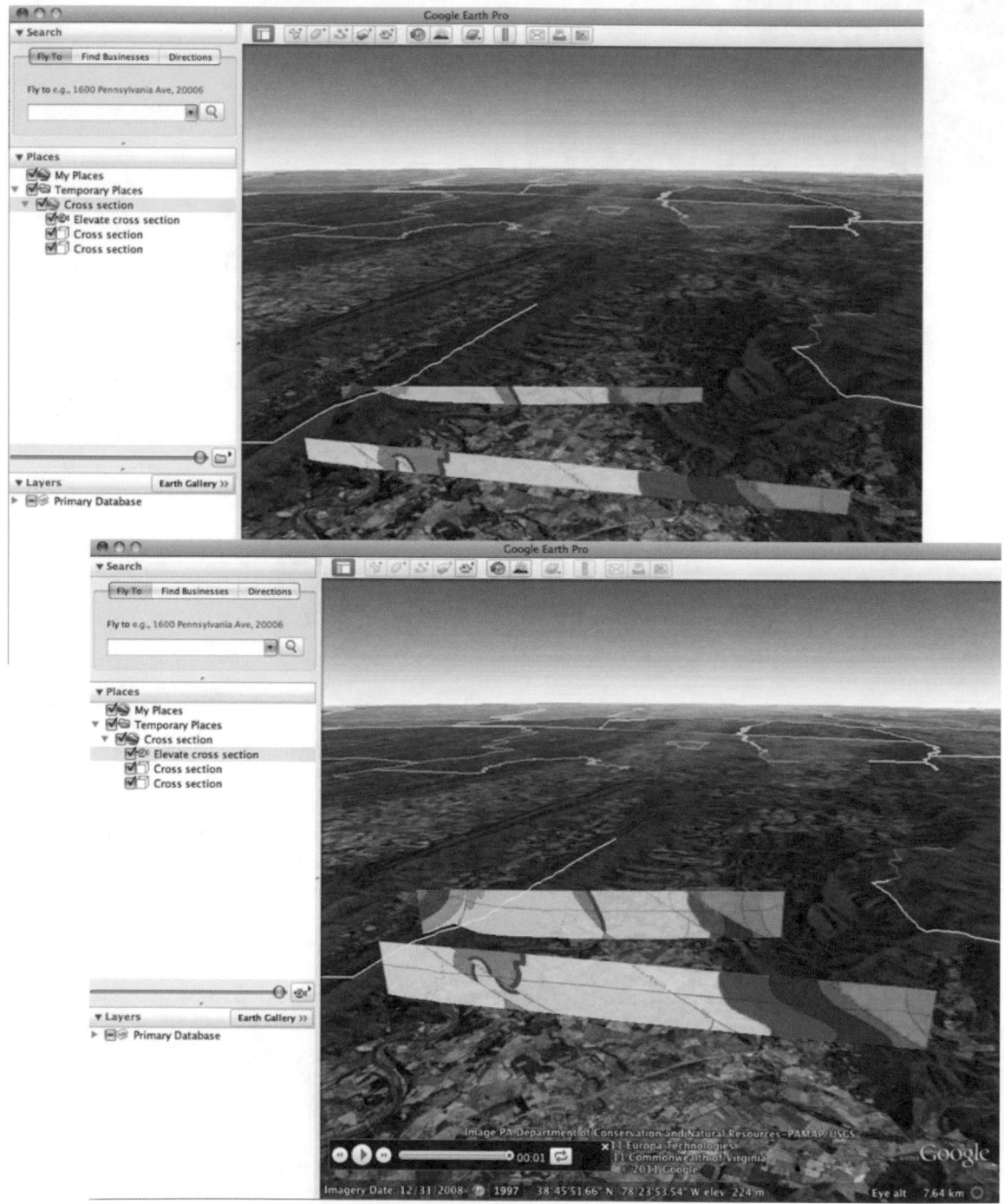

Figure 10. Cross-section emergence controlled by the Tour feature. © 2011 Google, Europa Technologies, Common-wealth of Virginia. Image courtesy of Penn. Department of Conservation and Natural Resources–PAMAP/USGS.

TABLE 13. MAKING A MODEL EMERGE WITH THE TOUR FUNCTION

```
<gx:Tour>
    <name>Elevate cross section</name>
    <gx:Playlist>
        <gx:AnimatedUpdate>
            <gx:duration>0.1</gx:duration>
            <Update>
                <targetHref></targetHref>
                <Change>
                    <Placemark targetId="AAmodel">
                        <Model>
                            <Location>
                                <altitude>-500</altitude>
                            </Location>
                        </Model>
                    </Placemark>
                </Change>
            </Update>
        </gx:AnimatedUpdate>
        <gx:Wait>
            <gx:duration>0.1</gx:duration>
        </gx:Wait>
        <gx:AnimatedUpdate>
            <gx:duration>0.1</gx:duration>
            <Update>
                <targetHref></targetHref>
                <Change>
                    <Placemark targetId="AAmodel">
                        <Model>
                            <Location>
                                <altitude>-450</altitude>
                            </Location>
                        </Model>
                    </Placemark>
                </Change>
            </Update>
        </gx:AnimatedUpdate>
        <gx:Wait>
            <gx:duration>0.1</gx:duration>
        </gx:Wait>
        ...
        ...
    </gx:Playlist>
</gx:Tour>
```

*Note:*This code is generated by client-side JavaScript.

Google Earth screen overlays, photo overlays, ground overlays, placemark icons, and balloons. The HTML file (which can be run either locally or remotely) contains a form that allows the user to choose a local PNG image file. When the user presses the submit button, the form contacts (posts to) the "upload.php" file which must be located on an Internet-connected server in a folder with read-write-execute permissions set. The image file is then uploaded and renamed "image.png," replacing any prior version. Any Google Earth stand-alone application or plug-in instance that includes a network link to the image file will show the new image provided the link in Google Earth is set to refresh at specific intervals (Fig. 13). Thus, a teacher can remotely change images—ground overlays, screen overlays, photo overlays, etc.—as they are being used by students. For security and/or performance purposes, it is best to limit the permitted image file size.

A HTML form can also be used to write or replace KML code in a file located on the server side (Fig. 14). Table 19 lists an example in which the user pastes KML into an HTML form's text area. Upon receiving the submit button event, the linked PHP file (Table 20) writes the text to a KML document. Two functions are particularly useful for dynamic KML editing (Table 21), namely the preformatted string denoted by a triple angle bracket ">>>" and the "str_replace" function that searches for a substring and replaces it with a specified string. For example, within a do-loop, it is possible to search for a placeholder string such as "replaceMe" and replace it repetitively with, say "string 1," "string 2," "string 3," etc.

SERVER-SIDE PYTHON SCRIPTS

A second server-side scripting language that is very popular and more sophisticated than PHP for KML generation and

TABLE 14. BARE-BONES HTML SOURCE FOR THE GOOGLE EARTH PLUG-IN

```
<html>
    <head>
        <script src="http://www.google.com/jsapi"></script>
        <script type="text/JavaScript">
            google.load("earth", "1");
            google.setOnLoadCallback(initialize);

            function initialize() {
                google.earth.createInstance('myWorld');
            }
        </script>
    </head>

    <body id="myWorld">
    </body>
</html>
```

Figure 11. The bare-bones Google Earth plug-in generated by the code in Table 12. © 2011 Google.

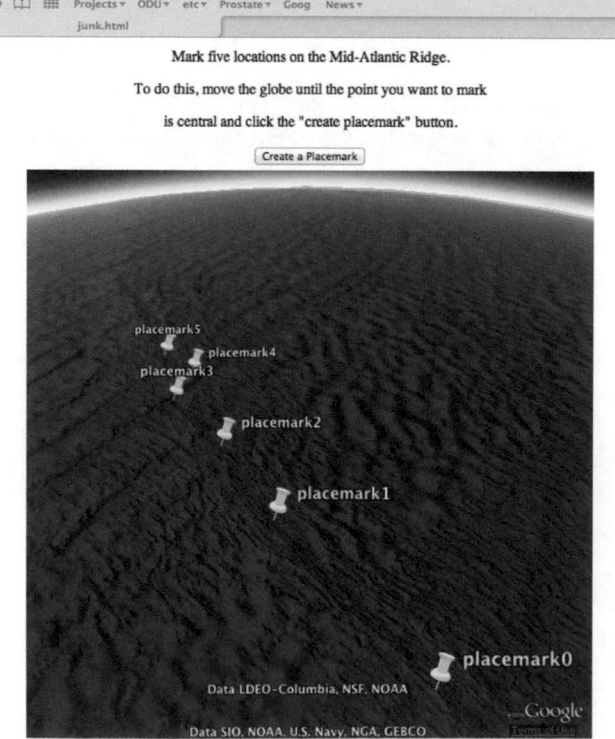

Figure 12. A plug-in that allows students to enter an enumerated series of placemarks.

modification is Python (http://www.python.org; see Beazley, 2009). Python presents several options for producing KML from code (Ballagh et al., 2011). As stated above, KML is an XML markup and native within Python is the ability to use traditional XML tools, such as Document Object Model (DOM) manipulation (http://docs.python.org/library/xml.dom.html), or the more Pythonic Element Tree XML API (http://docs.python.org/library/xml.etree.elementtree.html). These Python APIs allow the generation of KML directly in a pure Python environment. However, they require the developer to keep track of how KML is structured. There are several open source libraries for the production of KML. Three in particular stand out as beneficial when using Python to generate KML—they are GDAL/OGR, Libkml, and pyKML.

The Geographic Data Abstraction Library (GDAL, http://gdal.org) is an extensive set of utility libraries for working with GIS data files. With GDAL and its accompanying library OGR (note that these initials don't actually stand for anything; http://trac.osgeo.org/gdal/wiki/FAQGeneral#WhatdoesGDALstandsfor), geoscience content developers can start from most openly specified GIS data formats and produce KML. It is bundled with a lot of utility libraries. In particular, ogr2ogr.py allows you to implement data conversion between different file formats.

GDAL and OGR can be fairly heavyweight for use in many applications; consequently Google Inc. released the more light-weight libkml (http://libkml.googlecode.com). Both libkml and GDAL are written in C++ with Swig bindings (http://www.swig.org/) that allow you to use them in Python applications. Unlike GDAL, libkml works exclusively with KML. libkml is a high-performance library, but it can be hard to use and the default interface provided by Swig is not very Pythonic. The pylibkml library (http://pylibkml.googlecode.com) uses libkml, but provides a more Pythonic interface.

Finally, pyKML provides a lighter-weight method of reading and writing KML using Python. pyKML relies on Libxml2 (http://xmlsoft.org/), a standard C library for parsing XML files. It uses this library to parse the KML schema and to provide a Pythonic interface. Table 22 presents some sample code to generate a "Placemark" element with a "Point" element using pyKML. The "placemarkGeneration" function can be used to return an object that can be written out in a response. If the script is run by itself, it writes out a KML file to the file system.

CLIENT AND SERVER-SIDE RUBY SCRIPTS

A final scripting language that is very popular and useful for dynamic KML generation is called Ruby (e.g., Flanagan and Matsumoto 2008). For example, Schulze et al. (2009) used it to enable sketching or painting style annotation onto Google Earth, which has obvious applications to geological mapping.

Ruby comes built in with many modern operating systems. To test whether it is already installed on your computer, type the command line "ruby-v" which should return the current version details. If necessary, it may be downloaded from

TABLE 15. HTML SOURCE FOR A GOOGLE EARTH PLUG-IN
THAT ALLOWS USERS TO CREATE ENUMERATED PLACEMARKS

```
<html>
<head>
    <script src="http://www.google.com/jsapi?key=[my key]"></script>
    <script type="text/JavaScript">
        google.load("earth", "1");
        google.setOnLoadCallback(initialize);

        function initialize() {
            google.earth.createInstance('myWorld',success,failure);
            makeButton();
        }

        var ge = '';
        function success(instance){
            ge = instance;
        }

        function failure(errorCode){
            alert(errorCode);
        }

        function makeButton() {
            var btn = document.createElement('input');
            btn.type = 'button';
            btn.value = 'Create a Placemark';
            if (btn.attachEvent){
                btn.attachEvent('onclick', handleButton);
            }else{
                btn.addEventListener('click', handleButton, false);
            }
            document.getElementById('makeButton').appendChild(btn);
        }

        var counter = 0;
        function handleButton() {
            var nextPlace = ge.createPlacemark('');
            nextPlace.setName("placemark" + counter++);
            ge.getFeatures().appendChild(nextPlace);
            var la = ge.getView().copyAsLookAt(ge.ALTITUDE_ABSOLUTE);
            var point = ge.createPoint('');
            point.setLatitude(la.getLatitude());
            point.setLongitude(la.getLongitude());
            nextPlace.setGeometry(point);
        }
    </script>
</head>

<body>
    <table align="center">
        <tr>
            <td align="center">
                <p>Mark five locations on the Mid-Atlantic Ridge. </p>
                <p>To do this, move the globe until the point you want to mark </p>
                <p>is central and click the &quot;create placemark&quot; button.</p>
                <div id="makeButton"></div>
            </td>
        </tr>
        <tr>
            <td align="center">
                <div ALIGN=CENTER id="myWorld" style="width: 600px; height: 600px;">
                </div>
            </td>
        </tr>
    </table>
</body>
</html>
```

TABLE 16. HTML SOURCE FOR CONTROLLING A GOOGLE EARTH PLUG-IN

```
<!DOCTYPE html PUBLIC "-//W3C//DTD XHTML 1.0 Transitional//
   EN" "http://www.w3.org/TR/xhtml1/DTD/xhtml1-transitional.dtd">
   <html xmlns="http://www.w3.org/1999/xhtml" xml:lang="en">
      <head>
         <script src="http://www.google.com/jsapi?key=[my key]">
         </script>

         <script language="JavaScript"
src="http://www.lions.odu.edu/org/planetarium/steve/Tonga_API/
         tonga_roll_js/slider.js">
         </script>

         <script type="text/JavaScript"
src="http://www.lions.odu.edu/org/planetarium/steve/Tonga_API/
         tonga_roll_js/tonga_slab.js">
         </script>

         <script type="text/JavaScript"
src="http://www.lions.odu.edu/org/planetarium/steve/Tonga_API/
         tonga_roll_js/timer_tonga.js">
         </script>
      </head>

      <body>
         <form name="Form">

            <script type="text/JavaScript"
src="http://www.lions.odu.edu/org/planetarium/steve/Tonga_API/
      tonga_roll_js/tonga_data_slider.js">
            </script>

            <input name="sliderValue" id="sliderValue" value="0" type="Text"
size="3" onchange=
            "A_SLIDERS[0].f_setValue(this.value)">
            </input>

         </form>
      </body>
   </html>
```
Note: The symbol "↵" denotes a continuation line.

TABLE 17. AN HTML FILE THAT CALLS THE PHP FILE IN TABLE 18 TO ALLOW THE USER TO UPLOAD A PNG IMAGE

```
<!DOCTYPE html>
<html>
<head>
</head>
<body>
   <form enctype="multipart/form-data" action="http://www.digitalplanet.org/php/upload.php"
method="POST">
      <input type="hidden" name="MAX_FILE_SIZE" value="100000" />
      Choose a PNG file to upload: maximum size is 100 Kb.
      <input name="uploadedfile" type="file" /><br />
      <input type="submit" value="Upload File" />
   </form>
</body>
</html>
```

TABLE 18. LISTING OF THE PHP FILE STORED AT FILE http://www.digitalplanet.org/php/upload.php

```
<?php
   $target_path = "img/image.png";

   if(move_uploaded_file($_FILES['uploadedfile']['tmp_name'], $target_path)) {
      echo "The file image.png has been uploaded to the img folder";
   } else{
      echo "There was an error uploading the file, please try again!";
   }
?>
```

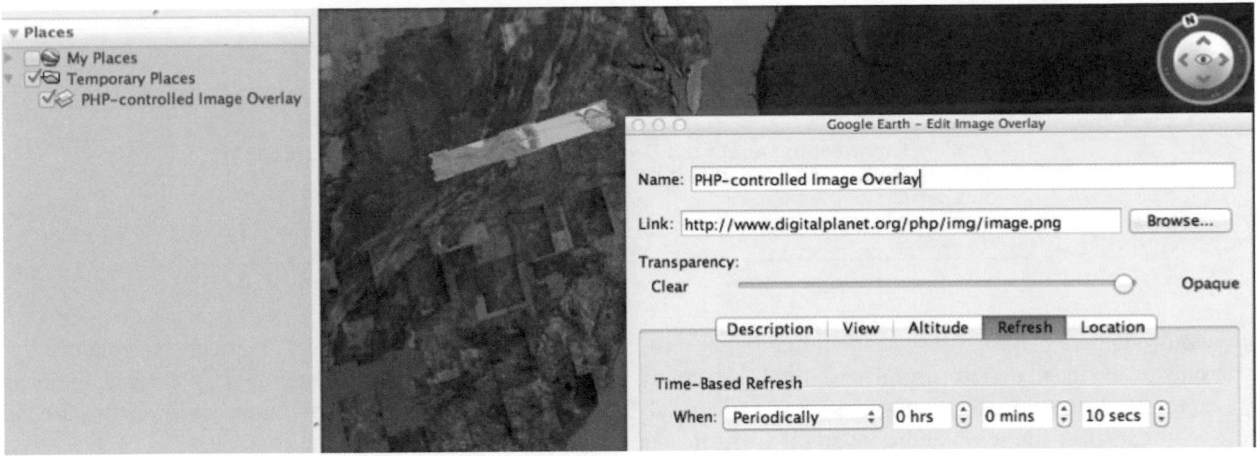

Figure 13. An image overlay in Google Earth set to refresh at 10-second intervals. © 2012 Google, TerraMetrics. Data courtesy of SIO, NOAA, U.S. Navy, NGA, GEBCO.

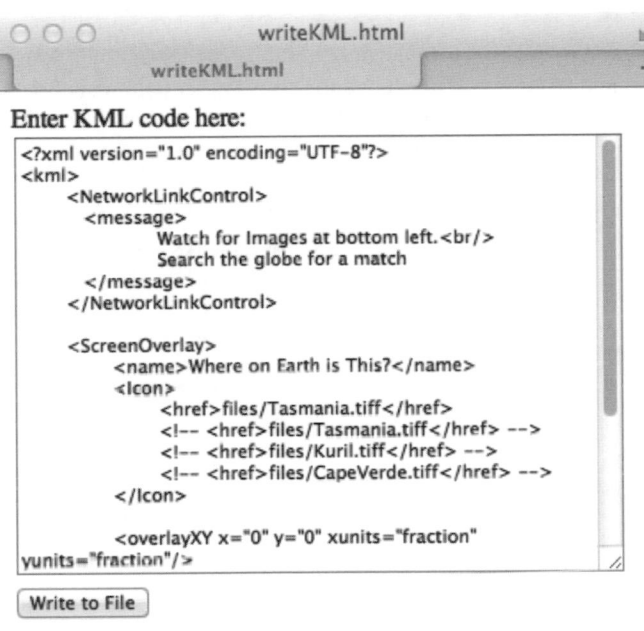

Figure 14. An HTML form linked to a PHP file can be used to write code to a KML file.

TABLE 19. LISTING OF AN HMTL FORM THAT CREATES OR REPLACES A KML FILE ON THE SERVER SIDE

```
<!DOCTYPE html>
<html>
    <body>
        <form
            enctype="multipart/form-data"
            action="http://www.digitalplanet.org/php/writeKML.php" method="POST">
            <input type="hidden" />
            Enter KML code here:
            <br />
             <TEXTAREA name="KML_Code" rows="20" cols="50"></TEXTAREA>
            <br />
            <input type="submit" value="Write to File" />
        </form>
    </body>
</html>
```

TABLE 20. LISTING OF THE PHP FILE CALLED IN TABLE 19

```php
<?php
$newMessage = $_POST['myTweet'];
    $messageFile = "KML_Code.kml";
    $messageCode =  $_POST['KML_Code'];
    $fh = fopen($messageFile, 'w') or die("can't open KML_Code");
    fwrite($fh, $messageCode);
    fclose($fh);
    echo "done!";
?>
```

http://www.ruby-lang.org. Ruby files are identified by the ".rb" suffix and may reside locally on the user's hard drive or remotely on a server. Table 23 illustrates an example of a script that reads data from a spreadsheet file (csv) and converts it to KML—in this case forming a set of "Placemark" elements. Of course, the same task can be completed in any of the previously discussed scripting languages (see, for example, http://code.google.com/apis/kml/articles/csvtokml.html). However, Ruby has particularly elegant file parsing and XML formatting features, as seen in the csvtokml function.

FUSION TABLES

Fusion Tables dramatically extend the functionality of Google Earth, transforming it from a geo-browser to a geospatial relational database (think ArcGIS without the learning curve).

Fusion Tables are spreadsheets that can be dynamically linked to Google Maps and/or Google Earth. If the purpose is to create and manage a set of geographical placemarks, then the linkage is simple. First, go to:

http://www.google.com/fusiontables/public/tour/index.html

and learn how to enter your own geospatial data into a Fusion Table (Fig. 15). The table must specify each location's longitude and latitude and may contain various other columns of data. Clicking on the "visualize map" option reveals a global map (Fig. 16) and a text area containing a URL such as:

http://www.google.com/fusiontables/exporttable?query=select+*+from+395795+&o=kmllink&g=col2

TABLE 21. EDITING AN EXISTING KML FILE
USING THE PHP PRE-FORMATTED STRING AND STR_REPLACE FUNCTIONS

```php
<?php
    if(isset($_POST['submit'])){
        $KML = '';
        $Headers =<<<Headers
        <?xml version="1.0" encoding="UTF-8"?>
        <kml>
            <Document>
        Headers;

        $PlaceHolder =<<<PlaceHolder
            <Placemark>
                <name>Change_Me</name>
            </Placemark>
        PlaceHolder;

        $Footers =<<<Footers
            </Document>
        </kml>
        Footers;

        $myName = $_POST['myName'];
        $PlaceHolder = str_replace("Change_Me", $myName, $PlaceHolder);
        $kml = $Headers + $PlaceHolder + $Footers;
        $kmlFile = fopen("PlaceName.kml", 'w') or die("can't open kml file");
        fwrite($kmlFile, $kml);
        fclose($kmlFile);
    }
?>
```

TABLE 22. A PYTHON SCRIPT FOR GENERATING A KML PLACEMARK

```python
from pykml.factory import KML_ElementMaker as K
from pykml.factory import ATOM_ElementMaker as ATOM
from lxml import etree
def placemarkGeneration(name,author,authorLink,longitude,latitude):
    doc = K.kml(
        K.Document(
          ATOM.author(
            ATOM.name(author)
          ),
          ATOM.link(href="{authorlink}".format(authorlink=authorLink)),
          K.Placemark(
            K.name(name),
            K.Point(
              K.coordinates("{longitude},{latitude}"
                .format(longitude=longitude,latitude=latitude))
            ))))
    return doc
def writeFile(document, filepath):
    docstring = etree.tostring(document)
    f=open(filepath,"w")
    f.write(docstring)
    f.close()

if __name__ == '__main__':
    kmldoc = placemarkGeneration("Null Island", "Mano Marks",
      "http://manomarks.net", "0.0", "0.0")
    writeFile(kmldoc,"/examplepath/documents/kml/example.kml")
```

TABLE 23. A RUBY SCRIPT FOR GENERATING KML PLACEMARKS
FROM A COMMA-DELIMITED SPREADSHEET (CSV) FILE

```ruby
def indent(data)
#function written by Paul Lutus
  tab = 0
  kml = []
  data.split("\n").each { |record|
    record.strip!
    if(record.size > 0)
      outc = record.scan(%r{(</|/>)}).length
      inc = record.scan(%r{<\w}).length
      net = inc - outc
      tab += (net < 0)?net:0
      kml << ("\t" * tab) + record
      tab += (net > 0)?net:0
    end
  }
  return kml.join("\n")
end

def wrap(tagName,data)
  return "<#{tagName}>\n#{data}</#{tagName}>\n"
end

def csv2kml(input)
  kml = []
  input.each do |record|
    record.strip!
    lng,lat,alt,name,desc = record.split(",")
    coords = "<coordinates>"+lng+","+lat+","+alt+"</coordinates>\n"
    output = []
    output << "<name>"+name+"</name>\n"
    output << "<description>"+desc+"</description>\n"
    output << "<Point>\n"+coords+"</Point>\n"
    kml << wrap("Placemark",output)
  end
  kml = wrap("Folder",kml)
  kml = wrap("kml",kml)
  puts "<?xml version=\"1.0\" encoding=\"UTF-8\"?>\n" + indent(kml)
end

csv2kml($stdin.readlines)
```

Note: Adapted from code by Paul Lutus, ©2008 http://www.arachnoid.com.

Google fusion tables **testSDdata.csv**

File   View   Edit   Visualize   Merge

Current view: **All** - Show options

ID ▾	Lon ▾	Lat ▾	◻ ▸	
1	-6.129135	53.302769	145	35
2	36.66236	-1.26361	30	45
3	36.840090	-1.13739	60	65
4	36.836050	-1.17129	90	85
5	36.6312	-0.9829	120	10
6	36.700972	-1.112857	150	20
7	36.6171	-1.08108	180	30
8	36.6896	-1.21635	210	40
9	36.7613	-1.1328	240	50
10	-76	36	123	45
11	-81	85	210	25

Figure 15. A Fusion Table with locations in decimal longitude and latitude and bedding orientations in integer strike (0° through 359°) and dip (0° to 90°, right-hand rule).

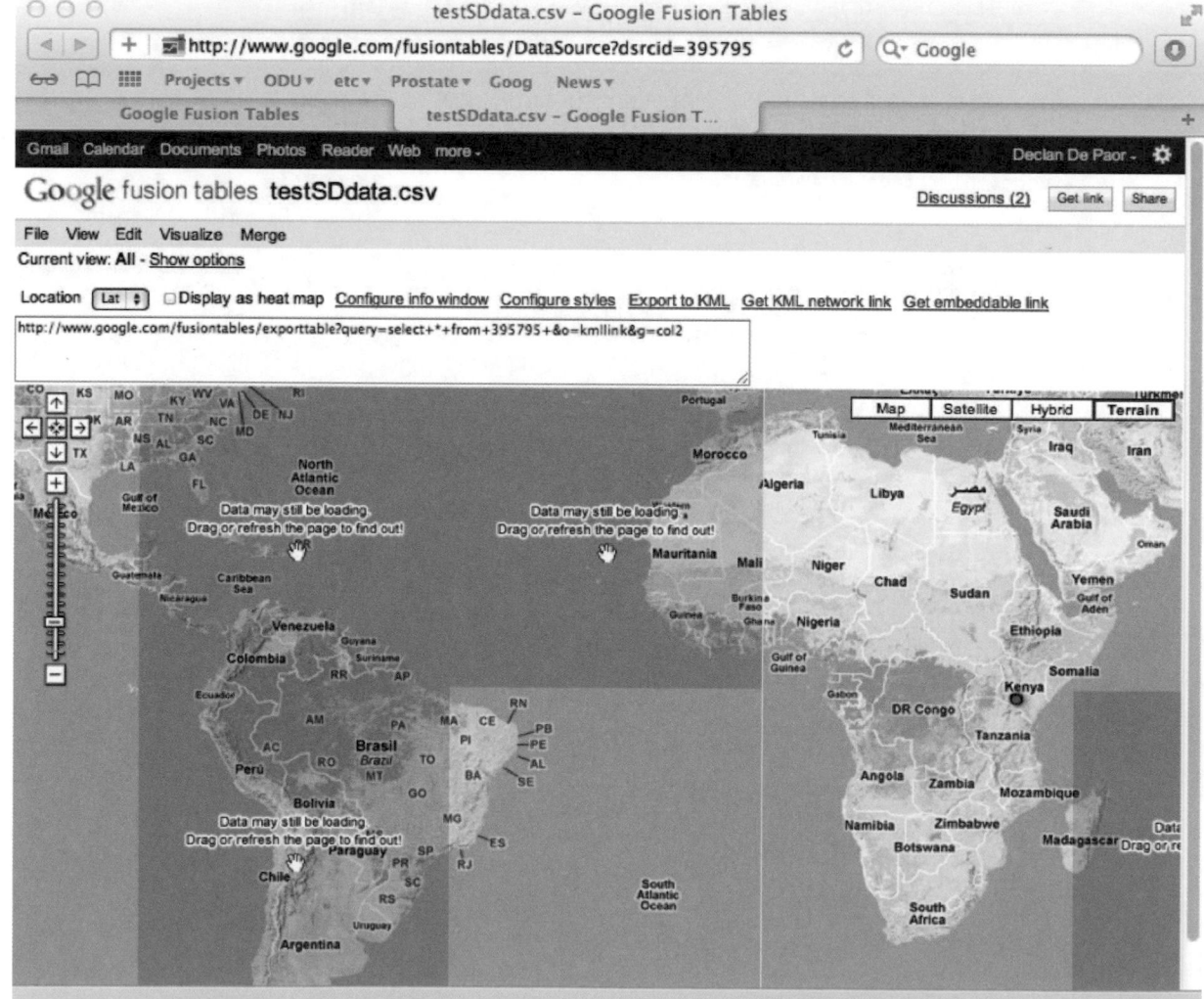

Figure 16. The network link for a Fusion Table is revealed using "visualize > map."

The view illustrated in Figure 17 results from pasting this URL into a new network link in the Google Earth desktop application, or writing it into the Google Earth API code for an instance of a plug-in. "Placemark" elements are automatically created for each longitude/latitude coordinate data pair and the other data columns are recorded in the placemarks' balloons.

In the geosciences, we often need to represent orientation data linked to geographical coordinates. For example, each "Placemark" element might represent a location where a geologist recorded a strike and dip reading specifying the orientation of a structural feature such as bedding or a fault. In this case the linkage is more convoluted but is made possible by a combination of the routines previously discussed, as explained next. (The logic of the linkage is illustrated in Figure 18.)

Instead of creating a network link from Google Earth directly to the Fusion Table, we create a network link to a KML file residing on a server, with a refresh rate set to an interval of several seconds, minutes, or hours, depending on the nature of the data.

A PHP, Python, or Ruby script also running on the server side accesses the Fusion Table and writes updates to the KML file whenever it receives new data (Table 24). The result is illustrated in Figure 19. Any changes in Fusion Table data (foreground window) are immediately reflected in the Google Earth application or API (background window).

DISCUSSION AND CONCLUSIONS

The Google Earth stand-alone application and web browser plug-in have already transformed how we teach and learn in the geospatially oriented sciences. However, the geoscience applications developed to date represent just the tip of an iceberg of future potential. In this manuscript, and in other contributions to this Special Paper, we and our colleagues have presented examples of KML, JavaScript, and other client- and server-side scripts that explore the current capabilities of Google Earth, and we have highlighted new

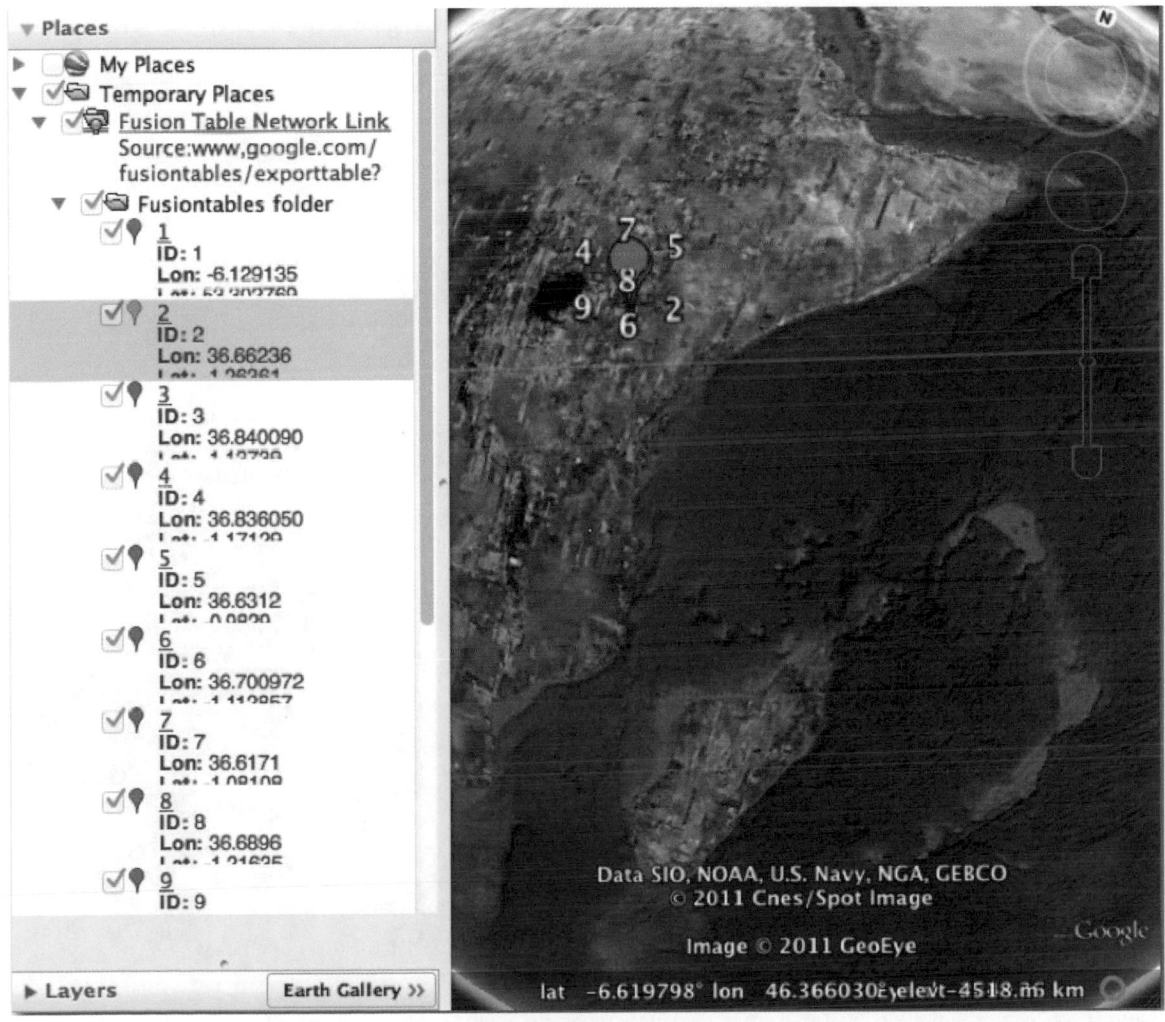

Figure 17. Placemarks in Google Earth linked to a Fusion Table.

tools such as Fusion Tables that hold considerable promise for future development.

Some programming experience is undoubtedly helpful for advanced applications but much can be achieved with a moderate investment of time and effort, and we have presented an array of tools and techniques using a wide variety of programming and scripting languages. Readers might be at a loss to choose amongst the various approaches described, therefore by way of an executive summary, we suggest the following:

(1) Readers who are familiar with any computer programming language with a current compiler for the Windows, Macintosh, or Linux operating systems can probably use it to generate and save KML without resorting to a new, unfamiliar development environment. This solution is ideal for outputting static KML files, especially files with dozens or hundreds of incrementally changing KML scripts that animate geological maps or models. The maps and models will change when the KML file is opened in the Google Earth application or plug-in but the KML file itself will not change. Languages such as FORTRAN or C cannot be used to change KML dynamically (in response to user actions) because they compile before they run, whereas scripts accessed by Google Earth via a network link or those included in an HTML document must be run line-by-line using an interpreter.

(2) Basic knowledge of JavaScript is essential and is sufficient for client-side applications. The JavaScript API is the most important tool for educators wishing to create engaging web pages with embedded instances of the Google Earth plug-in.

(3) Readers who wish to create dynamic server-side KML that is capable of responding to viewer queries or

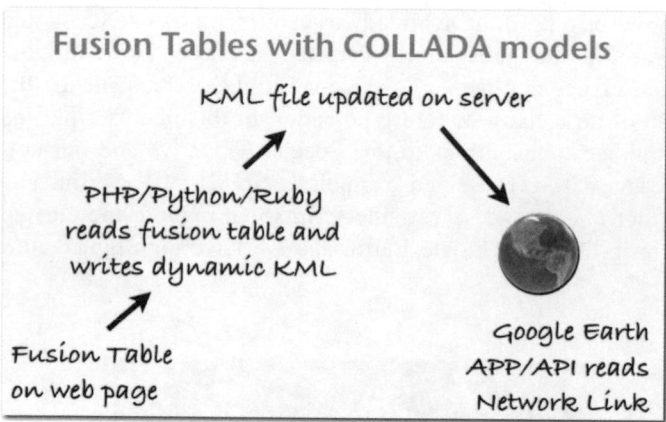

Figure 18. Dynamic linkage between Fusion Table and Google Earth.

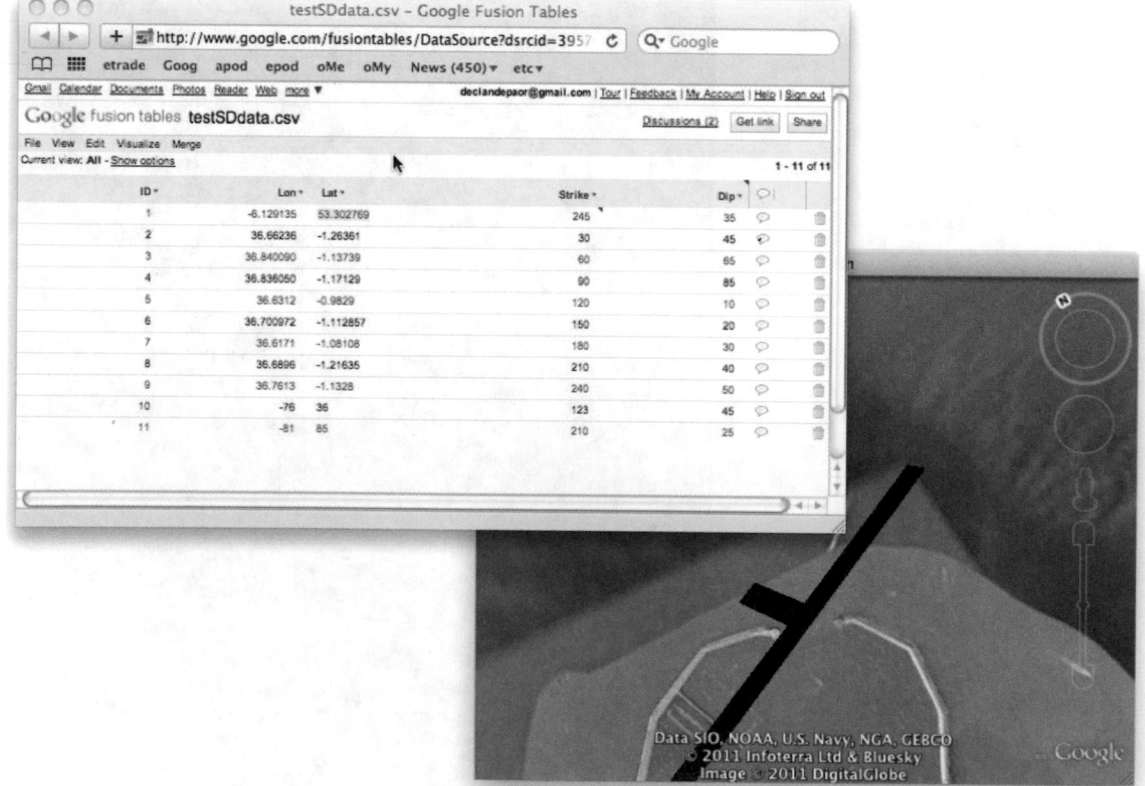

Figure 19. Changes in the Fusion Table are immediately reflected in any Google Earth application or plug-in that monitors the associated KML file.

TABLE 24. A PHP SCRIPT READS A FUSION TABLE AND WRITES KML THAT RENDERS GEO-ORIENTATION DATA SYMBOLS

```php
<?php
$KML_Start_String =<<<KML_Start
<?xml version="1.0" encoding="UTF-8"?>
<kml>
    <Document>
KML_Start;

$Placemark_String =<<<Placemark_shell
    <Placemark>
        <Model id="SDmodel">
            <Location>
                <longitude>lonHolder</longitude>
                <latitude>latHolder</latitude>
                <altitude>200</altitude>
            </Location>
            <Orientation>
                <heading>strikeHolder</heading>
                <tilt>0</tilt>
                <roll>-dipHolder</roll>
            </Orientation>
            <Link>

    <href>http://csmres.jmu.edu/Geollab/Whitmeyer/web/visuals/GoogleEarth/tools/SDsymbol.dae</href>
            </Link>
        </Model>
    </Placemark>
Placemark_shell;

$KML_End_String =<<<KML_End
    </Document>
</kml>
KML_End;

if(isset($_POST['submit'])){
    $fusionTable = $_POST['myFusionTable'];
    $fusionData = fopen($fusionTable, 'r') or die("can't open fusion table");
    if ($fusionData){
        //echo "Current table: $fusionTable";
        //echo "<br>";

        $i = 0; //count the fusion table rows
        while (($buffer = fgetcsv($fusionData, 4096)) !== false) {//put comma delimited data in array

            //insert values from fusion table in KML placemarks
            $placemark[$i] = $Placemark_String; //start with the placemark shell
            if ($buffer[0] != "ID"){// ignore the header
                $placemark[$i] = str_replace("ID_Shell", $buffer[0], $placemark[$i]);
            }
            if ($buffer[1] != "Lon"){// ignore the header
                $placemark[$i] = str_replace("lonHolder", $buffer[1], $placemark[$i]);
            }
            if ($buffer[2] != "Lat"){// ignore the header
                $placemark[$i] = str_replace("latHolder", $buffer[2], $placemark[$i]);
            }
            if ($buffer[3] != "Strike"){// ignore the header
                $placemark[$i] = str_replace("strikeHolder", $buffer[3], $placemark[$i]);
            }
            if ($buffer[4] != "Dip"){// ignore the header
                $placemark[$i] = str_replace("dipHolder", $buffer[4], $placemark[$i]);
            }
            $i++;
        }
        if (!feof($fusionData)) {
            echo "Failed to read data from fusion table - check format\n";
        }
        fclose($fusionData);

        $KML_Start_String = str_replace("docLonHolder", $buffer[1], $KML_Start_String);
        $KML_Start_String = str_replace("docLatHolder", $buffer[2], $KML_Start_String);

        $kmlFile = fopen("KMLfromFusionTable.kml", 'w') or die("can't open kml file");
        fwrite($kmlFile, $KML_Start_String);
        while ($i-- > 1){
            fwrite($kmlFile, $placemark[$i]);
        }
        fwrite($kmlFile, $KML_End_String);
        fclose($kmlFile);
    }
}?>
```

updating itself based on links to Fusion Table, for example, can choose between PHP, Python, and Ruby. There are numerous self-help books and websites for each, such as http://tryruby.org/levels/1/challenges/0. Python is arguably the most elegant and powerful solution, given the range of libraries designed for XML generally, and specifically for KML.

(4) Fusion Tables are potentially transformative. They are relatively easy to link to Google Earth if their data comprise location and placemark content only and can be linked in more complex ways for presentation of dynamic orientation data or real-time models.

The gold-rush style rewards for time invested in creating dynamic KML and other geo-browser resources remain to be staked and claimed. We trust that our colleagues and peers will stake these claims in fairly short order and render this volume semi-obsolete within the next year or two.

ACKNOWLEDGMENTS

We wish to thank Chris Condit for his editorial help. Because the authors included the volume editors, he served as special editor and arranged independent reviews. Reviews by Scott Matthews and Andy Bobyarchick helped to improve the final manuscript. This research was support in part by National Science Foundation (NSF) TUES 1022755, NSF GEO 1034643, and a 2010 Google Faculty Research Award. Any opinions, findings, and conclusions or recommendations expressed in this paper are those of the authors and do not necessarily reflect the views of the NSF or Google Inc.

REFERENCES CITED

Ballagh, L.M., Raup, B.H., Duerr, R.E., Jodh, S., Khalsa, S., Helm, C., Fowler, D., and Gupte, A., 2011, Representing scientific data sets in KML: Methods and challenges: Computers & Geosciences, v. 37, no. 1, p. 57–64, doi:10.1016/j.cageo.2010.05.004.

Beazley, D.M., 2009, Python Essential Reference, 4th edition: Addison-Wesley Professional, 717 p., ISBN 0672329786.

Brady, A., McDowell, R.C., and Schultz, K., 2002, JavaScript programming basics: a laboratory series for beginning programmers: Journal of Educational Resources in Computing, v. 2, no. 2, doi:10.1145/772938.772939.

Castagnetto, J., Rawat, H., Schumann, S., Scollo, C., and Veliath, D., 1999, Professional PHP Programming: Wrox Press, 909 p.

De Paor, D.G., 2008, Using Google SketchUp with Google Earth for Scientific Applications: Google Tech Talk, San Francisco, http://www.youtube.com/watch?v=6cVJqvsfxvo.

De Paor, D.G., and Pinan-Llamas, A., 2006, Application of novel presentation techniques to a structural and metamorphic map of the Pampean Orogenic Belt, Northwest Argentina: Geological Society of America Abstracts with Programs, v. 38, no. 7, p. 326, paper no. 131-12, http://gsa.confex.com/gsa/2006AM/finalprogram/abstract_112392.htm.

De Paor, D.G., and Williams, N.R., 2006, Solid modeling of Moment Tensor Solutions and Temporal Aftershock Sequences for the Kiholo Bay Earthquake using Google Earth with a surface bump-out: Eos (Transactions, American Geophysical Union), Fall Meeting, v. 87, no. 52, abstract S53E-05.

De Paor, D.G., Daniels, J., and Tyagi, I., 2007, Five geo-browser lesson plans: Eos (Transactions, American Geophysical Union), Fall Meeting, v. 88, no. 52, abstract IN43A-0904.

De Paor, D.G., Wild, S.C., and Dordevic, M.M., 2011, Emergent and animated COLLADA models of the Tonga Trench and Samoa Archipelago: Implications for geoscience modeling, education, and research: Geosphere, v. 8, p. 491–506, doi:10.1130/GES00758.1

Dordevic, M.M., 2012, this volume, Designing interactive screen overlays to enhance effectiveness of Google Earth geoscience resources, in Whitmeyer, S.J., Bailey, J.E., De Paor, D.G., and Ornduff, T., eds., Google Earth and Virtual Visualizations in Geoscience Education and Research: Geological Society of America Special Paper 492, doi:10.1130/2012.2492(07).

Flanagan, D., and Matsumoto, Y., 2008, The Ruby Programming Language: O'Reilly Media Inc., 429 p.

Schulze, C., Nienhaus, L., and Loviscach, J., 2009, Sketch-based annotations in Google Earth: Proceedings of SIGGRAPH '09: New York, New York, Association for Computing Machinery, doi:10.1145/1599301.1599354.

Wernecke, J., 2009, The KML Handbook: Geographic Visualization for the Web: Addison-Wesley, 368 p., ISBN: 9780321574404.

Yamagishi, Y., Yanaka, H., Suzuki, K., Tamura, H., Nagao, H., and Tsuboi, S., 2008, Visualization of geoscience data on Google Earth: Development of a data converter system for seismic tomography models, geochemical data of rocks, and geomagnetic field models: Eos (Transactions, American Geophysical Union), v. 89, no. 53, Fall Meeting Supplement, Abstract IN41B–1151.

Yamagishi, Y., Yanaka, H., Suzuki, K., Tsuboi, S., Isse, T., Obayashi, M., Tamura, H., and Nagao, H., 2010, Visualization of geoscience data on Google Earth: Development of a data converter system for seismic tomographic models: Computers & Geosciences, v. 36, p. 373–382, doi:10.1016/j.cageo.2009.08.007.

MANUSCRIPT ACCEPTED BY THE SOCIETY 16 APRIL 2012

The Geological Society of America
Special Paper 492
2012

Designing interactive screen overlays to enhance effectiveness of Google Earth geoscience resources

Mladen M. Dordevic*

Department of Physics, Old Dominion University, Norfolk, Virginia 23529, USA

ABSTRACT

The effectiveness of a computer application depends on, among other things, an efficient user interface. In order to visualize subsurface geologic phenomena using the Google Earth™ application, we initially employed the built-in Google Earth time slider. Dragging the slider's right thumb elevated a COLLADA model that initially loads at a subsurface altitude. However, the double-thumb feature of the time slide caused users some difficulties. It is not possible to turn off this feature when not required, so it can be misleading to users. Because of this and because of the need for more control, we transitioned from the stand-alone Google Earth application to the web-based Google Earth plug-in. To overcome some of the limitation for the existing user interface, such as the inability to make controls appear semitransparent, we designed and implemented a screen overlay using the plug-in's application programming interface. This approach opened new possibilities to build more customizable user interfaces. A demonstration of the approach and sample usage of JavaScript to create buttons, draggable images, slides, and slider controls is presented.

INTRODUCTION

One of the great challenges of using Google Earth for the geosciences is visualization of the subsurface. One technique employs 3-D COLLADA (collaborative design activity) models of the subsurface and exploits the build-in Google Earth time slider control for elevating models into view (De Paor and Williams, 2006; De Paor and Whitmeyer, 2010; Barne and Finch, 2008). The KML (Keyhole Markup Language) TimeSpan tag has the capability to store information about the position and appearance of models in specified time intervals. Selection of a particular time interval is achieved by dragging the time slider thumb (see Fig. 1). By defining different altitudes for the model in every time step using the "begin" and "end" tags, it is possible

Figure 1. Google Earth Time Slider with advanced functions. In the blue slider background, two gray thumbs can be dragged. Thumbs can be separated to select a time span. On left image only a single value selected 8/20/8 and on right image, time span from 2/27/9–12/29/11 is selected. It is not possible to show custom units; instead dates always appear.

*mdordevi@odu.edu

Dordevic, M.M., 2012, Designing interactive screen overlays to enhance effectiveness of Google Earth geoscience resources, *in* Whitmeyer, S.J., Bailey, J.E., De Paor, D.G., and Ornduff, T., eds., Google Earth and Virtual Visualizations in Geoscience Education and Research: Geological Society of America Special Paper 492, p. 105–111, doi:10.1130/2012.2492(07). For permission to copy, contact editing@geosociety.org. © 2012 The Geological Society of America. All rights reserved.

to elevate lithospheric blocks above the ground level and visualize them in cross section.

Problems were encountered using this approach during the user testing of the lab modules (Gobert et al., this volume). The inability to turn off advanced features of the Google Earth time slider created many opportunities for error in their usage. One of these features makes it possible to select a single time or a time interval by moving the right or left piece (thumb) of the slider (see Fig. 1). This created confusion with users as a result of unintentionally selecting time intervals. By doing so, models at different altitudes were seen to overlap. Furthermore, the display of appropriate units and values for elevation was not possible—the time slider might read "14 Jan 2001" when the desired output would be, say, 350 m elevation.

A solution for this problem was found by migrating from the Google Earth stand-alone application to the web-based Google Earth plug-in (Google c, Appendix 1). This transition offered more control over the models by using customizable interface elements on the web page containing the plug-in (Stenback et al., Appendix 1).

Before the release of the working draft of HTML 5 (Hickson, Appendix 1), a user interface element slider was usually available as a paid third-party add-on. Now, with the release of HTML5, the slider feature is natively supported in modern web browsers (THE HTML5 TEST, Appendix 1). Instead of having hundreds lines of code to create a slider, it is possible to do it with the single line:

```
<input type = "range" min = "0" max = "50" value = "10" />
```

This code creates a slider on a web page with the range 0–50 and with a current value of 10. It could possibly control the altitude of an elevating cross section or other COLLADA model. A problem could arise if the application that one builds has a lot of interface elements and the user is running it on a small laptop screen. There is a trade-off between the available space for a Google Earth container (Google b, Appendix 1) and space for other interface elements. Four cases will be discussed with various portions of the screen used for each component.

1. Make the size of the Google Earth container big enough to fit the Internet browser window and get the maximal screen usage. Sliders could be located on the web page beside the Google Earth container so they would exceed the physical size of the browser window resulting in the appearance of a scroll bar. When an action is to be performed, the user scrolls the screen, possibly losing partial visibility of the Google Earth window and therefore losing sight of the action performed by the slider (see Fig. 2A).

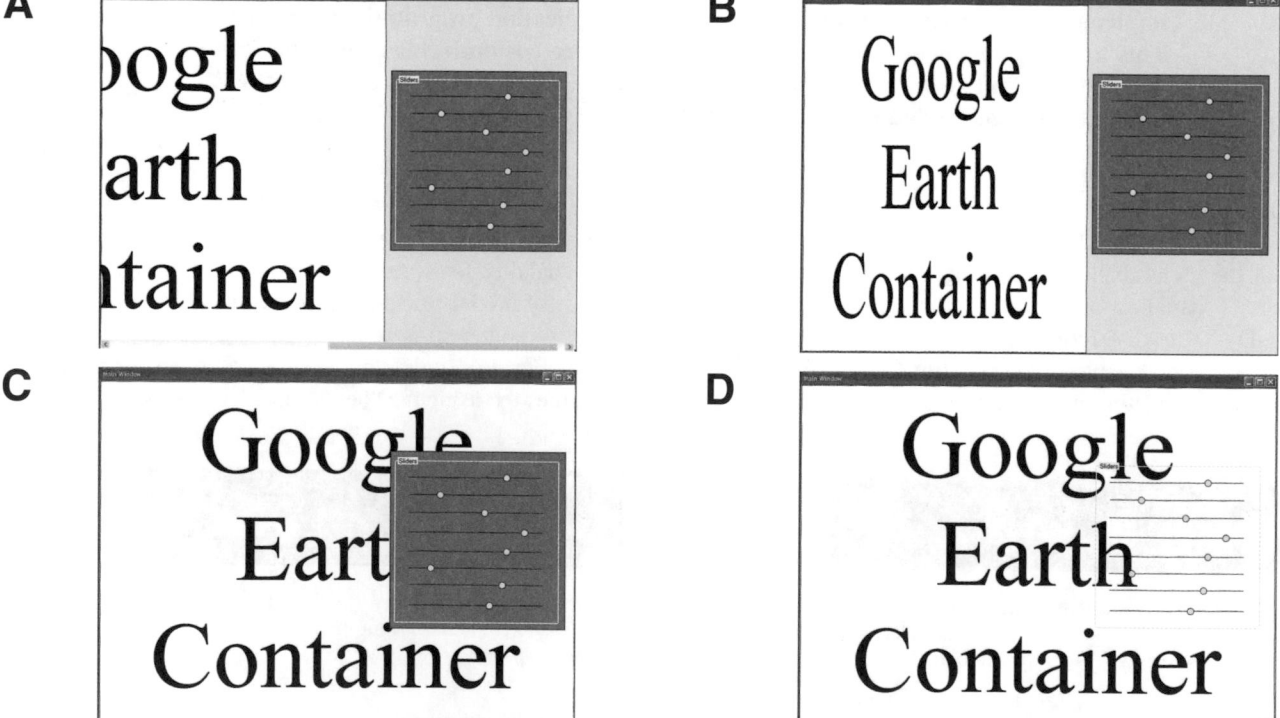

Figure 2. Possible arrangement of Google Earth container and sliders on smaller screen. (A) Size of Google Earth container fits browser window; sliders located on side and exceed physical size of browser window resulting in appearance of scroll bar. (B) Google Earth container and slider fitted to share browser window. (C) Slider on top of Google Earth container on hiding menu. (D) Slider on top of Google Earth container on semitransparent hiding menu.

2. Shorten the length or width of the Google Earth container relative to the size of the browser and position the sliders on the freed-up window space (see Fig. 2B). This removes the need to scroll and the action of the slider can be seen as it is performed. The downside is that the Google Earth container is smaller, which might not be wise in order to perform an action that only might be required only once.

3. Make the size of the Google Earth container big enough to fit the Internet browser window as in case one with the sliders located on a hiding menu bar (Microsoft, Appendix 1). The hiding menu bar appears on demand, automatically resizing the Google Earth container. Under this action, the Google Earth container changes its size to a predefined one and the end result is as in case two (see Fig. 2D).

4. The same setup as in case three, with the difference being the hiding menu extends over half of the Google Earth container (Williams, Appendix 1) and produces a similar result to case one (see Fig. 2C).

A good ratio of the Google Earth container to the slider area visibility would be that described in case four, with the possible addition of semitransparency to the hiding menu. This way, only parts of the Google Earth container that are directly under the sliders controls would be invisible and therefore minimizes the obstruction similar to native Google Earth navigation controls (see Fig. 3).

The current technique for making hiding menus uses IFRAME shims (King, Appendix 1) but with limited to no capability of making them semitransparent (Google d, Appendix 1). An advantage of this technique is that all the interface elements on the hiding menu are HTML-native, which is easy for implementation.

A different technique is suggested here to build semitransparent hiding menus. Using Google Earth screen overlays, it is possible to build some of the web page's main interface elements (such as a slider, button, or check box) and assign them a corresponding functionality.

The idea behind using screen overlays as part of the interface is to achieve customizable look, size, and position. They could be used to embed a PowerPoint-style presentation in a Google Earth Tour, as geological map keys, or as buttons, sliders, or check boxes controlling the viewing experience. In this paper the application and integration of such an interface is explained with a few examples.

CREATING LOW-LEVEL FUNCTIONALITY

In order to overcome the lack of functionality built into Google Earth's screen overlays (for example, the inability to add event listeners as is possible with Placemarks, Google g, Appendix 1), we designed virtual buttons. A virtual button is a predefined area of the Google Earth container (preferably square for simplicity) that is checked after every mouse-down event to determine if the mouse pointer is inside the area. If inside, a predefined action for that area is executed; if not, monitoring continues until the next mouse-down event is detected. (Mouse-down = user presses the mouse button on a page element; mouse-up = user releases the mouse button from an element.)

To make virtual button areas of the Google Earth container visible and recognizable by a user, a Google Earth screen overlay is used with a corresponding size and position. Consequently, the user is aware that clicking on the Google Earth screen overlay may result in some sort of action, pictographically or textually, represented by the image. It is also possible to simulate the pressed state of a button by changing the overlay image on receiving the mouse-down event and returning it to the original state on mouse-up.

The two examples that will be given are for a button and a slider. We start with the more basic example, the button.

Button

It is useful to add two methods (in the object-oriented programming sense of the word) to the window "Number" object. These methods are called toFractionX and toFractionY and they serve to convert from pixel to a fraction size based on the Google Earth container's width or height. They are written under the assumption that the instance of Google Earth is stored in an HTML DIV element with id = "map3d" and that a prototype framework is used (Prototype, Appendix 1):

```
Number.prototype.toFractionX = function(){
return this / $('map3d').offsetWidth
}
Number.prototype.toFractionY = function(){
return this / $('map3d').offsetHeight
}
```

Now we can add the screen overlay to Google Earth and set up its parameters as describe in the Google Code Playground (Google, Appendix 1):

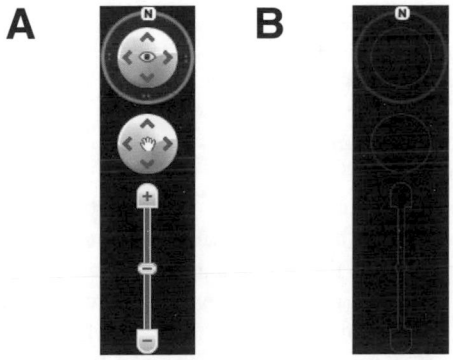

Figure 3. Native Google Earth navigation interface: active (A) and hidden (B).

```
// Create the ScreenOverlay
var button = ge.createScreenOverlay('');
```

We chose an image and set it to the top left corner of the Google Earth container, and we set its size to 60 × 60 pixels.

```
var buttonX = 0.1;
var buttonY = 0.1;
var buttonSizeX = 60;
var buttonSizeY = 60;
// Specify a path to the image and set as the icon
var icon = ge.createIcon('');
icon.setHref('http://www.google.com/intl/en_ALL/images/logo.
gif');
screenOverlay.setIcon(icon);
// Set the ScreenOverlay's position in the window
button .getOverlayXY().setXUnits(ge.UNITS_FRACTION);
button .getOverlayXY().setYUnits(ge.UNITS_FRACTION);
button .getOverlayXY().setX(buttonX);
button .getOverlayXY().setY(buttonY);
// Set the overlay's size in pixels
button .getSize().setXUnits(ge.UNITS_PIXELS);
button .getSize().setYUnits(ge.UNITS_PIXELS);
button .getSize().setX(buttonSizeX);
button .getSize().setY(buttonSizeY);
// Add the ScreenOverlay to Earth
ge.getFeatures().appendChild(button);
```

To achieve the effect of semitransparency, we use an image format such as PNG with a predefined level of transparency, or alpha layer. By adding an event listener that will monitor for a mouse-down event, this button will become interactive or clickable. An event listener is added using the Google Earth default function that accepts three arguments:

```
google.earth.addEventListener (object on which to add listener,
type of event, action upon event)
```

where the first argument identifies the Google Earth window, the second is "mouse-down" in this case, and the third argument will be executed upon the event. The triggered function needs to check whether the mouse pointer at the moment of an event is in the area defined by the button. This is done by checking four conditions: two for each x and y mouse coordinate. Since the position of the button in the Google Earth container is given in fractions and the mouse pointer is in pixels, it is necessary to convert them to the same units using the functions described above (toFractionX() and toFractionY()):

```
function inTheBox(event){
/*Check that mouse pointer X coordinate is to the right of the
left edge of the button*/
```

```
if((buttonX + buttonSizeX .toFractionX() / 2) > event.
getClientX().toFractionX()
/*and that mouse pointer X coordinate is to the left of the right
edge of the button*/
&& (buttonX - buttonSizeX .toFractionX() / 2) < event.
getClientX().toFractionX()
/*and that mouse pointer Y coordinate is to the above of the
bottom edge of the button*/
&& (buttonY + buttonSizeY .toFractionY() / 2) > 1 - event.
getClientY().toFractionY()
/*and that mouse pointer Y coordinate is to the below of the top
edge of the button*/
&& (buttonY - buttonSizeY .toFractionY() / 2) < 1 - event.
getClientY().toFractionY()){
/* if all conditions satisfied, mouse is inside of the area define
by the button*/
/*perform action on button click*/
}
}
/*add vent listener to Google Earth window that monitors for
mouse button down and triggers the function inTheBox defined
above*/
google.earth.addEventListener(ge.getWindow(), 'mousedown',
inTheBox)
```

If all conditions in the above "inTheBox" function return as true, we proceed with the execution of the code for the button action.

Sliders

Every slider consists of two elements: the draggable piece ("thumb") and the track along which the thumb can travel ("background"; see Fig. 1). Once those two elements are set and appended to the Google Earth window using the approach described above, listening for three consecutive events starts.

1. Event: on mouse-down
Response: the code checks the current mouse position. The function "inTheBox" could be reused here: is the mouse pointer in the box that is defined by the thumb? If so, we set a Boolean "drag" flag to indicate that the current slider is being dragged and return to listening for the next event.

2. Event: on mouse-move
Response: if the "drag" flag is true, we monitor the position of the mouse and, upon every horizontal change in position for the horizontal slider or vertical change for the vertical slider, we reposition the thumb and recalculate the thumb value of the slider based upon the new thumb versus background positions. JavaScript publishes this value change so that custom functions can respond to it. In this example, the response is to change the altitude attribute of the model representing the emergent cross section.

Figure 4. Draggable and resizable screen overlay in Google Earth with capability for user to enter image URL and see changes on-the-fly.

3. Event: on mouse-up

Response: The "drag" flag is set to false and the function waits for the next mouse-down event. It is also important to be aware of page resizing issues. If the size of the Google Earth container is set in percentages and a user resizes the page, the background is going to be shorter or longer so rescaling of the slider is also in order. To fix this, we listen for the HTML window .onresize event and restart sliders by removing them and adding them again. To conserve CPU usage on resize events, we put the action on resize in the function called "setTimeout(function() { },100)." In that way it is triggered at a slower rate.

INTERACTIVE SCREEN OVERLAYS

The first example is shown in Figure 4. It is possible to input the URL of any image from the web and click the "Post" but-

ton. This image immediately appears as a screen overlay in the Google Earth container. The screen overlay can be dragged and resized as shown in Figure 4.

It is not necessary for an input URL to be located on the same page as the Google Earth container. For example, a teacher could decide what image to show as the screen overlay and ask students to respond to a corresponding question by searching the globe. Alternatively, several small screen overlays could be floated over the Google Earth container and students could be instructed to click the correct one for a given geoscience lab exercise.

SCREEN OVERLAYS AS INTERACTIVE MAP KEYS

Previous authors have used both opaque and semitransparent screen overlays as map keys. Figure 5 shows an example from Whitmeyer (Whitmeyer, Appendix 1).

Figure 5. Screen overlay used as button. Legend fully interactive; it is possible to turn on and off map sections. In (B) Tertiary and Proterozoic Piedmont formations have been unchecked in key legend.

In Figure 5, the map key has been made interactive. Note how the Pennsylvanian formations have been unchecked in the key; consequently they disappear from the map. Showing and hiding individual formations can be done in the Google Earth desktop application in the Places sidebar, but the interactive map key opens up many possibilities that are not possible with built-in Google Earth functionality (for example, choice of colors or symbols).

SCREEN OVERLAYS AS POWERPOINT PRESENTATIONS

Google Earth cannot be embedded in Microsoft PowerPoint, Apple Keynote, or Google Doc presentations, and so presenters must either interrupt their slide show in order to give a live demonstration or pre-save screen grabs and movies. An interactive screen overlay could have forward and reverse buttons for navigating through a set of slides, thus embedding a PowerPoint-style presentation in Google Earth.

SCREEN OVERLAYS AS SLIDER CONTROLS

Previous authors have also used Google Earth's built-in time slider (De Paor et al., 2008) and tour slider to control features such as emergent cross sections. The time slider's two thumbs, as described, are difficult to move together and when they separate they create an overprinting of models. Figure 6 shows a pair of custom sliders that control emergent cross sections of the subsurface. Each slider consists of two screen overlays, a stationary bar (light blue) and a draggable thumb (dark blue). The orientations of these sliders are horizontal and vertical, but other angles are also possible.

Figure 6. Screen overlay used as slider in Google Earth. Two fully functional sliders used to control emergent cross sections.

CONCLUSION

As the usage of the Google Earth web plug-in becomes more diverse, the need for specialized user interface elements grows. We have described how one can use a screen overlay to make such interface elements. Basic examples of implementation have been given for button and slider controls.

Among the geoscience applications that immediately come to mind are: (*i*) interactive maps keys (screen overlays) linked to maps (ground overlays); (*ii*) PowerPoint-style presentations embedded in the Google Earth container, controlled by forward and reverse buttons; and (*iii*) remote control of the screen overlay content by an educator, enabling different tasks to be assigned to students both locally and in a distance education setting. We strongly believe that, until other advanced features are developed for Google Earth, interactive screen overlays will provide one more useful tool in applications writing, and we hope these examples motivate and inspire readers to implement this interface in their own creations.

APPENDIX 1. WEB REFERENCES

Google a. Adding Time as a Fourth Dimension: Google, http://earth .google.com/outreach/tutorial_time.html (2 December 2011).

Google b. Google Earth API Developer's Guide: Google, http://code. google.com/apis/earth/documentation/index.html#using_the _google_earth_api (accessed 2 December 2011).

Google c. Google Earth plug-in: Google, http://www.google.com/ earth/explore/products/plugin.html (accessed 2 December 2011).

Google d. KML Developer Support: Google, http://groups.google .com/group/google-earth-browser-plugin/browse_thread/thread/ f3a5ea13984fd8e2/b7f9b959f05dfa18.

Google e. KmlScreenOverlay Interface Reference, http://code.google .com/apis/earth/documentation/reference/interface_kml_screen _overlay.html (accessed 29 November 2011).

Google f. Other KML Features » Creating Screen Overlays: Google, http:// code.google.com/apis/ajax/playground/?exp=earth#creating _screen_overlays (accessed 2 December 2011).

Google g. Point Placemarks—Draggable Placemark, http://code .google.com/apis/ajax/playground/?exp=earth#draggable_place- mark (accessed 28 November 2011).

Google h. Time and Animation: Google, http://code.google.com/apis/ kml/documentation/time.html (accessed 2 December 2011).

Google i. What is Google Earth? Google, http://earth.google .com/support/bin/answer.py?hl=en&answer=176145 (accessed 2 December 2011).

Hickson, I., HTML5, http://dev.w3.org/html5/spec/Overview.html (accessed 2 December 2011).

King, J., How to cover an IE windowed control (Select Box, ActiveX Object, etc.) with a DHTML layer, http://www.macridesweb .com/oltest/IframeShim.html (accessed 2 December 2011).

Microsoft, Hiding menu bars: Microsoft, http://msdn.microsoft.com/ en-us/library/aa511502.aspx#hidingMenus (accessed 2 December 2011).

Prototype: Prototype, http://www.prototypejs.org/ (accessed 1 December 2011).

Stenback, J., Le Hégaret, P., and Le Hors, A., Interface HTMLMenuElement, http://www.w3.org/TR/DOM-Level-2-HTML/html.html#ID -72509186 (accessed 2 December 2011).

THE HTML5 TEST, http://html5test.com/results.html (accessed 2 December 2011).

Whitmeyer, S.J., Google Earth Labs and Exercises, http://csmres.jmu.edu/Geollab/Whitmeyer/web/visuals/exercises.html (accessed 2 December 2011).

Williams, M., Replay Routes, http://maps.myosotissp.com/ (accessed 2 December 2011).

ACKNOWLEDGMENTS

This research was support in part by National Science Foundation (NSF) TUES 1022755, NSF GEO 1034643, and a 2010 Google Faculty Research Award. Any opinions, findings, and conclusions or recommendations expressed in this paper are those of the authors and do not necessarily reflect the views of the NSF or Google Inc. The comments of two anonymous reviewers greatly improved the manuscript.

REFERENCES CITED

Barne, M., and Finch, E.L., 2008, COLLADA—Digital Asset Schema Release 1.5.0: Sony Computer Entertainment Inc.

De Paor, D.G., and Williams, N.R., 2006, Solid modeling of moment tensor solutions and temporal aftershock sequences for the Kiholo Bay earthquake using Google Earth with a surface bump-out: Eos (Transactions, American Geophysical Union), Fall Meeting, v. 87, no. 52, abstract S53E-05.

De Paor, D.G., and Whitmeyer, S.J., 2011, Geological and Geophysical Modeling on Virtual Globes Using KML, COLLADA, and Javascript: Computers & Geosciences, v. 37, p. 100–110, doi:10.1016/j.cageo.2010.05.003.

De Paor, D.G., Whitmeyer, S.J., and Gobert, J., 2008, Emergent models for teaching geology and geophysics using Google Earth: Eos, Transactions, American Geophysical Union, v. 89, no. 53, abstract ED31A-0599.

Gobert, J., Wild, S.C., and Rossi, L., 2012, this volume, Examining the learning gains from Google Earth activities designed for non-geology students, *in* Whitmeyer, S.J., Bailey, J.E., De Paor, D.G., and Ornduff, T., eds., Google Earth and Virtual Visualizations in Geoscience Education and Research: Geological Society of America Special Paper 492, doi:10.1130/2012.2492(35).

MANUSCRIPT ACCEPTED BY THE SOCIETY 16 APRIL 2012

The Geological Society of America
Special Paper 492
2012

Geomorphological analysis of coastal depositional systems in SE Brazil aided by Google Earth coupled with the integration of chronological and sedimentological data by means of a Google Fusion Table

André Zular*
Carlos C.F. Guedes
Vinícius R. Mendes
André O. Sawakuchi
Paulo C.F. Giannini
Ana P.B. Tanaka
Milene Fornari
Daniel R. Nascimento Jr.
Departamento de Geologia Sedimentar e Ambiental, Instituto de Geociências, Universidade de São Paulo, Brazil

ABSTRACT

Although the evolution of Brazilian coastal depositional systems in the Quaternary has been studied in past decades, it is only in the last couple of years that it has been possible to incorporate the latest remote sensing databases available to help understand their development. In comparison to other freely accessible imagery, high-resolution images available on Google Earth are advantageous when undertaking local coastal analysis. In some instances, it is possible to differentiate geomorphologic features such as tidal deltas, beach ridges, and dunes. Also, the monitoring of small-scale features allows evaluation of the sensitivity of coastal zones to high-frequency and low-intensity processes. Thus, the downscaling description of coastal zones is now easily accessible, permitting the analysis of the extensive Brazilian coastal depositional systems. On the regional scale, a quick glance of a coastal setting may help frame the sedimentary characteristics of the depositional system.

Coastal areas in the States of Santa Catarina and São Paulo are taken into consideration in this study. These areas illustrate representative prograded barrier formations from Middle to Late Holocene with dunes formed at a later development stage. A comparison is made in the use of Google Earth and its historic images with aerial photographs and Landsat images. In the past, small-scale features of these regions were evaluated in aerial photographs, while regional features were studied

*andrezular@gmail.com

Zular, A., Guedes, C.C.F., Mendes, V.R., Sawakuchi, A.O., Giannini, P.C.F., Tanaka, A.P.B., Fornari, M., and Nascimento, D.R, Jr., 2012, Geomorphological analysis of coastal depositional systems in SE Brazil aided by Google Earth coupled with the integration of chronological and sedimentological data by means of a Google Fusion Table, *in* Whitmeyer, S.J., Bailey, J.E., De Paor, D.G., and Ornduff, T., eds., Google Earth and Virtual Visualizations in Geoscience Education and Research: Geological Society of America Special Paper 492, p. 113–125, doi:10.1130/2012.2492(08). For permission to copy, contact editing@geosociety.org. ©

by low-resolution satellite images. Accordingly, integration of these two products was difficult. In this work, we show that Google Earth facilitates the analysis as a whole. Furthermore, comparison of Google Earth images with aerial photographs from 1938 onward allowed the study of short-term migration and deflation of the dunefields probably accelerated in recent years by human interference. In addition, Keyhole Markup Language (KML) files were saved from Google Earth placemarks to facilitate georeferencing raster images on GIS programs. Finally, information available from previous local studies, such as luminescence dating, geomorphology of the costal system, grain size, heavy minerals, pollen, and carbon isotope analyses, was gathered into a Google Fusion Table database making data retrieval and parsing easily accessible. This database provides information that can be shared with other researchers and may be used to address important questions about the development of Brazil's coastal system in the past, present, and future.

1. INTRODUCTION

Sea level has fluctuated periodically throughout geological time, encroaching on or retreating across continental margins (Lambeck and Chappell, 2001). The present physiography of the Earth's coastal zones resulted from the last flooding of the continental shelf by the sea-level rise (Flandrian Transgression) after the Last Glacial Maximum 20,000 yr B.P. and the Mid-Holocene sea-level highstand at around 6000 yr B.P. These events favored the development of transgressive and regressive depositional systems tracts, which include deltas, estuaries, tidal channels, lagoons, barriers, and dunefields (Morton et al., 2000; Woodroffe, 2002; Orford et al., 2003; Lessa and Masselink, 2006; Switzer et al., 2010). Also, Late Quaternary climate changes have influenced the development of coastal depositional systems through shifts in their morphology and dynamics. While estuaries and barriers are the most common geomorphologic features in the northern portion of the SE coast of Brazil, lagoons and dunefields occur frequently in the southern portion. A substantial amount of data from different coastal depositional systems have been collected over the past several decades in these areas (Angulo and Lessa, 1997; Martinho et al., 2006; Giannini et al., 2007; Nascimento et al., 2008; Sawakuchi et al., 2008; Guedes et al., 2011a). These data encompass information provided by grain size, heavy minerals, pollen, carbon isotope analyses, and optically stimulated luminescence ages using single-aliquot regenerative-dose (OSL-SAR) protocol dating supported by stratigraphic and geomorphologic surveys. However, only in the last couple of years has it been possible to couple these data with the latest imagery available to better understand the coastal evolution of SE Brazil in the Late Pleistocene and Holocene.

The advent of publicly available tools such as Google Earth has made high-resolution remote sensing images readily accessible. This has proven to be extremely useful in analyzing coastal depositional systems, where the influence of sea-level changes, storms, littoral drift, and availability of sediments patterns may affect geomorphologic features at different time and space scales. Recently, hundreds of barrier islands were newly identified in a global survey which included the Gurupi Islands, along the equatorial coast of Brazil (Stutz and Pilkey, 2011). This survey was conducted with the aid of the Earthsat Geocover Landsat mosaic data and Google Earth (Stutz and Pilkey, 2011). Livingstone et al. (2010) constructed a map of eolian dune types based on visual interpretation of Google Earth images. Google Earth images were used to survey and map the destruction and flooding caused by Hurricane Katrina (Klemas, 2009).

The aim of this research is to show different applications of Google Earth in the study of local coastal landforms and depositional systems and to bring forward new methods in displaying and sharing this information by means of a Google Fusion Table. In the first part of this study we show applications of Google Earth image analyses from two locations along Brazil's SE coast (Fig. 1).

The second part of this study emphasizes the functionality of Google Fusion Tables, a cloud-based georeferenced data management and integration tool that may be coupled with the Google Earth platform. This is pioneering work in establishing new directions to store and visualize records and build user-friendly interfaces with data from coastal depositional systems in SE Brazil. Google Fusion Tables support for collaboration among multiple users and institutions unlocked new ways of sharing data that appeal to a broader audience of users who are less technically skilled. Also, Google Fusion Tables provide a way of assisting regional interpretation in the future by compiling data from local studies as shown from grain size, heavy minerals, and OSL-SAR ages from Ilha Comprida (Nascimento et al., 2008; Sawakuchi et al., 2008; Giannini et al., 2009; Guedes et al., 2011a, 2011b) in the State of São Paulo and São Francisco do Sul Island (Zular, 2011) in the State of Santa Catarina (Fig. 1).

2. REGIONAL SETTINGS

The Quaternary evolution of SE Brazilian coastal depositional systems produced a large variety of landforms. These systems evolved in wave-dominated coastal regions characterized by a microtidal regime (0.6–1.2 m) and by storm events all year round that are more intense in the winter season (Stech and Lorenzzetti, 1992). Northeast trade winds act during fair

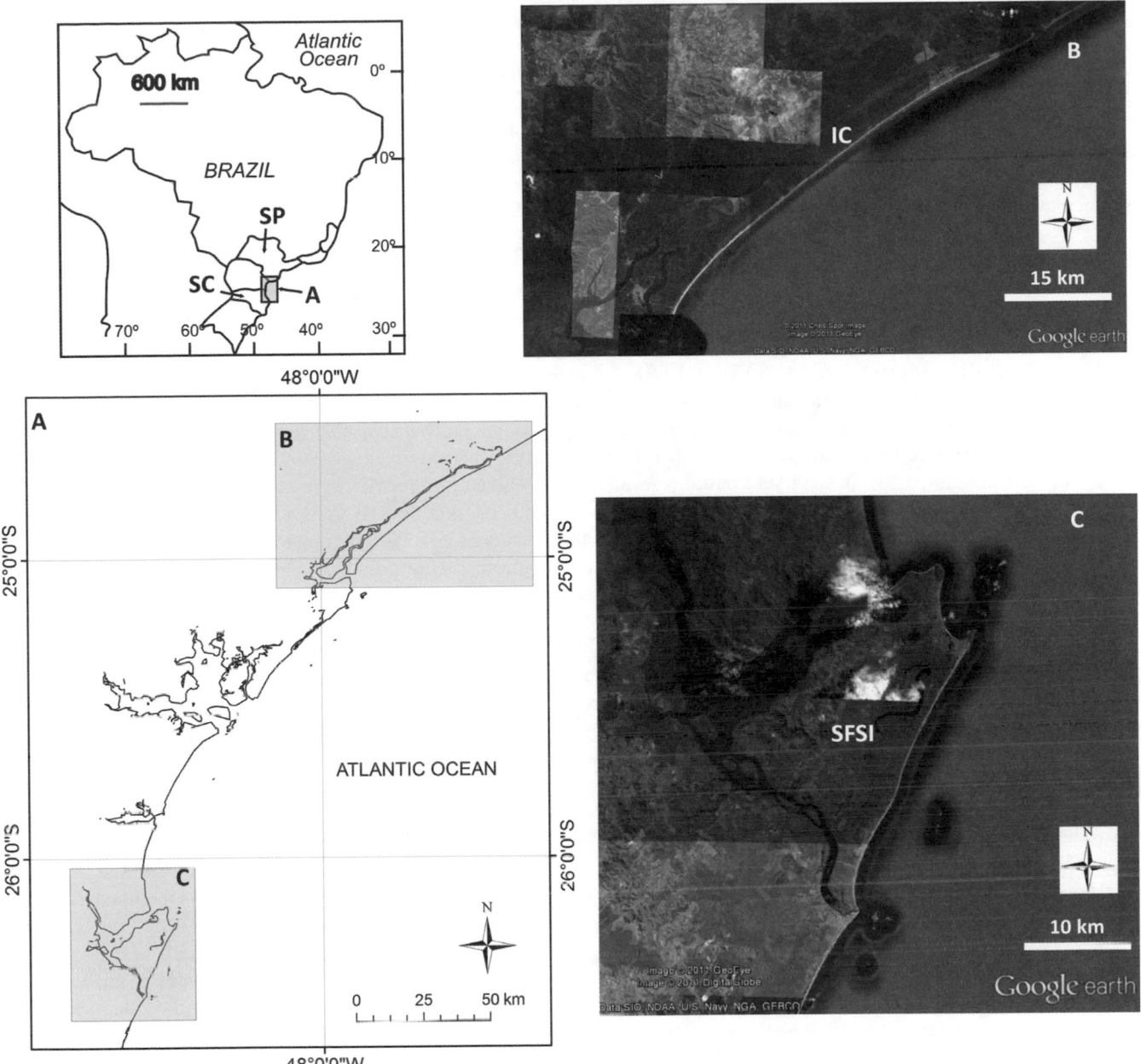

Figure 1. Maps and 2009 Google Earth images of Brazil's SE coast with locations discussed in the text. (A) States of São Paulo (SP) and Santa Catarina (SC). (B) Ilha Comprida (IC). (C) São Francisco do Sul Island (SFSI).

weather periods while the northward advance of polar air masses stimulates the action of S and SE winds (Stech and Lorenzzetti, 1992). Middle to Late Holocene barriers are conspicuous landforms in SE Brazilian regions (Giannini et al., 2009; Hesp et al., 2009). The barriers in the coastal zones of northern Santa Catarina and São Paulo States (Fig. 1) are formed mainly by beach ridge alignments, and many barriers present foredune and blowout alignments in their seaward portion (Sawakuchi et al., 2008; Angulo et al., 2009; Giannini et al., 2009; Hesp et al., 2009). A blowout is a saucer-, cup-, or trough-shaped depression or hollow formed by wind erosion on a preexisting sand deposit

(Hesp, 2002). Their formation may be related to natural and/or anthropogenic factors.

São Francisco do Sul Island is located on the northern coast of Santa Catarina State (Fig. 1). It covers an area of ~265 km^2 with an approximately 40-km-long coastline. The island was formed by coastal depositional systems that were active during the relative sea-level changes of the past 120,000 yr B.P. These depositional systems resulted from the trapping of sediment supply in the inner shelf and estuaries during relative sea-level highstands and by means of littoral drift. Holocene deposits with ages obtained by OSL-SAR protocol (Zular, 2011) located on the eastward side of the barrier consist primarily of prograded beach ridges and dune formations.

Ilha Comprida is a regressive barrier island located in the State of São Paulo in SE Brazil (Fig. 1). Sawakuchi et al. (2008) and Guedes et al. (2011a) obtained 36 luminescence ages (OSL-SAR) for the beach ridges, foredunes, and blowouts. This allowed the onset of a precise chronology sequencing of these landforms and calculation of rates of coastal progradation.

3. GOOGLE EARTH

3.1. Modeling

Although Landsat, SPOT-2, RADARSAT-1, CBERS-1, and other remote sensing images from satellites have been available in Brazil since 1973 and useful in the elaboration of multipurpose regional maps at scales 1:1,000,000, 1:500,000, or 1:250,000, they have been of limited use in detailed coastal research. In coastal studies when finer resolution is required, remote sensing images at scale fractions of 1:25,000, 1:20,000 to 1:45,000, or even 1:10,000 in special surveys can be advantageous. These scales can be obtained from aerial photography surveys. Coastal aerial photography surveys in coastal SE Brazil have been intermittent. Monochromatic surveys in the short wave spectrum have been conducted in the years of 1938 (1:20,000), 1957 (1:25,000), 1969 (1:20,000), and 1976/1978 (1:25,000; http://geo.pmf.sc.gov.br/historico.php). In past decades, these images were the primary resource used by researchers to further investigate small-range coastal environments despite the fact that most of them required exhaustive and time consuming rectification or even orthorectification. The process of acquiring high-resolution imagery by satellite or aerial photography and turning it into a continuously updated mosaic by Google Earth has given a better and easier option to work with for many coastal studies.

3.1.1. Geomorphological Divisions

Zular (2011) carried out a remote sensing survey in São Francisco do Sul Island based entirely on Google Earth images. These images were effective in framing the different depositional systems in the local and regional context, acquiring initial area measurements of dunefields or prograding terrains, identifying contrasting dune types, surveying terrain variation along elevation transects and in delineating geomorphologic divisions. Along the coastline, a narrow belt of 5–10-m-high foredune ridges associated with swales occurs. The foredunes are gradually replaced inland by 10–20-m-high vegetated parabolic or irregular dunes formed under the influence of SE winds whereas previously deposited beach ridges dominate further inland (Fig. 2).

3.1.2. Comparison to Other Remote Sensing Techniques

The synoptic views obtained from Google Earth of the SE coast of Brazil are a powerful tool for many aspects of geospatial analyses. Many coastal geomorphological features are identifiable at a quick glance. Figure 3 shows foredunes and blowouts from São Francisco do Sul Island displayed through different remote sensing techniques. Resolution provided by aerial photography and Landsat imagery was not sufficient to identify these coastal depositional systems in detail. Google Earth provided excellent elucidation at a scale fraction of 1:6000.

3.1.3. Georeferencing

The smallest scale available for most topographic maps in SE Brazil is the 1:50,000 fraction. In many instances geological maps are available only at a scale fraction of 1:1,000,000 or larger. In coastal studies small-scale geomorphological investigations are usually required, making it necessary to produce a map at the appropriate scale fraction. At São Francisco do Sul Island, this was done with ArcGIS 9.3 software associated with Google Earth images that were used for georeferencing. Google Earth images saved as JPG files at 1:6000 scale fraction with assigned placemarks were utilized. Placemarks were exported as KML files directly to ArcGIS software. Images with its associated geographic coordinates provided by placemarks, made georeferencing possible with the aid of ArcGIS 9.3 software (Fig. 4).

3.1.4. Blowout Evolution

Blowouts are common in the central and northern oceanic coast of São Francisco do Sul Island. Landmarks in past aerial photography surveys and still recognizable in the 2009 Google Earth image were used in georeferencing aerial photography, making appraisal of recent blowout evolution possible (Fig. 5).

3.1.5. Historical Images—Progradation and Erosion

Studies by Guedes et al. (2011a) showed variation in coastal progradation rates and sediment retention during the past 6000 yr B.P. in Ilha Comprida supported by OSL-SAR dating. Based on beach ridge alignments, six units of growth are identified with two growth directions, transverse and longitudinal. Rates of progradation with transverse growth vary from 0.13 to 4.6 m/yr. Rates of longitudinal growth to NE range from 5.2 to 30 m/yr. Alternation between longitudinal and transverse growth still occurs (Fig. 6).

4. GOOGLE FUSION TABLES

The ability to visualize spatial data distribution in maps or images in lieu of tables or graphs made geographic information systems (GIS) an essential tool for geoscience researchers, educators, and professionals alike. However, the high cost of programs, learning difficulties, and the lack of friendly interfaces have hindered its general acceptance. The emergence of new platforms such as Google Fusion Tables, providing several ways of visualizing and integrating data in a free, easy-to-use environment, appeals to a broader class of users. Google Fusion Tables is a georeferenced database tool available online that offers a platform for handling large amounts of information.

In addition to providing instant viewing with the aid of Google Earth or Google Maps, it also allows displaying data in tables or graphs. Google Fusion Tables offer tools to invite collaborators to view, edit, and merge data with other tables and handling online interaction between research groups. Overviews of the uses of Google Fusion Tables have been given by several authors (Gonzalez et al., 2010a, 2010b, 2010c; Lu et al., 2011) and detailed comparisons of GIS software and Google Earth are given by Goodchild (2008). Different examples of actual, ongoing uses of Google Fusion Tables can be found online at the Google Fusion Tables site (https://sites.google.com/site/fusiontablestalks/). In our study of coastal SE Brazil depositional systems, the use of Google Fusion Tables offered the possibility of regional integration and sharing of geomorphological, sedimentological (grain size and heavy minerals), chronological, and bibliographical data in a single, easily accessible platform. In order to explore the functionality of this tool, we utilized information on the sedimentary evolution of Ilha Comprida published by Sawakuchi et al. (2008) and Guedes et al. (2011a, 2011b) and recent data from São Francisco do Sul Island (Zular, 2011). São

Francisco do Sul Island and Ilha Comprida barriers are some 150 km apart, and in both locations there is an increase in the availability of sediments due to an intensification in longshore transport in SE Brazil starting at ~2000 yr B.P. to the present (Guedes et al., 2011b; Zular, 2011) with the onset of parabolic dune formation in São Francisco do Sul Island and intense progradation in Ilha Comprida. Google Fusion Tables coupled with Google Earth allowed easy comparison of OSL dating results, heavy minerals, and grain-size analyses from both locations.

Initially, structured data from a Microsoft Excel spreadsheet was uploaded into the Google Fusion Tables environment. Each record corresponded to a sediment sample. All pertinent data concerning different aspects of analyses and methods were included as a single field. In addition to these data, bibliographical references, photographs, overlays, and KML links were included in each record. The classification and arrangement of these data are shown in Table 1.

Google Fusion Tables may also be exported as KML to be viewed on Google Earth where all sample locations are placed as balloons. When the balloons are selected, relevant information,

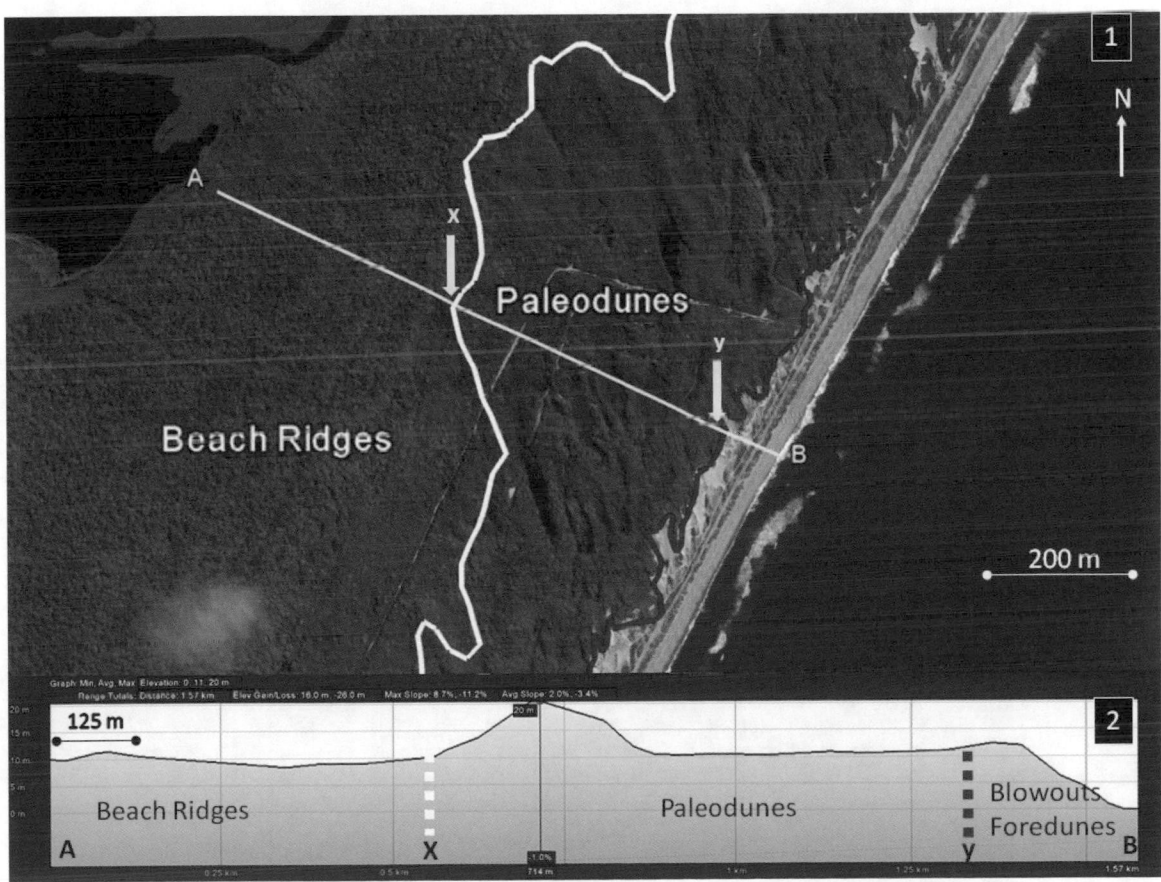

Figure 2. (1) 2009 Google Earth image showing different geomorphological units of São Francisco do Sul Island. Despite the cover of intense vegetation, well-preserved paleodune morphology is conspicuous and can be used to separate it from the beach ridges unit (white line). Red line shows boundary between the paleodunes and blowouts/foredunes units. (2) Elevation profile tool from Google Earth showing a slope between the beach ridges and paleodunes units and a decrease in altitude from the paleodunes and blowouts/foredunes units.

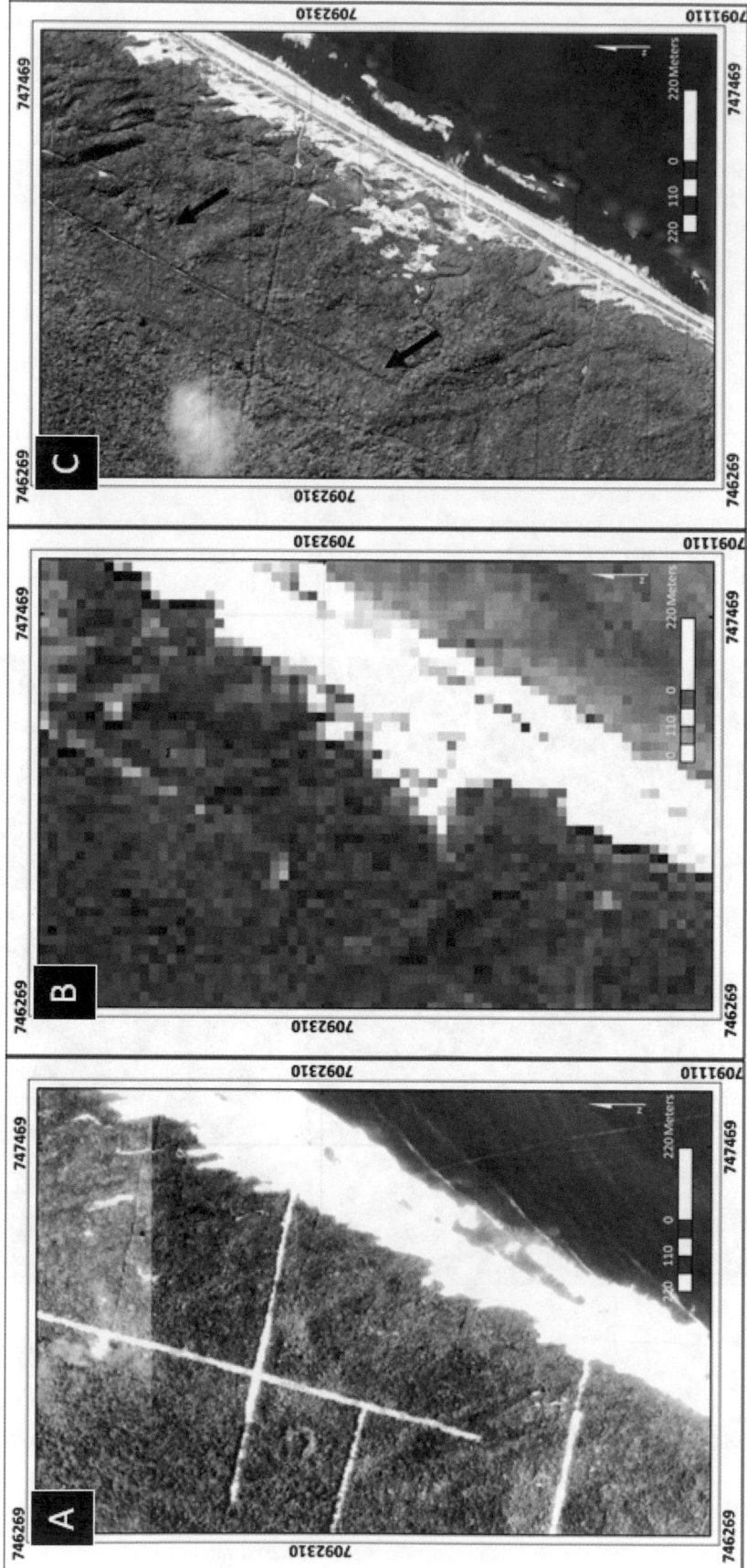

Figure 3. (A) 1957 aerial photography survey. (B) Landsat 7 image. (C) Scripps Institute of Oceanography, National Oceanic and Atmospheric Administration, U.S. Navy, National Geospatial Intelligence Agency, and General Bathymetric Chart of the Oceans. 2009 images available on Google Earth from São Francisco do Sul Island.

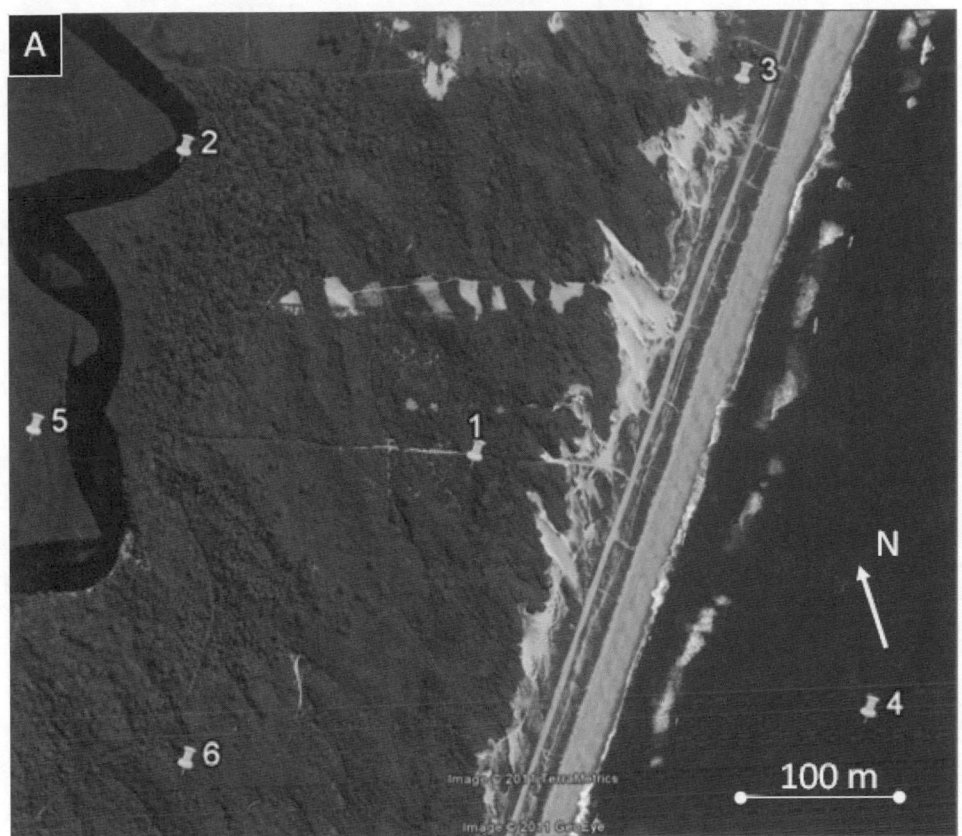

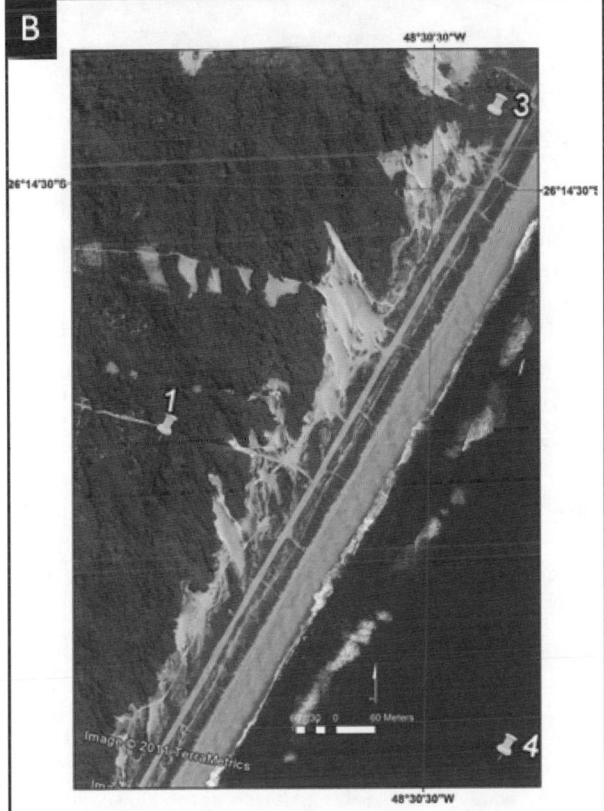

Figure 4. Placemarks on the 2009 Google Earth image (A) saved as KML files were used to georeference images with the aid of ArcGIS 9.3 software (B).

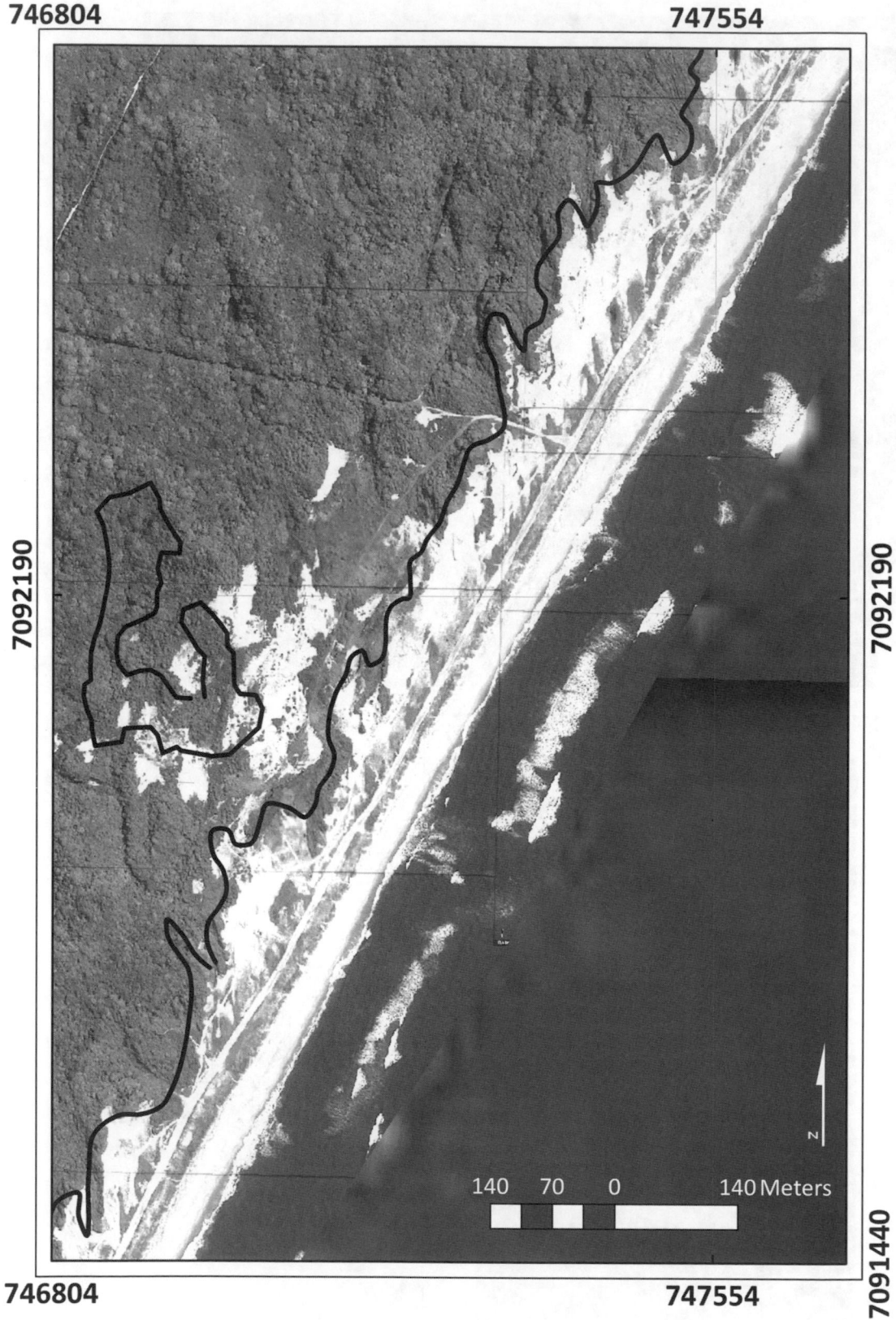

Figure 5. 2009 Google Earth image. Blowout evolution growth on the northern part of São Francisco do Sul Island. Solid line depicts blowout boundary obtained from the 1938 aerial photography survey. There has been little change in blowout coverage from the past 67 years except from the area located 150 m farther inland due to blowout migration and vegetation growth.

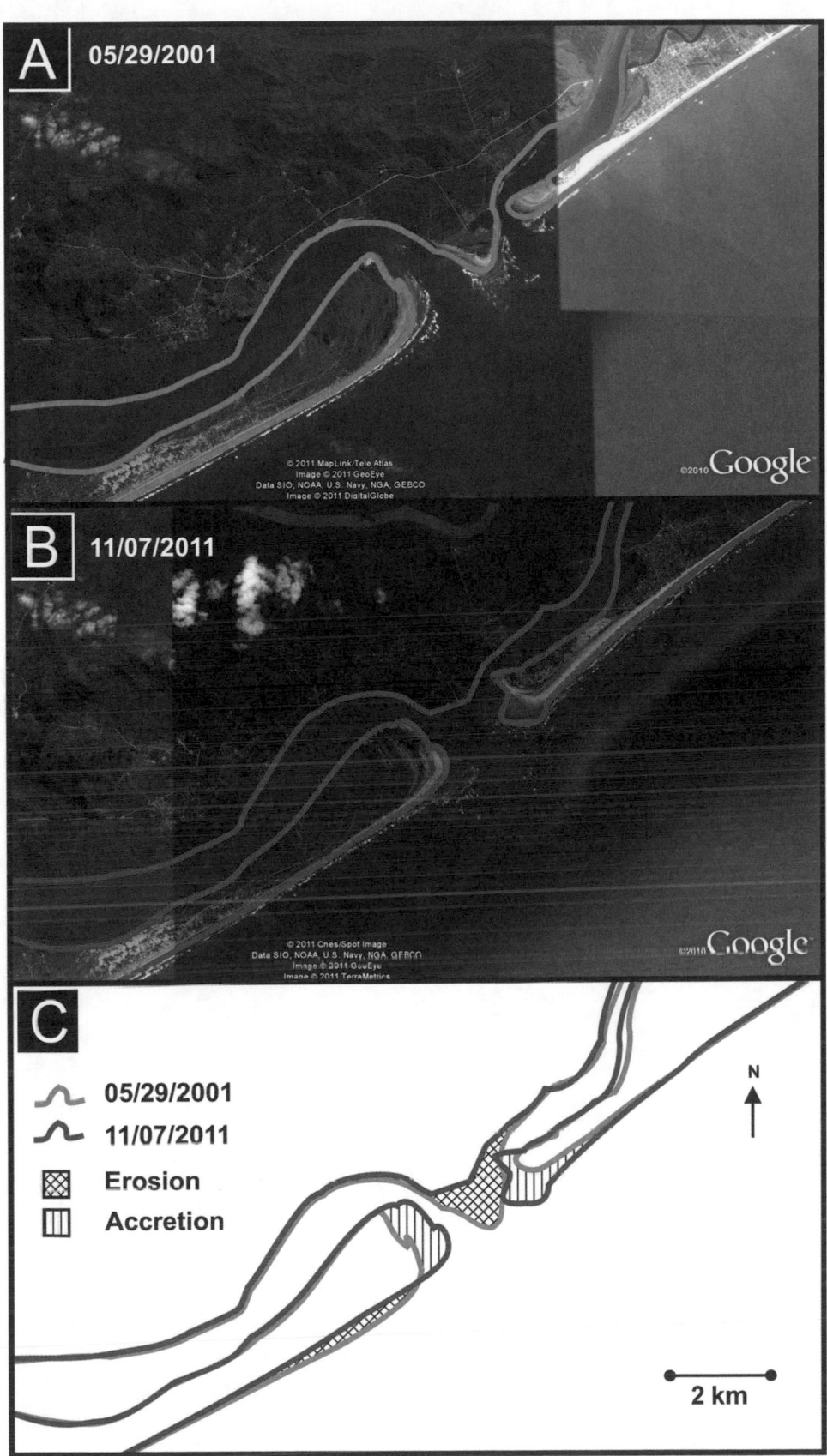

Figure 6. Google Earth historical images were utilized to elucidate sediment dynamics linking erosion and accretion in Ilha Comprida from 2001 to 2011. Ilha Comprida is the barrier island on the left. (A) 2001 Google Earth historical image. (B) 2011 Google Earth image. (C) Sediment budget between 2001 and 2011 in the Ilha Comprida area and vicinity.

TABLE 1. LIST OF FIELDS USED IN THE BRAZILIAN COASTAL LANDFORMS GOOGLE FUSION TABLE

Feature	Field	Remarks and Google Earth popup window contents
Geographic	UTM X, UTM Y, UTM Zone, Latitude, Longitude, Location	Local data
General	Date of sampling, Comments	Major comments regarding the local sample or study area
Bibliographic	Articles	URL link to relevant articles published on the study area
Geomorphologic	Distance from coastline (m), Elevation above RSL (m), Sample collection depth (m), Landform, Landform Unit	Data obtained by field trips or image analyses
Chronological and OSL parameters	Age determination method, Dose (mGy), Dose standard error (mGy), Gamma dose-rate (mGy/yr), Gamma dose-rate standard error (mGy/yr), Beta dose-rate (mGy/yr), Beta dose-rate standard error (mGy/yr), Cosmic dose-rate (mGy/yr), Cosmic dose-rate standard error (mGy/yr), Total dose-rate (mGy/yr), Total dose-rate standard error (mGy/yr), Age (yr), Age standard error (yr), OSL data analysis link	Age determination method, age and standard error shown on popup window. In the case of OSL method, URL link to data analysis table
Grain-size analyses (grain-size group)	Very Coarse Sand (%), Medium Sand (%), Fine Sand (%), Very Fine Sand (%), Mean Diameter (phi), Standard deviation (phi), Skewness (phi), Kurtosis (phi), Sorting Class, Comments	Bar chart displayed for the sand size class analysis. Statistical data shown as text.
Heavy minerals (heavy minerals group)	Opaques, Non-Opaques, Zircon, Tourmaline, Rutile, Hornblende, Epidote, Sillimanite, Kyanite, Staurolite, Apatite, Clinozoisite, Tremolite, Andaluzite, Actinolite, Enstatite, Monazite, Hypersthene, Diopside, Garnet, Topaz, Altered Minerals, % Opaques, % Non-Opaques, Total Non-Opaques, Comments, Heavy minerals table link	Total of heavy minerals counted is given in the "Total Non-Opaques" field. All individual numbers are absolute. Heavy minerals and % Opaques content shown in pie charts illustrating proportion. Heavy minerals table link.
Heavy minerals Indexes (heavy minerals group)	Zi-Ru, Tu-Hn, Tu-Zr	Text
Heavy minerals dedicated count (heavy minerals group)	Zi dedicated count Zi + Ru, Ru dedicated count Zi + Ru, Zi dedicated count Zi + Tu, Tu dedicated count Zi + Tu, Tu dedicated count Tu + Hn, Hn dedicated count Tu + Hn	URL link to server
Overlays	Overlay 1, Overlay 1 Description, Overlay 2, Overlay 2 Description, Overlay 3, Overlay 3 Description	URL link to server
Pictures	Image 1, Comments Image 1, Image 2, Comments Image 2, Image 3, Comments Image 3, Image 4, Comments Image 4	All pictures and comments displayed.
Evidences	Anthropological evidence, Paleontological evidence, Comments	Field or research evidences—Text

Note: RSL—relative sea level; OSL—optically stimulated luminescence.

graphs, links, and pictures are displayed in a customized window (Fig. 7). Customization was done with the help of KML language in the Google Fusion Tables environment (Appendix[1]). The framework of KML has been documented by Wernecke (2009). The KML code structure consisted of snippets of fixed text combined with dynamic elements provided by different fields from the Google Fusion Tables. Pop-up window displays of size class distribution and classification of sands into sorting classes based on standard deviations, heavy minerals content and indexes data as pie and bar charts were coded in KML. In the pop-up windows, links to overlays such as an interpolation map of the rutile-

[1]GSA Data Repository item 2012300, Supplementary KML code, is available at http://www.geosociety.org/pubs/ft2012.htm, or on request from editing@geosociety.org.

zircon index for the Island of São Francisco do Sul created with the help of ArcGIS 9.3 software may also be displayed (Fig. 8).

The Brazilian Coastal Landforms Google Fusion Table can be viewed at the site: https://www.google.com/fusiontables/DataSource?snapid=S467208frrJ. No sign-in is required.

5. CONCLUSIONS

This paper addresses how different uses of Google Earth can be applied to characterize coastal depositional systems and landforms in the SE Brazilian coast. We have illustrated features and methods using some examples that highlighted the role played by Google Earth in our coastal studies. Powerful visualization, interactivity, regional context and ease of use were important aspects of Google Earth utilized in our research. Moreover, as our

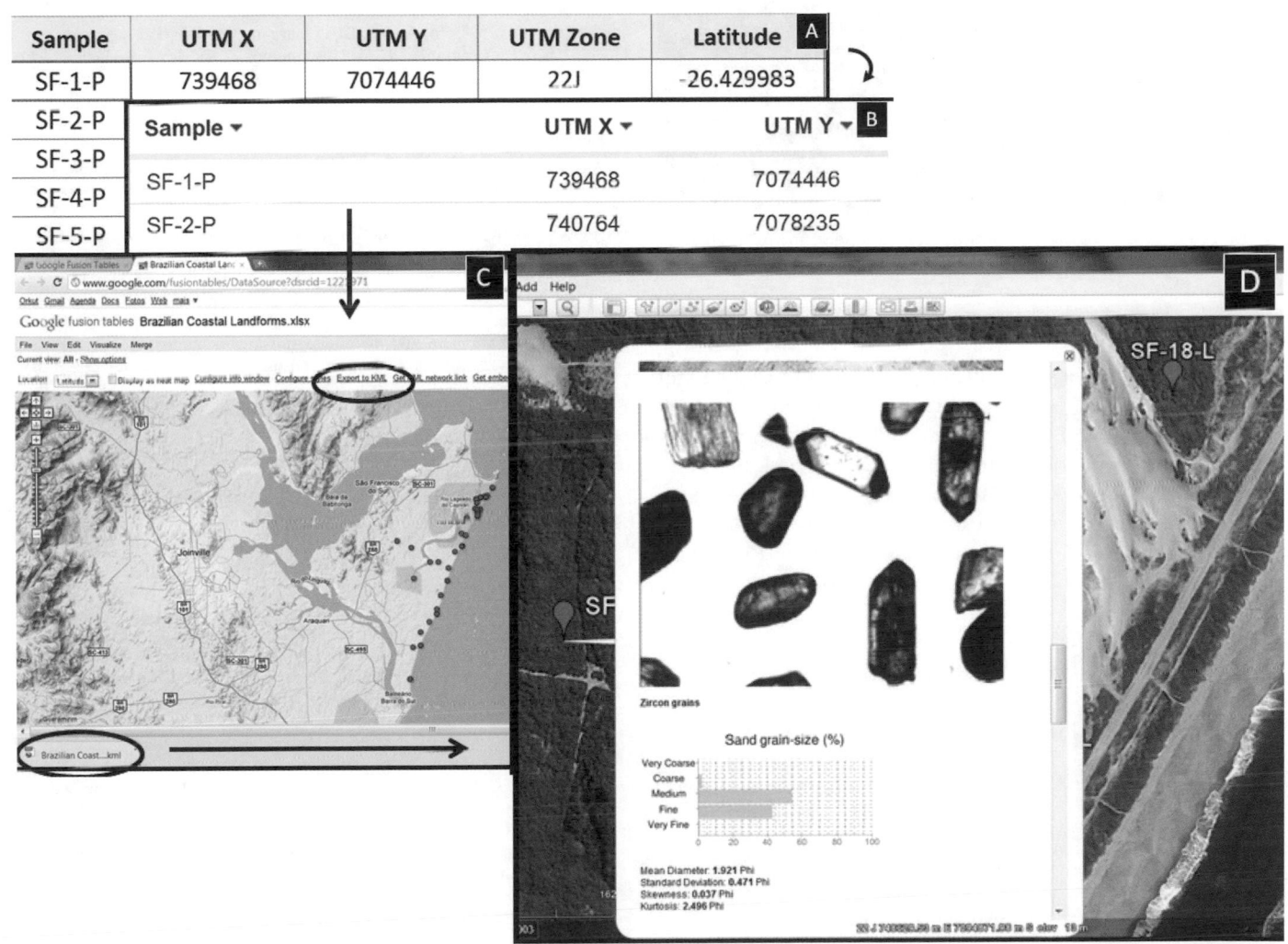

Figure 7. Screenshots illustrating different steps described in the text. (A) Excel spreadsheet with São Francisco do Sul Island and Ilha Comprida data. (B) Google Fusion Tables uploaded from the Excel spreadsheet. (C) Visualization with Google Maps in the Google Fusion Tables platform. Circle shows the Export to KML option to transfer data to Google Earth. (D) Sample site SF-9-L in São Francisco do Sul Island with customized pop-up window with links, sample site pictures and sedimentological and heavy minerals data displayed as graphs and charts.

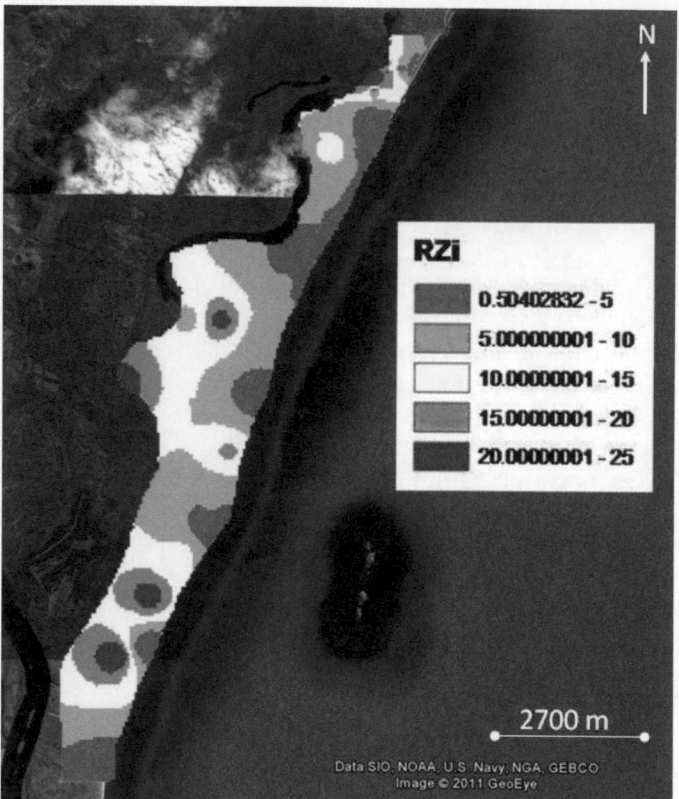

Figure 8. Rutile-zircon index interpolation uploaded from ArcGIS 9.3 software seen as an overlay in São Francisco do Sul Island Google Earth image.

examples suggest, there is a wide range of uses of Google Earth than can be enhanced when combined with a GIS application.

The integration of Google Earth and Google Fusion Tables brings forward some of the analytic, modeling, and inferential functionalities of typical GIS and database software to an easier to use and robust environment that can be shared by different collaborators for larger scale regional interpretation. This in turn can contribute to elucidating paleoclimate reconstructions, sediment provenance, and sea-level changes, among other studies. These tools promote easy dissemination of knowledge to fellow colleagues and the interested general public around the world.

REFERENCES CITED

Angulo, R.J., and Lessa, G.C., 1997, The Brazilian sea-level curves: a critical review with emphasis on the curves from the Paranaguá and Cananéia regions: Marine Geology, v. 140, p. 141–166, doi:10.1016/S0025-3227(97)00015-7.

Angulo, R.J., Lessa, G.C., and Souza, M.C., 2009. The Holocene barrier system of Paranaguá and northern Santa Catarina coasts, southern Brazil, *in* Dillenburg, S.R., and Hesp, P.A., eds., Geology and Geomorphology of Holocene Coastal Barriers of Brazil: Berlin–Heidelberg, Springer, p. 135–176.

Giannini, P.C.F., Sawakuchi, A.O., Martinho, C.T., and Tatumi, S.H., 2007, Eolian depositional episodes controlled by Late Quaternary relative sea level changes on the Imbituba-Laguna coast (southern Brazil): Marine Geology, v. 237, p. 143–168, doi:10.1016/j.margeo.2006.10.027.

Giannini, P.C.F., Guedes, C.C.F., Nascimento, D.R., Jr., Tanaka, A.P.B., Angulo, R.J., Assine, M.L., and Souza, M.C., 2009. Morphology and sed-

imentology of Ilha Comprida, southern São Paulo coast, *in* Dillenburg, S.R., and Hesp, P.A., eds., Geology and Geomorphology of Holocene Coastal Barriers of Brazil: Berlin–Heidelberg, Springer, p. 177–224.

Gonzalez, H., Halevy, A., Jensen, C.S., Langen, A., Madhavan, J., Shapley, R., and Shen, W., 2010a, Google fusion tables: data management, integration and collaboration in the cloud, *in* Proceedings of the 1st Association for Computing Machinery (ACM) Symposium on Cloud Computing, Indianapolis, Indiana, USA, 10–11 June 2010, p. 175–180.

Gonzalez, H., Halevy, A.Y., Jensen, C.S., Langen, A., Madhavan, J., Shapley, R., Shen, W., and Goldberg-Kidon, J., 2010b, Google Fusion Tables: Web-Centered Data Management and Collaboration, *in* Proceedings of the Association for Computing Machinery's Special Interest Group on Management of Data International Conference on the Management of Data, Indianapolis, Indiana, USA, 6–11 June 2010, Association for Computing Machinery, p. 1061–1066.

Gonzalez, H., Halevy, A.Y., Langen, A., Madhavan, J., McChesney, R., Shapley, R., Shen, W., and Goldberg-Kidon, J., 2010c, Socializing Data with Google Fusion Tables: IEEE Computer Society, Bulletin of the Technical Committee on Data Engineering, v. 33, no. 3, p. 25–32.

Goodchild, M.F., 2008, The use cases of digital earth: International Journal of Digital Earth, v. 1, p. 31–42, doi:10.1080/17538940701782528.

Guedes, C.C.F., Giannini, P.C.F., Sawakuchi, A.O., DeWitt, R., Nascimento, D.R., Jr., Aguiar, V.A.P., and Rossi, M.G., 2011a, Determination of controls on Holocene barrier progradation through application of OSL dating: The Ilha Comprida Barrier example, Southeastern Brazil: Marine Geology, v. 285, p. 1–16, doi:10.1016/j.margeo.2011.04.005.

Guedes, C.C.F., Giannini, P.C.F., Nascimento, D.R., Jr., Sawakuchi, A.O., Tanaka, A.P.B., and Rossi, M.G., 2011b, Controls of heavy minerals and grain size in a Holocene regressive barrier (Ilha Comprida, southeastern Brazil): Journal of South American Earth Sciences, v. 31, p. 110–123, doi:10.1016/j.jsames.2010.07.007.

Hesp, P.A., 2002, Foredunes and blowouts: initiation, geomorphology and dynamics: Geomorphology, v. 48, p. 245–268, doi:10.1016/S0169-555X(02)00184-8.

Hesp, P.A., Giannini, P.C.F., Martinho, C.T., Miot da Silva, G., and Asp Neto, N.E., 2009. The Holocene barrier system of the Santa Catarina coast, Southern Brazil, *in* Dillenburg, S.R., and Hesp, P.A., eds., Geology and Geomorphology of Holocene Coastal Barriers of Brazil: Berlin–Heidelberg, Springer, p. 93–134.

Klemas, V.V., 2009, The role of remote sensing in predicting and assessing coastal storm impacts: Journal of Coastal Research, v. 25, no. 6, p. 1264–1275, doi:10.2112/08-1146.1.

Lambeck, K., and Chappell, J., 2001, Sea Level Change through the Last Glacial Cycle: Science, v. 292, p. 679–686, doi:10.1126/science.1059549.

Lessa, G., and Masselink, G., 2006, Evidence of a mid-Holocene sea level highstand from the sedimentary record of a macrotidal barrier and paleoestuary system in northwestern Australia: Journal of Coastal Research, v. 22, no. 1, p. 100–112, doi:10.2112/05A-0009.1.

Livingstone, I., Bristow, I.C., Bryant, R.G., Bullard, J., White, K., Giles, F.S., Wiggs, G.F.S., Baas, A.C.W., Bateman, M.D., and Thomas, D.S.G., 2010, The Namib Sand Sea digital database of aeolian dunes and key forcing variables: Aeolian Research, v. 2, p. 93–104, doi:10.1016/j.aeolia.2010.08.001.

Lu, M., Agrawal, D., Dai, B.T., and Tung, A.K.H., 2011, Schema-as-you-go: On probabilistic tagging and querying of wide tables, *in* Proceedings of the Association for Computing Machinery's Special Interest Group on Management of Data, 12–16 June 2011, Athens, Greece, p. 181–192.

Martinho, C.T., Giannini, P.C.F., Sawakuchi, A.O., and Hesp, P.A., 2006, Morphological and depositional facies of transgressive dunefields of the Imbituba-Jaguaruna region, Santa Catarina State, Southern Brazil: Journal of Coastal Research, SI 39 (Proceedings of the 8th International Coastal Symposium), p. 673– 677.

Morton, R.A., Paine, J.G., and Blum, M.D., 2000, Responses of stable bay-margin and barrier-island systems to Holocene sea-level highstands, Western Gulf of Mexico: Journal of Sedimentary Research, v. 70, p. 478–490, doi:10.1306/2DC40921-0E47-11D7-8643000102C1865D.

Nascimento, D.R., Jr., Giannini, P.C.F., Tanaka, A.P.B., and Guedes, C.C.F., 2008, Mudanças morfologicas da extremidade NE da Ilha Comprida (SP) nos últimos dois séculos: Geologia-USP-Série Científica, v. 8, no. 1, p. 25–39.

Orford, J.D., Murdy, J.M., and Wintle, A.G., 2003, Prograded Holocene beach ridges with superimposed dunes in north-east Ireland: mechanisms and timescales of fine and coarse beach sediment decoupling and deposition: Marine Geology, v. 194, p. 47–64, doi:10.1016/S0025-3227(02)00698-9.

Sawakuchi, A.O., Kalchgruber, R., Gianinni, P.C.F., Nascimento, D.R., Jr., Guedes, C.C.F., and Umisedo, N.K., 2008, The development of blowouts and foredunes in the Ilha Comprida barrier (Southeastern Brazil): the influence of Late Holocene climate changes on coastal sedimentation: Quaternary Science Reviews, v. 27, p. 2076–2090, doi:10.1016/j.quascirev.2008.08.020.

Stech, J.L., and Lorenzzetti, J.A., 1992, The response of the South Brazil Bight to the passage of wintertime cold fronts: Journal of Geophysical Research, v. 97, C6, p. 9507–9520, doi:10.1029/92JC00486.

Stutz, M.L., and Pilkey, O.H., 2011, Open-Ocean Barrier Islands: Global Influence of Climatic, Oceanographic, and Depositional Settings: Journal of Coastal Research, v. 27, p. 207–222.

Switzer, A.D., Sloss, C.R., Jones, B.G., and Bristow, C.S., 2010, Geomorphic evidence for mid-late Holocene higher sea level from southeastern Australia: Quaternary International, v. 221, p. 13–22, doi:10.1016/j.quaint.2009.06.035.

Wernecke, J., 2009, The KML Handbook: Geographic Visualization for the Web: Upper Saddle River, New Jersey, Addison-Wesley, 368 p.

Woodroffe, C.D., 2002, Coasts, Form, Process and Evolution: Cambridge University Press, 623 p.

Zular, A., 2011, Sedimentologia e cronologia por luminescência da Ilha de São Francisco do Sul (SC): Considerações sobre a evolução holocênica de barreiras arenosas da costa sul e sudeste do Brasil [Master's thesis]: Universidade de São Paulo, Brazil, 93 p.

MANUSCRIPT ACCEPTED BY THE SOCIETY 16 APRIL 2012

The Geological Society of America
Special Paper 492
2012

Visualization of spatial and temporal trends in Louisiana water usage using Google Fusion Tables

Jeffrey A. Nunn*

Department of Geology & Geophysics, Louisiana State University, Baton Rouge, Louisiana, USA

Lauren Bentley

Department of Petroleum Engineering, Louisiana State University, Baton Rouge, Louisiana, USA

ABSTRACT

Data on ground-water and surface-water use in Louisiana are available online in tabular form from the U.S. Geological Survey. Data are categorized by parish and by type of usage (e.g., public supply, irrigation, industry, and power generation) from 1960 to 2005. Water usage in Louisiana has complicated spatial and temporal trends which are not readily apparent in static tables. For example, ground-water usage varies from more than 200 million gallons a day in some rice farming parishes to less than 40,000 gallons a day in coastal parishes where most ground water is not potable. Baton Rouge Parish uses mostly ground water even though it is on the Mississippi River because the ground water is high quality. Orleans Parish uses almost exclusively river water because most ground water is brackish. Significant temporal trends include the rapid rise of water use for power generation since 1960, a drop in overall water usage during the economic downturn following the oil bust of the 1980s, and the switch from surface to ground water in some areas due to decadal droughts or pollution of surface water. Google Fusion Tables represent a rapid and effective way to visualize water usage trends for K–16 education, research, and public policy. Using Google API (application programming interface), we have developed intensity maps that illustrate quantity and category of both surface-water and ground-water use by parish. Each parish within an intensity map has a pop-up bar chart that shows total water usage from 1960 to 2005 in five-year increments. We also have included versions of intensity maps that have pop-up pie/line charts that show the distribution of usage in each parish among public supply, agriculture, industry, and power generation. The dynamic feature of Fusion Tables allows students, researchers, and policy makers to clearly see temporal trends as well as illustrate connections among water usage and other factors. For examples, most students falsely assume that the steady rise in water use for public supply is related to population increase whereas it is primarily due to a substantial increase in per capita usage. These tables will be made available on the web.

*gljeff@lsu.edu

Nunn, J.A., and Bentley, L., 2012, Visualization of spatial and temporal trends in Louisiana water usage using Google Fusion Tables, *in* Whitmeyer, S.J., Bailey, J.E., De Paor, D.G., and Ornduff, T., eds., Google Earth and Virtual Visualizations in Geoscience Education and Research: Geological Society of America Special Paper 492, p. 127–138, doi:10.1130/2012.2492(09). For permission to copy, contact editing@geosociety.org. © 2012 The Geological Society of America. All rights reserved.

INTRODUCTION

Ground-water and surface-water withdrawal and use information for Louisiana on a five-year basis from 1960 to 2000 is available online from the U.S. Geological Survey. Data are categorized by parish and by type of usage (e.g., public supply, irrigation, industry, and power generation). While this information is valuable, it is not easy for many potential users to see spatial and/or temporal trends in water use or interrelations among users or outside factors such as population changes or decadal changes in rainfall in static tables.

As part of a Geological Society of America Penrose Conference held at Google's headquarters in Mountain View, California, in January 2011, Google engineers provided training to participants on the construction and potential use of Fusion Tables. Fusion Tables contain similar information to a standard table such as parish or county, year, and water use broken down by category. In addition, Fusion Tables have several other important features. The table can easily be made public. Data provided by a user can be merged with KML (Keyhole Markup Language) information from Google so that tabular data can be converted into intensity maps where viewers can see spatial variations (e.g., water use as color fields on a parish map of Louisiana). Fusion Tables can have custom pop-up info windows containing bar, pie, or line charts or text that contain information on water use within a parish. Finally, Fusion Tables can create embedded links so that intensity maps can be attached to web pages or blogs for easy access to a global audience.

In this study, we show how water use data available from the U.S. Geological Survey and polygon information for each parish in Louisiana from Google.com can be combined into Fusion Tables with associated intensity maps and pop-up bar, pie, and line charts and dynamic tables to make this information assessable to a broad audience of researchers, students, and layman for research, K–16 education, and public policy decisions.

DATA AND METHODS

Water use data in Louisiana has been collected and published by the U.S. Geological Survey in cooperation with the Louisiana Department of Transportation and Development. Water withdrawal and use information on a five-year basis from 1960 to 2000 is available online at http://la.water.usgs.gov/WaterUse/default.htm. Data are available in tabular form for all 64 Louisi-

ana parishes as total withdrawals, ground-water withdrawals, and surface-water withdrawals. Withdrawals are broken down into eight categories of use: aquaculture, general irrigation, industrial, livestock, power generation, public supply, rice irrigation, and rural domestic. The U.S. Geological Survey also provided similar information on 2005 withdrawals (J.K. Lovelace, 2011, personal commun.); 2010 data will be added when it becomes available. This site also contains downloadable pdf copies of Department of Transportation and Development Special Reports on water use in Louisiana for each five-year period (e.g., Sargent, 2007) and links to water use in the United States.

Several different Fusion Tables were created as part of this study. In each case, the rows are Louisiana parishes. The columns are parish name plus various categories of water use depending on the table. For example, the Fusion Table used to create Figure 2 has columns that represent the ground-water use in 1960 through 2005 in five-year increments plus "bin value" and "geometry" columns. The "bin value" column was created in this study to provide information used in either the map or the associated charts (e.g., maximum value for the y scale of the chart). The use of "bin value" is discussed in more detail in the results section. The "geometry" column contains the polygon information used to draw each parish in Louisiana on the map (Fig. 2). This information was acquired from Google by merging our Fusion Tables with its Fusion Table that has the outline of every county in the United States. Merging tables is done by having one column in common from each table that is used to correlate them. In this case, the parish name column was used for correlation. The KML file in each cell of the "geometry" column is a string of latitude and longitude values plus the following text string at both the beginning and end of the file: <Polygon><outerBoundaryIs> <LinearRing><coordinates>. Polygons outlining Louisiana parishes on Google maps are available from Google at http://www .google.com/fusiontables/DataSource?dsrcid=210217. The original source for the data is the U.S. Census Bureau.

Figure 1 shows a screen capture of the upper portion of the Google Fusion Table interface when it is in map mode. The lower portion of the interface is similar to Google maps. A few of the features used in this study are briefly described. A full discussion of Google Fusion Tables is available at http://www .google.com/fusiontables/public/tour/index.html. The "Visualize" pull-down menu allows the user to select the Fusion Table or create a map view, pie chart, line chart, etc., from all or part of the Fusion Table. Note that geometry must be selected in the

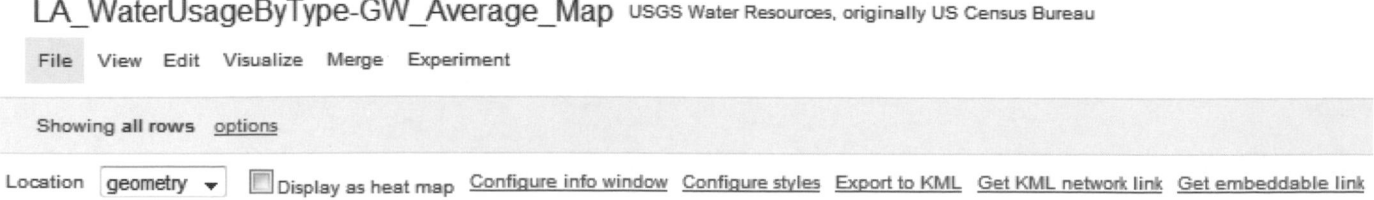

Figure 1. Screen dump of the upper portion of a Google Fusion Table in map visualization mode.

location box in order to fill polygons for maps similar to Figures 2–5. The "Merge" button brings up a window that allows the user to merge the Fusion Table with another table. This window was used to merge our water use data with polygon information for each parish. The "Configure info window" button allows the user to enter HTML that can produce pop-up bar, pie, or line charts for each parish (Figs. 3–7). The "Configure styles" button brings up a window that allows the user to specify background colors for polygon fills as well as line color and line thickness for the map. The buckets and gradient methods of polygon fill are used in this study and are discussed further in the results section below. In addition, the "Configure styles" interface allows the user to select which column is used to determine the polygon fill (e.g., which year or which type of water use). Finally, the "Get embeddable link" button provides a HTML link to the map that can be inserted into a blog or web page.

HTML code used to construct bar, pie, and line charts in this study (Figs. 3–7) are in Appendix 1. Important features for each type of chart are discussed in the results section. Additional information on Google charts as well as examples, tutorials, and a chart wizard are available at http://code.google.com/apis/chart/.

RESULTS

A series of maps and charts created with Google's Fusion Tables illustrating various aspects of water use in Louisiana are presented here. They are intended to be illustrative rather than exhaustive. In most instances, ground water and surface water are treated separately for two reasons. First, as is the case in most states, surface-water use is much higher than ground-water use, so combining results tends to obscure ground-water trends. Second, both the geographic distribution and type of water use in Louisiana is dramatically different between ground water and surface water. Figures are screen captures including information windows with bar, pie, and line charts. These maps and charts are available at http://www.geol.lsu.edu/LaWaterUse or in the GSA Data Repository[1].

The first set of maps shows ground-water and surface-water use by parish for each five-year increment from 1960 to 2005. Figure 2 shows ground-water use in 2005 as a bucket map. Inset in the upper right is the polygon background color pop up window which allows the user to select type of color fill (e.g., bucket or gradient), which column in the spreadsheet is mapped, and the colors and value ranges used. In Figure 2, only shades of blue are used. Lighter blue indicates low use and darker blue indicates high use. Each shade of blue represents a 20 million gallon/day increment. Figure 2 clearly shows that the heaviest ground-water use is restricted to parishes in southwest Louisiana plus East Baton Rouge Parish in the south-central part of the state. Calcasieu and Acadia are the top two parishes averaging

between 143 and 145 million gallons a day from 1960 to 2005. Ground-water use is lowest in the northwestern and southeast portions of the state. Plaquemines Parish has the lowest use with an average of 0.3 million gallons a day from 1960 to 2005 primarily because ground water in that part of Louisiana is brackish (Prakken, 2007).

Figure 3 shows surface-water use in 2005 as a gradient map. In a gradient map, minimum and maximum values are chosen as well as which column to use in assigning a color to each polygon (see inset in upper right of Fig. 3). In Figure 3, lighter red indicates low use and darker red indicates high use. Heavy surface-water use is also concentrated in a relatively small number of parishes. Most of these parishes are along the lower Mississippi river (Fig. 3). There are some outlier parishes in other parts of the state (Rapides, Ouachita, and Calcasieu). St. Charles Parish used more than 3 billion gallons of surface water a day for power generation or 35% of the state total in 2005, which is almost three times as much as the next largest user, Jefferson Parish, with 1.13 billion gallons a day. Despite the overall large numbers, some parishes used just a few hundred thousand gallons of surface water a day. In most instances, these parishes do not contain a large river and/or have a small population. For parishes within the Atchafalaya river basin to the south of the lower Mississippi River, the U.S. Army Corps of Engineers built the Old River Control Structure in 1963 to force most of the water down the Mississippi river to maintain New Orleans and Baton Rouge as deep water ports. Thus, the Atchafalaya and its tributaries are much smaller now than they were in 1960 (McPhee, 1987). Figure 3 contains an embedded chart for surface-water use in Jefferson Parish that illustrates the dramatic increase in water use in 1975 associated with power generation. HTML for this chart is contained in Appendix 1. It allows the user to specify the size of the chart as well as scales, colors, and line thicknesses. Columns from the Fusion Table used in the chart are enclosed by curly brackets ({}). In addition to water use in each five-year period, we created a column called "bin values" which was used in the HTML to control the maximum value used in the y-axis scale. This was necessary because of the wide variation in the amount of water used from parish to parish. In the pop-up chart for Jefferson Parish, the "bin value" is 1600.

Another way of analyzing water use is to look at the category of use. As noted above, the U.S. Geological Survey subdivided water use into eight categories: rural domestic, public supply, industrial, power generation, livestock, general irrigation, rice irrigation, and aquaculture. Figures 4 and 5 are bucket maps of Louisiana showing each parish color-coded by which category is the largest use of ground water and surface water in the parish, respectively. While a few parishes may have their water use partitioned among two or more significant uses, in most cases, the primary use dominates. For ground water, the biggest use category exceeds all other uses combined in 49 of 64 parishes and is at least 40% of total water use in all but six parishes. For surface water, the biggest use category exceeds all other uses combined in 54 out of 64 parishes and is at least 40% of total water use in

[1]GSA Data Repository item 2012301, Louisiana water use maps and charts, is available at http://www.geosociety.org/pubs/ft2012.htm, or on request from editing@geosociety.org.

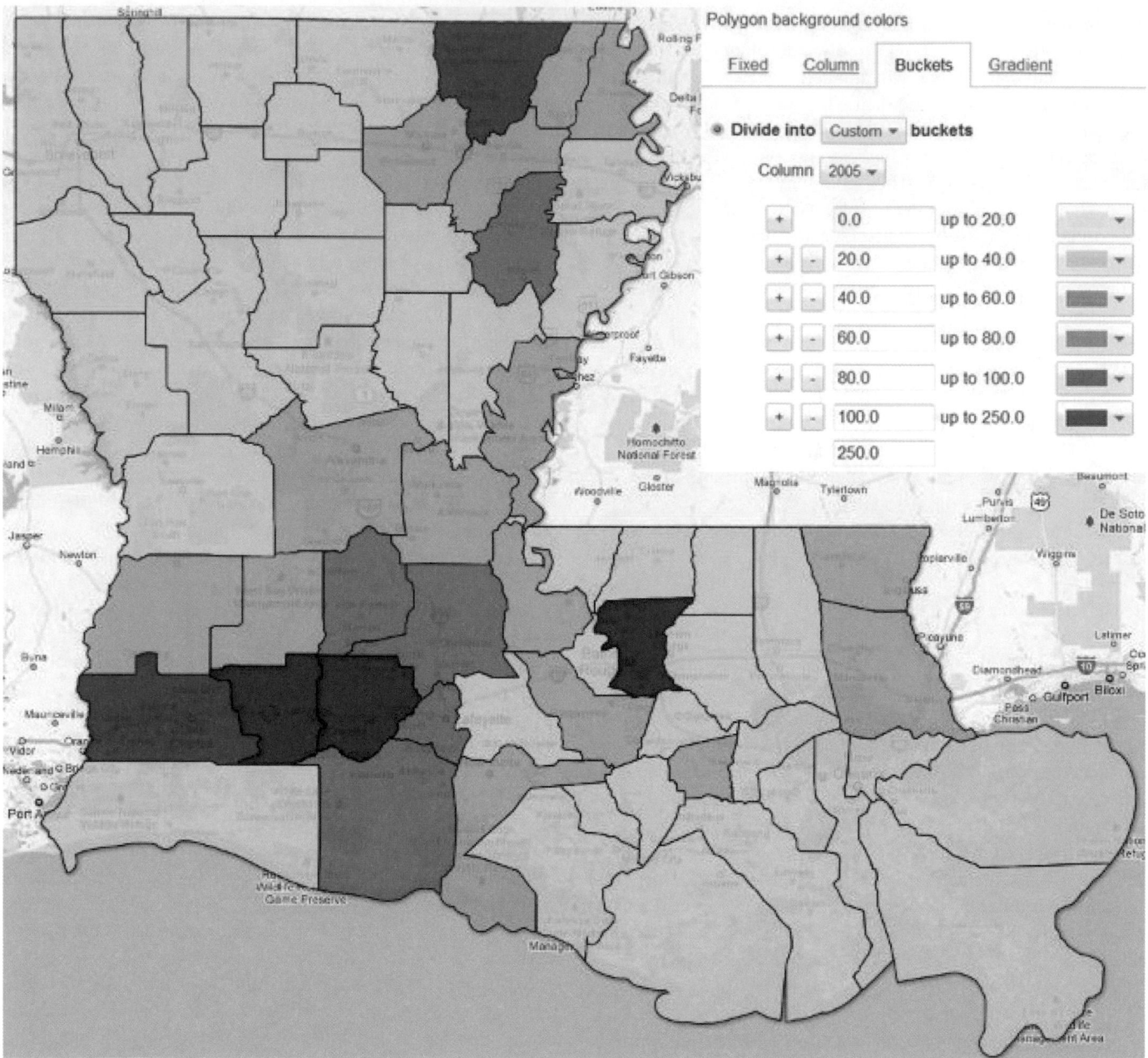

Figure 2. Bucket map showing Louisiana ground-water use in 2005 by parish. Range is from 0 to 150 million gallons per day with a bucket interval of 20 million gallons per day (see inset). Darker blue means higher ground-water use. Parishes with 100 million gallons per day or more are dark blue.

all but four parishes. Thus, these maps give a fairly robust characterization of how water is used in each parish. Bucket maps were created using a column in the Fusion Table called "MaxDigit." The MaxDigit column contains an integer value associated with each type of use (e.g., if industrial was the largest use of water that parish would be assigned a MaxDigit of five). The bucket map contains eight buckets, one bucket for each category of use. In the maps in Figures 4 and 5, a distinct color was used for each bucket rather than shades of blue or red.

Intensity maps of ground water and surface water by dominant category of use for every five-year increment from 1960 to 2005 were created. Intensity maps have a pop-up pie chart for each parish showing the distribution of water use by category (Figs. 4 and 5).

In a broad, roughly north-south and continuous belt from East Carroll Parish in the north and Cameron Parish in the south, ground water is primarily used for rice farming (Fig. 4). This agricultural area is clearly visible on satellite photos of the state.

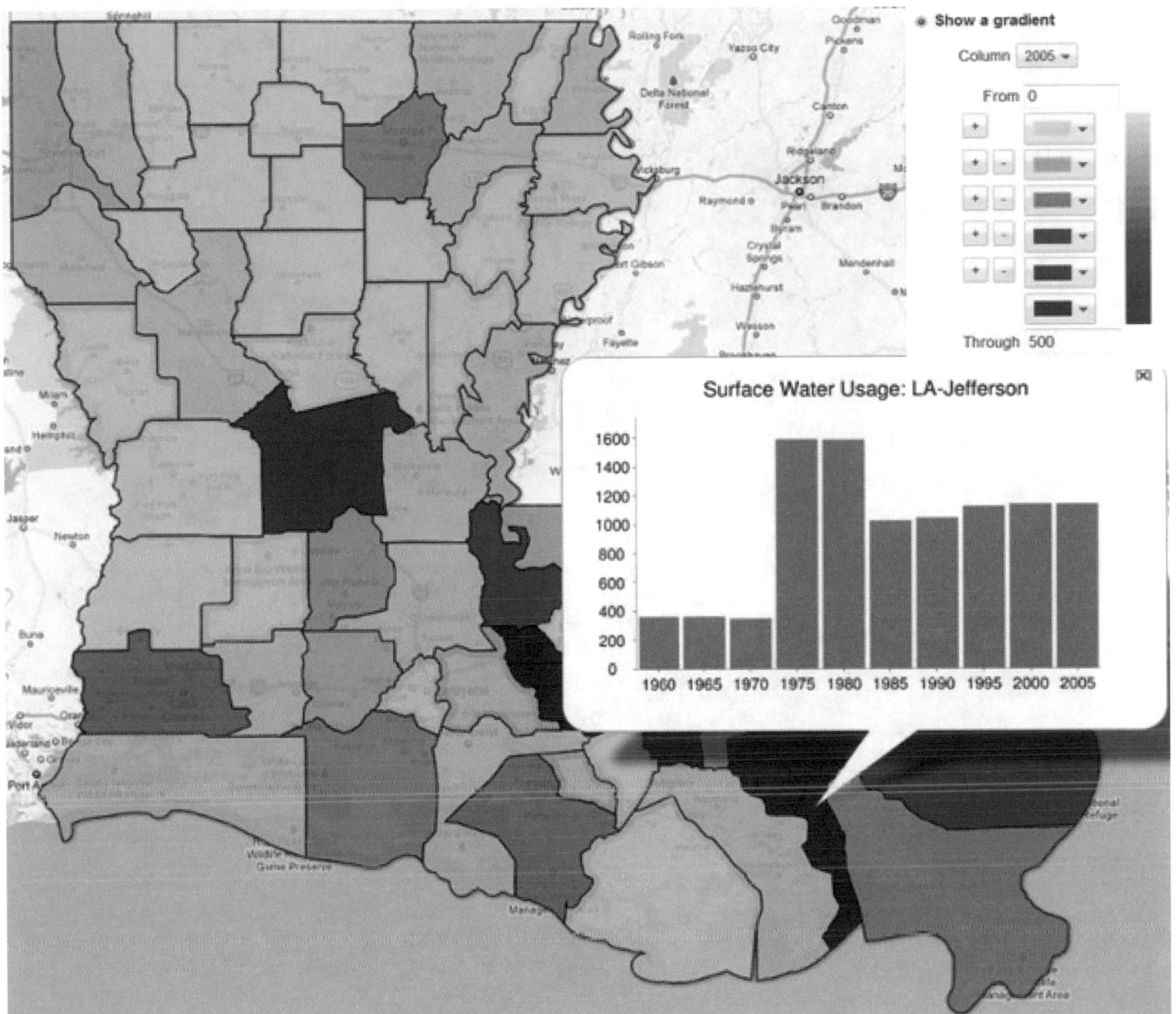

Figure 3. Gradient map showing Louisiana surface-water use in 2005 by parish. Range is from 0 to 500 million gallons per day. Darker red means higher ground-water use (see inset). Parishes with 415 million gallons per day or more are dark red. Embedded bar charts show surface-water use from 1960 to 2005 for St. Bernard Parish.

There is also a large industrial corridor along the lower Mississippi River as well as industrial use centered in Lake Charles in the southwest and from Monroe to Shreveport along Interstate 20 in the northern part of the state. Most other parishes primarily use ground water for public supply or rural domestic usage. There are also two parishes in southeast Louisiana where aquaculture is the dominant use (note St. Martin Parish is bisected into two separate regions by Iberia Parish). The embedded pie chart for Livingston Parish shows that public supply and rural domestic are the dominant uses but also indicates that overall ground-water use in that parish is small (less than ½ percent of the state total). The west-

ern portion of the parish is primarily bedroom communities for Baton Rouge and the eastern portion is sparsely populated farm land or marsh. HTML for this chart is contained in Appendix 1. It also contains instructions for scales, colors, and line thickness. In addition, the appendix contains information on colors and names for the legend (Fig. 4).

Surface-water use from 1960 to 2005 also shows the north to south agricultural belt seen in ground-water use except the rice irrigation area is not as wide or continuous. Surface water is also used for livestock and general irrigation in many rural areas of north Louisiana. A few parishes have public supply or aquaculture

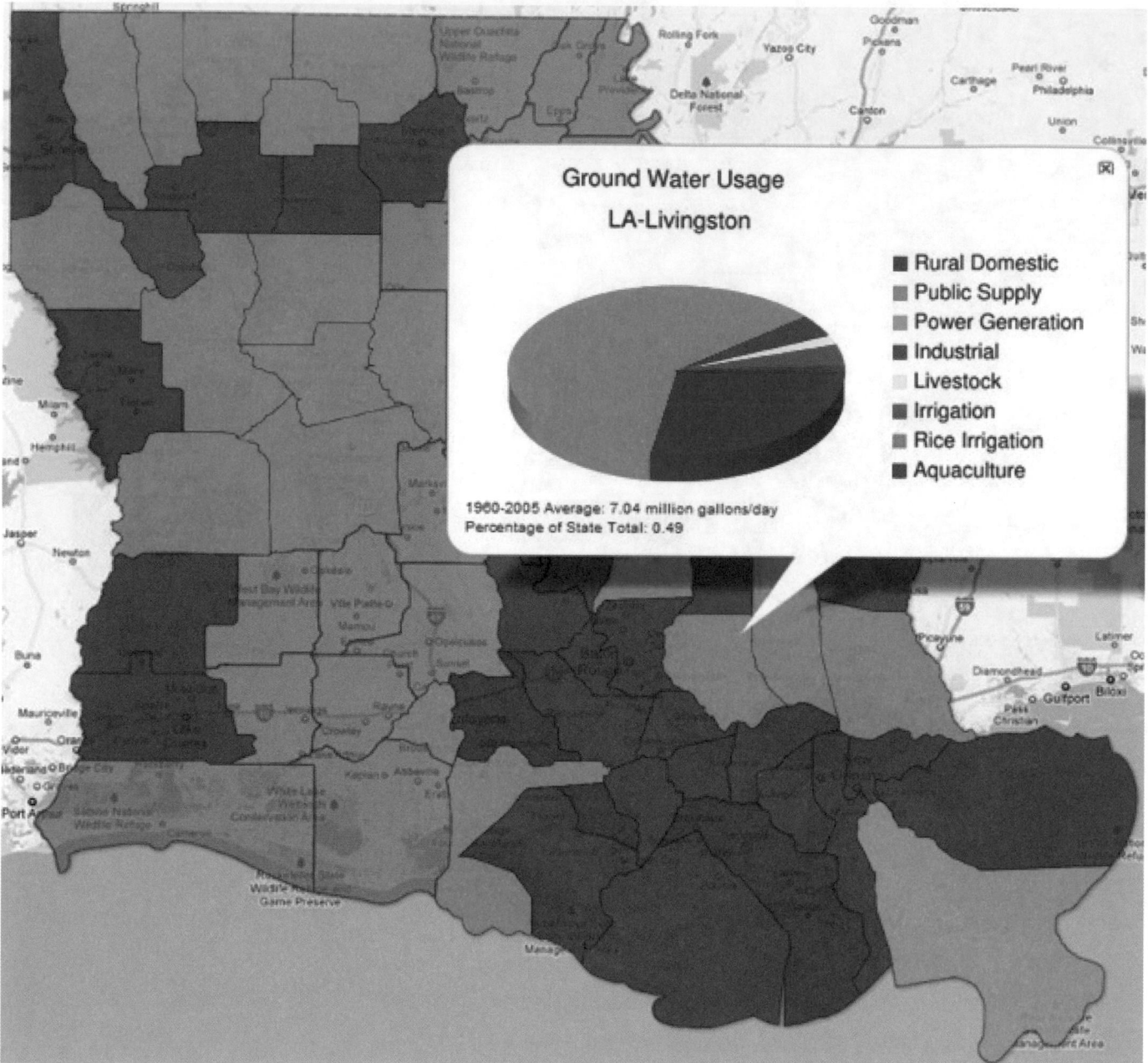

Figure 4. Intensity map showing maximum ground-water use by type in each parish averaged over 1960–2005. Each color represents a different category of use (e.g., public supply or rice irrigation). Embedded pie chart shows the distribution of water use by type for Livingston Parish.

as the dominant use of surface water. In general, higher quality ground water is used for people and surface water is used for animals. As with ground-water use, there is heavy industrial use of water along the lower Mississippi River as well as Lake Charles and several parishes in north Louisiana. Power generation is the dominant use of surface water in eleven parishes adjacent to the Mississippi, Red, Atchafalaya, and Sabine rivers. Averaged over 1960–2005, more than half of all surface-water use in Louisiana is for power generation. More than 35% of all surface water used in Louisiana from 1960 to 2005 was for power generation

in the three parishes near New Orleans (St. Charles, Jefferson, and Orleans; see Fig. 5). The embedded pie chart in Figure 5 for Tangipahoa Parish shows use of surface water for livestock, irrigation, and industrial purposes. HTML for this chart is very similar to the HMTL for the chart shown in Figure 4 and is listed in Appendix 1.

A second set of intensity maps of ground-water and surface-water use by dominant category were created with pop-up line charts that show water use by each category in five-year increments from 1960 to 2005 (Figs. 6 and 7). The line charts illustrate

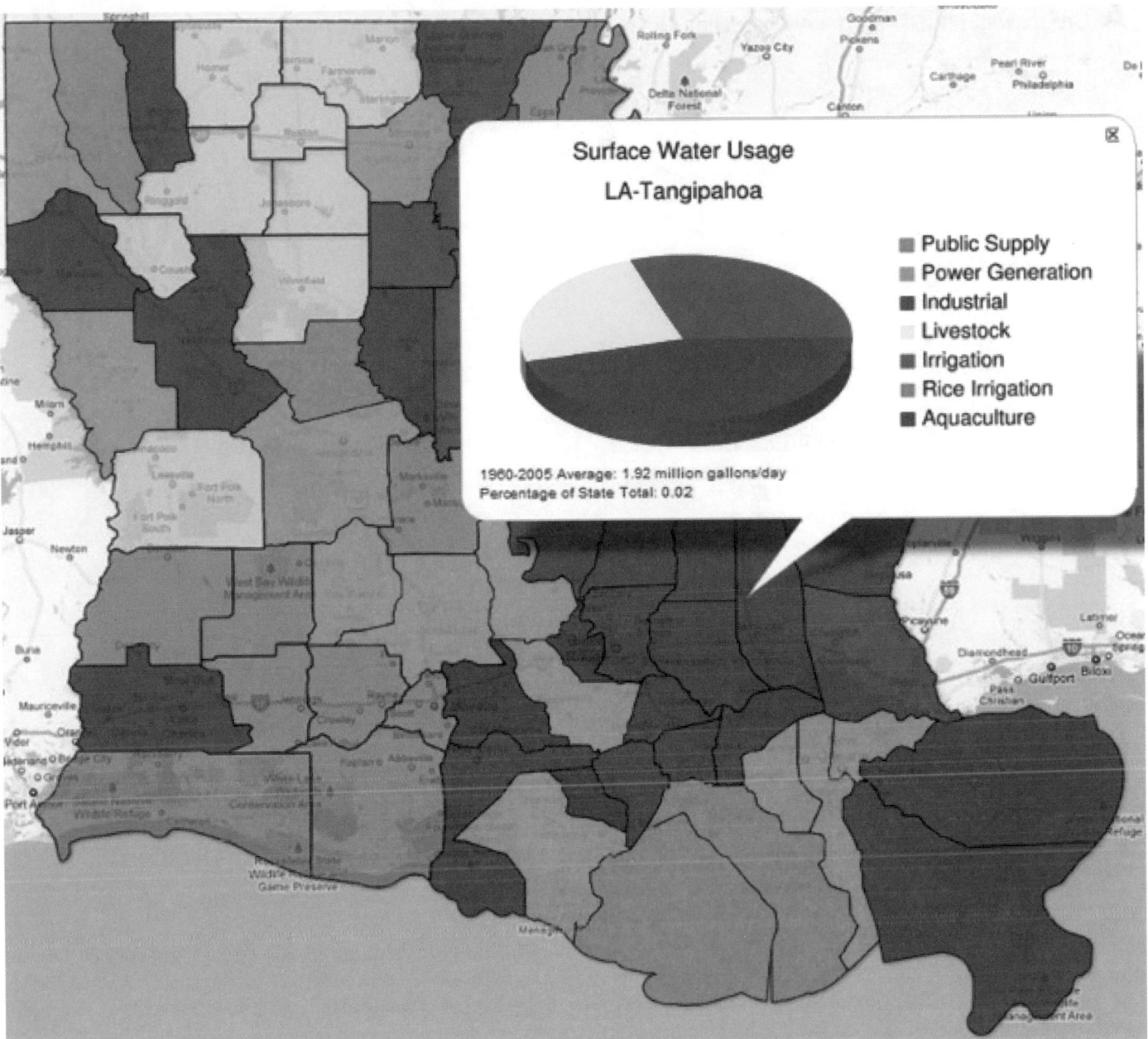

Figure 5. Intensity map showing maximum surface-water use by type in each parish averaged over 1960–2005. Each color represents a different category of use (e.g., public supply or rice irrigation). Embedded pie chart show the distribution of water use by type for Tangipahoa Parish.

temporal trends in water use in terms of quantity and type of use. HTML for these charts is contained in Appendix 1. The HTML for this chart is different from the previous ones in two respects. These charts use many more columns in the Fusion Table. There must be a column for each year and each category of use (72 columns = 9 five-year increments and 8 use categories). Also, the HTML must specify the color, symbol type, symbol size, line thickness, and line type used in the chart.

Figure 6 illustrates ground-water and surface-water use by category from 1960 to 2005 for the two largest metropolitan areas in Louisiana: Baton Rouge and New Orleans. The two largest ground-water uses in East Baton Rouge Parish are industrial and public supply. A small amount of water is used for power generation (~6 million gallons a day) and all other uses are minimal (<0.5 million gallons a day). Industrial use is the dominant category. There are significant fluctuations from year to year with notable declines in 1965 and 1985. It is important to note that one large-diameter well can pump 10–20 million gallons a day, so temporal changes of that magnitude might represent the opening or closing of a single paper mill or some other industrial plant.

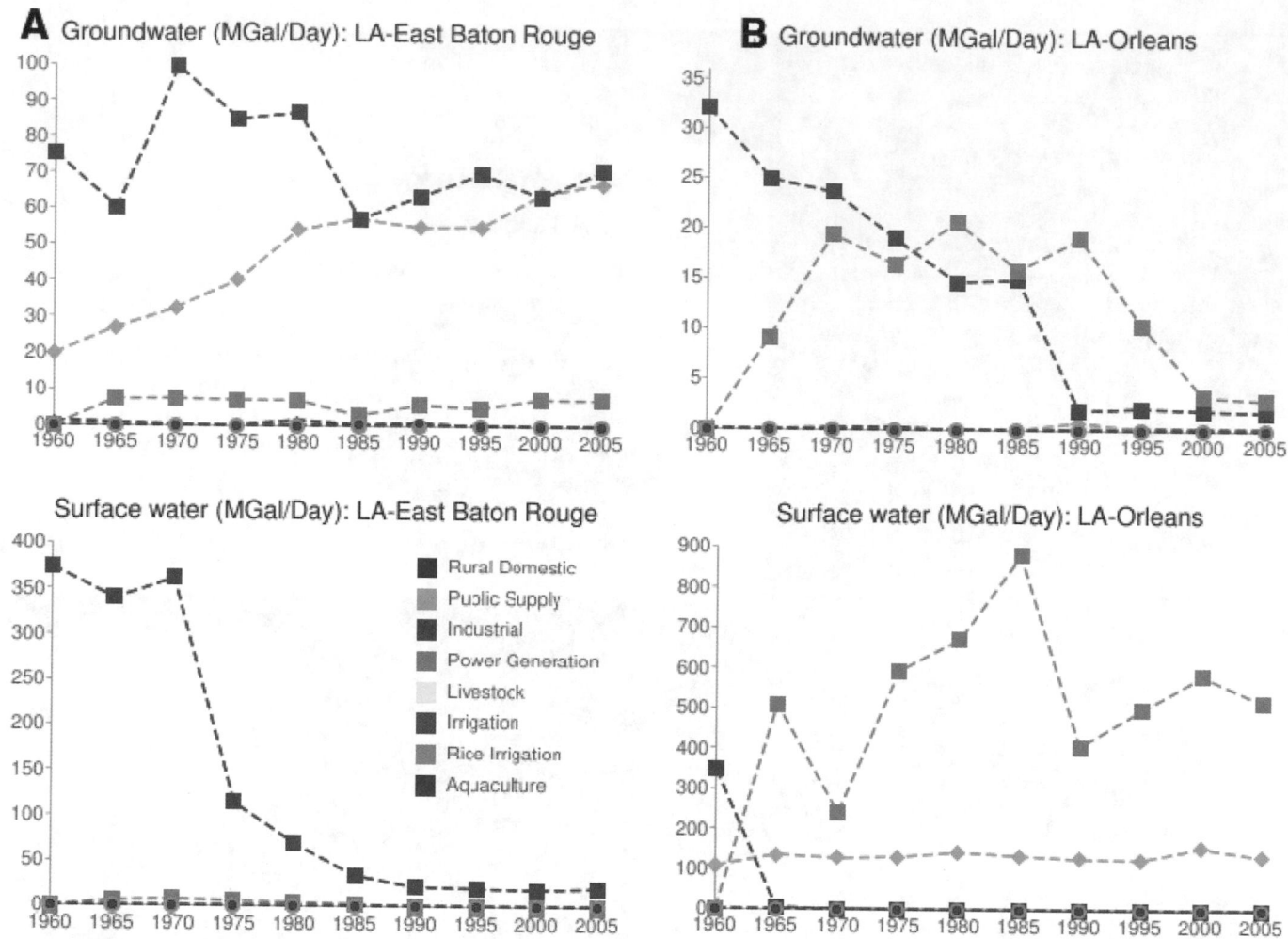

Figure 6. Line charts showing water use by type from 1960 to 2005 for East Baton Rouge (A) and Orleans (B) Parishes. Top row is ground water and bottom row is surface water.

Public supply has steadily risen since 1960 with the exception of slower growth from 1980 to 1995 when Louisiana's population was stagnant. Surface-water use in East Baton Rouge Parish is almost exclusively industrial; it has declined dramatically since the 1960s from around 350 million gallons a day to around 20 million gallons a day. This is somewhat surprising given that East Baton Rouge Parish is bounded on the west by the Mississippi River. River water contains a lot of mud and silt and must first go through a settling tank for most purposes, which increases costs, whereas ground water in East Baton Rouge Parish is very high quality with low amounts of dissolved solids. In addition, the aquifers underneath East Baton Rouge Parish were initially under high enough pressure that water flowed to the surface. In contrast, New Orleans uses relatively little ground water because of poor water quality (Fig. 6). Primary uses of ground water are industrial and power generation. Industrial use has declined substantially since 1960 with an especially large drop between 1985 and 1990 associated with the oil bust. Use of ground water for power generation ramped up in the 1960s, peaked in the 1980s,

and has declined in the last decade. Surface water in New Orleans is used for power generation and public supply. Use of surface water for power generation grew from 1960 to 1985 except for a significant drop in 1970. There was a large drop in use in 1990 followed by an increase and then a leveling off of use over the last decade. Public supply use is fairly steady even though the Orleans Parish population has declined significantly (627,525 in 1960 versus 483,674 in 2000; U.S. Census Bureau).

Figure 7 shows ground-water and surface-water use by category from 1960 to 2005 for two parishes in southwest Louisiana: Calcasieu, which contains the industrial city of Lake Charles with a population of 183,000 in 2000, and Vermilion, which is a largely rural parish with a farm-based economy and a population of 53,000 in 2000 (U.S. Census Bureau). Calcasieu Parish shows a more diverse use of ground water than many other parishes, with significant use at various times for industrial, rice farming, public supply, and power generation as well as smaller amounts for general irrigation and aquaculture. As with most areas, public supply increases over time. Industrial and rice irrigation use

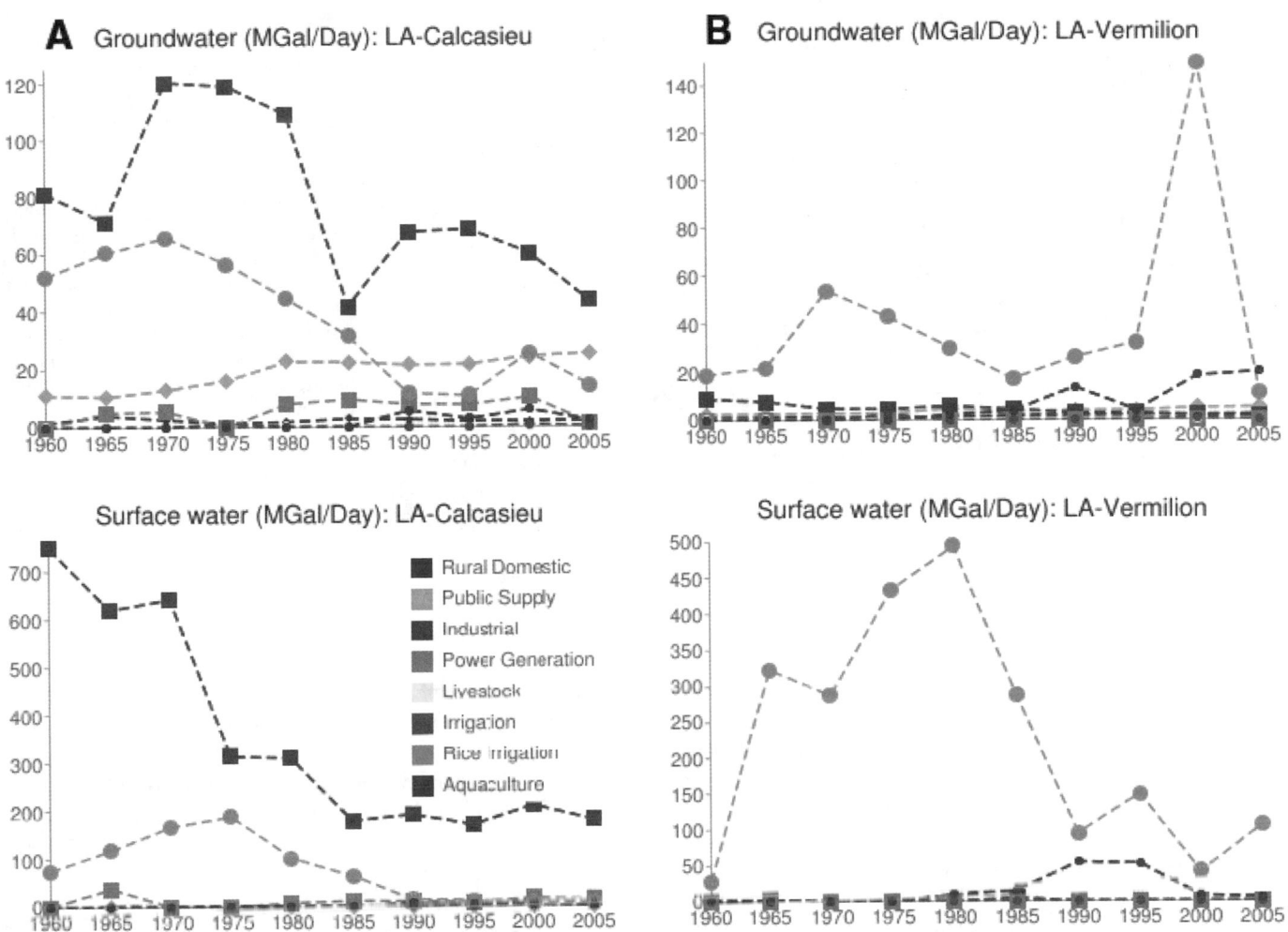

Figure 7. Line charts showing water use by type from 1960 to 2005 for Calcasieu (A) and Vermilion (B) Parishes. Top row is ground water and bottom row is surface water.

of ground water declines significantly over time. Industrial use of ground water shows dramatic changes over five-year intervals with a spectacular drop in 1985. Surface-water use in Calcasieu Parish shows a dramatic decline over time. Rice irrigation use increased from 1960 to 1975 with a peak of 188 million gallons a day and then declines to less than 10 million gallons a day by 2005. Both ground water and surface-water use in Vermilion Parish are dominated by rice irrigation (Fig. 7). At certain stages of germination, rice fields are flooded and thus rice irrigation uses large amounts of water. Rice irrigation also shows large fluctuations over time. Averaged over 1960–2005, Vermilion Parish uses ~40 million gallons of ground water a day for rice irrigation. However, in 2000, rice irrigation use spiked to almost 150 million gallons a day from 32 million gallons a day in 1995 and then dropped to 11 million gallons a day in 2005. How much ground water is used for rice irrigation depends on the number of acres planted, rainfall, and amount of surface water available. In 2000, drought conditions required more irrigation, plus surface water supplies were becoming contaminated by saltwater intrusion

(McClain, 2003). Rice is very sensitive to water salinity. Figure 7 clearly shows the corresponding drop in surface-water use to the upward spike in ground-water use. The only other significant use of both ground water and surface water in Vermilion Parish is aquaculture, which also shows fluctuations from year to year.

Finally, a Google dynamic table showing statewide ground-water and surface-water use versus time was created. Dynamic tables, when viewed online, can show an animated bar chart of either ground water or surface water where the bars representing each category of use grow or shrink with time. Dynamic tables can also show an animated cross-plot of ground water and surface water where bubbles representing each category move with time (Fig. 8). Dynamic tables have the option to track and/or label categories of interest. This feature makes it easy to see correlations between ground-water and surface-water use. Dynamic tables can also produce line plots of use versus time.

Figure 8 shows a cross-plot bubble chart of surface-water versus ground-water use for major categories from 1960 to 2000. Rice irrigation over the past 45 years has been the largest use

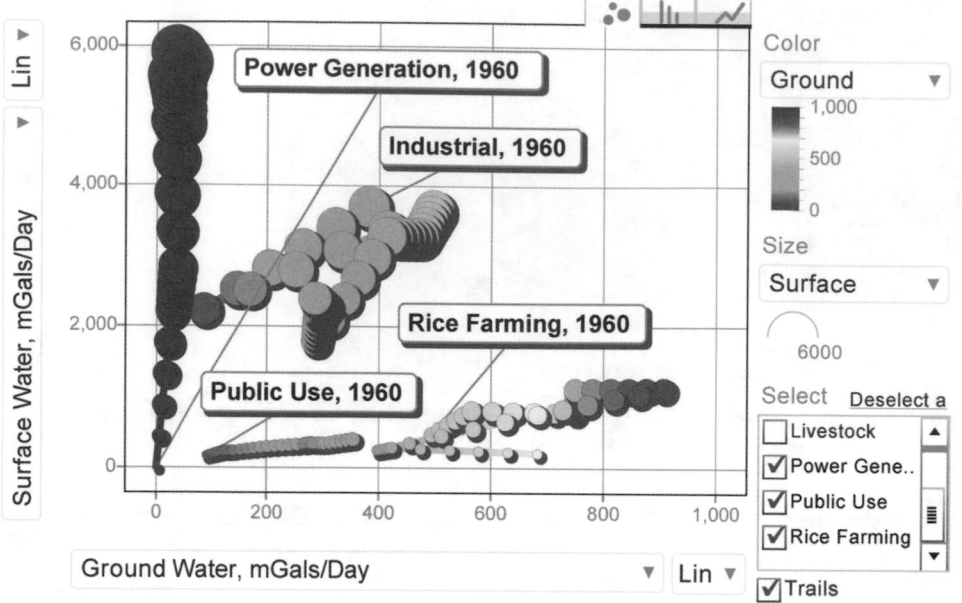

Figure 8. Snapshot of bubble motion chart showing Louisiana water use by category from 1960 to 2000. Size of bubbles is proportional to surface-water use. Color of bubbles is proportional to ground-water use. Curves for the four smallest categories (irrigation, rural domestic, livestock and aquaculture) have been removed for clarity.

of ground water in Louisiana, followed by industrial and public use. Both rice irrigation and industrial use reached maximum values in the late 1970s followed by dramatic drops in the 1980s and then a recovery in the late 1990s. Industrial use also had a spectacular decline in 1965. In contrast, public supply has shown continuous growth even though population growth has been slow since the 1985 oil bust. Surface-water use is dominated by power generation and industrial. Power generation shows spectacular growth in the 1960s and has remained high through 2005 with the exception of a dip in 1990. Industrial use has shown a slow decline over the past 45 years with dramatic declines in 1965 and 1985. Rice irrigation use peaked in 1980 and has declined substantially since then, especially in surface-water use (Fig. 8).

APPLICATIONS

As with most states, there are numerous research programs regarding water resources in Louisiana. These include studies of declining water levels in aquifers from over use (Griffith and Lovelace, 2003; Lovelace et al., 2004; Sargent, 2007); water quality (Lovelace, 1999; Prakken, 2007; Welch and Hanor, 2011); saltwater intrusion both in surface water and ground water (Tomaszewski, 1996; Milner and Van Biersel, 2006; Servan-Camas and Tsai, 2009); and surface subsidence (Nunn, 2003, 2010). These potential problems are occurring statewide, not just in coastal regions, as saline fluids exist in the subsurface throughout the state, including in the Sparta aquifer in North Louisiana (Carlson and Van Biersel, 2009). While the Fusion Tables described above do not represent new data available to researchers, the ability to quickly and easily visualize data allows researchers to see trends or interconnections that might otherwise not be obvious. An example is the hydrogeology of Ver-

milion Parish, where extensive pumping of both surface water and ground water for rice farming is occurring in a low-lying coastal region that is highly sensitive to saltwater intrusion, subsidence, and decadal variations in rainfall. Easy access to this type of information also is useful to policy makers as well as the average citizen as Louisiana attempts to wisely manage its water resources in the twenty-first century.

Most Louisiana State University (LSU) students have no idea of the source of their water, how much they use, or for what purposes. A simple water resources exercise used in freshman geology for a number of years asks students to identify on a map the top five parishes for surface-water and ground-water use, what category of use constitutes the dominant use, and statewide temporal trends for farming, industrial/power generation, and individual use using the U.S. Geological Survey data. The students are then asked to discuss temporal trends in water use in Louisiana. Note that much of the effort in this exercise is in the students converting tabular data into maps and charts rather than in higher-order thinking skills. The intensity maps described above will allow students to spend more time thinking about trends, causes, and interconnections. This exercise has also been used as part of a National Science Foundation–sponsored outreach program to Louisiana high school science teachers. It will now be easier to share this information with K–12 teachers. We plan to develop an online lesson plan so all teachers will have to do is distribute the URL to their students. Plans including having college students create their own Fusion Tables. For example, a class at LSU will have students create Fusion Tables for water use by aquifer. Other possibilities would be for students to compile information on climate information such as rainfall or other resources such as oil, gas, and coal. A valuable feature of Fusion Tables is that they are shared data that can be merged with new

HTML for Bar Chart in Figure 3

```
<div style="width:500px; height:300px">
<img src="http://chart.apis.google.com/chart
?chxl=0:|1960|1965|1970|1975|1980|1985|1990|1995|2000|2005
&chxp=1960,1965,1970,1975,1980,1985,1990,1995,2000,2005
&chxr=1,0,{Bin Values}|0,1960,2005
&chxs=0,000000,14.5,0,lt,000000|1,000000,14.5,0,lt,000000
&chxt=x,y
&chs=475x285
&cht=bvg&chco=4D89F9,C6D9FD
&chds=0,{Bin Values},1960,2005
&chd=t1:{1960},{1965},{1970},{1975},{1980},{1985},{1990},{1995},{2000},{2000}|1960,1965,1970,1975,19
80,1985,1990,1995,2000,2005
&chbh=a
&chdlp=b
&chls=2,4,1
&chma=50,5,5,5&
&chtt=Surface+Water+Usage: {Location}
&chts=000000,17.5">
</div>
```

HTML for Pie Chart in Figures 4 and 5

```
<div style="width:550px; height:300px">
<img src="http://chart.apis.google.com/chart
?chxt=Rural Domestic,Public Supply,Power Generation,Industrial,Livestock,Irrigation,Rice Irrigation,
Aquaculture
&chs=530x280
&cht=p3
&chco=FF0000|00FF00|00FFFF|9900FF|FFFF00|FF00FF|FF9900|0000FF
&chd=t:{Rural Domestic},{Public Supply},{Power Generation},{Industrial},{Livestock},
{Irrigation},{Rice Irrigation},{Aquaculture}
& chl=Rural Domestic|Public Supply|Power Generation|Industrial|Livestock|Irrigation|Rice
Irrigation|Aquaculture
&chdl=Rural Domestic|Public Supply|Power Generation|Industrial|Livestock|Irrigation|Rice
Irrigation|Aquaculture
&chdls=000000,18.5
&chls=2,4,1
&chma=5,5,5,5
&chtt=Ground+Water+Usage | {Location}
&chts=000000,20.5">
1960-2005 Average: {Total} million gallons/day
</div>
Percentage of State Total: {Percentage}
```

HTML for Line Charts in Figures 6 and 7

```
<img src="http://chart.apis.google.com/chart
?chxt=x,y
&chs=450x250
&cht=lxy
&chls=1.5,6,3|1.5,6,3|1.5,6,3|1.5,6,3|1.5,6,3|1.5,6,3|1.5,6,3| 1.5,0,3
&chma=0,0,0,0
&cht=c,s,FFFFFF
&chm=d,FF0000,0,-1,5,0|d,00FF00,1,-1,9,0|s,9900FF,2,-1,9,0| s,0099FF,3,-1,9,0|o,FFFF00,4,-
1,5,0|o,FF00FF,5,-1,5,0| o,FF9900,6,-1,9,0|o,0000FF,7,-1,5,0
&chco=FF0000,00FF00,9900FF,0099FF,FFFF00,FF00FF,FF9900,0000FF
&chdl=Rural Domestic|Public Supply|Industrial|Power Generation| Livestock|Irrigation|Rice
Irrigation|Aquaculture
&chxr=1,0,{BinValue}|0,1960,2005,5
&chds=1960,2005,0,{BinValue}
&chtt=Groundwater+(MGal/Day): {Location}
&chd=t:1960,1965,1970,1975,1980,1985,1990,1995,2000,2005|
{RD_1960},{RD_1965},{RD_1970},{RD_1975},{RD_1980},{RD_1985},
{RD_1990},{RD_1995},{RD_2000},{RD_2005}|-1|{PS_1960},{PS_1965},
{PS_1970},{PS_1975},{PS_1980},{PS_1985},{PS_1990},{PS_1995}, {PS_2000},{PS_2005}|-
1|{ID_1960},{ID_1965},{ID_1970},{ID_1975},
{ID_1980},{ID_1985},{ID_1990},{ID_1995},{ID_2000},{ID_2005}
|-1|{PG_1960},{PG_1965},{PG_1970},{PG_1975},{PG_1980},{PG_1985},
{PG_1990},{PG_1995},{PG_2000},{PG_2005}|-1|{LS_1960},{LS_1965},
{LS_1970},{LS_1975},{LS_1980},{LS_1985},{LS_1990},{LS_1995}, {LS_2000},{LS_2005}|-
1|{IR_1960},{IR_1965},{IR_1970},{IR_1975},
{IR_1980},{IR_1985},{IR_1990},{IR_1995},{IR_2000},{IR_2005}|-1|{RI_1960},{RI_1965},
{RI_1970},{RI_1975},{RI_1980},{RI_1985}, {RI_1990},{RI_1995},{RI_2000},{RI_2005}|-
1|{AQ_1960},{AQ_1965}, {AQ_1970},{AQ_1975},{AQ_1980},{AQ_1985},{AQ_1990},{AQ_1995},
{AQ_2000},{AQ_2005}
&" width="450" height="250" alt="" />
```

information. Thus, a compilation of rainfall information could be merged with the Fusion Tables described above to create maps or dynamic tables that relate rainfall to ground-water use on a parish by parish basis or versus time. Fusion tables created by college students could also be shared with K–12 teachers and students.

SUMMARY AND CONCLUSIONS

Google Fusion Tables represent a rapid and effective way to visualize water usage trends for K–16 education, research, and public policy.

Water usage in Louisiana has complicated spatial and temporal trends which are not readily apparent in static tables. For example, ground-water usage varies from more than 200 million gallons a day in some rice farming parishes to less than 40,000 gallons a day in coastal parishes where most ground water is not potable. East Baton Rouge Parish uses mostly ground water even though it is on the Mississippi River because the ground water is high quality and less expensive than river water. Orleans Parish uses almost exclusively river water because most ground water is brackish.

Significant temporal trends include the rapid rise of water use for power generation since 1960, the drop in overall water usage during the economic downturn following the oil bust of the mid-1980s, and the switch from surface to ground water in some areas due to decadal droughts or pollution of surface water.

The dynamic feature of Fusion Tables allows students, researchers, and policy makers to clearly see spatial or temporal trends as well as illustrate connections among water usage and other factors. For examples, most students falsely assume that the steady rise in water use for public supply is related to population increase whereas it is primarily due to an increase in per capita usage. Vermilion Parishes ground-water use varies significantly from year to year depending on saltwater intrusion of surface water and rainfall.

These tables are available at http://www.geol.lsu.edu/LaWaterUse or in the GSA Data Repository (see footnote 1).

ACKNOWLEDGMENTS

The authors thank John Lovelace of the U.S. Geological Survey for providing data on Louisiana water usage in 2005. Andre Zular and Ioannis Georgiou provided useful comments and corrections which greatly improved the clarity and quality of the manuscript. Nunn would like to thank John Bailey, Steven Whitmeyer, Declan De Paor, and Tina Ornduff for organizing the Penrose Conference. Our work has been supported in part by National Science Foundation Grant ERA-0557555 (Nunn and Hanor). Bentley received support from the LA-STEM research scholars program funded by NSF and the Louisiana Board of Regents.

REFERENCES CITED

Carlson, D., and Van Biersel, T., 2009, Is Chloride Concentration Increasing in the Sparta Aquifer of North-Central Louisiana?: Gulf Coast Association of Geological Societies Transactions, v. 59, p. 171–180.

Griffith, J.M., and Lovelace, J.K., 2003, Louisiana Ground-Water Map No. 15: U.S. Geological Survey Water-Resources Investigations Report 03-3020.

Lovelace, J.K., 1999, Distribution of saltwater in the Chicot aquifer system in the Calcasieu Parish area, Louisiana 1995–1996: State of Louisiana, Water Resources Technical Report No. 66, 61 p.

Lovelace, J.K., Fontenot, J.W., and Frederick, C.P., 2004, Withdrawals, water levels, and specific conductance in the Chicot aquifer system in southwestern Louisiana, 2000–2003: U.S. Geological Survey Water-Resources Investigations Report 04-5212, 56 p..

McClain, R., 2003, In Louisiana rice: Rain helps fight saltwater intrusion: Delta Farm Press, 20 June 2003.

McPhee, J., 1987, The control of nature Atchafalaya: New Yorker, 23 February 1987.

Milner, R., and Van Biersel, T., 2006, Updated geology and saltwater intrusion for the Chicot Aquifer of Southwestern Louisiana, *in* Proceedings, 25th Anniversary Meeting and International Conference, Challenge in Coastal Hydrology and Water Quality, p. 141–152.

Nunn, J.A., 2003, Land surface subsidence caused by ground-water withdrawal in southeastern Louisiana: Gulf Coast Association of Geological Societies Transactions, v. 53, p. 630–638.

Nunn, J.A., 2010, Seasonal Groundwater Withdrawal in Southwestern Louisiana: Implications for Land Subsidence and Resource Management: Gulf Coast Association of Geological Societies Transactions, v. 60, p. 515–524.

Prakken, L.B., 2007, Chloride Concentrations in the Southern Hills Regional Aquifer system: State of Louisiana, Department of Transportation and Development, Water Resources Technical Report No. 76, 30 p.

Sargent, B.P., 2007, Water Use in Louisiana, 2005: State of Louisiana, Department of Transportation and Development, Water Resources Special Report No. 16, 133 p.

Servan-Camas, B., and Tsai, F.T.-C., 2009, Saltwater intrusion modeling in heterogeneous confined aquifers using two-relaxation-time lattice Boltzmann method: Advances in Water Resources, v. 32, no. 4, p. 620–631, doi:10.1016/j.advwatres.2009.02.001.

Tomaszewski, D. J., 1996, Distribution and movement of saltwater in aquifers in the Baton Rouge area, Louisiana, 1990–1992: State of Louisiana, Office of Public Works, Water Resources Technical Publication No. 59.

Welch, S.E., and Hanor, J.S., 2011, Sources of elevated salinity in the Mississippi River Alluvial Aquifer, south-central Louisiana, U.S.A.: Applied Geochemistry, v. 26, p. 1446–1451.

MANUSCRIPT ACCEPTED BY THE SOCIETY 16 APRIL 2012

The Geological Society of America
Special Paper 492
2012

Extreme dynamic mapping: Animals map themselves on the "Cloud"

Eugene Potapov*
Bryn Athyn College, College Drive, Bryn Athyn, Pennsylvania 19009, USA

Valery Hronusov*
"Inform++ Ltd," Shatrova St, 20-25, Perm 614064, Russia

ABSTRACT

Animal tracking data are routinely delivered in the form of e-mail messages with an attachment or in the main text of an e-mail that includes satellite-telemetry data provided by Argos services. Downloading these data onto a computer, transferring them into shapefiles, filtering, processing, and displaying them consumes considerable end-user time and energy. In this paper, we demonstrate that freely available "Cloud"-based services are sufficient to take over this workload and fast enough to deliver spatial data to an end-user without a considerable investment of time. The animal-generated spatial data we present come in two forms: satellite data from the Argos service and GPS data delivered as text messages using a Short Message Service (SMS). We suggest a simple mail-to-map system, which automatically archives data (coordinates, time, telemetry) and displays it dynamically on various Internet applications such as Google Maps/Google Earth or Google Graphs. We use the Gmail service to filter messages, a free blog service (e.g., blogger.com or wordpress.com) for unlimited-time data storage and the Google spreadsheets to dynamically assemble the KML (Keyhole Markup Language) files. To demonstrate the utility of our mail-to-map system, we apply the approach to two contrasting wildlife case studies—the highly endangered Steller's Sea Eagle (*Haliaetus pelagicus*) of northeast Asia and White-tailed Deer (*Odolescens virginianus*), which is ubiquitous in the eastern United States—and discuss conservation implications of the near-real-time data publication opportunities that our system provides.

INTRODUCTION

The conventional methods of obtaining geospatial data from various PTT/GPS (Platform Transmitter Terminal/Global Positioning System) data loggers involve downloading the data from a company provider's server (e.g., Argos website, Followit, North-Star Telemetry) or extracting the data from provider-generated e-mails. Once downloaded, extensive processing on the client's computer is necessary to make these data available on an Internet web page. Failure to download the data in time generates serious gaps in the data flow, as some providers do not store the data forever. For example, to this day, the Argos system, which has a worldwide monopoly on satellite-relayed wildlife data, keeps only 10 days of the data available for online access (Argos User

---

*Eugene.Potapov@brynathyn.edu; xbbster@gmail.com.

Potapov, E., and Hronusov, V., 2012, Extreme dynamic mapping: Animals map themselves on the "Cloud," *in* Whitmeyer, S.J., Bailey, J.E., De Paor, D.G., and Ornduff, T., eds., Google Earth and Virtual Visualizations in Geoscience Education and Research: Geological Society of America Special Paper 492, p. 139–145, doi:10.1130/2012.2492(10). For permission to copy, contact editing@geosociety.org. © 2012 The Geological Society of America. All rights reserved.

Manual, 6.3.3, p. 38). Data downloads that have to be done every 10 days require the effort and time of a researcher and often are incompatible with fieldwork. These problems can be overcome by using automated scripts on dedicated corporate servers. However, this requires substantial overhead costs, including system administrator salaries and software and hardware expenses. In this paper, we suggest a simple and cost-efficient alternative to the conventional (human-to-computer intensive) and corporate (high cost) models. We suggest accumulating and storing data on the theoretically indestructible and free "Google-cloud"–based web service, which allows near-real-time data processing for generating outputs on Google Earth/Google Map services.

Prior to 2004, none of the existing wildlife tracking systems was designed to provide continual real-time data access (Clark et al., 2006). In 2004, there was an attempt to design a real-time tracking system for moose in Sweden (Dettki et al., 2004), which was based on ArcIMS server (ESRI, 2004). A new philosophy of data storage and access in the form of "software as an online service" arrived with the launch of Google Maps and Google Earth in 2005 and Google Spreadsheets in 2006. The availability of these highly integrative products has opened new possibilities for the presentation of the dynamically changing geolocations of animals.

METHODS

Configuration of E-mail Clients

The majority of tracking systems (Argos or GPS/GSM [Global System for Mobile Communications]) offer e-mail as a vehicle to deliver the data, a convenient choice since e-mail systems can be easily automated. We recommended Google's Gmail service for mailing telemetry data since it is not affected by corporate e-mail server rules and has sufficient capacity to store the data. It is also possible to set individual e-mail accounts for an Argos program or subprogram, or even for an individual animal in the case of high-density GPS/GSM data streams. Gmail accounts are an attractive option since these can be configured with ease and have virtually no size limits. The size of an incoming e-mail associated with telemetry data can jeopardize data flow in the corporate model because servers in these environments might use unpredictable firewall settings and require a third party to make changes. A filter should be set up to forward e-mail messages to the user's data-storage system. It is recommended to filter the messages by the sender e-mail address and by subject line.

Data Storage

We suggest storing the data in a "blog," originally an online journal for sharing diary entries (Fellbaum, 1998). Most blogging software provides the option to send posts via e-mail. The free, cloud-based, blog service run by Google (www.blogger.com) is appealing, since it is virtually indestructible and has no size limit. The data is forwarded, as described above, from Gmail (or another mailer) to the blog by automated filters. For GPS/

GSM monitoring it is recommended to have one blog per animal (http://goo.gl/EnKCl). For the Argos system, filters should be set-up to send diagnostic data (DIA; http://wp.me/2egoY) and telemetry data (DS; http://wp.me/2efLy) files to separate blogs (Fig. 1). The configuration of the blog is critical, as the data stored there will be accessed by spreadsheets for data import. The amount of the data displayed on the front page of the blog determines the amount of data subsequently displayed on the associated map. A blog does not set limits on data storage time (can be indefinite), and it also allows for the exporting of the entire blog, i.e., the entire data set. This is a significant improvement over other systems, for example, the Argos system, which limits the online data access to "one current and past 9 days" (Argos User Manual, 2011, chapter 6.3.3, p. 38).

Data Processing

Once the data arrive at the designated blog(s), they become available for automated reading. This can be achieved by the import command within online spreadsheets (e.g., Google Documents or editgrid.com). Since Google spreadsheets are linked to Gmail, we highly recommend using this service for its convenience. The data from the blog gets imported into the Google spreadsheet by the command:

=importxml("http://*yourblog*.blogspot.com/","//p[1]/text()")

The values of "//p[1]/text()" depend on the way the blog is configured. In our case studies we deployed "Tic-Tac" and "Simple" templates of the www.blogger.com service. The *importxml* function reads the data into a spreadsheet with a periodicity of ~30 min (Google spreadsheet function list, 2011). The read data are broken up into values using "split()" command, and are cleaned up of unnecessary bits of text using the command "filter." The data are sorted chronologically, checked by simple error-tests that eliminate erroneous fixes, and are assembled into dynamic Keyhole Markup Language (KML) file(s) for the near-real-time display of the data on Google Maps/Google Earth/Google Graphs (see example at http://goo.gl/EnKCl). For the GPS/GSM service the error test is a simple elimination of the lines with the GPS time-out error, the elimination of duplicates, and optionally removing noisy fixes, whereas for Argos the error test includes not only the elimination of duplicates, but also a switch between solution 1 and solution 2 (see Argos User Manual, 2011, for terminology and example spreadsheet http://goo.gl/gGVPU). When using Argos-based PTT (Platform Transmitter Terminal, also known as satellite tag), it is very important to have the readings of telemetry sensors (battery voltage, temperature, etc.). This can be achieved automatically by reading the blog containing DS data (http://wp.me/2efLy), and decoding the sensors using the formula supplied by the device's manufacturer (http://goo.gl/gGVPU). The end result is a graph with the telemetry readings (Fig. 2), which can be embedded into KML file.

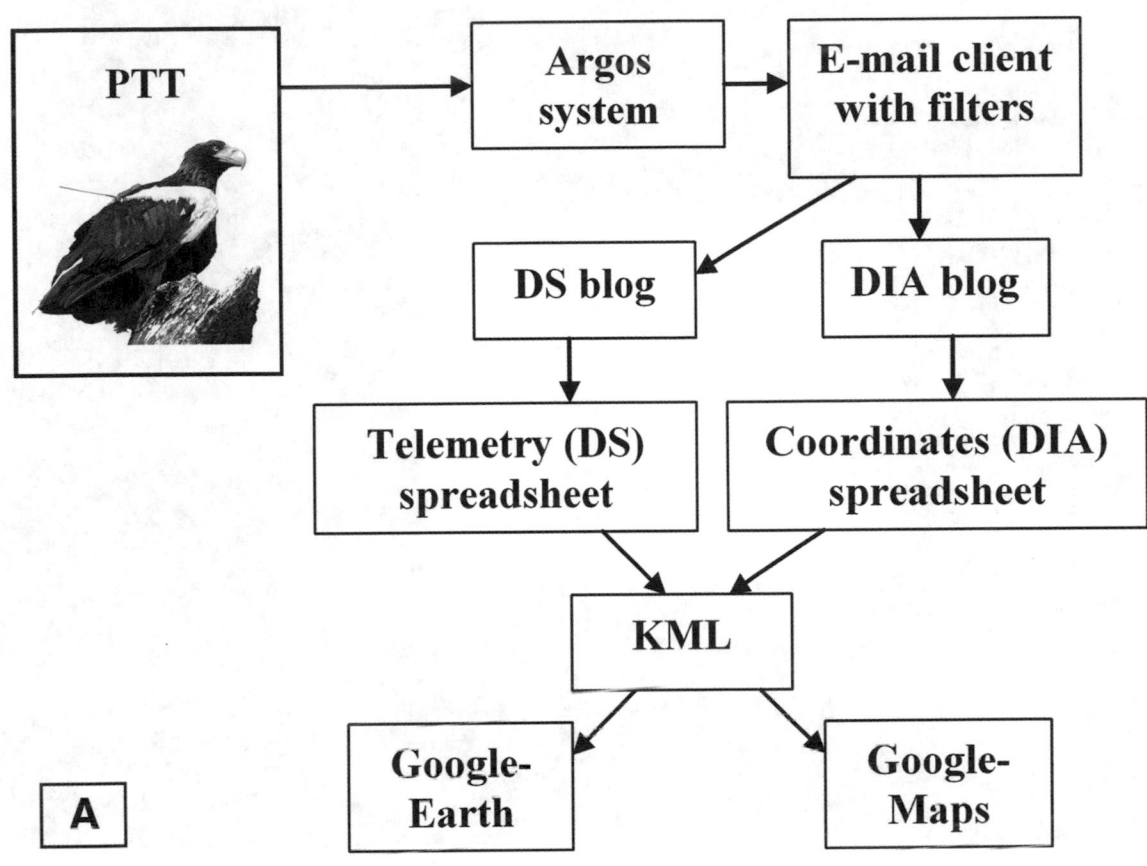

Figure 1. (A) Block diagram of the automated decoding of the Argos-based system. (B) Screenshot of the telemetry DS blog (http://wp.me/2efLy). (C) Screenshot of the DIA data blog which provides coordinates and time-stamp (http://wp.me/2egoY).

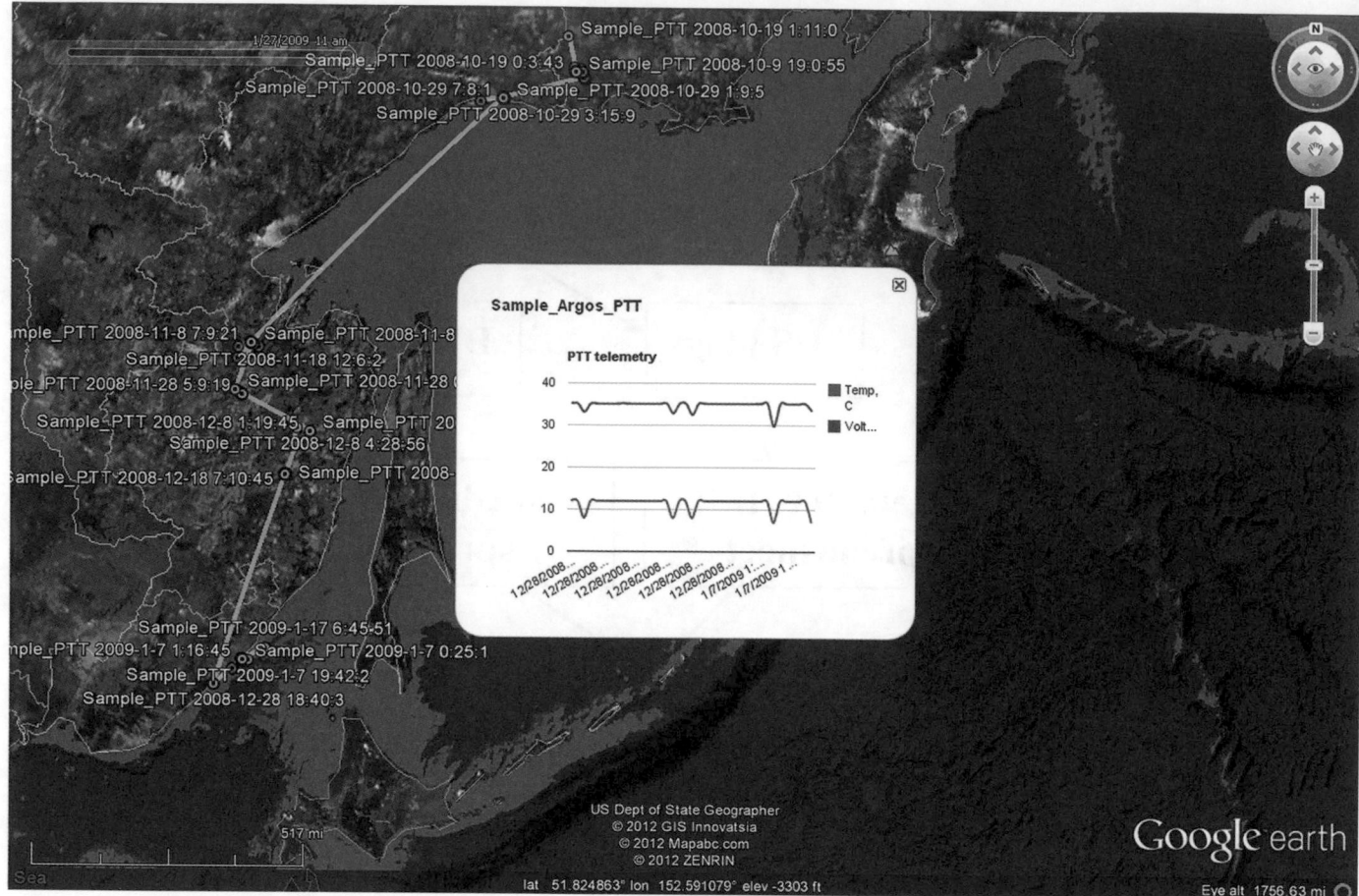

Figure 2. The trajectory of young Steller's Sea Eagle instrumented with Argos-based PTT migrating in the fall along coasts of Sea of Okhotsk, northeast Siberia. The telemetry data are decoded using spreadsheet, which reads DS data (http://goo.gl/r2OeG), the coordinates are decoded into dynamic KML using separate spreadsheet (http://goo.gl/gGVPU). Google-Earth file can be downloaded here (http://goo.gl/Kkg5d).

Such a service is exceptionally efficient when used on web-enabled phone devices as it does not involve downloading, importing, sorting, and displaying of all of the data, which would require additional computer resources and time. Once set up, the system runs unattended and is not susceptible to server shut-downs. Other advantages include free server access for data storage and accessibility from any internet connection.

CASE STUDIES

Case Study 1: GPS/GSM Biologging Collars on White-Tailed Deer

The white-tailed deer is widespread in the United States, especially in densely populated suburbs of the northeast where human/deer conflicts are common (DeNicola et al., 2000). We tracked the deer in a suburban area of Philadelphia in order to understand their spatial movements in areas of potential conflict. Tracking collars (Tellus, Followit, Sweden) transmitted a GPS location every five minutes. Once eight fixes accumulated in the collar, it transmitted the data via a cell phone network in

form of a Short Message Service (SMS) text message. The text messages were then e-mailed to the end-user. The e-mail system (Gmail) has filters configured to send the e-mail messages to a blog which was set-up within the same Google account. The blog (http://goo.gl/0Kwi2) was configured to display the 15 latest messages on the front page and to archive older messages. Next, we used the import command within Google spreadsheets to read and process the data, and assemble the KML file (http://goo.gl/EnKCl). We automated our system so that each time a new e-mail is received by the Gmail account, the data appear on the blog, and this new content is read by the spreadsheet. Thus the generated KML file is continually updated with the latest data available for display in Google Earth and Google Maps (http://goo.gl/Nze33).

The high frequency of location fixes, and intuitive graphical representation of the resulting track (that can be viewed on web-enabled phones) accommodated highly sophisticated behavioral experiments. This system has been running successfully since 2008, and has been used to track 32 individual deer. A block diagram of the system is given in Figure 3 and the resulting map is given in Figure 4.

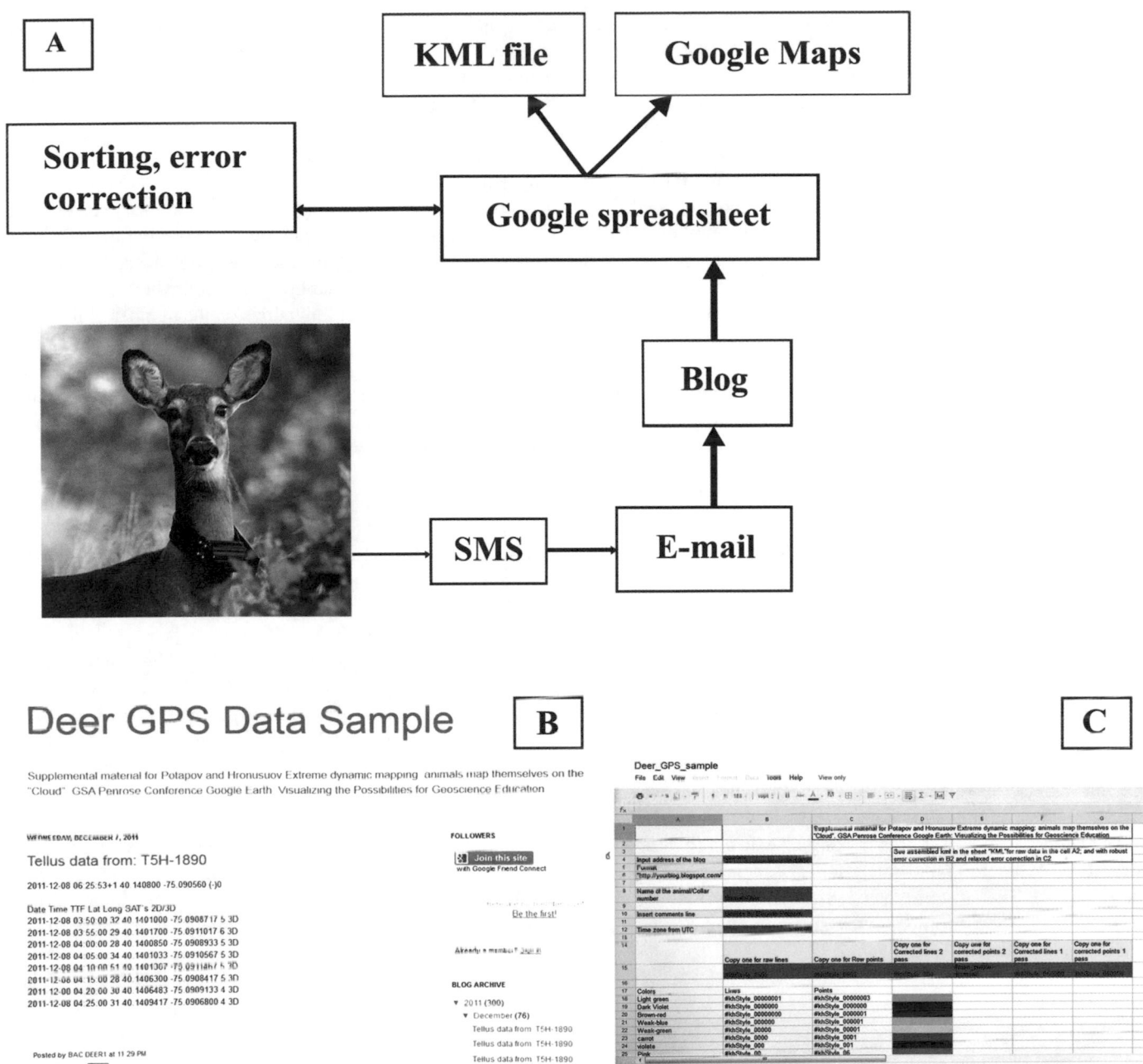

Figure 3. (A) Block diagram of the GPS/GSM-based system. (B) Screenshot of the blog which stores the data (http://goo.gl/0Kwi2). (C) Screenshot of the spreadsheet (http://goo.gl/EnKCl), which reads the blog, runs error correction, and dynamically assembles the KML file.

Case Study 2: Steller's Sea Eagles with Argos Doppler PTTs

The Steller's Sea Eagle is an endangered species, which breeds exclusively in northeast Siberia, Russia, and winters in the Russian far east, Kamchatka, and Japan. It is protected in both countries, nonetheless population numbers are on the decline. Since the breeding grounds are located in areas not yet affected by industrial society, the problems are thought to be concentrated within wintering areas and along migration routes (McGrady et al., 2003). To study the migration of the eagles, Argos-serviced PTTs were deployed. The Argos-based system is in principle no different from the GPS/GSM-based system presented above. Although the format of the Argos files is notoriously complicated, the files can be decoded with relative ease. The challenge is to decode observations that are presented in several lines.

The most widespread option of obtaining files from the Argos provider is via e-mail. The provider sends two data files to the end user: one in diagnostic format (DIA) and one in full format (DS). The first one is best to use for extracting coordinates, location class (LC), and date and time for Doppler locations. The second one is needed to extract telemetry information. The system in place for the eagle project is based on a Google account. The Gmail account receives Argos-distributed e-mails and, using filters, forwards these to separate blogs: one for DS data (http://wp.me/2efLy) and another for DIA (http://wp.me/2egoY), a practice that is highly recommended. The length of the blog's front page depends on the number of PTTs in the program and the number of e-mail messages received from the Argos system per day. If the number of PTTs in a program is 10, it is sufficient to set the number of posts displayed on the front page at the value of 12.

In this example, two spreadsheets read the DS (http://goo.gl/r2OeG) and DIA (http://goo.gl/gGVPU) blogs (Fig. 1). Each filters only one PTT at a time using the command "IMPORTXML." In the next step, the spreadsheet splits the lines of the messages into meaningful data using the command:

```
= FILTER(INDIRECT("A1:A"&B1) ;
(left(INDIRECT("A1:A"&B1),5) = "YOUR
PTT NUMBER")+SIGN(IFERROR(FIND("
"&ROW(INDIRECT("A1:A"&B1))&" " ; " "&CON
CATENATE(ARRAYFORMULA(IF(left(INDIR
ECT("A1:A"&B1),5) = "YOUR PTT NUMBER"
; (ROW(INDIRECT("A1:A"&B1))+1)&" " ;
IFERROR(1/0)))));0)))
```

This complicated formula reads two consecutive lines that are relevant to the selected PTT. Processed data are sorted by date, and filtered for desired location class (LC) accuracy. Duplicates are removed and either a dynamic KML file (http://goo.gl/Kkg5d) or dynamic graph with telemetry parameters is assembled (Fig. 2). This system has been running successfully since 2008, and has been used to track six individual Steller's Sea Eagles and also ten Tundra Swans.

DISCUSSION

As with any of the systems built on the new cloud-based architecture, the suggested automatic mapping system has some constraints. If the data stream exceeds a certain level, the blog treats the dataflow as spam and automatically evokes the Internet bot protection called "CAPTCHA." CAPTCHA is usually a distorted image of letters and numbers, which is used to prevent automated use of blogs. It is based on the fact that a computer cannot recognize distorted letters and numbers, whereas humans can recognize them with ease. CAPTCHA activation requires human intervention, and thus can potentially lead to a loss of data messages. With a low frequency of messages this is not a problem. However, if the data provider sends hundreds of messages simultaneously (say after emergency reboot), this can potentially cause a data gap, which has to be filled by manually resending the messages. As a workaround we suggest users funnel the data into

a backup blog for 24 hours, allowing the original blog to recover from the "data flood." The spreadsheet has to be reconfigured manually for a new blog address. Such "overflow" conditions occurred twice in three years of our continuous monitoring of the white-tailed deer.

Other problems are built into the Google spreadsheets. Currently there is a limit of 20,000 formulas per spreadsheet (with the current settings that means 20,000 cells, as every cell contains some sort of formula), which might be a potential limit for large Argos programs. For large data sets the problem can be solved by using the alternative service EditGrid.com, which offers similar spreadsheets, but without limitations on the number of formulas. There are some other alternatives, which are rapidly developing, e.g., dabbledb.com or zoho.com. The system that we outlined above suggests that the blog displays only the latest locations, which may be a constraint for some users. The length of the time period for these displayed fixes depends on the number of PTTs in the program and varies from 2 weeks to 1 month. Thus, it does not completely eliminate the necessity to download from the blog, which is capable of archiving all the data and processes the data in a conventional manner (GIS, filters, etc.). The speed of updates of the Google spreadsheets is currently once per hour or so, which is fast enough for both Argos-based satellite telemetry data and for GPS/GSM data.

Google spreadsheet-based maps also have limits on the number of points shown in Google Maps. Currently it is 100 elements. A larger number of points can make the map difficult to read. There are no such limits in Google Earth. A prominent advantage of the suggested system is that both the blog and the final spreadsheet-generated KML file are "web-ready" and can be used within various web clients. Although the blogs are potentially accessible to all Internet users the exact address is known only to informed users. Thus, the Internet address plays the role of a password. Of course, there is a potential for concern when displaying the current locations of endangered species on the web. One possible solution is to display only delayed locations. This is easy to achieve in the spreadsheet by simply unselecting the latest filtered position, and proved to be a very effective measure. In more than a year of continuous tracking of the white-tailed deer in southeastern Pennsylvania, including the legal hunting season, a 10-hour delay was implemented, and there were no attempts to harass the instrumented animals. For an endangered species, such as the Steller's Sea Eagle, a delay of a few days is considered sufficient.

Another important feature of our system is that it can generate output for web-enabled phones, and this near-real-time output can be brought to the field for experiments and rapid checks of the tagged animals without downloads and data processing. The near-real-time mapping is especially useful when used on mobile Internet devices (e.g., iPhones) as the latter allows researchers to get close to the tracked animal during field experiments. Once set up, the system runs unattended, is unaffected by corporate-based or other server shutdowns, firewalls, etc., and is based on free servers which are accessible from any Internet connection.

Figure 4. Snapshot of the Google Earth page with the track of the animal shown in near-real time. The dynamic KML can be downloaded here (http://goo.gl/Nze33).

ACKNOWLEDGMENTS

We thank Dr. M. McGrady of Natural Research, Ltd., Scotland and Dr. D. Rimlinger of San Diego Zoo for their support throughout this work, and Bryn Athyn College faculty for their help and encouragement. The method was conceived while one of the authors was chasing a Tundra Swan at the Colville River, Alaska, when working for the Swans Research Program funded by CONOCO Philips. We are grateful to Dr. W. Sladen and John Whissel, Ecological Studies at Airlie, Virginia, for supporting this study. We also thank André H. Banen (aka Ahab) for assistance with tuning Google spreadsheets and Dr. J. Bailey, Matt Heavner, and an anonymous reviewer for their suggestions to improve the manuscript, and the editorial team of this Special Paper for their hard work.

REFERENCES CITED

Argos User Manual, 2011, http://www.argos-system.org/files/pmedia/public/r363_9_argos_manual_en.pdf (accessed November 2011).

Clark, P.E., Johnson, D.E., Kniep, M.A., Jermann, P., Huttash, B., Wood, A., Johnson, M., McGillivan, C., and Titus, K., 2006, An advanced, low-cost, GPS-based animal tracking system: Rangeland Ecology and Management, v. 59, p. 334–340, doi:10.2111/05-162R.1.

DeNicola, A.J., VerCauteren, K.C., Curtis, P.D., and Hygnstrom, S.E., 2000, Managing white-tailed deer in suburban environments: A technical guide: Ithaca, New York, Cornell Cooperative Extension, Cornell University, 52 p.

Dettki, H., Ericsson, G., and Edenius, L., 2004, Real-time moose tracking: an internet based mapping application using GPS/GSM-collars: Alces (Thunder Bay, Ont.), v. 40, p. 13–21.

ESRI, 2004, ArcGIS Engine developer guide: Redlands, California, ESRI Press.

Fellbaum, C., ed., 1998, WordNet: An electronic lexical database: Cambridge, Massachusetts, MIT Press, http://wordnet.princeton.edu (accessed January, 2009).

Google spreadsheets function list, 2011, https://support.google.com/docs/bin/static.py?hl=en&topic=25273&page=table.cs (accessed November 2011).

McGrady, M., Ueta, M., Potapov, E.R., Utekhina, I.G., Masterov, V., Ladyguine, A., Zykov, V., Cibor, J., Fuller, M., and Seegar, W.S., 2003, Movements by juvenile and immature Steller's Sea Eagles *Haliaeetus pelagicus* tracked by satellite: The Ibis, v. 145, no. 2, p. 318–328, doi:10.1046/j.1474-919X.2003.00153.x.

MANUSCRIPT ACCEPTED BY THE SOCIETY 16 APRIL 2012

Printed in the USA

The Geological Society of America
Special Paper 492
2012

The new frontier of interactive, digital geologic maps: Google Earth–based multi-level maps of Virginia geology

Owen P. Shufeldt
Steven J. Whitmeyer*
Department of Geology & Environmental Science, James Madison University, Memorial Hall MSC 6903, Harrisonburg, Virginia 22807, USA

Christopher M. Bailey
Geology Department, College of William & Mary, McGlothlin-Street Hall 215, Williamsburg, Virginia 23187, USA

ABSTRACT

Digital geologic maps that use a virtual globe interface, like Google Earth (GE), are a relatively new medium for presenting geologic data and interpretations. This format incorporates significant advantages over traditional paper geologic maps and cross sections, including:

• A user-friendly and intuitive interface for novice users, which enhances the utility of geologic information for students and the general public;

• The ability to view multiple maps simultaneously and seamlessly transition between maps by zooming or panning;

• The option of displaying cross sections in situ on geologic maps as vertical interpretations of above ground or subsurface geology; and

• A facility for integrating map interpretations with individual outcrop and field data, which traditionally has been relegated to field books.

This paper outlines a digital maps package, composed of geologic maps of regions of Virginia, as a proof of concept and template for possible future expansion beyond state boundaries or into the realm of soils, geomorphological, or hydrological maps. Through collaboration among universities, state agencies, and federal organizations we have assembled a multi-layered, fully interactive map accessible through two portals: the stand-alone Google Earth application, and as a web page using the GE web browser plug-in (GE API). All maps within this package have selectable polygons, polylines ("paths"), and points ("placemarks"), many of which contain associated metadata, such as lithologic descriptions, fault information, outcrop orientation data, etc. At the smallest scale, a generalized geologic map of Virginia is displayed with a selectable overlay of regional physiographic provinces. As users pan and zoom, the maps automatically transition from generalized statewide maps to more refined

*Corresponding author: whitmesj@jmu.edu.

Shufeldt, O.P., Whitmeyer, S.J., and Bailey, C.M., 2012, The new frontier of interactive, digital geologic maps: Google Earth–based multi-level maps of Virginia geology, *in* Whitmeyer, S.J., Bailey, J.E., De Paor, D.G., and Ornduff, T., eds., Google Earth and Virtual Visualizations in Geoscience Education and Research: Geological Society of America Special Paper 492, p. 147–163, doi:10.1130/2012.2492(11). For permission to copy, contact editing@geosociety.org. © 2012 The Geological Society of America. All rights reserved.

regional maps and 1:24,000 scale quadrangle maps. Many of the map components (cross sections, explanations, and orientation symbols) cannot be created directly in GE but are added to the digital maps using KML scripts derived from an HTML-based toolkit.

Challenges related to the method of digital map development described herein include: effective importation of vector data from other GIS databases, style limitations inherent in GE, and time-consuming labor associated with the digitization of polygons and polylines in GE. There are also conceptual challenges at the user interface level, including possible misconceptions with the display of vertical cross sections due to the inability to look below the GE digital elevation model and associated surface imagery.

INTRODUCTION

Geoscience educators and professionals are often challenged by using 2-D static tools to present and teach 3-D spatial concepts like rock orientations and folded structures. Studies have discussed this difficulty while describing challenges faced when performing Euclidean or projective spatial tasks (Kastens and Ishikawa, 2006). The advent of virtual globes has provided a new medium for addressing these challenges (Butler, 2006; Lisle, 2006; Goodchild, 2008) through their facility in displaying spatial geological data in a virtual 3-D, dynamic environment (Hennessy and Feely, 2008; De Paor and Whitmeyer, 2011). Techniques of displaying geological data on virtual globes include overlaying map images, creating individual selectable polygons, importing data from a geographic information system (GIS) (Whitmeyer et al., 2010), and modifying Keyhole Markup Language scripts (KML) to link maps with associated Collaborative Design Activity (COLLADA) models of structural orientation symbols and cross sections (De Paor and Whitmeyer, 2011). While many of these topics have been previously presented and discussed (e.g., articles in Chen and Bailey, 2011), the literature lacks a thorough manual for creating a complete, multi-level, interactive maps package for virtual globes, specifically Google Earth.

This paper discusses the collaborative development of a prototype maps package for Google Earth (GE) displaying Virginia geology and containing a full range of geological data, including maps at multiple scales that automatically transition as users zoom in and out. The maps contain the most important features of geologic maps: lithologic units and faults with descriptions (metadata), structural data (orientation symbols of planar features), cross sections, and outcrop photos and notes. Other components include a layer that highlights the 7.5 min quadrangles that currently are included in the maps package with hyperlinks to state geological survey (Virginia Department of Mines, Minerals and Energy) publications and reference material for each quadrangle. This paper will also briefly discuss the GE web browser plug-in (GE API) and the advantages and disadvantages of displaying this maps package in such a platform. Finally, potential uses of digital geologic maps packages for professionals, educators, and the general public will be addressed.

OVERVIEW OF THE VIRGINIA GEOLOGIC MAPS PACKAGE

The Virginia geologic maps package is a GE-based, integrated set of maps, cross sections, and outcrop-scale data designed as an intuitive interface for a broad spectrum of users, including geology professionals, educators, students, and the general public. To access the maps package, either download the master KML file at:

http://csmres.jmu.edu/Geollab/Whitmeyer/web/visuals/GoogleEarth/VirginiaGeologicMaps.kml

or view it as a web page by utilizing the GE API plugin at:

http://csmres.jmu.edu/Geollab/Whitmeyer/web/visuals/GoogleEarth/Virginia/VirginiaMaps.html

Fast, seamless transitions between statewide maps, regional 1:50,000 scale geologic maps, and detailed 7.5 min quadrangle (1:24,000 scale) geologic maps are key features of the maps package. The user controls what maps and features are displayed by zooming and panning in the GE window. Similarly, detailed outcrop data, such as orientation symbols and photos, are only visible when the user is zoomed in close to the ground surface. This design prevents voluminous point data from cluttering the user's field of view when zoomed out to smaller scales.

At the smallest scale, where the field of view encompasses Virginia (or more of the eastern seaboard) a map of the physiographic provinces of Virginia (Fig. 1A) is visible, with the option to show the extensions of these physiographic provinces to the north (Fig. 1B) and south (Fig. 1C) by toggling check boxes in the GE Layers menu. These maps, and others in this maps package, were digitized from a variety of sources, the citations for which can be found in the description pop-up bubbles associated with each colored polygon.

As the user zooms closer to the ground surface, the physiographic province maps automatically transition to a generalized geologic map of Virginia (Fig. 2A; C.M. Bailey's 1999 "Simplified Geologic Map of Virginia," http://web.wm.edu/geology/virginia/provinces/pdf/va_geology.pdf). A west-to-east cross

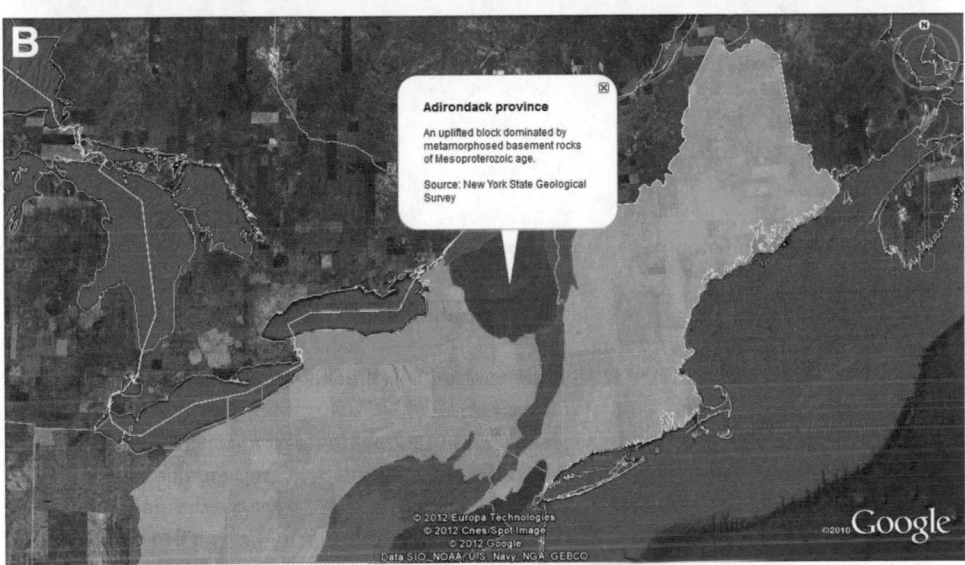

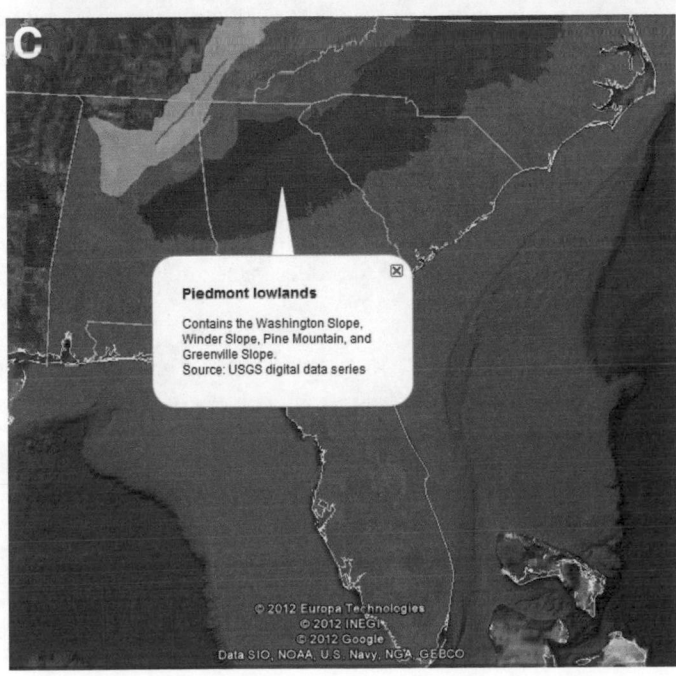

Figure 1. (A) Physiographic province map of Virginia with the provinces labeled. (B) Physiographic province map of the northeastern United States showing a general description of the selected polygon in the pop-up balloon. (C) Physiographic province map of the southeastern United States showing a general description of the selected polygon in the pop-up balloon.

Figure 2. (A) Simplified geologic map of Virginia as polygons with information in pop-up balloons (after C.M. Bailey's 1999 "Simplified Geologic Map of Virginia," http://web .wm.edu/geology/virginia/provinces/pdf/va_geology.pdf). (*Continued*).

Figure 2 (*continued*). (B) Northwest-southeast cross section interpreting subsurface geology spanning Virginia (after Williams et al., 2005). The cross section is elevated out of the subsurface by clicking on the "Elevate I-64 Cross Section" Tour highlighted in blue.

section interpretation of the subsurface geology can be viewed by tilting the view and running the "Elevate I-64 Cross Section" tour in the Virginia I-64 Cross Section folder (Fig. 2B). Continued zooming in the central Blue Ridge region displays a 1:50,000 scale map of the Blue Ridge to Valley and Ridge transition in the southern Shenandoah National Park area (Fig. 3A; Bailey, C.M., Gattuso, A.P., and Tadlock, E.D., 2008, Digital Geologic Map of the southern Shenandoah National Park region, Virginia, 1:50,000 scale, unpublished). With continued zooming in the Shenandoah Valley region, the largest scale (1:24,000) geologic maps appear (Fig. 3B). At present, the coverage of intermediate- to large-scale maps is restricted to northwestern Virginia. Figure 3B shows a typical region of partial coverage in the vicinity of Harrisonburg, Virginia. Work to expand map coverage at the 1:50,000 and 1:24,000 scales is ongoing.

Below we briefly discuss the geologic setting for Virginia and the Shenandoah Valley/Blue Ridge region, after which we discuss design components of the maps package in detail.

GEOLOGIC OVERVIEW OF VIRGINIA

The geology of Virginia is quite diverse, ranging from sediments currently being deposited in marshes and barrier islands along the Atlantic coast to Eocene volcanic rocks to metamorphic rocks formed over a billion years ago. At present, Virginia is located well within the North American tectonic plate along a passive margin setting. Virginia's geology is the result of two supercontinent cycles during the last billion years: (1) the formation of Rodinia to the opening of the Iapetus ocean, and (2) the formation of Pangaea to the opening of the Atlantic ocean.

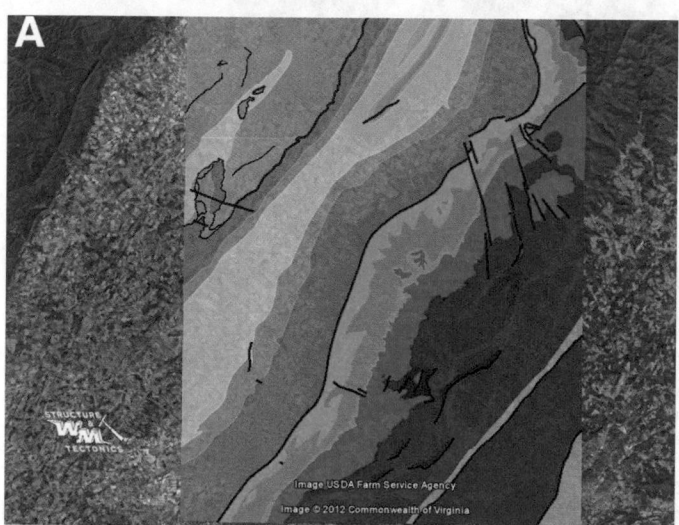

Figure 3. (A) 1:50,000 scale geologic map of the southern region of Shenandoah National Park and surrounding areas (after Bailey, C.M., Gattuso, A.P., and Tadlock, E.D., 2008, Digital Geologic Map of the southern Shenandoah National Park region, Virginia, 1:50,000 scale, unpublished). (B) Collection of 1:24,000 scale 7.5 min quadrangle geologic maps covering much of the same area as (A) (after Forte et al., 2005; Campbell et al., 2006, and references therein), with gaps showing quadrangles that have not yet been included in this map compilation.

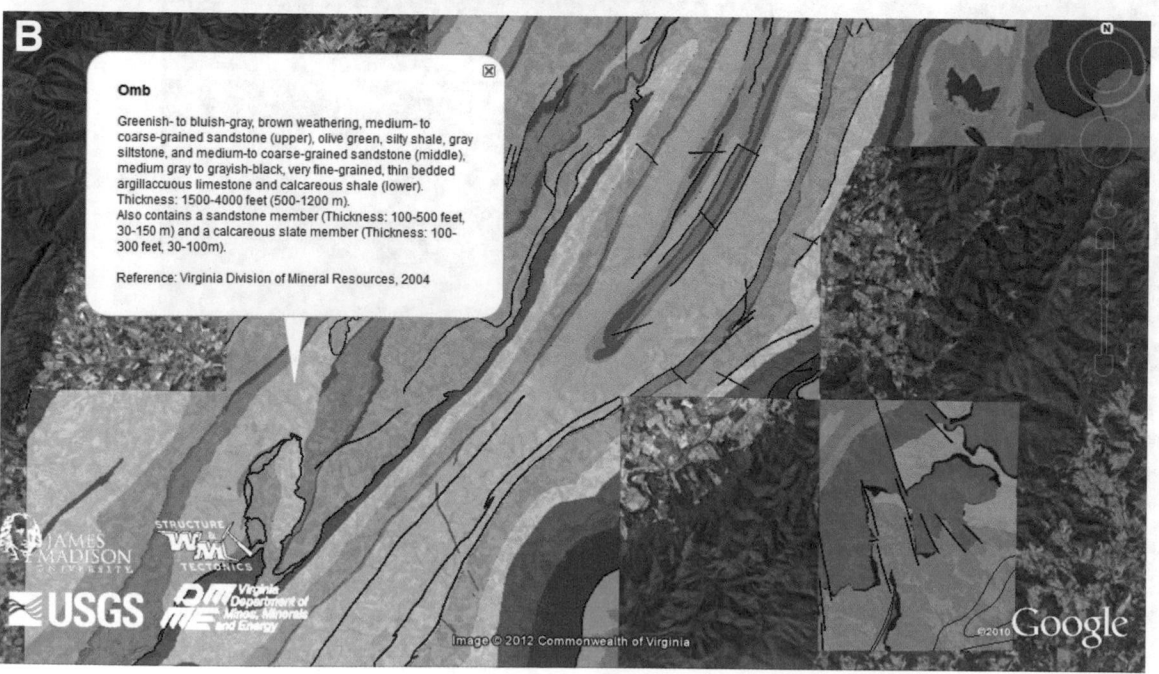

Traditionally, Virginia is divided into five physiographic provinces, each with its own unique topographic character that reflects the underlying materials and geologic structure of the province (Fig. 1A; Fenneman and Johnson, 1946; Bingham, 1991). From west to east, the provinces include the Allegheny Plateau, Valley and Ridge, Blue Ridge, Piedmont, and Coastal Plain (Fig. 1A). The Allegheny Plateau and Valley and Ridge provinces are part of the Appalachian foreland basin and are underlain by Paleozoic sedimentary rocks. Rocks on the Plateau are generally flat-lying, while rocks in the Valley and Ridge were folded and faulted during the late Paleozoic Alleghanian Orogeny (Fig. 2; Virginia Division of Mineral Resources, 1993, 2003). Differential erosion in the Valley and Ridge has produced the distinctive topography of the province. The Blue Ridge province encompasses Virginia's highest peaks and is characterized by relatively high relief. Rocks of the Blue Ridge province include 1.0–1.2 Ga granitoid gneisses and an overlying Neoproterozoic–Early Paleozoic cover sequence of metamorphosed volcanic and sedimentary rocks (Fig. 2; Bailey et al., 2006). A complex suite of Proterozoic to Paleozoic metamorphic and igneous rocks underlies the Piedmont province, forming the hinterland of the Appalachian Orogen, sensu lato (Horton et al., 1989). Major ductile fault zones and faults bound Piedmont terranes, and the province is cut by a series of early Mesozoic rift basins. Early Jurassic magmatism produced dikes, sills, and flows in the Culpeper basin, and mafic dikes throughout much of the Piedmont (Virginia Division of Mineral Resources, 1993, 2003). The Coastal Plain consists of a gently inclined sequence of Cretaceous to Recent sedimentary rocks and sediments deposited under shallow marine, estuarine, and fluvial conditions associated with changing sea levels during the past 100 million years (Fig. 2; Mixon et al., 1989; Virginia Division of Mineral Resources, 1993, 2003).

MAP DEVELOPMENT IN GOOGLE EARTH

The integrated geologic maps package described in detail below was developed specifically for presentation using GE, allowing for features to be completely controlled and displayed using the tools and controls within the program. However, the geologic maps displayed were, almost without exception, originally developed using other graphics platforms, such as ArcGIS and Adobe Illustrator. Thus, GE map development typically begins in one of two ways: by manually creating polygons and paths (poly-lines) through tracing over an overlay imported as a raster image (JPEG, PNG, TIFF); or by importing shapefile data from ArcGIS and modifying the vector data (polygon colors, line weights, etc.) to be consistent with the rest of the maps package. Individual polygons and paths are used for lithologic units and linear features, respectively, to enable users to access descriptions and metadata via pop-up bubbles by clicking on the features. This can also be accomplished by creating image overlays for each unit as a separate PNG (portable networks graphics) file, but GE renders raster overlays much more slowly than vector polygons. Fast and efficient rendering of map images is a constant challenge for any maps package that displays data on a quadrangle or larger scale.

Image Import Method

As discussed above, many digital maps are created in programs like Adobe Illustrator, where the final map can be exported as a JPEG or PNG file. If the map is a quadrangle, with known bounding coordinates, importing these files into GE simply requires using the "Add Image Overlay" tool and specifying the coordinates for the north, south, east, and west boundaries of the map. Once this is done, polygons can be created using the Add Polygon feature to manually trace the outline of each lithologic unit. To enhance interactivity, adding unit descriptions into the "Description" field of the polygon allows users to click anywhere within the polygon and obtain a pop-up balloon with the lithologic information.

Shapefile Import Method

For maps constructed in ArcGIS, the latest versions of Google Earth Pro allow the user to import unit polygons or linework from a shapefile (.shp). However, Google Earth Pro currently limits the number of unique colors that can be assigned to shapefile elements, and thus the shapefile sometimes is necessarily imported with a random color scheme, and all polygon colors have to be manually reassigned. Our currently preferred method is to export shapefiles or layers as KML files out of ArcGIS using Arc Toolbox, which does a better job of preserving polygon styles following the shapefile to KML conversion. Linework, however, is still problematic when imported into GE from ArcGIS, as individual lines in ArcGIS often become segmented into separate lines when converted to KML code. Thus, we typically have to manually redigitize all of the linework within GE. Another issue with lines in GE is that there is no option for creating ornamentations. Therefore, adding ornaments like barbs on thrust faults or tick marks on normal faults is a non-trivial task.

Adding Structural Symbols

At the largest map scale (1:24,000), structural data often is necessary to get a thorough understanding of the surface and subsurface geology. However, since this GE maps package is designed for a wide range of users, our goal is to display structural data in a manner that attempts to preclude typical confusion regarding strike and dip of planar surfaces (Kastens and Ishikawa, 2006). As discussed in Whitmeyer et al. (2010), we have found that the most accurate and intuitive way to display outcrop orientation symbols is as 3-D models hovering slightly above the ground surface (Fig. 4A).

The construction of these models previously required some knowledge of SketchUp, COLLADA models, and KML scripting (De Paor and Whitmeyer, 2011). Some basic KML scripting is included in Tables 1 and 2 and within other papers in this volume (e.g., De Paor et al., this volume, Chapter 6), but many

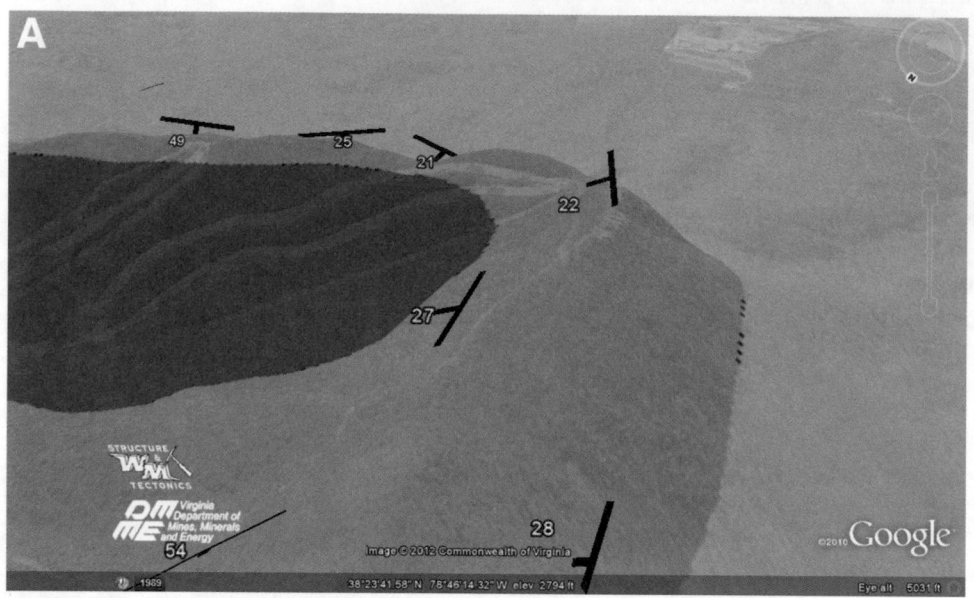

ORIENTATION SYMBOL GENERATOR   **B**

To generate and position orientation symbols in Google Earth:

1. *Complete the form below and click on "Enter data for a symbol" to generate the necessary KML code in a popup window. Repeat this step to enter data for as many symbols as you desire.*

2. *Click on "Finish data entry". The popup window with KML code for all of your symbols will appear.*

3. *Copy and paste the code from the popup window (use "Paste Special" - unformatted text) into a blank document and save that document (as plain text) with a .kml extension.*
   IMPORTANT: make sure you include the .kml extension at the end of your document name.

   Recommended: *Firefox users can "Save Page As" the popup window as a text file with a .kml extension.*

4. *In Google Earth, open the .kml file you created (or double-click on the icon).*

Figure 4. (A) Oblique view of Massanutten Mountain, Virginia, looking northwest with oriented, 3-D strike and dip symbols hovering above topography. (B) Orientation Symbol Generator web page. (*Continued.*)

Latitude:	⊙ Decimal Degrees [ ]	or ○ Degrees [ ]	Minutes [ ]	Seconds [ ]
Longitude:	⊙ Decimal Degrees [ ]	or ○ Degrees [ ]	Minutes [ ]	Seconds [ ]

Unit / Formation: [   ]

Feature measured: ⊙ Bedding   ○ Overturned bedding   ○ Foliation   ○ Lineation
Strike / trend (azimuth, 0-359): [000]
Dip / plunge amount (0-90): [00]
        ⊙ Right hand rule   or   Dip Direction: ○ North   ○ South   ○ East   ○ West
Display dip number next to   ○ Yes   ⊙ No
  symbol in Google Earth?

Symbol color: ⊙   ○   ○   ○   ○   ○   ○   ○

Notes: [   ]

(Enter data for a symbol)

(Finish data entry)

geologists will find it easier to use a web-based tool we have developed that partially automates the creation and positioning of orientation symbols. The Orientation Symbol Generator (Fig. 4B) is available at:

http://csmres.jmu.edu/Geollab/Whitmeyer/web/visuals/
maptools.html

To use this tool, the user supplies location (latitude and longitude) coordinates, formation name, azimuth of strike (or trend), dip (or plunge) angle and direction, and any additional notes about the outcrop. This generates code in a pop-up window that is saved as a KML file (Table 1), which, once opened in GE, displays a series of oriented 3-D strike and dip symbols (Fig. 4A) or lineation arrows. While all structural symbols created with this tool display basic outcrop data (lithologic unit, structural orientation) when selected, symbols can also include pictures of, and notes about, the outcrop at which the measurement was taken (Fig. 4C). The description and images can be added to the text balloon using basic KML code (see Table 1).

Adding Cross Sections

In order to display cross sections in GE, COLLADA models of transparent vertical rectangles are necessary, on which PNG images of the geologic interpretation are superimposed. Methods for doing this using SketchUp and the model import feature in GE have been described elsewhere (De Paor and Whitmeyer, 2011; Hill and Harrison, this volume). To simplify this process and remove the necessity of using SketchUp, a web-based tool has been created to position a cross section in the correct location and orientation in GE. This tool (Fig. 5A) is available at:

http://csmres.jmu.edu/Geollab/Whitmeyer/web/visuals/
GoogleEarth/tools/XS.html.

The cross section image must first be constructed in a graphics program like Adobe Illustrator. The image file must have a transparent background and be saved in the .png file format to the user's desktop. A reference .dae (digital asset exchange) file containing the COLLADA model for the vertical cross section rectangle must also be downloaded from the website and saved

Figure 4 (*continued*). (C) View of 3-D strike and dip model with a text balloon containing metadata and an outcrop photo. See Table 1 for the relevant KML code.

TABLE 1. KML CODE GENERATED BY STRIKE AND DIP SYMBOL GENERATOR WEB PAGE AT
http://csmres.jmu.edu/geollab/whitmeyer/web/visuals/googleearth/tools/sd.html

```
<Placemark>
        <name>Symbol3</name>
        <Model id="SDmodel">
                <altitudeMode>relativeToGround</altitudeMode>
                <Location>
                        <longitude>-78.74285500000001</longitude>
                        <latitude>38.406964</latitude>
                        <altitude>40</altitude>
                </Location>
                <Orientation>
                        <heading>240</heading>
                        <tilt>0</tilt>
                        <roll>-89</roll>
                </Orientation>
                <Scale>
                        <x>40</x>
                        <y>50</y>
                        <z>50</z>
                </Scale>
                <Link>
                        <href>http://csmres.jmu.edu/Geollab/Whitmeyer/web/visuals/
                                GoogleEarth/tools/SDsymbol.dae</href>
                </Link>
                <ResourceMap>
                </ResourceMap>
        </Model>
</Placemark>

<Placemark>
        <name>Massanutten Fm.3</name>
        <description><![CDATA[Unit: <b>Massanutten Fm.</b><br>
strike & dip of bedding: 240, 89 N<br>
<br>
Notes:<br>
Subvertical sandstones offset at Harshberger Gap<br>
<br>
<img
src="http://csmres.jmu.edu/Geollab/Whitmeyer/web/visuals/GoogleEarth/Virginia/outcropphotos/SmHarshberger.jpg">
<br>
View looking west from Massanutten Rd. at Harshberger Gap<br>
(4" wide compass case for scale)]]></description>
        <styleUrl>#sn_shaded_dot30</styleUrl>
        <Point>
                <coordinates>-78.74285500000001,38.406964,0</coordinates>
        </Point>
</Placemark>

<Placemark>
        <name>89</name>
        <styleUrl>#sn_no_icon</styleUrl>
        <Point>
                <coordinates>-78.74335500000001,38.407364,0</coordinates>
        </Point>
</Placemark>
```

Note: Three Placemarks are generated for each orientation symbol. The first one listed displays the 3-D model of the symbol positioned with the <Location tags> and aligned using the <Orientation> tags. Note that the symbol model is stored on an external server and referenced by the <Link> <href> tags. The second Placemark contains outcrop information that appears in a pop-up balloon. In this case it includes a photo stored on an external server and referenced with the <img src="…"> tag. The third Placemark displays the dip number next to the symbol. This is an option that can be selected on the Strike and Dip Symbol Generator web page.

to the user's desktop. Once these files are in place, the user enters latitude and longitude data, desired altitude, orientation (azimuthal strike of cross section), length, and height (in meters) of the cross section image into the input fields on the web page. This generates (in a pop-up window) the KML code for positioning the cross section model that is saved with a .kml file extension and opened in GE (Fig. 5B). If desired, all of the elements of the cross section can be packaged together as a stand-alone file by saving the cross section folder as a KMZ file from within GE.

As is evident from the Figure 5B, there are some drawbacks to cross section display in GE. First, GE does not easily or accurately display images below the ground surface. One way to get around this problem is to add a feature to elevate or "drag" the cross sections up out of the ground to show that they are vertical subsurface interpretations of the geology shown on the map. This can be done by creating a GE Tour, which is an option on the web page shown in Figure 5A (detailed in De Paor et al., this volume, Chapter 6) or by using the time slider feature (upper left corner of Fig. 5B) as detailed in De Paor and Whitmeyer (2011). The time slider feature can also be used to display a series of cross sections above the terrain, which can illustrate along-strike features such as a down-plunge projection (i.e., the syncline shown in Figure 5B).

Citing Authors and Copyrights

Building an integrated digital maps package necessitates assembling existing geologic maps from many sources. This prompts the question of how to display source information for the authors and organizations that created the individual maps. We list references at the bottoms of pop-up text balloons for material cited from publications, but users will not see these references if the balloons aren't opened. Users can be forced to view a pop-up window when individual maps are first loaded (e.g., the White Hall quadrangle; Fig. 6) by adding a <Description> to the Network Link folder (see third <Folder> in Table 2). For basic source acknowledgment we decided to use screen overlays of logos for the authoring institutions and have the logos appear or disappear depending on whether the specific map is visible to the user (bottom left corner of Figures 3B, 4C, 5B). Screen overlays cannot be created from within the GE application; they require external KML coding. Screen Overlays can be created using the web page at

http://csmres.jmu.edu/Geollab/Whitmeyer/web/visuals/ GoogleEarth/tools/SO.html

or see the KML tutorial at

http://code.google.com/apis/kml/documentation/kml _tut.html.

Integrating the Maps

The most significant concern for creating a multi-layer, integrated system of geologic maps is speed of rendering of the images. Any digital maps package becomes less convenient for the general public to use if it takes a significant amount of time for individual maps to load. Early versions of our maps package were developed with image overlays for lithologic units, which caused GE to quickly grind to a halt as the number of maps on display increased. Similarly, our early attempts to display the full aerial extent of every map at all times, regardless of the viewpoint of the user, slowed GE considerably. Thus, we settled on vector polygons and paths for lithologic units and linework, and we developed a Network Linked system of individual map files controlled through a downloadable master KML file.

Network Links is the mechanism that allows KML files to load other KML files based on certain criteria. When used with <Region>, <LatLonAltBox>, and <Lod> tags the master KML file can control when maps (as individual KML or KMZ files) are displayed based on the viewpoint of the user. Table 2 shows KML code from the VirginiaGeologicMaps.kml master control file, downloadable from:

http://csmres.jmu.edu/Geollab/Whitmeyer/web/visuals/ GoogleEarth/VirginiaGeologicMaps.kml

The <Region> tag defines the view area of the GE window within which a given map will be displayed. The children of the <Region> tag include <LatLonAltBox> which sets the boundaries of the viewable area, and <Lod> (Level of detail) which sets the zoom altitude at which the map will be displayed. The use of these tags greatly enhances the speed of the maps package by only loading a given map when users are looking at areas and zoom levels appropriate for viewing specific map features. We also use these tags to display photos and orientation symbols only when users are zoomed in close to the ground surface (e.g., Fig. 4A). This alleviates the problem of unnecessary clutter in the viewing window when zoomed out to a wider field of view. Note that the individual map KML files are loaded with the <NetworkLink><Link><href> tags that are coded after the <Region> tags (Table 2).

The Maps Package Viewed within a Web Browser

KML and KMZ files can also be viewed in a web page by installing the GE API in your web browser of choice. With knowledge of some basic HTML, JavaScript, and a free license key from Google (http://code.google.com/apis/maps/ signup.html), a web designer can include a GE window in a web page and load any desired KML or KMZ files. All of the functions described previously in this paper will behave exactly the same within the GE API. See De Paor et al. (this volume, Chapter 6) for a more complete description of similarities and differences between the GE application and the GE API.

One distinct advantage to using the GE API is the capability to restrict a user's ability to wander off task while using custom KML and KMZ files in GE. With some JavaScript coding,

CROSS SECTION MODEL GENERATOR

A

To generate and position a vertical cross section model in Google Earth:

1. *Draw your cross section in a graphics program (like Adobe Illustrator) and save it to your desktop as a PNG file (with a transparent background) named "image.png" - or type the file name in the box below.*

2. *Download the file* **Xsection.dae** *and save it to your desktop (right-click, "Save Link As...")*

3. *Complete the form below and click on "Enter information" to generate the necessary KML code in a popup window.*

4. *Copy and paste the code from the popup window (use "Paste Special" - unformatted text) into a blank document, and save that document (as plain text) with a .kml extension to your desktop. IMPORTANT: make sure you include the .kml extension at the end of your document name.*

   Recommended: *Firefox users can "Save Page As" the popup window as a text file with a .kml extension.*

5. *In Google Earth, open the .kml file you created (or double-click on the icon).*

6. *In Google Earth, save ("Save Place as...") the "Cross Section" folder as a KMZ file.*

Filename for cross section image (png format): image.png

Longitude (in decimal degrees) for NW corner of cross section: -78.8095
Latitude (in decimal degrees) for NW corner of cross section: 38.3817
Altitude relative to ground (for base of cross section): 0
Orientation (i.e. strike: 0 - 180) of cross section: 000
Length of cross section (in meters): 10000
Height of cross section (in meters): 1000

Create Tour to elevate cross section?   ○ Yes ⊙ No
   If Yes, enter amount to elevate cross section (in meters): 0
   (Recommended: to elevate the base of the cross section to the ground surface, set the Altitude to the negative value of the elevation amount entered above.)

(Enter information)

Figure 5. (A) Cross Section Model Generator web page. (B) A series of cross sections displayed above the geologic maps of the Massanutten Synclinorium.

Figure 6. View of the White Hall quadrangle geologic map in Google Earth showing source information displayed in a pop-up balloon when the quadrangle map is first viewed. From Doctor et al. (2010).

a menu of radio button selections can be displayed on the same web page as the GE API, so that users can display only those KML files that they desire. Similarly, the JavaScript programmer can make obvious which content is available to users. The current prototype web page for the Virginia geologic maps package: (http://csmres.jmu.edu/Geollab/Whitmeyer/web/visuals/ GoogleEarth/Virginia/VirginiaMaps.html) has toggle buttons to turn on/off the specific geologic maps and cross sections (Fig. 7). Other less relevant items that are displayed in the Layers and Places windows of the stand-alone GE application are not visible to users, which streamlines and simplifies the maps functionality. Initial informal feedback suggests that novice users prefer the direct, but restricted, GE API interface, while more experienced users of Google Earth prefer to have the full functionality of the stand-alone GE application.

APPLICATIONS OF THE MAPS PACKAGE

We envision the Virginia maps package as a useful compilation of geologic information for geology professionals and novices alike. The intuitive design of the maps interface, coupled with GE's search function, should provide quick and easy access to rock types and other geologic data at multiple levels of detail. As examples: (1) industry professionals, such as construction or environmental engineers, could quickly find information on subsurface rock type in a construction or wetland site; (2) professional geologists or advanced students could examine bedding orientations of rock outcrops at locations like the southern end of the Massanutten Synclinorium to determine regional structure (Fig. 4A); or (3) teachers could design a GE Tour to virtually visit outcrops as a preview for an upcoming school field trip.

TABLE 2. KML CODE SNIPPETS FROM THE MASTER VIRGINIAGEOLOGICMAPS.KML FILE THAT USERS DOWNLOAD TO OPEN THE MAPS PACKAGE IN GOOGLE EARTH

```
<Document>
        <name>Virginia Geologic Maps</name>
        <open>1</open>
        <Folder>
                <name>Physiographic Provinces</name>
                <Region>
                        <LatLonAltBox>
                                <north>42.00</north>
                                <south>34.00</south>
                                <east>-73.00</east>
                                <west>-86.00</west>
                                <minAltitude>0</minAltitude>
                                <maxAltitude>0</maxAltitude>
                        </LatLonAltBox>
                        <Lod>
                                <minLodPixels>100</minLodPixels>
                                <maxLodPixels>2000</maxLodPixels>
                                <minFadeExtent>0</minFadeExtent>
                                <maxFadeExtent>0</maxFadeExtent>
                        </Lod>
                </Region>
                <NetworkLink>
                        <Link>
                                <href>http://csmres.jmu.edu/Geollab/Whitmeyer/web/
                                        visuals/GoogleEarth/Virginia/
                                        PhysiographicProvinces.kml</href>
                                <viewRefreshMode>onRegion</viewRefreshMode>
                        </Link>
                </NetworkLink>
        </Folder>

        <Folder>
                <name>Virginia Geologic Map and Cross-section</name>
                <open>1</open>
                        <Region>
                                <LatLonAltBox>
                                        <north>42.00</north>
                                        <south>34.00</south>
                                        <east>-73.00</east>
                                        <west>-86.00</west>
                                        <minAltitude>0</minAltitude>
                                        <maxAltitude>0</maxAltitude>
                                </LatLonAltBox>
                                <Lod>
                                        <minLodPixels>1500</minLodPixels>
                                        <maxLodPixels>10000</maxLodPixels>
                                        <minFadeExtent>0</minFadeExtent>
                                        <maxFadeExtent>0</maxFadeExtent>
                                </Lod>
                        </Region>
                <Folder>
                        <name>Virginia Geologic Map</name>
                        <NetworkLink>
                                <Link>
                                        <href>http://csmres.jmu.edu/Geollab/Whitmeyer/
                                                web/visuals/GoogleEarth/Virginia/
                                                VirginiaGeologicMap.kml</href>
                                        <viewRefreshMode>onRegion</viewRefreshMode>
                                </Link>
                        </NetworkLink>
                </Folder>
```

Note: This master code loads each of the maps in turn depending on the users viewpoint within Google Earth. The <Region><LatLonAltBox><Lod> tags set the viewing bounds for each map and the <NetworkLink><Link><href> tags load the individual maps from an external server. The <Description> tag contains source information for the original map.

(*continued*)

TABLE 2. (*continued*)

```
<Folder>
            <name>Virginia I-64 Cross Section</name>
            <NetworkLink>
            <visibility>0</visibility>
                    <Link>
                            <href>http://csmres.jmu.edu/Geollab/Whitmeyer/
                                    web/visuals/GoogleEarth/Virginia/
                                    I64XS.kmz</href>
                            <viewRefreshMode>onRegion</viewRefreshMode>
                    </Link>
            </NetworkLink>
        </Folder>
    </Folder>

    <Folder>
            <name>White Hall Geologic Map</name>
            <description>SOURCE: Doctor, D.H., Orndorff, R.C., Parker, R.A., Weary, D.J., and Repetski, J.E., 2010,
Geologic map of the White Hall quadrangle, Frederick County, Virginia, and Berkeley County, West Virginia: U.S. Geological
Survey Open-File Report 2010–1265, 1 pl., scale 1:24,000, available only at http://pubs.usgs.gov/of/2010/1265</description>
            <Region>
                    <LatLonAltBox>
                            <north>39.375</north>
                            <south>39.25</south>
                            <east>-78.125</east>
                            <west>-78.25</west>
                    </LatLonAltBox>
                    <Lod>
                            <minLodPixels>192</minLodPixels>
                    </Lod>
            </Region>
            <NetworkLink>
                    <Link>
            <href>http://csmres.jmu.edu/Geollab/Whitmeyer/web/visuals/GoogleEarth/Virginia/WhiteHall.kmz</href>
                            <viewRefreshMode>onRegion</viewRefreshMode>
                    </Link>
            </NetworkLink>
        </Folder>
        ...
        ...
</Document>
```

Note: This master code loads each of the maps in turn depending on the users viewpoint within Google Earth. The <Region><LatLonAltBox><Lod> tags set the viewing bounds for each map and the <NetworkLink><Link><href> tags load the individual maps from an external server. The <Description> tag contains source information for the original map.

The touring capability of GE can be a powerful tool for illustrating geology to students and the general public. Scripted tours, in which the user starts the tour and the program automatically flies over a terrain, are intuitively appealing for students raised with digital devices. GE Tours can be created with relative ease, incorporating many of the map elements discussed in previous sections of this manuscript, and then saved as KMZ files. GE Tours can include a narrated audio track as well as text/image balloons that turn on and off as required during the tour. See Treves and Bailey (this volume) for details on creating effective tours in GE. In the geology department at the College of William and Mary tours of the Blue Ridge and Great Valley regions (Fig. 3A) are used to introduce students to the topography and geology of the areas prior to visiting the region on a field trip. Tours can highlight salient topographic features and illustrate the linkage between the topography and bedrock geology. Additionally,

tours serve as a springboard for individual student inquiry into a region. Finally, in situations where student access to outcrops is limited by landowner restrictions or individual mobility issues, virtual tours may provide the only way to examine geology in a setting that, in many respects, emulates the natural one.

CONCLUSIONS

The geologic maps package presented here provides a new, intuitive, digital interface for the investigation of geologic maps and data. The maps package is currently a prototype that focuses on Virginia, and specifically the Shenandoah Valley region of northwestern Virginia. However, it is hoped that this design can be exported to other states and regions to create an integrated continental-scale digital geologic maps package. Applications for this sort of publicly available maps package include quick

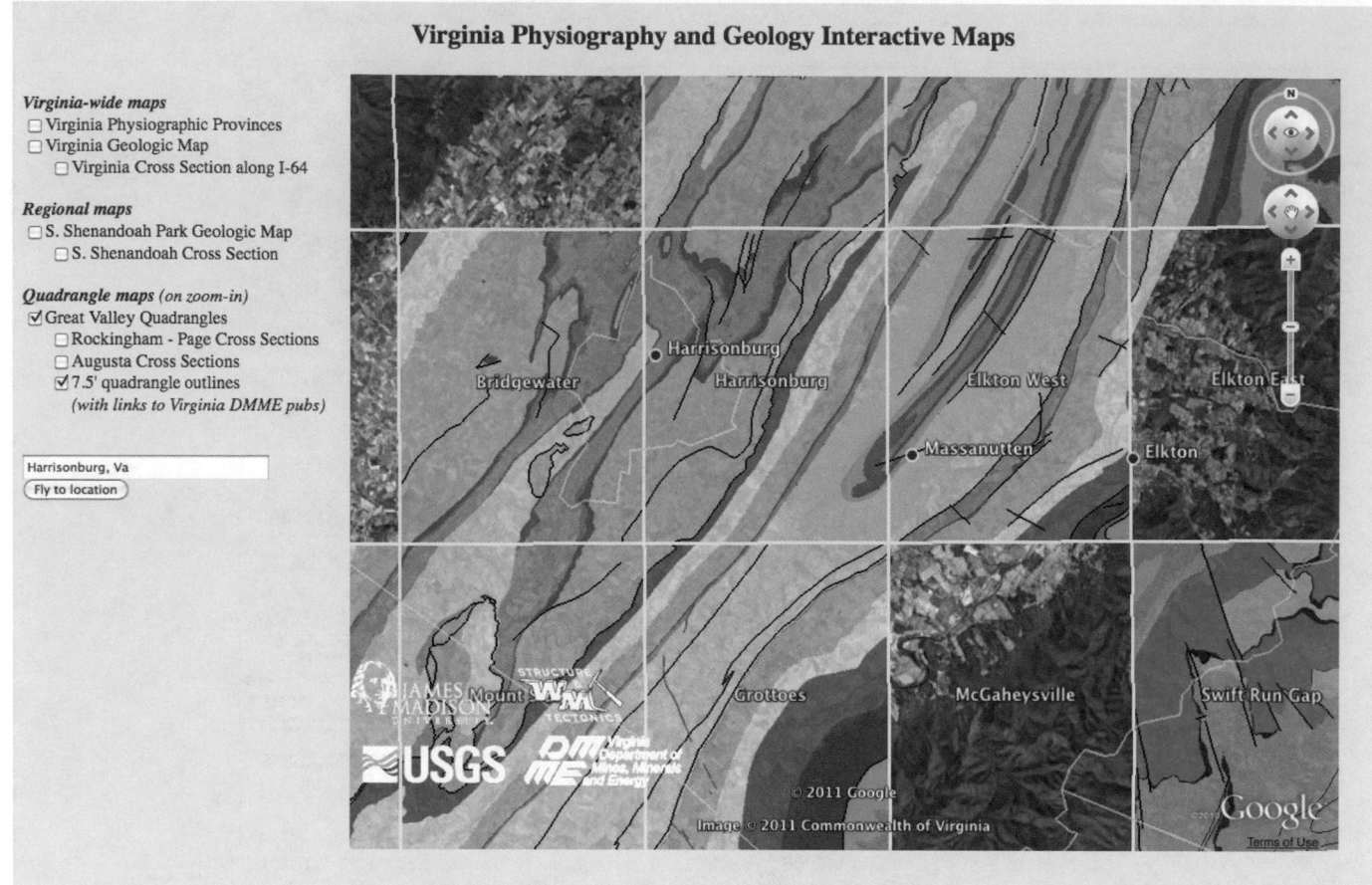

Figure 7. The Virginia geologic maps package viewed in a web browser using the Google Earth API. The maps displayed are toggled by the buttons to the left of the Google Earth window, so that several of the Great Valley quadrangle maps are displayed along with yellow 7.5′ quadrangle outlines. The region shown is similar to the region in Figure 3B. This web page can be accessed at: http://csmres.jmu.edu/Geollab/Whitmeyer/web/visuals/GoogleEarth/Virginia/VirginiaMaps.html.

access to geologic information for geology novices as well as professionals, and custom tours for presentations and inquiry-based investigations by students and the general public. This prototype multi-level maps package was developed with collaboration among universities (James Madison University, College of William and Mary), state agencies (Virginia Department of Mines, Minerals and Energy, Department of Geology and Mineral Resources) and federal organizations (U.S. Geological Survey). We envision the expansion of this effort to other states and regions by similar collaborative teams, ultimately resulting in a continent-wide package of digital geologic maps that are universally accessible.

ACKNOWLEDGMENTS

This manuscript was improved by comments from Andy Bobyarchick, Jesse Hill, and John Bailey. Early versions of this maps package were improved though comments and suggestions by Declan De Paor, Dan Doctor, Randy Orndorff, and Matt Heller, among others. This work has been partially supported by USGS EDMAP, Virginia DMME STATEMAP, and NSF grants (0837049, 1022782, 1034660). Any opinions, findings, and conclusions or recommendations expressed in this paper are those of the authors and do not necessarily reflect the views of the USGS, Virginia DMME, National Science Foundation, or Google Inc.

REFERENCES CITED

Bailey, C.M., Southworth, S., and Tollo, R.P., 2006, Tectonic history of the Blue Ridge north-central Virginia, *in* Pazzaglia, F.J., ed., Excursions in Geology and History: Field Trips in the Middle Atlantic States: Geological Society of America Field Guide 8, p. 113–134, doi:10.1130/2006.fld008(07).

Bingham, E., 1991, Physiographic diagram of Virginia: Virginia Division of Mineral Resources Publication 105, scale ~1:685,000.

Butler, D., 2006, Virtual globes: The web-wide world: Nature, v. 439, p. 776–778, doi:10.1038/439776a.

Campbell, E.V., Hibbitts, H.A., Williams, S.T., Duncan, I.J., Reis, J.S., Floyd, J.M., and Wilkes, G.P., 2006, Interstate 81 Corridor Digital Geologic Compilation, Virginia DMME Open-File Report 06-01.

Chen, A., and Bailey, J., eds., 2011, Virtual Globes in Science: Computers & Geosciences, v. 37, no. 1, p. 1–110.

De Paor, D.G., and Whitmeyer, S.J., 2011, Geological and geophysical modeling on virtual globes using KML, COLLADA, and Javascript: Computers & Geosciences, v. 37, no. 1, p. 100–110, doi:10.1016/j.cageo.2010.05.003.

De Paor, D.G., Whitmeyer, S.J., Marks, M., and Bailey, J.E., 2012, this volume, Geoscience applications of client/server scripts, Google Fusion Tables, and dynamic KML, *in* Whitmeyer, S.J., Bailey, J.E., De Paor, D.G., and Ornduff, T., eds., Google Earth and Virtual Visualizations in Geoscience Education and Research: Geological Society of America Special Paper 492, doi:10.1130/2012.2492(06).

Doctor, D.H., Orndorff, R.C., Parker, R.A., Weary, D.J., and Repetski, J.E., 2010, Geologic map of the White Hall quadrangle, Frederick County, Virginia, and Berkeley County, West Virginia: U.S. Geological Survey Open-File Report 2010-1265, 1 pl., scale 1:24,000.

Fenneman, N.M., and Johnson, D.W., 1946, Physical divisions of the United States: Washington, D.C., U.S. Geological Survey map, scale 1:7,000,000.

Forte, A.M., Wooton, K.M., Hasty, B.A., and Bailey, C.M., 2005, Bedrock geology of the Swift Run Gap 7.5′ Quadrangle, Blue Ridge Province, Virginia, Geological Society of America Abstracts with Programs, v. 37, no. 2, p. 34.

Goodchild, M.F., 2008, The use cases of digital earth: International Journal of Digital Earth, v. 1, no. 1, p. 31–42, doi:10.1080/17538940701782528.

Hennessy, R., and Feely, M., 2008, Visualization of Magmatic Emplacement Sequences and Radioelement Distribution Patterns in a Granite Batholith: An Innovative approach using Google Earth, *in* De Paor, D., ed., Google Earth Science: Journal of the Virtual Explorer, Electronic Edition, v. 30, paper 2.

Hill, J.S., and Harrison, M.J., 2012, this volume, Terrain modification in Google Earth using Google SketchUp: An example from the Western Blue Ridge of Tennessee, *in* Whitmeyer, S.J., Bailey, J.E., De Paor, D.G., and Ornduff, T., eds., Google Earth and Virtual Visualizations in Geoscience Education and Research: Geological Society of America Special Paper 492, doi:10.1130/2012.2492(18).

Horton, J.W., Jr., Avery, A.D., Jr., and Rankin, D.W., 1989, Tectonostratigraphic terranes and their Paleozoic boundaries in the central and southern Appalachians, *in* Dallmeyer, R.D., ed., Terranes in the Circum-Atlantic Paleozoic Orogens: Geological Society of America Special Paper 230, p. 213–245.

Kastens, K., and Ishikawa, T., 2006, Spatial thinking in the geosciences and cognitive sciences: A cross-disciplinary look at the intersection of the two fields, *in* Manduca, C., and Mogk, D., eds., Earth and Mind: How Geologists Think and Learn about the Earth: Geological Society of America Special Paper 413, p. 53–76, doi:10.1130/2006.2413(05).

Lisle, R.J., 2006, Google Earth: A new geological resource: Geology Today, v. 22, p. 29–32, doi:10.1111/j.1365-2451.2006.00546.x.

Mixon, R.B., Berquist, C.R., Jr., Newell, W.L., Johnson, G.H., Powars, D.S., and Schindler, J.S., and Rader, E.K., 1989, Geologic map and generalized cross sections of the Coastal Plain and adjacent parts of the Piedmont, Virginia: U.S. Geological Survey Miscellaneous Investigations, Map I-2033, 1:250,000 scale.

Treves, R., and Bailey, J.E., 2012, this volume, Best practices on how to design Google Earth tours for education, *in* Whitmeyer, S.J., Bailey, J.E., De Paor, D.G., and Ornduff, T., eds., Google Earth and Virtual Visualizations in Geoscience Education and Research: Geological Society of America Special Paper 492, doi:10.1130/2012.2492(28).

Virginia Division of Mineral Resources, 1993, Geologic Map of Virginia: Virginia Division of Mineral Resources, scale 1:500,000.

Virginia Division of Mineral Resources, 2003, Digital representation of the 1993 geologic map of Virginia: Virginia Division of Mineral Resources Publication 174 [CD-ROM; 2003, December 31]. Adapted from Virginia Division of Mineral Resources, 1993, Geologic map of Virginia and Expanded Explanation: Virginia Division of Mineral Resources, scale 1:500,000.

Whitmeyer, S.J., Nicoletti, J., and De Paor, D.G., 2010, The digital revolution in geologic mapping: GSA Today, v. 20, no. 4–5, p. 4–10, doi:10.1130/GSATG70A.1.

Williams, S.T., Bleick, H., Carter, M.W., and Berquist, C.R., 2005, Generalized Geologic Cross Section of Virginia: unpublished Virginia Division of Mineral Resources data.

MANUSCRIPT ACCEPTED BY THE SOCIETY 16 APRIL 2012

The Geological Society of America
Special Paper 492
2012

Automated export of GIS maps to Google Earth: Tool for research and teaching

Peter L. Guth*

Department of Oceanography, U.S. Naval Academy, Annapolis, Maryland 21402, USA

ABSTRACT

Google Earth offers an excellent example of software design balancing enough power while retaining a simple and intuitive interface, and provides tremendous capabilities both for teaching and research interaction. It displays data with a well-documented, standard keyhole markup language (KML) format on generally high resolution base imagery, with roads, borders and other relevant layers which can be turned on and off, and allows easy combination of multiple data sets. Google Earth lacks the analysis capabilities of geographical information system (GIS) software, but is outstanding for visualization and dissemination of results. Users can zoom in and out and view animations or 3-D displays of the data. The freeware MICRODEM program allows easy export of GIS data to KML and linking of text and graphics to an icon on the map, and it facilitates registration of maps. Examples of this usage include student projects, animations used for teaching, sharing data among research groups, and interactive display of published maps. For the earth sciences, where almost all data has a geographic component, virtual globes provide an integrated way to interact with information.

INTRODUCTION

The geologic map has held a central place in scientific studies almost from the start of geology as a science at the end of the eighteenth century. Smith's 1815 map of England, Wales, and Scotland (Winchester, 2001) used a large oversized map which could show both a large area and significant detail. Lyell's *Principles of Geology* (1830–1833, three volumes) has ~10 maps, with a large foldout in each volume and several smaller maps that occupied part of a page in the text. Maps accompanied papers in publications of the Geological Society of America from its founding in 1888. While technology changed incrementally during the 1800s and into the first part of the 1900s, the basic nature of the map did not. The geologist, perhaps assisted by a draftsman or cartographer, prepared a map, it was printed, and the audience interpreted the geology through the eye of the geologist. The user of the map had extremely limited ability to add to or change the content on the map.

The development of geographical information systems (GIS) in the 1980s began to change this paradigm. Originally, the benefits of GIS only benefited the map makers, who could change scales, add and subtract data layers, and generally create tailored versions of a map relatively easily. Files could be shared with other users, but the large file sizes, expensive and proprietary GIS software, and steep learning curves for the software limited the process. Educators experimented with exposing students to the

---

*pguth@usna.edu

Guth, P.L., 2012, Automated export of GIS maps to Google Earth: Tool for research and teaching, *in* Whitmeyer, S.J., Bailey, J.E., De Paor, D.G., and Ornduff, T., eds., Google Earth and Virtual Visualizations in Geoscience Education and Research: Geological Society of America Special Paper 492, p. 165–182, doi:10.1130/2012.2492(12). For permission to copy, contact editing@geosociety.org. © 2012 The Geological Society of America. All rights reserved.

complex GIS systems, devoting a lot of time to software training. The end product remained a static map, whether on a printed page or the screen, and only another user with GIS software and training could manipulate the actual map data. Traditional GIS software offers tremendous versatility for the trained user to visualize and explore data, but it cannot readily replicate that experience for others.

The arrival of Google Earth in 2005 changed the scene (Butler, 2006). Three variables (price, ease of use, and capabilities) often define the success of computer software, and Google Earth found a sweet spot. Free for educational and personal use, it has a simple and intuitive interface that lets most users rapidly and easily do most of what they would like. Both free and commercial GIS programs such as GRASS (GRASS, 2011), Quantum GIS (QGIS, 2011), and ESRI's ArcGIS (ESRI, 2011a) offer much more capability, but at the cost of significant training to learn how to operate the myriad program options and functions. With the adoption of the Keyhole Markup Language (KML) as an open standard by the Open Geospatial Consortium (OGC, 2011), the data files for Google Earth can also be viewed in other 3-D Earth browser software such as the free ArcGIS Explorer (ESRI, 2011b). For simplicity in this paper, I will refer only to the Google Earth browser, but the standard KML format means the options are not restricted to a single viewing platform if other programs can redefine Google's sweet spot or extend it in a useful manner. It should be noted that KMZ files are compressed KML files, which will be smaller, and perhaps more importantly, place all of the images in the same file so they cannot be separated from the KML file and then fail to display.

KEY ELEMENTS OF KML AND GOOGLE EARTH

KML allows only a single choice for the geodetic datum, the World Geodetic System 1984 (WGS84), and data must be in a geographic projection. Google Earth offers no choice for the map projection, although the display changes from an orthographic display viewing the entire globe to a cylindrical, conformal projection when zoomed in to large-scale maps. This is a much better choice than the default in ArcGIS, which will display Shuttle Radar Topography Mission (STRM; Farr et al., 2007) digital elevation models (DEMs) or U.S. Geological Survey National Elevation Data (NED; Gesch et al., 2002) in a Plat Caree projection, with severe distortion (Snyder, 1987), unless the user manually sets a better projection. There might be reasons to want another projection, such as a preference for either equal area or conformal properties, but the Google Earth projections have few significant drawbacks for the intended audience. Allowing user selection of the projection would increase the complexity of the program and required user training.

KML files include both data and symbolization in a single file. This is not ideal from a database standards and management perspective, and the common, industry standard GIS formats like shapefiles (ESRI, 1998) and GeoTIFF (Ritter and Ruth, 2000) do not include symbolization. Short of going to proprietary formats

like ESRI layers or layer packages, most readily available GIS data requires user effort to set the symbolization. In contrast, a KML file opens with the symbolization chosen by its creator.

Compared to full capability GIS programs, Google Earth has several key limitations: (1) once in Google Earth, unless the data were exported as multiple layers, you cannot filter to restrict the display to a subset having specified characteristics; (2) you cannot change the display based on different attributes of the data; (3) you can only change the labels by manually retyping them for every record, which is tedious for large data sets; and (4) you cannot see the entire world at one time. The strength of Google Earth lies in its ability to display and manipulate visualizations of data, giving the end user an interactive experience to go between a detailed view of the data and an overview to assess its context, and to rapidly combine the data with other layers. Rather than having to juggle multiple maps with different scales, extents, and coverages, the user can see each layer in its correct relationship with the other layers.

PREVIOUS WORK

Ballagh et al. (2011) described several ways to create KML: (1) proprietary software like ArcGIS with a steep learning curve and limited ability to generate KML on the fly; (2) custom code, with complete control but variable level of effort depending on the required scripts; and (3) open source software, which provides extreme flexibility and a learning curve less severe than with proprietary software. While they did not directly address freeware like MICRODEM (Guth, 2009), it shares many characteristics with open source software. Chiang et al. (2009) described a Fortran library for creating KML output. They noted that KML files are self-contained and can be emailed or posted on the web, and recipients can easily view the data without need for additional format conversions. However, only Fortran programmers benefit from their WKML library; this is custom code that must be integrated with other software. De Paor and Whitmeyer (2011) described several sophisticated techniques for geologic data representation, especially data tiling for huge raster data sets, and animating COLLADA (Collaborative Design Activity) models for 3-D depictions of geology. These tools require a high level of user sophistication and training, and the goal of the work in this paper is to automate the generation of KML to put it within the reach of introductory earth science students.

TOOLS

The free version of Google Earth has no ability to create databases or import standard GIS data like GeoTIFF grids or shapefiles. I have created a simple graphical interface within the MICRODEM freeware GIS (Guth, 2009) to export raster and vector data to display in Google Earth. This does not exploit all the capabilities of Google Earth, many of which are described in other papers in this volume, but concentrates on the following capabilities: (1) basic display of raster and vector GIS data;

(2) displaying the attributes of a vector record by clicking on the map display; (3) displaying text and images linked to locations on the map; (4) time animations; (5) placing legends in a corner of the screen; and (6) using icon symbols for point features.

MICRODEM is a freeware GIS program that started as a viewer for DEMs, and has evolved into a complete GIS with the ability to manipulate imagery and vector data (Guth et al., 1987; Guth, 2009). The program retains a more geological and geomorphometric flavor compared to other GIS software (Wood, 2009) and has a number of optimizations to provide easy operation for students and other casual users.

The most useful capabilities of MICRODEM include:

(1) fast export of common raster and vector data sets to KML, with an easy to use graphical user interface (GUI);

(2) easy export of data sets with multiple filters and different symbology, allowing the simulation of sophisticated GIS options in Google Earth to change the display;

(3) placing documents, including text, graphics, and web links with icons on the virtual globe and allowing them to be opened while navigating the globe;

(4) easy export of earthquake focal mechanisms;

(5) automatic recognition and smart display for Census Bureau TIGER files or OpenStreetMap vector data with roads and streams;

(6) suitable for student use with minimal training; and

(7) freeware with no limitations on use.

MICRODEM has a standard graphical user interface allowing multiple open map windows, controlled by the mouse, icons, and menus. MICRODEM reads most of the common GIS raster formats; for formats it does not understand, it can use the Geospatial Data Abstraction Library (GDAL, 2011) to convert the data to GeoTIFF format (Ritter and Ruth, 2000). For vector data, like Google Earth, MICRODEM only supports geographic coordinates on the WGS84 datum, but provides conversion routines that can be run before using the data.

Figure 1 shows MICRODEM with two data sets open. The map on the left side of the screen overlays the geological map of the former Nevada Test Site (Slate et al., 2000) in vector shapefile format, on the NED in raster GeoTIFF format. The attribute table for the shapefile appears on the right side of the screen. Right clicking on the map brings up the pop-up menu shown in the center of the screen. Selection of the option to "Export, quick map to Google Earth" automatically performs the following steps: (1) reprojects the map to WGS84, if required; (2) creates a properly formatted KML file; and (3) opens the KML file in Google Earth or another viewer associated by the operating system with the KML extension. The export will include any map elements such as scale bars or grids, which could be turned off prior to the export. Figure 2 shows the map display in Google Earth, rotated for a 3D view.

The quick export is designed to create a usable map with no user interaction, and is particularly useful for students in introductory courses. Sophisticated users can select a deliberate instead of a quick export, which can convert the entire map, or

each layer can be exported separately, with significant control over the export. Figure 3 shows two graphical forms that customize the export. MICRODEM remembers the selections, and will apply them to future exports. Google Earth provides easy tools to iteratively adjust the display, renaming layers, tweaking symbology, organizing layers into folders, and saving the file when finished.

EXAMPLES

I will discuss three uses of MICRODEM to export data for Google Earth, with a metaphor of using the virtual globe to explore data. Students can write a paper, collect or analyze data, and display it in its geographic context. In addition, because they use Google Earth in their life outside the classroom, the work acquires greater relevance for many of them.

Simple Student Projects

The simplest projects involve students preparing a paper and then displaying the paper in Google Earth, with text, graphics, and web links. Their papers could be on any subject that refers to a point on the earth. We have done this in a non-laboratory, introductory physical geography course with minimal GIS instruction, largely confined to creating some maps to illustrate a paper on the physical geography of a foreign province. The students save their paper in hypertext markup language (HTML) format, a single operation in both Microsoft Word or OpenOffice. They then prepare a simple comma separated variable (CSV) file in a spreadsheet describing where and how the HTML document will display. MICRODEM opens and converts the CSV file into a standard GIS database, and the user can export it to a KML file which includes the HTML document.

As an example, in spring 2011 my Honors Oceanography Research Methods students did a lab report on the March 2011 Tohuku earthquake and tsunami. This course primarily teaches Matlab programming, so the students only get one or two labs on GIS. Each student analyzed a tide gauge record and a buoy record, and computed travel paths to try to predict arrival times around the Pacific based on shallow wave theory. They presented their results in Google Earth. After collecting data from the web and performing the analysis, they pasted GIS maps and graphs into Microsoft Word, OpenOffice Writer, or any HTML editor, added text for a complete paper, and saved it as an HTML file. They must keep the formatting simple, since the files will go through a number of programs before the final display in Google Earth. Both Word and OpenOffice Writer produce verbose HTML code, but MICRODEM strips out the worst of the excessive HTML code. To create a database, students use Excel, OpenOffice Calc, or even Notepad to create a CSV file with the location of the buoy or tide gauge, the name of an icon to display, text to display next to the icon, and the name of the HTML file to display when clicking at the point (Fig. 4). MICRODEM imports this file, creates a database, and then exports it to Google Earth. Figure 5 shows the

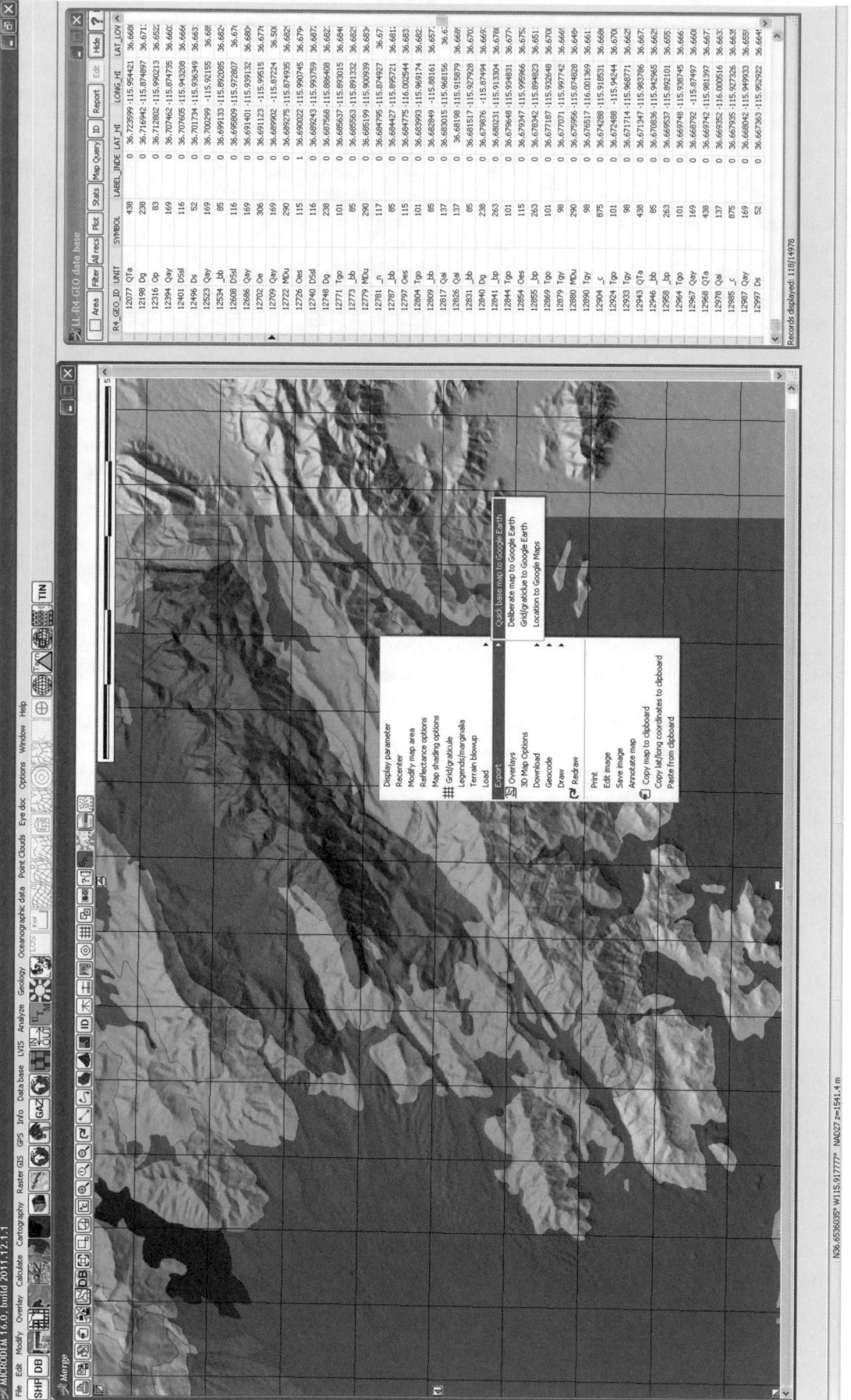

Figure 1. Graphical user interface of MICRODEM showing multiple windows, menus, and icons.

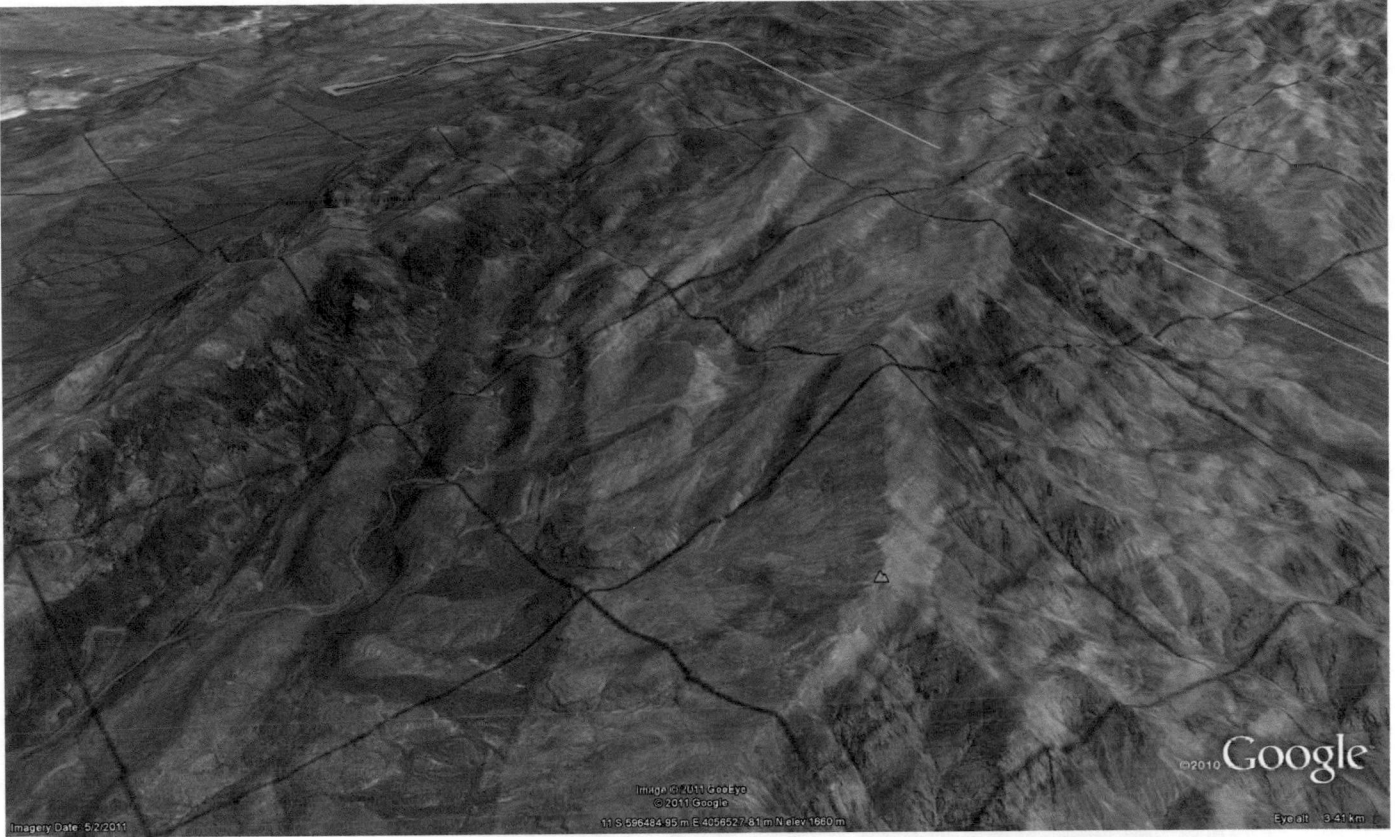

Figure 2. Map exported from MICRODEM and displayed in 3-D by Google Earth

Figure 3. Forms for selection of the deliberate KML export options for (A) a database and (B) a map.

LAT	LONG	NAME	ICON	TEXT
38.322	142.369	Epicenter	BMW95.ico	epicent.htm
30.515	152.117	Station 21413	BMW95.ico	dart_buoy.htm
41.745	-124.1816667	Tide Gauge	BMW95.ico	tide_gauge.htm

Figure 4. Simple database table used to display icons on the virtual globe, linked to documents with text and images.

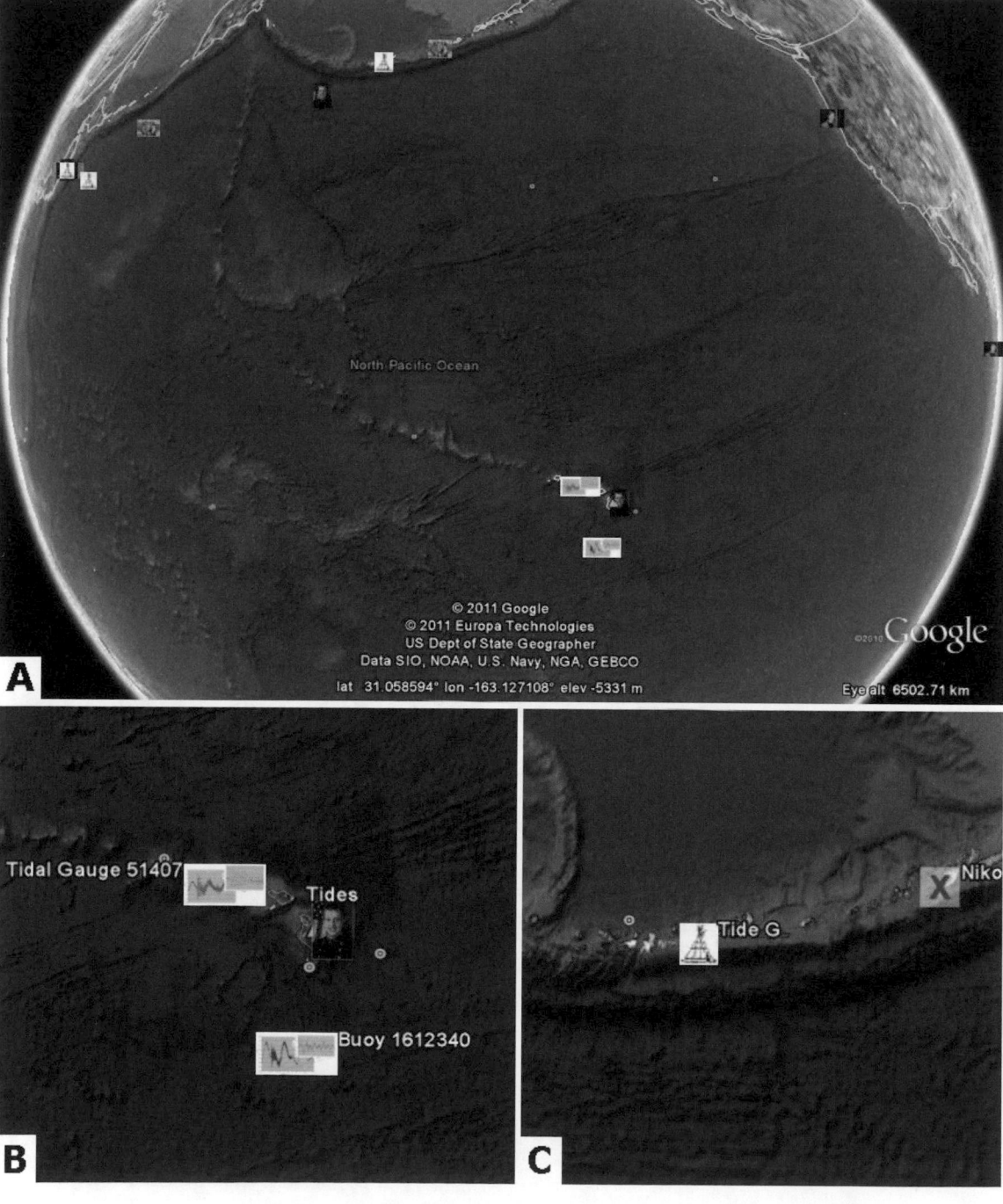

Figure 5. Google Earth display combining several student lab assignments showing icons and text labels.

combined exports from the class. Students have the flexibility to choose an icon, but lack of attention to detail in entering the icon name, or in the correct storage location for the icon file lead to the red X icon (Fig. 5C) (and missing images in the HTML document). Clicking on the icon in Google Earth opens the student's report (Fig. 6).

Combining the projects from the class produces an interactive display in Google Earth which allows selection of data via its geographical location, and immediately displays the student's report. As the students collected the data, they observed that scientific data delivery is moving in this direction. The download sites for the buoys (NDBC, 2011) and the tides (CO-OPS, 2011) both have graphical interfaces to retrieve the data, although neither uses Google Earth or allows combination with other data on their websites. Google Earth provides links to other material, such as photos, again via the geographical context, and students can see details of the landscape in the region of their project. Students see that with a little effort they can create effective displays in Google Earth, and it helps them understand and appreciate the technology.

Advanced Student Projects and Class Demonstrations

Earthquake locations can showcase the time animation capabilities of Google Earth. After the 2011 Japan Tohoku earthquake, I prepared demonstrations for the sequence of fore and aftershocks. This highlights Google Earth's ability to display databases, and how careful planning for the GIS data export can compensate for some of the limits of Google Earth.

The USGS allows selection of earthquake data by using a web form to specify a desired search region and time period, and exporting with the CSV format (USGS, 2011). The resulting list can be copied to a spreadsheet or text editor (Fig. 7A). Minor edits must be done here: a title line must be removed, the field names cleaned up, and any summaries at the bottom of the listing removed (Fig. 7B). It can then be saved as a CSV file for import into MICRODEM as a GIS database (Fig. 7C).

Once the database has been imported in the GIS, two steps are required to export it to Google Earth. To symbolize the earthquakes, I created a database table with 7 categories (Fig. 8). For each category, the FILTER is applied to the database. Three fields define symbolization in the GIS (SYM_COLOR, SYM_TYPE, and SYM_SIZE, with the symbol in the NAME column displaying them graphically); and the ICON and ICON_SCALE fields apply for the export to Google Earth. Using these rules, MICRODEM creates the required icons and copies the appropriate ICON and ICON_SCALE values into the database for each earthquake (Fig. 7B).

Time fields must be created for the time animation. These have a format that would be very hard to create manually (shown in the rightmost two columns of Fig. 9A), but MICRODEM can easily create them from YEAR, MONTH, DAY, and TIME_STR (time as a character string) fields. In the case of the earthquake data, one command converts the time in HHMMSS.SS format

supplied by USGS into HH:MM:SS (Google Earth cannot use fractions of a second, and it would be overkill for animations). Then MICRODEM creates the START_DATE and END_DATE for the display of each earthquake. At this point the database can be exported automatically, displayed in Google Earth (Fig. 10), and run as a time animation. The printed page provides only a limited demonstration of the real power of the interactive display where the time and geographic coverage can be adjusted.

Earthquake focal mechanisms from the Global Centroid-Moment-Tensor (CMT) Catalog (Ekström and Nettles, 2011) provide additional insight into the earthquake. This can be done in MICRODEM in just four steps: (1) download a file with the CMT data converted to a GIS table, (2) open the file, (3) filter the database to the region of interest, and (4) select the focal mechanism KML export option. The focal mechanisms will plot at the epicenter, and clicking on one will bring up the full record for that event. Figure 11 shows two different exports of the historical record which starts in 1975, and includes only events with a focal mechanism. The rupture area of the 2011 earthquake has ~700 focal mechanisms. This recoloring of the display, and turning labels on and off, are easy to do in a GIS (provided, in this particular case, that the GIS can plot focal mechanisms), but is not possible in Google Earth. Multiple exports provide an easy workaround, and as the table of contents in Figure 9A shows, the user can easily select which layers to display. Multiple exports increase file storage somewhat, but these are not particularly large files and multiple versions increase display flexibility with the simple user interface.

Published Maps

The maps published in journals and other publications represent a potentially underutilized resource, limited by the difficulty in combining them with other data. The digital versions of the maps available online, such as those in Geological Society of America (GSA) publications, prove to be only slightly more useful than their paper counterparts. Since the vast majority of the maps were probably generated in a GIS, the publication process greatly reduces their usability for users whose first instinct is to combine layers in a GIS.

Google Earth allows image overlays, and for some maps this process can provide a very easy display. The map should be in a cylindrical projection such as Universal Transverse Mercator (UTM) or Mercator (Snyder, 1987), with either the UTM grid or the graticule having parallel lines which intersect in right angles. Other maps can be imported, but they must be registered and reprojected in the GIS before exporting to KML. The key lies in having something with which to register the map such as a geographic feature like the coastline, or the geographic or UTM grid or ticks. MICRODEM makes the process very easy, as illustrated by several maps from recent papers in the *GSA Bulletin* (Centeno-García et al., 2011; Oner and Dilek, 2011).

While Google Earth has an option ("View grid") to display a graticule, it has several disadvantages: (1) unlike most printed

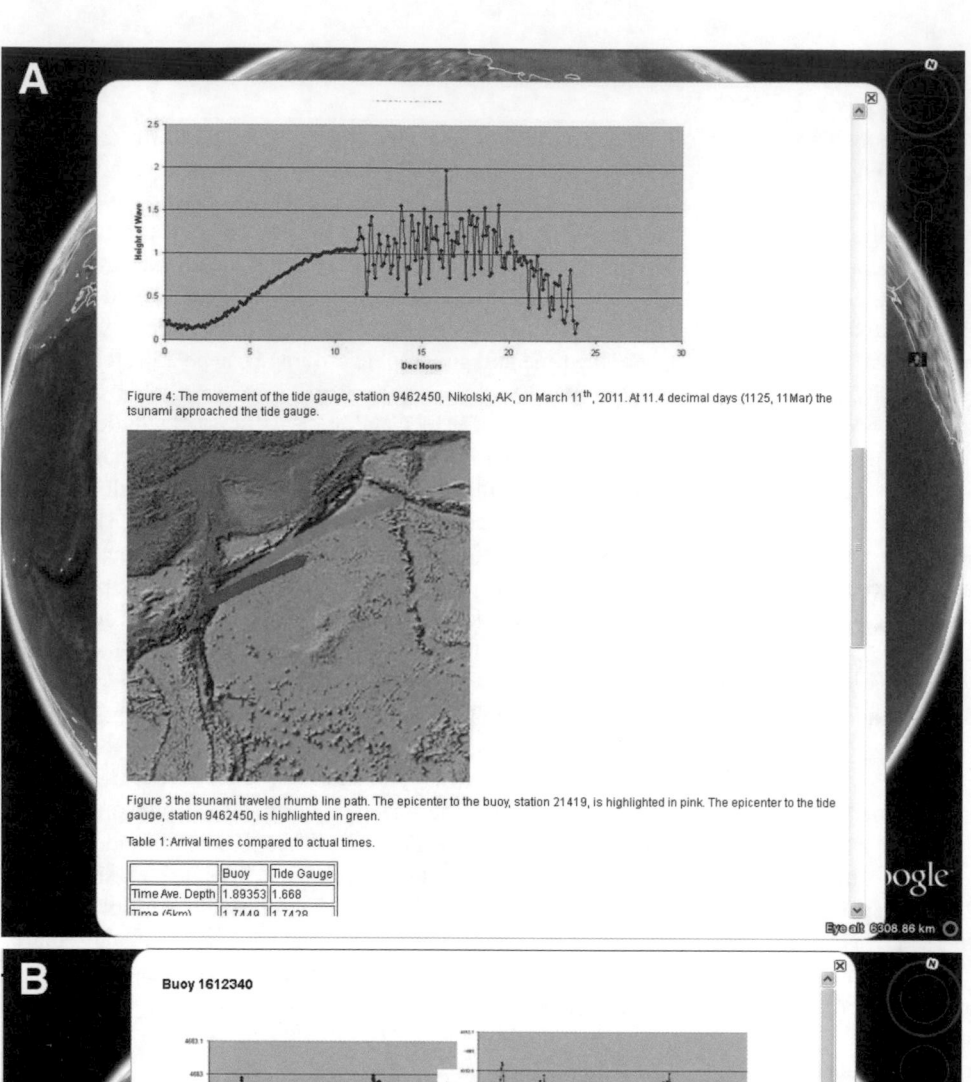

A

Figure 4: The movement of the tide gauge, station 9462450, Nikolski, AK, on March 11th, 2011. At 11.4 decimal days (1125, 11Mar) the tsunami approached the tide gauge.

Figure 3 the tsunami traveled rhumb line path. The epicenter to the buoy, station 21419, is highlighted in pink. The epicenter to the tide gauge, station 9462450, is highlighted in green.

Table 1: Arrival times compared to actual times.

	Buoy	Tide Gauge
Time Ave. Depth	1.89353	1.668
Time (5km)	1.7449	1.7429

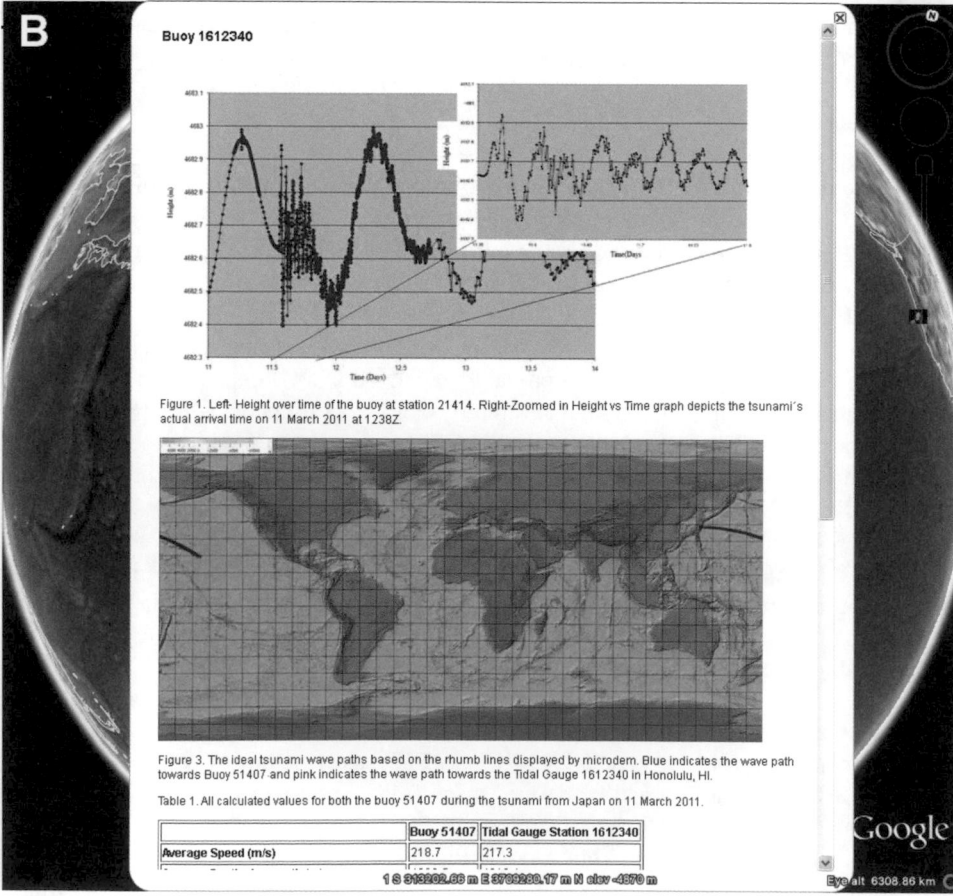

B

Buoy 1612340

Figure 1. Left- Height over time of the buoy at station 21414. Right-Zoomed in Height vs Time graph depicts the tsunami's actual arrival time on 11 March 2011 at 1238Z.

Figure 3. The ideal tsunami wave paths based on the rhumb lines displayed by microdem. Blue indicates the wave path towards Buoy 51407 and pink indicates the wave path towards the Tidal Gauge 1612340 in Honolulu, HI.

Table 1. All calculated values for both the buoy 51407 during the tsunami from Japan on 11 March 2011.

	Buoy 51407	Tidal Gauge Station 1612340
Average Speed (m/s)	218.7	217.3

Figure 6. Extract of student report, opened by clicking on the icon plotted at the location of a tide gauge. (A) Report includes a graph of the tide record showing the arrival of the tsunami. (B) Report shows the record from a DART buoy. Both include a map with likely paths for the tsunami.

```
Link to USGS Home Page
NEIC: Earthquake Search Results

Year,Month,Day,Time(hhmmss.mm)UTC,Latitude,Longitude,Magnitude,Depth,Catalog
  2011,01,01,153434.49, 37.802, 138.918,4.5,173,PDE-W
  2011,01,02,191023.54, 36.769, 141.321,4.5, 39,PDE-W
  2011,01,03,044515.28, 38.567, 139.099,4.8, 39,PDE-W
  2011,01,05,005732.22, 31.545, 142.177,5.6, 21,PDE-W
  2011,01,06,002641.42, 31.527, 142.282,4.7, 10,PDE-W
  2011,01,06,034208.63, 31.506, 142.344,4.4, 10,PDE-W
  2011,01,06,044904.27, 31.543, 142.288,4.5, 10,PDE-W
  2011,01,06,091224.10, 30.790, 141.525,4.8, 17,PDE-W
  2011,01,07,174514.24, 35.122, 141.351,4.4, 10,PDE-W
  2011,01,08,074748.30, 41.986, 141.401,4.8,116,PDE-W
  2011,01,08,182152.03, 42.594, 142.162,4.7,124,PDE-W


USGS National Earthquake Information Center
USGS Privacy Statement | Disclaimer      FirstGov
```

A

```
Year,Month,Day,Time_UTC,Latitude,Longitude,Magnitude,Depth,Catalog
  2011,01,01,153434.49, 37.802, 138.918,4.5,173,PDE-W
  2011,01,02,191023.54, 36.769, 141.321,4.5, 39,PDE-W
  2011,01,03,044515.28, 38.567, 139.099,4.8, 39,PDE-W
  2011,01,05,005732.22, 31.545, 142.177,5.6, 21,PDE-W
  2011,01,06,002641.42, 31.527, 142.282,4.7, 10,PDE-W
  2011,01,06,034208.63, 31.506, 142.344,4.4, 10,PDE-W
  2011,01,06,044904.27, 31.543, 142.288,4.5, 10,PDE-W
  2011,01,06,091224.10, 30.790, 141.525,4.8, 17,PDE-W
  2011,01,07,174514.24, 35.122, 141.351,4.4, 10,PDE-W
  2011,01,08,074748.30, 41.986, 141.401,4.8,116,PDE-W
  2011,01,08,182152.03, 42.594, 142.162,4.7,124,PDE-W
```

B

LAT	LONG	YEAR	MONTH	DAY	TIME_UTC	MAGNITUDE	DEPTH	CATALOG
37.802	138.918	2011	01	01	153434.49	4.5	173	PDE-W
36.769	141.321	2011	01	02	191023.54	4.5	39	PDE-W
38.567	139.099	2011	01	03	44515.28	4.8	39	PDE-W
31.545	142.177	2011	01	05	5732.22	5.6	21	PDE-W
31.527	142.282	2011	01	06	2641.42	4.7	10	PDE-W
31.506	142.344	2011	01	06	34208.63	4.4	10	PDE-W
31.543	142.288	2011	01	06	44904.27	4.5	10	PDE-W
30.79	141.525	2011	01	06	91224.1	4.8	17	PDE-W
35.122	141.351	2011	01	07	174514.24	4.4	10	PDE-W

C

Figure 7. Steps in the import process of earthquake data downloaded from USGS site (2011). (A) Data saved from the browser window. (B) File after removing extraneous lines at the beginning and end, and renaming the time field to comply with the rules for database field names. (C) Data displayed in the GIS database.

NAME	FILTER	SYM_COLOR	SYM_TYPE	SYM_SIZE	ICON	ICON_SCALE
M < 3	MAGNITUDE < 3	8421376	13	3	MAG_3.GIF	0.4
M >=3 AND M <4	MAGNITUDE >=3 AND MAGNITUDE <4	12615680	13	4	MAG_3_4.GIF	0.5
M >=4 AND M <5	MAGNITUDE >=4 AND MAGNITUDE <5	16744576	13	5	MAG_4_5.GIF	0.6
M >=5 AND M <6	MAGNITUDE >=5 AND MAGNITUDE <6	33023	13	6	MAG_5_6.GIF	0.7
M >=6 AND M <7	MAGNITUDE >=6 AND MAGNITUDE <7	65280	13	7	MAG_6_7.GIF	1.1
M >=7 AND M <8	MAGNITUDE >=7 AND MAGNITUDE <8	16711935	13	9	MAG_7_8.GIF	1.1
M >=8	MAGNITUDE >=8	255	13	11	MAG_8.GIF	1.6

Figure 8. Display rules for the earthquake data, stored in a database.

LAT	LONG	YEAR	TIME_UTC	TIME_STR	START_DATE	END_DATE
37.802	138.918	2011	153434.49	15:34:34	2011-01-01T15:34:34Z	2011-01-02T19:10:23Z
36.769	141.321	2011	191023.54	19:10:23	2011-01-02T19:10:23Z	2011-01-03T04:45:15Z
38.567	139.099	2011	44515.28	04:45:15	2011-01-03T04:45:15Z	2011-01-05T00:57:32Z
31.545	142.177	2011	5732.22	00:57:32	2011-01-05T00:57:32Z	2011-01-06T00:26:41Z
31.527	142.282	2011	2641.42	00:26:41	2011-01-06T00:26:41Z	2011-01-06T03:42:08Z
31.506	142.344	2011	34208.63	03:42:08	2011-01-06T03:42:08Z	2011-01-06T04:49:04Z
31.543	142.288	2011	44904.27	04:49:04	2011-01-06T04:49:04Z	2011-01-06T00:91:22Z
30.79	141.525	2011	91224.1	00:91:22	2011-01-06T00:91:22Z	2011-01-07T17:45:14Z
35.122	141.351	2011	174514.24	17:45:14	2011-01-07T17:45:14Z	2011-01-08T00:74:74Z

A

LAT	LONG	TIME_UTC	MAGNITUDE	ICON_SCALE	ICON
37.93	143.802	61523.12	6.3	1.1	MAG_6_7.GIF
37.704	143.833	83801.51	5.4	0.7	MAG_5_6.GIF
37.556	143.642	85105.36	5.3	0.7	MAG_5_6.GIF
37.683	143.311	90914.68	5.5	0.7	MAG_5_6.GIF
37.572	143.889	93535.51	4.9	0.6	MAG_4_5.GIF
37.961	143.571	102821.04	5	0.7	MAG_5_6.GIF
37.588	144.053	120752.81	4.6	0.6	MAG_4_5.GIF
37.882	143.321	131310.36	4.8	0.6	MAG_4_5.GIF

B

Figure 9. Changes to the database for export to Google Earth. The database allows hiding or rearranging fields to improve the display. (A) Two steps are required for time animation, done with two commands in MICRODEM: convert the time obtained from USGS to a string, and then convert that, with the rest of the date information, into the correct format for KML. (B) Application of the display rules from Figure 6 into two fields for the icon display used in Google Earth.

maps, the spacing in the latitude and longitude directions will probably be different; (2) the spacing of the graticule cannot be set and might not match that on the geologic map to be registered; (3) the graticule spacing will change as you zoom the map; and (4) the labels are in the center of the map, whereas they might be more useful on the edges.

MICRODEM exports a graticule, or a UTM grid, with complete control over its parameters (Fig. 12). In this case I set the extent of the graticule to be slightly larger than the map to be registered, and made the grid spacing 3′ (arc minutes) to match the map. The graticule has been recolored so that it will contrast with the white background of the map. The graticule consists of two layers, which can be turned on and off separately, for the lines and the labels. Label placement follows the internal rules of Google Earth, like many display options, and can move in seemingly random ways, but is only there to assist in matching grid lines and is not necessarily meant to be part of the final display.

The map can now be downloaded from the GSA journals website and saved on the user's hard disk. From there the user selects the "Add Image Overlay" icon in Google Earth, and picks the file with the map. Three commands size and register the map and are active when the cursor becomes a pointing finger (Fig. 13A): (1) the large green cross in the center of the map moves the entire map; (2) the green marks in each corner, and the center of each side, allow adjusting the size of the map, either in one direction at a time or both simultaneously; and (3) the green diamond

along the left center rotates the overlay about the center point. Rotation should generally only be required for a UTM map. Figure 13A shows the results, with the gold graticule exported by MICRODEM aligning with the ticks on the map in the journal paper. I have left the cross section and the legend on the map, although they could easily be removed. This registration required only correct placement and resizing, and after a few trials can be done easily and iteratively.

If the user had adjacent maps and wanted to see how they compared, the map marginalia can be clipped in any image editor, and only the main portion of the map imported (Fig. 13B). The legend can be cut from the original map and put in a corner of the screen, again as a simple option in MICRODEM. The legend can be toggled on and off from the map's table of contents. The scale

Figure 10. Four displays of the great Tohoku megathrust, foreshocks, and aftershocks in early 2011. Figure (A) shows 856 earthquakes from 1 January until 2 March, clearly showing the huge increase in seismicity associated with the magnitude 9 earthquake on 11 March compared to other subduction zone segments in the western Pacific. Figures (B) and (C) show the ability to zoom in, and change the dates shown using the slider bar in the top left of the screen. Figure (B) shows the small number of earthquakes in the first two and a half months of 2011, and (C) shows the huge increase during the three-day period following the main event. Figure (D) shows zooming in to better see individual events.

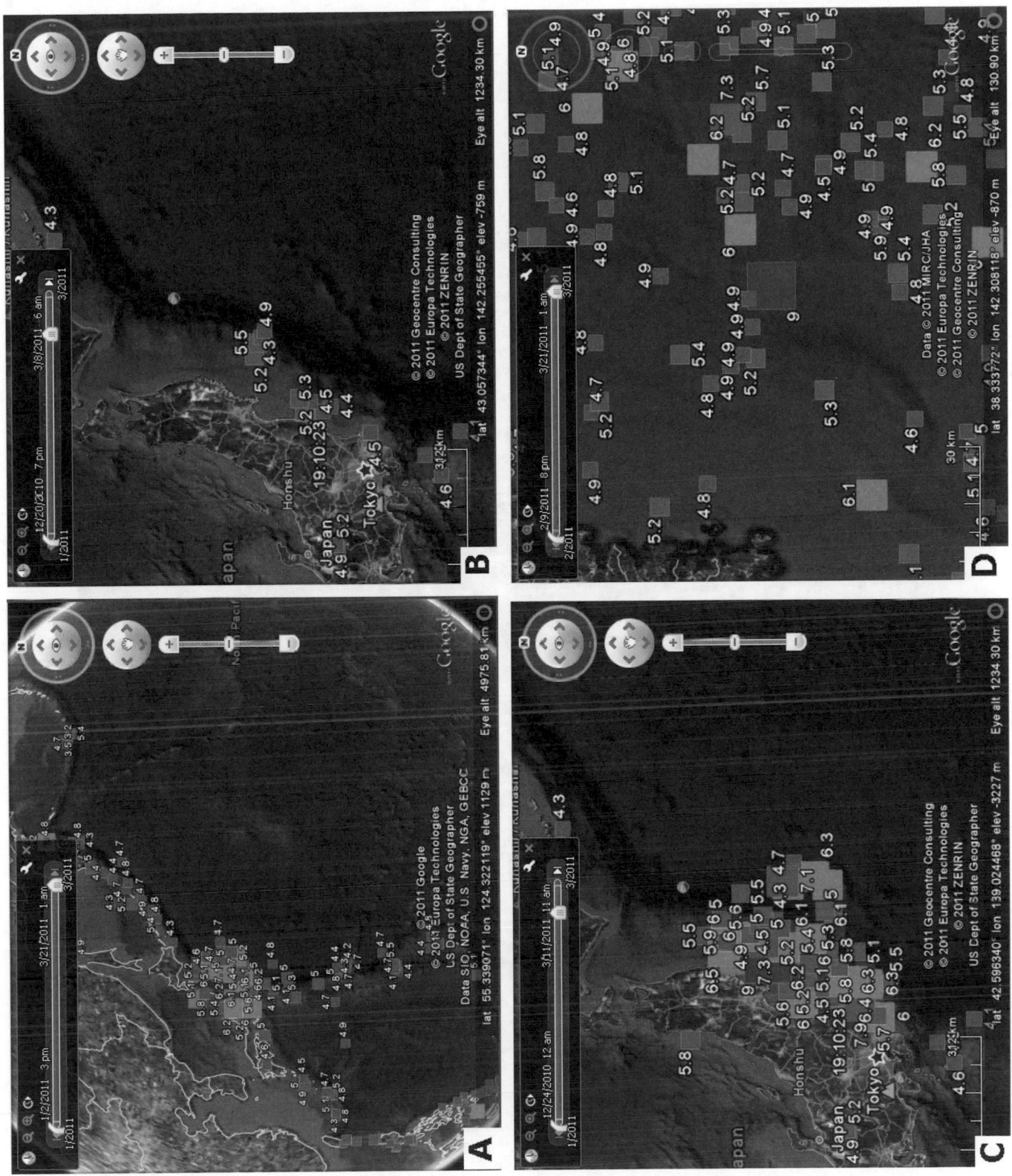

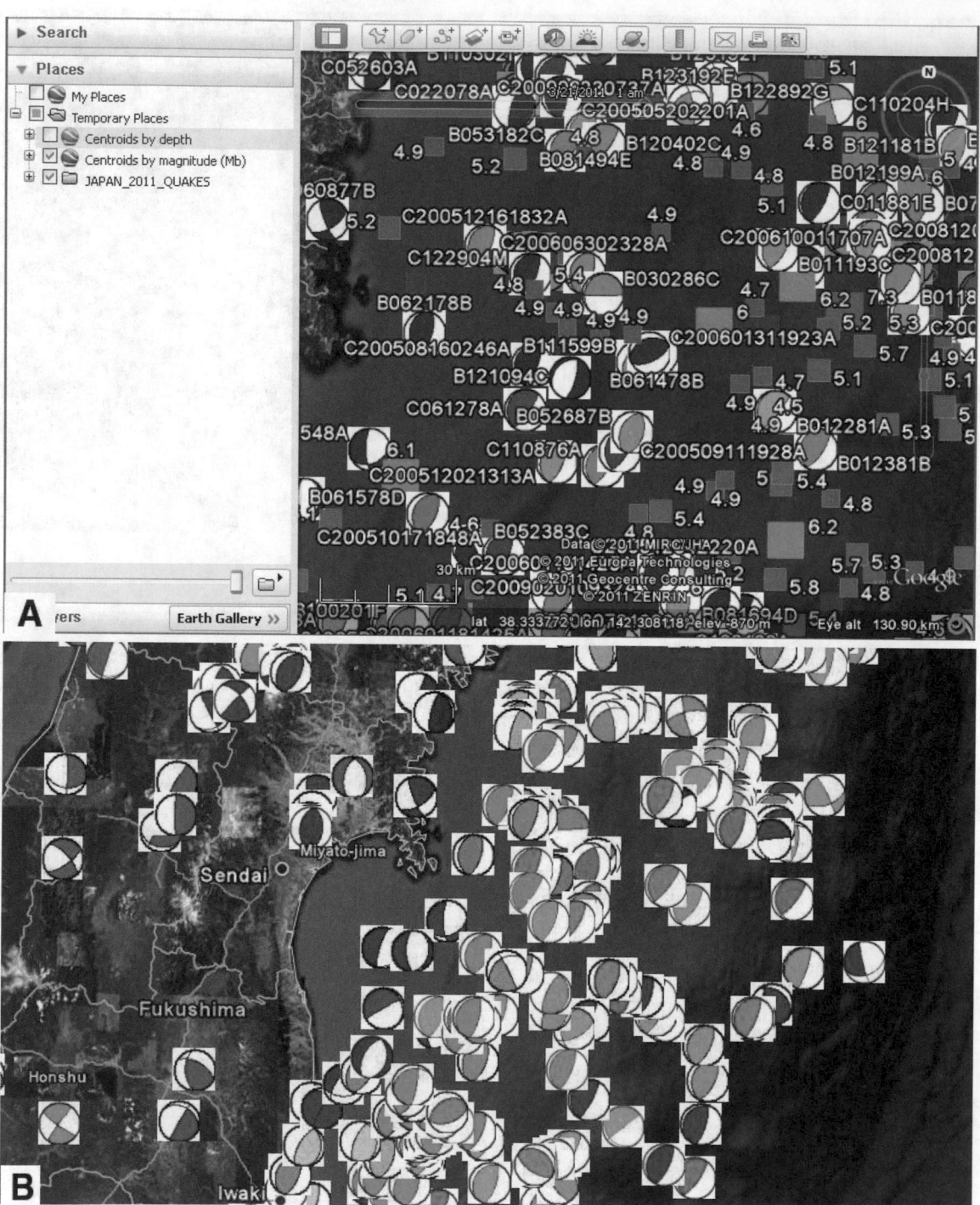

Figure 11. Two views of the Global CMT data (Ekström and Nettles, M., 2011). Figure (A) shows color coding by magnitude, with event labels, and shows the Goggle Earth table of contents which allows easily turning layers on and off. Figure (B) shows a second export from MICRODEM with the labels turned off and color coding by depth, showing the dip of the Benioff zone underneath Japan.

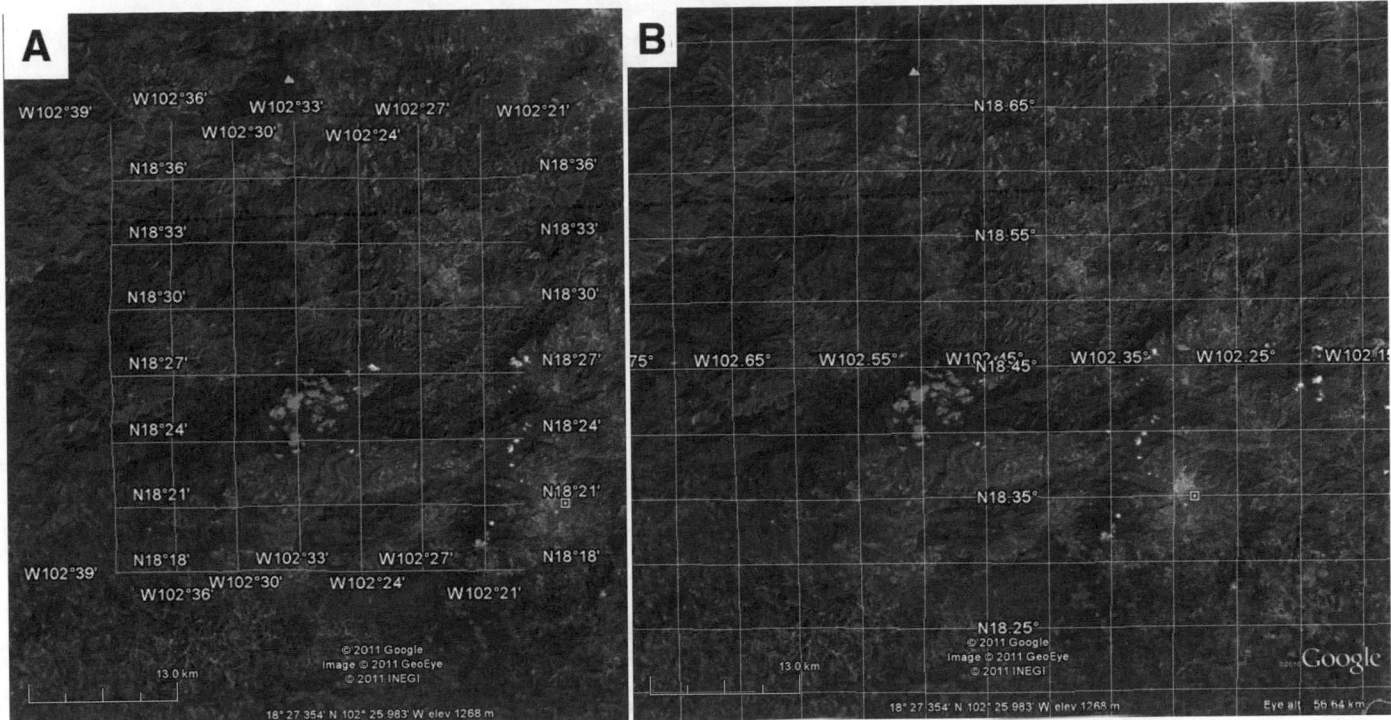

Figure 12. Lat/long graticules exported from MICRODEM which allows full customization (A) and the default display from Google Earth (B) which cannot be customized.

bar was clipped from the legend before the import, because the legend will maintain a constant size as the user zooms the map, and the static scale would rarely be correct. In contrast the scale bar in Figure 14 will correctly scale as the map zooms, but scale bars are not one of the cartographic strengths of Google Earth. "View scale legend" does put a legend in the lower left corner, but it can be hard to see, and the 71 km total length of the scale bar makes the ticks 17.75 km apart when a cartographer would use a 20 km spacing.

Figure 14 shows a second map (after Centeno-García et al., 2011, their fig. 2), with the earthquake focal mechanisms from the Global CMT Catalog (Ekström and Nettles, 2011) overlaid. The focal mechanisms plot at the epicenter, and clicking on one will bring up the information about that event (Fig. 14A). This allows easy comparison of recent seismicity with the published map. Similar combinations could be done with any other GIS data, including the larger earthquake catalogs (USGS, 2011). Figure 14B shows the map with its transparency adjusted in Google Earth, showing how the mapped geology correlates with the regional satellite imagery shown at this scale.

Figure 15 shows a second map from the *GSA Bulletin* (after Oner and Dilek, 2011, their fig. 3). This shows the green controls used to adjust the overlay, and the white UTM grid exported from MICRODEM which was used to register the map. The legend of the map has been cut and pinned in the lower left corner of the map, with the scale bar and north arrow removed since they will not be accurate as the map is zoomed, and rotated to face south

in this view. The figure also shows a KML file generated by the USGS Earthquake Search page (USGS, 2011) which comes with a legend on the left side of the screen, and a USGS logo on the bottom right. This area has no focal mechanisms (Ekström and Nettles, 2011).

DISCUSSION

Students today have grown up in the digital age, and arrive in geoscience classrooms used to electronic media and Google Earth, even if their most memorable experiences have only been using the flight simulator as a game. Publishers of introductory textbooks provide KML files to guide students to relevant locations to observe features for particular lessons. Helping students create content in this familiar medium adds a particular relevance to geoscience courses which students appreciate—it's cool to make an interesting interactive product, instead of just another paper or lab report.

As the delivery of data in KML format increases, the options for combining data sets on the virtual globe will increase. USGS (2011) now provides direct KML export of earthquake data (Fig. 15), but until the web services provide full selection of symbology online, there will still be occasions for more sophisticated display option and time animation available only with manipulation in the GIS environment (Fig. 8).

The limitations of the KML format, and Google Earth, which is not a full GIS, must be clearly understood. Google

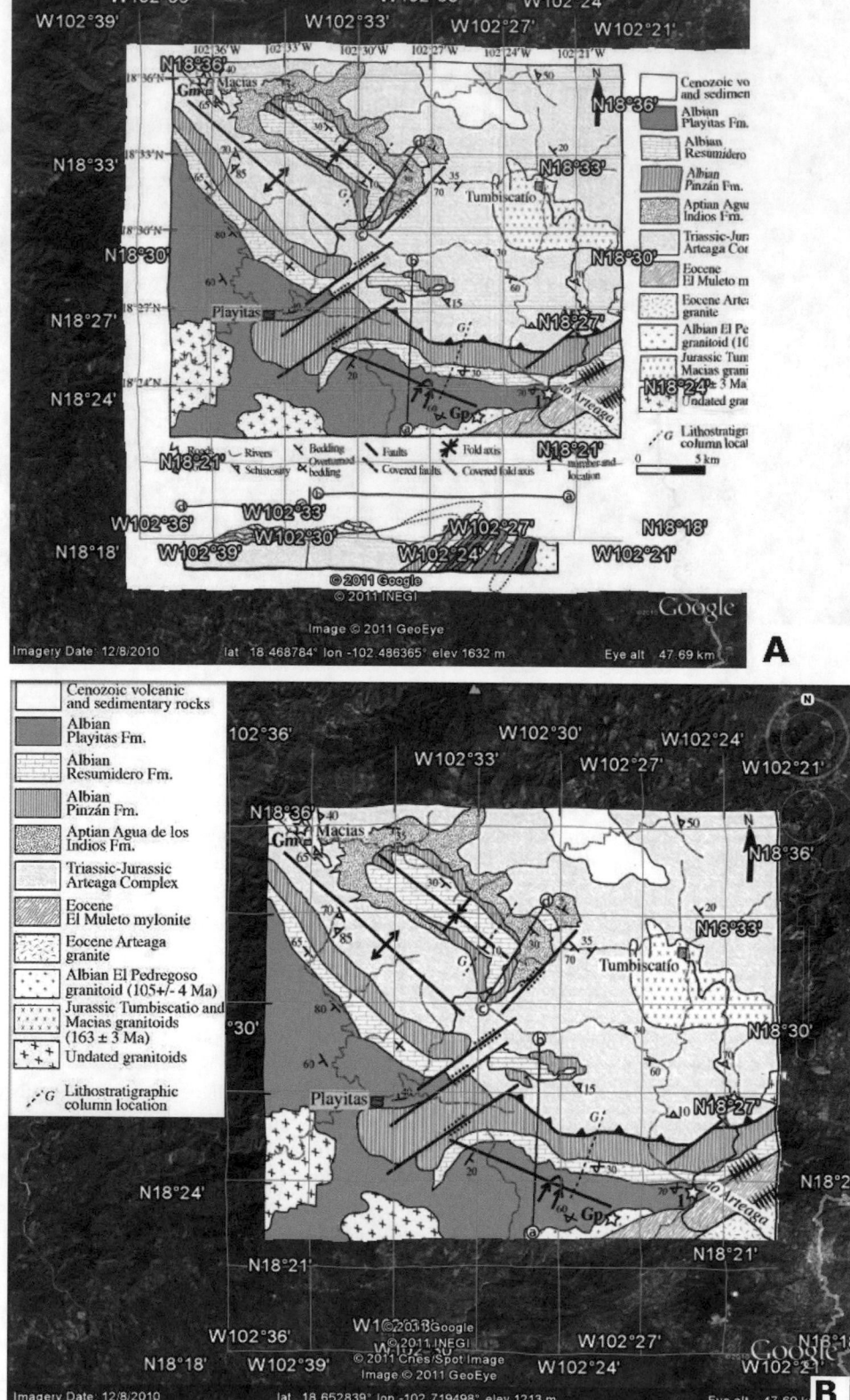

Figure 13. (A) Map (after Centeno-García et al., 2011, their fig. 4) displayed in Google Earth with the graticule exported from MICRODEM. (B) Same map displayed in Google Earth with the legend and marginalia removed. The legend remains fixed in the upper left, but can be hidden by unchecking its entry in the table of contents.

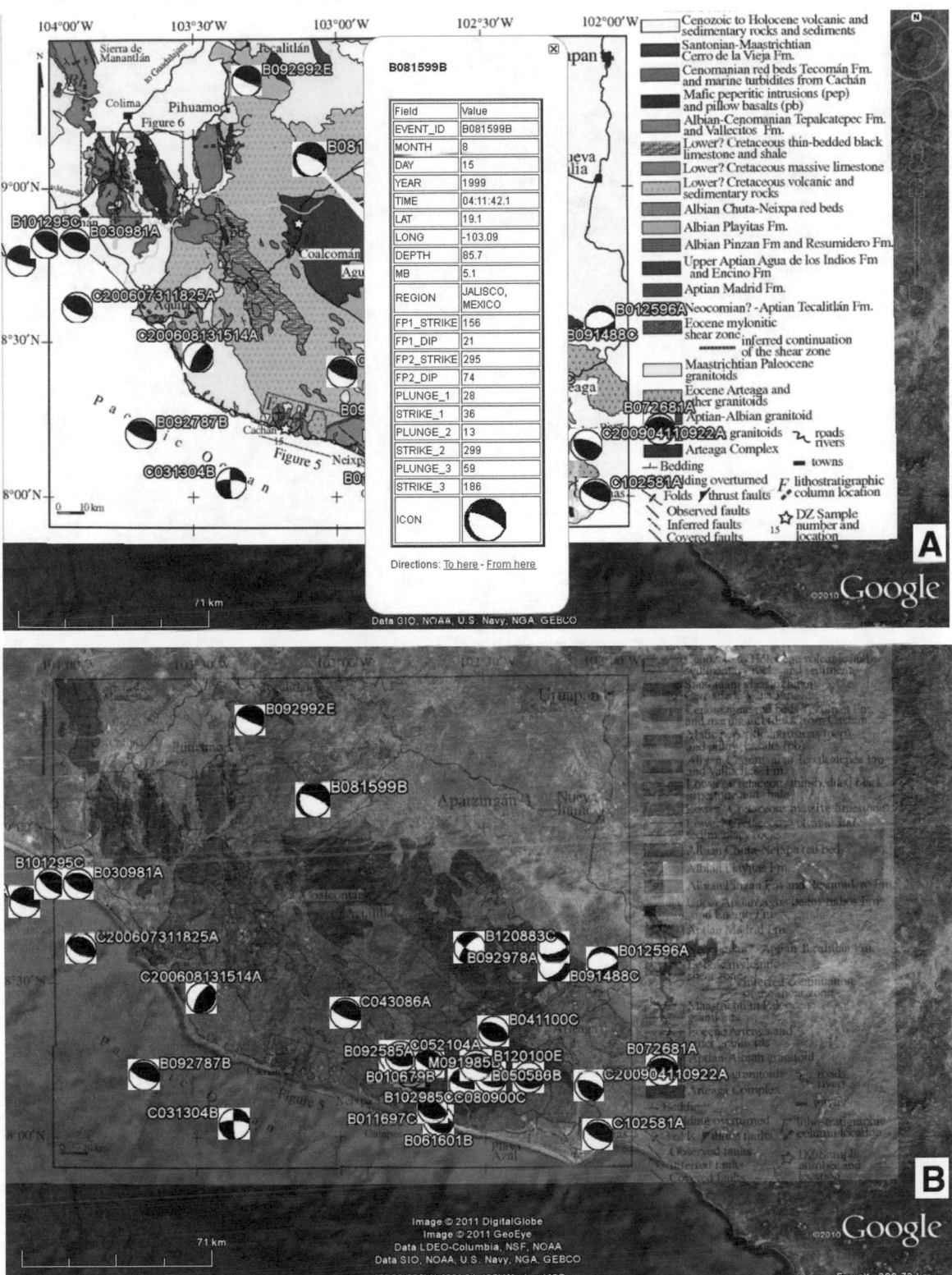

Figure 14. (A) Map (after Centeno-García et al., 2011, their fig. 2) displayed in Google Earth with earthquake focal mechanisms from the Global CMT data (Ekström and Nettles, 2011) overlaid. The data for one earthquake appears after clicking on the beach ball. (B) Same map with the transparency increased to emphasize the satellite image base map.

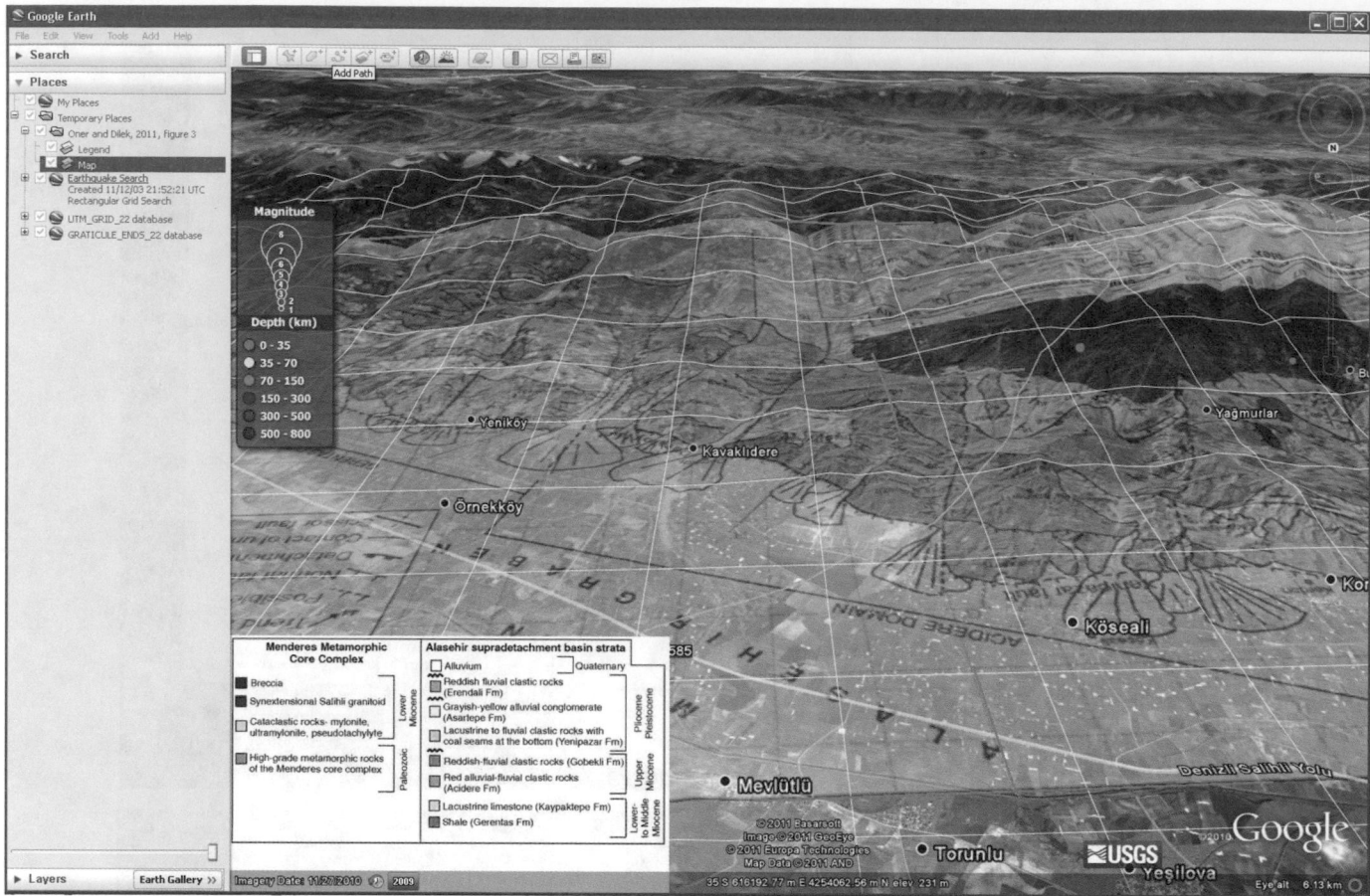

Figure 15. Map (after Oner and Dilek, 2011, their fig. 3) in a 3-D oblique view in Google Earth. This shows the green controls used to align the image to the white UTM grid exported from MICRODEM, and earthquake epicenters from USGS.

Earth cannot filter a database, to hide or display only the records matching selected attributes in the table, or easily change the display symbolization. The user can toggle on/off, or change the display characteristics, for an entire map layer or individual records, but cannot perform the sophisticated and rapid changes or analysis that characterize GIS operations. For large data sets selection of individual records proves impractical. Google Earth can best be employed as a means to disseminate analysis results done in a full-featured GIS program, and preparation of the KML files can be done interactively using both the GIS program and Google Earth.

Users must understand the limitation of scale in digital maps, as Google Earth makes it easy to zoom data and effortlessly supply appropriate image data. Table 1 shows the sizes of the maps from the *GSA Bulletin* shown in Figures 13–15. None

of these are large-scale maps, particularly compared to the imagery in Google Earth which frequently has sub-meter pixel resolution. Map and GIS literacy in understanding resolution will be important as mapping data moves from the GIS lab to the smart phone. Users will expect to download higher resolution imagery on demand, but must clearly understand that scientific maps will have only a single scale at which they were compiled, which will often be significantly different than the imagery.

The map projection and datum will also limit results when trying to register maps. For maps on WGS84 or NAD83, now becoming the standard, the datum shift will not make a difference, and the simple registration provided by Google Earth could be sufficiently accurate (Figs. 13–15). In other cases, a more detailed analysis will be required, and registration might have to be done in a GIS. For small-scale maps, any datum shift problems will be subsumed in the several pixel limit to how close you can register the map, and in the map pixel size relative to a possible datum shift. For example, with a map pixel size of 258 m, the map shown in Figure 14 would have a datum shift of less than one pixel whether it had been digitized on NAD27 or NAD83. Most maps in recent issues of the *GSA Bulletin* appear to have a cylindrical map projection and import easily into Google Earth.

TABLE 1. MAP RESOLUTION FOR MAPS FROM
GEOLOGICAL SOCIETY OF AMERICA BULLETIN

Figure	Map width		Pixel size	Source
	Pixels	km	(m)	
13	833	32.7	37	Centeno-García et al. (2011, fig. 4)
14	828	214	258	Centeno-García et al. (2011, fig. 2)
15	925	22	42	Oner and Dilek (2011, fig. 3)

A few conical projections of very large areas would require a more complex import; Guth and Jacks (2011) showed how to use MICRODEM to do a Delaunay triangulation with the map grid and provided an example using historic World War II maps of the Normandy invasion. While not as easy as a registration in Google Earth, students can accomplish the task.

Google Earth has limits on data file sizes, and may not be able to handle data sets that conventional GIS software displays, or may slow down significantly. The limits apply to the sizes of individual rasters, which may require tiling (De Paor and Whitmeyer, 2011). MICRODEM can tile large rasters before export, but in my experience the need for tiling single map exports has been decreasing in newer versions of Google Earth unless the raster layer is truly huge. The overall size and complexity of a project creates a larger challenge, and may best be accomplished by managing a map project outside Google Earth. I have used HTML files, stored both on a web server for public access and on a personal computer for individual access. A table can hold links to the layers, a thumbnail image of the layer, and notes (e.g., Guth, 2011). Users can even use the browser search feature to augment the logical arrangement in the HTML file to find layers, and clicking on a link opens the layer in Google Earth. In this way I have managed a research collaboration with over 150 KML layers and almost 600 MB of data which could never open simultaneously in Google Earth.

While it proves relatively easy to import most of the maps currently available with journal papers online into Google Earth, the greatest value would occur if the maps were posted online by the journals already formatted as KML or KMZ files. Authors could prepare the files in their GIS software, ensuring correct registration, and with full control over map content. The actual map data need not be published, only the published figures with the information needed to correctly display them in a virtual globe. The additional registration information would add a trivial amount to the file size, so this would only require storing a second version of the figure with each map. Photos could also be linked to the map, and potentially cross sections, and graphs which would be reasonable to associate with a particular place. Tables, such as radiometric dates, could also be included as a data layer, and would be more useful than the current PDF files in an electronic supplement. Ultimately, as map data becomes much more common with the proliferation of geographically aware smart phones, standard browsers may display the images in KML files and the journals could provide just the KML files.

Unlike some earlier forays into new technology, like the microfiche experiment in the early 1980s, web delivery for journals appears to be unstoppable. Building on their heritage with maps, the geological journals should lead a move to deliver content in a fully geographic context. Instead of just static maps on the computer monitor, maps should be fully interactive in a virtual globe. The technology currently exists for that delivery. While the intellectual property issues must be addressed, the experience of the music and movie industries shows that users will find a way to get the content into the format they want, and the publish-

ers must adapt. I anticipate the day when the download page for figures from a scientific paper will include a button to download a KMZ file (Fig. 16), and that clicking on the button would lead straight to Google Earth with the map properly registered. All the maps for a paper could download with a single click, and could contain links to other maps and data. In addition to display and manipulation of data, the virtual globe could provide the search mechanism for additional resources.

CONCLUSION

The freeware MICRODEM program provides a simple graphical interface for the conversion of GIS maps and databases into KML files for display in Google Earth or comparable viewers. Users can filter data sets, select display symbology, and prepare time animations that can be widely shared. Users with no GIS training, and no access to expensive commercial software, can view and manipulate the data. MICRODEM can register existing maps, such as the digital figures from online journals, and provide a tailored grid to assist in simple registration done in Google Earth.

The earth sciences have a rich tradition of using maps to disseminate information. Google Earth and similar virtual globes allow users to adjust the map display, and combine information

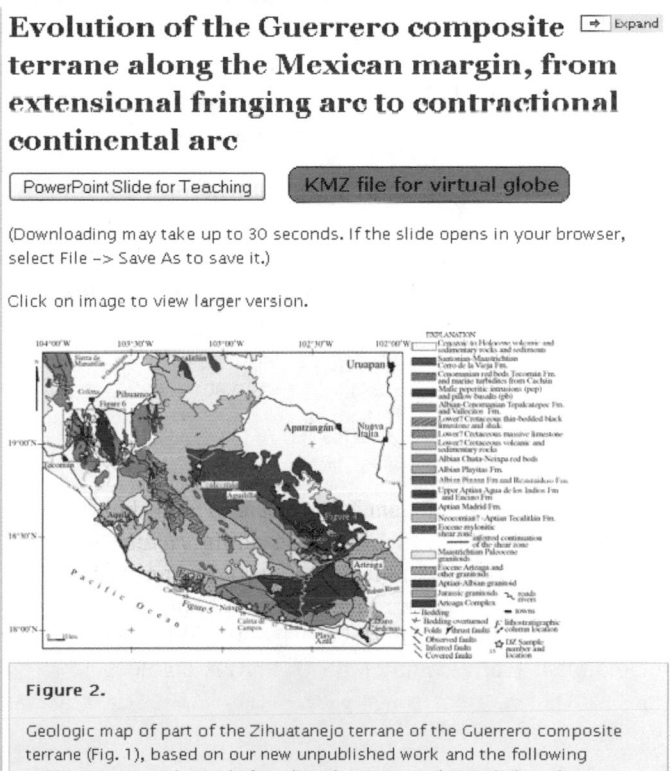

Figure 16. Download for a map from a journal paper displayed by the Geological Society of America website, with the suggested addition of a button to download a KMZ version.

from multiple sources. GIS software enhances the capabilities of map users, who are no longer restricted by the filter of a trained cartographer, but requires significant training and often expensive and proprietary software. KML and virtual globes offer a significant extension to a much wider audience, who can use free viewers to visualize map data, changing scales and turning layers on and off. This offers exciting possibilities for teaching, research collaboration, and viewing and combining data.

ACKNOWLEDGMENTS

Complete directions, along with screen captures for how to perform the operations discussed, are included in the MICRODEM help. MICRODEM is freeware (no restrictions on use, educational or commercial) and can be downloaded from http://www.usna.edu/Users/oceano/pguth/website/microdem/microdemdown.htm. KMZ files for most of the examples shown in this paper, and links to additional KMZ exports from MICRODEM, are located at http://www.usna.edu/Users/oceano/pguth/website/kml/gsa_special_paper/guth_chapter.htm.

I thank students Midshipman Hardie, Milian, Minker, Pritchard, Testino, and Thompson for sharing their labs on the March 2011 tsunami. SEAP intern Olivia Goldberg provided the impetus for automating legend export to Google Earth. Finally, I thank the editor and anonymous referees for very thorough reviews that significantly helped to focus and improve this paper.

REFERENCES CITED

Ballagh, L.M., Raup, B.H., Duerr, R.E., Khalsa, S.J.S., Helm, C., Fowler, D., and Gupte, A., 2011, Representing scientific data sets in KML; methods and challenges: Computers & Geosciences, v. 37, p. 57–64, doi:10.1016/j.cageo.2010.05.004.
Butler, D., 2006, Virtual globes: The web-wide world: Nature, v. 439, p. 776–778, doi:10.1038/439776a.
Centeno-García, E., Busby, C., Busby, M., and Gehrels, G., 2011, Evolution of the Guerrero composite terrane along the Mexican margin, from extensional fringing arc to contractional continental arc: Geological Society of America Bulletin, v. 123, p. 1776–1797, doi:10.1130/B30057.1.
Chiang, G.-T., White, T.O.H., and Dove, M.T., 2009, Geospatial visualization tool kit for scientists using Fortran: Eos (Transactions, American Geophysical Union), v. 90, p. 250–251, doi:10.1029/2009EO290001.
CO-OPS (Center for Operational Products and Services), 2011, Tides and currents: http://tidesandcurrents.noaa.gov/ (accessed 4 December 2011).
De Paor, D.G., and Whitmeyer, S.J., 2011, Geological and geophysical modeling on virtual globes using KML, COLLADA, and Javascript: Computers & Geosciences, v. 37, p. 100–110, doi:10.1016/j.cageo.2010.05.003.
Ekström, G., and Nettles, M., 2011, Global CMT web page: http://www.globalcmt.org/ (accessed 2 December 2011).
ESRI, 1998, ESRI shapefile technical description: ESRI White Paper, 34 p., http://www.esri.com/library/whitepapers/pdfs/shapefile.pdf (accessed 5 December 2011).
ESRI, 2011a, ArcGIS: http://www.esri.com/software/arcgis/index.html (accessed 5 December 2011).
ESRI, 2011b, ArcGIS Explorer Desktop: http://www.esri.com/software/arcgis/explorer/index.html (accessed 5 December 2011).
Farr, T.G., Rosen, P.A., Caro, E., Crippen, R., Duren, R., Hensley, S., Kobrick, M., Paller, M., Rodriguez, E., Roth, L., Seal, D., Shaffer, S., Shimada, J., Umland, J., Werner, M., Oksin, M., Burbank, D., and Alsdorf, D., 2007, The shuttle radar topography mission: Reviews of Geophysics, v. 45, p. RG2004, doi:10.1029/2005RG000183.
GDAL, 2011, GDAL—Geospatial data abstraction library, http://www.gdal.org/ (accessed 5 December 2011).
Gesch, D.B., Oimoen, M., Greenlee, S., Nelson, C., Steuck, M., and Tyler, D., 2002, The national elevation dataset: Photogrammetric Engineering and Remote Sensing, v. 68, p. 5–11.
GRASS, 2011, Welcome to GRASS GIS: Geographic Resources Analysis Support System, http://grass.osgeo.org/ (accessed 5 December 2011).
Guth, P.L., 2009, Geomorphometry in MICRODEM, in Hengl, T., and Reuter, H.I., eds., Geomorphometry: concepts, software, applications: Developments in Soil Science Series, v. 33, Elsevier, p. 351–366.
Guth, P.L., 2011, Automated export of GIS maps to Google Earth: Tool for Research and Teaching: http://www.usna.edu/Users/oceano/pguth/website/kml/gsa_special_paper/guth_chapter.htm (accessed 7 December 2011).
Guth, P., and Jacks, J., 2011, Using KML to disseminate military geography: in International Handbook Military Geography, Proceedings 8th International Conference on Military Geosciences, Vienna Austria, June 2009, Redaktion Truppendienst, p. 220–232.
Guth, P.L., Ressler, E.K., and Bacastow, T.S., 1987, Microcomputer program for manipulating large digital terrain models: Computers & Geosciences, v. 13, p. 209–213, doi:10.1016/0098-3004(87)90041-0.
Lyell, C., 1830–1833, Principles of geology, being an attempt to explain the former changes of the Earth's surface, by reference to causes now in operation: John Murray, London, http://darwin-online.org.uk/content/frameset?viewtype=text&itemID=A505.1&pageseq=1, http://darwin-online.org.uk/content/frameset?viewtype=text&itemID=A505.2&pageseq=1, and http://darwin-online.org.uk/content/frameset?viewtype=text&itemID=A505.3&pageseq=1.
NDBC, 2011, National data buoy center: http://www.ndbc.noaa.gov/dart.shtml (accessed 7 December 2011).
OGC, 2011, KML: http://www.opengeospatial.org/standards/kml (accessed 7 December 2011).
Oner, Z., and Dilek, Y., 2011, Supradetachment basin evolution during continental extension: The Aegean province of western Anatolia, Turkey: Geological Society of America Bulletin, v. 123, p. 2115–2141, doi:10.1130/B30468.1.
QGIS, 2011, Quantum GIS: http://www.qgis.org/ (accessed 5 December 2011).
Ritter, N., and Ruth, M., 2000, GeoTIFF format specification GeoTIFF Revision 1.0: http://www.remotesensing.org/GeoTIFF/spec/GeoTIFFhome.html (accessed 5 December 2011).
Slate, J.L., Berry, M.E., Rowley, P.D., Fridrich, C.J., Morgan, K.S., Workman, J.B., Young, O.D., Dixon, G.L., Williams, V.S., McKee, E.H., Ponce, D.A., Hildenbrand, T.G., Swadley, W.C., Lundstrom, S.C., Ekren, E.B., Warren, R.G., Cole, J.C., Fleck, R.J., Lanphere, M.A., Sawyer, D.A., Minor, S.A., Grunwald, D.J., Laczniak, R.J., Menges, C.M., Yount, J.C., and Jayko, A.S., 2000, PART A. Digital Geologic Map of the Nevada Test Site and Vicinity, Nye, Lincoln, and Clark Counties, Nevada, and Inyo County, California, Revision 4: U.S. Geological Survey Open-File Report 99-554, Online Version 1.1: http://pubs.usgs.gov/of/2000/ofr-00-0554 (accessed 4 December 2000).
Snyder, J.P., 1987, Map projections—a working manual: U.S. Geological Survey Professional Paper 1395, 383 p.
USGS, 2011, Rectangular Area Earthquake Search: http://earthquake.usgs.gov/earthquakes/eqarchives/epic/epic_rect.php (accessed 3 December 2011).
Winchester, S., 2001, The map that changed the world: William Smith and the birth of modern geology: Harper Williams, 329 p.
Wood, J., 2009, Overview of software packages used in geomorphometry, in Hengl, T., and Reuter, H.I., eds., Geomorphometry: concepts, software, applications, Developments in Soil Science Series, v. 33, Elsevier, p. 257–267.

MANUSCRIPT ACCEPTED BY THE SOCIETY 16 APRIL 2012

The Geological Society of America
Special Paper 492
2012

Transferring maps and data from pre-digital era theses to Google Earth: A case study from the Vredefort Dome, South Africa

C. Simpson*
D.G. De Paor
M.R. Beebe
J.M. Strand
Old Dominion University, Norfolk, Virginia 23529, USA

ABSTRACT

College geoscience departments keep archives of student research ranging from senior theses to master's and Ph.D. dissertations. In field geology, these archives often include maps, cross sections, stereographic projections, field notes and photographs, hand specimens, and thin sections. Subsequent publications may result from the thesis work, but much of this valuable legacy data is difficult to access and assess. Here we describe the conversion of a pre–digital-era thesis on the Vredefort Rim Synclinorium in South Africa from hard copy to digital format using Keyhole Markup Language (KML) to drape maps and inset photographs, and COLLADA (COLLAborative Design Activity) models to create stereographic projections, emergent cross sections, and virtual specimens. In addition to using the Google Earth terrain to fine-tune draped map locations, errors in field locations arising from pace and compass or bearing methods of geo-location that preceded the availability of Global Positioning Systems (GPS) were recognized and corrected.

At 2.023 billion years in age and an estimated 300 km in original diameter, the Vredefort Dome is the world's oldest and largest known impact structure. The Vredefort region has been designated a World Heritage Site and specimen collection is prohibited. Only a few geologists are ever likely to visit the region, so geo-referenced field photography, specimens, and structural data are irreplaceable. An interpretative center is being planned for the Vredefort structure by South African authorities and our interactive Google Earth resources will be made available to the visiting public as well as those browsing over the Internet. Thus draped maps and scanned models provide an invaluable opportunity for enhanced instruction, continued research, and public outreach.

*CSimpson@odu.edu

Simpson, C., De Paor, D.G., Beebe, M.R., and Strand, J.M., 2012, Transferring maps and data from pre-digital era theses to Google Earth: A case study from the Vredefort Dome, South Africa, *in* Whitmeyer, S.J., Bailey, J.E., De Paor, D.G., and Ornduff, T., eds., Google Earth and Virtual Visualizations in Geoscience Education and Research: Geological Society of America Special Paper 492, p. 183–197, doi:10.1130/2012.2492(13). For permission to copy, contact editing@ geosociety.org. © 2012 The Geological Society of America. All rights reserved.

INTRODUCTION

The worlds of academic publishing and library science are currently undergoing radical change as the Portable Document Format (PDF) and Digital Object Identifier (DOI) replace the paper page and bound volume. As researchers rely ever more on electronic media, access to a rich source of geological legacy data is becoming increasingly difficult. That resource includes the stacks of bound B.S., M.S., and Ph.D. theses and post-doctoral dissertations with accompanying geological maps that are housed in academic libraries. Few theses are in sufficient demand to warrant a Google Books–style conversion to digital format by use of optical character recognition software or even by scanning as image files. Consequently, readership is extremely low. It would not be surprising if many university libraries soon ceased their Inter-Library Loan services for unpublished works. In some fields—for example, genetic engineering or stem cell research—this is not a major problem as older information becomes obsolete and little data preceded the digital era. However, in the geosciences, a huge wealth of important and still-relevant field data is stored in theses and memoirs. Remapping from scratch using digital technology (Global Positioning Systems, GPS, and geographic information systems, GIS) is difficult to fund and time-consuming to implement. The solution is for those with access to the theses to collaborate with digital data technologists in order to convert the work, especially maps and other geospatial data, to accessible digital format before they are lost.

In this paper, we use the example of a field-based M.S. thesis (Simpson, 1977) on the topic of the Rim Synclinorium surrounding the Vredefort Dome, South Africa. The thesis reported on months of mapping and map-making using traditional methods including pace and compass, clinometer readings, town-lands fence maps, and scarce oblique aerial photographs. In 1996, Simpson and De Paor both visited the region while offering a University of Witwatersrand short course entitled "New Directions in Structural Geology" and were able to collect additional specimens and take new field photographs. In 2005, Google Earth™ was released to the general public and it became possible to view the structure and the legacy data from an entirely new perspective.

Building upon the methods employed in Simpson and De Paor's (2010) virtual field trip (VFT) to the Assynt region of Scotland, we scanned maps and figures from Simpson's (1977) thesis, together with photographs from our subsequent visit, and imported them to Google Earth as ground-overlays and photo-overlays. Using additional recently developed techniques (De Paor, 2009; De Paor and Whitmeyer, 2011; De Paor et al., 2012a, this volume), we constructed a series of emergent cross sections in their correct field locations and orientations. We also scanned rock specimens using a NextEngine™ 3D scanner in order to create virtual specimens as COLLADA (COLLAborative Design Activity) models geo-referenced to their original collection locations (Beebe and De Paor, 2010; Simpson et al., 2010). These measures are particularly important to the current field area as it is now a World Heritage Site (Fleminger, 2008) and is protected from specimen collection. We took structural orientation data that had been plotted on paper maps and stereographic projections and made them available in Keyhole Markup Language (KML) documents using Whitmeyer's map symbol generator (http://csmres.jmu.edu/Geollab/Whitmeyer/web/visuals/GoogleEarth/tools/SD.html). Finally we created a presentation designed to focus students' or museum visitors' attention on the Kaapvaal region and to highlight the differences between the Archean terrain 2 billion years ago and the surface imagery of today's Google Earth.

THE AREA OF INTEREST

At 2.023 billion years old (Gibson and Reimold, 2000), the Vredefort Dome is the world's oldest generally accepted giant impact structure, or astrobleme. The present geomorphological expression is a ring of high terrain ~70 km in diameter, but the original structure is estimated at 300 km diameter, making it also the largest preserved astrobleme on Earth—larger than the Sudbury impact crater in Ontario, Canada, or the Chicxulub structure in Yucatan, Mexico. Figure 1 shows the location of the Vredefort Dome around the towns of Vredefort and Parys, ~120 km southwest of Johannesburg, South Africa. Clearly visible on the Google Earth image are the central domal uplift, composed of Archean granite gneiss, surrounded by a prominent ring structure of steeply dipping to overturned Witwatersrand Supergroup sedimentary beds of Late Archean age (Catuneanu,

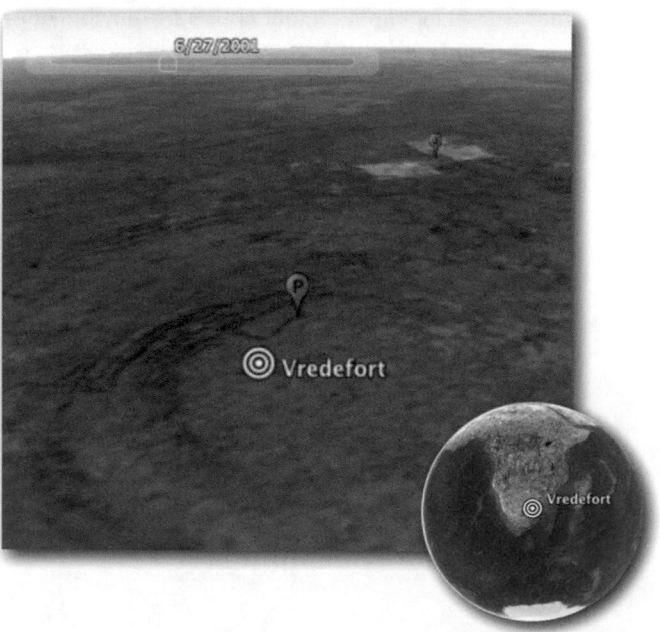

Figure 1. Outline of the Vredefort Structure in Google Earth. P—Town of Parys, J—City of Johannesburg. The May 2001 historic imagery is blurred, but free of distracting agricultural patterns.

2001), and a surrounding rim synclinorium of overlying Transvaal Sequence sedimentary and volcanic rocks (Simpson, 1978) of latest Archean to early Proterozoic age, spanning ~2.65–2.05 Ga (Moore et al., 2001). The feature is prominent in geological, magnetic, and gravity maps, and concentric bedding strikes may be easily recognized and traced on the 3D Google Earth terrain (Fig. 2), especially in the vicinity of the River Vaal in the north and northwest of the structure; to the southeast, the surface expression is overprinted by Late Carboniferous to Early Jurassic sedimentary and volcanic rocks of the Karoo Supergroup (Johnson et al., 1996), but the buried geophysical structure of the astrobleme is still clear.

Today, the Vredefort Dome is generally recognized as the central rebound mound of a giant impact structure (Wolf and Gibson, 2010). However, in the late twentieth century, less was known about impact tectonics on terrestrial bodies, least of all on Earth, and there was still controversy between those who supported an incoming asteroid impact hypothesis (beginning with Daly, 1947) and those who favored an outward-directed crypto-explosion (e.g., Nicolaysen et al., 1963). Simpson (1977) concentrated research efforts on structural field mapping, analysis of rock specimens, and construction of cross sections in the rim synclinorium surrounding the central uplift. The tools and techniques of the time were strictly analog.

FIELDWORK THEN AND NOW

In the 1970s, field geologists located themselves using a combination of compass bearings and estimating distances by pacing. Geological mapping was carried out by penciling data onto printed field slips or air photos during the day, and then "inking-in" the data onto fair-copy maps perhaps by the light of a Coleman™ Camp Lantern in the evening. Back in the laboratory, individual map sheets were laboriously combined onto larger, poster-sized Mylar sheets using fine-nib Rotring™ pens and stencils or Letraset™ transfer lettering, and these sheets were stored in data repositories or occasionally converted to printed paper or canvas maps (see Simpson and De Paor, 2010).

Today's geologists, like everyone else, can determine their position to the nearest couple of meters using their cellphones, and they can track their every move in the field with dedicated GPS devices. Hand-held computers such as the Trimble™ running ArcPad™ or the iPad™ running the GIS Roam™ or iGIS™ mobile application allow geologists to plot data directly onto digitized maps. Digital cameras can geo-reference images in preparation for automatic display in the correct geographic location after uploading to the Internet (note however, that cameras still do not automatically record the lens-axis azimuth and inclination). It would be nice to be able to turn back the clock and redo old field projects with modern methods. However, there is little time or funding available for such exercises. We therefore undertook the task of "map inversion" (a term coined by De Paor and Sharma, 2007) in order to make legacy data ready available in digital format.

RETRIEVING LEGACY DATA

Since the submission of Simpson's (1977) thesis, its base maps, original figures, notebooks, specimens, and photographs have been transported across continents and from one campus laboratory to another and have been supplemented by specimens and photographs from a subsequent field trip. In the not-too-distant future, their journey will end with the author's retirement. Were it not for this study, it is likely that much of these materials would end up in an academic catacomb, at best. The methods described below are applicable to numerous other legacy field mapping collections that otherwise would never again see the light of day. We hope that many others will be encouraged to undertake the admittedly tedious, but highly rewarding process of map inversion using our methods.

SCANNING, TILING, AND DRAPING MAPS

The first task was to make maps from the thesis available as ground-overlays on the Google Earth terrain. The basic concept is simple: any digital image such as a JPEG, GIF, PNG, or TIFF file can be draped over the Google Earth terrain using the "add image overlay" menu option (Fig. 3A). The image can be positioned, scaled, and rotated using draggable green handles or the bounding latitudes and longitudes may be entered at the keyboard. Strictly, only Plate Carrée (also known as Simple Cylindrical) projection will fit the Google Earth terrain perfectly, but on a deca-kilometer scale other projections are often close enough for practical purposes. A little-appreciated feature of the current version of Google Earth is the "Convert to LatLonQuad" option (Fig. 3B). This feature, which was previously confined to Google Earth Pro, allows a map to be "rubber-sheeted" by a quadrilateral transformation (see De Paor, 1994), thus improving the fit to the terrain.

Two practical problems arise with the creation of ground overlay maps. First, maps are often larger than U.S. letter or A4 paper size. They have to be either cut or folded in order to fit a standard flat-bed scanner or else fed through a large format scanner. Many poster printers are also drum-scanners, where care has to be taken when feeding fragile, sometimes brittle, maps through such large format scanners. Second, hand-drawn maps generally contain very fine handwriting, equivalent to a 9 point font or smaller, therefore maps have to be scanned at high resolution resulting in multi-megabyte or even gigabyte file sizes. If several of these scanned images are added to Google Earth, the application's responsiveness will grind to a halt. Indeed there are many examples of geological maps available for download from the World Wide Web where the act of loading the file causes Google Earth to halt. The solution is to create image tiles arranged in image pyramids (the Google term is Super-Overlays). The free, beta version of the application "MapTiler" was used in this study in order to generate tile pyramids automatically. However, several maps had to be pre-processed in Photoshop in order to ensure that they were strictly orthogonal, because MapTiler does not handle rotated orientations or quadrilateral distortions.

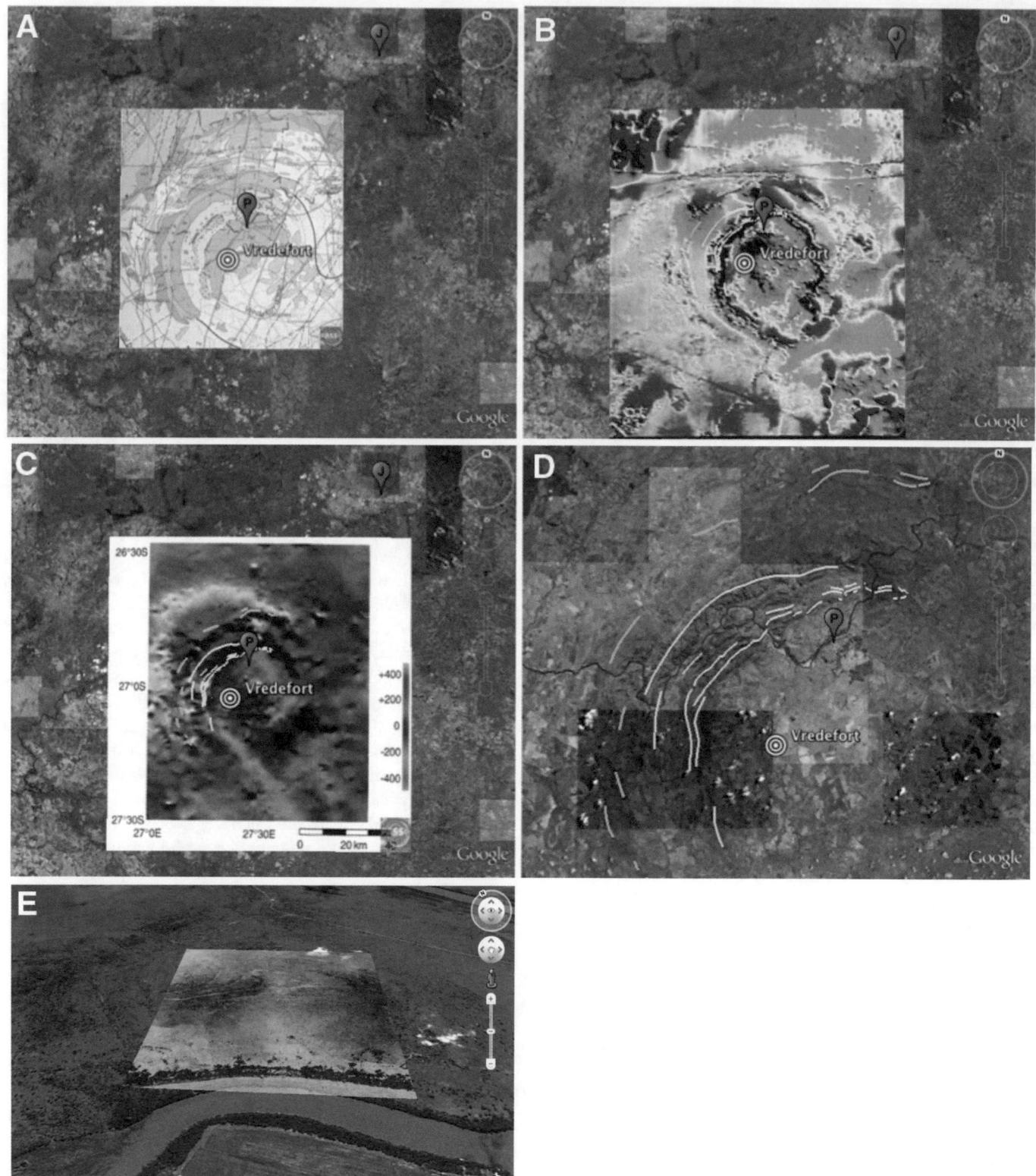

Figure 2. The Vredefort structure on (A) geologic map by Martin Tuchscherer, (B) Aeromagnetics, (C) Bouguer gravity image (B and C both courtesy of South African Council for Geoscience), (D) bedding traces on the Google Earth terrain (source: http://www.unb.ca/passc/ ImpactDatabase/images/vredefort.htm). (E) Oblique aerial photos were rare treats ca. 1970. This one was used in the thesis mapping. Here projected as an inclined photo overlay over Google Earth.

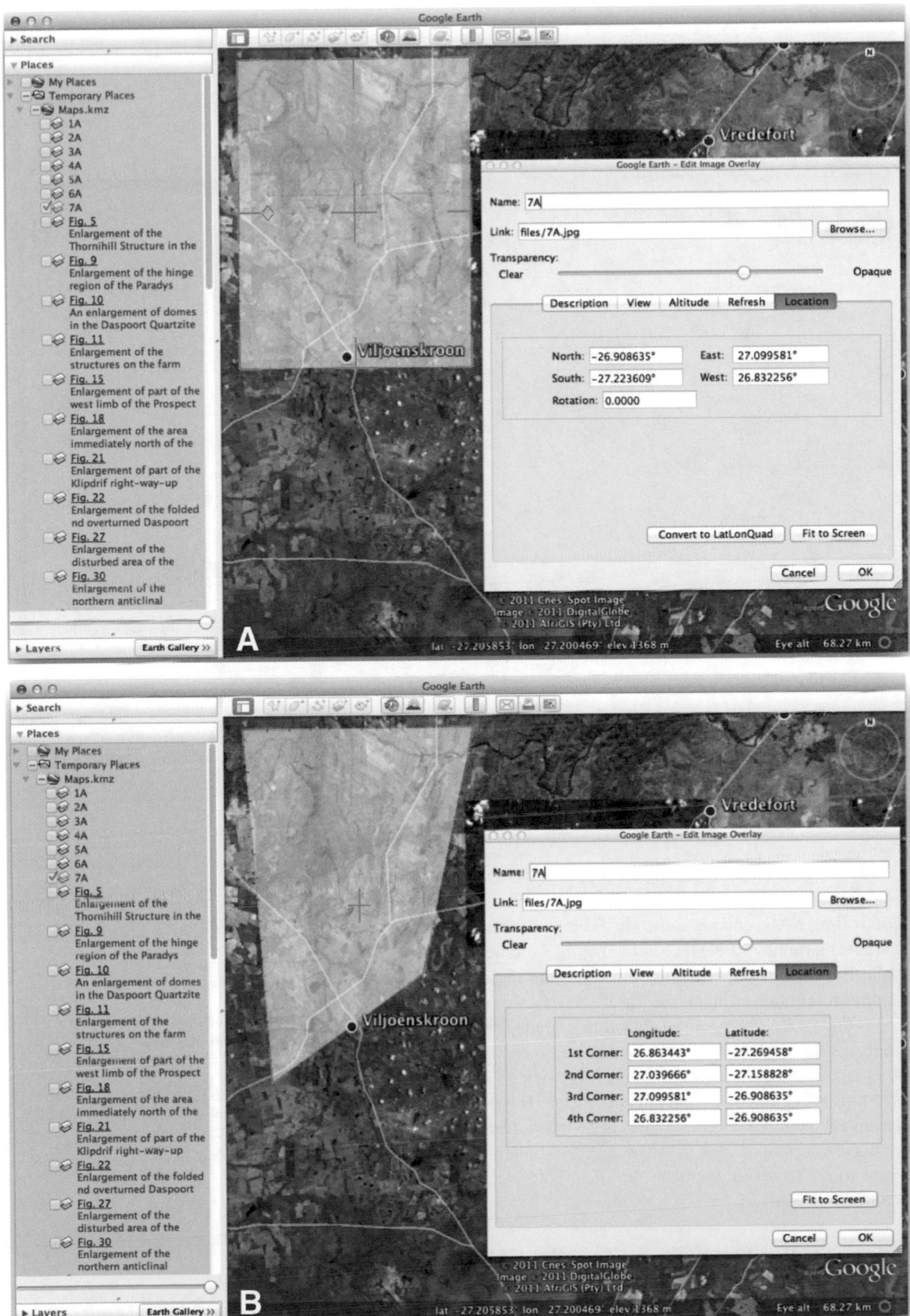

Figure 3. Adding maps to Google Earth as ground overlays.

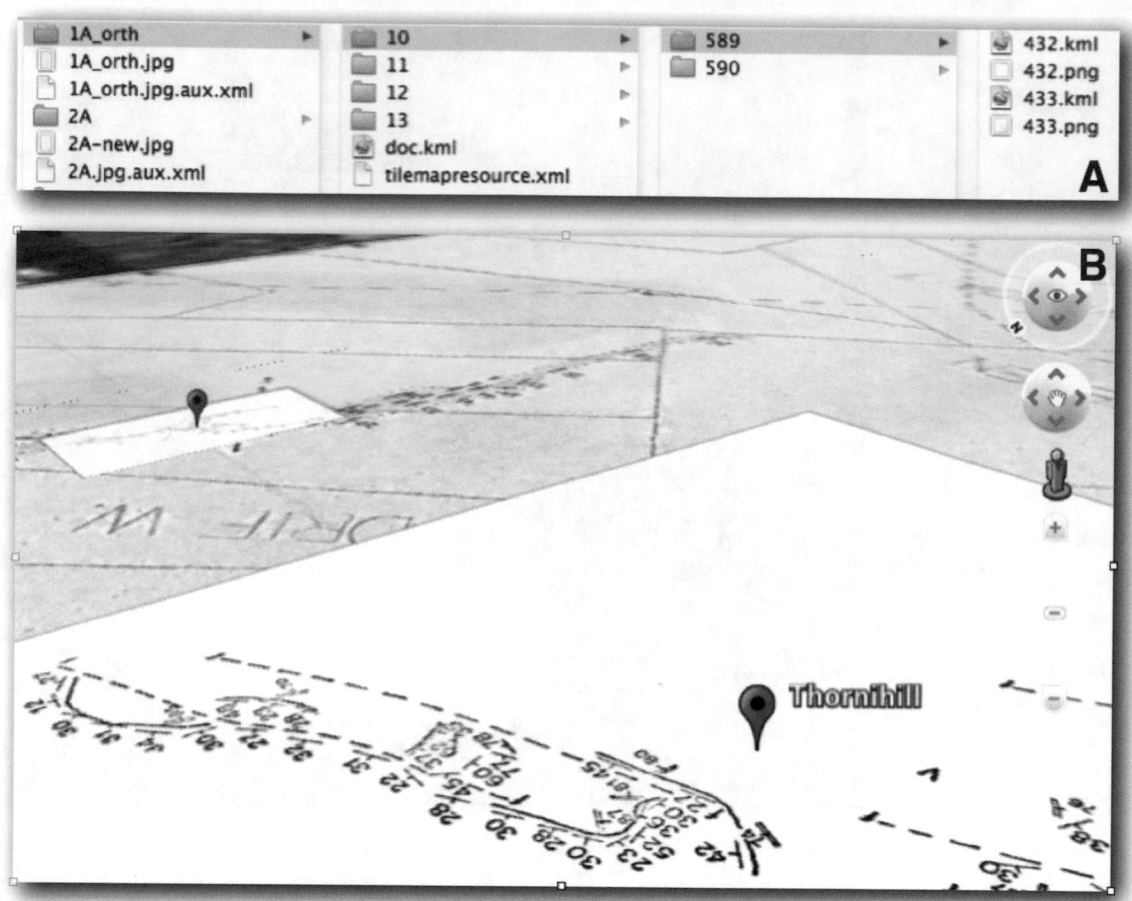

Figure 4. Maximizing foreground image resolution and file size. (A) Image pyramid file structure. (B) Foreground detail on map has higher resolution than background region.

The file structure of an image pyramid is shown in Figure 4A, and the controlling KML code is shown in Table 1. At every level of the pyramid structure, a KML "NetworkLink" element checks the visibility of each region of the image. If the region occupies a minimum number of pixels, an "href" link to a more detailed image is invoked. The result is that foreground detail is shown at high resolution but lower resolution (and therefore smaller file size) is used for background regions (Fig. 4B).

Image pyramids are central to the way in which Google Earth renders the terrain efficiently. However, issues arise when different image tile sets are superimposed, as when a map inset shows detail of a larger map. To avoid conflicts, the tile "drawOrder" must be set manually in KML.

Reverse-Engineering Structural Data

In addition to formation identification and contacts, the thesis maps abound in structural data. Strikes and dips of bedding, foliation, faults, etc., were drawn from field notebook records. To

TABLE 1. A TYPICAL NETWORK LINK FROM AN IMAGE PYRAMID (ALSO KNOW AS A SUPER-OVERLAY)

```
<NetworkLink>
  <name>10/589/432.png</name>
  <Region>
    <Lod>
      <minLodPixels>128</minLodPixels>
      <maxLodPixels>-1</maxLodPixels>
    </Lod>
    <LatLonAltBox>
      <north>-26.74561038219902</north>
      <south>-27.05912578437405</south>
      <east>27.42187499999999</east>
      <west>27.07031250000000</west>
    </LatLonAltBox>
  </Region>
  <Link>
    <href>10/589/432.kml</href>
    <viewRefreshMode>onRegion</viewRefreshMode>
  </Link>
</NetworkLink>
```

Note: The "Region" element checks whether the level of detail ("Lod") of an image tile covering the latitudes and longitudes listed is sufficient to occupy 128 pixels of the screen at the current zoom level. The "Link" element applies the same test to the next level of the image pyramid.

fully leverage the advantages of a digital globe such as Google Earth we converted orientation data into 3D COLLADA models (Fig. 5). This can be done using the symbol generator from Whitmeyer (http://csmres.jmu.edu/Geollab/Whitmeyer/web/visuals/GoogleEarth/tools/SD.html). In this case, we substituted a custom strike and dip symbol that incorporates the dip amount as seen in red (Table 2). The KML Placemark contains a COLLADA model that links to a Digital Asset Exchange file called "17°.dae." There are 90 such models on the server, one for every degree of dip. The model is rotated into the correct strike orientation using the KML "heading" tag and is inclined by the dip amount using the "roll" tag (the negative sign reflects COLLADA's z-up reference frame in contrast to field geologists' positive-down convention for dip or plunge). The result is that all structural data is query-able with the click of a computer mouse and structures can be more easily visualized when viewed obliquely (Fig. 5B). Future plans include use of linked Fusion Tables for added database management. In some cases, pace and compass errors in the original mapping were evident. We repositioned data collection sites using the Google Earth terrain imagery as a guide; however, as with any archival restoration work, we were careful to distinguish between original and corrected data.

Some data were too numerous or complex to convert to individual digital objects. For example, some original tables of structural measurements such as en echelon vein arrays (Fig. 5C) were simply scanned as JPEG images. If access to this data warrants further work in the future, material can be rescanned with optical character recognition software.

EMERGENT CROSS SECTIONS

Since the first geological mapping by William Smith (Smith, 1815) cross sections have played a vital role in helping viewers visualize 3D structures. On paper maps, cross sections are often added to the end of a map sheet or printed on a separate sheet. Google Earth allows us to view these sections in their correct position and orientation on the virtual globe. Furthermore, using the approach of De Paor (2007), cross sections can be made to emerge from the Google Earth terrain when a slider control is dragged. A representative snippet of the controlling KML code is shown in Table 3. A series of placemarks are created with incremental altitude tags and a "Timespan" element bounded by corresponding begin and end times. When the user drags the time slider thumb, the cross section emerges from the terrain. (Since

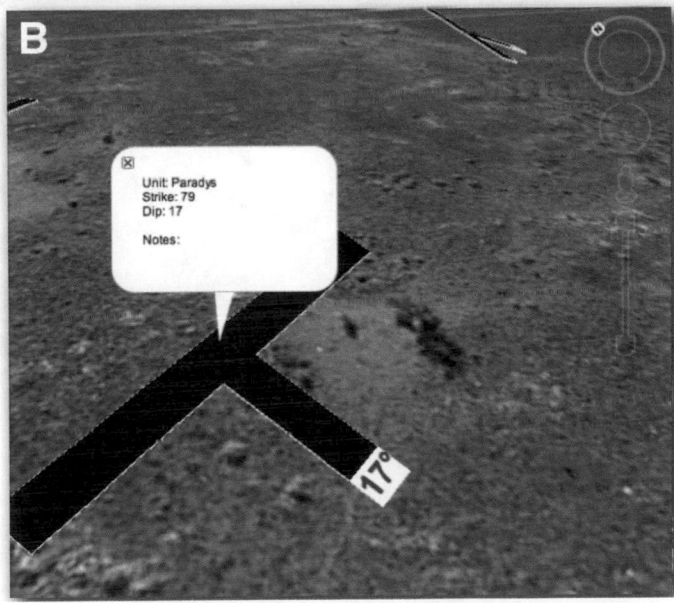

Figure 5. Converting structural orientation data to 3D models. (A) COLLADA model of a dip and strike symbol hovers over the paper equivalent on the draped map. (B) The model data is query-able; placemark balloon reveals metadata. (*Continued.*)

TABLE 2. KML CODE FOR INSERTING A DIP AND STRIKE SYMBOL AT A LOCATION IN ITS CORRECT GEOGRAPHICAL ORIENTATION

```
<Placemark>
        <name>Symbol8</name>
        <Model id="SDmodel">
                <altitudeMode>relativeToGround</altitudeMode>
                <Location>

        <longitude>26.89451363309894</longitude>
                        <latitude>-26.9996947209925</latitude>
                        <altitude>15</altitude>
                </Location>
                <Orientation>
                        <heading>79</heading>
                        <tilt>0</tilt>
                        <roll>-17</roll>
                </Orientation>
                <Link>
                        <href>17°.dae</href>
                </Link>
        </Model>
</Placemark>
```

Note: "heading" corresponds to strike, but "roll" is the negative of dip.

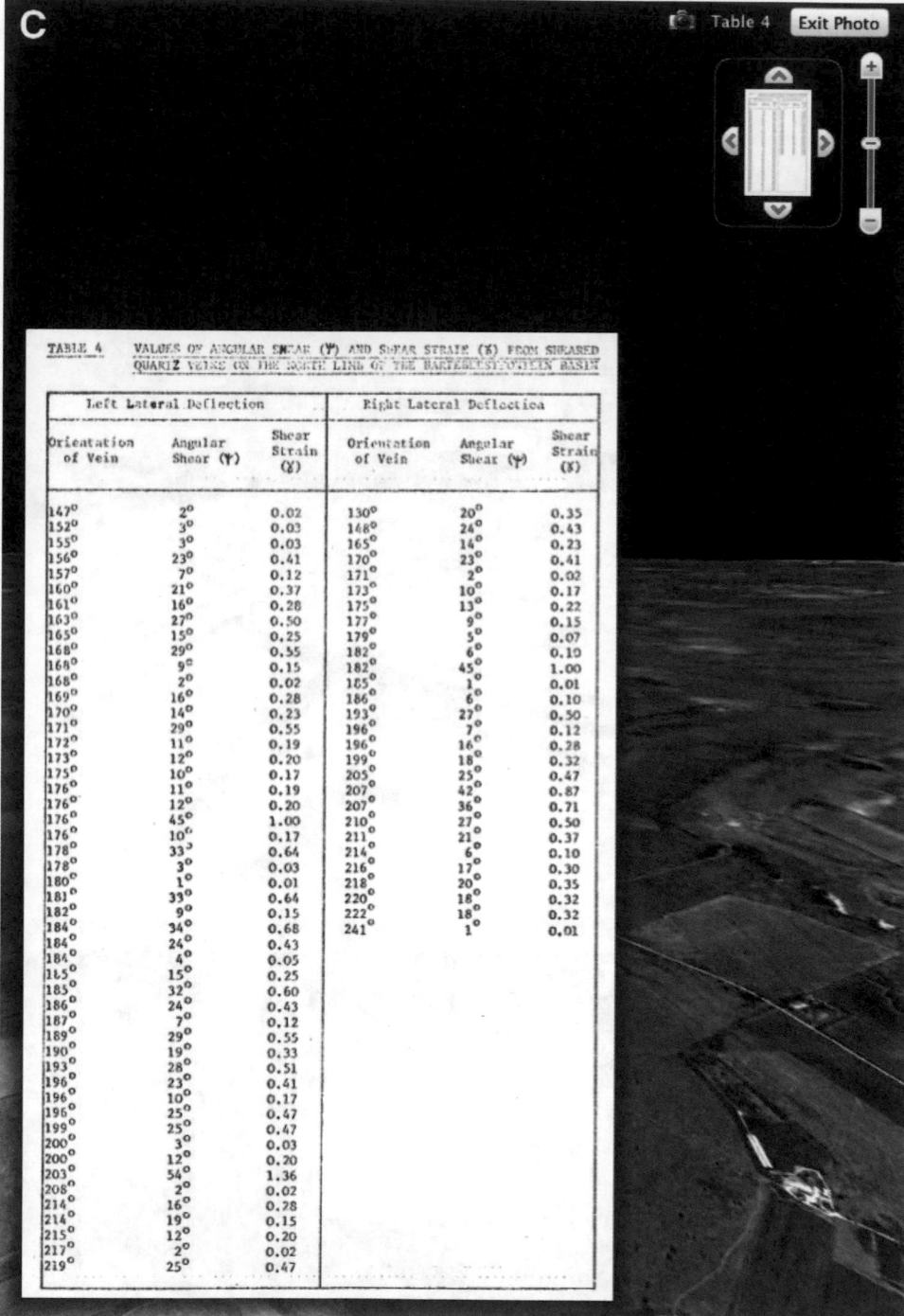

Figure 5 (*Continued*). (C) Some thesis data was scanned and geolocated as JPEG images. If they become important in the future, these data may be retrieved by optical character recognition.

submission of this manuscript, other methods of elevating cross sections using the Google Earth tour feature have been developed by Steve Whitmeyer. See De Paor et al., this volume.)

Simpson (1977) constructed a series of radial sections to represent the rim synclinorium of the Vredefort structure. We scanned each section individually and saved two mirror image versions for use as textures on opposite sides of rectangular COLLADA models. The models were scaled and oriented to fit along the lines of section drawn on the draped maps. We animated these sections in unison as illustrated in Figure 6.

PHOTO-OVERLAYS AND MONSTER BILLBOARDS

In addition to maps and cross sections, Simpson's (1977) thesis, just like most of the era, contained a large collection of figures, including sketches and photographic prints physically

TABLE 3. KML CODE SNIPPET FOR EMERGENT
CROSS SECTIONS SHOWING TWO OF A LONG SERIES OF
"PLACEMARK" ELEMENTS THAT DIFFER ONLY
IN THE BEGINNING AND ENDING TIMES OF THE TIMESPAN
AND THE ALTITUDE OF THE CROSS SECTION MODEL

```
<Placemark>
  <TimeSpan>
          <begin>2300</begin>
          <end>2400</end>
  </TimeSpan>
  <Model id="model_192">
          <Location>
                  <longitude>27.35364</longitude>
                  <latitude>-26.65267</latitude>
                  <altitude>2400</altitude>
          </Location>
          <Link>
                  <href>files/A-B.dae</href>
          </Link>
  </Model>
</Placemark>

<Placemark>
  <TimeSpan>
          <begin>2400</begin>
          <end>2500</end>
  </TimeSpan>
  <Model id="model_1912">
          <Location>
                  <longitude>27.35364</longitude>
                  <latitude>-26.65267</latitude>
                  <altitude>2500</altitude>
          </Location>
          <Link>
                  <href>files/A-B.dae</href>
          </Link>
  </Model>
</Placemark>
```

pasted to the page. Associated field records include Kodachrome transparencies. We scanned all of these images, added subsequent field trip photos, and saved them as PNGs and high-resolution JPEG files. We then used two methods to present thesis figures on the Google Earth terrain. First, we imported images as photo-overlays using the "Add Photo" menu item (Fig. 7A). During the adding process, Google Earth enables the content creator to position and orient the image to match the background view. Of course, the original images were taken with a chemical film camera that lacked today's automatic geo-referencing so some locations and orientations are approximate. Nevertheless, Figure 7A shows a photograph of a bench of the Leeukop Quarry, famous for its exposure of pseudotachylite (Reimold and Colliston, 1994; Gibson and Reimold, 2001), aligned against the bench which is identifiable in Google Earth imagery. A KML code snippet for a typical photo-overlay is shown in Table 4. Photo-overlays have the advantage that they handle large file tiling automatically; however, they have a disadvantage in that the viewer becomes locked into the photo view until they choose to "Exit Photo." This limits browsing options, as Google Earth becomes a Zoomify-type viewer (www.zoomify.com). Therefore we also created so-called "monster billboards" for viewing photos such as in Figure 7B. These are COLLADA models and are essentially the same as

cross sections. Monster billboards allow oblique viewing without locking into the camera view, which can be important when the user wishes to compare a photograph with the background vista.

SCANNING AND GEO-REFERENCING VIRTUAL SPECIMENS

One of the most important forms of legacy data associated with academic theses, field expeditions, and even casual conference-related field trips, are the hand specimens that geoscientists collect. In the case of Vredefort, rock specimens containing shocked quartz, pseudotachylite, and shatter cones are critical for interpreting the tectonic origin of the structure. Since the region was declared a World Heritage Site in 2005, specimen collection is no longer permitted. In any case, the total number of people who get to travel to Vredefort and to examine these outcrops in person is miniscule.

To enable scientists and students across the globe to examine Vredefort rocks, we scanned specimens that had been collected during thesis fieldwork and subsequent short course trips using a NextEngine 3D scanner (Fig. 8). This scanner combines a high resolution digital camera, custom multi-scan laser technology, and a revolving specimen stage. The instrument's software, called ScanStudio™, creates a 3D wireframe model of each specimen face and drapes a photographic image over it. The user must stitch models of the different faces together as is commonly done when creating a panorama photo, for example. Parts of the specimen may not scan well because they are touching the stage or because of their intricate topography, including holes, however these are easily recognized on the final model.

To present virtual rock specimens on the Google Earth terrain, we exported the models from ScanStudio as 3D object files (dot-obj) and imported them to an intermediate program called MeshLab. No changes were made to the imported models; rather they were immediately exported with change of file type. This allowed file type conversion from object to Digital Asset Exchange format (dot-dae). The latter can be imported into Google Earth using the "add…model" menu option. Associated KML code was used to position, orientate, and scale the models to match their original collection sites. Figure 9A shows three specimens collected from the Leeukop Quarry. Zooming in on the specimens (Fig. 9B) allows the viewer to clearly identify coarse granitic minerals and dark fine-grained shock-induced glass. In the background of Figure 9B one can make out a specimen hovering over the rim synclinorium on the horizon. Zooming in on this (Fig. 9C) reveals a shatter cone structure indicative of impact.

ARCHEO–GOOGLE EARTH

The Earth was very different two billion years ago. Continents were not green because plants had not yet evolved. The only extensive green surfaces were greenstone belt rocks and

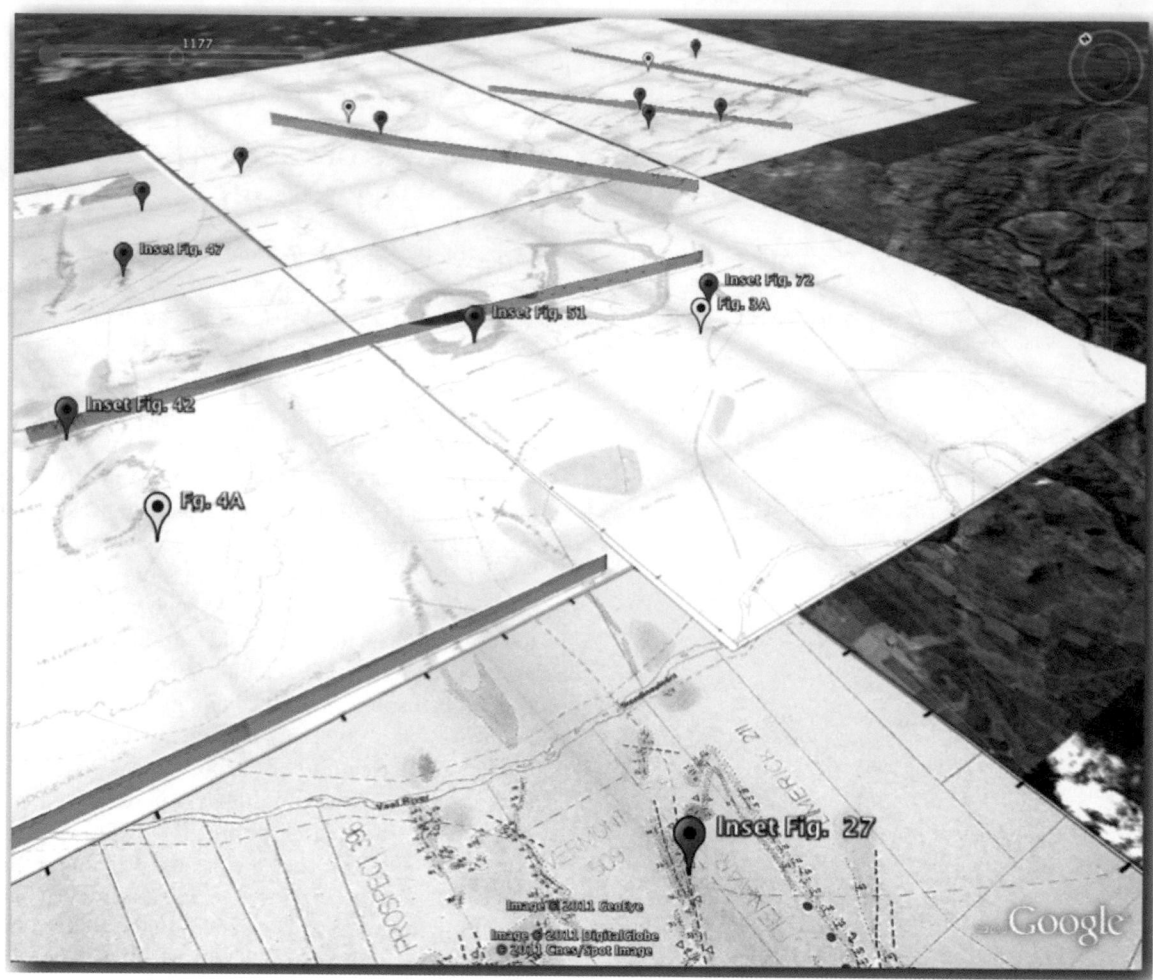

Figure 6. Five emergent cross-sections positioned in correct orientation and location above geological map in Google Earth. Placemark pins and pin labels indicate locations of more detailed maps and figures from original thesis.

pond scum due to cyanobacteria. Free atmospheric oxygen had just begun to build up, but there was probably little rusty iron oxide coloration even in deserts. No ozone layer protected the surface from bombardment by UV sunlight. The Earth's inner core had not yet solidified so there was little or no magnetic field protecting the surface from the solar wind. In short, it was not a nice place to live unless you were a submarine microbe. Only the oldest Archean cores of continents and the oldest Protero-zoic platform strata existed—the rest of today's mobile belts had not yet accreted to the micro-continental margins. However, although the micro-continents may have resembled today's des-erts, they were not all arid. Except for latitudes 30N and 30S and high plateaus, the desert-like surface of present-day southern Africa had plenty of rivers, lakes, and inland seas as represented in the stratigraphic column (e.g., Minter, 2006).

Earth's current vegetated and developed landscape interferes with our visualization of the Archean world and may constitute a hindrance to learning. In Figure 10, we illustrate the barren surface of the Archean Kaapvaal Craton, with a large inland sea

TABLE 4. KML CODE SNIPPET FOR A TYPICAL PHOTO-OVERLAY

```
<PhotoOverlay>
        <name>Leeukop Quarry</name>
        <Camera>...</Camera>
        <Style>...</Style>
        <Icon>
                <href>photos/Leeukop.jpg</href>
        </Icon>
        <ViewVolume>
                <leftFov>-11.88</leftFov>
                <rightFov>11.88</rightFov>
                <bottomFov>-7.865</bottomFov>
                <topFov>7.865</topFov>
                <near>0.884024</near>
        </ViewVolume>
        <Point>
           <coordinates>
                27.40739939435396,
                -26.89660771652535,
                2.525783803827311
           </coordinates>
        </Point>
</PhotoOverlay>
```

Note: This code is automatically generated when a photograph is added using the Google Earth "Add Photo" menu item.

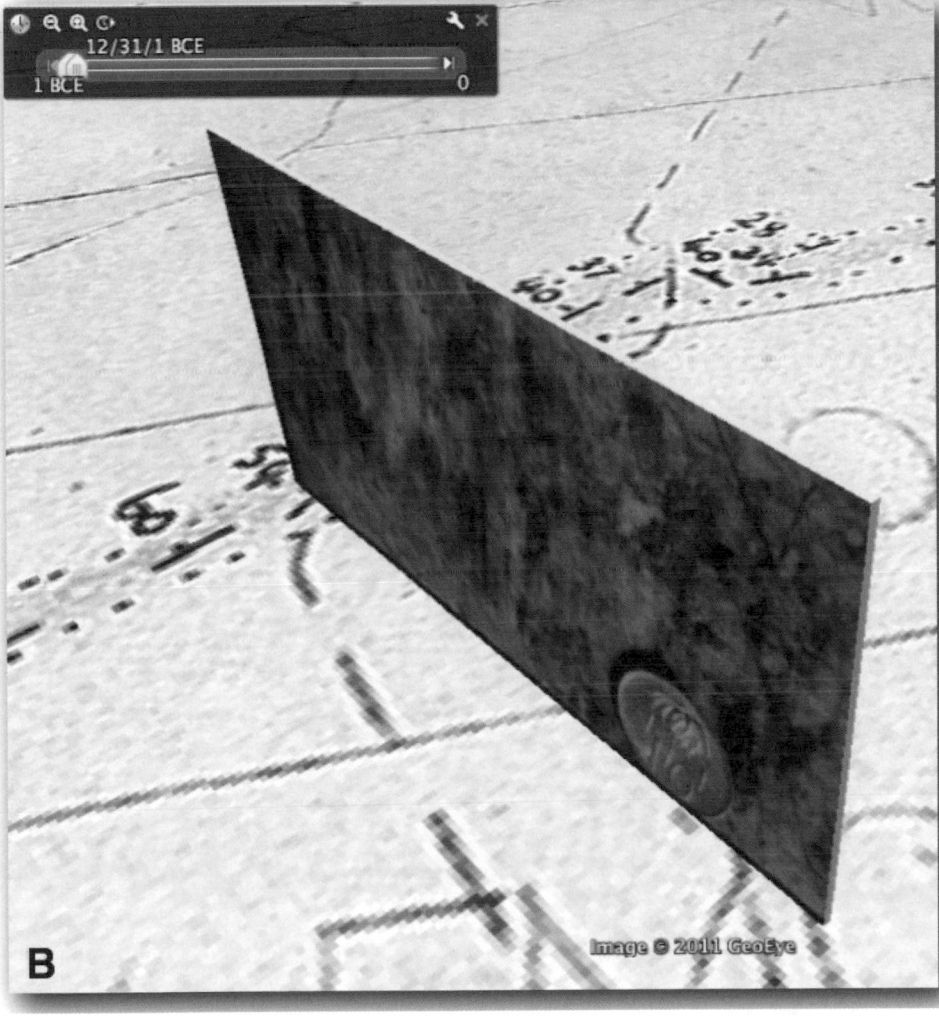

Figure 7. Outcrop photography incorporated into Google Earth in appropriate location and orientation as (A) photo overlay, (B) a "monster billboard."

representing the Witwatersrand Basin. Zooming in on the sedimentary basin reveals a number of dome structures in red on the Frimmel et al. (2005) map. The Vredefort Dome is the large red area near the center of the Witwatersrand Basin.

To create this visualization, we first covered the entire present Google Earth terrain—from 90° north to 90° south and 180° east to 180° west—with a plain gray ground overlay. Table 5 shows the required KML script. Then we superimposed ground overlays representing the Kaapvaal Craton and Witwatersrand Basin (Table 6). Note the "drawOrder" tag which ensures the correct sequence of image superposition. The images accessed by the

Figure 8. Scanning a pseudotachylite specimen using NextEngine, Inc.'s 3D Scanner and ScanStudio™ software. Orange spot shows stage rotation.

Figure 9. Virtual Specimens. (A) Three pseudotachylite samples from Leeukop Quarry. (B) Close-up of specimen on left in (A). (C) Shatter cone segment in quartzite in field orientation, positioned above sample location.

"href" element in Table 6 are in PNG format saved with an alpha layer so that their background is transparent. Finally the geologic map was overlain and brought into view based on pixel level of detail using the image pyramid technique discussed previously.

One aspect of impact tectonics that students and other novices find difficult to visualize is the relative size of impactor and impact crater (e.g., Koeberl and Henkel, 2005; see also http://www.hartrao.ac.za/other/vredefort/vredefort.html). Earth's largest craters, such as Vredefort, Sudbury, and Chicxulub, were made by relatively modest-sized impactors, each with a diameter of ~10 km. To convey a sense of relative size, we created a model of the Vredefort bolide (Fig. 11) and animated its flight through the Earth's atmosphere. The bolide is a simple COLLADA model of a sphere draped with a plain texture and was created in SketchUp. It was exported as a ".dae" file and linked to a series of "placemark" elements with "TimeSpan" child elements. The technique is identical to that used to elevate emergent cross sections except that the longitude, latitude, and altitude data were all incremented in each time step. Needless to say, we have no data regarding the azimuth or steepness of the trajectory, but the visualization may help students realize how small the bolide was in comparison to the crater it created.

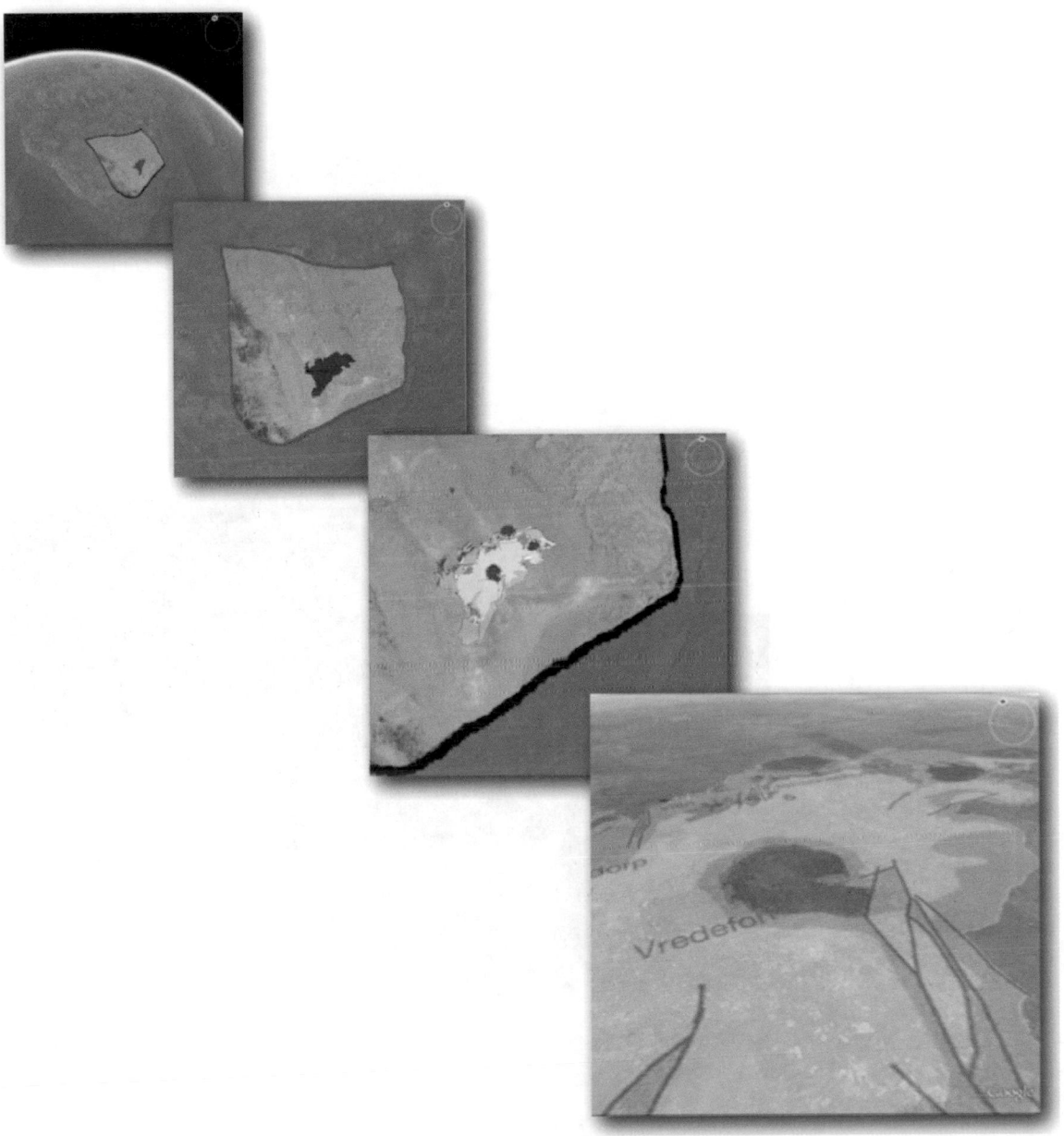

Figure 10. Graying out the current landscape to convey a sense of the Archean world. Images progressively zoom in to view of geological map (source: http://www.min.tu-clausthal.de/www/lager/Exc2005/bilder/klein/sa003.jpg) overlain onto representative Archean surface.

TABLE 5. KML CODE SNIPPET FOR COVERING THE GOOGLE
EARTH TERRAIN IMAGERY WITH A PLAIN GRAY IMAGE

```
<GroundOverlay id="go1">
        <name>gray</name>
        <Icon>
                <href>gray.png</href>
        </Icon>
        <LatLonBox>
                <north>90</north>
                <south>-90</south>
                <east>180</east>
                <west>-180</west>
        </LatLonBox>
</GroundOverlay>
```

TABLE 6. KML CODE SNIPPET FOR OVERLAYING PNG IMAGES
OF THE KAAPVAAL CRATON AND WITWATERSRAND BASIN

```
<GroundOverlay id="go2">
        <name>Kaapvaal Craton</name>
        <drawOrder>1</drawOrder>
        <Icon>
                <href>Kaapvaal.png</href>
        </Icon>
        <LatLonBox>
                <north>-19.14779864224844</north>
                <south>-31.96711203738477</south>
                <east>34.12108293991445</east>
                <west>19.68170347819974</west>
        </LatLonBox>
</GroundOverlay>

<GroundOverlay id="go3">
        <name>Lake</name>
        <drawOrder>2</drawOrder>
        <Icon>
                <href>lake.png</href>
        </Icon>
        <LatLonBox>
                <north>-25.65356696439984</north>
                <south>-28.84456462466237</south>
                <east>31.19291374126456</east>
                <west>23.94207711714848</west>
        </LatLonBox>
</GroundOverlay>
```

Note: The "drawOrder" element which determines overprinting
sequence.

Figure 11. Simulation of Vredefort impactor bolide to scale.

CONCLUSIONS

The conversion of legacy data to digital format opens up many opportunities for advanced visualizations and engaging learning experiences. In this example, we imported detailed maps, photographs, specimens, cross sections and orientation data into Google Earth. We also constructed an animation of the impact event that is thought to have created the Vredefort Dome 2.03 billion years ago. The example demonstrates that data stored in long-unopened dissertations on shelves and in map and specimen drawers can be made readily available to today's and future geological communities in a format that is both informative and useful. Inevitable errors in precise location on the ground that occurred when mapping in areas that lacked accurate survey maps without the assistance of GPS can now be corrected by comparing with the land surface images in Google Earth. As new tools are developed to allow incorporation of historical time sliders, COLLADA models, photo "billboards," and the reverse engineering of structural data, the interactivity and accuracy of geological maps will only improve. In addition to uses in research and in undergraduate and graduate education, these resources will contribute to the informal education and public outreach activities of the Vredefort and any studied site on the globe.

ACKNOWLEDGMENTS

The manuscript was improved by reviews by Gary Solar and Katherine Boggs. This research was support in part by NSF-TUES 1022755, NSF-GEO 1034643, and a 2010 Google Faculty Research Award. Any opinions, findings, and conclusions or recommendations expressed in this paper are those of the authors and do not necessarily reflect the views of the National Science Foundation or Google Inc.

REFERENCES CITED

Beebe, M.R., and De Paor, D.G., 2010, Virtual Specimens for geophysical modeling: Southeast Conference for Undergraduate Women in Physics, Duke University, http://www.physics.ncsu.edu/scuwp.

Catuneanu, O., 2001, Flexural partitioning of the Late Archaean Witwatersrand foreland system, South Africa: Sedimentary Geology, v. 141–142, p. 95–112, doi:10.1016/S0037-0738(01)00070-7.

Daly, R.A., 1947, The Vredefort ring structure of South Africa: The Journal of Geology, v. 55, p. 125–145, doi:10.1086/625423.

De Paor, D.G., 1994, A parametric representation of ellipses and ellipsoids: Journal of Structural Geology, v. 16, p. 1331–1333, doi:10.1016/0191-8141(94)90074-4.

De Paor, D.G., 2007, Embedding Collada models in geo-browser visualizations: a powerful tool for geological research and teaching: Eos (Transactions, American Geophysical Union), Fall Meeting Supplement Abstract, v. 88, p. 2277.

De Paor, D.G., 2009, Virtual Specimens: Eos (Transactions, American Geophysical Union), Fall Meeting Supplement Abstract, v. 90, p. 52.

De Paor, D.G., and Sharma, A., 2007, Map inversion: Geological Society of America Abstracts with Programs, v. 39, no. 1, p. 41.

De Paor, D.G., and Whitmeyer, S.J., 2011, Geological and Geophysical Modeling on Virtual Globes Using KML, COLLADA, and Javascript: Computers & Geosciences, v. 37, p. 100–110, doi:10.1016/j.cageo.2010.05.003.

De Paor, D.G., Wild, S.C., and Dordevic, M.M., 2012a, Emergent and animated COLLADA models of the Tonga Trench and Samoa Archipelago: Implications for geoscience modeling, education, and research: Geosphere, v. 8, p. 491–506, doi:10.1130/GES00758.1

De Paor, D.G., Whitmeyer, S.J., Marks, M., and Bailey, J.E., 2012b, this volume, Geoscience applications of client/server scripts, Google Fusion Tables, and dynamic KML, *in* Whitmeyer, S.J., Bailey, J.E., De Paor, D.G., and Ornduff, T., eds., Google Earth and Virtual Visualizations in Geoscience Education and Research: Geological Society of America Special Paper 492, doi:10.1130/2012.2492(06).

Fleminger, D., 2008, Vredefort Dome: World Heritage Sites of South Africa: 30 Degrees South Publishers, 144 p., ISBN-10: 0958489149.

Frimmel, H.E., Groves, D.I., Kirk, J., Ruiz, J., Chesley, J., and Minter, W.E.L., 2005, The formation and preservation of the Witwatersrand goldfields, the world's largest gold province: Economic Geology, 100th Anniversary Volume, p. 769–797.

Gibson, R.L. and Reimold, W.U., 2000, ^{40}Ar/^{39}Ar constraints on the age of metamorphism in the Witwatersrand Supergroup, Vredefort dome (South Africa): South African Journal of Geology, v. 103, no. 3-4, p. 175–190, doi:10.2113/1030175.

Gibson, R.L., and Reimold, U.W., 2001, The Vredefort Impact Structure, South Africa: The scientific evidence and a two-day excursion guide: Council for Geosciences (Geological Survey), Pretoria, Memoir 92, 111 p.

Johnson, M.R., Van Vuuren, C.J., Hegenberger, W.F., Key, R., and Show, U., 1996, Stratigraphy of the Karoo Supergroup in southern Africa: an overview: Journal of African Earth Sciences, v. 23, p. 3–15, doi:10.1016/S0899-5362(96)00048-6.

Koeberl, C., and Henkel, H., editors, 2005, Impact tectonics: Springer, 572 p., ISBN-10: 3540241817.

Minter, W.E.L., 2006, The sedimentary setting of Witwatersrand placer mineral deposits in an Archean atmosphere, *in* Kesler, S.E., and Ohmoto, H., eds., Evolution of Early Earth's Atmosphere, Hydrosphere, and Biosphere—Constraints from Ore Deposits: Geological Society of America Memoir 198, p. 105–119, doi:10.1130/2006.1198(06).

Moore, J.M., Tsikos, H., and Poltea, S., 2001, Deconstructing the Transvaal Supergroup: Implications for Paleoproterozoic paleoclimate models, South Africa: Journal of African Earth Sciences, v. 33, p. 437–444, doi:10.1016/S0899-5362(01)00084-7.

Nicolaysen, L.O., Burger, A.J., and Van Niekerk, C.B., 1963, The origin of the Vredefort dome structure in the light of new isotopic data: International Union of Geology and Geophysics Abstract, 13th General Assembly, Berkeley, California.

Reimold, W.U., and Colliston, W.P., 1994, Pseudotachylites of the Vredefort Dome and the surrounding Witwatersrand Basin, South Africa, *in* Dressler, B.O., Grieve, R.A.F., and Sharpton, V.L., eds., Large meteorite impacts and planetary evolution: Geological Society of America Special Paper 293, p. 177–196.

Simpson, C., 1977, A structural analysis of the rim synclinorium of the Vredefort Dome [unpublished M.Sc. thesis]: Johannesburg, South Africa, University of Witwatersrand, 257 p.

Simpson, C., 1978, The structure of the rim synclinorium of the Vredefort Dome: Transactions, Geological Society of South Africa, v. 81, p. 115–121.

Simpson, C., and De Paor, D.G., 2010, Restoring maps and memoirs to four-dimensional space using virtual globe technology: a case study from the Scottish Highlands, *in* Law, R.D., Butler, R.W.H., Holdsworth, R.E., Krabbendam, M. and Strachan, R.A., eds., Continental Tectonics and Mountain Building: The Legacy of Peach and Horne: Geological Society of London Special Publication 335, p.427–439, doi:10.1144/SP335.20.

Simpson, C., De Paor, D.G., Beebe, M.R., and Strand, J.M., 2010, A self-drive virtual field trip to the Vredefort Impact Structure, South Africa, incorporating virtual specimens and digitized legacy mapping data: Geological Society of America Abstracts with Programs, v. 42, no. 5, p. 422.

Smith, W., 1815, A Delineation of the Strata of England and Wales with Part of Scotland: Series 53 (reproduction), British Geological Survey, London.

Wolf, U.R., and Gibson, R.L., 2010, Meteorite Impact! The Danger from Space and South Africa's Mega-Impact—The Vredefort Structure: Springer, 3rd edition, 326 p. ISBN-10: 9783642104633.

MANUSCRIPT ACCEPTED BY THE SOCIETY 16 APRIL 2012

The Geological Society of America
Special Paper 492
2012

A test of the three-point vector method to determine strike and dip utilizing digital aerial imagery and topography

Leslie E. Hasbargen*

Department of Earth and Atmospheric Sciences, State University of New York, College at Oneonta,
Oneonta, New York 13820, USA

ABSTRACT

This paper focuses on the potential for developing geologic maps using high-resolution aerial imagery and digital elevation data. A key component of geologic mapping discussed herein addresses the determination of geologic layer orientation using a planar approximation (also known as the three-point method). An analytical solution is presented which can be readily implemented in a spreadsheet. The path from initial data collection in a GIS to spreadsheet computation and back to GIS is outlined. The Mecca Hills in southern California serve as a test case, as there is good coverage with high-resolution aerial imagery, a variety of elevation data including airborne LiDAR, and a published geologic map for the area. The comparison between vector-derived strike and dip and traditional field measured strike and dip suggests that remote mapping can successfully capture the regional aspects of the geology in an area. Estimates of rock orientation can be made if the elevation data is accurate and sufficient visual contrast exists in aerial imagery to define mappable features. A comparison shows that United States Geological Survey's National Elevation Data sets and airborne LiDAR can identify regional geologic structures, while Shuttle Radar Topography Mission and ASTER GDEM data yield results that diverge substantially from higher resolution elevation data. Remote mapping using the vector form of the three-point method offers a promising tool for geologic exploration as data sets continue to improve in quality and resolution.

INTRODUCTION

Geologists routinely utilize topographic and aerial imagery data to assist with location and mapping of rock features, including rock type, orientation, and geologic structures, such as fractures, folds, and faults. Air photos and topographic maps are often perused prior to a season of mapping in the field to provide a preliminary glimpse of geologic contacts, overall structural patterns, and exposures. The digitization of air photos and topography has made a preliminary perusal much easier than before. Once in the field, geologists measure the strike and dip angles of geologic surfaces with a compass and inclinometer.

*Leslie.Hasbargen@oneonta.edu

Hasbargen, L.E., 2012, A test of the three-point vector method to determine strike and dip utilizing digital aerial imagery and topography, *in* Whitmeyer, S.J., Bailey, J.E., De Paor, D.G., and Ornduff, T., eds., Google Earth and Virtual Visualizations in Geoscience Education and Research: Geological Society of America Special Paper 492, p. 199–208, doi:10.1130/2012.2492(14). For permission to copy, contact editing@geosociety.org. © 2012 The Geological Society of America. All rights reserved.

This information is crucial for developing three-dimensional visualizations of rock layers at depth beneath Earth's surface, as well as characterizing the style of deformation in a region. Thus, rock orientation is of fundamental importance for a broad range of applied aspects of geology.

We tend to trust that measurements made by a geologist in the field with compass and inclinometer are "ground truth," but there are limitations to field investigations. First, a surface that is measured by hand is typically on the order of a few cm square in dimension, and thus may not be truly representative of the orientation over a longer length of 10s to 100s m. Poorly defined bedding surfaces can make it difficult to find a suitable plane to characterize strike and dip angles. A diverse range of problems from rough terrain to political turmoil can limit access to significant portions of an area. Time constraints can also restrict investigations on the ground. Remotely sensed data can remove several of these hindrances.

The combination of high-resolution aerial imagery and digital elevation data provides a basis for remote geologic mapping. Contacts can be mapped based on topographic changes and visual contrast. These types of data are in a format which permits easy access to georeferenced coordinates, particularly when they are in a Universal Transverse Mercator projection, with units of length. Coordinates are the key to the methods presented below, as they can be directly entered into formulas to compute layer orientation. A numeric approach can lead to a significant reduction in data collection time and data processing, and thus facilitate remote mapping. With some practice, an individual can collect a few dozen strikes and dips in an hour over a fairly significant area.

The three-point method to characterize a plane and find strike and dip has deep roots in structural geology. The implementation of the method is usually a blend of graphical and numeric techniques where one finds the strike from two points on a dipping feature that intersects a contour line, and the dip from the elevation difference from the contact to the strike line (Marshak and Mitra, 1988). This method works well with paper geologic maps with elevation contours, that is, where the data is graphical in nature. Maps are increasingly digital, with location of features stored as coordinates. The coordinate nature of the point data facilitates vector operations, and there have been significant efforts to extract geologic structures and render the objects in perspective views in a purely numerical framework. The advent of topographic scanning with densely spaced point data has led to routine usage of vectors to estimate rock orientation (Haneberg, 1990; Schetselaar, 1995; de Kemp, 1998; Slob and Hack, 2004; Bellian et al., 2005; Ferrero et al., 2009; Jones et al., 2009; Lapponi et al., 2011; Pearce et al., 2011; Pavlis and Bruhn, 2011; Wilson et al., 2011). The vector method tends to get subsumed into proprietary or expensive software. Herein the focus rests squarely on the usage of vectors in determining the orientation of a surface where the user selects points manually, and the solution is worked out in a spreadsheet. This simpler approach opens up the opportunity for remote geologic investigation using readily available software without the need of high-end computing power. The method is sufficiently tractable for introduction to undergraduates, and powerful enough to characterize large-scale structures by researchers.

This paper has largely been motivated by the wide variety of freely available online data. Remote imagery used in this study includes the U.S. Geological Survey (USGS) National Elevation Data (NED) at 10 m and 30 m spacing; Shuttle Radar Topography Mission data (SRTM) at 90 m spacing; Advanced Spaceborne Thermal Emission and Reflection radiometer Global Digital Elevation Model (ASTER GDEM) at 30 m spacing; light detection and ranging (LiDAR) at 1 m spacing; and aerial imagery from the National Agricultural Imagery Program (NAIP) at 1 m spacing. Such a data rich digital environment begs for quantitative methods to extract geologic data.

The mathematical steps needed to extract strike and dip from visible layers and topography are iterated here, along with an analytical solution to the vector form of the three-point method to characterize a plane, with the goal of providing a tool for those without access to expensive software. Well-exposed sedimentary rocks along the southern section of the San Andreas fault in the Mecca Hills Wilderness in California provide a test bed for the method. The paper then provides an assessment of the quality of geologic orientations derived from remotely sensed data using a comparison to a published geologic map created by expert field geologists and an analysis of the effect of data resolution on computed strike and dip.

DESCRIPTION OF THE THREE-POINT METHOD

Overview of the Process to Extract Geologic Information from Digital Data

This section introduces the basic operations to extract rock layer orientation using the vector form of the three-point method, and proceeds with a test of the quality of the derived orientations based on different sources and resolution of topographic data. It also discusses necessary conditions for successful collection of remote geologic data.

The following sequence of operations provides a road map to extract the orientation of geologic features from remotely sensed data: (1) obtain aerial imagery for an area of interest; (2) obtain elevation data at the highest resolution available; (3) convert the data to an orthogonal coordinate system, such as Universal Transverse Mercator, or state plane; (4) collect coordinates for three points on geologic surfaces in a spatial data software; (5) extract the point data and implement the three-point vector calculations in a spreadsheet to determine strike and dip; and (6) export the strike and dip data from the spreadsheet back to the spatial data software. Note that many spatial data sets are in geographic coordinates of latitude and longitude (units of degree), and are not in an orthogonal reference frame. The data must be in an orthogonal projection to work in the formulas below.

The Three-Point Method to Determine Strike and Dip

The basic problem is to determine the orientation of a geologic surface, which we can approximate with a general form of the equation of a plane over some reasonable area:

$$Ax + By + Cz + D = 0 \qquad (1)$$

The parameters A, B, and C define the vector normal to the plane, also known as the pole to the plane, starting somewhere on the plane and extending as a directed line segment in 3-D space (x-y-z) orthogonal to the plane. The derivation below follows the vector methods portrayed in standard calculus textbooks (see Larson and Hostetler, 1986 for an example). The pole to the plane can be used to compute the strike and dip angles. In order to calculate the pole, we first construct two vectors using the three points on the plane, with each vector starting at the same point. Thus, these two vectors lie in the plane of interest. The cross product of the two vectors yields another vector that is orthogonal to both, and this is the pole to the plane.

Given three points: P_1 (x_1, y_1, z_1), P_2 (x_2, y_2, z_2), and P_3 (x_3, y_3, z_3), let the vector from P_2 to P_1 be $\mathbf{V}_1$, and the vector from P_2 to P_3 be $\mathbf{V}_2$, depicted in Figure 1. Note that boldface indicates a vector. We find $\mathbf{V}_1$ and $\mathbf{V}_2$ from:

$$\mathbf{V}_1 = (x_1 - x_2, y_1 - y_2, z_1 - z_2) = (a, b, c) \qquad (2)$$

$$\mathbf{V}_2 = (x_3 - x_2, y_3 - y_2, z_3 - z_2) = (d, e, f) \qquad (3)$$

The normal vector $\mathbf{U}$ (the pole to the plane containing $\mathbf{V}_1$ and $\mathbf{V}_2$) is then given as:

$$\mathbf{U} = \mathbf{V}_1 \times \mathbf{V}_2 = (b * f - e * c)\mathbf{i} - (a * f - d * c)\mathbf{j} + (a * e - d * b)\mathbf{k} = U_1\mathbf{i} - U_2\mathbf{j} + U_3\mathbf{k} \qquad (4)$$

where $U_1\mathbf{i}$ is the vector component of $\mathbf{U}$ in the x-direction, $U_2\mathbf{j}$ is the vector component of $\mathbf{U}$ in the y-direction, $U_3\mathbf{k}$ is the vector component of $\mathbf{U}$ in the z-direction, and $\mathbf{i}$, $\mathbf{j}$, and $\mathbf{k}$ represent unit vectors in the X, Y, and Z directions respectively (see Fig. 2). Written out with coordinates, the equations for the components of $\mathbf{U}$ are as follows:

$$U_1\mathbf{i} = \big((y_1 - y_2) * (z_3 - z_2) - (y_3 - y_2) * (z_1 - z_2)\big)\mathbf{i} \qquad (5)$$

$$U_2\mathbf{j} = -\big((x_1 - x_2) * (z_3 - z_2) - (x_3 - x_2) * (z_1 - z_2)\big)\mathbf{j} \qquad (6)$$

$$U_3\mathbf{k} = \big((x_1 - x_2) * (y_3 - y_2) - (x_3 - x_2) * (y_1 - y_2)\big)\mathbf{k} \qquad (7)$$

The components of $\mathbf{U}$, that is (U_1, U_2, U_3), represent the pole to the plane. Note that in Equations 5 through 7 the components of $\mathbf{U}$ are solely a function of the coordinates of three points on the plane. In our case, we assume that points are collected in a reference frame with units of length, such as Universal Transverse Mercator (UTM), in units of m. Thus, X corresponds to UTM easting (referred to subsequently as E), Y corresponds to UTM northing (referred to subsequently as N), and Z corresponds to height in m above datum.

The strike line is defined as the intersection of a horizontal plane with a dipping plane. The cross product can be used to find the line of intersection between two planes. Given the normal components (U_1, U_2, U_3) for the pole to the dipping plane, and the pole to an arbitrary horizontal plane $P_{XYZ} = (0,0,1)$, we compute the cross product to find our strike vector $\mathbf{S}$:

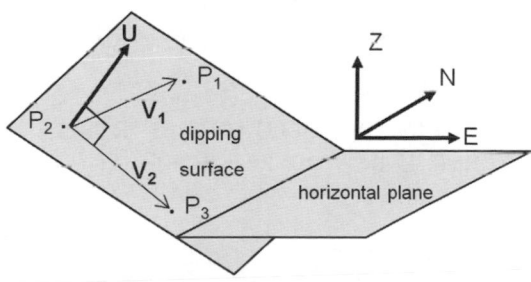

Figure 1. Geometry of points P_1 (x_1, y_1, z_1), P_2 (x_2, y_2, z_2), and P_3 (x_3, y_3, z_3) on a dipping geologic surface. $\mathbf{V}_1$ is constructed from P_2 to P_1 and $\mathbf{V}_2$ from P_2 to P_3. $\mathbf{U}$ is the pole to the dipping plane. E, N, Z indicate easting, northing, and elevation, respectively, in a right-handed reference frame, such as Universal Transverse Mercator.

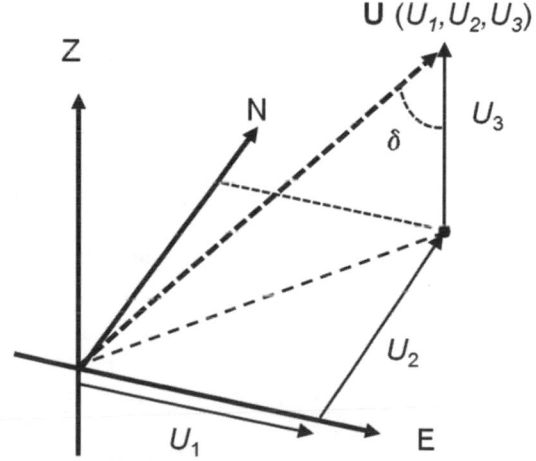

Figure 2. Components (U_1, U_2, U_3) of the pole ($\mathbf{U}$) to the dipping geologic surface, with the underlying coordinate directions easting (E), northing (N), and elevation (Z). δ denotes the angle between $\mathbf{U}$ and U_3, and is equivalent in magnitude to the dip of the geologic surface.

$$\mathbf{S} = \{(U_2 * 1 - 0 * U_3), -(U_1 * 1 - 0 * U_3), \quad (8)$$
$$(U_1 * 0 - 0 * U_2)\}$$

which reduces to

$$\mathbf{S} = \{U_2, -U_1, 0\} = \{E, N, 0\} \quad (9)$$

where E is easting and N is northing in units of length (m, in our case), and the asterisk represents multiplication. $\mathbf{S}$ represents a directed line segment from the origin to a point in the horizontal plane, with coordinate axes aligned east-west, and north-south (E and N respectively). Note that we must check for the condition where

$$U_2 = U_1 = 0 \quad (10)$$

If Equation (10) is true, then the geologic surface is horizontal, the two planes are coplanar, and an infinite number of strike lines are possible.

We define azimuth α as the angle from geographic north to the strike vector in a clockwise direction. Thus, the azimuth of the strike line is:

$$\alpha = \arccos\left(\frac{N}{\sqrt{E^2+N^2}}\right) \quad (11)$$

The inverse cosine function resolves problems where the strike is east-west, and is preferable to using the tangent function. The azimuth we derive from this expression is in π-*radians*. We can convert to degrees by multiplying by $180/\pi$.

There are two ambiguities to resolve: one concerning the polarity of the pole (above or below the plane) and the other with the geographic convention which places the geographic North Pole at $0°$. For polarity, there can be two pole orientations to a given plane, one extending above and one extending below the plane. Polarity affects the dip direction, and thus the strike. Here we follow the convention that the strike is $90°$ counterclockwise from the dip direction, which yields a strike consistent with the right hand rule (looking in the strike direction, dip direction is always "to the right"). The sign of U_3 determines whether the pole extends above or below the plane. If the sign is positive, we need to change the direction of the strike line, and we do so by changing the sign of (E,N) to $(-E,-N)$.

In addition, we require the bearing to be in azimuthal form, where values increase in a clockwise direction, with north at 0. The cosine function returns a value which ranges from -1 to 1, and needs to be converted into geographic directional degrees. This can be done on a quadrant by quadrant basis, but simplifies to an adjustment for negative easting values. Table 1 provides a guide.

The dip magnitude δ can be determined using the components of $\mathbf{U}$:

$$\delta = \arcsin\left(\frac{\sqrt{U_1^2+U_2^2}}{\sqrt{U_1^2+U_2^2+U_3^2}}\right) \quad (12)$$

Note the dip angle is in units of π-*radians*, and can be converted to degrees, if convention requires.

Implementing the Three-Point Vector Equations in a Spreadsheet

Solving the vector equations for numerous strike and dip locations is intractable for manual calculations. Spreadsheets can handle these types of repetitive operations easily. A spreadsheet which performs these operations is provided in the GSA Data Repository[1]. The section that follows demonstrates a procedure for implementing Equations 5 through 12 in a spreadsheet.

One method for managing the three-point data is to place all point coordinate data for an individual planar surface measurement into the same row in the spreadsheet. Implementing the calculations for a single layer in the same row facilitates copying and pasting formulas, if one has numerous locations at which to determine strike and dip. This means there will be 10 columns of input data—a location identifier, and an x-y-z coordinate for each of the three points. In the next three columns, we compute the vector components of the pole to the plane, and determine the strike and dip of the feature. Equations 5, 6, and 7 for $U_1\mathbf{i}$, $U_2\mathbf{j}$, $U_3\mathbf{k}$ can be represented in a spreadsheet with the following formulas respectively. The notation is for row-column cell naming convention in Excel.

$$= ((RC[-8]-RC[-5])*(RC[-1]-RC[-4])- \quad (RC[-2]-RC[-5])*(RC[-7]-RC[-4])) \quad (13)$$

$$= -((RC[-10]-RC[-7])*(RC[-2]-RC[-5])- \quad (RC[-4]-RC[-7])*(RC[-8]-RC[-5])) \quad (14)$$

$$= ((RC[-11]-RC[-8])*(RC[-4]-RC[-7])- \quad (RC[-5]-RC[-8])*(RC[-10]-RC[-7])) \quad (15)$$

TABLE 1. CONVERSION OF BEARING TO AZIMUTHAL FORM

Easting	Northing	Quadrant	Formula to convert bearing to azimuth (in degrees)
E > 0	N > 0	0-90	$= \alpha * 180/\pi$
E > 0	N < 0	90-180	$= \alpha * 180/\pi$
E < 0	N < 0	180-270	$= 360 - \alpha * 180/\pi$
E < 0	N > 0	270-360	$= 360 - \alpha * 180/\pi$

[1]GSA Data Repository item 2012302, strike and dip spreadsheet and training video, is available at http://www.geosociety.org/pubs/ft2012.htm, or on request from editing@geosociety.org.

The strike vector is readily obtained from Equations 13–15, once sign conventions are enforced. Equations 16 and 17 enforce the sign convention and yield the strike vector easting and northing respectively, assuming these are implemented in columns adjacent to the vector components of the pole to the plane.

$$= \text{IF(RC[}-1]<0,\text{RC[}-2],-\text{RC[}-2]) \qquad (16)$$

$$= \text{IF(RC[}-2]>0,\text{RC[}-4],-\text{RC[}-4]) \qquad (17)$$

The azimuth of the strike vector can be computed with the following formula.

$$= \text{DEGREES(IF(RC[}-2]> =0,$$
$$\text{ACOS(RC[}-1]/(\text{RC[}-2]^2+\text{RC[}-1]^2)^0.5), \qquad (18)$$
$$2*\text{PI()}-\text{ACOS(RC[}-1]/(\text{RC[}-2]^2+\text{RC[}-1]^2)^0.5)))$$

Finally, the dip angle can be computed from the following formula.

$$= \text{DEGREES(ASIN((RC[}-6]^2+\text{RC[}-5]^2)^0.5/$$
$$(\text{RC[}-6]^2+\text{RC[}-5]^2+\text{RC[}-4]^2)^0.5)) \qquad (19)$$

Table 2 shows the arrangement of data in rows and columns for a strike and dip determination. Equations 13–19 are valid for columns arranged in the fashion of Table 2. Numbers in the columns for calculated values in Table 2 refer to the formulas in Equations 13–19. As provided above, a person can copy and paste the formulas directly into the corresponding cells, and have a calculator for computing strike and dip from three-point coordinates.

If one collects numerous point data from a region of interest, the transfer of information from the GIS to a spreadsheet and back to the GIS can be onerous. A VisualBasic® macro contained within the spreadsheet (see footnote 1) reads text files with three-point coordinate information, computes the strike and dip, and writes the strike-dip data both to the spreadsheet and to a file readable by Global Mapper, a GIS software. This functionality makes data transfer relatively simple, and thus facilitates the extraction of geologic orientation data.

Considerations for Data Collection in GIS

The process begins with the collection of elevation coverage for the region of interest. There are a variety of data sources for worldwide elevation. In our case, we chose NED (both 10 m and 30 m spacing); SRTM (90 m spacing); ASTER GDEM (30 m spacing) and LiDAR DEM (supplied by B4 at 1 m spacing). Many GIS applications take advantage of web mapping services (WMS), which provide direct links from the GIS software to online data sources for a region of interest, and this greatly simplifies data access and collection.

The next step is to obtain aerial imagery of the region of interest. This kind of data is available for the United States from the USGS Seamless Data Server, which also has a WMS available for National Agricultural Imagery Program (NAIP) color photographs at 1 m spacing. Google Earth provides a suite of aerial imagery, and perusing these images can help identify those photos which best capture the features of interest. If one chooses to use Google Earth, access to elevation data is somewhat restrictive. Point coordinates (easting and northing) can readily be collected in Google Earth and exported to GIS where the horizontal coordinates can be combined with elevation data, and subsequently exported as x-y-z points to a spreadsheet for computation of strike and dip.

Identifying suitable locations for a strike and dip determination in remote imagery takes some practice. GIS software provides perspective views with imagery wrapped on to topography, and this greatly aids the identification of surfaces. Google Earth is an outstanding tool for this part of the process. Points on the dipping surfaces should be selected from the highest and lowest locations on the surface, and points must not lie in the same line. Layers which can be traced over a ridge represent ideal features to select for the three-point method. Dip slopes, if present, offer excellent choices for planes as well. Lines which connect the three selected points on the surface can be viewed in perspective, and rotated until the view is along strike. This functionality greatly assists with selection of points, and can help eliminate spurious data.

It is not always obvious when picking points whether the geometry is sufficiently robust for a determination of strike and dip. If the points fall on the same line, that is, they are collinear, then an indeterminate solution results. The suitability of the points can be tested while calculating strike and dip. Recall that two vectors on the surface are constructed from the three points. The dot product of the two vectors can be computed as a check that the points are non-collinear. As the angle between vectors approaches zero or 180 degrees, the points approach collinearity and the solution becomes indeterminate. Thus, poor solutions can be eliminated with a filter for points exhibiting approximate collinearity.

The spacing of the underlying elevation data is a very important consideration when viewing aerial imagery and selecting points along a surface. It is unreasonable to make a determination of strike and dip on a surface with points separated by a

TABLE 2. EXAMPLE SPREADSHEET SETUP FOR COMPUTING STRIKE AND DIP

	Point 1			Point 2			Point 3			Calculated Values						
ID	E, m	N, m	Z, m	E, m	N, m	Z, m	E, m	N, m	Z, m	U_1 i	$-U_2$ j	U_3 k	E	N	*Strike*	*Dip*
A	x_1	y_1	z_1	x_2	y_2	z_2	x_3	y_3	z_3	13	14	15	16	17	18	19

distance less than the posting of the elevation data. Elevation data at 30 m spacing, for instance, places a lower limit on the separation distance between sample locations of 30 m. A much longer separation distance than the data posting should be used, at least two to three times the spacing. Clearly, tight folds with wavelengths of twice the data spacing or less would be below the detection limit. Strike and dip determinations over longer distances can reduce errors from the fixed spacing of points in digital elevation grids, so one might be tempted to make determinations over longer distances, but it must be noted that structural complexity at short length scales can make remote determination of strike and dip intractable.

An Application of the Three-Point Vector Method

The Mecca Hills in southern California, USA, provide a rich area for geologic investigation. Along this plate boundary, several fault strands, including the San Andreas fault, transect a transpressive tectonic setting, and lead to complex structures (Sylvester and Smith, 1976; Sylvester, 1999). A large portion of the area is underlain by Miocene and younger sedimentary rocks (Sylvester and Smith, 1976; Sylvester, 1999). Rock erodibility varies between lithologic units and indeed between individual sedimentary layers in the area. Tilted sedimentary layers in this arid nearly vegetation-free environment provide numerous exposures for determining layer orientation. LiDAR data, high resolution aerial imagery, and detailed geologic maps provide good coverage and thus a means to compare the effects of digital

data resolution on rock orientations derived from the three-point method as outlined above. A training video is included in the supplemental material for this paper (see footnote 1), and demonstrates data collection and computation of strike and dip.

As an overview of regional structures and to gain experience with virtual mapping, strike and dip determinations were made at locations where layers could readily be traced across the landscape (area = 27 km²). This was in essence a training exercise in remote mapping utilizing NED 10 m grid spacing elevation data and NAIP imagery. Google Earth, with access to time series aerial imagery, greatly aided in visualizing the landscape, and in identification of suitable bedding planes for measurement. Average length of each three-point line was 103 m, which greatly exceeds the grid spacing of the elevation data. Of 178 measurements, 139 yielded points which formed a robust geometry, where a filter was implemented with the dot product between vectors arbitrarily set such that the vectors never approach closer than 25° of collinearity. Thus, 78% of the point selections resulted in clearly non-collinear geometry. The rather high number of nearly collinear vectors documents the need for a filter. A stereoplot of the poles to bedding surfaces (not shown here) reveals a regional fold axis trend at 300° and plunging 12°.

How well do rock layer orientations derived from the three-point method compare to on the ground measurements by skilled professional geologists? The answer of course depends on the quality of the digital data and the choice of point locations. Figure 3 depicts an overlay of digitally derived orientations onto a published geologic map (from fig. 10, Sylvester,

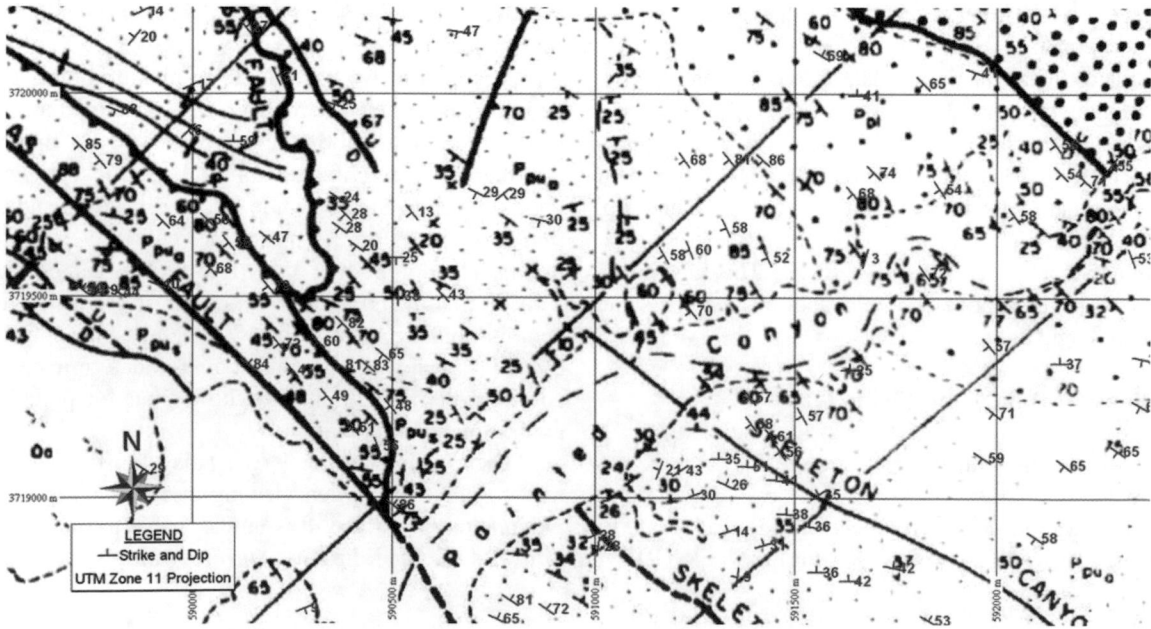

Figure 3. Remote (red) and field (black) determined strikes and dips for Painted Canyon in Mecca Hills, California. Underlying geologic map modified from Sylvester (1999). Remotely determined orientations mimic field measurements, and correctly identify a regional syncline (Skeleton Canyon syncline) in southeast quadrant of the map. Thick black lines delineate faults, including the San Andreas fault in the southwest corner.

1999, and originally published in Sylvester and Smith, 1976). There is moderately good correlation between both strike and dip, though there are clearly a few that diverge greatly from the map of Sylvester and Smith. Discordance could result from poor point selection, poor quality of underlying data, misalignment of data layers in the GIS, or from small wavelength folds. The remotely derived strikes and dips accurately identify the Skeleton Canyon syncline in the southeast quadrant of the map. In areas of very rough and well-exposed topography such as Mecca Hills, remote data provide access and can take advantage of the both the relief and exposure of layers. The overall approximate agreement between field and digital data collection is very encouraging.

Quantitative Comparison between Topographic Data Sets

How much of an effect does elevation data resolution have on strike and dip measurements? The analysis below compares strike and dip between data derived from airborne LiDAR and data from mapping conducted on the ground, and then follows with a comparison between elevation data sets of varying resolution, including B4 LiDAR data gridded at 1 m spacing (http://www .earthsciences.osu.edu/b4, data retrieved 27 December 2010); the National Elevation Data set at 1 arc second and 1/3 arc second spacing (30 m and 10 m data grid spacing, respectively); Space Shuttle Radar Topography Mission (SRTM) data at 3 arc second spacing (90 m), and Version 1 of the Advanced Spaceborne Thermal Emission and Reflection Radiometer Global Digital Elevation Model (ASTER GDEM) data at 30 m spacing.

The strongest comparison is between the geologic map and the best remote data—B4 LiDAR data. Strike and dip measures (n = 50) were extracted from a published geologic map of Painted Canyon (Sylvester and Smith, 1976; Sylvester, 1999) which had been scanned and rectified in a GIS software. A dedicated effort was made to extract strikes and dips from aerial imagery and LiDAR B4 topography at the same sample locations as the geologic map. Not every location provided easily identifiable bedding surfaces; ~30 locations yielded strike and dip determinations within 50 m of the published field data. It is worth noting that geologic attitudes can vary over this scale in Mecca Hills. Figure 4 shows the stereographic projection of the poles to the bedding planes from the geologic map (see Fig. 4A), and from the vector method utilizing LiDAR B4 topography (see Fig. 4B). There is good agreement between the two data sets. Both data indicate the presence of a fold axis trend at 306°, and plunging 12°.

A scatter plot of remotely mapped strike and dip plotted against field mapped data show that the two methods are generally congruent (see Fig. 5). Best fit linear equations with intercepts set to 0 provide a convenient way to measure the congruency of solutions, as the slope of the best fit line approaches unity when the two sets consistently yield the same measure. Further, correlation coefficients (i.e., R^2 values) provide an estimate of the scatter around the trend. Strike determinations for the study here with a relatively small sample size (n = 30) yield a slope of the trendline close to unity (slope = 0.97, R^2 = 0.87), while dip determinations based on LiDAR topography tend to underestimate dip (slope = 0.85, R^2 = 0.26). Large scatter for dip determination could be due to location selection problems. Points collected

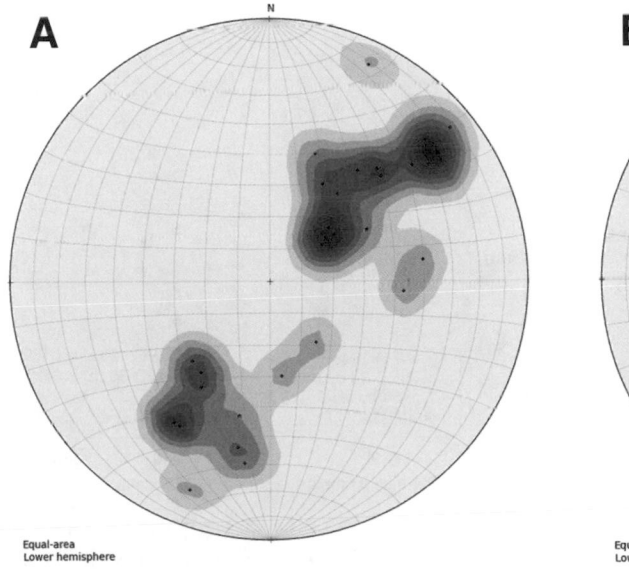
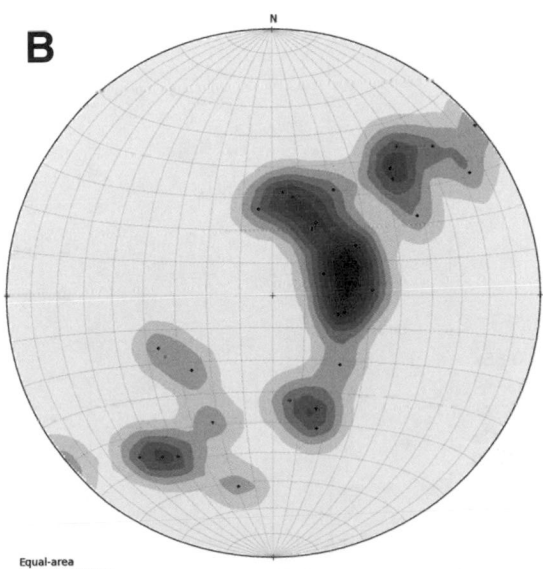

Figure 4. Lower hemisphere projection of poles to strike and dip surfaces for data derived from (A) field-based measurements (Sylvester, 1999), and (B) remote mapping with aerial imagery and LiDAR B4 data. Shaded contours show very similar patterns, suggesting good correspondence between field-derived and remotely mapped strike and dip. Schmidt net projection, grid spacing = 10°.

along a dipping layer as it transects topography appear to yield better solutions than a uniformly sloping surface, which can be deceitful in appearing as the dip slope of a bedding surface. There may also be some errors introduced by processing of LiDAR data in the generation of gridded data files (Liu, 2008).

To ascertain the differences, if any, between the various topographic data sets (LiDAR, NED, SRTM, and ASTER GDEM), a set of 25 measurement sites was selected based on easily identifiable sedimentary layers in the imagery. The easting, northing, and elevation coordinates of three points for each measurement site were extracted for each topographic data set. To isolate the effect of elevation, only the elevation coordinate varies between the three-point data sets. Namely, the GIS software can write the coordinate data to a file, and the active elevation layer provides the data for the elevation coordinate. The data were written repeatedly with a different topographic data layer (B4 LiDAR, NED, etc.) activated for each. Strikes and dips were then computed for the selected features. The LiDAR data was used as the comparative standard for each data set. The published geologic map could have been used, but since that map did not always have data at ideal imaging sites, LiDAR was used as the best proxy for "ground truth."

Of the 25 sites, four yielded collinear or nearly collinear solutions and were expunged from the following comparison of strikes and dips. Thus, the use of the vector dot product to filter site selections during the computation of strike and dip is a very useful and necessary part of the process, as the data were selected with due care and diligence. The remaining valid strikes and dips were then plotted against the LiDAR derived strikes and dips as scatter plot charts.

The strike and dip values derived from these various data sets, when compared to the result for the LiDAR data set, exhibit a moderate convergence in solutions. Strikes show less variability than dips. Strikes of NED 10 m and 30 m data compare favorably with B4 LiDAR data. Linear scatter plot trends between NED strikes and B4 LiDAR strikes have a trendline slope close to unity (0.97 and 0.95 for NED 10 m and NED 30 m, respectively; see Fig. 6). Linear scatter plot trends for dip angles show a weaker relation, and typically underestimate the dip angle determined from LiDAR (trendline slope of 0.82 and 0.79 for NED 10 m and 30 m respectively; see Fig. 7). Strike and dip derived from ASTER GDEM and SRTM data exhibit much weaker correlations to LiDAR data (0.84 and 0.69 trendline slopes for strike, and 0.21 and 0.18 trendline slopes for dip for ASTER GDEM and SRTM respectively). Indeed, dip angles show almost no correspondence with LiDAR data.

Discussion

Given the broad coverage of NED data for the United States, the above results are very encouraging. Moderately detailed topography with data postings from 10 to 30 m can capture the regional patterns of rock orientation, so long as the plane being measured is much larger than the spacing of the underlying topographic data set. This study measured surfaces over lengths on the order of 100 m, and obtained good results. Many geologic structures, such as drag folds or parasitic folds, are much smaller than this length, and so are below the detection limit of the data used in this study.

The divergence of the computed strike and dips shown by ASTER GDEM and SRTM data from the LiDAR data is disconcerting, and suggests that these data can be useful for regional

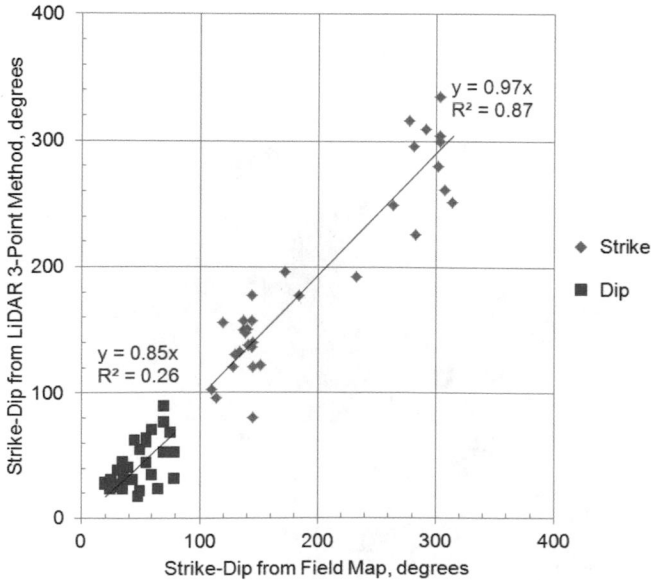

Figure 5. Remotely mapped data plotted against field derived data (strike and dip). Strike determinations based on LiDAR topography are generally congruent with field determinations, while dip determinations based on LiDAR topography tend to underestimate dip.

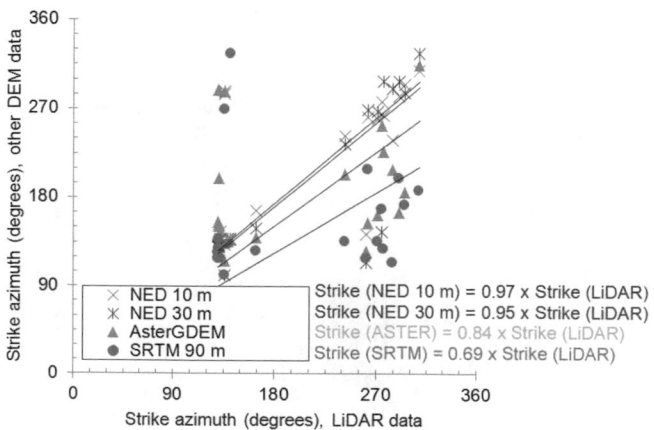

Figure 6. Strikes derived from NED 10 m, NED 30 m, ASTER GDEM, and SRTM data plotted against B4 LiDAR strikes. Best fit linear trend lines demonstrate that NED strikes compare favorably to LiDAR (trend line slope of 0.97 and 0.95 and R^2 of 0.62 and 0.52 for NED 10 m and NED 30 m, respectively), while SRTM and to a lesser extent ASTER data diverge (trend line slope of 0.84 and 0.68 and R^2 of 0.63 and 0.81 for ASTER GDEM and SRTM respectively).

visualizations, but cannot be trusted to adequately capture geologic orientation. Elevations in ASTER GDEM have been shown to vary systematically from NED data (ASTER GDEM Validation Team, 2009), and this is a likely source of the poor correspondence between dip angles in this study. Additionally, part of the problem with larger distances between postings is that landscapes are rougher at smaller scales, and this information cannot be adequately captured with sparse elevation data. The global coverage offered by these data sets is exciting and provides unprecedented access to the extraordinary structures in otherwise inaccessible regions, such as the Near East, for instance. Accurate remote mapping of the orientation of those structures with the current data, however, is still not entirely trustworthy. This result thus calls for higher precision and more detailed elevation coverage to facilitate remote mapping efforts.

The analytical solution to the vector form of the three-point method is a very potent tool for geologists and has not received enough attention. Researchers utilizing point clouds derived from ground-based LiDAR scanning have greatly extended vector solutions to estimate planes and automatically extract fracture orientations, for instance. However, working with point clouds requires somewhat specialized and often proprietary software. The method as outlined here does not require expensive software. It is a method which can be taught at the undergraduate level, and implemented in commonly available software. The method clearly can work for aerial imagery and topography. It can also work if contacts between geologic units and topography are available. The author has estimated dip of Devonian sedimentary rocks in upstate New York based on a combination of the New York digital geologic map (based on Fisher et al., 1970) and NED 10 m data. Three points extracted along a contact yielded a dip angle around 1 degree. This dip angle was verified with a geologic well log which encountered the contact at depth. The method can also be used on a very local scale if the data are sufficiently precise. Total station surveys, particularly those where a reflecting prism is not necessary, can collect three-dimensional coordinates to the nearest mm, and they represent another means of determining orientations for layers, contacts, faults, fractures, and the like.

SUMMARY

We live in data-rich times, as topography and visual imagery of Earth's surface is available to anyone with access to the World Wide Web. This form of data is ideal for geologic exploration, which has traditionally utilized topography and aerial imagery to assist with reconnaissance mapping. The easy access to three-dimensional coordinates of such data open the door to using vector math to determine orientations of planar features, as outlined above. Strike and dip extraction by selecting three points on a surface can be accomplished reliably for data sets with sufficiently high resolution. The comparison of strike and dip derived from various topographic data sets shows that LiDAR layer orientations (strike and dip) compare favorably with data collected on the ground by experienced geologists. Strike orientations derived from lower resolution NED at both 10 and 30 m spacing compare well with LiDAR, and NED-derived dips show moderate correlations with LiDAR. This is a very encouraging result, and opens the door for additional digital exploration with these data. SRTM and ASTER GDEM data yielded poor results for strike and dip, and thus should probably not be used for strike and dip determination.

As a side note, Google Earth inspired this project, as it provides an outstanding portal to aerial imagery. Google Earth can serve as a GIS tool to collect two-dimensional point coordinates of geologic surfaces. However, elevation data is not exported with the point coordinates, and so the data exported from Google Earth must be combined with topography in separate GIS software at the time this article has gone to print. Contacts between rock types and faults can be traced as lines, and this can greatly assist with mapping other aspects of geologic maps in addition to strike and dip of planar features.

In summary, the analytical solution to the vector form of the three-point method is a powerful tool in a geologist's tool kit. Reconnaissance mapping can be conducted in the office in advance of field efforts to provide a broader view of the geology, and highlight regions which warrant close inspection. While remote mapping will never supplant field mapping, it certainly offers a complementary view from above.

ACKNOWLEDGMENTS

Comments by reviewers Eric de Kemp and Colin Shaw improved this manuscript. The author is extremely grateful for

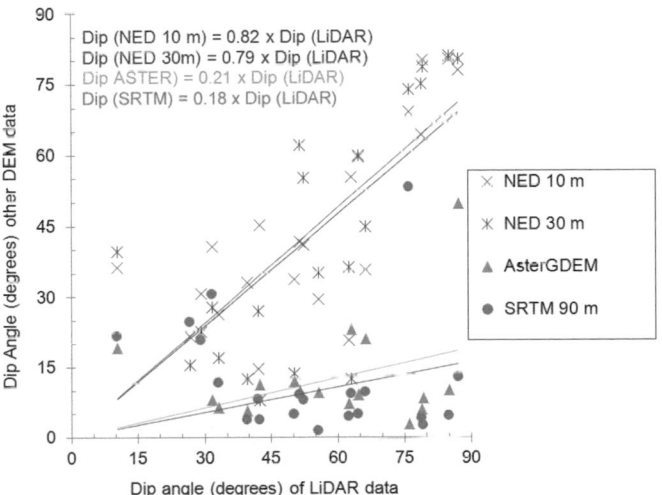

Figure 7. Dips derived from NED 10 m, NED 30 m, ASTER GDEM, and SRTM data plotted against B4 LiDAR dips. Best fit linear trend lines demonstrate that NED dips consistently underestimate LiDAR dips (trend line slope of 0.82 and 0.79 and R^2 of 0.56 and 0.53 for NED 10 m and NED 30 m, respectively). Both SRTM and ASTER data consistently find dips of ~20% of the B4 LiDAR dips (trend line slope of 0.21 and 0.18 and R^2 of 0.01 and 0.25 for ASTER GDEM and SRTM respectively).

the open and free access to spatial data from the following providers: LiDAR topographic data from the B4 website, http://www.earthsciences.osu.edu/b4; ASTER GDEM, the property of METI and NASA (http://www.science.aster.ersdac.or.jp/en/index.html); and NED and NAIP data in the USGS National Seamless Data Warehouse: http://seamless.usgs.gov/index.php and http://seamless.usgs.gov/. The Instituto de Geociencias, Universidade de São Paulo has graciously provided open access to OpenStereo, a software to plot geologic structural data (http://www.igc.usp.br/index.php?id=391). This project relied heavily on web mapping service access and point selection in Global Mapper, a low-cost GIS. Finally, the author expresses appreciation to Professor Arthur G. Sylvester for providing the geologic ground truth for this project, and introducing the author to the extraordinary geology in Mecca Hills.

REFERENCES CITED

ASTER GDEM Validation Team: METI/ERSDAC, NASA/LPDAAC, USGS/EROSASTER, 2009, Global DEM Validation Summary Report, interagency report.

Bellian, J., Kerans, C., and Jennette, D., 2005, Digital Outcrop Models: Applications of Terrestrial Scanning Lidar Technology in Stratigraphic Modeling: Journal of Sedimentary Research, v. 75, no. 2, p. 166–176, doi:10.2110/jsr.2005.013.

de Kemp, E.A., 1998, Three-dimensional projection of curvilinear Geological features through direction cosine Interpolation of structural field observations: Computers & Geosciences, v. 24, no. 3, p. 269–284, doi:10.1016/S0098-3004(97)00066-6.

Ferrero, A., Forlani, G., Roncella, R., and Voyat, H.C., 2009, Advanced Geo-structural Survey Methods Applied to Rock Mass Characterization: Rock Mechanics and Rock Engineering, v. 42, no. 4, p. 631–665, doi:10.1007/s00603-008-0010-4.

Haneberg, W., 1990, A Lagrangian interpolation method for three-point problems: Journal of Structural Geology, v. 12, no. 7, p. 945–947, doi:10.1016/0191-8141(90)90069-B.

Jones, R., Kokkalas, S., and McCaffrey, K., 2009, Quantitative analysis and visualization of nonplanar fault surfaces using terrestrial laser scanning (LIDAR)–The Arkitsa fault, central Greece, as a case study: Geosphere, v. 5, no. 6, p. 465–482, doi:10.1130/GES00216.1.

Fisher, D.W., Isachsen, Y.W., and Rickard, L.V., 1970, Geologic Map of New York State, 1:250,000: New York State Museum, Map and Chart Series No. 15.

Lapponi, F., Casini, G., Sharp, I., Blendinger, W., Fernández, N., Romaire, I., and Hunt, D., 2011, From outcrop to 3D modeling: a case study of a dolomitized carbonate reservoir, Zagros Mountains, Iran: Petroleum Geoscience, v. 17, no. 3, p. 283–307, doi:10.1144/1354-079310-040.

Larson, R.E., and Hostetler, R.P., 1986, Calculus with Analytic Geometry, 3rd edition: Lexington, Massachusetts, D. C. Heath and Company.

Liu, X., 2008, Airborne lidar for DEM generation; some critical issues: SAGE Publications, Progress in Physical Geography, v. 32, no. 1, p. 31–49.

Marshak, S., and Mitra, G., 1988, Basic Methods of Structural Geology: Englewood Cliffs, New Jersey, Prentice Hall, 446 p.

Pavlis, T.L., and Bruhn, R.L., 2011, Application of LIDAR to resolving bedrock structure in areas of poor exposure: An example from the STEEP study area, southern Alaska: Geological Society of America Bulletin, v. 123, no. 1-2, p. 206–217, doi:10.1130/B30132.1.

Pearce, M.A., Jones, R.R., Smith, S.A.F., and McCaffrey, K.J.W., 2011, Quantification of fold curvature and fracturing using terrestrial laser scanning: The American Association of Petroleum Geologists Bulletin, v. 95, no. 5, p. 771–794.

Schetselaar, E., 1995, Computerized field-data capture and GIS analysis for generation of cross sections in 3-D perspective views: Computers & Geosciences, v. 21, no. 5, p. 687–701, doi:10.1016/0098-3004(94)00104-3.

Slob, S., and Hack, R., 2004, 3D Terrestrial Laser Scanning as a New Field Measurement and Monitoring Technique, *in* Hack, R.A., and Charlier, R., eds., Engineering Geology for Infrastructure Planning in Europe, Lecture Notes in Earth Sciences: Berlin-Heidelberg-New York, Springer-Verlag, International (III), v. 104, p. 179–189.

Sylvester, A.G., 1999, Rifting, transpression, and neotectonics in the Central Mecca Hills, Salton Trough, California, Field Trip Guide Book: Pacific Section: Society of Economic Paleontologists and Mineralogists, v. 85, p. 1–11.

Sylvester, A.G., and Smith, R.R., 1976, Tectonic transpression and basement-controlled deformation in San Andreas fault zone, Salton trough, California: American Society of Petroleum Geologists Bulletin, v. 60, no. 12, p. 2081–2102.

Wilson, C.E., Aydin, A., Karimi-Fard, M., Durlofsky, L.J., Amir, S., Brodsky, E.E., Kreylos, O., and Kellogg, L.H., 2011, From outcrop to flow simulation: Constructing discrete fracture models from a LIDAR survey: The American Association of Petroleum Geologists Bulletin, v. 95, no. 11, p. 1883–1905.

MANUSCRIPT ACCEPTED BY THE SOCIETY 16 APRIL 2012

The Geological Society of America
Special Paper 492
2012

Applications of Google Earth Pro to fracture and fault studies of Laramide anticlines in the Rocky Mountain foreland

David R. Lageson*
Martin C. Larsen
Helen B. Lynn
Whitney A. Treadway
Department of Earth Sciences, Montana State University, Bozeman, Montana 59717, USA

ABSTRACT

Google Earth Pro imagery was used by graduate students for a course project to identify, describe, and interpret lineament patterns on two oil-producing anticlines in Wyoming, one in the northwest Wind River Basin and the other in the southern Bighorn Basin (Maverick Springs and Thermopolis anticlines, respectively). These anticlines lie on opposite sides of the east-west–trending Owl Creek arch, which is a sinistral, transpressive array of en echelon, basement-involved thrust blocks. Both anticlines are well-exposed and display extensive near-surface fracturing and faulting, making them ideal candidates for a study of fold-related lineament patterns. Google Earth Pro was used to map and measure the orientation of lineaments and faults in a digital format. The lineaments identified include a set parallel to dip (A–C), a set parallel to strike (B–C), and two sets oblique to strike. Lineament orientation data were analyzed using length-weighted rose diagrams, whereas fold geometry and plunge were evaluated using equal-area (lower hemisphere) stereonets. Although the study was limited in scope to a computer-based geometric analysis and did not include outcrop-based kinematic data, the lineament/fracture data derived from Google Earth mapping are nevertheless compatible with published studies that demonstrate regional NE-SW shortening along the western Owl Creek transpressive zone during the Laramide orogeny. Google Earth Pro proved to be a highly effective tool for gathering lineament orientation and spatial distribution data across these well-exposed anticlines.

*lageson@montana.edu

Lageson, D.R., Larsen, M.C., Lynn, H.B., and Treadway, W.A., 2012, Applications of Google Earth Pro to fracture and fault studies of Laramide anticlines in the Rocky Mountain foreland, *in* Whitmeyer, S.J., Bailey, J.E., De Paor, D.G., and Ornduff, T., eds., Google Earth and Virtual Visualizations in Geoscience Education and Research: Geological Society of America Special Paper 492, p. 209–220, doi:10.1130/2012.2492(15). For permission to copy, contact editing@geosociety.org. © 2012 The Geological Society of America. All rights reserved.

INTRODUCTION

Hydrocarbons have been produced from anticlines and domes since the "anticlinal theory of oil and gas entrapment" was conceptualized in the mid-nineteenth century by Thomas Hunt (1861) and others, eventually formalized by I.C. White (1885). In addition, basic fracture patterns associated with inclined rock layers (limbs of anticlines) have been well understood and geometrically classified for at least a century, with "strike joints" and "dip joints" being the most common fracture sets (e.g., Willis and Willis, 1929, p. 58, and Price, 1966, p. 113). While the significance of "fracture porosity" to enhanced oil and gas recovery from the crest and limbs of anticlines has long been understood by petroleum geologists and reservoir engineers (Selley, 1998, p. 244), the current domestic exploration emphasis on developing unconventional oil and gas resources through directional drilling in fractured mudrocks and chalks has inspired a new generation of research into the origin, geometry, kinematics, and structural diagenesis of fractured reservoir rocks. In addition, there is considerable ongoing research into the structural integrity of seal rocks (relative to surface leakage potential) at structural domes under consideration for long-term CO_2 sequestration, as well as the role of fractures and faults in governing CO_2 migration pathways in subsurface reservoir rocks (Grobe et al., 2009). There-

fore, an overarching goal of this research project was to introduce graduate students to the modern literature on fold-related fracture sets and then have them apply this knowledge to the identification, description and interpretation of lineament patterns on two oil-producing anticlines in a structurally complex part of the Central Rocky Mountains of Wyoming. This project was conducted as part of a graduate course in structural geology and the results were presented in two research colloquia at Montana State University (Burnham et al., 2010). The two anticlines chosen for this study are Maverick Springs anticline on the south side of the east-west–trending Owl Creek arch in the northwestern Wind River Basin, and Thermopolis anticline on the north side of the arch in the southern Big Horn Basin (Fig. 1). Google Earth Pro (GEP) proved to be a highly effective platform for the identification and mapping of lineament patterns on the crest and limbs of these anticlines, as well as for other fold-related structural applications as discussed later in this paper.

Detailed structural analysis (Davis et al., 2012, p. 29) is normally conducted in three sequential steps: (1) descriptive geometric analysis, (2) kinematic analysis, and (3) dynamic analysis. Geometric analysis is largely descriptive and seeks to understand the 2D and 3D shape, size, and orientation of naturally deformed rocks, as well as the relationship between structures at various scales of observation. GEP is recommended as a first step for

Figure 1. Google Earth Pro image of a portion of central Wyoming showing the Owl Creek Mountains with Thermopolis anticline in the southern Bighorn Basin, and Maverick anticline in the northwest Wind River Basin. The east-west–trending Owl Creek arch has been interpreted to be a distributed zone of sinistral transpression (represented by bold red arrows) comprising several en echelon, northwest-trending, basement-cored thrust blocks (see Paylor and Yin, 1993, and figures therein).

the collection of basic orientation data on geometric features and is highly useful for reconnaissance (pre–field season) geologic mapping of a region, especially in semi-arid or arid regions. What GEP cannot do is provide 3D (i.e., dip) information on mesoscopic fracture sets that have dimensions of meters or tens of meters, nor can it provide kinematic data on fractures (e.g., differentiation of Mode I, II, and III fractures; Gudmundsson, 2011, p. 260); these data must be collected from the outcrop. Therefore, GEP is not a replacement for boots-on-the-ground fieldwork.

In summary, the objectives of this research project were to (1) introduce students to the literature on fold-related fracture sets, (2) have them apply this knowledge to the GEP mapping and interpretation of lineament patterns on two oil-producing anticlines in the Central Rocky Mountains of Wyoming, (3) become familiar with the mapping tools of GEP and compatible software applications for descriptive structural analysis, and (4) explore the possibilities of GEP for other applications in structural geology.

GEOLOGIC SETTING

Structural Setting

Maverick Springs anticline is located along the northwest margin of the Wind River Basin, along the south margin of the western Owl Creek Mountains (T. 6 N., R. 2 W., Fremont County, Wyoming). The western Owl Creek Mountains are a world-class example of a mountain-scale, basement-involved, transpressive duplex fault system that accommodated regional sinistral-shear shortening during the Laramide orogeny (Brown, 1993; Stone, 1993; Paylor and Yin, 1993). This transpressive system comprises an array of en echelon, northwest-striking, northeast-dipping thrust faults and hanging wall anticlines, with east-west–striking, high-angle faults.

Maverick Springs anticline is, in many respects, a classic Laramide-style, basement-involved, tri-shear anticline (Fig. 2A) (Stone, 1993). The generic model of a composite, basin-margin, foreland anticline (Brown, 1993, p. 336) is remarkably similar to Maverick Springs in terms of the geometry of basement deformation, details of the tri-shear zone, and mechanical response of the sedimentary cover to basement shortening. The anticline is markedly asymmetric and verges to the southwest, as verified by looking at outcrop widths of equivalent formations on the southwest and northeast flanks of the fold using GEP (Fig. 3). The subtending, northeast-dipping Maverick Springs thrust fault (Fig. 2A) has a reported vertical separation of ~3048 m (~10,000 ft) and a horizontal separation of ~2590 m (~8500 ft) at the top of the Precambrian basement; net slip on the fault is ~3658 m (~12,000 ft) (Stone, 1993, p. 284). The Maverick Springs thrust can be traced ~56 km to the southwest where it dies out beneath the Eocene Wind River Formation.

Thermopolis anticline is located on the north flank of the Owl Creek arch in T. 43–44 N., R. 93–96 W., Hot Springs County, Wyoming. Thermopolis anticline is also markedly asym-

metric with southwest-vergence, facing the gently north-dipping flank of the Owl Creek arch; the south limb of Thermopolis anticline dips ~60° SSW whereas the north limb dips ~15–20° NNE. The fold is subtended by a NNE-dipping thrust fault (Fig. 2B) (Blackstone, 1986, p. 18). Thermopolis anticline is characterized by four small-amplitude domes with intervening saddles along its hinge line (~50 km in overall length); from northwest-to-southeast, these are King, Rose, Cedar Ridge and Warm Springs Domes (Zahm, 2002). The anticline has been eroded to the top of the Permian Phosphoria Formation at Rose and Cedar Ridge Domes, providing outcrop access to a major reservoir rock in this region. The Bighorn River is superposed over Thermopolis anticline between Cedar Ridge and Warm Spring Domes in an area where the anticline changes trend from WNW-ESE to NW-SE; this is also an area of large-volume hot spring discharge and extensive active and non-active travertine deposits. This "kink" in the overall trend of the anticline could mark the position of a tear fault that slightly offsets the subtending thrust fault, or a relay ramp between two en echelon subtending thrust faults, or simply a zone of fractures perpendicular to the hinge line that formed in response to a change-in-strike in the underlying thrust.

Maverick Springs and Thermopolis anticlines are extremely well-exposed in the sense of the quintessential Wyoming "sheep-herder" anticline (McPhee, 1986, p. 17) and, therefore, are well suited for Google Earth applications. Although there have been many stories about the origin of the term "sheepherder anticline," suffice it to say that these are extremely well-exposed, basin-margin or intra-basin anticlines in the Central Rockies with a topographically high rim formed, in many cases, by resistant sandstones in the Upper Cretaceous Mesa Verde Group.

Stratigraphic Framework

Details of the Phanerozoic stratigraphic section in central Wyoming are discussed by Keefer and Van Lieu (1966) and Paylor and Yin (1993) and will not be reiterated herein. Similarly, the stratigraphic elements of the petroleum system in central Wyoming are summarized by Glaser (1989), Johnson (2005), and Kirschbaum et al. (2007). The stratigraphy at both localities is basically the same, although Maverick Springs is ~50 km farther west than Thermopolis anticline and therefore has facies that are slightly more "western" within equivalent Phanerozoic units. As previously mentioned, Thermopolis anticline exposes the Permian Phosphoria Formation along its crest, whereas the deepest level of erosion at Maverick Springs is the upper part of the Triassic red-bed succession. Lineament sets are easily seen in GEP across the crest and limbs of both anticlines, particularly in resistant dip-slopes formed by the Triassic Chugwater, Jurassic Sundance, and Cretaceous Cloverly and Muddy Sandstones; resistant sandstones of the Frontier and Mesa Verde Formations are also highly fractured higher in the section. Lineament distribution and density are not uniform throughout the exposed stratigraphic section, reflecting the importance of mechanical stratigraphy in the length, distribution, and spacing of fracture sets.

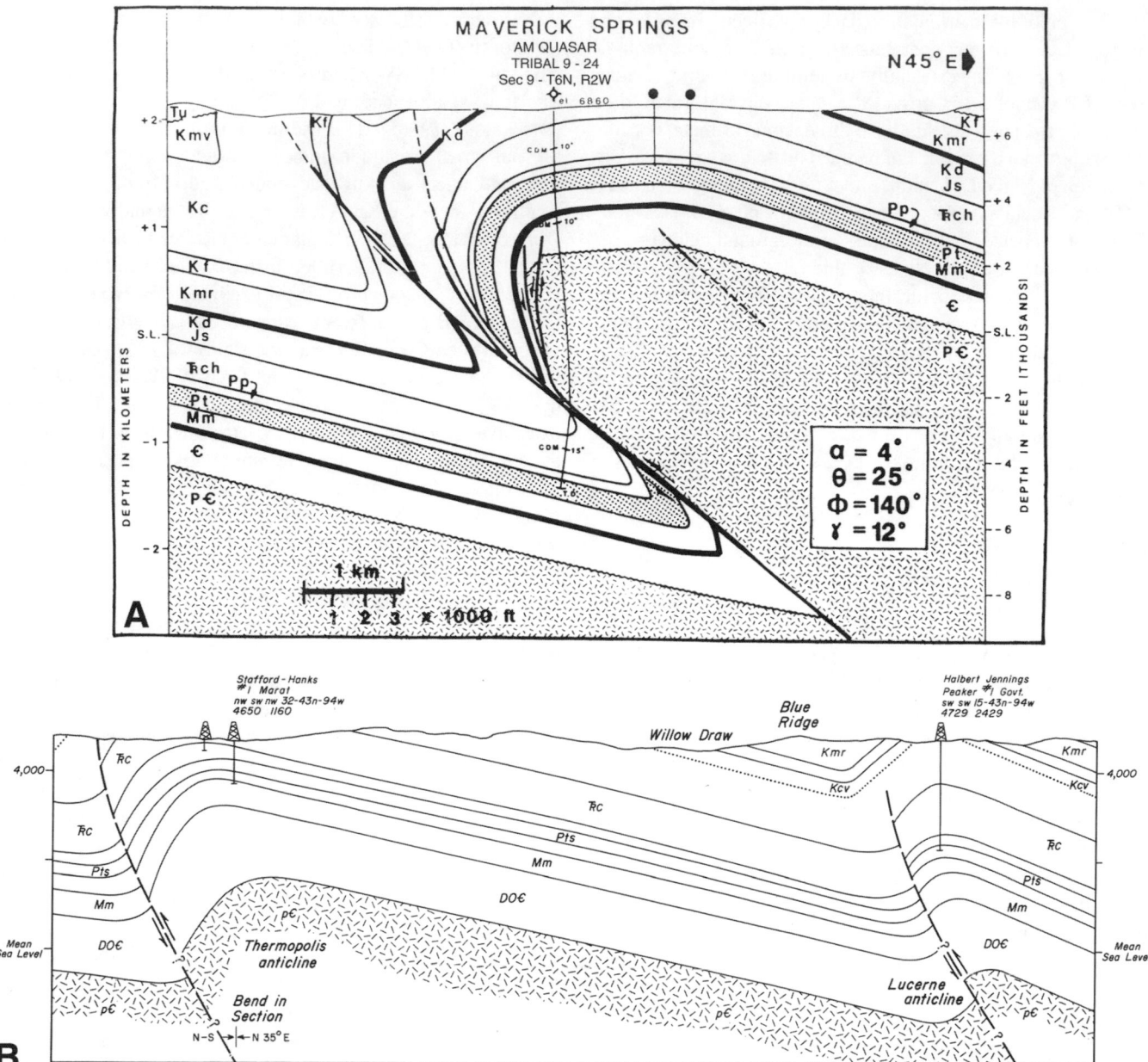

Figure 2. (A) Structural cross section across Maverick Springs anticline showing overall geometry of the fold (from Stone, 1993). Line-of-section is oriented NE-SW across the anticline. The inset box showing degrees (lower right) pertains geometric details of the fold (α = back-limb rotation angle; θ = initial thrust angle with top of basement; ϕ = fault cutoff angle at top of hanging wall basement; γ = regional dip in the footwall), as discussed in Stone's paper. (B) Structural cross section across Thermopolis anticline showing overall geometry of the fold (from Blackstone, 1986). Line-of-section is oriented NE-SW across the anticline.

METHODS

Maverick Springs and Thermopolis anticlines were selected for this study because of their structural position on opposite sides of the east-west–trending Owl Creek arch, the bedrock detail displayed in GEP coupled with little or no vegetation, and readily mappable linear features. GEP was considered essential for this project because of the measurement tools it has avail-

able, but mostly because of the ability to save "premium quality" images for subsequent use.

The terms lineament, linear, line, and lineation have been used in geologic literature for over a hundred years, sometimes with confusing or overlapping meaning. O'Leary et al. (1976) provide a good summary of the nomenclature problems, with recommendations that are still useful and valid today. For the sake of clarity, we measured *lineaments* on GEP images in this project, in accord with the definitions of O'Leary et al. (1976) and Fossen (2010, p. 260), being linear [*adjective*] features at the scale of topographic maps, aerial photos, digital elevation models, satellite images, and Google Earth images. Our a priori assumption is that many, if not most lineaments on images of Maverick Springs and Thermopolis anticlines correspond to four dominant fracture sets and related small faults on the crest and limbs of these folds, based on their repeating geometry and spatial distribution across the folds. This assumption should be tested in the field at the outcrop-scale of observation, but this was beyond the scope of this computer-based project. However, the interpretation of lineaments (mapped from air photos or other imagery) as related to fracture patterns on folds is not without precedence based on decades of field work on fold-related fracture patterns by scores of researchers (e.g., Willis and Willis, 1929; Price, 1966; Stephenson, et al., 2007; Wennberg et al., 2007). The orientation of these fracture-controlled lineaments relative to the hinge lines at Maverick Springs and Thermopolis anticlines will be discussed later in this paper. In addition, the majority of lineaments measured for this project are expressed as *lines* on bedrock dip-slopes and do not represent topographic features like ridgelines or stream channel terraces, although some joints are certainly exploited and widened by geomorphic processes. Anthropogenic features were also avoided in lineament mapping. Faults were identified by small offsets in bedding or topography. Fault recognition was likely biased toward those faults at high-angles to fold hinge lines (i.e., sub-perpendicular to bedding strike).

Google Earth Pro Methods

Determining the scale of observation for lineament measurements is the first task that must be completed. This can be done either qualitatively or quantitatively, depending on the objectives of the study. Premium quality, poster-size GEP images can be printed using a plotter at a predetermined scale of observation that allows the entire structure to be viewed on a single map, for example at 1:5,000. The orientation and length of every lineament can then be manually measured (with a protractor and scale) across the fold using a grid system or a reference line; this is called the *inventory* method of analysis and is useful for determining lineament or fracture density in a rock body (van der Pluijm and Marshak, 2004, p. 151). Alternatively, the *selection* method of lineament mapping is conducted by visually scanning a GEP image at a scale that encompasses the entire structure, subjectively determining the dominant lineament sets based on repeating patterns across the fold with respect to the hinge line, and then measuring the orientation and length of major lineaments that stand out and are representative of those sets. The inventory and selection methods can also be applied at the outcrop-scale of observation for measuring fractures (van der Pluijm and Marshak, 2004, p. 151). Both measurement techniques have their advantages and limitations, again depending on the objectives of the study.

For the purposes of this project (see introduction), the selection method was better suited for digital analysis with GEP and ArcGIS. The first step in the selection process involves viewing the entire anticline to see major lineaments and faults, as well as the expression of stratigraphy around the fold as correlated with available geologic maps. After measuring the major lineaments at this scale (as described in detail below), a GEP image is "zoomed in" to approximately ½ of the fractional scale that encompasses the entire structure for detailed mapping, using the scale bar in the lower-left corner of the screen which is activated under the view tab. For example, if the entire structure can be viewed at a scale of 1:5,000, lineaments should first be mapped at this scale; following this, the GEP image should be zoomed in to 1:2500 for more detailed lineament mapping. The degree to which one decides to zoom in is entirely dependent on the objectives of the study and the quality of outcrops as displayed in GEP; students are encouraged to experiment with measuring lineaments at different scales of observation before deciding on which scale is best. Along the same lines, it is a judgment call for students to *stop* drawing lineament features on the GEP image at some point; it is possible to get carried away and draw lines on every tiny straight feature, so one needs to have "size filter" in mind and always be mindful of the geologic context of every lineament and the scale of observation appropriate to the analysis. It is wise to have a group meeting and discuss lineaments, the criteria for distinguishing various fracture sets, and the issue of scale before opening GEP and starting to draw lines.

Lineaments are mapped on GEP images using the "path" tool; the "ruler" tool can then be used to determine orientation and length of each lineament. The "line" tab of the ruler dialogue box displays the length and heading (azimuth) of the lineament that was just measured; the units for length can be changed in the ruler dialogue box. The "ruler" tool can be used to measure lines, paths, and polygons. The difference between a line and a path is important to understand because a line has only two end points, whereas a path can have more than two end points (if the path being drawn is not a straight line). When using GEP to map and measure lineaments, it is important to only draw lines with beginning and end points while using the "path" tool; this will aid in measurements using either the GEP "ruler" tool or after transferring to ArcGIS. Lineaments can be color-coded to distinguish different sets by editing the properties of individual paths (edit>properties>style/color). Lineament attributes can then be recorded on a spreadsheet and imported into a structural analysis program, like RockWare StereoStat, for statistical and geometric evaluation.

ArcGIS Methods

Exporting GEP lineament data to ArcGIS is a more efficient way of digitally managing lineament data. Specifically, ArcGIS can simultaneously measure all of the mapped lineaments and generate a table that includes the length and orientation of each lineament. This method can save a great deal of time that would have been spent laboriously measuring and recording the orientations and lengths of each lineament. Obviously, this method is useful only if ArcGIS is available and students have the requisite

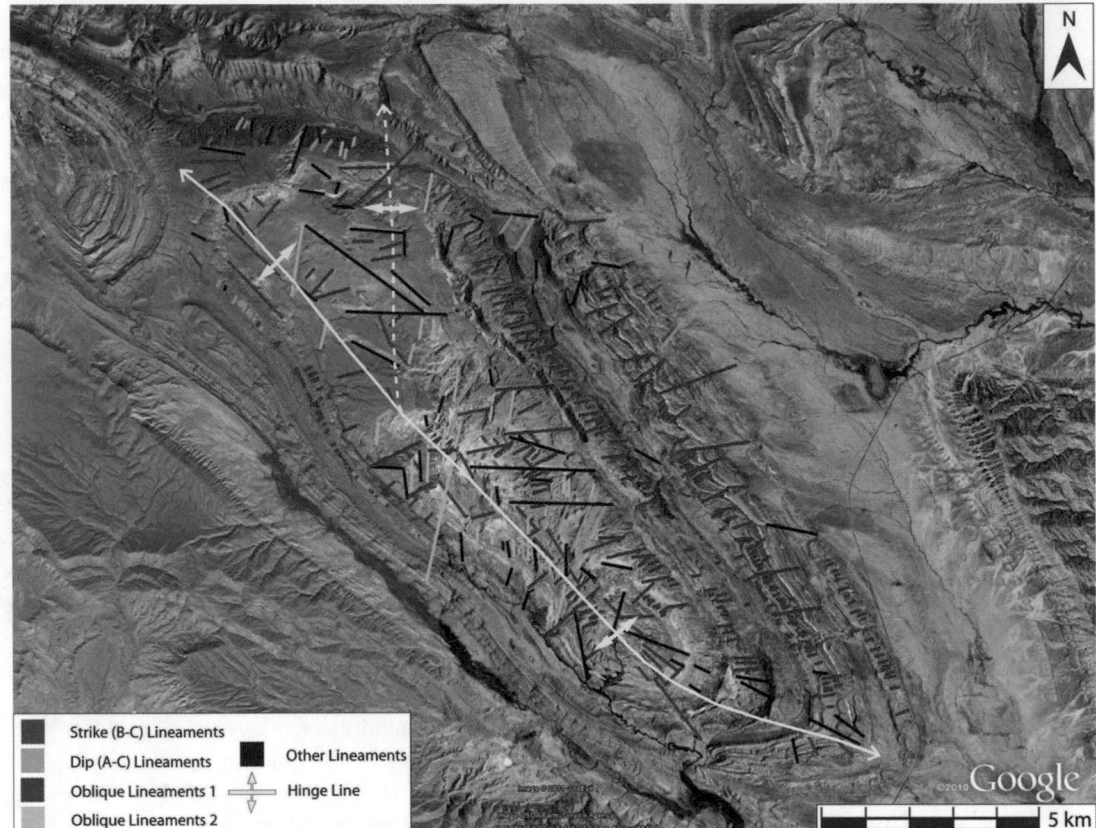

Figure 3. Map of Maverick Springs anticline on a Google Earth Pro base showing five sets of lineaments. Lineaments are color-coded according to their geometry with respect to the trend of the axial surface trace (hinge line) of the fold. See inset key and text for explanation. Bold yellow line is the approximate location of the hinge line.

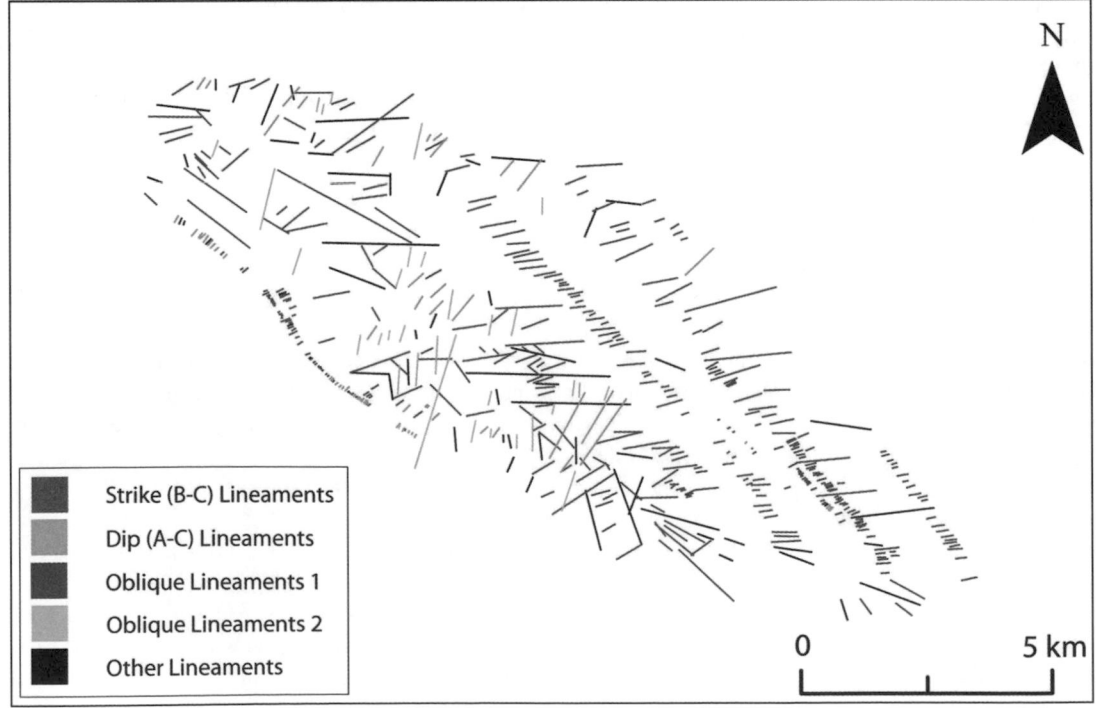

Figure 4. Map of Maverick Springs anticline emphasizing the five lineament sets, without the Google Earth Pro image in the background.

training and experience in GIS. After mapping lineaments in GEP with the "path" tool, the lineament map can be saved as a KML file, imported into ArcGIS and converted into a shapefile to easily calculate the azimuth of each lineament. This is done using the arc toolbox conversion tools' "KML-to-Layer" tool which creates a geodatabase containing a layer file (.lyr) representing the mapped lineaments; this layer file can also be exported as a shapefile. Arc-Catalog is then used to create coordinate geometry fields for the shapefile using the "Create COGO Fields" command. These fields represent geometric attributes including the length and orientation for each of the mapped features. Once these fields are created, they can be populated using the "update COGO attributes" command in ArcMap, which calculates the length and orientation of each lineament and adds it to the attribute table of the shapefile. This attribute table can then be exported as a text file and opened in most stereonet programs which are used to create rose diagrams for examination of lineament orientations.

In addition, GEP was also used for construction of down-plunge projections of the two anticlines. The method of down-plunge viewing (Mackin, 1950) is an extremely powerful technique for viewing "true" cross-section profiles of plunging folds (i.e., profiles perpendicular to the fold axis) in irregular or rough terrain by simply aligning one's line-of-sight parallel to the plunging fold axis. The down-plunge view allows one to see the variable geometry of stratigraphic units (mechanical stratigraphy) that lies above the line of cross section on a geologic map, thereby allowing students to project those beds up-dip into the line-of-section above the level of erosion. Down-plunge viewing is not just a visualization trick employed by old structural geologists; rather, it is the only way to correctly view the geometry of deformed rock units on a geologic map. The trend-and-plunge of the fold axis can be easily determined from a β-point or π-point plot on an equal-area stereonet of strike-and-dip values from the limbs and hinge area of a fold; strike-and-dips can be tabulated from published geologic maps for this purpose (Warlow, 1985). The trend-and-plunge of the fold axis can then be entered into GEP by editing the properties of the screen image (edit>properties>view); trend of the fold axis should be entered in the "heading" box and plunge in the "tilt" box. Profiles may be constructed freehand while looking down-plunge (best for introducing students to the down-plunge visualization technique), or by using the graphical "grid" technique, or through the use of computer algorithms that rely on digitized points from a geologic map (Marshak and Mitra, 1988, p. 287–297). GEP is a wonderful way to teach the down-plunge projection technique, particularly using folds that are well-exposed in moderate terrain with little or no vegetative cover, such as those found in Wyoming or the Zagros Mountains.

LINEAMENT INTERPRETATIONS

Maverick Springs Anticline

Four lineament orientations were identified on GEP images of Maverick Springs anticline (Figs. 3 and 4); note that the axial

surface trace bifurcates at the northwest end of Maverick Springs where the fold becomes more "box-like." These lineament orientations are: (1) within 15° from a line orthogonal to the average trend of the hinge line, interpreted to be dip-fractures (A-C) shown in green; (2) parallel to (or within 15°) of the hinge line, interpreted as strike-fractures (B-C) shown in dark blue; (3) oblique to the hinge line, trending NNE shown in light blue; and (4) oblique to the hinge line, trending ENE shown in red. In addition, the lineament orientation data were plotted on a length-weighted rose diagram (Fig. 5). The dominant lineament set trends ENE (red), being oblique to hinge line of the fold; the significance of this trend will be discussed later. The NNE-trending oblique set (light blue) is also distinct on the rose diagram. Lineaments trending

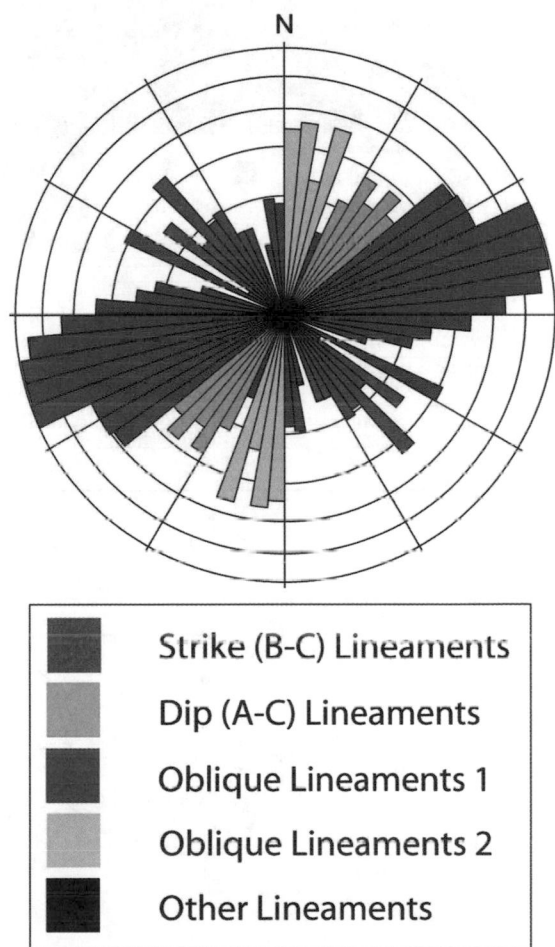

Figure 5. Length-weighted rose diagram representing 519 lineaments measured a Google Earth Pro image of Maverick Springs anticline. Each petal of the diagram equals 5°. Northwest-trending lineaments are strike-lineaments (B-C) and northeast-trending lineaments are dip-lineaments (A-C). The acute angle between the two oblique lineament sets (NNE in light blue and ENE in red) is approximately bisected by the trend of dip-lineaments (green), suggesting a possible conjugate interpretation for the oblique lineaments. See text for more explanation.

216

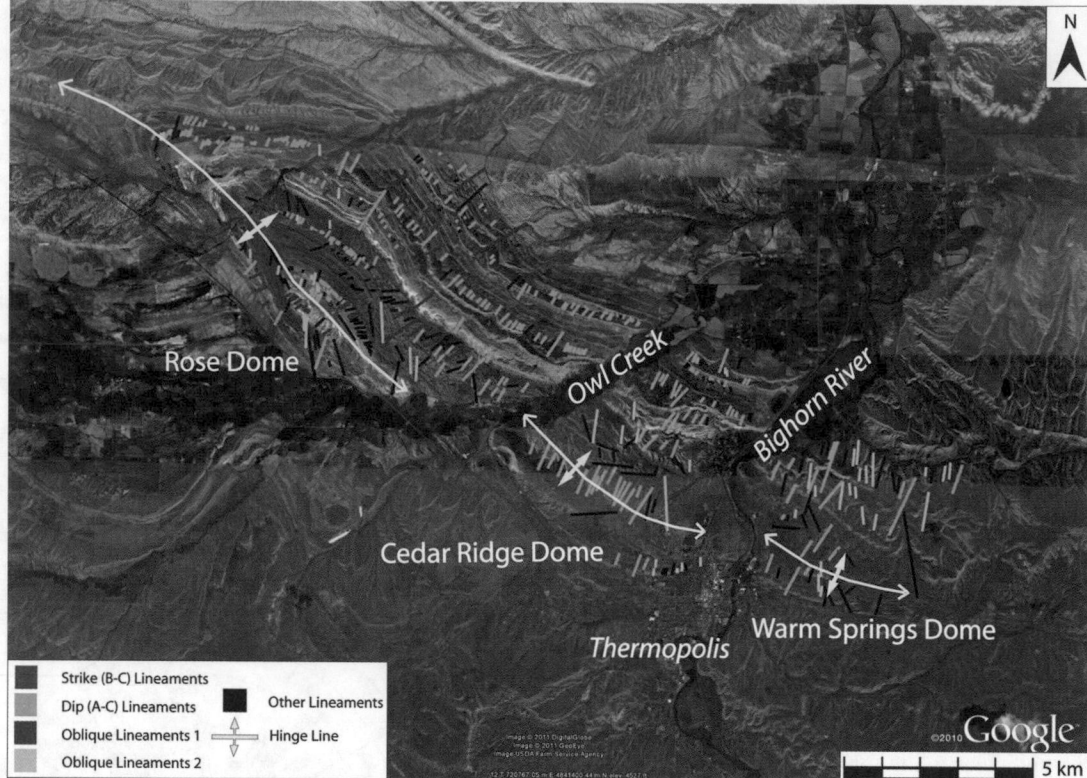

Figure 6. Map of Thermopolis anticline on a Google Earth Pro base showing five sets of lineaments. Lineaments are color-coded according to their geometry with respect to the trend of the axial surface trace (hinge line) of the fold. See inset key and text for explanation. Bold yellow lines show the approximate location of the hinge line over low-amplitude structural culminations along the crest of the anticline, namely Rose, Cedar Ridge, and Warm Springs Domes. The Bighorn River is superposed over the east end of the anticline between Cedar Ridge and Warm Springs Domes and flows through the city of Thermopolis, Wyoming (lower right).

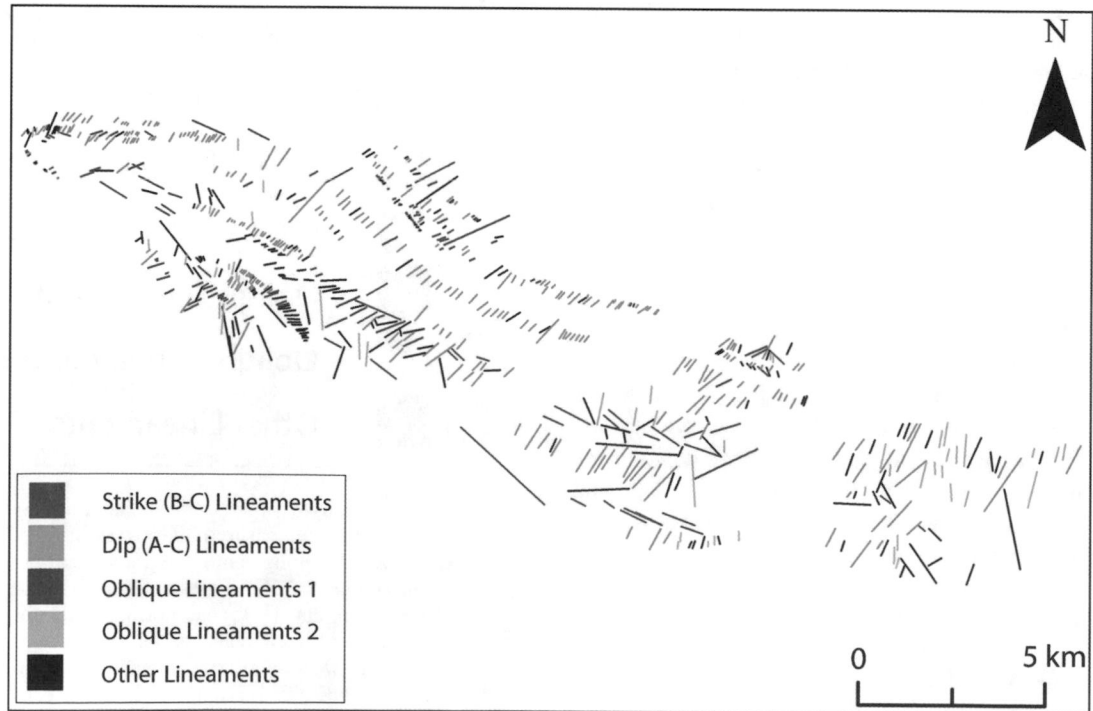

Figure 7. Map of Thermopolis anticline emphasizing the five lineament sets, without the Google Earth Pro image in the background.

sub-parallel to the hinge line (interpreted as strike fractures, dark blue) also form a distinct population, as do those trending sub-normal to the hinge line (dip fractures, green).

Thermopolis Anticline

Figure 6 is a GEP image of Thermopolis anticline showing Rose Dome on the northwest end, Cedar Ridge Dome west of the Bighorn River and Warm Springs Dome to the east, as well as several sets of mapped lineaments; the general trend of the hinge line across the three domes is shown in yellow. As before, lineaments sub-orthogonal to the hinge line (A-C) are shown in green, those sub-parallel to the hinge (B-C) are in dark blue, and oblique lineaments are shown in light blue and red. The highest frequency of lineaments appears to be orthogonal to the hinge (A-C, green) on the flanks as well as the crest of the anticline (Figs. 7 and 8). It is clear that most of the hinge-parallel fractures

(B-C, dark blue) are located either near the crest of the anticline or along the steeply dipping south limb where flexural rotation of beds was the greatest.

At both anticlines, B-C and A-C lineaments are interpreted to be mode-I extensional joints based on their based on their repeating geometry and spatial distribution with respect to the axial surface trace (hinge line) of the fold (Price, 1966, p. 113; Ramsay and Huber, 1987, p. 647). B-C joints tend to be concentrated in areas of maximum curvature in the hinge area of a fold and typically form by outer-arc extension of a bed during flexural-slip folding; A-C joints are common across an entire fold (hinge area and limbs) and typically form due to plunge-parallel extension (Fig. 9). In some cases, pairs of oblique lineaments (NNE and ENE) with acute angles close to 60° could be interpreted as conjugate fractures, but this of course is interpretational and would need to be verified by the collection of kinematic data in the field.

DISCUSSION

The scope of this project was to test the applicability of GEP for the collection and preliminary interpretation of orientation (azimuthal) data for lineaments, given the limitations stated previously. Because the project lacked a fieldwork component to gather kinematic data, we were limited in our ability to interpret the results in a more rigorous fashion. With a follow-on field component, one could employ several techniques that combine geometric and kinematic data from fault and fracture arrays to constrain the orientation of the paleostress field, such as slip-linear plots (Marshak and Mitra, 1988), fault-slip (stress) inversion and tangent-lineation (M-plane) diagrams (Angelier, 1984; Fossen, 2010). For example, based on extensive fracture and stress inversion studies along the east-west–trending Owl Creek and Casper Mountain arches by Molzer and Erslev (1995), the direction of the principal maximum stress direction (σ_1) in central Wyoming during the Laramide orogeny was interpreted to be N48°E to N65°E

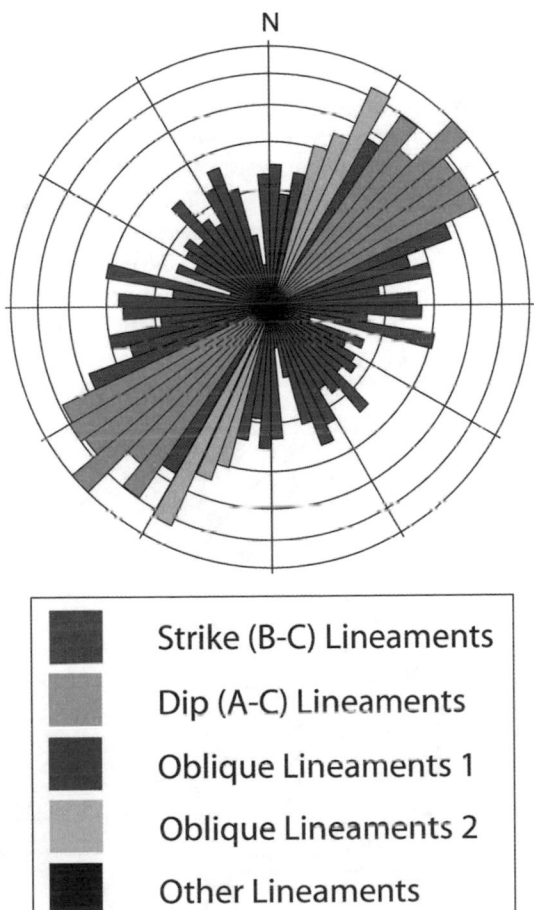

Figure 8. Length-weighted rose diagram representing 667 lineaments measured a Google Earth Pro image of Maverick Springs anticline. Each petal of the diagram equals 5°. Northwest-trending lineaments are hinge-parallel (B–C) and northeast-trending lineaments (green), which are the most prominent set at Thermopolis, are dip-lineaments (A–C). See text for more explanation.

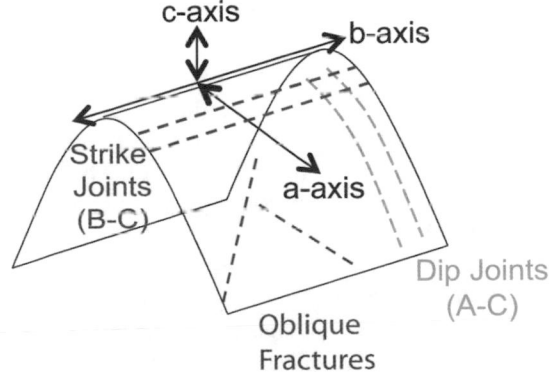

Figure 9. Diagram of an upright anticline showing the conventional a-b-c coordinate system used to classify fold-related lineaments and fractures; the b-axis is always parallel to the hinge line of the fold (Price, 1966, p. 113; Ramsay and Huber, 1987, p. 647).

(northeast-southwest). Blackstone (1990) and Paylor and Yin (1993) also concluded that the principal maximum stress direction was oriented NNE-SSW, with a distributed component of significant sinistral transpression along the Owl Creek trend.

To a first approximation, the lineament analyses from Maverick and Thermopolis anticlines presented herein appear to have overall geometric compatibility with the more rigorous, previously published studies in the area. The NE-SW direction of Laramide shortening derived by other workers is perpendicular to the hinge lines of both anticlines and the hinge-parallel (B-C) lineament set. Also, this shortening direction bisects the acute angle of several intersecting oblique lineaments on the flanks of Maverick and Thermopolis anticlines that may be interpreted to be conjugate fractures, assuming they formed at the same time. Of course, there are many assumptions and pitfalls associated with even the most rigorous paleostress calculations and/or interpretations for shear fractures and fault arrays, not the least of which is structural control by preexisting zones of weakness in the basement and anisotropies within cover rocks (e.g., Blackstone, 1990; Brown, 1984, 1988).

The most prominent lineament set at Maverick Springs trends ENE (red) and is highly oblique to the hinge line of the fold. It is tempting to classify all the lineaments in this set as A-C dip-fractures along the southwest flank of the anticline, and a few do indeed just fall within the A-C range on the weighted rose diagram (i.e., within 15° of a line orthogonal to the average trend of the hinge line) (Fig. 4). However, this oblique trend is clearly persistent across the crest and northeast limb of the anticline and must represent a more complex process than that normally ascribed to mode-I dip-joints. We propose that progressive sinistral shear along the western Owl Creek transpressive zone resulted in a strong over-print of joints normal to the local NNW extension direction and sub-parallel to the local shortening direction. This same phenomenon has been interpreted for lineament/fracture patterns associated with folds in the Zagros Mountains (Stephenson et al., 2007), whereby early, nearly orthogonal fracture fabrics were reactivated and rotated by the diffuse shear (Fig. 10). The lineament orientations at Thermopolis anticline seem to be more consistent with the regional Laramide stress orientations derived by Blackstone (1990), Paylor and Yin (1993), and Molzer and Erslev (1995). The NE-SW–shortening direction is also perpendicular to the subtending thrust fault at Thermopolis and sub-perpendicular to B-C lineaments. Furthermore, A-C lineaments appear to "track" the curvature of the hinge line from dome to dome and do not appear as consistently oriented oblique lineaments, as at Maverick Springs.

Lastly, another GEP application that proved to be exceptionally useful in the analysis of folds is the ability to quickly and easily obtain a down-plunge projection view of the anticlines (Mackin, 1950). Down-plunge viewing is a major aid in the construction of cross sections over plunging folds because folds appear in their true geometric attitude (as seen in a projection plane normal to the fold axis) and relative fault separation is observed directly. Experience has shown that students sometimes have difficulty "seeing" a down-plunge view on a 2D map (such

as a USGS 7.5 min quadrangle), but with GEP the down-plunge learning curve is virtually instantaneous (Fig. 11). The ideal scenario is to couple geologic maps with a GEP image of the same area, so students can see the stratigraphic units that create the topography in a down-plunge-view.

CONCLUSIONS

This project demonstrates that Google Earth Pro is an excellent platform for remote mapping of structural lineaments, joint sets and faults as the first step in a structural analysis of an area. The mapping tools available in GEP and ArcGIS can quickly help students understand the orientation and distribution of lineaments across a host structure like Maverick Springs anticline, while integrating other graphical techniques like rose diagrams, equal-area stereonets and down-plunge projections. It is recommended that the results from a GEP study of the type described herein be followed up with fieldwork, particularly with kinematic data on fracture and fault surfaces. However, even in the absence of field work, stand-alone orientation data derived from GEP imagery is nevertheless exceedingly useful for understanding the spatial orientation, distribution with respect to the host structure, and stratigraphic compartmentalization of lineament sets in the study of exhumed petroleum reservoir rocks and/or fractured seal rocks. Thus, we feel that this graduate course project demonstrated that Google Earth Pro is a powerful mapping tool that can be effectively used to describe and analyze fracture networks.

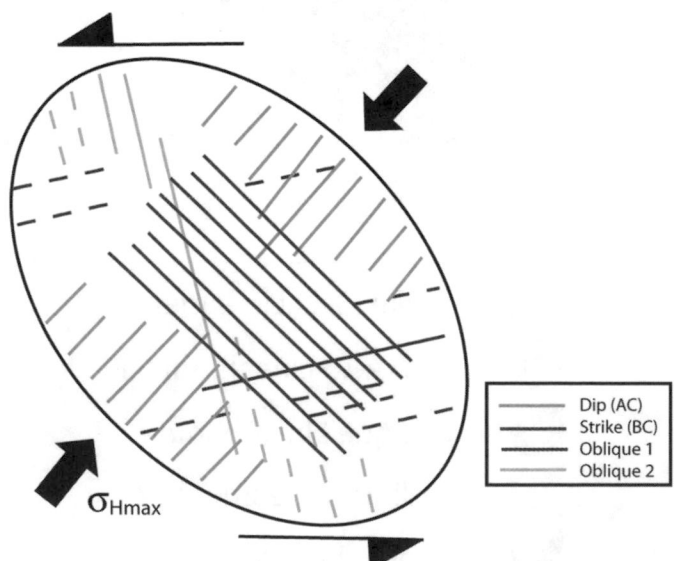

Figure 10. Interpretation of fracture patterns associated with folds in the Zagros Mountains, in which early, nearly orthogonal fracture fabrics were reactivated and rotated by the diffuse shear (modified from Stephenson et al., 2007). The four sets are color-coded to correspond to lineament sets mapped at Maverick Springs and Thermopolis anticlines and include hinge-parallel lineaments (dark blue), hinge-orthogonal lineaments (green), and two sets of oblique lineaments (light blue and red).

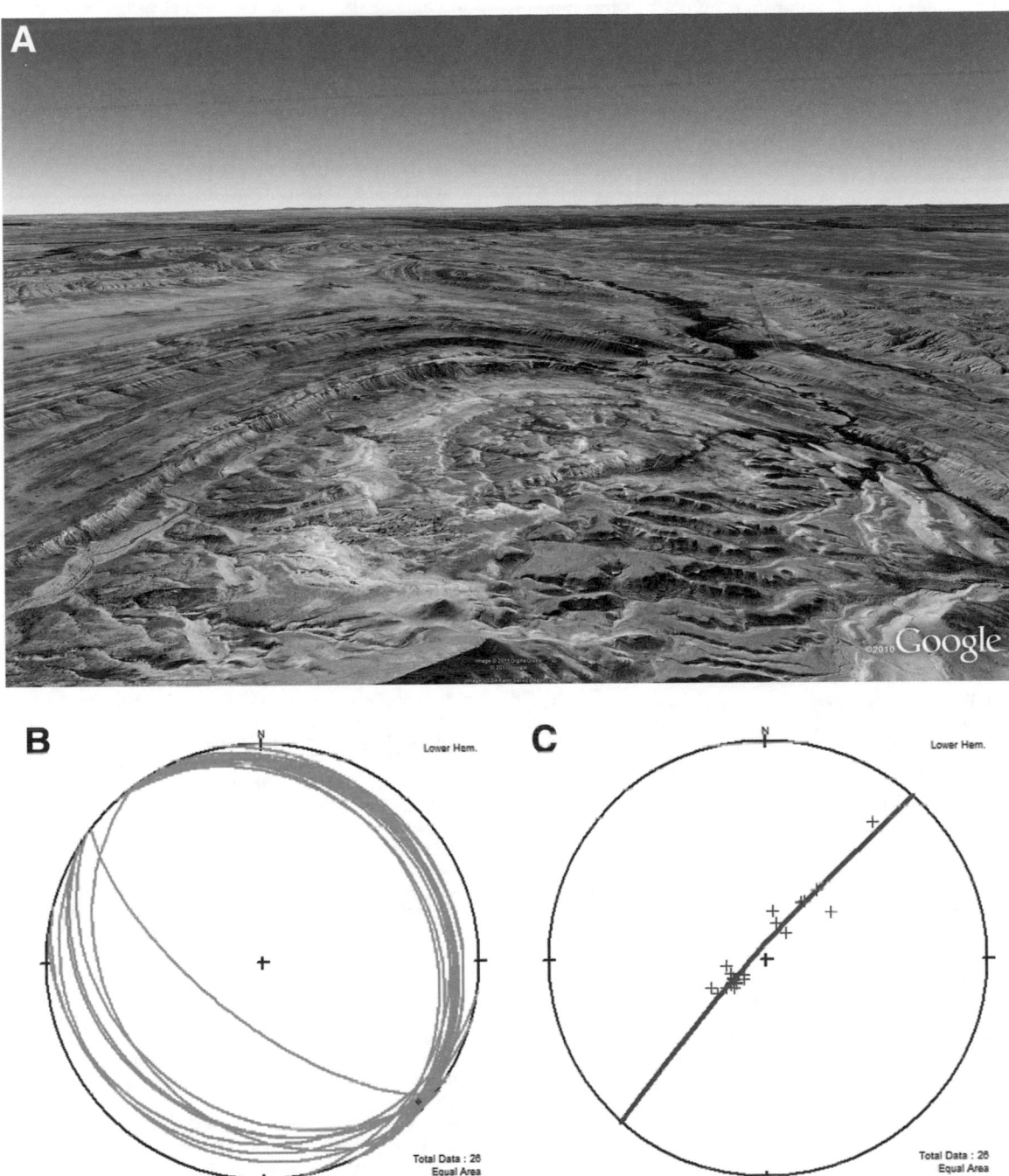

Figure 11. (A) Approximate down-plunge view over the southeast end of Maverick Springs anticline; Little Dome anticline is visible in the middle distance below the horizon. (B) The most accurate way to view down-plunge structure is to construct an equal-area plot of the trend-and-plunge of the fold-axis from strike-and-dip values gathered on the limbs and crest of the fold from geologic maps (β-point diagram), thus allowing a more accurate direction and angle of plunge to be incorporated into your down-plunge view. (C) Another way to achieve this is to plot poles to bedding and then find the best-fit great circle, which is the pole to the fold axis (π-point diagram). Strike-and-dip data for stereonet plots were gathered from Warlow (1985).

ACKNOWLEDGMENTS

The authors are grateful for the thorough reviews of this manuscript provided by two reviewers. Google Earth Pro licenses were donated to the Department of Earth Sciences at Montana State University, for which we are grateful. Portions of this research (Thermopolis anticline) were supported by the Department of Energy Award Number DE-FE0000397, Zero Emissions Research and Technology (ZERT II), awarded to the Energy Research Institute at Montana State University. This report was prepared as an account of work sponsored by an agency of the United States Government. Neither the United States Government nor any agency thereof, nor any of their employees, makes any warranty, express or implied, or assumes any legal liability or responsibility for the accuracy, completeness, or usefulness of any information, apparatus, product, or process disclosed, or represents that its use would not infringe privately owned rights. Reference herein to any specific commercial product, process, or service by trade name, trademark, manufacturer, or otherwise does not necessarily constitute or imply its endorsement, recommendation, or favoring by the United States Government or any agency thereof. The views and opinions of authors expressed herein do not necessarily state or reflect those of the United States Government or any agency thereof.

REFERENCES CITED

Angelier, J., 1984, Tectonic analysis of fault slip data sets: Journal of Geophysical Research, v. 89, p. 5835–5848, doi:10.1029/JB089iB07p05835.

Blackstone, D.L., Jr., 1986, Foreland compressional tectonics: southern Bighorn Basin and adjacent areas, Wyoming: The Geological Survey of Wyoming, Report of Investigations, no. 34, 32 p.

Blackstone, D.L., Jr., 1990, Rocky Mountain foreland structure exemplified by the Owl Creek Mountains, Bridger Range and Casper arch, central Wyoming, *in* Specht, R., ed., Wyoming tectonics and sedimentation: Wyoming Geological Association forty-first annual field conference guidebook, p. 151–166.

Brown, W.G., 1984, Basement involved tectonics in foreland areas: American Association of Petroleum Geologists, Continuing Education Course Note Series 26, 92 p.

Brown, W.G., 1988, Deformational style of Laramide uplifts in the Wyoming foreland, *in* Schmidt, C.J., and Perry, W.J., eds., Interaction of the Rocky Mountain foreland and Cordilleran thrust belt: Geological Society of America Memoir 171, p. 1–25.

Brown, W.G., 1993, Structural style of Laramide basement-cored uplifts and associated folds, *in* Snoke, A.W., Steidtmann, J.R., and Roberts, S.M., eds., Geology of Wyoming: Laramie, Wyoming, The Geological Survey of Wyoming Memoir no. 5, v. 1, p. 312–371.

Burnham, D., Larsen, M., Lynn, H., Porter, J., Treadway, W., and Lageson, D., 2010, Fracture analysis of two hydrocarbon-producing anticlines in the Wind River Basin, Wyoming: comparison with similar structures in the Zagros Mountains, Iran: 2010 Montana State University Student Research Celebration and 5th Annual Earth Sciences Colloquium, unpublished abstracts with program.

Davis, G.H., Reynolds, S.J., and Kluth, C.F., 2012, Structural geology of rocks and regions (3rd edition): John Wiley and Sons, Inc., 839 p.

Fossen, H., 2010, Structural geology: New York, Cambridge University Press, 463 p.

Glaser, T., 1989, Maverick Springs, Wyoming oil and gas fields, Bighorn and Wind River Basins: Casper, Wyoming, Wyoming Geological Association, p. 302–303.

Grobe, M., Pashin, J.C., and Dodge, R.L., eds., 2009, Carbon dioxide sequestration in geological media—state of the science: Tulsa, Oklahoma, American Association of Petroleum Geologists, Studies in Geology 59, 715 p.

Gudmundsson, A., 2011, Rock fractures in geological processes: Cambridge, UK, Cambridge University Press, 578 p.

Hunt, T.S., 1861, Bitumens and mineral oils: Montreal Gazette, March 1.

Johnson, E.A., 2005, Geologic assessment of undiscovered oil and gas resources in the Phosphoria total petroleum system, southwestern Wyoming province, Wyoming, Colorado, and Utah, *in* USGS southwestern Wyoming province assessment team, petroleum systems and geologic assessment of oil and gas in the southwestern Wyoming province, Wyoming, Colorado, and Utah: U.S. Geological Survey Digital Data Series DDS-69-D, Chapter 4, 46 p., CD-ROM.

Keefer, W.R., and Van Lieu, J.A., 1966, Paleozoic formations of the Wind River Basin, Wyoming: U.S. Geological Survey Professional Paper 495-B, 60 p.

Kirschbaum, M.A., Lillis, P.G., and Roberts, L.N.R., 2007, Geologic assessment of undiscovered oil and gas resources in the Phosphoria total petroleum system of the Wind River Basin province, Wyoming, *in* USGS southwestern Wyoming province assessment team, petroleum systems and geologic assessment of oil and gas in the southwestern Wyoming province, Wyoming, Colorado, and Utah: U.S. Geological Survey Digital Data Series DDS-69-J, Chapter 3, 27 p., CD-ROM.

Mackin, J.H., 1950, The down-structure method of viewing geologic maps: The Journal of Geology, v. 58, p. 55–72, doi:10.1086/625695.

Marshak, S., and Mitra, G., 1988, Basic methods of structural geology: Englewood Cliffs, New Jersey, Prentice Hall Publishing, 446 p.

McPhee, J., 1986, Rising from the plains: New York, Farrar, Straus and Giroux, 214 p.

Molzer, P.C., and Erslev, E.A., 1995, Oblique convergence during northeast-southwest Laramide compression along the east-west Owl Creek and Casper Mountain arches, central Wyoming: The American Association of Petroleum Geologists Bulletin, v. 79, no. 9, p. 1377–1394.

O'Leary, D.W., Friedman, J.D., and Pohn, H.A., 1976, Lineament, linear, lineation: Some proposed new standards for old terms: Geological Society of America Bulletin, v. 87, p. 1463–1469, doi:10.1130/0016-7606(1976)87<1463:LLLSPN>2.0.CO;2.

Paylor, E.D., and Yin, A., 1993, Left-slip evolution of the North Owl Creek fault system, Wyoming, during Laramide shortening, *in* Schmidt, C.J., Chase, R.B., and Erslev, E.A., eds., Laramide basement deformation in the Rocky Mountain foreland of the western United States: Geological Society of America Special Paper 280, p. 229–242.

Price, N.J., 1966, Fault and joint development in brittle and semi-brittle rock: Oxford, Pergamon Press, 176 p.

Ramsay, J.G., and Huber, M.I., 1987, The techniques of modern structural geology, volume 2: folds and fractures: London, Academic Press, 700 p.

Selley, R.C., 1998, Elements of petroleum geology (2nd edition): San Diego, Academic Press, 470 p.

Stephenson, B.J., Koopman, A., Hillgartner, H., Mcquillan, H., Bourne, S., Noad, J.J., and Rawnsley, K., 2007, Structural and stratigraphic controls on fold-related fractures in the Zagros Mountains, Iran: implications for reservoir development, *in* Lonergan et al., eds., Fractured Reservoirs: Geological Society of London Special Publication 270, p. 1–21.

Stone, D.S., 1993, Basement-involved thrust-generated folds as seismically imaged in the subsurface of the central Rocky Mountain foreland, *in* Schmidt, C.J., Chase, R.B., and Erslev, E.A., eds., Laramide basement deformation in the Rocky Mountain foreland of the western United States: Geological Society of America Special Paper 280, p. 271–318.

van der Pluijm, B.A., and Marshak, S., 2004, Earth structure: An introduction to structural geology and tectonics (2nd edition): New York, W.W. Norton and Company, 656 p.

Warlow, R.C., 1985, Geologic map and coal resources of the Maverick Spring quadrangle, Fremont County, Wyoming: U.S. Geological Survey Coal Investigations Map, C-91, 1:24,000.

Wennberg, O.P., Azizzadeh, M., Aqrawi, A.A.M., Blanc, E., Brockbank, P., Lyslo, K.B., Pickard, N., Salem, L.D., and Svana, T., 2007, *in* Lonergan et al., eds., Fractured Reservoirs: Geological Society of London, Special Publication 270, p. 23–42.

White, I.C., 1885, The geology of natural gas: Science (June and July).

Willis, B., and Willis, R., 1929, Geologic structures (2nd edition): New York, McGraw-Hill, 518 p.

Zahm, C.K., 2002, 3D strain and basement-involved folding, Thermopolis anticline, Wyoming [doctoral thesis]: Golden, Colorado, Colorado School of Mines, 242 p.

MANUSCRIPT ACCEPTED BY THE SOCIETY 16 APRIL 2012

The Geological Society of America
Special Paper 492
2012

Geology from real field to 3D modeling and Google Earth virtual environments: Methods and goals from the Apennines (Furlo Gorge, Italy)

Mauro De Donatis*
Sara Susini
Marco Foi
DiSTeVA—Department of Earth, Life and Environmental Sciences, Università degli Studi di Urbino "Carlo Bo," Urbino, Italy

ABSTRACT

A workflow for digital geological mapping, from fieldwork to multidimensional digital map, has been designed and tested in a sector of the Northern Apennines (Furlo Anticline). Using digital tools, a field mapping campaign was conducted after the organization of conceptual schemas for data capture, storage, and management. Once the data and schemas had been tested to enable a map to be drawn using GIS tools, they were almost ready to be imported into modeling tools for maps, sections, and 3D models. Moreover, we believe that web visualization and distribution using Google Earth is a further step in the direction of knowledge transfer to a greater number of people.

INTRODUCTION

The Importance of Digital Mapping

Over the last few years, field digital mapping, geographic information systems (GIS) and multidimensional modeling have been some of the new ways of producing innovative geological cartography (Bonham-Carter, 1994; McCaffrey et al., 2005; Wawrzyniec et al., 2007).

The use of portable hardware in the field (Tablet PCs, PDAs, and recently also smartphones and iPads) enables the more accurate georeferencing (Global Positioning System receivers) and digital storage (database) of data, information, hypotheses, and interpretations. Their availability in digital format has led to new types of analysis, final map interpretation, and drawing. Moreover, this use of information technology not only considerably boosts the ability of private companies and public agencies to reduce the time and money they spend, but also enables them to reduce uncertainty (Jones et al., 2004) and improve their management of knowledge (Bond et al., 2007). Accordingly, academic research could become increasingly free to explore new methods and tools for digital mapping, ranging from *off-the-shelf* products to *homemade* systems. Many research teams have thus worked on finding their own paths from fieldwork to map elaboration (Briner et al., 1999; Walker and Black, 2000; Akciz et al., 2002; Howard, 2002; Brodaric, 2004; De Donatis et al., 2005).

The next step in the process is knowledge transfer. In the digital era, colored 2D paper maps and sections (geologists are

*mauro.dedonatis@uniurb.it.

De Donatis, M., Susini, S., and Foi, M., 2012, Geology from real field to 3D modeling and Google Earth virtual environments: Methods and goals from the Apennines (Furlo Gorge, Italy), *in* Whitmeyer, S.J., Bailey, J.E., De Paor, D.G., and Ornduff, T., eds., Google Earth and Virtual Visualizations in Geoscience Education and Research: Geological Society of America Special Paper 492, p. 221–233, doi:10.1130/2012.2492(16). For permission to copy, contact editing@geosociety .org. © 2012 The Geological Society of America. All rights reserved.

still fascinated by them) are unable to reach much of the population, including administrators and politicians (Jackson, 2009), but 3D models are frequently self-explanatory and can be visualized on almost any PC and/or TV screen. In the 1990s, 3D modeling techniques in virtual environments were confined almost exclusively to a few industry sectors (cinema, medics, petroleum), because of the expensive hardware and software requirements. Nowadays, a large number of people can use free, online, advanced and 3D augmented geographical reality tools such as Google Earth (Ballagh et al., 2011). As a consequence, new ways of producing maps can be achieved using web-distributed tools, as this paper proposes.

Scope

This project was designed to integrate the main lines of research on mapping techniques adopted by LINEE (Laboratory of Information Technology for Earth and Environmental Sciences of Urbino University) since 1997. Consequently, the purpose of this work is to delineate a form of workflow from field survey to map elaboration in a digital and innovative manner. Digital field mapping is a somewhat recent phenomenon, but in just a few years has become a common approach, despite the number of skeptics and the different ways of working (Athey, 2011). Model building in 3D is a more detailed form of geological cartography. Accordingly, a framework with which to facilitate the delivery of knowledge with multi-dimensional digital tools was created, covering the early stages of data collection to the final results. Finally, the availability of information carriers, such as Google's geographical tools, leads this work in the direction of a much wider form of web distribution (Bailey and Chen, 2011).

Location and Geological Settings

The Furlo Mountains (Monte Pietralata and Monte Paganuccio) in the northern part of the Marche Region in Italy define a small anticline which is geologically located in the Northern Apennines (Fig. 1). The mountains are separated by a narrow gorge by a combination of antecedence and superposition of the cross-cutting Candigliano River (Mazzanti and Trevisan, 1978; Alvarez, 1999). Spectacular outcrops reveal mainly carbonate rocks from a Jurassic rifted pelagic succession to Cretaceous passive margin units up to Miocene foredeep deposits belonging to the early stages of Alpine orogeny affecting this area. The relatively small anticline is detached at the regional Triassic evaporite décollement level, while other important décollement levels in the upper parts of the succession accommodate major shortening, especially in the periclinal areas of the major fold. Normal and strike-slip faults dissect sectors of the main anticline and are related to both the rifting and the shortening stages of the regional structural evolution.

Many authors have worked in this geologically well-known site (Alvarez, 2008, and references therein), and, apart from the

close proximity of our laboratory thereto, this was one of the main reasons that this 40 km² area was chosen for this project.

An additional reason for our choice of site was the fact that our team has previously worked to survey the official national geological map of the area at a scale of 1:50,000 (Cecca et al., 1996b). We were also responsible for building the still unique prototype (for the official national map) of the 3D model of the entire sheet 280-Fossombrone, where the Furlo anticline is located in the southern sector (De Donatis et al., 2002; Borraccini et al., 2004; De Donatis et al., 2009).

Tools: Hardware and Software

LINEE's digital equipment (Fig. 2) includes workstations, Tablet PCs, desktop PCs, and open source and proprietary software exploited in different ways in both the laboratory and fieldwork (Table 1). As a consequence of the rapid development of the digital world, these were not the most up-to-date tools available, but they were nevertheless efficient enough to support the entire range of the study.

Field Tools

With respect to field instruments, a **rugged slate Tablet PC** (Fig. 2A, Xplore ix104, with a Mobile Intel Pentium 866MHz CPU, 504 Mb of RAM and Windows XP Tablet PC Edition Operating System) was utilized with a GPS receiver and connected using a USB Bluetooth dongle or an internal Bluetooth PCMCIA card.

The proprietary **software MapIT** is installed in this type of Tablet PC. This is a mobile GIS which has been available

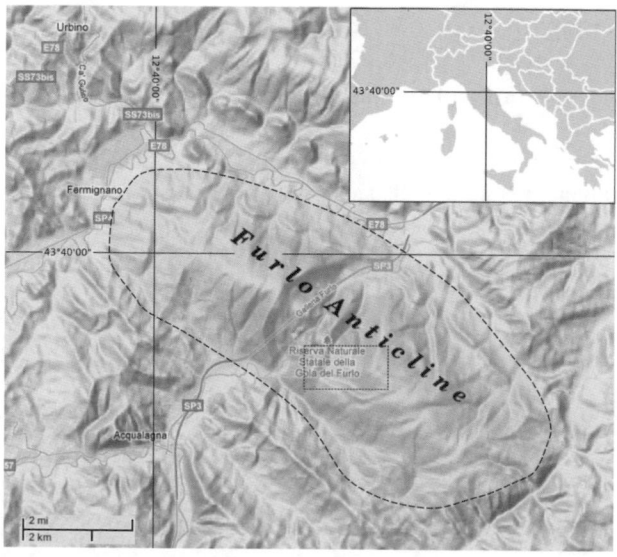

Figure 1. Relief map of the study sector of the Furlo Anticline located inside the dashed line. See also on Google Maps at http://g.co/maps/r3424. The dotted rectangle shows the location of the map in Figure 9.

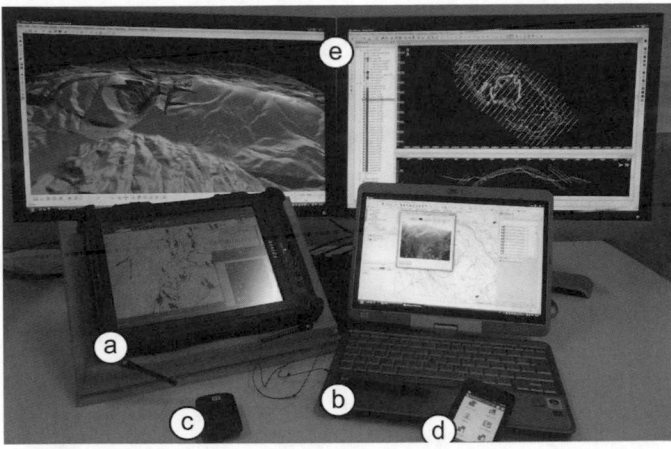

Figure 2. Hardware and software tools: a—rugged Xplore Tablet PC with MapIT; b—HP Tablet PC with BeeGIS; c—GPS receiver; d—Android smartphone with Geopaparazzi; e—workstation with 3DMove model (on the left monitor) and 2DMove (on the right monitor).

commercially since 2004 as a result of a collaboration between LINEE and a software company (De Donatis and Bruciatelli, 2006). It has the following key characteristics:

- A clean graphical user interface which has been carefully designed to provide direct access to the functions that are the most useful in the field;
- Editing tools that operate like those in all desktop GIS but slightly modified to be operated with a stylus;
- The exploitation of ink-technology with the capacity to write and draw with the stylus directly on the screen;
- Several spatial analysis functions with which to perform the most common post-processing tasks (buffering, intersections, etc.);
- Georeferencing tools to assign the correct geographic position to scanned paper maps;

- Reliable GPS support for both positioning and data capturing;
- A user friendly form-editor to easily organize forms, which is helpful for collecting data in the field; and
- "Easy-Note" a sort of sticky note which enables small sketches to be hand drawn, links to be made to larger external documents (spreadsheets, databases, voice recordings, etc.), notes to be taken on embedded pictures and all of this information to be related to a geographic element inside the map.

Unfortunately, the software company suspended the collaboration with LINEE, meaning that software is no longer available. Accordingly, during our research, a parallel project on the development of other mobile GIS software was conducted in our laboratory. To continue our experience of a mobile GIS, we decided to place our trust in the free, open-source, geographic software world by producing **BeeGIS** (De Donatis et al., 2008a; De Donatis et al., 2010), which is compatible with the main operating systems (Windows, Linux, and MacOS). To avoid "reinventing the wheel" we started with uDig, which is open source software that was already available. All of its functionalities (e.g., online map visualization) have been exploited, and a number of plug-ins developed, enabling the main features of its predecessor, MapIT, to be retained and improved upon. Later, when the software was reliable enough to be used in the field, all of the elements of the project were migrated from MapIT to BeeGIS without any major problems.

Due to the increased CPU and RAM requirements when migrating from the old to the newer software, more recent **convertible Tablet PCs** were utilized (Fig. 2B, HP Compaq 2710p, CPU Intel Core Duo 1, 20 GHz, 2GB of RAM, WinVista Business). Although they are not rugged, we used those Tablet PCs because in the study area dust storms do not occur and we do not work in the field during rainfalls. Moreover the screen visibility is largely improved.

TABLE 1. SUMMARY OF DIGITAL TOOLS UTILIZED AND THEIR PURPOSES AND/OR FUNCTIONALITIES

	Hardware	Software	Functionality
Field tools	Rugged Tablet PC (Win XP)	MapIT	Mobile GIS mapping (older, proprietary software)
	Convertible Tablet PC (Win Vista)	BeeGIS	Mobile GIS mapping (newer, open source software)
	Smartphone (Android 2.1)	Geopaparazzi	GPS tracks, geotagged pictures and notes
		eGEO Compass	Digital geological compass/clinometer
	Bluetooth GPS receiver	*firmware*	NMEA signal acquisition
	Digital cameras	*firmware*	Exif JPEG pictures
Lab tools	Desktop PCs (Win XP and Ubuntu)	Desktop GIS (i.e., ESRI ArcGIS)	Standard GIS functionalities
	Workstation PC (Win Vista)	Move 2010 (by M.V.E. ltd)	Modeling package

Note: NMEA—National Marine Electronics Association; Exif—exchangeable image format.

A number of different **GPS receivers** were also used. All of the tested antennas relied on the NMEA 0183 Standard communication protocol, and a 51 channels Bluetooth WASS+EGNOS GPS receiver (Fig. 2C) was employed (high sensitivity receiver, 25 hours of operation).

The **digital photo camera** is an additional useful tool. Compared to old film-based reflex cameras which were used until a few years ago, this device can import (by USB cable or SD card reader) the project images so as to analyze and interpret beds, structures, etc., directly in the field.

In the final period of the study, **Android smartphones** (Fig. 2D, HTC Hero and HTC Magic) became available and our laboratory was able to develop two applications for this device that were useful for surveying purposes:

- **Geopaparazzi** is a lightweight solution for the rapid acquisition of GPS traces, with written and vocal notes and photos which are displayed on an online map (Open Street Map or Google Map). Its files can also be exported in KML (Keyhole Markup Language) files to Google Earth and can likewise be easily imported into BeeGIS, where notes and pictures are immediately georeferenced.
- **eGEO Compass** is a primitive but promising version of the digital compass/clinometer for Android smartphones. By placing the phone on a surface and pressing a finger on the screen, the georeferenced bed orientation data can be collected and exported to any software editing .csv file format.

Laboratory Tools

Although the Tablet PC can be used as a laptop (for the slate version an external USB keyboard and mouse were used), we primarily employed an ordinary desktop computer (operating system: Windows XP or Vista, Linux Ubuntu) in the laboratory.

In particular, and due to the excellent performance of its video card (Nvidia Quadro FX1700), which is frequently used with heavy files, a desktop workstation (Fig. 2E, Dell Precision T7400; CPU Intel Xeon E5430; RAM 4GB) and two wide screen 22″ monitors were utilized to produce a 3D model.

An important software tool for 2D maps, cross-sections, and 3D model drawing was the Move suite by Midland Valley Exploration Ltd. (MVE; see http://www.mve.com/software/move). The choice of this software collection was related (apart from 15 years of experience thereof) to its working method, which adopts the approach of a geologist rather than that of a GIS or CAD operator. This meant that we could easily organize both the collection of data in the field and the database for the next stage of importing this information directly into this modeler software (e.g., dip data).

Other tools used to export the projects to a Google Earth world were the following.

- **Shp2Kml**: a very small piece of portable software that is available with a freeware license (http://www.zonums.com/) and can be used, copied, and distributed although no source code is provided. This tool does not require installation and can be launched directly.
- **Export to KML** for ArcGIS 9.x: a script that is freely available to download from the ESRI support site; it can even be run on a limited ArcGIS 9.x license (http://arcscripts .esri.com/details.asp?dbid=14273).

METHOD

The purpose of this paper is to describe a number of methodological proposals for new types of geological mapping work (De Paor and Whitmeyer, 2011). The article describes our efforts in trying to explain the method we followed in four possible phases, even though the entire process of the workflow design was conducted using a trial-and-error approach. As set out in Table 2,

TABLE 2. WORKFLOW CHART THAT SUMMARISES THE FOUR PHASES
OF THE PROJECT AND THE DIFFERENT STEPS TAKEN

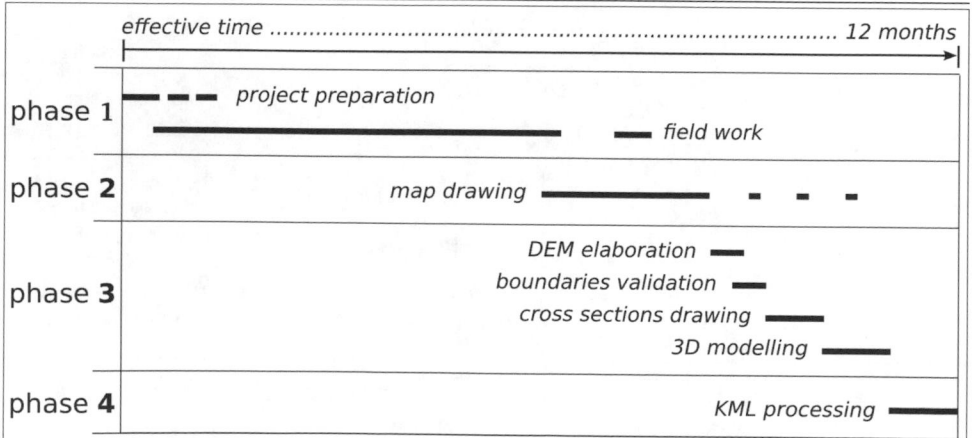

Note: The relative chronology (the effective working time was almost one year) of the single steps reveals some overlap among the different steps and phases.

these steps, reported as elements of a systematic progression, have also been subject to a number of revisions.

First Phase: Project Preparation and Fieldwork

Laboratory Preparation

As geology students, we were always taught to prepare any mapping project very carefully before going out into the field; photocopying a topographic map, gluing a canvas onto the back, checking pencils, and looking for aerial photos and other geological maps of the area were all tasks that were undertaken on a regular basis. In the "digital era" however, this material needs to be particularly well organized. Accordingly, we imported regional 1:10,000 scale topographic vector maps and subsequently transformed them from DWG ("drawing" format from CAD) to Shape files. A scanned geological map (Cecca et al., 1996a) and some aerial photos were also imported and georeferenced as raster (normally TIFF format) files. However, the most important aspect of our preparation was the customization of the mobile GIS project using MapIT. Our initial database organization consisted of defining the main elements to be mapped, including outcrops, bed attitudes, structural surfaces, and unit boundaries, along with their attributes and topological features.

Forms (Fig. 3) for the elements to be mapped were prepared to enable us to easily collect data and information in the field. Symbols and styles for points, lines, and areas were also chosen. Some attributes of the mapping elements were used to automatically draw the field map. For example, stratigraphic unit attributes were employed to color the polygons of the out-

crops and the dip direction in order to define the rotation of the bed attitude symbols.

Moreover, because we are aware that field geologists need to be free to include information, hypotheses and thoughts (De Donatis et al., 2008b), pre-organized forms are not enough for ensuring the correct interpretation of the geology of the studied area. In paper-based mapping, many notes and sketches are reported in the field book and/or directly on the map. Accordingly, to retain this traditional way of working, the ink technology of the Tablet PC was exploited by using its stylus to write and draw on the screen thanks to software such as Windows Journal. Furthermore, a sort of georeferenced sticky note, known as *Easynote* in MapIT (Fig. 4) and *Geonote* in BeeGIS, was utilized to write notes (also using the screen keyboard), draw sketches and import digital photos and draw on them. Moreover, other application files could also be imported, such as spreadsheets or stereoplots where several pieces of data can be easily collected in a structural measurement location.

In the field

During the initial running phase, some corrections were driven onto the database by including attributes that were not taken into account when the project was set up in the laboratory. For example, the distinction of the members of the lithostratigraphic units had to be reorganized as a result of the observations carried out in the first period of fieldwork; this is a typical geological approach that must be taken into account in the digital forms and database.

The GPS acquisition has been an important tool for good quality approximation (less than two meters, on average),

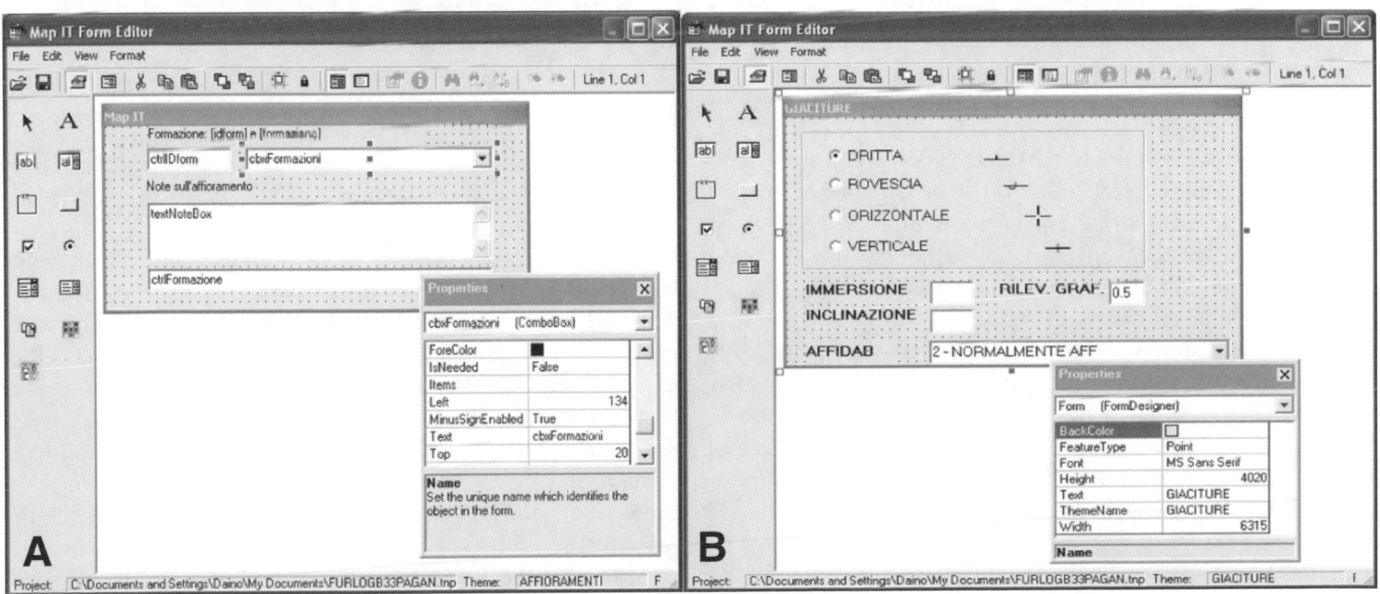

Figure 3. Forms for the field survey created and edited in MapIT: (A) The "AFFIORAMENTI" (outcrops) polygon layer contains a ComboBox control named "cbxFormazioni," the content of which is filled dynamically from a Microsoft Access database table; (B) The custom form created to handle data acquisition for the "GIACITURE" (dip attitudes) point layer. The most notable aspect is the set of four option buttons designed to graphically help the user while selecting the type of measurement being recorded.

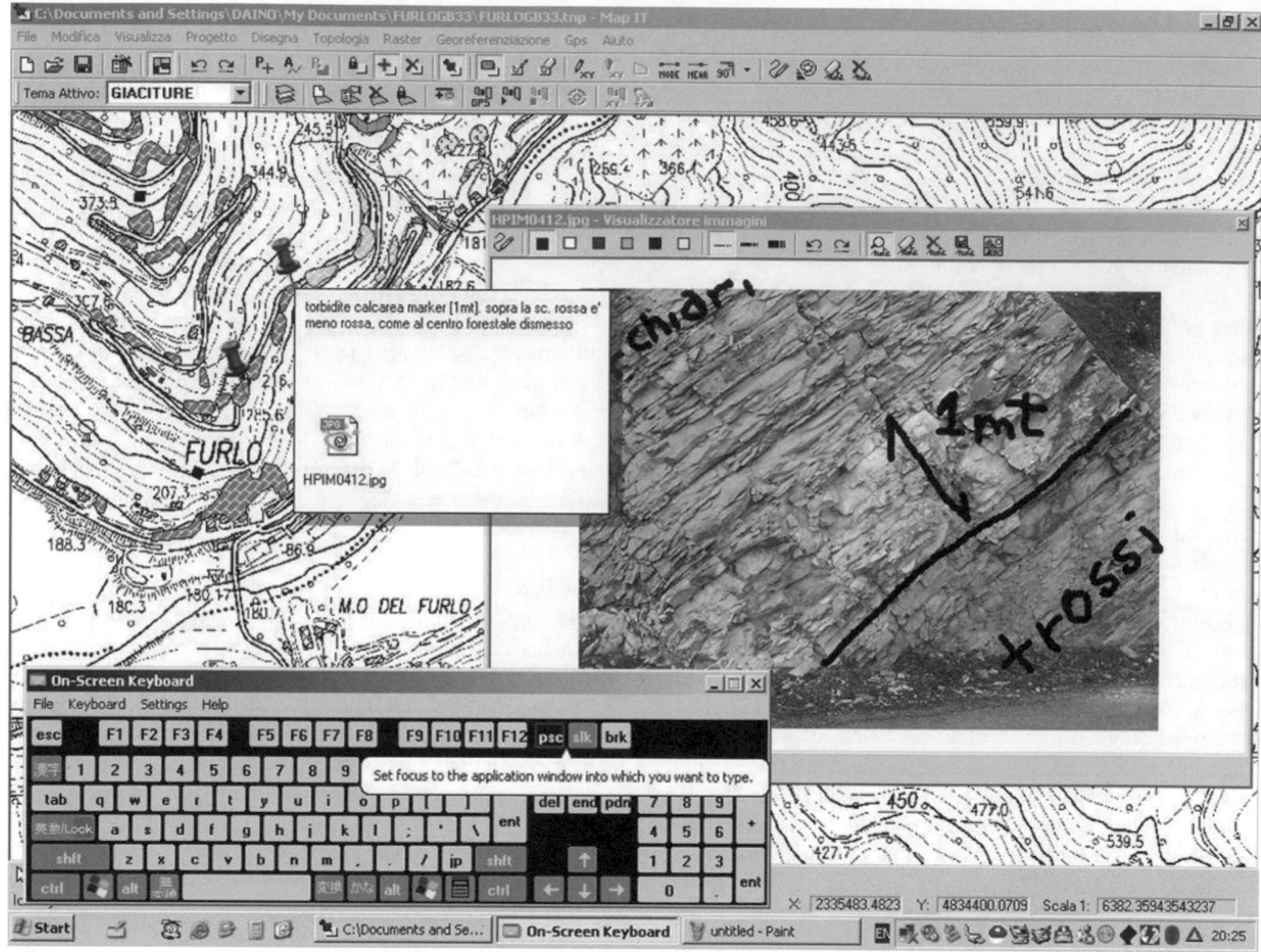

Figure 4. Easy Note is a type of georeferenced sticky note in MapIT for rapid data and information capture.

although there were some remarkable errors which were probably due to satellite constellation on some days, or to position when the work was carried out in the middle of the gorge or under trees. In the areas of the field with poor GPS reception, classical positioning techniques were employed. GPS was frequently used to insert Geonotes where notes, sketches, and digital pictures were stored, and the opportunity to sketch directly on the pictures was also taken. Occasionally, the USB connection of the camera in the field was awkward, meaning that a georeferencing picture tool (with a wizard in BeeGIS) was used in the laboratory, enabling us to georeference a folder of the day pictures in ad-hoc Geonotes by synchronizing the time of the GPS with that one of the photo camera.

In the final period of this study, the applications for the Android smartphones (with camera, GPS, and digital compass), such as Geopaparazzi (http://code.google.com/p/geopaparazzi/), were also used. This enabled us to easily export into BeeGIS both a trace of the journey and georeferenced pictures visualizing also

the shot direction. These Geonotes were collected into the Fieldbook tool of BeeGIS where they could be managed, selected, and edited (Fig. 5).

Because of the specificity of the Geonotes, which relate only to BeeGIS, not all of the elements (notes, sketches, etc.) stored by this tool can be exported to other GIS software. Accordingly, to allow us to share all of this information, a tool called *Dump Geonote* was created. This enabled us to export into a new folder all of the Geonote elements that can be opened with a text editor (.txt), a drawing software (.svg) or an image visualizer (.jpg or other raster formats). When drawing the field map, points, lines, and areas were usually digitized in the appropriate layers (i.e. bedding attitude, unit boundaries, outcrops) using the Tablet PC's stylus. In the case of the faster or tentative drawing of lines (for example, a hypothetical fault), the *Annotation tool* could be employed, which enables the surveyor to draw lines on the map onto an automatically created layer. This layer does not, however, correspond to similar tools

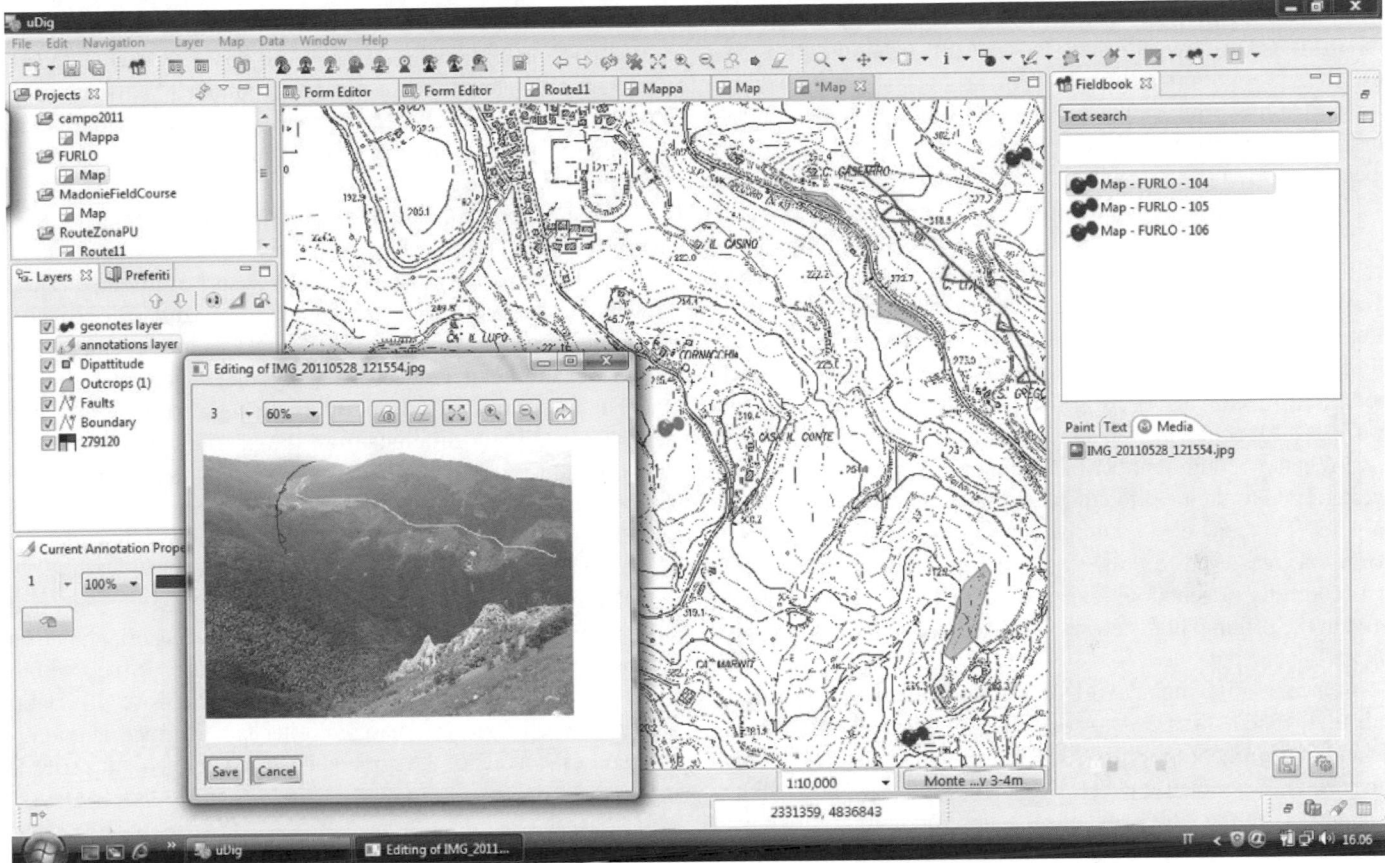

Figure 5. The Fieldbook (on the right side of the BeeGIS window) is the easily accessible collection of Geonotes. The selected Geonote shows the media collector and a picture interpreted from it.

in other GIS software and cannot therefore be exported. Nevertheless, the annotations were very useful during the interpretation phase.

Second Phase: Map Elaboration in GIS Environment

One of the final outcomes of the first working phase was the availability of conceptual schemas for data collection and physical implementation within a software environment. Thereafter, the main purpose of the second phase was to compile a geological map of the surveyed area by employing the tools developed to perform field surveys and post-processing operations.

During the first stages of the fieldwork, the practical use of the functionalities implemented in the mobile GIS platforms highlighted the existence of a number of problems due to either the software tools or the data schemas.

Once the entire area had been surveyed, a significant amount of data was available in digital format inside the mobile GIS platform. The majority of data were already in the standard ESRI Shapefile format (.shp) and were thus ready to be transferred to other tools for subsequent activities. Despite this, both of the adopted pieces of software were tested to push the

workflow further towards the production of a two-dimensional geological map.

Specifically, the process of map creation from the field data required two further steps: field data validation and map data structure implementation.

All information was checked against null attribution errors, with the purpose of this key data validation control being to spot the features that completely lacked their main attributes. In this way, outcrops with the formation attribute set to null, or boundaries with null type, could be easily located and repaired.

Next, during the data representation structure implementation stage, layers which would be filled with geometric elements for the final 2D map were formulated in the GIS platform. The production of such data structures involved the creation of a set of corresponding Shapefiles. This task was accomplished by relying on the basic facilities for layer creation and management that are available in GIS platforms, and usually starts with doubling the schema of a few available field data structures. Since the creation of a geological map requires at least one layer for depicting the geologic formations (polygons) and one for the lithologic/tectonic boundaries (polylines), these data structures were copied from the outcrop layer and the field boundary layer, respectively.

This was possible because the geometries were the same type, and the reuse of some attributes was feasible, whereas the useless attributes were omitted.

Since this process was performed using the same hardware and software that was utilized in the field, it was possible to define specific areas of the map when the available data were inadequate for acceptable interpretation purposes.

Before redrawing the map in the laboratory, we return to the field and check directly for inconsistent or missing data. There were two main benefits of doing so: (1) mistakes could be reinterpreted and corrected directly in the field with all of the necessary digital tools; and (2) previous inconsistent interpretations could be stored in GIS layers as a history of the geologist's thought processes for achieving the synthesis of the final map.

When a certain number of inconsistencies were recognized, we returned to the field to collect further data and resolve uncertainties. This new data could be collected in the same GIS platform that was used to draw the final map, and this information was carefully attached with specific notes. Any mistake discovered in the original field dataset was fixed and noted in the appropriate log attribute.

Once a satisfying network of boundaries had been produced, the final extent of the map area was defined using a dummy boundary to encompass the entire area within a closed polyline. Next, a geometric check was carried out on the polylines. Thereafter, in the polygonization stage, polygons were created which depicted the extent of the rock formation across holes in the boundary network.

The final outcome of the second phase was a digital 2D geological map composed of a formation layer beneath a network of stratigraphic and tectonic boundaries, as well as a set of field layers that could be queried.

Third Phase: *n*-Dimensional Modeling

In this phase, the *structural modeling and analysis tool kit* of Move suite (version 2010.1) was used to build a volumetric 3D model. The work began with the digital collection of a georeferenced dataset of dip data (point element) and geological boundaries, limits, and faults (linear elements). The dataset was thus suitable for importation into the Move environment.

The model building procedure consisted of four steps:
1. Field data importation and conversion from the GIS .shp to the .mve format (4DMove and 2DMove).
2. Field data projection onto a digital elevation model (DEM) and validation of geological boundaries (3DMove).
3. Drawing geological cross-sections (3DMove and 2DMove).
4. 3D model building using limits and cross-sections (3DMove).

Importing Data

The dip.shp file is loaded into a 4D Move environment where all dip data become available as a single element "vertex cloud."

The vertex cloud is saved as a .csv file and then converted to a proprietary .mve file using 4DMove. In this step, each data column in the .csv text file is assigned an "internal" attribute (e.g., dip, azimuth). A further step is necessary if a data set contains overturned beds, as younging information must be edited manually in Move.

By launching 2DMove and importing the .mve file, all measurements are automatically displayed as dip measurements. An algorithm enables to automatically scan the different values available in the "Horizon" column and to assign a color to each of them. With the "Update Model" tool, all measurements are colored according to the geological unit they represent.

The procedure for importing geologic boundaries with 4DMove is slightly different from the previous process, but is more streamlined because values such as Dip and Azimuth do not need to be defined. Any kind of polylines drawn in the GIS environment (e.g. unit boundaries, faults, or outcrop boundaries) are treated just like linear objects.

During the import phase in 4DMove, GIS files are converted from .shp to .mve format, while attributes such as horizon names can be exported since they have been previously defined within the GIS. At this point, lines can be edited in 2DMove. If a database is created from the map, coloring is performed automatically after the name of the horizon has been defined, thus speeding up the process.

DEM, Field Data, and Geological Boundaries Validation

Once the field data have been loaded in the modeling framework, the next step is to import the base map and DEM in 3DMove. This software generates the topographic surfaces as triangular irregular networks (TIN) from the elevation contour lines. With the tessellation algorithm, the TIN is created starting from the nodes of the contour lines, which are assigned X, Y, and Z attributes. Then, georeferenced dip data are inserted onto DEM, acquiring Z attributes by shifting them onto the surface with the "Unfold to Target" tool.

During the construction of the 3D model, it is possible that some minor errors could have been introduced and a check for these can thus be conducted. 3DMove also permits there to be a check of the drawing accuracy of the geological boundaries using an algorithm that interpolates the DEM (surface) with stratigraphic contacts and faults (lines). The geometry of any linear element plunging into the ground is transformed into a "ribbon" which is a long and narrow surface displaying the geometry of unit boundaries under the topographic surface. Where this analysis indicates rapid changes in ribbon orientation, this often means that either the line of contact is probably incorrectly drawn compared to the dip data, or that the tectonic structures are not recognized. At the end of this step, the geological boundaries must be redrawn in 2DMove and checked again in 3DMove.

All of these operations are facilitated by the Move suite, where any edited element can be easily and quickly transferred between a 2D and 3D environment.

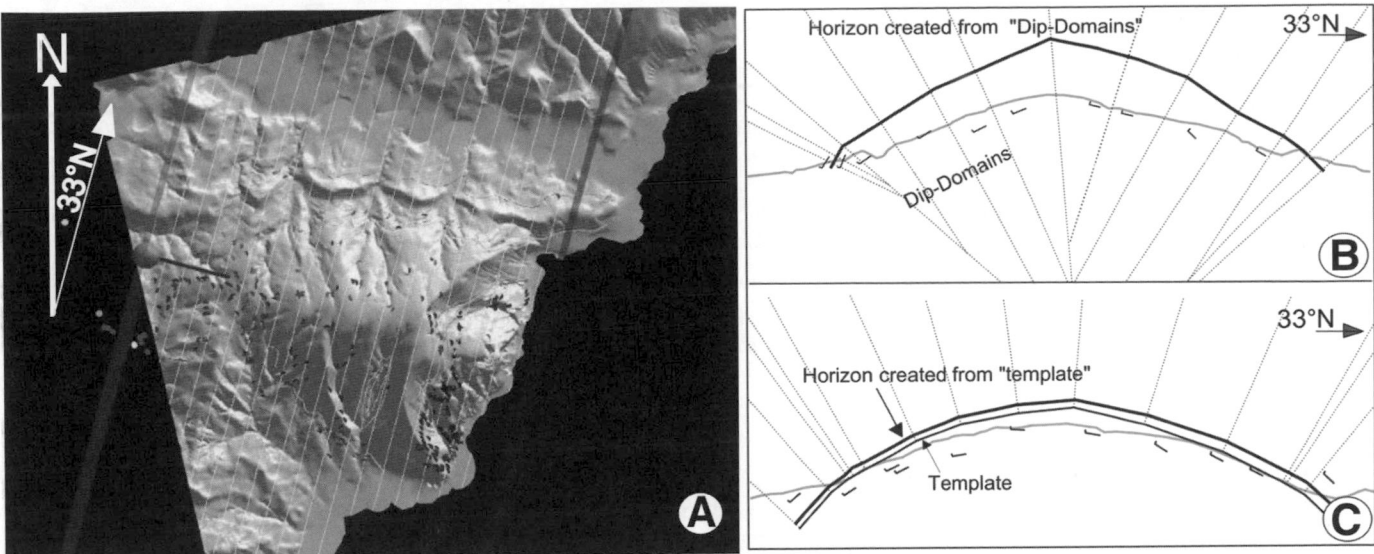

Figure 6. Cross-sections normal to the Furlo Anticline axis (parallel view mode). (A) Map view (in 3DMove) of northwestern part of the anticline where the cross-section series defined by the operator are traced. (B) 2DMove section view; the template limit (black line) is automatically generated taking into account the dip-domains. (C) 2DMove section view; the white horizon is generated starting from a template limit (grey line), based on parallel folding (with shear-angle = 0°) algorithm.

Cross-Sections from 3D

By using the section drawing tool in 3DMove, the geologist can create geological cross-sections from the 3D model. The number, spacing, and orientation of the traces of cross-sections can be chosen depending on the orientation of the geological structure (Fig. 6A). A dozen sections were drawn normally to the axis of the Furlo Anticline. Once visualized in the 2D map mode, the dip data can then be projected onto sections according to the algorithm chosen by the operator. Template limits were automatically generated by the software in the 2D section mode and took into account the "dip-domains method" (Fig. 6B).

Horizons were generated starting from template limits with a parallel folding algorithm (0° shear-angle) which creates horizons from a "custom list." This list is a table where the layer heights are defined (Fig. 6C). The thickness of the rock formations was considered to be constant in the model unless evidence of tectonically driven thickness changes was provided.

Surfaces and 3D Model Construction

Each geological surface was individually built by correlating between cross-sections and observing the limits of the boundaries on the DEM (Fig. 7A). When a stratigraphic surface was complete, it was used to build the next surface when the unit thicknesses were known (Fig. 7B). Bounding surfaces, such as faults, were built independently by stratigraphic surfaces using the tessellation algorithm.

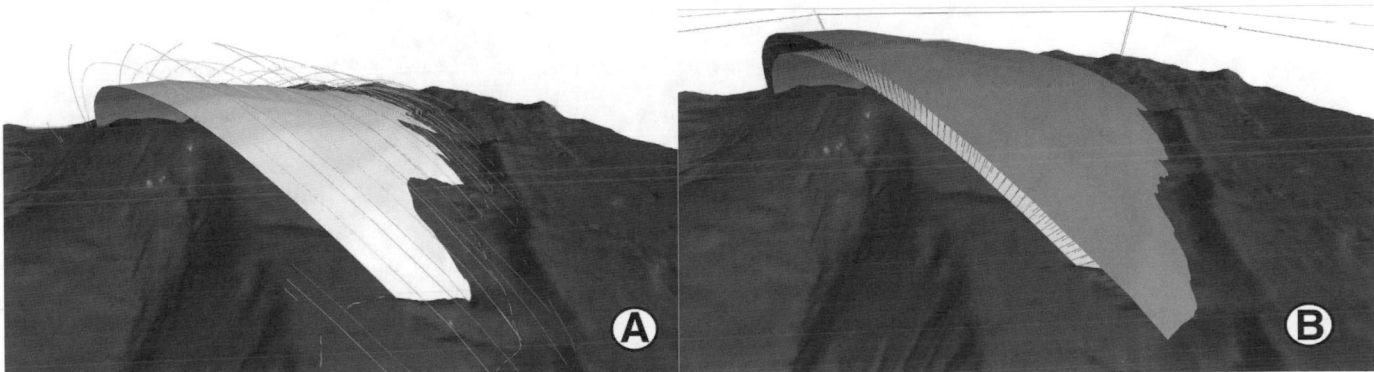

Figure 7. Monte Pietralata anticline structure construction in 3DMove (digital elevation model is dark grey irregular surface). (A) The whitish surface is generated from two or more lines representing the same element in the stratigraphic sequence with the tessellation algorithm. (B) The upper grey horizon is created starting from the previous surface with the "Construct fold surface from" tool with a parallel fold geometry.

The surfaces were drawn following two different algorithms. The first method involves the construction of a surface from two or more lines representing the same element in the stratigraphic sequence. The second produces the creation of a new from a pre-existing surface (constructed with the previous algorithm). The construction of fold surface from the template surface tool generates a plane that is parallel to the template surface once the distance between the two surfaces is set and the operator has decided between parallel or similar fold geometry and the shear angle.

Accordingly, during the development of the model, the one-way path process (from a 2D data field to a 2D geological map and a 3D geological model) was not followed. Meanwhile, the tools and results were tested moving from a 2D to a 3D environment and vice versa.

The resulting 3D model (Fig. 8) produced at this stage can now be used for further Move analysis or for visualization using the MoveViewer, a free download of which is available on the website www.mve.com.

Fourth Phase: Web Publication

In a traditional form of workflow, field maps are passed to a professional designer to produce an accurate map from the available vector geometries. Although the workflow could come to an end at this stage, our aim was nevertheless to take the process one step further by making our products more suitable for digital distribution. Various different technologies like web mapping services could have been adopted to pursue this aim. However,

to reduce the complexity of the workflow and loosen the ties to enterprise software frameworks, a file-based dissemination approach was chosen. The two file formats we considered were the .vrlm and the .kml, both of which are open standards that use regular ASCII text files. The main difference between the two formats is in the intrinsic 3D information delivered by the .vrml format, which allows for a true representation of physical surfaces in a defined three-dimensional space. Unfortunately, little information was available on the file formats used by the MidlandValley Move suite to store the 3D geological models from which the .vrml files should have been derived. This forced us to revert to the .kml file format and scale down from the notion of disseminating true 3D data.

The adoption of this format provided a simple way to merge custom information with the high-resolution satellite imagery obtained from Google web services. Despite being based on 90 m cells, the terrain model available in the application proved to be adequate for delivering an effective "2.5-dimensional" environment for exploring the geological data. However, because Google Earth is a surface viewing tool with no facility to display subsurface perspectives, it is not a true 3D viewing device (Jones et al., 2008). It is nevertheless adequate for both observing the stratigraphic boundaries running on the surface from a new perspective and correlating their course with the bedding data. More generally, it also greatly helps with comprehension of geological maps, especially in the case of inexperienced users, as it does not require the ability to mentally rebuild topographic relief from contour lines.

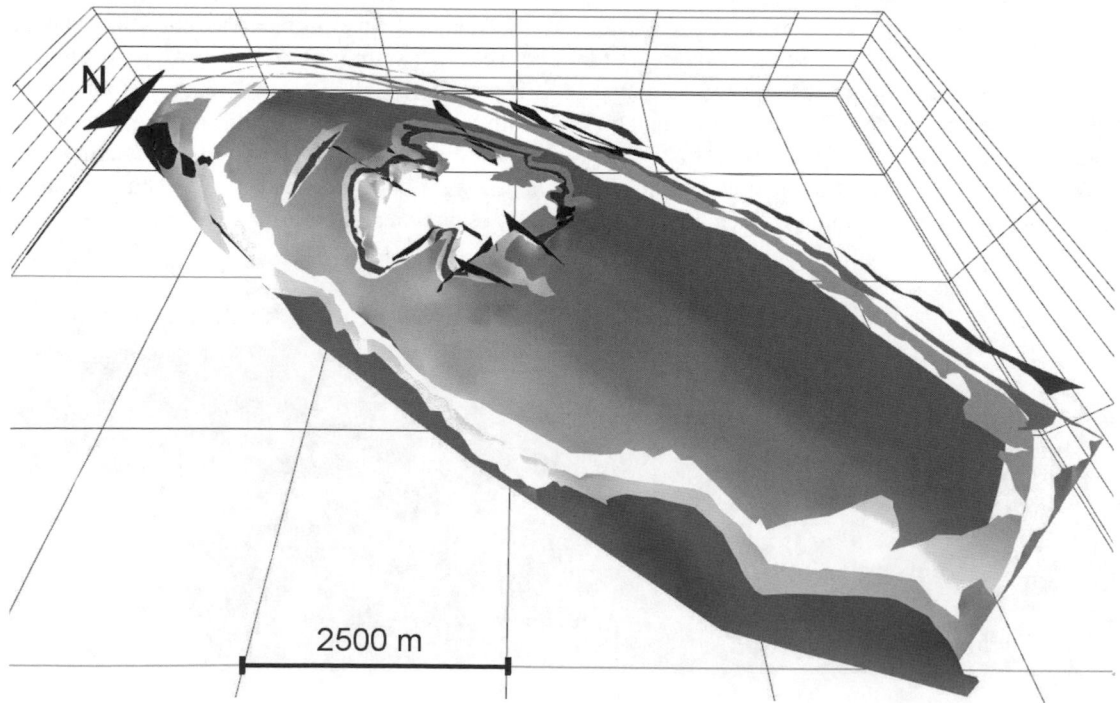

Figure 8. Furlo Anticline structure in 3DMove (parallel view mode) within the UTM box. The surfaces with different grey tones represent contact between geological units of the outcropping succession. Black planes are faults.

Creation of the KML Map

The creation of the complete map as a KML file was subdivided into the conversion of the single input layers arising from the second phase of this work (map elaboration in GIS environment).

All of the 2D maps from previous stages were represented in the local Italian Spatial Reference System (SRS) named "Gauss-Boaga." This system had to be transformed because Google Earth can only handle KML containing geometries in the WGS84 geocentric coordinate system.

We again employed BeeGIS for this purpose. This is because it provides a standard facility for SRS management since it relies on the European Petroleum Surveying Group (EPSG) database to ensure the most complete and up to date SRS list.

Once all data were finally available in the WGS84 SRS, the files were converted to KML format using "Shp2Kml" freeware.

This software has a simple interface that guides the user through the process of KML creation, from the selection of the input Shapefile to the setup of all of the rendering parameters. It also allows for the creation of custom styling that is based on unique values, which is the most appropriate method for rendering both geologic boundaries (polylines) and formations (polygons).

We were able to perform this task quickly, and the resulting KML could even be spatially queried in Google Earth. Clicks on a geographic element triggered the appearance of an information window containing a pre-formatted table of all of its textual attributes. This avoided the need to create obtrusive legends that float on the digital-globe window and prevent the full use of the display area, which has great benefits when it comes to smaller screen sizes.

The only major issue with the use of "Shp2KML" was the erroneous rendering of holes in formation polygons, resulting in the production of odd, overlapping shapes. This problem had to be hand-fixed by editing the wrongly encoded KML file using a text editor to convert "MultiGeometry" entities into holed polygons.

The Shp2KML tool had to be abandoned when the need for icon rotation support arose when trying to convert dip data to KML. As such data is basically "point data" depicted by simple icons rotated according to an angle (azimuth angle/strike angle), the lack of a "rotate based on attribute" feature is unacceptable.

The most affordable solution was the "Export to KML" ArcObject script. This tool succeeded in replicating, in the KML output, the custom styling applied within ArcGIS to the dip data layer. Despite the complexity of the classification conditions, the script was able to perform quite well. The different kinds of customized icons were correctly applied to the various dip data classes (vertical, horizontal, overturned, etc.). The rotation angle was also preserved, resulting in a KML with a satisfying rendering in Google Earth (Fig. 9).

Only one problem could not be resolved: the generated KML was indeed a 2D layer and the dip data symbols did not own the real 3D orientation in space of the surfaces they represented.

To resolve this issue, some more complex tools should had been introduced into the workflow to render the dip data as 3D models with effective spatial orientation. Although this was possible, the decision was made to stop at this stage and continue with the "best-effort" approach that characterized the entire workflow.

After each layer had been created as a stand-alone KML document, everything was assembled in a single file. This was done directly within the Google Earth environment. Each file was simply loaded in the viewer and the vertical order of the layers was decided by dragging them to compose the definitive arrangement. Some titles were changed to enhance immediate understanding of the content and some layers were even turned off to initially display only the most relevant ones to users. This stack of layers was finally saved as a KMZ file to benefit from the compression it performs on data. It was later made available for download (http://www.mcfoi.it/furlo _geology_kmz/).

CONCLUDING REMARKS

The procedure explained in this paper outlines an integrated approach to up-to-date cartography for geology. The traditional way of surveying an area, ranging from the use of a pen/stylus to keep "the feel of a pencil on the map" to a visualization of appropriate symbols on a colored map, have been taken into account. However, the new digital methods and available tools allow for more accurate positioning, the easier management of large quantities of data and a wider dissemination of knowledge. By using digital ink technology with a pen-based PC and a GPS, a field geologist is able to not only acquire data more accurately, but also store interpretations and keep track of working hypotheses.

The process of synthesizing and drawing a "final" map is not only facilitated, but also enhanced, by using digital tools; different GIS layers can be visualized, managed, and merged with others to gather and add more information to the classic geological sheet.

Moreover, n-D modeling (from two dimensions of maps and cross-sections, to a 3D geometrical construction, up to a 4D time dependent evolutionary model) can exploit the data that have been digitally acquired. In addition to using these models for advanced analysis (ex. fracture modeling of fluid reservoir), they are often also self-explanatory objects that can be demonstrated to a larger number of people who are unfamiliar with classical geological maps.

This goal is more easily achieved if web-based visualization tools are employed. For this purpose Google Earth is a powerful engine that enables us to share geological and territorial knowledge beyond simply earth sciences scientists and professionals.

Throughout the entire sequence of work, open-source applications were preferred to commercially available software. The reason for this related to the structured customization of the tools by users and the worldwide collaborative system of development. Moreover, open-source codes and software often have free public distribution, and the economic advantages thereof are evident.

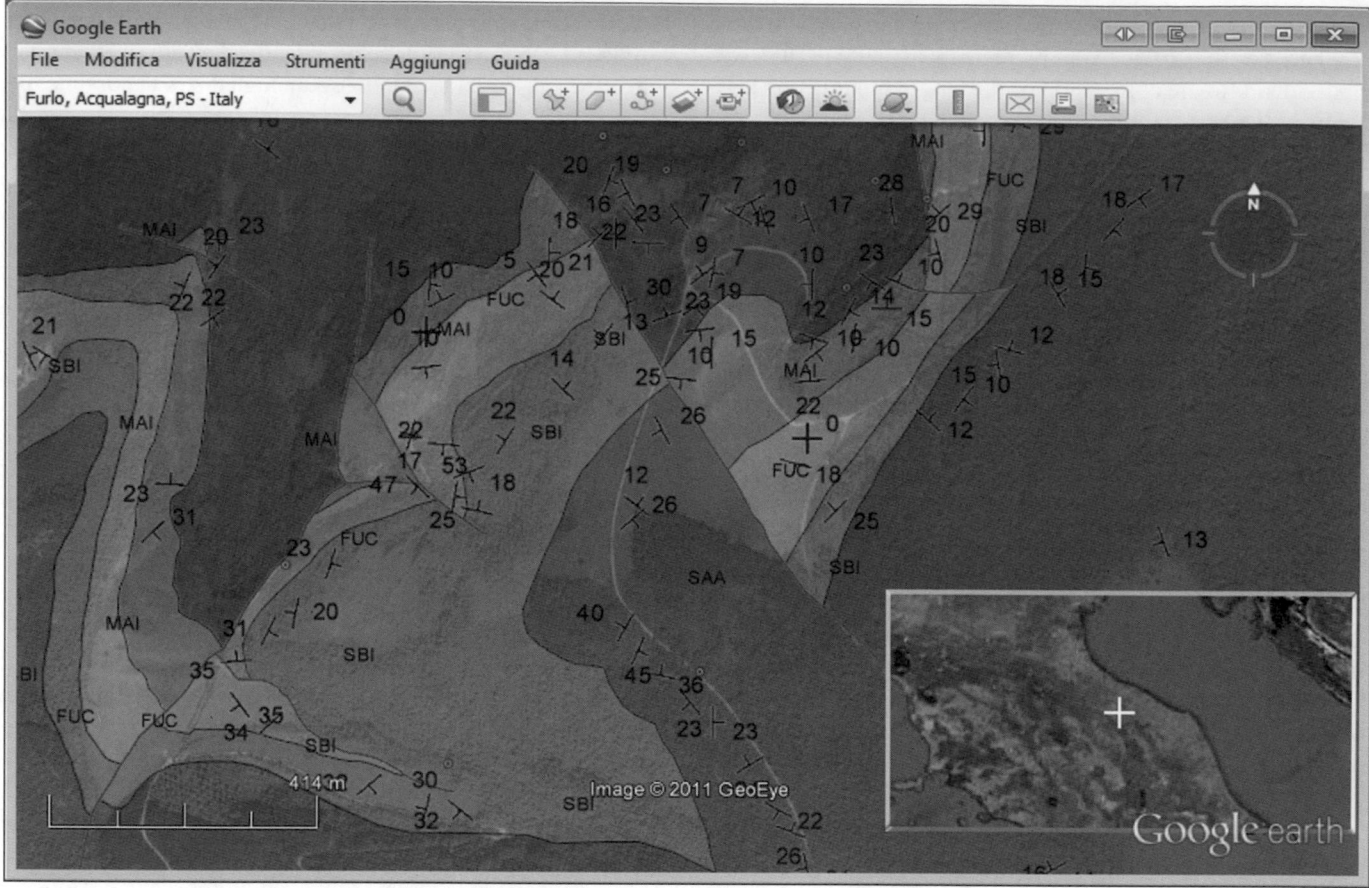

Figure 9. The "GIACITURE" (dip attitudes) layer created by the "Export to KML" free script for ArcGIS 9.x and reproduced in Google Earth 5.1.

On the other hand, the ceaseless, ongoing development of the tools requires knowledge of them to be kept up to date. For the same reason, the software employed in this research was replaced by newer versions due to its continuous development.

Another limitation is the fragmentation of the tools that are available in the different software, which forced us to look for assistance in a number of applications. On the assumption that both hardware (e.g., weight of rugged tablet PCs) and software will become more integrated and user-friendly in due course, our team will continue to build and develop new concepts and relative procedures for geological mapping.

ACKNOWLEDGMENTS

We are grateful to the editors and the reviewers who helped us to improve the quality of the text. We acknowledge the 2008 HP Teaching Grant Program which awarded us with the Tablet PCs utilized in this work. We also acknowledge MVE Ltd (Glasgow-UK) for the availability of the MOVE package and the founder and leader, Dr. Alan Gibbs, for the friendly collaboration.

REFERENCES CITED

Akciz, S.O., Sheehan, D.D., Niemi, N.A., Nguyen, H., Hutchinson, W.E., Carr, C.E., Hodges, K.V., Burchfiel, B.C., and Fuller, E., 2002, What does it take to collect GIS data in the field?: Geological Society of America Abstracts with Programs, v. 34, no. 6, p. 419, abstract 185-29, http://gsa .confex.com/gsa/2002AM/finalprogram/abstract_43554.htm.

Alvarez, W., 1999, Drainage on evolving fold-thrust belts: a study of transverse canyons in the Apennines: Basin Research, v. 11, p. 267–284, doi:10.1046/j .1365-2117.1999.00100.x.

Alvarez, W., 2008, The Mountains of Saint Francis: Discovering the Geologic Event that shaped the Earth: New York, Norton W.W. & Company Inc., 413 p.

Athey, J.E., 2011, Final Results from 2010 Digital Field Mapping Technology Survey: Digital Mapping Techniques '09, poster presentations, U.S. Geological Survey Open-File Report at http://ngmdb.usgs.gov/Info/dmt/ DMT11presentations.html.

Bailey, J.E., and Chen, A., 2011, The role of Virtual Globes in geoscience: Computers & Geosciences, v. 37, p. 1–2, doi:10.1016/j.cageo.2010.06.001.

Ballagh, L.M., Raup, B.H., Duerr, R.E., Khalsa, S.J.S., Helm, C., Fowler, D., and Gupte, A., 2011, Representing scientific data sets in KML; methods and challenges: Computers & Geosciences, v. 37, p. 57–64, doi:10.1016/j .cageo.2010.05.004.

Bond, C.E., Shipton, Z.K., Jones, R.R., Butler, R.W.H., and Gibbs, A.D., 2007, Knowledge transfer in a digital world: Field data acquisition, uncertainty, visualization and data management: Geosphere, v. 3, p. 568–576, doi:10.1130/GES00094.1.

Bonham-Carter, G.R., 1994, Geographic information systems for geoscientists modeling with GIS: Computer methods in the geosciences: Oxford, Pergamon Press, Merriam D.F., v. 13, 398 p.

Borraccini, F., De Donatis, M., Pantaloni, M., and D'Ambrogi, C., 2004, Il Foglio 280—Fossombrone 3D: un progetto pilota per la cartografia geologica nazionale in tre dimensioni: Bollettino della Società Geologica Italiana, v. 123, p. 319–331.

Briner, A.P., Kronenberg, H., Mazurek, M., Horn, H., Engi, M., and Peters, T., 1999, Fieldbook and geodatabase-tools for field data acquisition and analysis: Computers & Geosciences, v. 25, p. 1101–1111, doi:10.1016/S0098-3004(99)00078-3.

Brodaric, B., 2004, The design of GSC fieldLog: ontology-based software for computer aided geological field mapping: Computers & Geosciences, v. 30, p. 5–20, doi:10.1016/j.cageo.2003.08.009.

Cecca, F., Catenacci, V., Conte, G., Cresta, S., D'Andrea, M., Graziano, R., Menichetti, M., Molinari, V., Pampaloni, M.L., Pantaloni, M., Pichezzi, R.M., and Rossi, M.G., 1996a, Carta geologica dell'area dell'Anticlinale di M. Pietralata–M. Paganuccio (Furlo): Istituto Poligrafico e Zecca dello Stato—Settore Cartografia, Roma, scale 1:10.000, 1 sheet.

Cecca, F., Catenacci, V., Conte, G., Cresta, S., D'Andrea, M., Graziano, R., Menichetti, M., Molinari, V., Pampaloni, M.L., Pantaloni, M., Pichezzi, R.M., and Rossi, M.G., 1996b, Preliminary results of the survey on the South-Western area of the Sheet n. 280 Fossombrone of the Geological Map of Italy: Bollettino del Servizio Geologico d'Italia, v. 115, p. 3–70.

De Donatis, M., and Bruciatelli, L., 2006, MAP IT: The GIS software for field mapping with tablet pc: Computers & Geosciences, v. 32, p. 673–680, doi:10.1016/j.cageo.2005.09.003.

De Donatis, M., Jones, S., Pantaloni, M., Bonora, M., Borraccini, F., Gallerini, G., and D'Ambrogi, C., 2002, A National Project on Three-Dimensional Geology of Italy: Sheet 280—Fossombrone in 3D: Episodes, v. 25, p. 29–32.

De Donatis, M., Bruciatelli, L., and Susini, S., 2005, MAP IT: a GIS/GPS Software Solution for Digital Mapping, *in* Proceedings, Digital Mapping Techniques '05—Workshop, Baton Rouge, Louisiana: U.S. Geological Survey, p. 97–101.

De Donatis, M., Antonello, A., Foi, M., Foresto, C., Franceschi, S., and Susini, S., 2008a, BeeGIS: a new open source tool for mobile GIS applications: Rendiconti online della Società Geologica Italiana, v. 8, p. 46–49.

De Donatis, M., Susini, S., and Delmonaco, G., 2008b, Digital geologic mapping methods: from field to 3D model: International Journal of Geology, v. 2, p. 47–52.

De Donatis, M., Borraccini, F., and Susini, S., 2009, Sheet 280—Fossombrone 3D: a study project for the new geological map of Italy at the scale of 1:50,000 in three dimensions: Computers & Geosciences, v. 35, p. 19–32.

De Donatis, M., Antonello, A., Lanteri, L., Susini, S., and Foi, M., 2010, BeeGIS: a new open source a multiplatform mobile GIS, *in* Proceedings, Digital Mapping Techniques—Workshop: U.S. Geological Survey, http://pubs.usgs.gov/of/2010/1335/pdf/usgs_of2010-1335_DeDonatis.pdf.

De Paor, D.G., and Whitmeyer, S.J., 2011, Geological and geophysical modeling on virtual globes using KML, COLLADA, and JavaScript: Computers & Geosciences, v. 37, p. 100–110, doi:10.1016/j.cageo.2010.05.003.

Howard, A.S., 2002, Capturing digital data in the field—The British Geological Survey's SIGMA project: digital field data capture in a corporate context, *in* Proceedings, Capturing digital data in the field Workshop 2002: British Geological Survey, http://www.bgs.ac.uk/science/dfdc/ahoward.html.

Jackson, I., 2009, 174 years and you still haven't finished?—Do geological surveys have a role in the 21st century knowledge economy? *in* Proceedings, 6th Euregeo 2009: Bayerisches Landesamt für Umwelt, v. 1, p. 3–6.

Jones, R.R., McCaffrey, K.J.W., Wilson, R.W., and Holdsworth, R.E., 2004, Digital field data acquisition: Towards increased quantification of uncertainty during geological mapping, *in* Curtis, A., and Wood, R., eds., Geological prior information: Informing science and engineering: Geological Society of London Special Publication 239, p. 43–56.

Jones, R.R., Wawrzyniec, T.F., Holliman, N.S., McCaffrey, K.J.W., Imber, J., and Holdsworth, R.E., 2008, Describing the dimensionality of geospatial data in the earth sciences—recommendations for nomenclature: Geosphere, v. 4, p. 354–359, doi:10.1130/GES00158.1.

Mazzanti, R., and Trevisan, L., 1978, Evoluzione della rete idrografica nell'Appennino centro-settentrionale: Geografia Fisica e Dinamica Quaternaria, v. 1, p. 55–62.

McCaffrey, K.J.W., Jones, R.R., Holdsworth, R.E., Wilson, R.W., Clegg, P., Imber, J., Holliman, N., and Trinks, I., 2005, Unlocking the spatial dimension: Digital technologies and the future of geosciences fieldwork: Journal of the Geological Society, v. 162, p. 927–938, doi:10.1144/0016-764905-017.

Walker, J.D., and Black, R.A., 2000, Mapping the outcrop: Geotimes, v. 45, p. 28–31.

Wawrzyniec, T.F., Jones, R.R., McCaffrey, K.J.W., Imer, J., Holliman, N., and Holdsworth, R.E., 2007, Introduction: Unlocking 3D earth system-harnessing new digital technologies to revolutionize multi-scale geological models: Geosphere, v. 3, p. 406–407, doi:10.1130/GES00156.1.

MANUSCRIPT ACCEPTED BY THE SOCIETY 16 APRIL 2012

The Geological Society of America
Special Paper 492
2012

Creating interactive 3-D block diagrams from geologic maps and cross-sections

Paul Karabinos
Department of Geosciences, Williams College, Williamstown, Massachusetts 01267, USA

ABSTRACT

Geologic maps and cross-sections effectively summarize the structural geology of a region, but they can be difficult for non-geologists to interpret. Textbooks and interpretive guides commonly integrate maps and cross-sections into static perspective block diagrams to help novices visualize basic concepts in geology. The inherent power of block diagrams, however, is dramatically increased by software such as Google SketchUp, a free downloadable program, which can create interactive 3-D models of a region. The stand-alone models can be Rotated, Panned, and Zoomed by the user and exported for animations. An efficient way to create block diagrams is to combine the individual strengths of dedicated GIS software with SketchUp, and merge the results into a single 3-D model.

Effective 3-D block diagrams drape a geologic map on a digital elevation model and show how the map and cross-sections connect at the topographic surface. Creating block diagrams in such a way that portions of the map between cross-section planes are independent segments gives the user flexibility to make portions of the map invisible. By "turning off" parts of the surface, it is possible to sequentially reveal multiple cross-sections. 3-D block diagrams help students and non-specialists visualize geologic structures. Once created, the 3-D block diagrams can be quickly edited by substituting alternate images of geologic maps and cross-sections. Thus they provide an elegant approach for comparing different interpretations of a region. Combined with tools available in SketchUp, they also provide geologists with a valuable resource for assessing the geometric plausibility of geologic cross-sections.

INTRODUCTION

Traditional block diagrams use one- or two-point perspective and hand drafting expertise to combine a map and one or two vertical cross-section views of a region (Lobeck, 1924). Geologists typically combine geologic maps and cross-sections into block diagrams to portray the structure of a region. Geographic information system (GIS) programs and 3-D modeling software now make it possible to construct accurate interactive digital block diagrams that can be widely disseminated.

Geologic maps and cross-sections are fundamental to many aspects of earth science, and they compactly summarize many observations and interpretations. Yet geologists commonly fail to appreciate how difficult it is for non-specialists to understand how to use them, or even what they represent (Kastens et al., 2009). One of the most valuable functions of block diagrams

Karabinos, P., 2012, Creating interactive 3-D block diagrams from geologic maps and cross-sections, *in* Whitmeyer, S.J., Bailey, J.E., De Paor, D.G., and Ornduff, T., eds., Google Earth and Virtual Visualizations in Geoscience Education and Research: Geological Society of America Special Paper 492, p. 235–251, doi:10.1130/2012.2492(17). For permission to copy, contact editing@geosociety.org. © 2012 The Geological Society of America. All rights reserved.

is to illustrate what geologic maps and cross-sections represent to non-specialists. In this regard, the value of a map and cross-sections integrated into a 3-D model far surpasses what can be explained with separate 2-D images. The power of block diagrams increases significantly when they are constructed in SketchUp, a free downloadable program, as 3-D animations or interactive models, which can be **Rotated**, **Panned**, and **Zoomed**. They can be used for teaching, museum displays, and presentations to policy makers with limited geologic background. They are also valuable for presentations to specialists because structural interpretations can be absorbed more quickly from block diagrams than is typically possible with separate 2-D images of geologic maps and cross-sections. They can also help researchers test the geometric plausibility of interpretations shown in cross-sections. Finally, 3-D block diagrams created with SketchUp can be used to assist in cross-section construction and revision.

The most effective 3-D block diagrams drape geologic maps on a digital elevation model (DEM) and show how the maps and cross-sections connect at the topographic surface. Creating block diagrams in such a way that portions of the map between adjacent cross-sections are individual segments gives the user flexibility to make parts of the map invisible, and to sequentially reveal multiple cross-sections.

Lobeck's (1924) classic work on block diagrams set out all the rules and tricks for drafting them by hand. He also succinctly explains their value (Lobeck, 1924, p. 2):

A block diagram owes its value to the ease with which it can be understood. The reader makes immediate acquaintance with it because there is nothing unusual about it. No conventions are needed to represent the topography. No explanation is needed to indicate the position of the geological cross-section. It carries its message directly to the eye and leaves a visual impression unencumbered by lengthy descriptions.

Most introductory and advanced textbooks take advantage of these characteristics of block diagrams to illustrate important geologic concepts to students. Block diagrams are also valuable for presenting research results, both orally and in print. The use of GIS programs to produce maps for presentations and publications is becoming increasingly common, yet combining these carefully crafted and georeferenced maps with cross-sections to create 3-D block diagrams and thereby leverage their value remains rare.

Google Earth, a free downloadable program, offers geologists a tool for integrating geologic maps and cross-sections (Ishikawa and Kastens, 2005; Kastens and Ishikawa, 2006), but cross-sections are typically elevated above the ground surface completely (De Paor et al., 2008; Whitmeyer and De Paor, 2008), or only high enough to show that part of the cross-section that lies above the erosion surface (Walsh, 2009). Other programs can drape a map onto a DEM and combine it with cross-sections to create block diagrams, such as Rockworks and Midland Valley's Move software packages.

I outline here a process for creating 3-D block diagrams using SketchUp, GIS software, an image-processing program, a DEM, and digital images of geologic maps and cross-sections.

The block diagram created by this process is an interactive stand-alone model that can be used for presentations, exported for animations, and disseminated as a digital file. It is also possible to export such models to Google Earth to add greater geographic context. Finally, 3-D block diagrams can be used to construct models of geologic structures shown in maps and cross-sections.

METHOD FOR CREATING 3-D BLOCK DIAGRAMS

Detailed instructions for constructing 3-D block diagrams using SketchUp can be downloaded from Karabinos (2011). Here I give an overview of the process to illustrate how 3-D block diagrams are created, but I do not describe the detailed steps for all the programs used. This is because there are many GIS and image processing programs that could be used with SketchUp to produce 3-D block diagrams, and it is impossible to describe them all. Furthermore, as new versions of these programs are introduced, the details of performing specific tasks change, making detailed instructions obsolete. The online detailed instructions will be updated as necessary.

Required Files and Software

- A digital elevation model (DEM), which can be downloaded from http://seamless.usgs.gov/.
- A digital geologic map and the accompanying cross-sections. These can be scanned from paper maps or downloaded from http://ngmdb.usgs.gov/ngmdb/ngm_catalog.ora.html.
- The free version of SketchUp.
- A geographic information system (GIS) program (e.g., ArcMap or Global Mapper) for preparing the DEM, geologic map, and cross-sections for import into SketchUp.
- An image-processing program (e.g., Photoshop) for cropping borders from the map and cross-sections.

Preparing and Importing the DEM

The first step is to obtain a DEM for the map area and prepare it for import into SketchUp. If a DEM is not available it can be downloaded from the U.S. Geological Survey's Seamless Server site (http://seamless.usgs.gov/) as a zipped file. A GIS program is necessary to open the DEM file, to define the projection and datum to match those of the geologic map to be used, and to save the DEM in a file format recognized by SketchUp.

The DEM can be imported into SketchUp using the **File/Import** menu. It is necessary to specify the import file type as DEM. The **Options** button in the **File/Import** dialog makes it possible to select the number of points used in the import. For example, a 7.5′ quadrangle located at 43° latitude is ~10 km wide in the east to west direction and 14 km long in the north to south direction, thus a 30 m DEM contains almost 1.6×10^5 pixels. Importing a DEM for a 7.5′ quadrangle at full resolution would create a file too large for SketchUp to handle on most computers. Using 10,000–20,000 points for the DEM import is a reasonable compromise between file size and resolution, and computers designed for graphics-intensive work can process such a file in a reasonable amount of time.

SketchUp converts the DEM into a triangulated irregular network (TIN) made up of triangular faces separated by edges,

and combines them into a single SketchUp component, or entity. To improve the appearance of the DEM, which is now really a TIN, it is necessary to edit the component to erase unwanted vertical faces along the borders, and to soften and smooth the triangles' edges (Fig. 1).

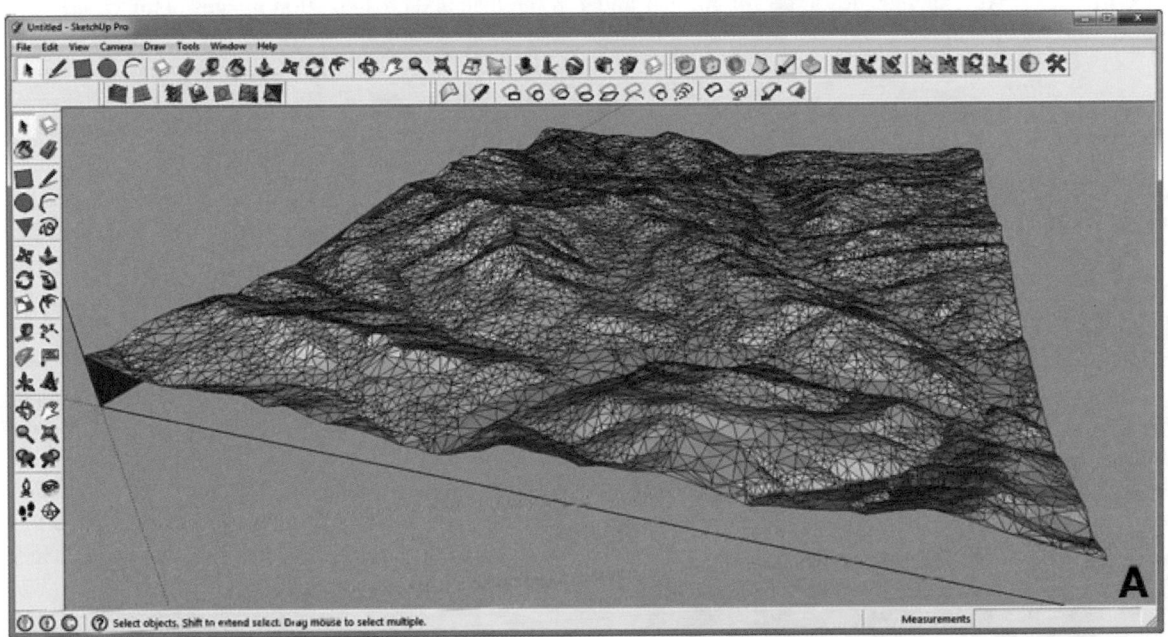

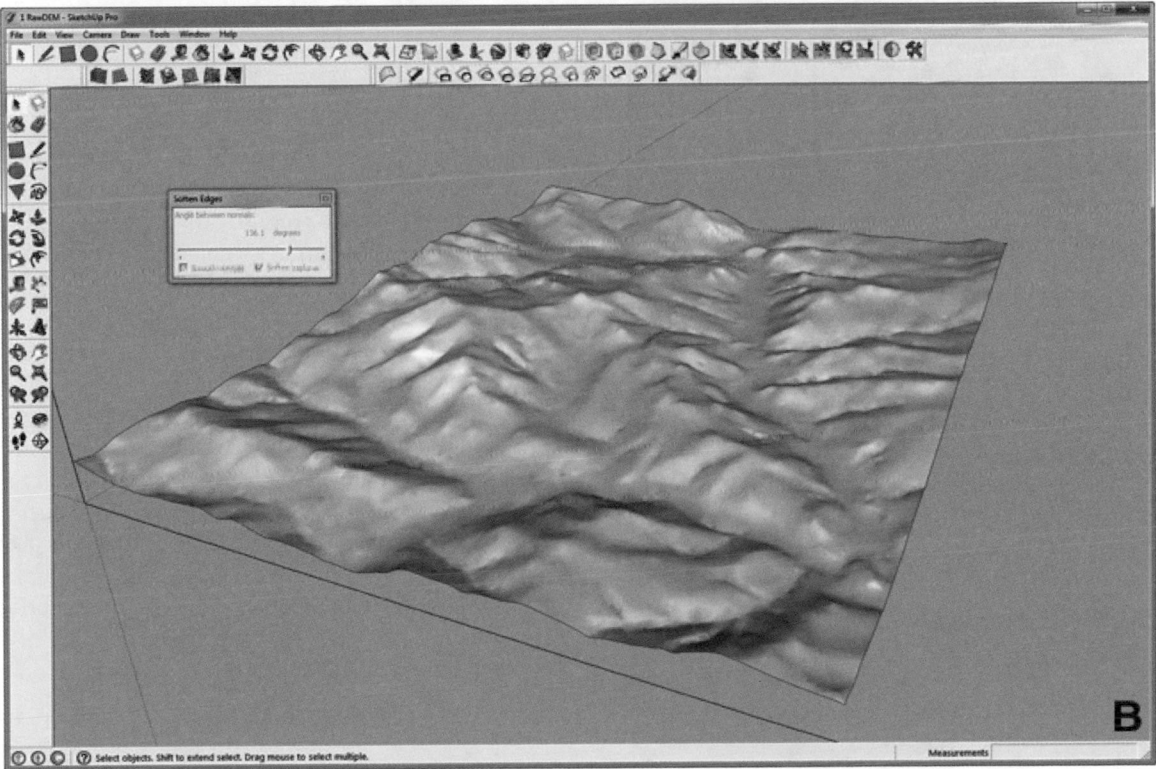

Figure 1. Screen captures of the Rochester, Vermont, digital elevation model (DEM) in SketchUp. (A) DEM shows individual triangular faces of the triangulated irregular network. Note vertical edges on left side of DEM. (B) DEM after vertical edges deleted and surface edges softened and smoothed in SketchUp.

Preparing and Importing the Map and Cross-Sections

Digital versions of a geologic map and the accompanying cross-sections are also needed. Paper copies of maps and cross-sections can be scanned. However, high-quality digital versions of many published maps and sections can be downloaded, eliminating the need to scan. An excellent resource for finding and downloading digital maps and sections is http://ngmdb.usgs

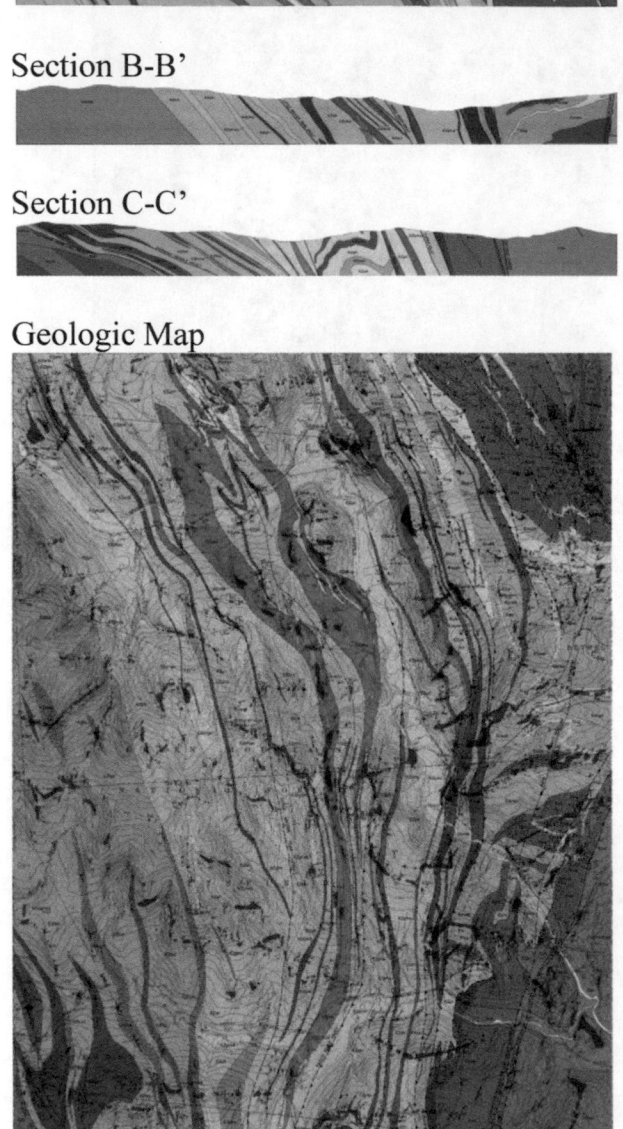

Section A-A'

Section B-B'

Section C-C'

Geologic Map

Figure 2. Cropped images of cross-sections and geologic map of the Rochester, Vermont, 7.5′ quadrangle (Walsh and Falta, 2001). Note cross-sections cropped to show only subsurface geology.

.gov/ngmdb/ngm_catalog.ora.html. Once the digital copies are obtained, commonly in multi-resolution seamless image database file format (MrSID), the map and sections must be cropped and saved in a file format that is recognized by SketchUp and that supports transparent backgrounds, such as portable network graphics (png). GIS software may be required for file conversion if the map and cross-section images obtained are MrSID files. If that is the case, Global Mapper or ArcMap can be used. Cropping can be done in ArcMap or an image-processing program such as Photoshop. A GIS program is also used to georeference the geologic map if the downloaded file does not already contain spatial information.

Cropping the map eliminates the white border, or collar, around the colored portion of the map. This step is essential to permit the map to overlay the DEM correctly. The cross-sections are cropped to eliminate the white border along the sides and bottom. Also, the portion of the cross-section above the topographic surface needs to be cropped (Fig. 2). Some potentially valuable information is lost in this step, but cropping that part of the cross-section that interprets the geology above the erosion surface allows the map and cross-section to meet seamlessly at the topographic surface, and form an edge of the block diagram. The image-processing program can also be used to ensure that the bases of the cross-sections are horizontal, which will help later to align the cross-sections with the map.

Drape the Map onto the DEM

The image file of the geologic map cannot be imported directly onto the DEM representation of the topography in SketchUp; a flat surface is required. To ensure that the map will drape conformably onto the DEM, a rectangle of the same horizontal dimensions as the DEM must be created. This rectangle serves as the flat surface for importing the image file of the geologic map in the **File/Import** menu, and it is convenient to position the rectangle directly over the DEM (Fig. 3A) to simplify some of the steps described below. Once the map is imported as an image and scaled to fit the rectangle, the underlying rectangle should be erased, leaving just the map, which will be a **component** in SketchUp. Because the imported map image is a **component**, it must be **exploded** using the **Context** menu before it can be selected as a **material** with the **Eyedropper** tool in the **Window/Materials** menu. Once the map image becomes a **material**, specifically a **texture**, it can be applied to a non-flat surface. If the rectangle and DEM have the same horizontal dimensions, the map is used exactly once to tile the entire DEM surface, and the geologic units will be positioned correctly (Fig. 3B). The flat, hovering imported map image can be **hidden** or, preferably,

Figure 3. Screen captures of the Rochester, Vermont, digital elevation model (DEM) and geologic map. (A) Map imported onto a rectangle above the DEM. (B) Geologic map used as a **material** in SketchUp and transferred to the DEM surface.

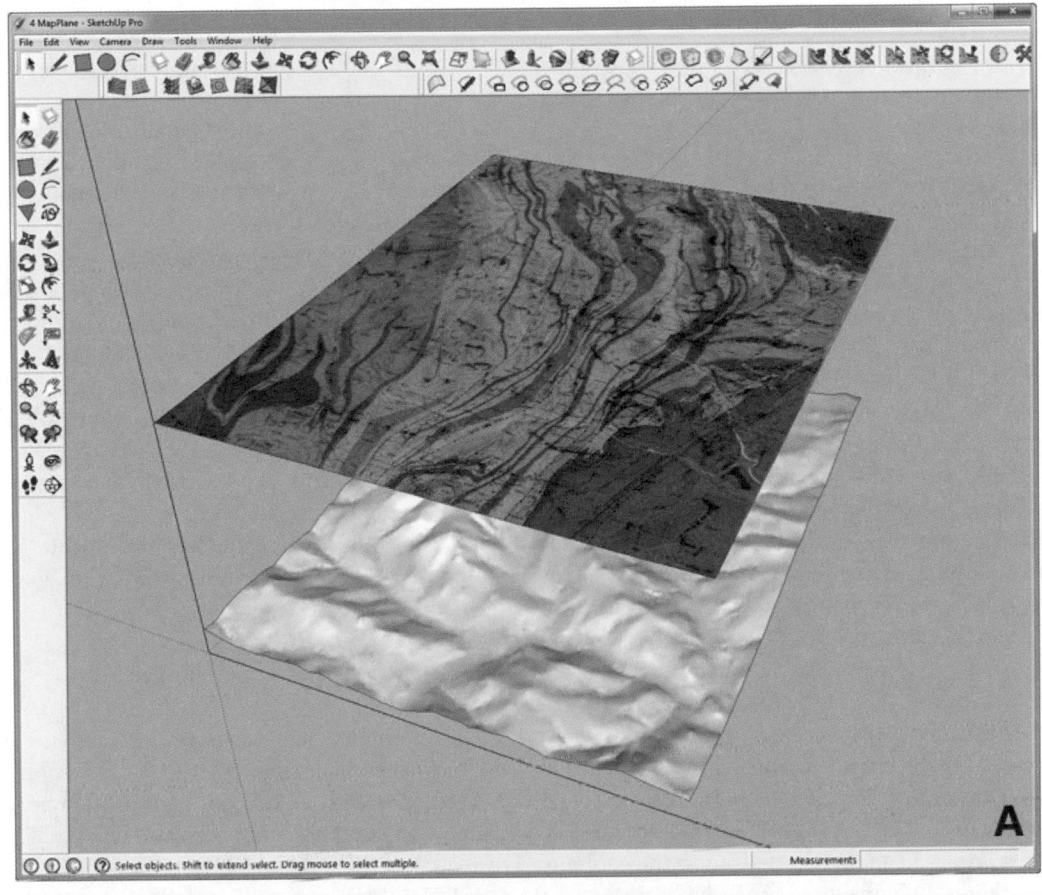

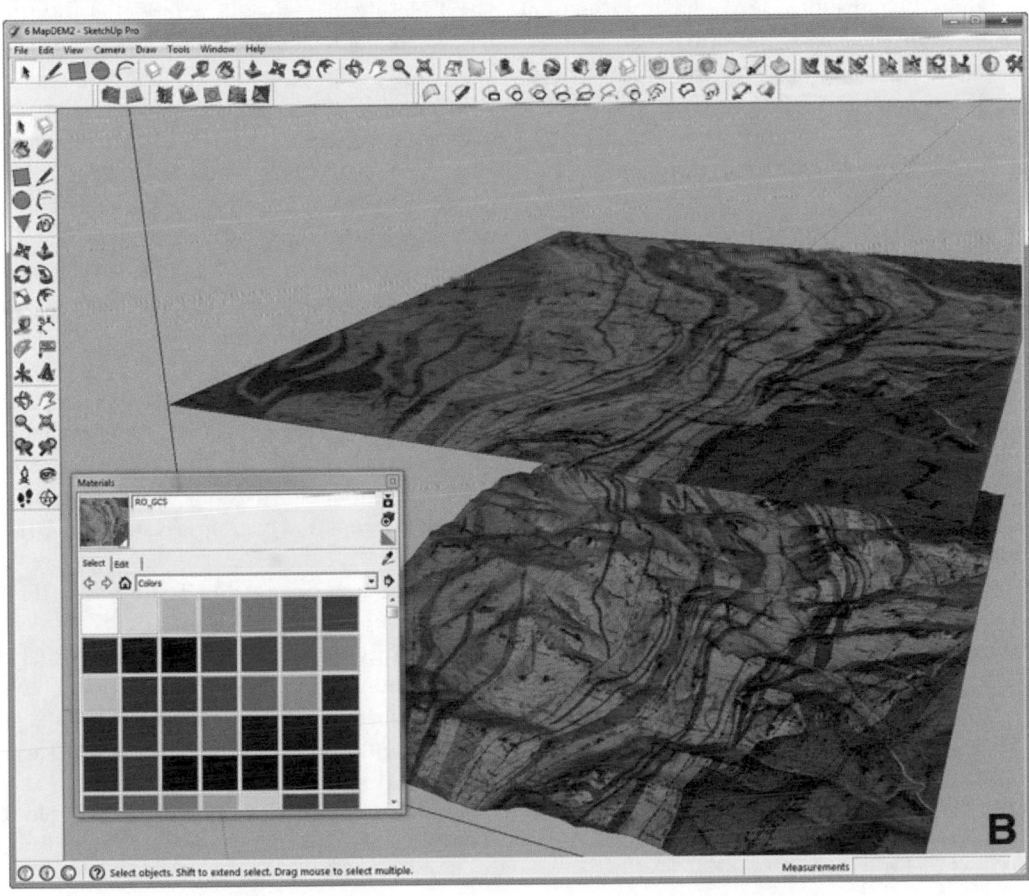

moved to a **layer** that can be made invisible, but it should not be erased because it will be used as a template to create the vertical cross-section planes.

Create Vertical Planes for Cross-Sections

Accurately aligning the cross-sections with the map and DEM in SketchUp can be challenging. To achieve consistently good results in a reasonable amount of time, it is essential to create the vertical planes used to import the cross-section images in the correct location relative to the map. To do this, the cross-section lines on the flat map from the hovering plane can be traced, and each pair of adjacent section lines joined to form quadrilaterals (Fig. 4A). The quadrilaterals are then turned into 3-D objects with the **Push-Pull** tool in SketchUp (Fig. 4B). For each quadrilateral, the two vertical rectangles above the cross-section lines will serve as the import planes for the cross-sections, and everything else can be erased (Fig. 4C). The horizontal lengths of the vertical rectangles need to match those of the cross-sections exactly, but the rectangle heights are not critical because when the cross-section images are imported, they can be scaled using only the lengths. Once the cropped cross-section images are imported using the cross-section planes, the rectangle backgrounds should be erased, leaving only the cross-section images visible (Fig. 4D).

Because the hovering rectangle used to import the geologic map was positioned directly above the DEM, the cross-section planes will be vertically aligned with the cross-section lines on the map draped on the DEM. Copies of the cross-section planes can be used to segment the DEM to create individual map portions bounded by the cross-section lines. In this way, the independent segments can be hidden to reveal successive cross-sections.

Segment the DEM

To take full advantage of the 3-D block diagram, it is necessary to segment the DEM so that parts of the map can be hidden to reveal the subsurface cross-sections. An easy way to do this is to (1) make a copy of each cross-section, (2) lower it vertically to intersect the DEM along the section line making sure it extends above and below the DEM, (3) use the **Scale** tool to make the sections long enough to extend beyond the edges of the DEM, and (4) use the **Edit/Intersect Faces/With Model** command to split the DEM into multiple segments (Fig. 5). For a map with n cross-sections, there should be n + 1 segments. Each segment becomes an independent feature that can be made invisible. The copies of the cross-sections are then erased.

Copies of the cross-sections are essentially "knives" used to cut the DEM. They are well suited for this task because, as components, they remain parallel with the section lines on the map when they are scaled to extend beyond the DEM. The only requirement is that the copies of the cross-section planes *completely* divide the DEM—above, below, and on both edges—

otherwise, the **Edit/Intersect Faces/With Model** command will not work properly.

If a cross-section line does not extend from one map edge to another, it is still possible to scale the cross-section plane, as described, above to cut the DEM and segment it. When the map segment is hidden to reveal the underlying cross-section, however, the map will extend beyond the cross-section plane. Another strategy is to use a quadrilateral, one edge of which coincides with the cross-section line, to create a "keyhole" in the map that can be hidden to reveal the cross-section plane.

The best way to portray a cross-section that is composed of two or more non-colinear segments is to crop the published cross-section into multiple pieces and to create a separate image for each segment. A vertical cross-section plane for each segment is created and used to import the individual images, and the images are adjusted with the **Move** and **Rotate** tools to optimize their registry. Finally, the cross-section images are aligned with the map as described below.

Align the Cross-Sections with the Map

The original cross-sections can be moved vertically to align them with the map (Fig. 6). The cross-sections may need to be **Moved**, **Rotated**, or **Scaled** slightly to improve the alignment with the topography of the DEM. Ideally, the topographic profile of the cross-sections will match the topographic surface of the DEM closely, but several factors contribute to deviations. The DEM may not match the topography of the geologic map, especially if it was published on an old base map. Also, because the DEM must be downsampled before it is imported to SketchUp, due to file size constraints, it is less accurate than the original topographic map on which it may have been based. Finally, the topographic profile used for the cross-section may have been drawn with some inaccuracies. Beyond these factors, errors in block diagram construction may produce discrepancies that prevent the map and cross-sections from aligning properly. For example, the aspect ratio of the cross-section may have been accidentally altered during image processing.

If the discrepancies between the map and cross-sections are acceptable, there are several tools available in SketchUp to edit the surface of the DEM that make it possible to improve the *visual appearance* of the join between the map and sections by minimizing gaps and overlaps (Karabinos, 2011).

Use Layers and Scenes to Create Animations

The **Layers** window in SketchUp makes it possible to assign features to an individual **Layer**, which can then easily be made invisible as needed. Each map segment and cross-section should be assigned to a separate **Layer** for maximum flexibility (Fig. 7). The **Scenes** window in SketchUp is used to create a sequence of images of the 3-D block diagram. Each **Scene** preserves the perspective and visibility of the layers that existed when the **Scene** was created (Fig. 8). **Scenes** can be

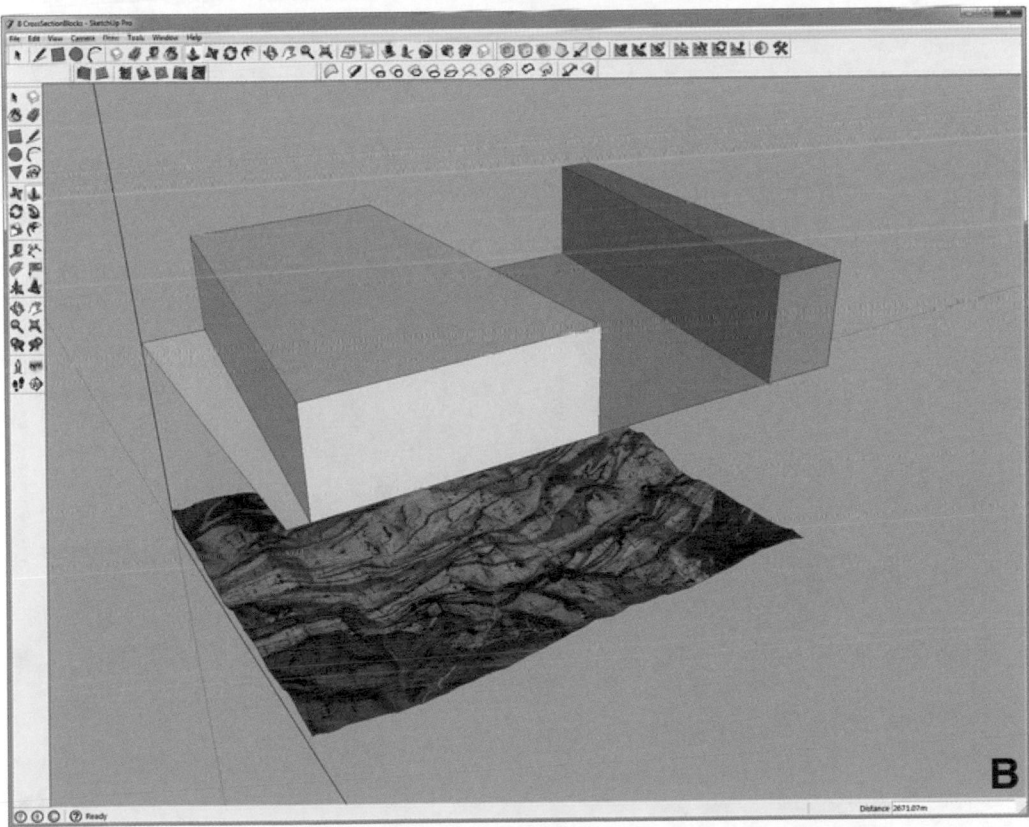

Figure 4. Screen captures showing how to create vertical cross-section planes. (A) Cross-section lines traced from geologic map on the hovering rectangle and connected to form quadrilaterals. For clarity, the map is replaced with gray quadrilaterals. (B) Quadrilaterals converted into 3-D objects with the **Push-Pull** tool in SketchUp. (*Continued*.)

242

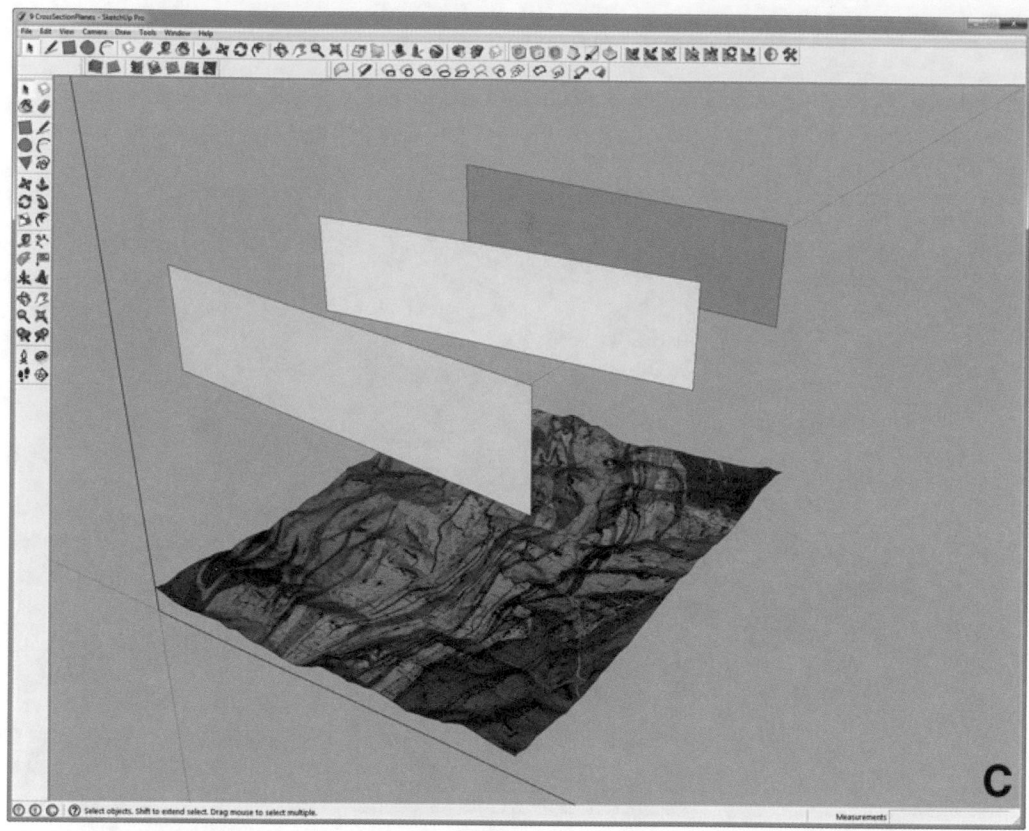

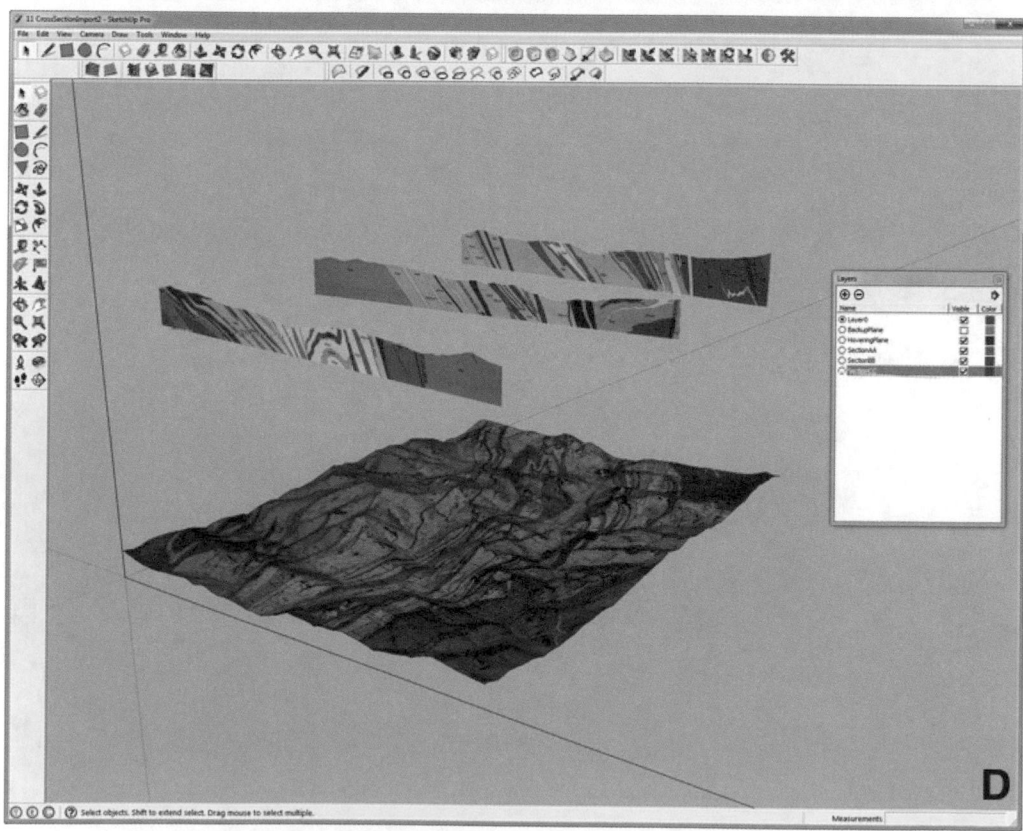

Figure 4 (*continued*). (C) Everything but the cross-section planes erased on the hovering rectangle. (D) Cross-section images imported to the cross-section planes; note that the rectangles used for importing the images have been erased.

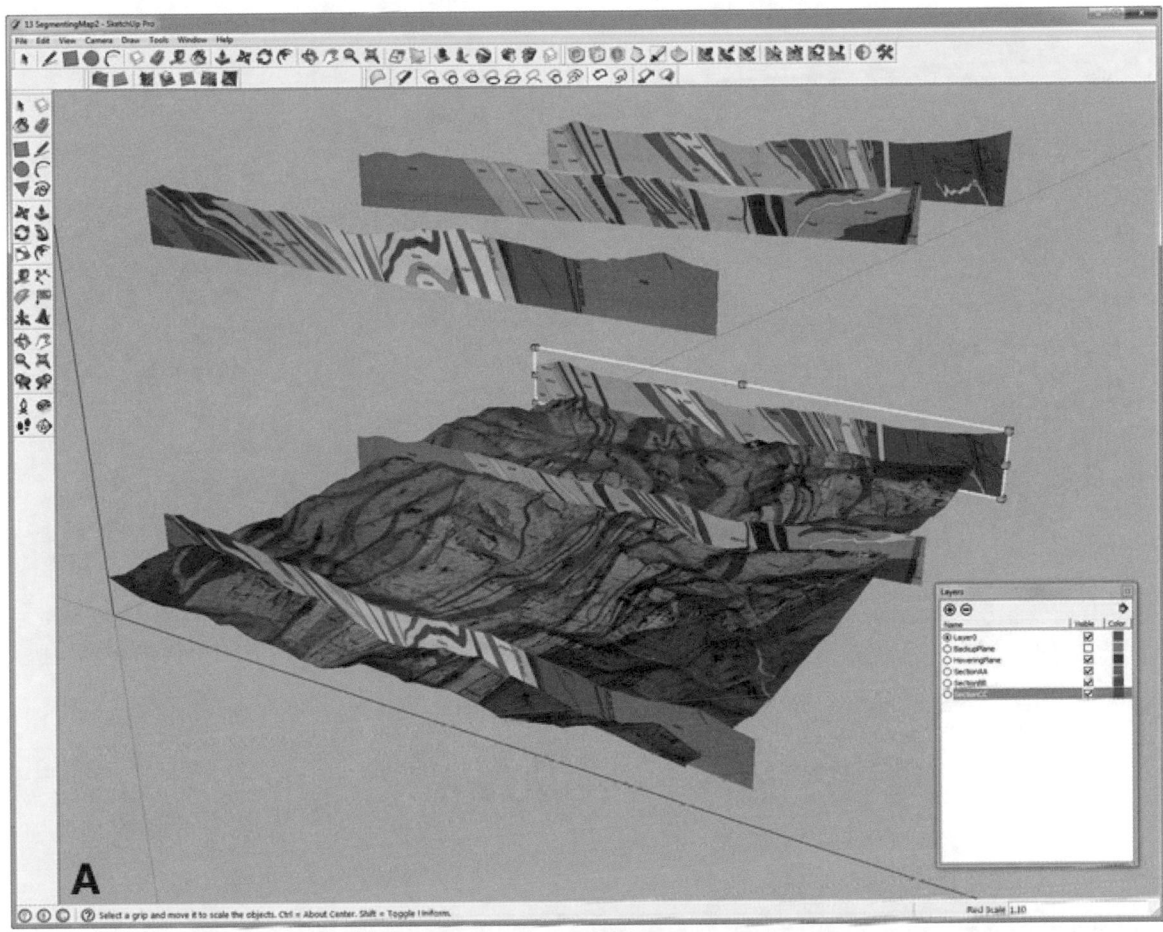

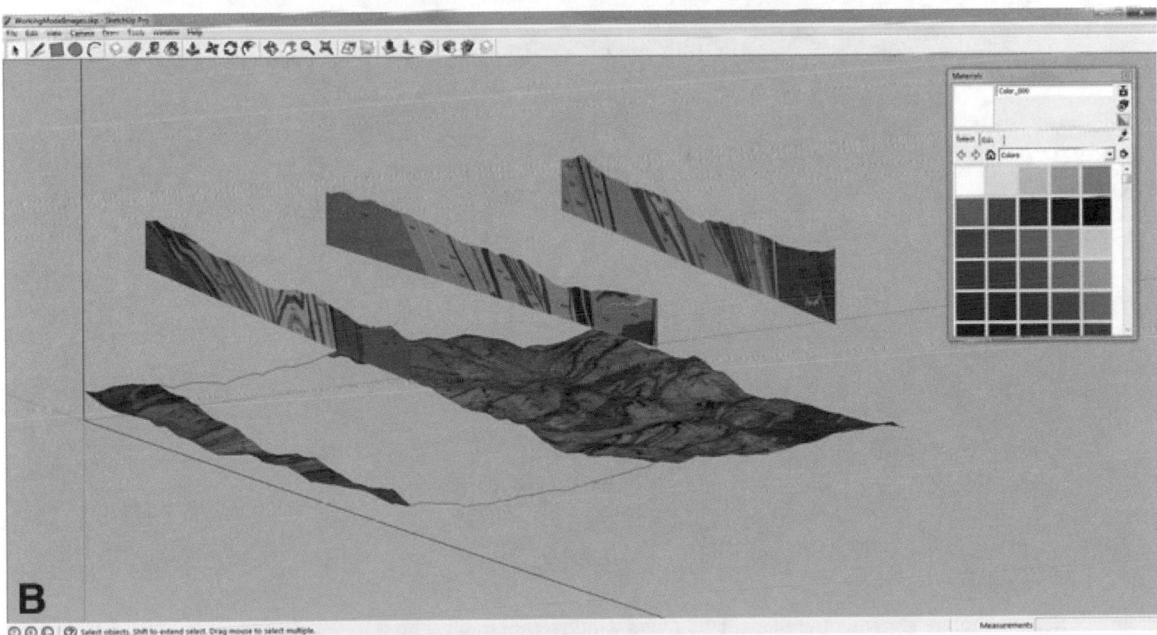

Figure 5. Screen captures showing how the digital elevation model (DEM) is segmented. (A) Copies of the cross-section plane are moved vertically to intersect the DEM and scaled to extend beyond it. The **Edit/Intersect/With Model** command segments the DEM along the cross-section lines. (B) One portion of the DEM is made invisible to show that the DEM is segmented parallel to the cross-section lines.

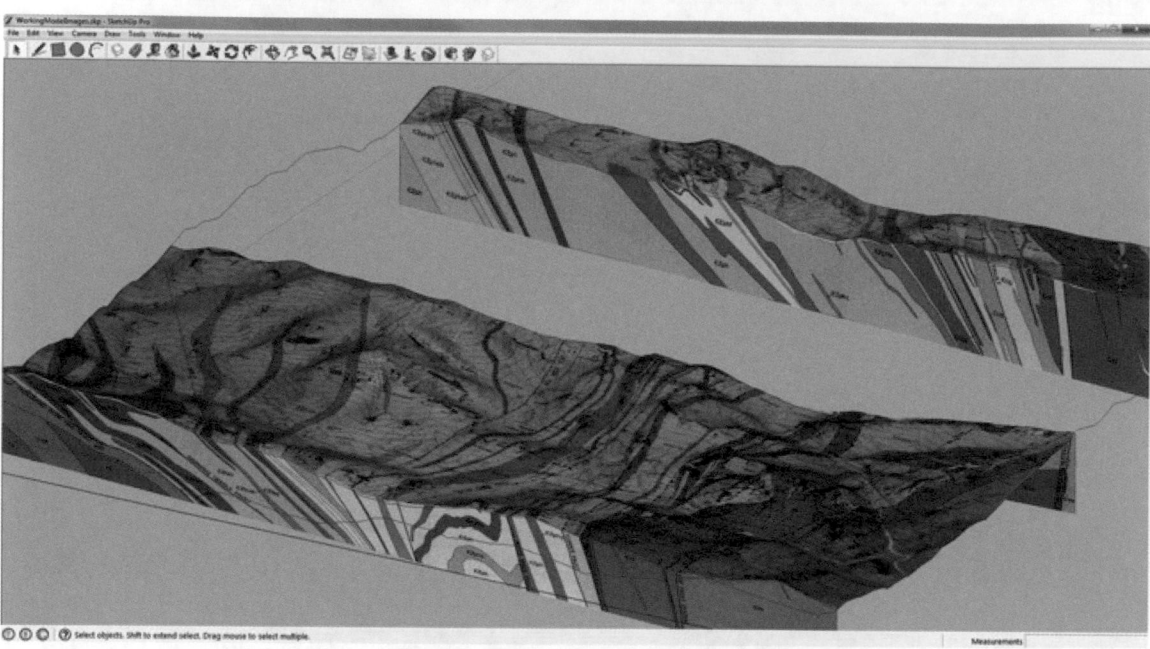

Figure 6. Screen capture showing two cross-sections integrated with the geologic map of the Rochester, Vermont, 7.5′ quadrangle.

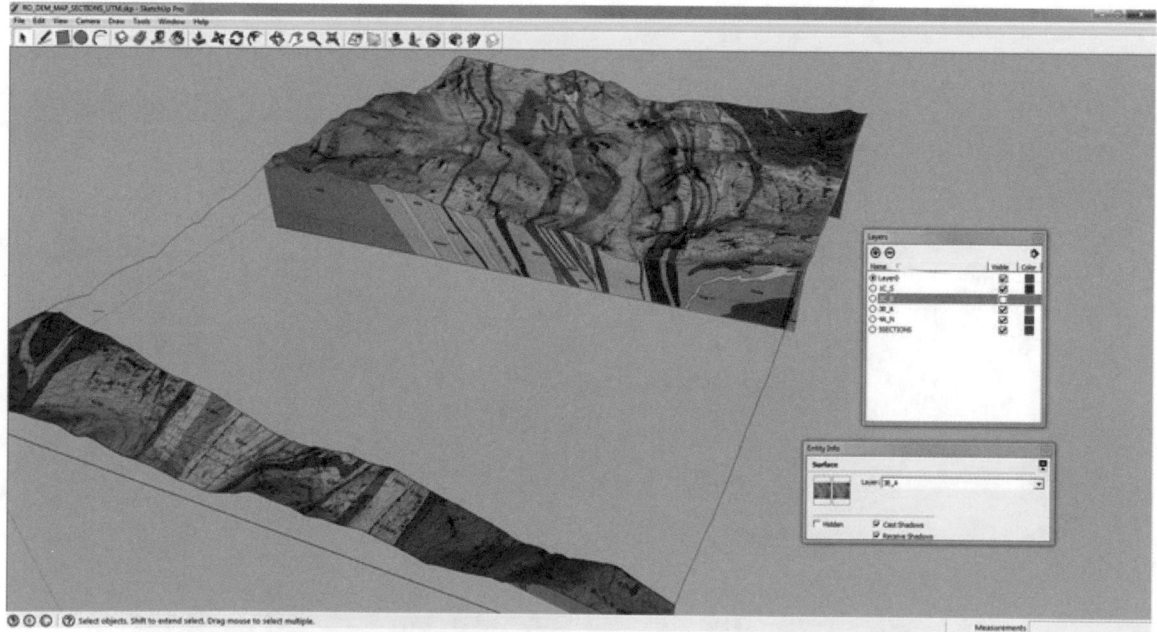

Figure 7. Screen capture showing the SketchUp Layers window (upper inset box). Each element of the SketchUp model is assigned to its own layer so that it can be made invisible as needed. Here the second segment from the south end of the map is "turned off" by unchecking the corresponding box in the Layers window, highlighted in blue.

used to quickly recreate a particular view of the model. Even more valuable is the ability to create **Animations** with **Scenes** to move smoothly through a well-planned sequence of views. The **Animations** can be shown in SketchUp or exported to commonly used file formats.

Export the 3-D Block Diagram to Google Earth

Once the 3-D block diagram has been created in SketchUp, it can be **Exported** to Google Earth. Showing the model in Google Earth gives it more geographic and geologic context. It can also be integrated with a Google Earth **Tour**. The model can be viewed by making the **Primary Database** invisible in Google Earth and turning off map segments to reveal the cross-sections (Fig. 9A). It is also possible to elevate the model above the surface to make the cross-sections visible (Fig. 9B). The detailed description of exporting a 3-D block diagram to Google Earth can be found in Karabinos (2011).

APPLICATIONS OF 3-D BLOCK DIAGRAMS

Teaching and Exhibits for Non-Specialists

Experienced geologists commonly overlook the fact that students and non-professionals do not understand how to extract information summarized on a geologic map, and that cross-sections are even more difficult to decipher (Kastens et al., 2009). The fundamental concept that cross-sections illustrate an interpretation of the distribution of rocks in the subsurface and, in many cases, rocks above the current erosion surface, is not widely appreciated by even well-educated students and adults. Furthermore, the ability to mentally merge information shown separately on a 2-D map and cross-section when one must look back and forth between the map and cross-section on a large sheet of paper varies widely, even among experienced geologists.

Block diagrams constructed with SketchUp provide a powerful tool for demonstrating how geologic maps and cross-sections *function* to non-professionals. The value of a map and cross-section integrated into a 3-D block diagram far surpasses that of separate 2-D images for explaining what they represent. Beyond this important function, they can be used to explain complex geology to novices. Particularly in regions where bedrock geology exerts an important control on topography, a geologic map draped on a DEM and an accurately positioned vertical cross-section clarifies the relationship between the color-coded

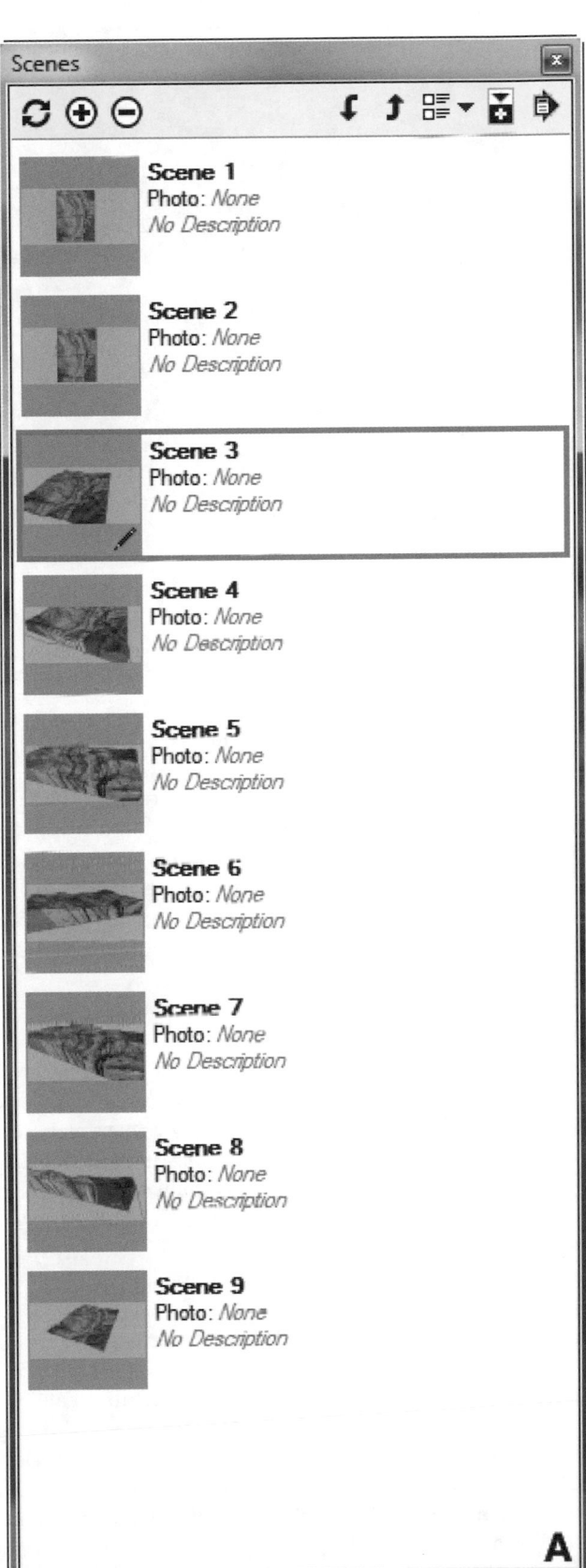

Figure 8. Screen captures showing the Scenes window (A) and one scene of the Rochester, Vermont, 7.5′ quadrangle (B). Each scene records the perspective and the visible layers of the SketchUp model when the scene was created. Scenes can be combined to create animations. (*Continued on following page.*)

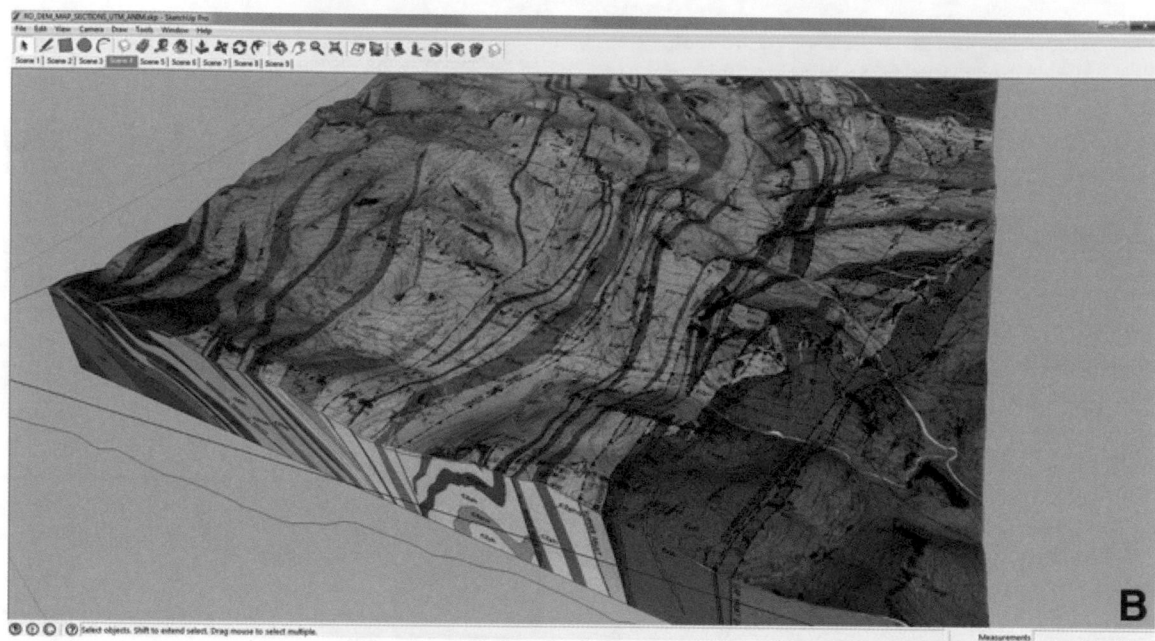

Figure 8 (*continued*).

polygons on maps and cross-sections and geologic units on the ground and in the subsurface.

One of the advantages of 3-D block diagrams created with SketchUp is that they can be made easily accessible, giving others the opportunity to **Pan**, **Zoom**, and **Rotate** the models. Interacting with block diagrams is a particularly effective way for students to develop an appreciation for maps and cross-sections as abstract concepts, as well as for teaching the geology of a specific area.

Professional Presentations

Even experienced geologists need time to visualize the geometric relationships portrayed in 2-D geologic maps and cross-sections. During professional presentations it is difficult for experts in the audience to assess even skillfully drafted maps and cross-sections in the time available. Well-constructed block diagrams help illustrate geologic interpretations clearly, and give the audience a tool to quickly assess proposed models.

Comparing Different Styles and Interpretations

A large percentage of the work it takes to create 3-D block diagrams with SketchUp is constructing the DEM, segmenting it, and aligning the cross-section planes with it. However, once the block diagram is made, it is easy to replace the draped map and cross-section images with others that emphasize different aspects of the geology, or show different interpretations. For example, the block diagram shown in Figure 10A integrates the published geologic map and cross-sections

(Ratcliffe et al., 1993) for three 7.5′ quadrangles located in the northwestern corner of Massachusetts into a 3-D block diagram. It shows the published interpretation, but the details portrayed on the map and sections are difficult to discern in the block diagram. This is because the geology is complex, most geologic formations are subdivided into numerous members by Ratcliffe et al. (1993), and the color scheme does not highlight the difference between the hanging-wall and footwall rocks of the major thrusts. In Figure 10B, I used the same DEM and cross-section planes to show the geology, but I replaced the original images with digitized versions of the map and cross-sections that I made using ArcMap. The new images combine and simplify some of the geologic units and, more importantly, highlight the difference between hanging-wall rocks, shown in gray, and footwall rocks, shown in bright shades of yellow and green. It seems likely that the simplified version of the block diagram would be more readily understood by students and non-specialists, and easier for geologists to absorb during an oral presentation.

SketchUp 3-D block diagrams also help compare different geologic interpretations. One of the units shown as part of the footwall of a recumbently folded thrust in Figures 10A and 10B is a graphite-rich schist (Ratcliffe et al., 1993). However, based on structural and petrologic evidence, Karabinos et al. (2010) argued that the graphite-rich schist is part of the hanging-wall of the thrust, and this interpretation is shown in Figure 10C where it is shown as dark gray. The important point here is that 3-D block diagrams, once created, can be used to compare different geologic interpretations and assess their relative merits.

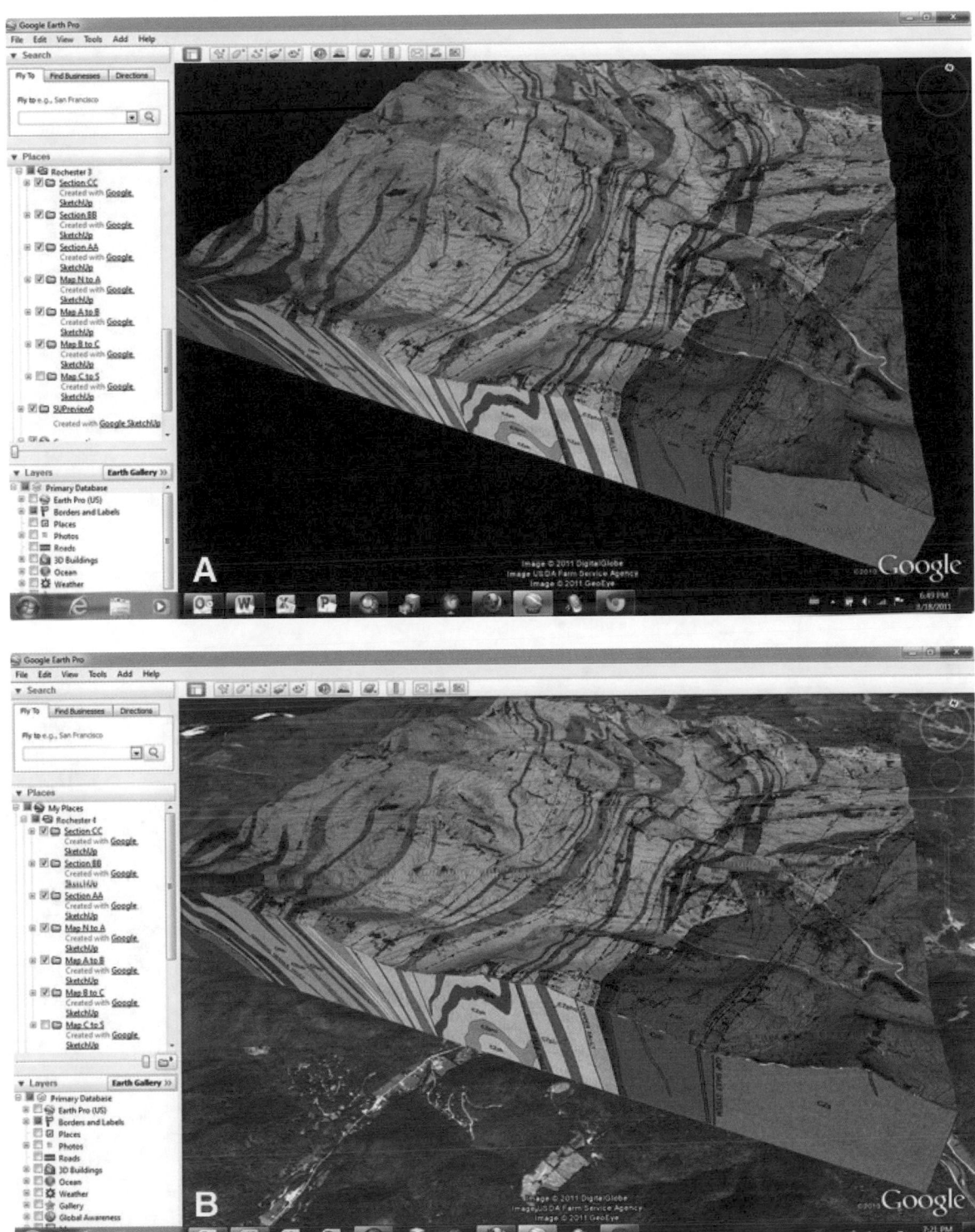

Figure 9. Screen captures showing the Rochester, Vermont, SketchUp model exported to Google Earth. (A) The digital elevation map of the SketchUp model is at the same elevation as the ground surface in Google Earth. The Primary Database has been made invisible to reveal the subsurface cross-section in the model. (B) The Primary Database is visible, and the entire SketchUp model is elevated above the ground surface in Google Earth to show the map and subsurface cross-section.

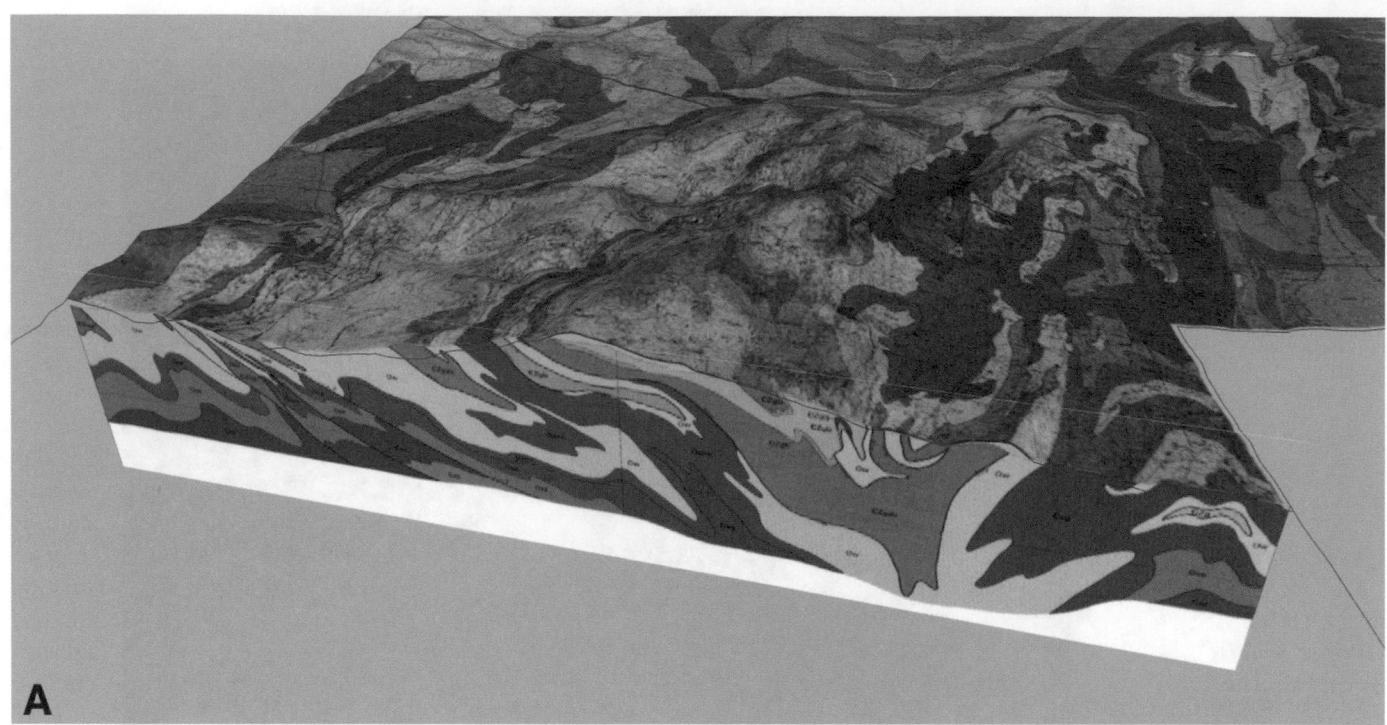

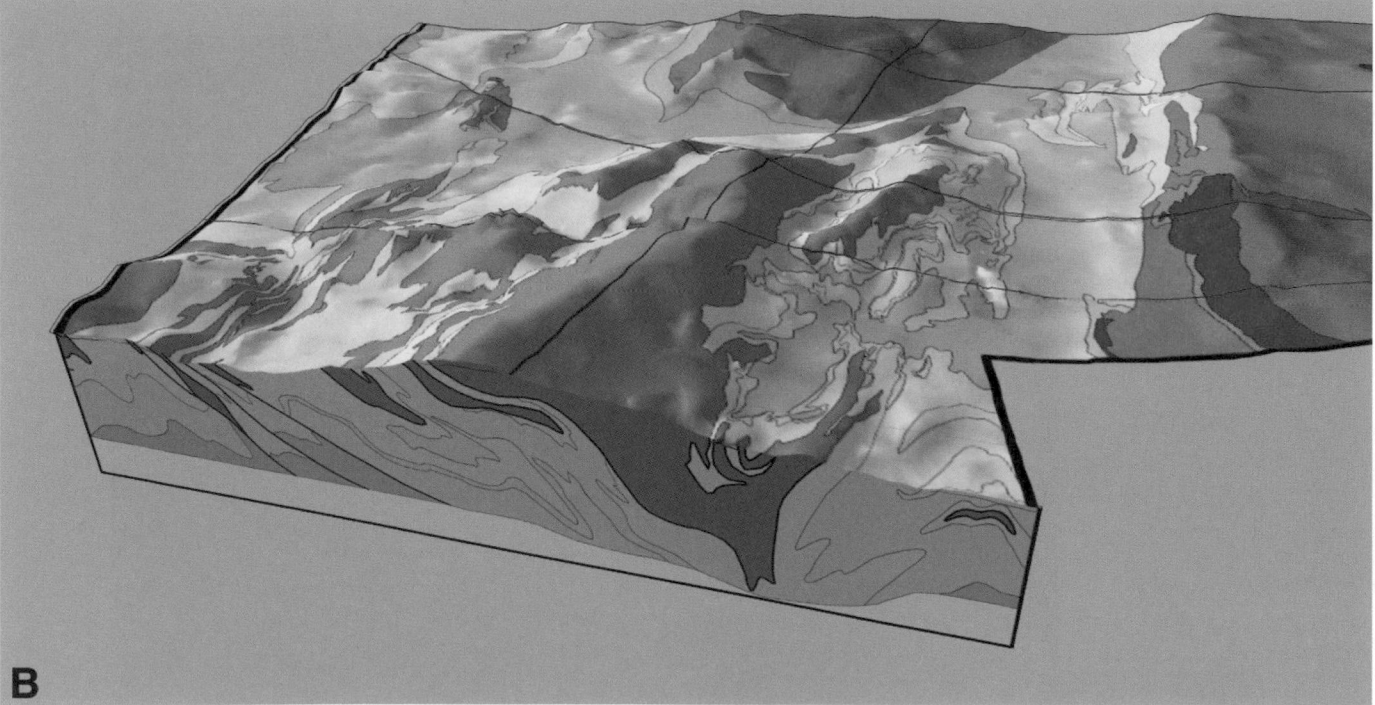

Figure 10. Block diagrams of the Williamstown, North Adams, and Cheshire, Massachusetts, 7.5′ quadrangles. Images exported from SketchUp showing three different sets of maps and cross-sections. Geology from Ratcliffe et al. (1993). (A) Cropped images from the published map and cross-sections. (B) Images used for block diagram were digitized, simplified, and recolored using the published geologic map. Hanging-wall rocks of major thrust shown in gray; footwall rocks shown in yellow and green. (*Continued.*)

Figure 10 (*continued*). (C) As in B, but a graphitic schist, shown in green in B, and in dark gray in C, is shown as part of the hanging-wall of the major thrust.

Visualizing and Modeling Structures

Interactive 3-D block diagrams created with SketchUp offer the possibility of enhancing our visualization of structures portrayed in maps and cross-sections, and can facilitate construction of models that approximate complex structures to help assess their geometric plausibility. Figure 11A shows a block diagram of Mount Greylock in northwestern Massachusetts (the same images as used in Fig. 10B) with a series of evenly spaced horizontal planes intersecting the map and sections. The structures are too complicated to be modeled exactly, but the enveloping surface of the folded basal thrust, as interpreted by Ratcliffe et al. (1993) can be approximately located on each horizontal sectioning plane. By combining information from each horizontal section, it is possible to model the surface, as shown in Figure 11B. The cross-sections clearly show the interpretation of Ratcliffe et al. (1993)—that the basal thrust was recumbently folded and then refolded about nearly vertical axial planes. What is less clear from the map and cross-sections is the implicit 3-D geometry of the structure. Because Mount Greylock is a klippe, the hanging-wall rocks (schist shown in gray) are completely surrounded by footwall rocks (marble shown in yellow). Figure 11B shows that the folded thrust surface is a large-scale sheath fold elongated in the north to south direction, and refolded with a steep north to south–striking axial plane. Thus, the SketchUp model gives a 3-D approximation of the structure implied by the map and cross-sections and can be used to assess the geometric plausibility of the proposed interpretation and compare it with thrust systems in other mountain belts.

Because 3-D block diagrams created with SketchUp can be rotated and viewed from any perspective, they are ideally suited to the down-plunge projection method (Groshong, 2008). When structural domains can be approximated as cylindrical folds, an oblique view of the map parallel to the fold axis should coincide with the cross-section for that domain. Thus, 3-D block diagrams have potential to illustrate how the down-plunge method works to students, and to create cross-sections of geometrically compatible regions.

CONCLUSIONS

Interactive 3-D block diagrams of geologic structures can be created from DEMs and published geologic maps and cross-sections using SketchUp, GIS software, and image-processing programs. They help students and non-professional to understand how geologic maps and cross-sections portray observations and interpretations and to visualize geologic structures. 3-D block diagrams can also be used in professional presentations to clarify interpretations proposed by authors, and to help audience members assess the geometric plausibility of those interpretations.

Once a block diagram has been constructed with SketchUp, it is possible to replace the original map and cross-section images

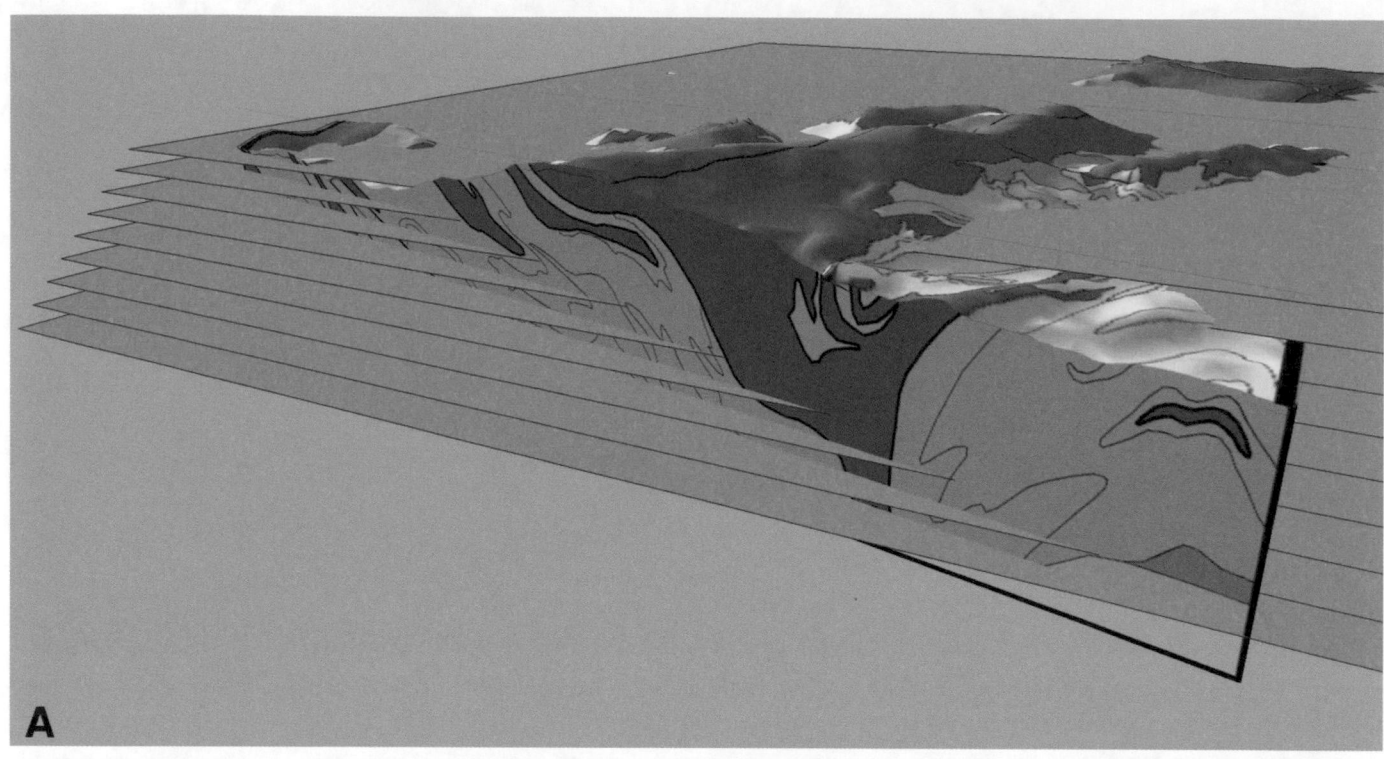

Figure 11. (A) Horizontal planes, shown in blue, intersect map and cross-sections and were used to define the enveloping surface of the folded thrust. (B) Approximate model of the folded thrust surface reconstructed from the interpretation shown on the map and cross-sections. The outer part of the thrust surface is red, and the inner part is orange.

with others to improve the visual appearance of the model and to test different interpretations. Furthermore, SketchUp can be used to construct 3-D models that approximate the structures portrayed in the map and cross-sections.

ACKNOWLEDGMENTS

Thanks to Jack Loveless and David Lageson for their helpful reviews of this manuscript. Thanks to Evan Dethier and Caleb Lucey for help digitizing geologic maps and cross-sections. Thanks to Caleb for help thinking about the problem of making 3-D block diagrams when we had no idea what to do.

REFERENCES CITED

De Paor, D.G., Simpson, C., and Whitmeyer, S., 2008, Deconstructing classical geologic maps using Google Earth's keyhole markup language: Geological Society of America Abstracts with Programs, v. 40, no. 6, p. 348.

Ishikawa, T., and Kastens, K., 2005, Why some students have trouble with maps and other spatial representations: Journal of Geoscience Education, v. 53, no. 2, p. 184–197.

Groshong, R.H., Jr., 2008, 3-D Structural Geology, a Practical Guide to Quantitative Surface and Subsurface Map Interpretation, 2nd edition: Springer, Berlin, 400 p.

Karabinos, P., Aronoff, R.F., and Nemser, E.S., 2010, Evidence for kilometer-scale fluid-controlled redistribution of graphite in pelctic schist in the Taconic thrust belt on Mount Greylock, Massachusetts: Geological Society of America Abstracts with Programs, v. 42, no. 1, p. 97.

Karabinos, P., 2011, Integrating a geologic map and cross-sections into an interactive 3-D block diagram with Google SketchUp, online tutorial: http://web.williams.edu/wp-etc/geosciences/facultypages/Paul/SketchUp/sketchup.html.

Kastens, K.A., and Ishikawa, T., 2006, Spatial Thinking in the Geosciences and Cognitive Sciences, *in* Manduca, C., and Mogk, D.W., eds., Earth and Mind: How Geoscientists Think and Learn about the Complex Earth: Geological Society of America Special Paper 413, p. 53–76, doi:10.1130/2006.2413(05).

Kastens, K.A., Agrawal, S., and Liben, L.S., 2009, How students and field geologists reason in integrating spatial observations from outcrops to visualize a 3-D geological structure, *in* Ramadas, J., and Gilbert, J., eds., Visual and Spatial Modes of Learning: International Journal of Science Education, Special Issue, v. 3, p. 365–393.

Lobeck, A.K., 1924, Block Diagrams and Other Graphic Methods used in Geology and Geography: New York, John Wiley and Sons, 206 p.

Ratcliffe, N.M., Potter, D.B., and Staley, R.S., 1993, Bedrock geologic map of the Williamstown and North Adams quadrangles, Massachusetts and Vermont, and part of the Cheshire Quadrangle, Massachusetts: U.S. Geological Survey Miscellaneous Investigations Series I-2369, scale 1:24,000, 2 sheets, 13 p.

Walsh, G.J., 2009, A method for creating a three dimensional model from published geologic maps and cross sections: U.S. Geological Survey Open-File Report 2009–1229, 16 p.

Walsh, G.J., and Falta, C.K., 2001, Bedrock geologic map of the Rochester quadrangle, Rutland, Windsor, and Addison Counties, Vermont: U.S. Geological Survey Geologic Investigations Series Map I-2626, 14 p., scale 1:24,000.

Whitmeyer, S.J., and De Paor, D.G., 2008, Large-scale emergent cross sections of crustal structures in Google Earth: Geological Society of America Abstracts with Programs, v. 40, no. 6, p. 189.

MANUSCRIPT ACCEPTED BY THE SOCIETY 16 APRIL 2012

The Geological Society of America
Special Paper 492
2012

Terrain modification in Google Earth using SketchUp: An example from the Western Blue Ridge of Tennessee

Jesse S. Hill*
Michael J. Harrison
Department of Earth Sciences, Tennessee Tech University, Cookeville, Tennessee 38506, USA

ABSTRACT

The creation of new outcrops through construction is an important source of field data for geologists, especially in parts of the Appalachians with limited rock exposure. Users of Google Earth for field research often encounter disparities between the digital topography and the current-day Earth's surface, as newly formed outcrops may not be represented in the topography. Such is the case along sections of the I-26 corridor in Unicoi County, northeastern Tennessee. Twenty-four kilometers of U.S. 23 (future I-26) was widened to four lanes from Sams Gap at the North Carolina–Tennessee line to the Nolichucky River near Erwin, Tennessee, in the early 1990s. The series of outcrops created along the corridor provide an exceptional traverse through Grenvillian-age basement and cover strata which contain numerous stacked Alleghanian thrust sheets and shear zones. Near mile marker 44 along I-26, an ~250 m-long and 65 m-high outcrop was formed as part of the early 1990s construction. Google Earth satellite and Street View images show the outcrop, but the digital terrain in Google Earth does not reflect the approximate 150,000 m³ of rock removed to form this roadcut. To correct for this, terrain modifications were made with SketchUp by copying and virtually excavating the landscape. The SketchUp model was then imported into Google Earth to show the outcrop and interstate as it looks today, with the interstate passing uninterrupted through a ridge rather than draping over hilly topography. This technique can be applied to any area in Google Earth where a mismatch exists between real and virtual topography.

INTRODUCTION

Portable computers and applications such as Google Earth have sparked a revolution in digital mapping (e.g., Knoop and van der Pluijm, 2006; De Paor and Whitmeyer, 2009; Whitmeyer et al., 2010). Google Earth is widely available and it can be utilized with GPS for real-time, in-the-field mapping. Moreover, the user can add geologic maps, cross sections, field photos, and struc-

tural data to the Google Earth virtual globe without the support of ArcGIS (Whitmeyer and Beard, 2010). This facilitates the building of a virtual field site in Google Earth because unlike when using ArcGIS, users do not have to manipulate the code used for Google Earth's application programming interface (API).

Although Google regularly adds and updates satellite imagery in Google Earth, elevation data may be outdated and may not accurately represent the current-day topography. Such is the case

*Hilljs@live.unc.edu

Hill, J.S., and Harrison, M.J., 2012, Terrain modification in Google Earth using SketchUp: An example from the Western Blue Ridge of Tennessee, *in* Whitmeyer, S.J., Bailey, J.E., De Paor, D.G., and Ornduff, T., eds., Google Earth and Virtual Visualizations in Geoscience Education and Research: Geological Society of America Special Paper 492, p. 253–262, doi:10.1130/2012.2492(18). For permission to copy, contact editing@geosociety.org. © 2012 The Geological Society of America. All rights reserved.

in this study, where 1990s construction of a four-lane interstate in eastern Tennessee created numerous outcrops that are seen in Google Earth satellite and Street View photographs, but are not represented in the Google Earth terrain. This creates a disparity between the current-day topography and the Google Earth virtual terrain (Fig. 1). In order to use Google Earth for accurate outcrop-scale mapping and data representation, the virtual terrain has to be modified. In this paper, we demonstrate how to use

SketchUp to reshape the Google Earth terrain. The techniques described in this paper use the default drop-down menu commands included with the free online versions of Google Earth version 6 and SketchUp version 8 for Windows. The techniques may differ slightly for Google Earth Pro or Mac users.

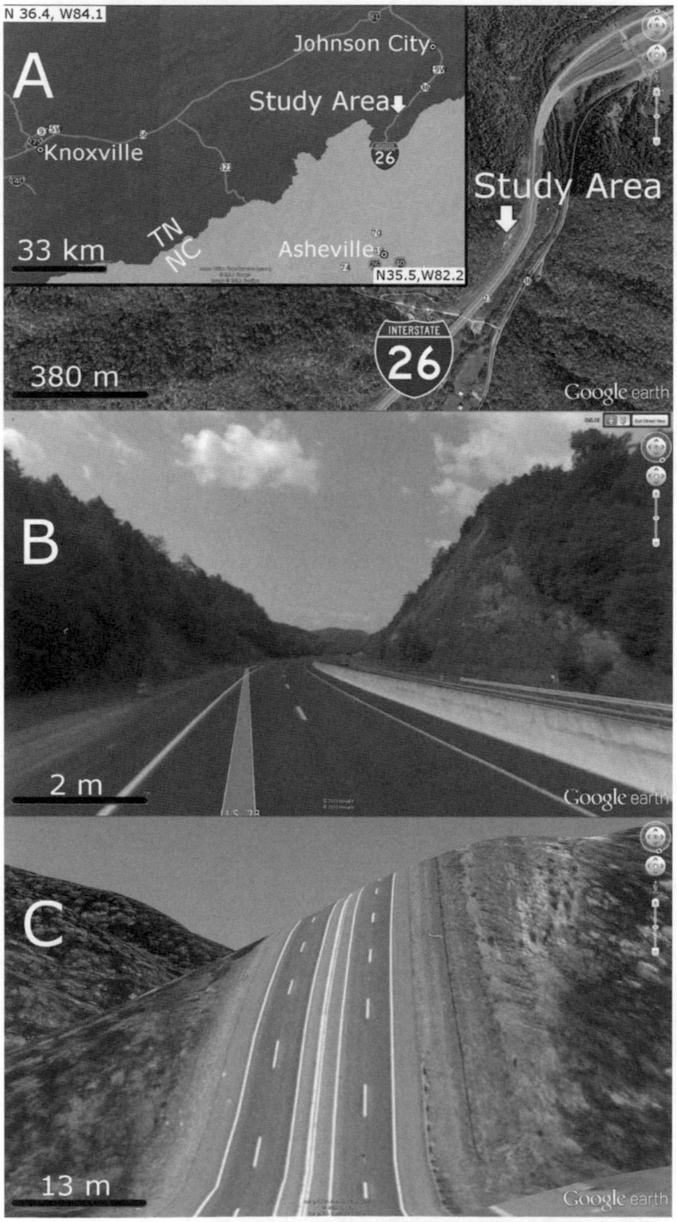

RESHAPING THE GOOGLE EARTH TERRAIN USING SKETCHUP

SketchUp has been used to model geologic structures and create block diagrams for Google Earth (e.g., Karabinos, 2010). In this study, SketchUp was used to modify the Google Earth terrain to reflect a new I-26 roadcut that was created during interstate construction in the 1990s. The outcrop is located ~13 km south of Erwin, Tennessee, on the NW side of I-26 near mile marker 44 (36.080472°N/82.495874°W) (Fig. 1). Here, 1.1 Ga Cranberry Mine Gneiss is unconformably overlain by Neoproterozoic Ocoee Supergroup metasedimentary rocks (see Hatcher et al., 2006, and references therein for a complete geologic description[1]). During excavation, more than 150,000 m^3 of rock was removed from this roadcut, creating a notch through a NW-SE–trending ridge (Moore, 1993). Currently, Google Earth satellite and Street View images show the outcrop, although it is not reflected in the Google Earth terrain (Fig. 1). This mismatch between the actual topography and the Google Earth virtual topography causes the interstate to rise sharply over the ridge that existed before the construction of the interstate.

Techniques for Creating a Terrain Model for Google Earth

To correct for the terrain disparities in Google Earth, SketchUp was opened and the coordinates of the outcrop were entered into the dialog box after clicking the "add location" command. Then, the "search" button was clicked followed by "select region." The area to be modified was defined by dragging the box corners and then clicking "grab." A copy of the Google Earth terrain was now imported into SketchUp.

In SketchUp, a rectangle was created in the shape of the roadcut using the "rectangle" tool (Fig. 2) and the terrain was turned on with the "toggle terrain" button (Fig. 3). Next, the rectangle was extruded into a box using the "push/pull" tool (Fig. 4). Using the "rotation" tool, the box was rotated parallel to the roadcut; and with the "scale" tool, the height, width, and length were adjusted to intersect the surface of the roadcut (Fig. 5). The sides of the rectangle were then tilted to match the outcrop faces in the satellite image (Fig. 6).

Next, the volume to be excavated was removed from the SketchUp model. This was done by right-clicking the terrain and then clicking "unlock." The terrain was right-clicked again and then "explode" was selected. This caused the surface to break

Figure 1. Google Earth satellite view (A), Street View (B), and terrain rendering (C) of the study outcrop. The outcrop is ~35 km south of Johnson City, Tennessee, along I-26. Note how the outcrop is depicted in the satellite (A) and Street View (B) but not in the Google Earth terrain (C). In (B) and (C), view is to the southwest; in (C) there is no vertical exaggeration.

---

[1]GSA Data Repository item 2012303, Hatcher et al. field guide, is available at http://www.geosociety.org/pubs/ft2012.htm, or on request from editing@ geosociety.org.

into many polygons. The terrain was right-clicked once more and then "intersect faces" → "with model" was selected (Fig. 7). The unwanted material was removed by selecting each volume and clicking "delete" or by using the "eraser" tool (Figs. 8 and 9).

After building the terrain model in SketchUp, photomosaics were added to the face of the virtual outcrop in SketchUp (Fig. 10). Field photos were added by clicking "file" → "import." "Use as image" must be selected in the "import" window. Google Earth Street View photos can be used instead of field photos by using the "add photo texture," but their availability is localized and their resolution is generally poor.

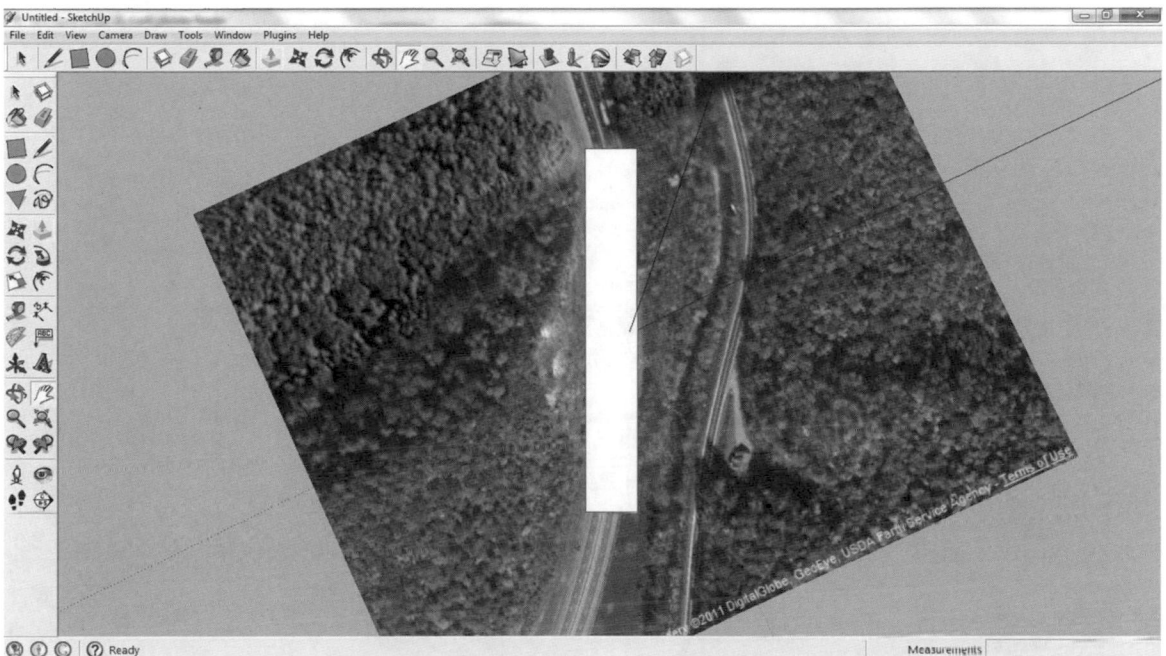

Figure 2. A rectangle was constructed in SketchUp and rotated parallel to the road.

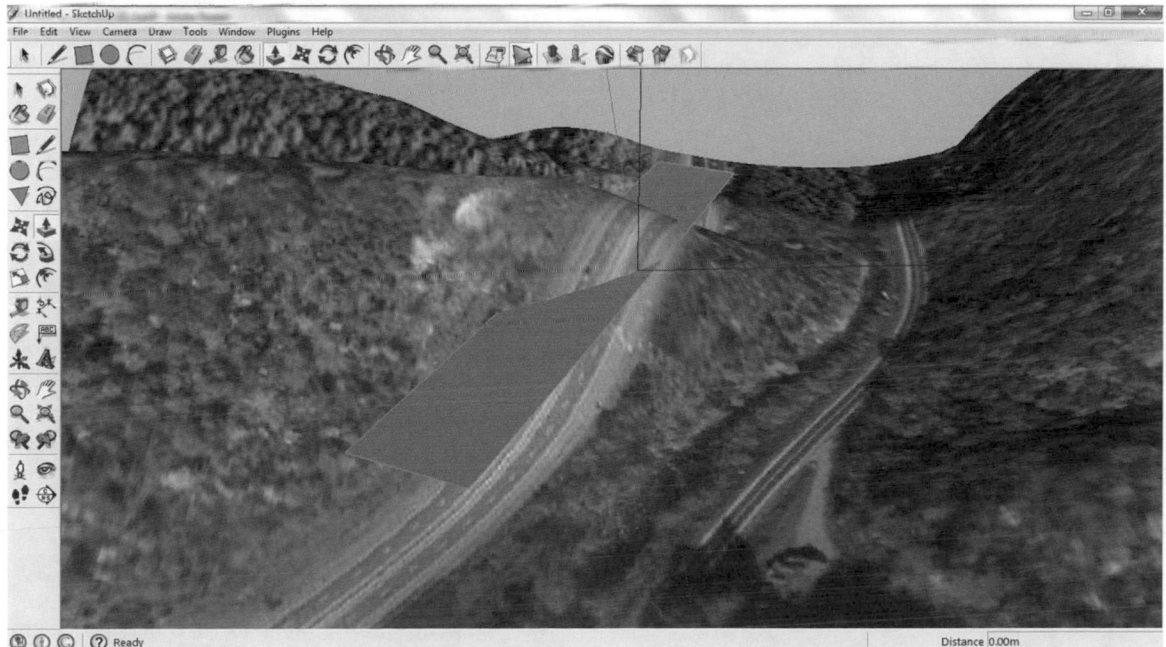

Figure 3. The terrain was turned on by clicking the "toggle terrain" button.

Once the photomosaics were added in SketchUp, the new terrain model was exported as a .kmz file (a zipped Keyhole Markup Language [KML] file) and then opened in Google Earth. Once opened, the .kmz file was automatically georeferenced. The model was displayed in Google Earth by unchecking the "show terrain" box in the "options" menu under the "tools" drop-down tab (Fig. 11). If the Google Earth terrain is not deactivated, the SketchUp terrain model will not appear. Lastly, when the model was added to Google Earth, it was not level with the Earth's surface. The altitude was adjusted to align the bottom of the model to the Earth's surface (Fig. 12); the terrain modification was now complete (Figs. 13 and 14).

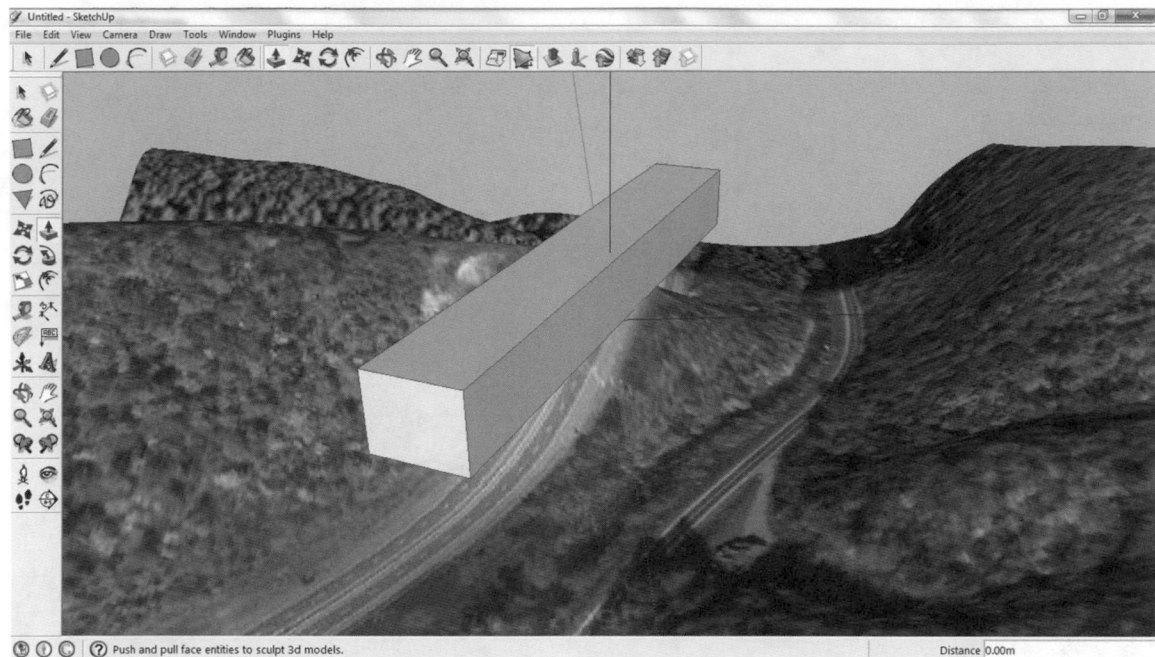

Figure 4. Using the "push/pull" tool, the rectangle was extruded into a box.

Figure 5. With the "scale" tool, the sides of the box were adjusted to the height, width, and length of the outcrop.

Issues Modifying the Google Earth Terrain Model

A serious limitation of this terrain-modification technique is the necessity by the user to flatten the terrain in Google Earth in order to display the model constructed in SketchUp. The imported terrain models from SketchUp have topography but they sit atop a flat Google Earth surface without topography (Figs. 12 and 13). This problem arises because, to the best of our knowledge, a user cannot manipulate the elevation data in Google Earth. However, if the user was able to manipulate the topographic data or

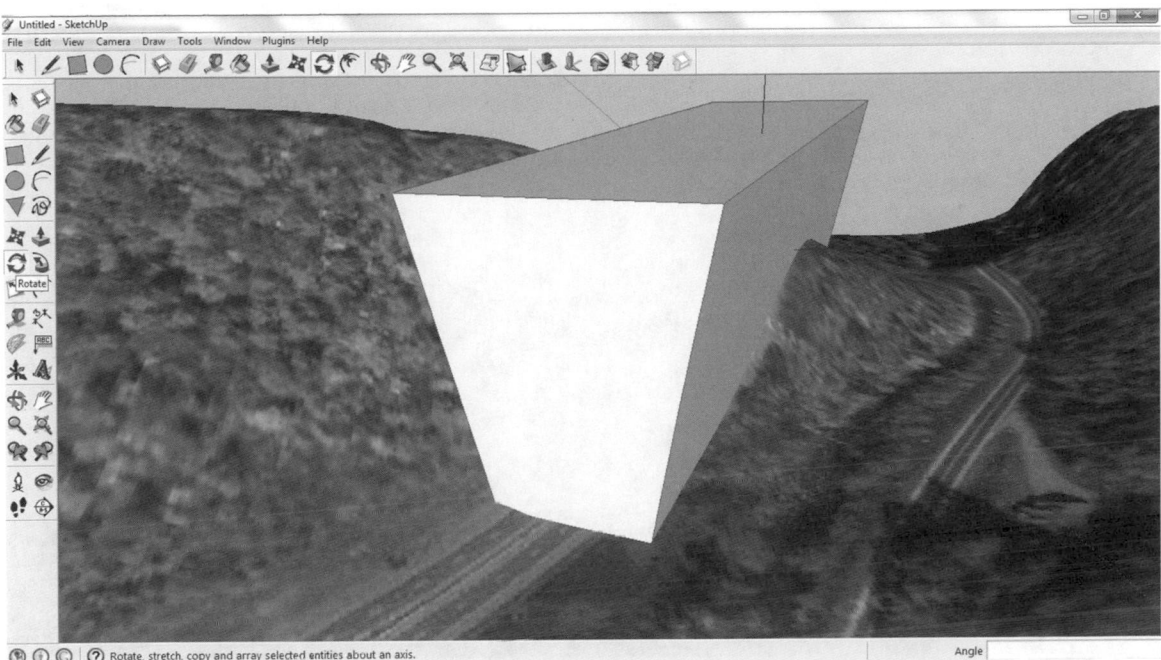

Figure 6. The sides of the box were rotated to align parallel to the outcrop faces.

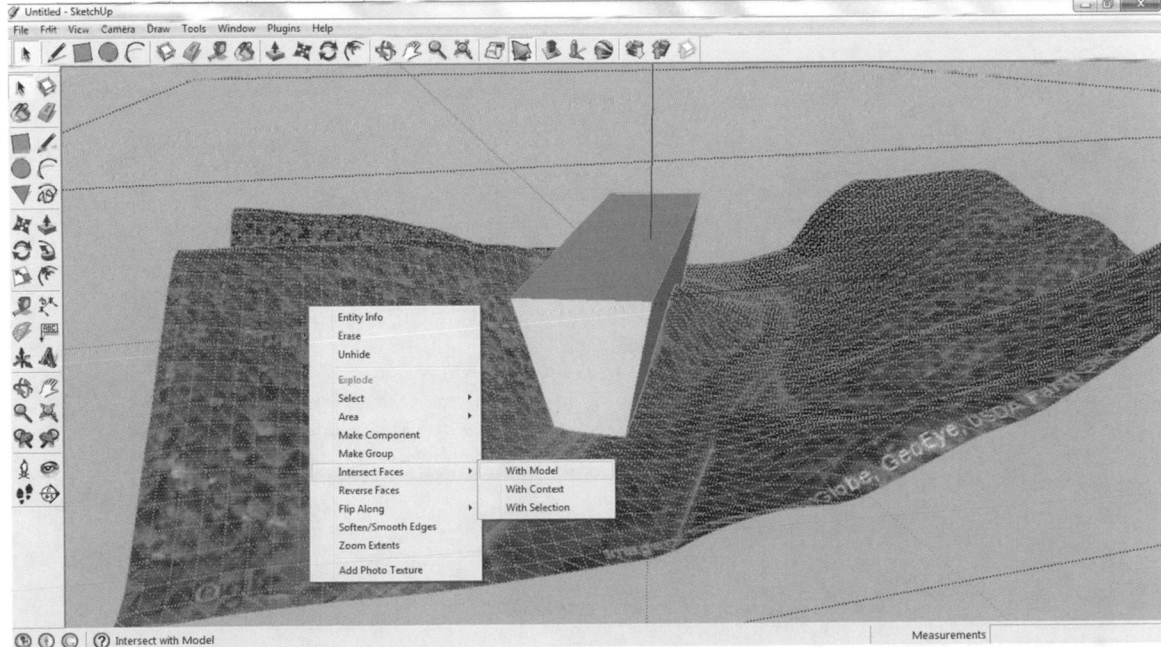

Figure 7. After "unlocking" and "exploding" the surface, the surface was combined with the box by right-clicking and selecting "intersect faces" → "with model."

delineate zones of transparency, one could modify the landscape without having to turn off the terrain.

Another issue with modifying the Google Earth terrain is that large areas (>4 km²) cannot be copied into SketchUp. This can be overcome by copying the terrain piecemeal until the desired area is achieved (Fig. 15). This technique is useful for creating topographic context around the SketchUp model (Fig. 16). Also, other than changing location, models produced in SketchUp cannot be modified in Google Earth. Any changes to the model must be performed in SketchUp and the process of exporting and importing into Google Earth must be repeated.

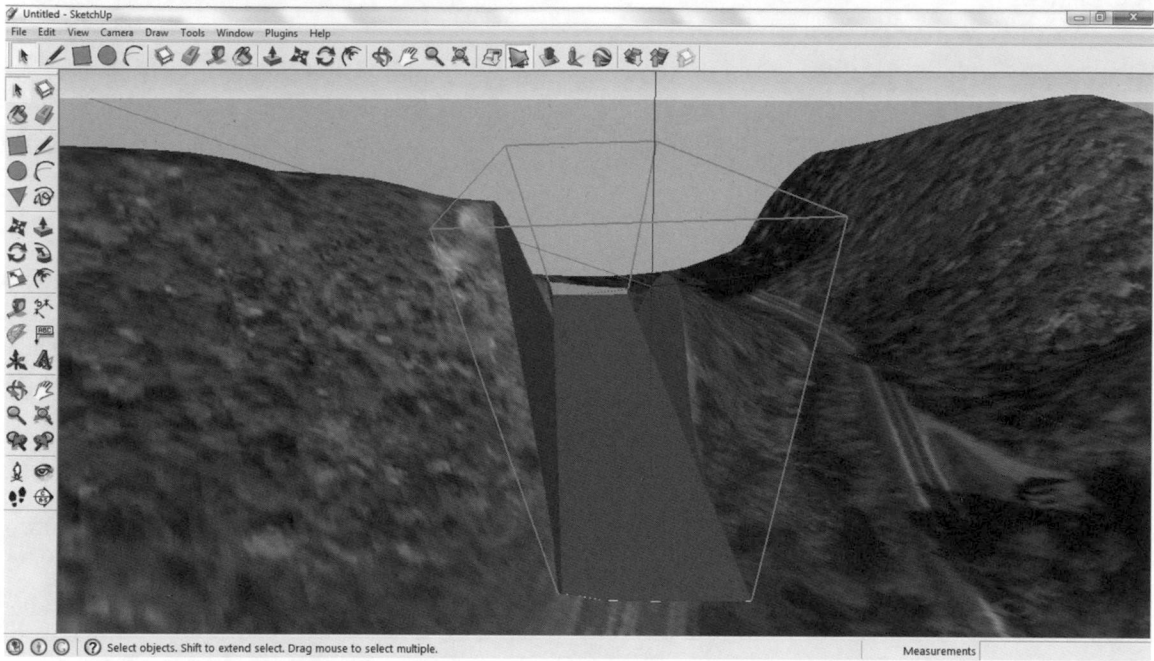

Figure 8. The volume to be excavated was deleted with the eraser tool. This removed the portion of the copied Google Earth surface that intersected the box.

Figure 9. The lines of the box were deleted and the bottom surface was removed.

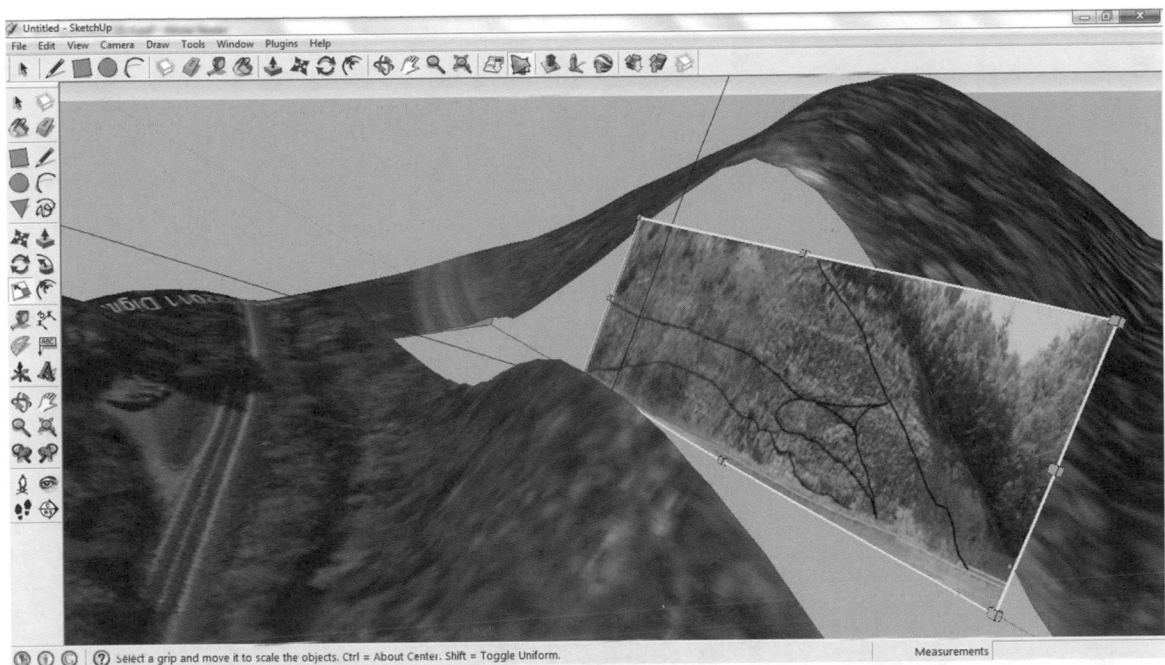

Figure 10. Once the excavation was complete, a photomosaic was added to the face of the terrain model. Lines were added to the photomosaic to highlight outcrop-scale structures. The model was then exported as a .kmz file using "export" → "3-D model."

Figure 11. In Google Earth, the .kmz file was opened in the "File" menu. This added the model in a georeferenced location. To view the model, the terrain must be turned off. This is found under "Tools" → "Options." The check box next to "show terrain" was deselected.

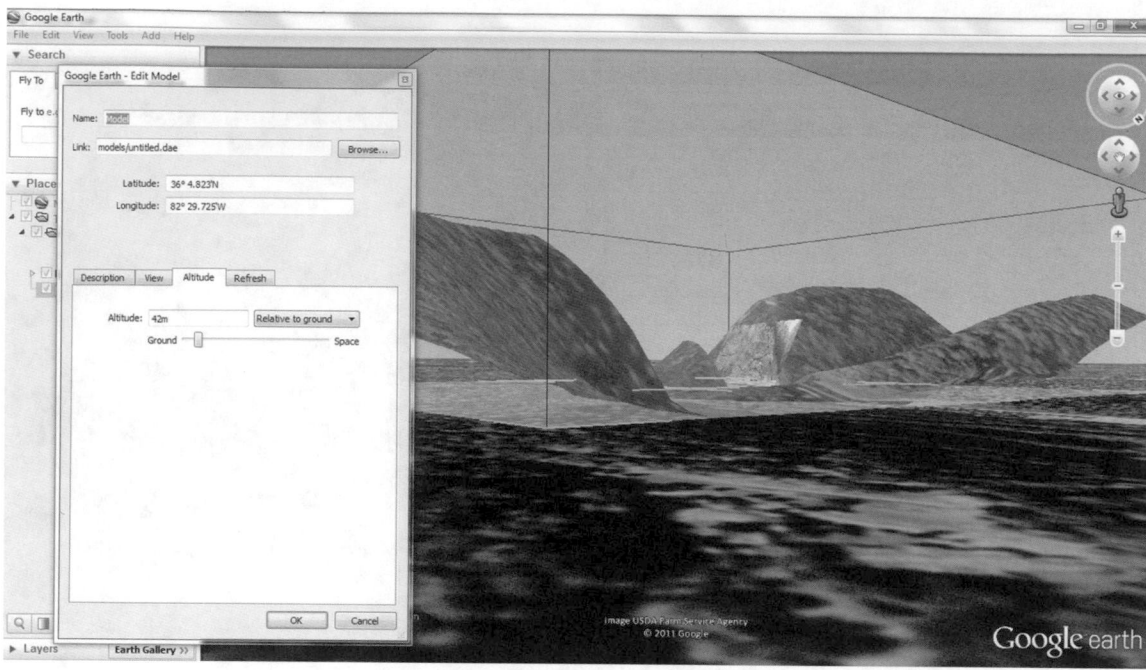

Figure 12. Adjustments were made in Google Earth to level the bottom of the terrain model to the Earth's surface. This was done with the "altitude" slider bar.

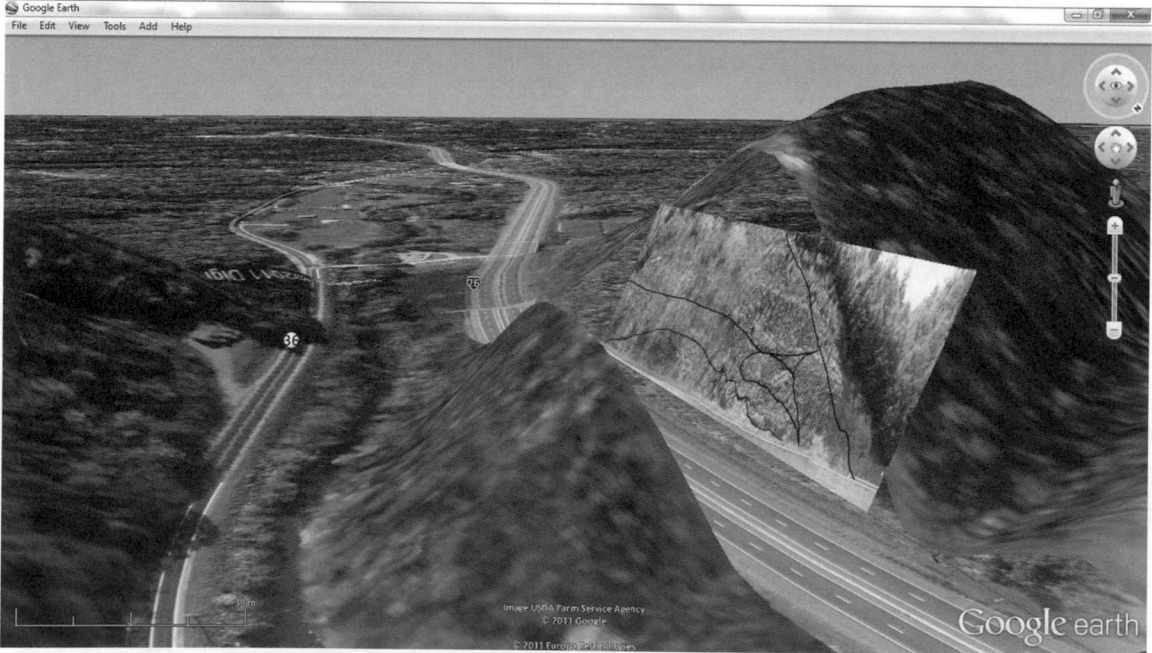

Figure 13. The SketchUp terrain model shows topography whereas the surrounding area in Google Earth is flat. The terrain must be deactivated to show the model.

Figure 14. Terrain modification before and after images. The before image shows I-26 rising over a ridge that is no longer present. The after image shows the modified terrain with photomosaic that was geo-referenced in Google Earth. Lines on the mosaic show representative outcrop structures.

Figure 15. The maximum area that can be imported into SketchUp at one time with the "add location" command is 4 km². However, larger areas can be constructed piecemeal in Google Earth SketchUp by repeating the import process.

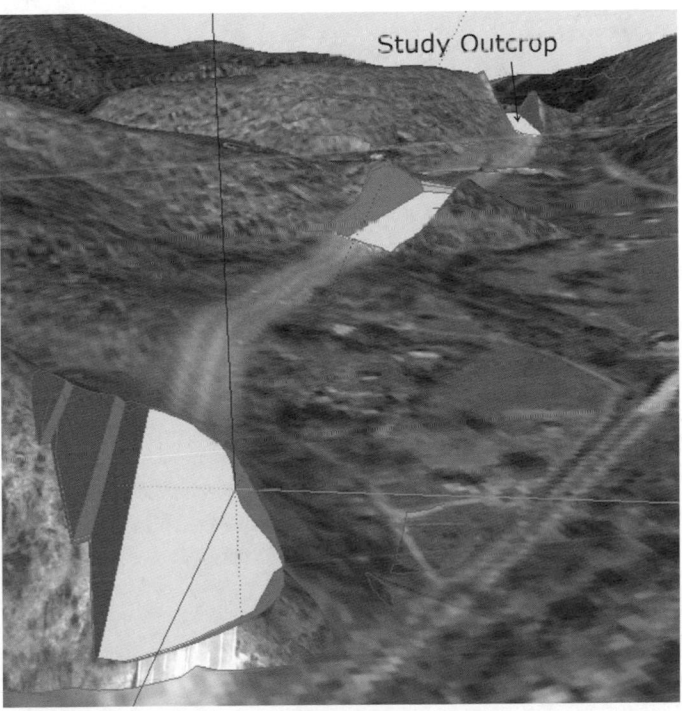

Figure 16. SketchUp terrain model showing multiple surface copies imported from Google Earth and pieced together. Once this larger area was constructed in SketchUp, multiple excavations were completed along I-26. This created topographic context around our study outcrop (upper right of image).

CONCLUSIONS

Google Earth is a free and dynamic virtual globe that is revolutionizing geoscience research and how that research is conceptualized and communicated. However, Google Earth elevation data may be outdated and may not accurately represent the current-day topography. In this paper, we show how to build a terrain model in SketchUp that can be used to modify the Google Earth terrain. These techniques can be applied to any region in Google Earth where disparities exist between the Google Earth terrain and the real world.

ACKNOWLEDGMENTS

The authors wish to thank Paul Karabinos, Crystal Wilson, and Steve Whitmeyer for providing detailed reviews and comments on an earlier version of the manuscript.

REFERENCES CITED

De Paor, D.G., and Whitmeyer, S.J., 2009, Innovations and obsolescence in geoscience field courses: Past experiences and proposals for the future, *in* Whitmeyer, S.J., Mogk, D.W., and Pyle, E.J., eds., Field Geology Education: Historical Perspectives and Modern Approaches: Geological Society of America Special Paper 461, p. 45–56, doi:10.1130/2009.2461(05).

Hatcher, R.D., Jr., Merschat, A.J., and Raymond, L.A., 2006, Geotraverse: Geology of northeastern Tennessee and the Grandfather Mountain region, *in* Labotka, T.C., and Hatcher, R.D., Jr., eds., Geological Society of America 2006 Southeastern Section Meeting Field Trip Guidebook, p. 129–184. (This paper is available as item 2012303 in the GSA Data Repository at http://www.geosociety.org/pubs/ft2012.htm, or on request from editing@geosociety.org.)

Karabinos, P., 2010, Adding structures to 3-D geologic maps: Geological Society of America Abstracts with Programs, v. 42, no. 5, p. 421.

Knoop, P.A., and van der Pluijm, B., 2006, GeoPad: Tablet PC-enabled field science education, *in* Berque, D., Prey, J., and Reed, R., eds., The Impact of Pen-based Technology of Education: Vignettes, Evaluations, and Future Directions: West Lafayette, Indiana, Purdue University Press, p. 103–114.

Moore, H., 1993, Engineering Geology of I-26: Engineering Geology Field Trip Guide, Tennessee Department of Transportation.

Whitmeyer, S.J., and Beard, D.L., 2010, CIT sandbox workshop: Creating content in Google Earth: http://csmres.jmu.edu/Geollab/Whitmeyer/web/visuals/GoogleEarth/CITworkshops.html (accessed 22 Nov. 2011).

Whitmeyer, S.J., Nicoletti, J., and De Paor, D.G., 2010, The digital revolution in geologic mapping: GSA Today, v. 20, no. 4/5, p. 4–10, doi:10.1130/GSATG70A.1.

MANUSCRIPT ACCEPTED BY THE SOCIETY 16 APRIL 2012

The Geological Society of America
Special Paper 492
2012

Interacting with existing 3D photorealistic outcrop models on site and in the lab or classroom, facilitated with an iPad and a PC

Miao Wang*
M. Iris Rodriguez-Gomez
Carlos L.V. Aiken
Department of Geosciences, University of Texas at Dallas, 800 West Campbell Road, Richardson, Texas 75080-3021, USA

ABSTRACT

Utilizing 3D photorealistic outcrop models for research and education is becoming more and more common in industry and academia. A system is presented for annotating on preloaded distortion-free photos used for model construction while in front of a real outcrop or a virtual 3D model in the lab or classroom. This interaction is accomplished through tracing geologic boundaries, writing notes and recording locations using an applet on an iPad. These geologic tracings are sent to a PC where 3D geometric information is extracted with a provided MatLab program. Annotations on the photo can then be visualized on the 3D model because control points from the model building process link 2D pixels on the photo with the 3D mesh by six transformation coefficients. The MatLab Program separates meaningful geological information based on pixel value defined by geologists for further analysis (strike/dip, rock type, etc.). Therefore geological features can be attributed, annotated, quantitatively analyzed and visualized in 3D or 3D stereo with available software. The University of Texas at Dallas as well as other groups have been creating such 3D models with laser scanning, GNSS (Global Navigation Satellite Systems) and digital photography since 1998. This approach can be used on any photorealistic model built based on the transformation correlation between imagery and 3D geometry. The website www.utdallas .edu/iGeology provides access to several such models of road cuts across the Arbuckle Anticline, Oklahoma. An instructor or project leader can access and set up the environment for students or users. This is not a field geologic logging/mapping system but is designed for interacting with and extracting quantitative from existing virtual models for research and education. Computations taking place on the PC are transparent to the user, and therefore the system can be readily used in academia and industry at many different levels of expertise. The tablet's portability enables users to interact with the outcrop in the field or with a 3D display in the lab. Similar applications can be built on android tablets. The system is provided in detail in three appendices.

---

*miao.wang@utdallas.edu

Wang, M., Rodriguez-Gomez, M.I., and Aiken, C.L.V., 2012, Interacting with existing 3D photorealistic outcrop models on site and in the lab or classroom, facilitated with an iPad and a PC, *in* Whitmeyer, S.J., Bailey, J.E., De Paor, D.G., and Ornduff, T., eds., Google Earth and Virtual Visualizations in Geoscience Education and Research: Geological Society of America Special Paper 492, p. 263–283, doi:10.1130/2012.2492(19). For permission to copy, contact editing@geosociety. org. © 2012 The Geological Society of America. All rights reserved.

INTRODUCTION

Photorealistic 3D outcrop modeling has been developed using 3D point clouds captured by terrestrial laser scanners with some draping digital photos resulting in colored point clouds. In others, the points are fit with a triangulated mesh upon which are draped the digital photos. The models can be integrated and geo-referenced, preferably by accurate survey grade (cm accuracy) GNSS. Each pixel contains spatial information (x, y, and z), intensity, and if available, color (RGB). The resulting virtual model provides a detailed and accurate (at the cm to mm level, depending on the equipment used for their creation) view of the actual outcrop, which can be displayed and interpreted with a variety of different software and hardware platforms. Such virtual models have been created at the University of Texas at Dallas in a procedure as defined by a patent (C.L.V. Aiken and X. Xu, Method and Apparatus for 3D Feature Mapping, U.S. Patent No. 6590640, 8 July 2003) and allow any photo draped on any feature to be utilized for this paper's iPad annotation/digitization scenario. The general solu-

tion is discussed and presented in Xu (2000), Xu et al. (2000), and Aiken et al. (2004) and is further developed and discussed in Thurmond et al. (2005), Oldow et al. (2006), Alfarhan et al. (2008), Olariu et al. (2008, 2012), White (2010), and Alfarhan (2010). Other examples of LiDAR (light detection and ranging) combined with photo work by other groups include Jones et al.(2004), Pringle et al. (2004), McCaffrey et al. (2005), Bellian et al. (2005), Clegg et al. (2005), Trinks et al. (2005), Enge et al. (2007), Labourdette and Jones (2007), Buckley et al. (2008), Fabuel-Perez et al. (2009), Fabuel-Perez et al. (2010), and Pringle et al. (2010). Detailed virtual models with surfaces can recreate the actual geology very realistically and, if displayed at great detail and in 3D stereo, allow a recreation of the field experience—effectively a virtual field trip. Using such models in teaching and research enhances archiving of thoughts and speculations and information extraction based on shape and color and on a 3D model. We illustrate the workflow for building a 3D photorealistic model with a case study of one of several existing outcrops of the Pennsylvanian Arbuckle Anticline along Interstate 35 in south-central Oklahoma (Fig. 1).

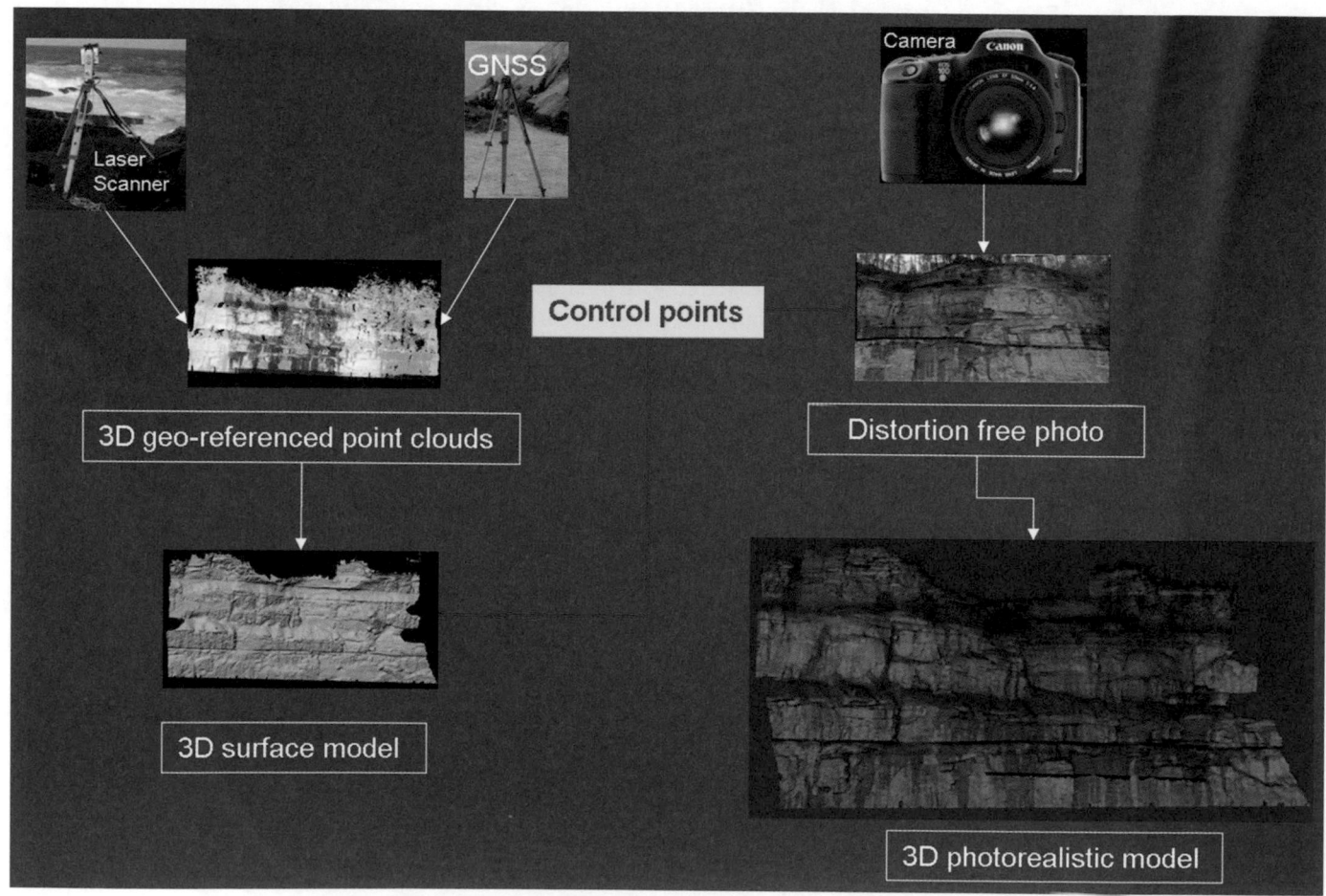

Figure 1. Workflow of generating a photorealistic model. Left: Field work using laser scanner and GNSS-generated geo-referenced surface model from point clouds. Right: Using control points, surface models are linked with distortion-free photos to produce 3D photorealistic models.

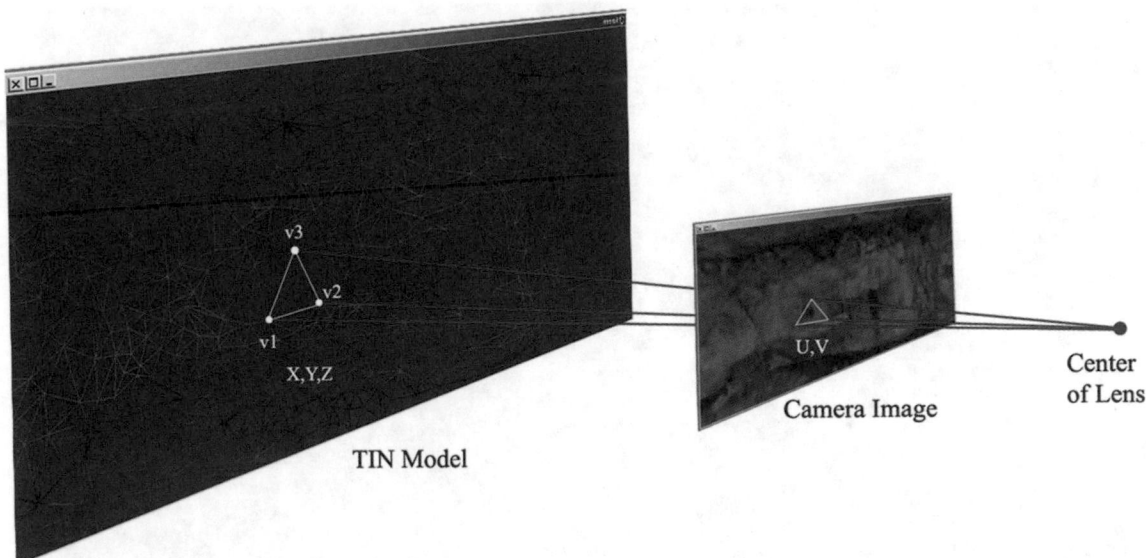

Figure 2. The vertices in the triangulated irregular network (TIN) model are mapped to corresponding pixels in the photographs (White, 2010).

DIGITIZING FEATURES AND ANNOTATING INFORMATION

Photorealistic modeling is accomplished by aligning the photograph to the model in approximately the same relationship (six transformation parameters) that the camera was aligned to the outcrop. Then, each of the vertices of the triangles is projected toward the photograph (Fig. 2). The mapping of 2D pixels into 3D involves five mathematical concepts—collinearity equations (Mikhail et al., 2001), image coordinates conversion, barycentric coordinates (Dunn and Parberry, 2002), k-dimensional (KD) Tree and k-dimensional nearest neighborhood (KNN) search. The collinearity equations are used for projecting 3D vertices into the 2D plane (charge coupled device, or CCD, sensor). The image coordinates conversion converts between different image coordinates, in this case, UV coordinates are transformed into Wavefront® OBJ coordinates. The barycentric coordinates are used for verifying if the center of one pixel is inside a triangle or not. KDTree and KNN do an efficient search of all the vertices to find the nearest vertex to each pixel defining the features to be extracted (Fig. 3).

Geological Example

One of several models of outcrops of the Pennsylvanian Arbuckle Anticline is of a fold exposed in a road cut ~5 m tall and 100 m long, along Interstate 35 in south-central Oklahoma (34°22′37.90″N, 97°08′37.81″W), available for downloading at www.utdallas.edu/igeology. This is an outcrop visited by academic organizations throughout the Midwest because of its well-exposed structures, in this case, a local anticline and syn-cline (a fold). The Arbuckle Anticline runs northwest to southeast, consisting of Paleozoic dolomite, limestone, sandstone subsequently folded and faulted during the Middle and Late Pennsylvanian (Fig. 4). The site is regularly visited by fieldtrips from undergraduate and graduate courses at University of Texas at Dallas.

The photorealistic model consists of five 2048 × 2048 pixel photos draped on a surface triangulated irregular network (TIN) model, an obj file, (Fig. 5), which can be downloaded at www.utdallas.edu/igeology).

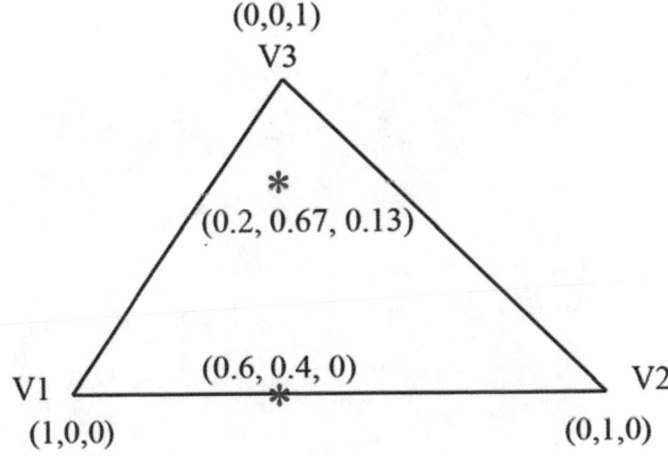

Figure 3. A triangle showing an example of barycentric coordinates (red) and vertices of the triangle (black) (White, 2010).

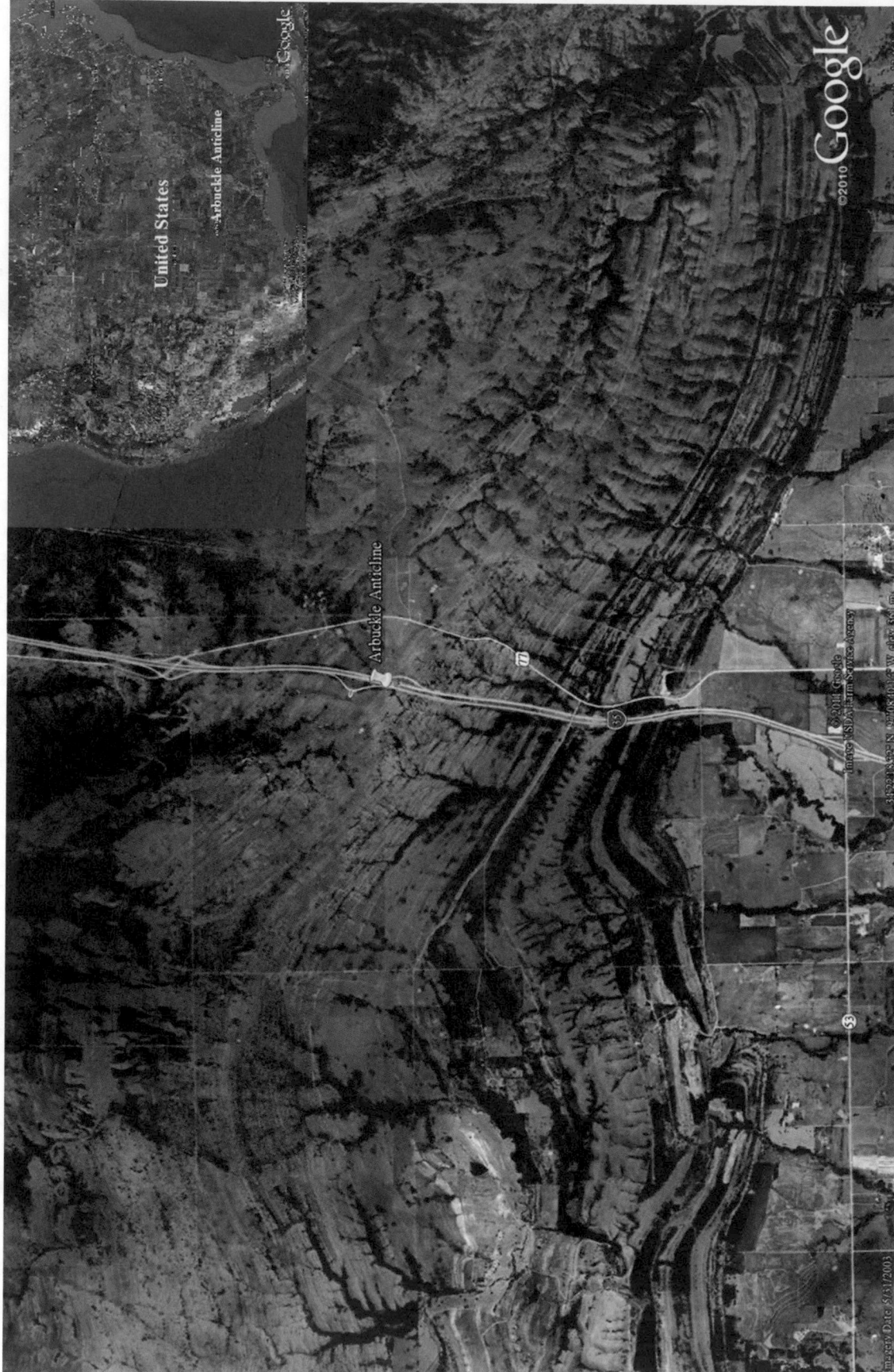

Figure 4. Arbuckle Anticline, Oklahoma, viewed with Google Earth. The yellow pin locates the anticline/syncline outcrop along I-35 and the inset is its location in North America.

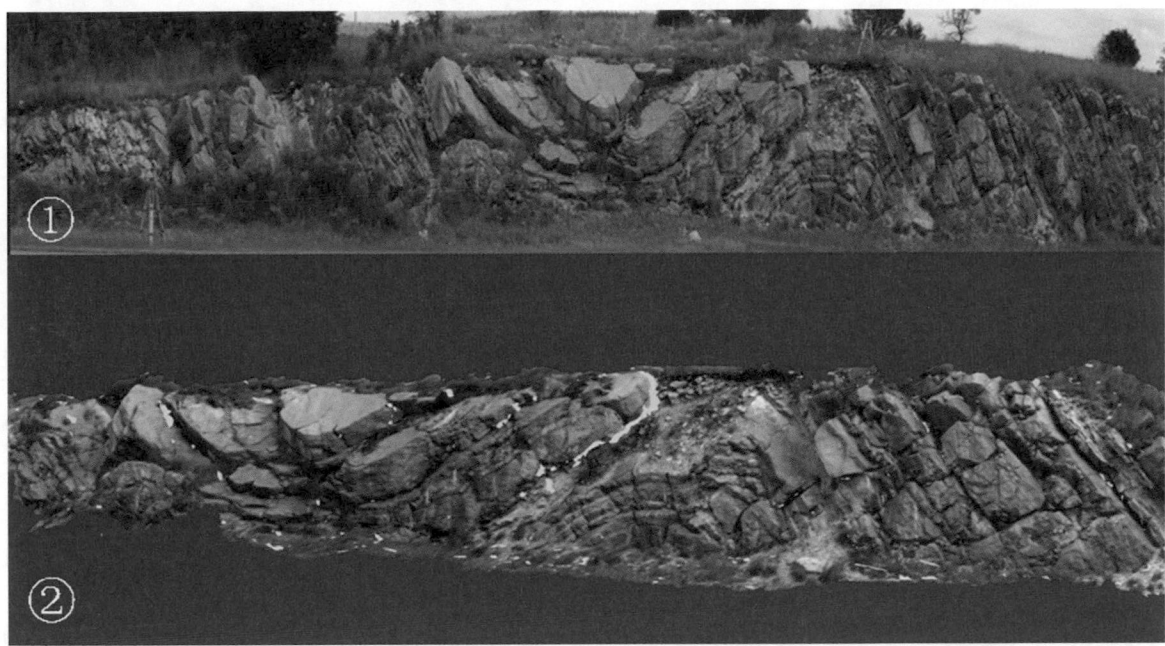

Figure 5. (1) Photomosaic of Arbuckle anticline/syncline (using Nikon D200 with 50 mm focal length) and 5 photos. (2) Photorealistic model of Arbuckle anticline/syncline. Yellow areas are those lacking in appropriate photo pixels (White, 2010).

INTERACTION WITH THE 3D PHOTOREALISTIC MODEL

Hardware and Software Requirements

To utilize this interaction system, it's necessary to have an iPad (1 or 2; for this application it make no difference) and a PC (512 MB graphic card or better, memory card no less than 1 GB; very modest requirements). On the iPad, we use a free applet named PaperPad as an example of an app that has the ability to trace lines on photos, and is available to be downloaded through the App Store. On the PC, OpenSceneGraph is one example of many available programs that can be used for visualizing 3D photorealistic models and is an open source software available free for download at http://www.openscenegraph.org/projects/osg (the version currently used in this paper can be downloaded at www.utdallas.edu/igeology) (Appendix I). PDF Converter is necessary if an app like PaperPad is used where the output is in a .pdf format and therefore must be converted into a .jpg format in order to use our processing software on the PC. The software can be downloaded at http://www.pdftiger.com/, or at www.utdallas.edu/igeology.

System Initialization by Instructor—Setting up Data Sets and Software

We present a system that allows a user to utilize available 3D photorealistic models by using a tablet like an iPad and an applet for tracing upon its screen annotations and comments and indicating locations with values of measurements very efficiently (see Appendix II). This information is then displayed in 3D (or even 3D stereo) on the 3D model on a PC. Digitized geometric contacts and other geologic features can be traced and are defined as pixels. These can be subsequently sent to a PC and transformed into 3D points for 3D geometric analyses and interpretations to be carried out (Appendix III). This entire presented system does not require the instructor/project director or the user to be a programmer or a specialist in LiDAR or virtual mapping and model building to be effective. Once the data set and software is prepared there is no need to come back and redo it to operate at a specific site.

A work folder is created on a PC. At the website (www.utdallas.edu/igeology) the instructor can download the folder "data.zip" which includes the 3D photorealistic model (in .obj file format or if preferred in multipatch format used for digitizing the 3D photorealistic model in ArcGIS ArcScene software), the original photo folder and software folder (follow the instructions inside or refer to the appendices of this paper) (Fig. 6).

Annotating in the Field and Analyzing and Visualizing in the Lab

The user begins by opening a paint applet (like PaperPad used for this example), loading a photograph from the 3D model covering of the area of interest on the outcrop, annotating by defining features based on color (for instance, red can indicate faults,

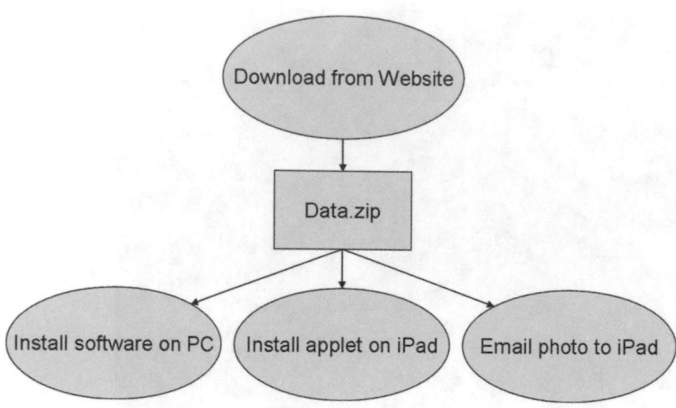

Figure 6. Flowchart of system initialization by instructor setting up data sets and software. Details are provided in Appendices I, II and III.

blue for text), then saving the photo and emailing it back to the PC with the provided programs. While either in the lab or while still in the field, the user can download photos from the email and use a PDF converter to convert the emailed photo from .pdf to .jpg as discussed earlier. It is important to remember to change the photo name given to the file sent from PaperPad to the name written on the corner of the photo. The user can drag the renamed photo into the model folder and replace the original photo (save the original photo in some other folder). Geologists can use the iPad to trace geological features of interest and write down measurements such as manually determined strike/dip, bedding thickness, or other notes on the preloaded distortion-free photos distinguished by using different colors and thicknesses of line (Fig. 7). The YouTube video, which is available at http://www.youtube .com/watch?v=GsK3kUe1IWg, demonstrates this concept.

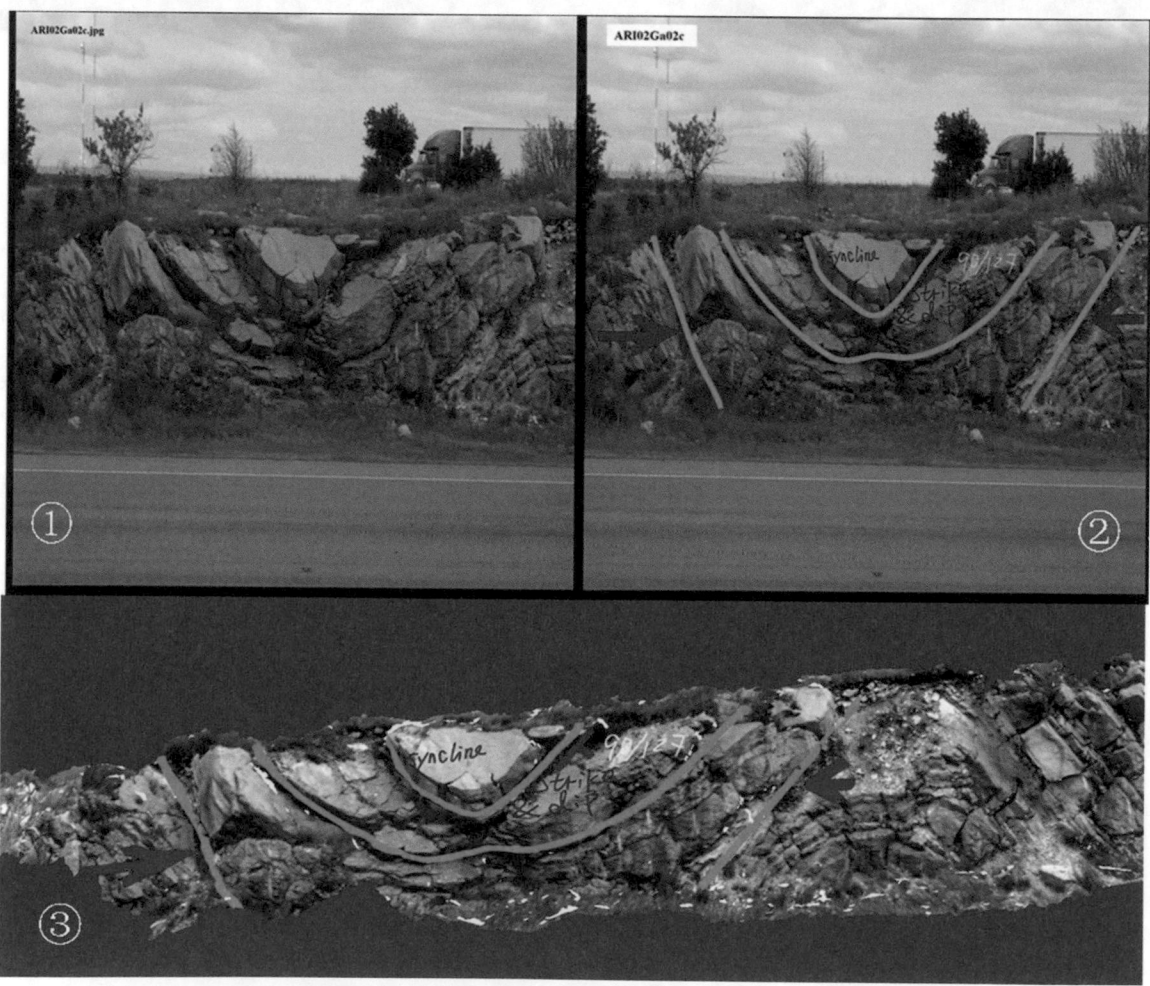

Figure 7. Example of a variety of annotations on the model. (1) Distortion-free photos preloaded on the iPad (photo ID is manually typed at the corner of the photo because iPad email system does not keep the original name of the photo). (2) Annotations on the photos with applet tracing (green lines—contacts; red—strike/dip location and interpreted deformation direction; blue—miscellaneous comments). (3) All annotations (contacts, descriptions, etc.) from (2) now displayed in 3D when overlain onto the original photorealistic model.

"Automatically" Extracting Geological Features on the PC

Original Matlab code is created for automatically extracting geological features annotated on the photo (Appendix III). Two different sets of contacts, the syncline in green and anticline in yellow, have been identified in the outcrop (Figs. 8, 9, and 10). Note that the choice of where to trace a contact and the thicknesses of the tracings is a factor to consider because an analysis is based on the pixels positions as defined by the tracings. The accuracy of the extracted geometries is dependent on how contacts are defined. For example, in this case the edges of the dip slopes are a more accurate indication of the ultimate orientation of interest than the contact between the lithologies. In using a compass for such measurements, the decision of where the compass is placed on the rock to best measure the orientation is important, as is the placing of the tracings of the contacts. The uncertainty of the computations due to the pixel distribution must be considered. This raster approach to digitizing geometries is quick and expeditious as a reconnaissance tool but is not as accurate as digitizing on a 3D surface model with a mouse.

Analysis of Extracted 3D Points

The extracted 3D points are saved as a .txt file in XYZ format and can be loaded into and analyzed by any software such as RockWorks®, GoCAD®, Petrel®, etc. This paper shows the strike/dip calculation tool applied in RockWorks® (Fig. 11).

Educational Applications

While working in the field during a project or as part of a field trip, having the ability to digitize and annotate one's observations

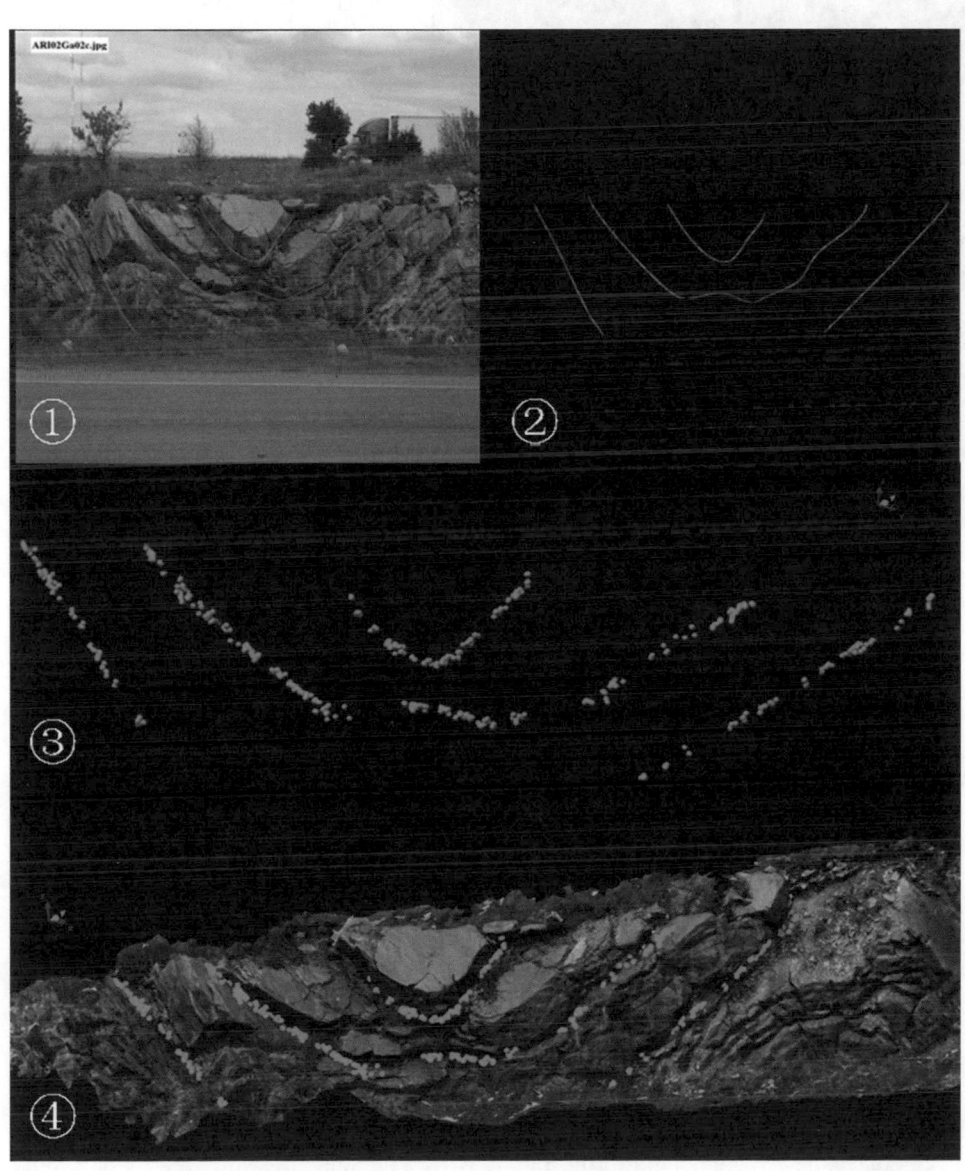

Figure 8. (1) Tracings are seen on the distortion-free photo preloaded into iPad using a stylus pen or finger (green). (2) Tracings are separated from the photo in UV coordinates based on their RGB value. (3) Extracted points in 3D space after their 3D transformation. (4) 3D extracted points mapped back onto the photorealistic mode for comparison.

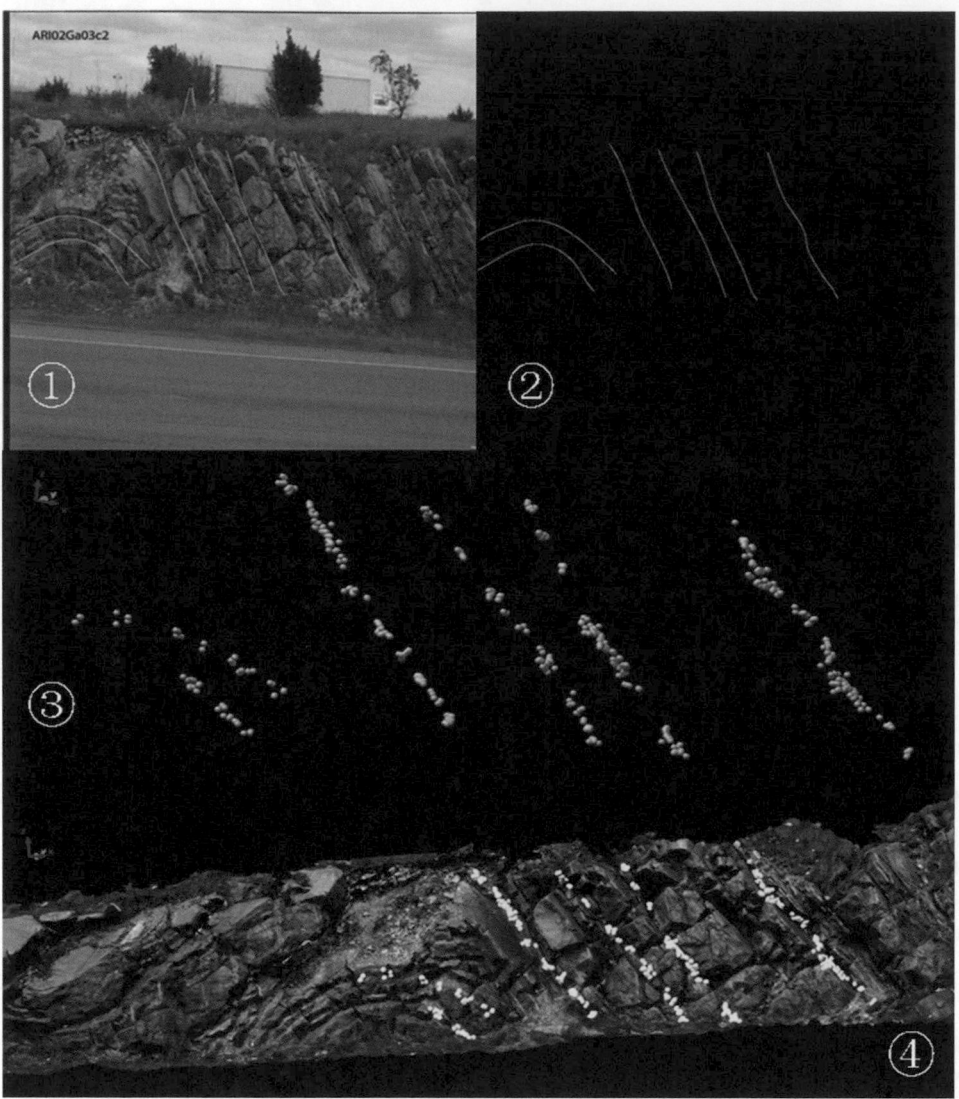

Figure 9. (1) The anticline contacts drawn onto the iPad using stylus pen or finger (thin yellow lines). (2) Extracted geometry tracings using feature extraction code. (3) Extracted anticline points in 3D space by transformation code. (4) 3D extracted points mapped back onto the photorealistic model. Different thicknesses of lines drawn by the applet on the photo effects the resultant density (number of pixels) that are available to be transformed into 3D to geometrically define features of interest and also accuracy of the computations.

directly onto a photograph that correlates to a 3D model of the same area can open up the user to further analysis and deeper understanding of the geology being observed. The user can zoom in or out of areas of interest on a photograph in order to investigate the geology more closely (Fig. 12). "Teachable moments" in the field are opportunities for the professor and student to dissect the geology at hand. Currently, discussion takes place and students write notes and collect samples. The iPad—essentially a digital field notebook–enables both professor and student to go a step further by providing a digital copy of observations, notes, and findings. Using the iPad and 3D model in classroom scenarios before (virtual field trips), during (expanding teachable moments) and after (review and analyze) create an opportunity for the students to dig deeper into the area being studied and its related geology (Fig. 13).

Utilizing familiar technologies such as an iPad in field work and classroom scenarios is a concept that is receiving more attention and consideration by the geosciences community. During field work (field trips), groups can utilize the iPad to digitize the geology and make annotations directly onto the photograph that correlates to that section of the outcrop. These photographs can be saved and emailed to a server for use later in the lab (or classroom). The iPad can also be used in conjunction with a low cost document camera for classroom scenarios. Placing the iPad under the document camera enables the professor or student to display real time digitization and annotations of the geology for the other students to see and comment on. Recent iPad developments include the capability to project onto a screen. Students who have their own iPad can digitize photographs in class or field, email them to the server and have them available for the professor or other students to critique or comment on. Anything being emailed from and to the iPad can also be received by iPhones or other smartphones monitoring the activity. The student-owned iPad would only

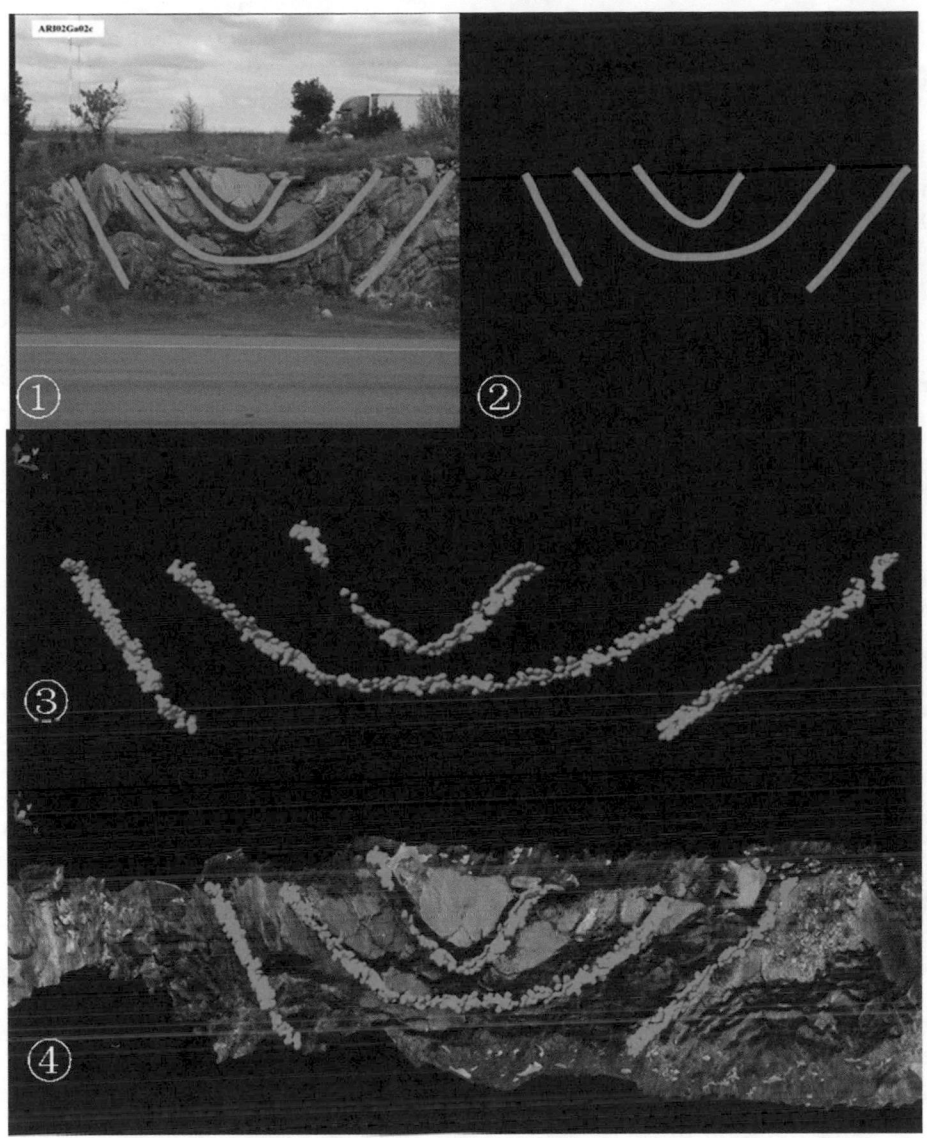

Figure 10. Example of consequences of thicker lines traced on photo: (1) Anticline contacts drawn on the iPad using stylus pen. (2) Extracted anticline tracings using feature extraction code. (3) Extracted anticline points in 3D space by transformation code. (4) 3D extracted points mapped back onto the photorealistic model for comparison.

need the photograph folder and the PaperPad app (or similar app). Fly-through capabilities are being developed for use in educational and research settings. The fly-through will enable the user to visit the outcrop sites and interact with the model in a familiar environment such as Google Earth or ArcGlobe (Fig. 14).

CONCLUSIONS

Once the instructor sets up the environment (software install and system initialization), the user will be working with simple steps—annotate a photo by drawing on a screen with a stylus or finger; send annotated photo to a PC and click one button in Mat-Lab. The mathematics and their computations are handled in the background, and the user does not need programming or other special training and training would be expeditious.

The photographs used for digitization are photographs which have been taken with a calibrated camera and have been mathematically related for 3D model development (embedded and draped onto the 3D model). Taking photographs with an iPad2 or other mobile device with internal camera and any uncalibrated camera will be provide a useful record but cannot be digitally or spatially related to the images on a 3D model; that is the purpose of this system. An instructor could immediately review in 3D the work collected in the field, but this real time iteration assumes an Internet connection in the field is available. In the Arbuckle case and in many models in north Texas, Internet access is in fact available.

Using the iPad allows the user to quickly, easily, and cost effectively interact with the photographs used in the 3D models. Repeated exposure to the same geology leads to deeper understanding and the ability to apply learning in new scenarios. The

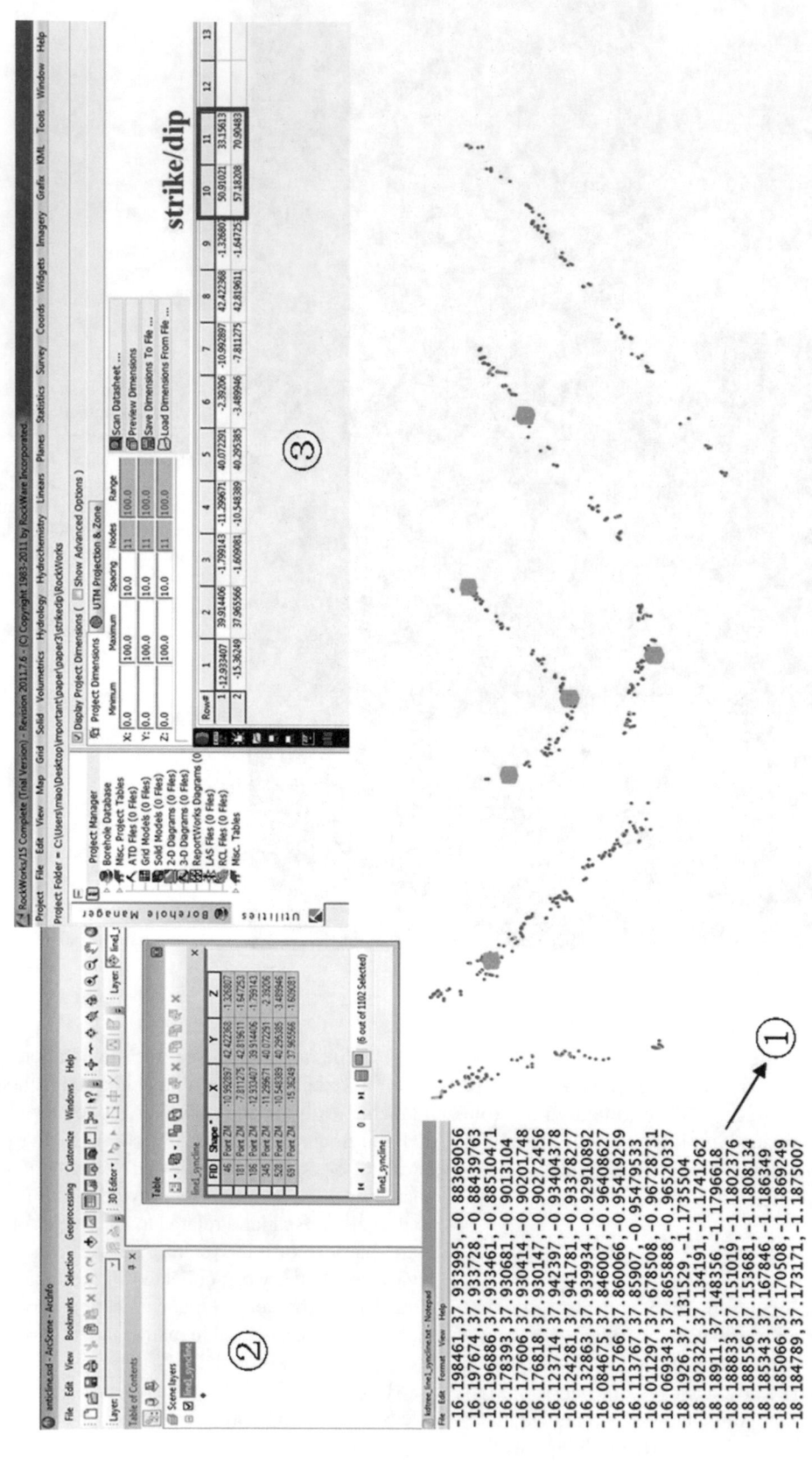

Figure 11. Analysis of extracted 3D points using RockWorks®: (1) xyz point clouds extracted from original MatLab codes in .txt file; (2) three points are picked of two contacts in ArcScene; (3) xyz values of three points in two sets are loaded into RockWorks. Strike and dips of three points for the upper contact is 50.91/33.15 while the other of three points for the lower contact is 57.18/70.90.

Figure 12. Students use an iPad to annotate a photo in field at the Preston Railroad cut, Plano, Texas (Preston model built by Xu et al., 2000).

Figure 13. Visualizing annotations on 3D photorealistic model in the lab with a 3D stereo TV.

Figure 14. Fly-through screen shot of Arbuckle Anticline. (Full movie can be downloaded at www.utdallas.edu/igeology.)

iPad has been considered to have a limitation in not having a USB port for data import and export but we have used it to communicate between the field and the lab and between systems in the field when the Internet is available, such as in urban areas, noting the mapping results in 3D stereo on a 3D TV.

This has been a useful approach, taking into account the technology of the present and the future. We are exploring the use of Android tablets, which may provide more flexibility.

Being able to interact individually or in a group in 3D or even 3D stereo as shown in Figure 13 can facilitate communication and understanding in teaching or in any other interaction as for example discussing opinions on geologic features in reservoir characterization and not requiring an immersive visualization system, but this is a simpler venue. A record of the annotations and observations on the photos is made and archived. As an example, this approach is discussed and being implemented in The Digital Integrated Stratigraphy Project (DISP) (Munnecke et al., 2012). This project aims to eliminate stratigraphic ambiguity associated with sample position within a stratigraphic section by utilizing 3D photorealistic models of GSSPs sites (Global Boundary Stratotype Sections and Points) for the International Stratigraphic Commission, and will be incorporating this raster approach for interaction with 3D models through a new website with accessible outcrops in a similar format that is worldwide in scope.

ACKNOWLEDGMENTS

The study was partially supported by National Science Foundation (NSF Projects 632050, 632102, 651529, 632402). We would like to thank Lionel White (Geological & Historical Virtual Models, LLC); Chander Ahuja; John Ferguson; Georgia Fotopoulos (who also provided the iPad); and Martin Orlob, Brian Burnham, Graham Mills, and Ranyah Kharwat (University of Texas at Dallas). Their contributions are gratefully acknowledged.

APPENDIX I. INSTRUCTIONS FOR INSTALLING OPENSCENEGRAPH ON THE COMPUTER

1. Run osgviewer.exe, follow OpenSceneGraph Setup Wizard, remember where you put installed file, such as C:\Program Files (x86)\OpenSceneGraph
2. Set up the environment variables. (Depending on your version of Windows, you may need to get there differently. Below is an example for Windows 7 users.)

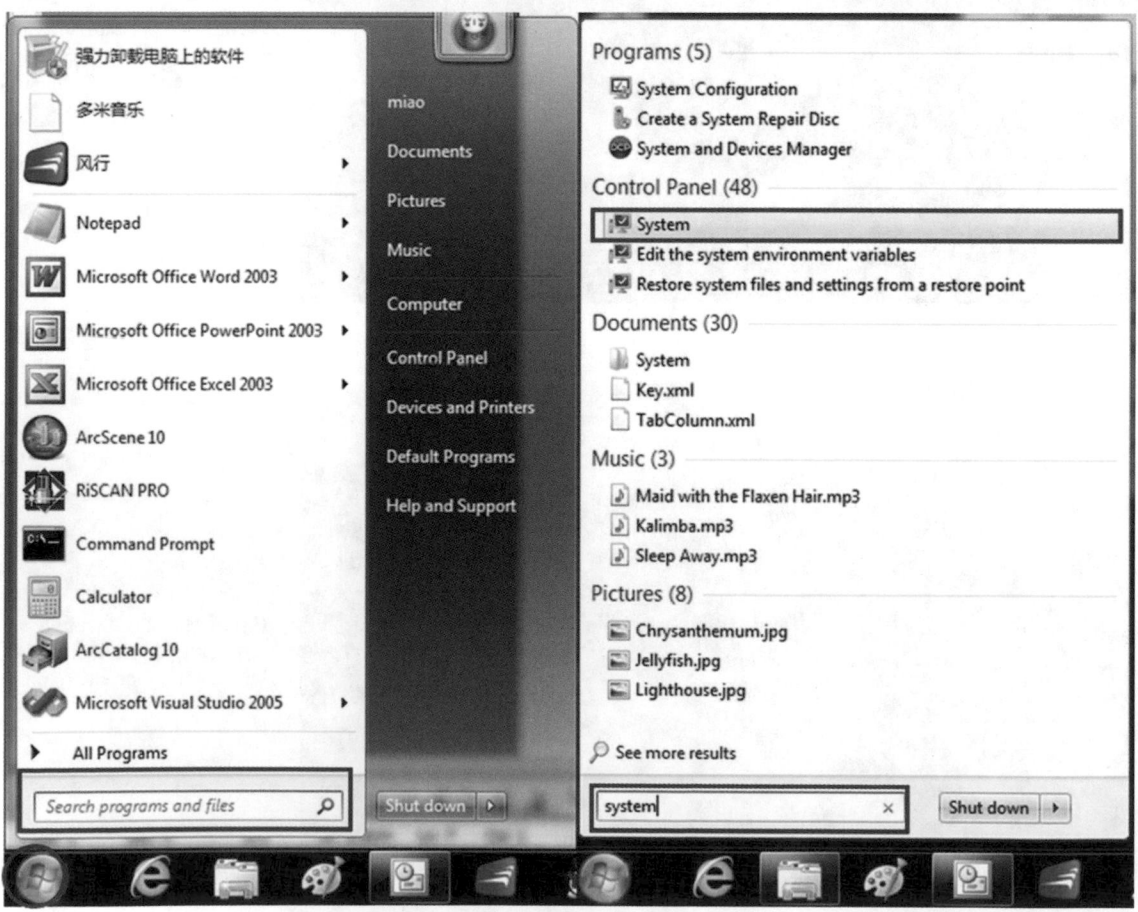

Figure AI_1.

Start → Search "System" → Choose "System" under Control Panel → on the left choose "advanced system setting" (Fig. AI_1, Fig. AI_2)

3. Here you need to tell Windows to let you access osgviewer. exe as a program from the command-prompt so you can type things like "osgviewer model.obj" in command-window no matter what directory you are in, or you can also use bat files. Add the bin folder containing osgviewer.exe to the Windows PATH environment variable. Edit the Path variable by first copying the full path to your bin folder which contains osgviewer.exe; you can copy the folder path from the address bar using Windows Explorer if you want. Then, you can add this folder path to the end of the list of folders specified in the Path variable. The folders listed in the Path variable are separated by semicolons, so you may need to add a semicolon at the end and then paste the folder path you are adding. See below:

Variable Name: PATH
Variable Value: C:\Program Files (x86)\OpenSceneGraph\bin to the end of variable value (Fig. AI_3)

New variables need to be added:
Variable Name: OSG_PATH
Variable Value: C:\Program Files (x86)\OpenSceneGraph\bin

Variable Name: OSG_ROOT
Variable Value: C:\Program Files (x86)\OpenSceneGraph
Variable Name: OSG_FILE_PATH
Variable Value: C:\Program Files (x86)\OpenSceneGraph\data;C:\Program Files (x86)\OpenSceneGraph\data\Images;C:\Program Files (x86)\OpenSceneGraph\data\fonts;C:\Program Files (x86)\OpenSceneGraph\data

APPENDIX II. INSTRUCTIONS FOR ANNOTATING ON 3D PHOTOREALISTIC MODEL

Data Preparation

1. Download Zip Tools and data.zip at http://www.utdallas.edu/igeology/data.html (Fig. AII_1).
2. Unzip the data.zip.
3. In software folder on your computer, install OpenSceneGraph and PDFTiger. After installing OpenSceneGraph, please review at the ReadMe.doc in the folder to learn how to set up environment parameters in your computer.
4. Install PaperPad on your iPad (Fig, AII_2).
5. Email the photo in original photo folder on your computer to the iPad and save it.

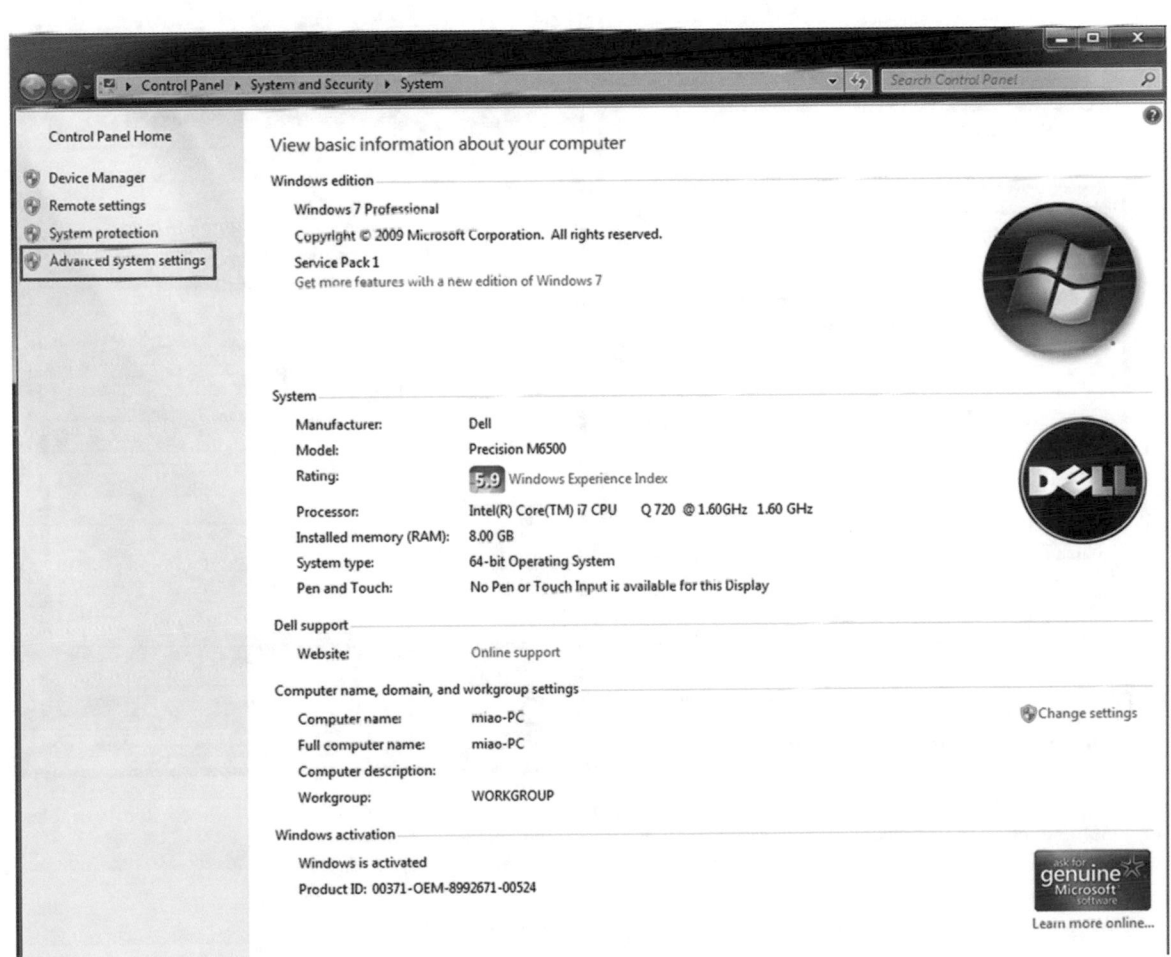

Figure AI_2.

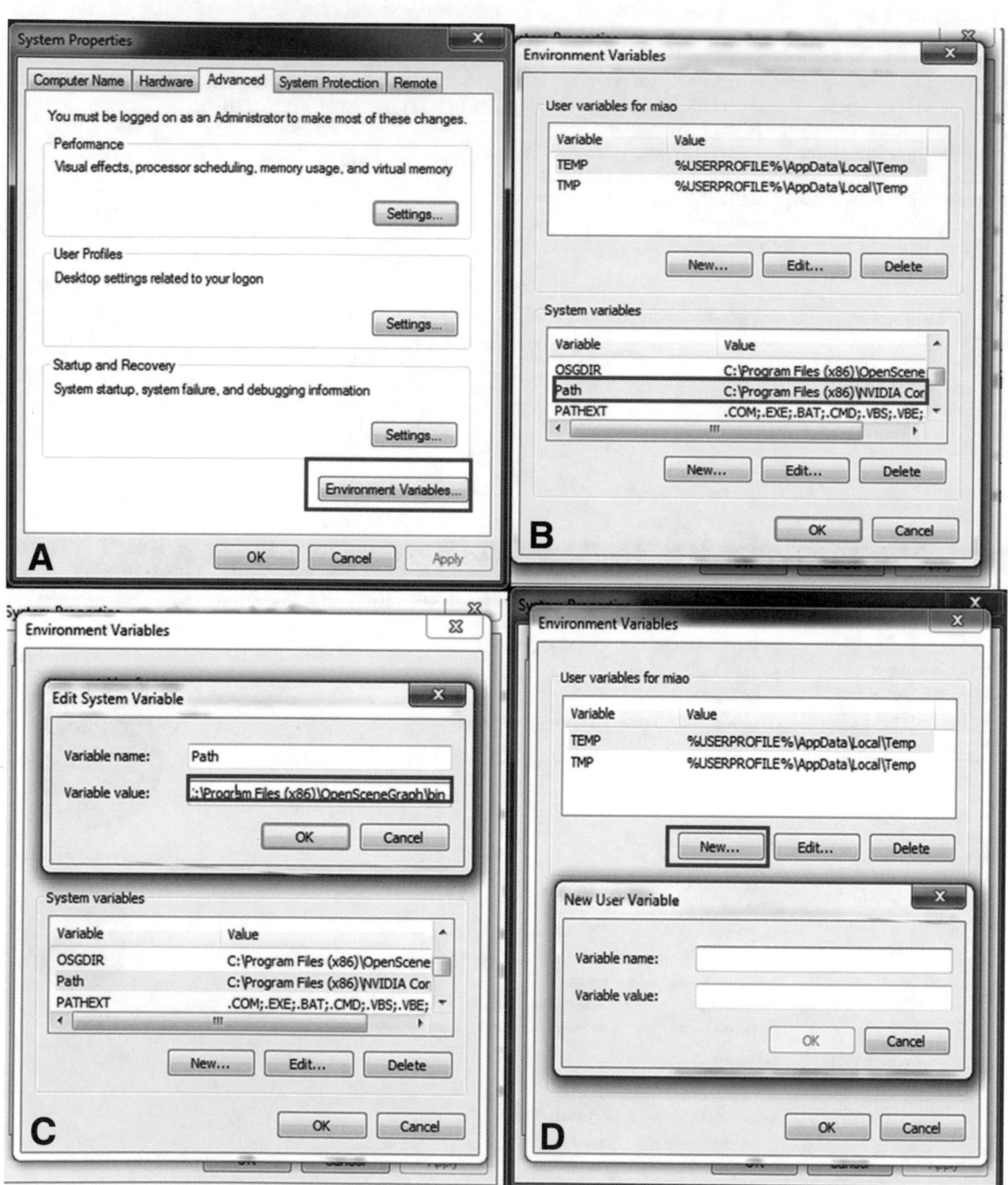

Figure AI_3. (A) Advanced system setting, choose environment variables. (B) Choose "Path" in System variables. (C) Double click it and edit as explain above. (D) Click "New" and add new variable.

Annotate in the Field or Lab

1. PaperPad on iPad (Fig. AII_3, Fig. AII_4).
2. Annotate geological features distinguished by color and thickness using finger or stylus (Fig. AII_5).
3. Email the annotated photo back to your email address. PaperPad generates .pdf file and will create a unique name for it (for this reason we need to include the photo name in the SUBJECT line of your email) (Fig. AII_6).
4. Download the .pdf file and use PDFTiger to convert it into .jpg file (Fig. AII_7).
5. Open folder which contains the .jpg file generated by PDFTiger, change the photo name to the original name i.e. ARI02Ga02c. Copy and paste this photo to the folder which contains 3D model (replace the photo with same name in that folder, better to save original photo without annotate into original photo folder).

6. Double click AntiSynClineFinal.bat file in the folder and visualize the annotations on 3D model.

APPENDIX III. INTRODUCTION FOR EXTRACTING GEOLOGICAL INFORMATION FROM PHOTOS WITH ANNOTATIONS

1. Copy photo with annotations into the code folder and change the name to ARI02Ga02c.jpg.
2. Open Matlab code depends on color you want to extract (Fig. AIII_1).
3. Click on green arrow button on the top to run the code (Fig. AIII_2).
4. Results are in the same folder of code, name by color for instance, red.txt (Fig. AIII_3).
5. Load .txt file into Petrel, GoCAD, ArcGIS, RockWorks and etc. to get strike/dip, surface fitting.

Figure AII_1. (A) Open www.utdallas.edu/igeology and click on "Data" at the bottom of the page.

Figure AII_2. (A) Open AppStore on the desktop. (B) Type PaperPad Lite in the search column. (C) Click "Free" on the right side. (D) Click "Install App" button. (E) Type in the iTunes Account. (F) PaperPad application show on the desktop.

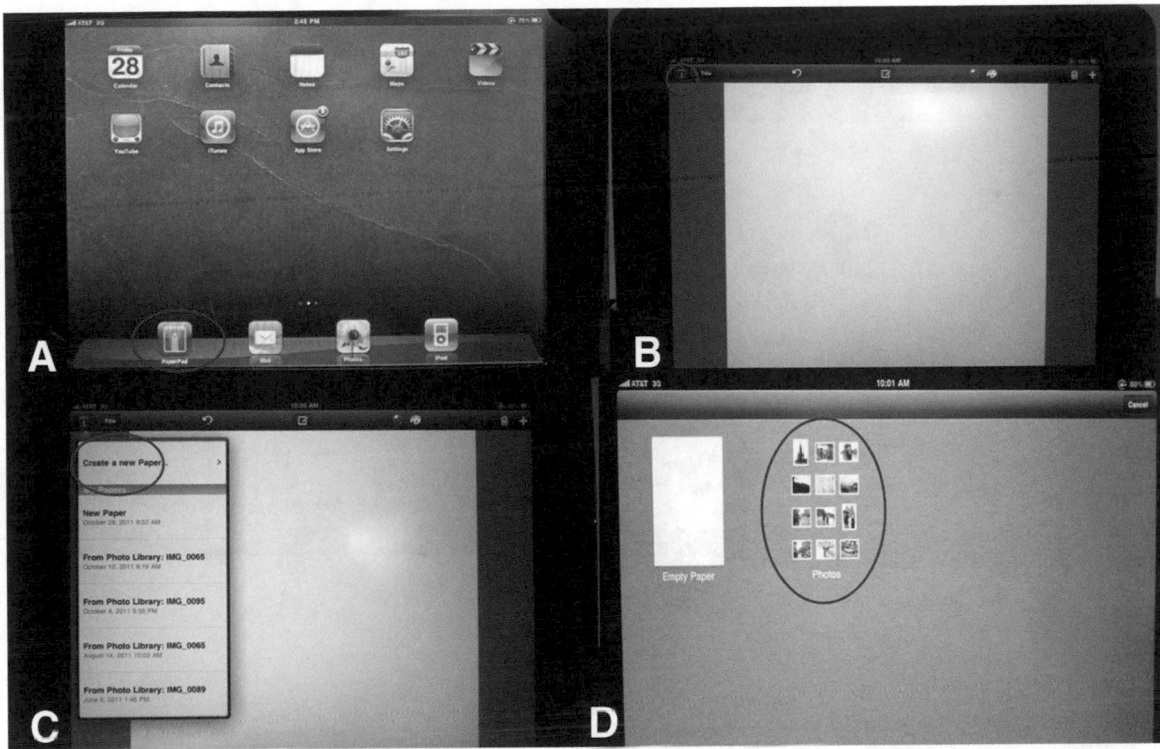

Figure AII_3. (A) Start PaperPad. (B) Click the button on the left corner. (C) Click on "create a new paper." (D) Click on photos.

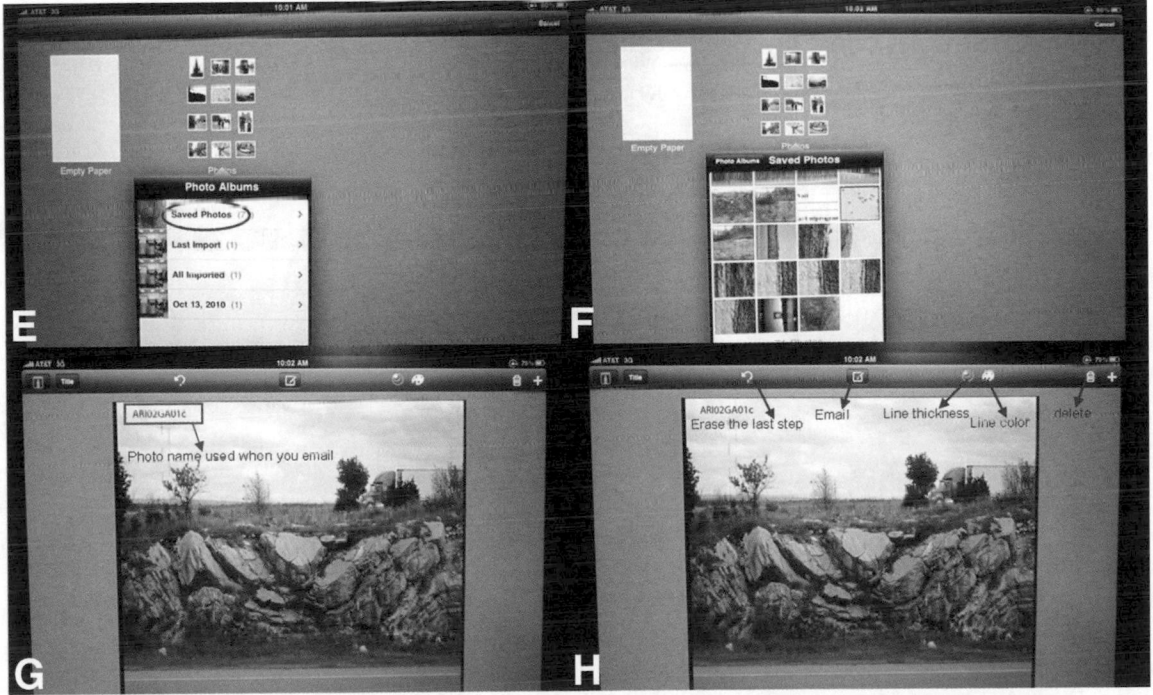

Figure AII_4. (E) Click saved photos. (F) Choose the photo you want to annotate. (G) Name on the left corner will be the information you type in the SUBJECT line in the email you send back to your email address. (It's better to write down or type the photo names on notepad to prevent confusion if you are working on several different photos.) (H) Explains the function of different icons seen on the annotation screen.

Figure AII_5.

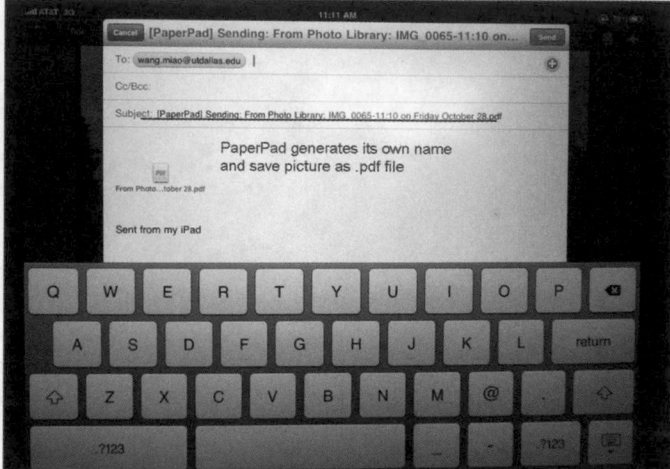

Figure AII_6.

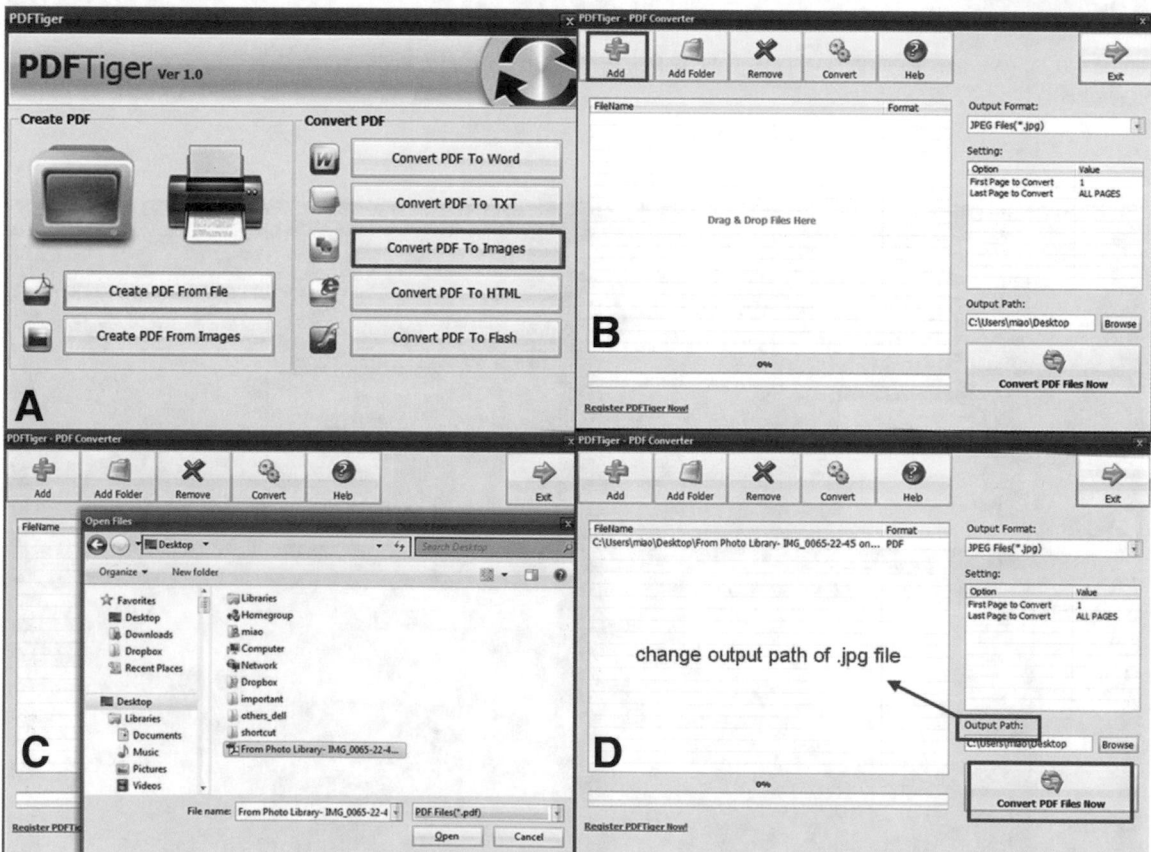

Figure AII_7. (A) Open PDFTiger, choose "convert PDF to Images." (B) Click "Add" on the top left corner. (C) Locate your .pdf file is and click open. (D) Browse the output path and choose where to put the .jpg file, then click "convert PDF file now."

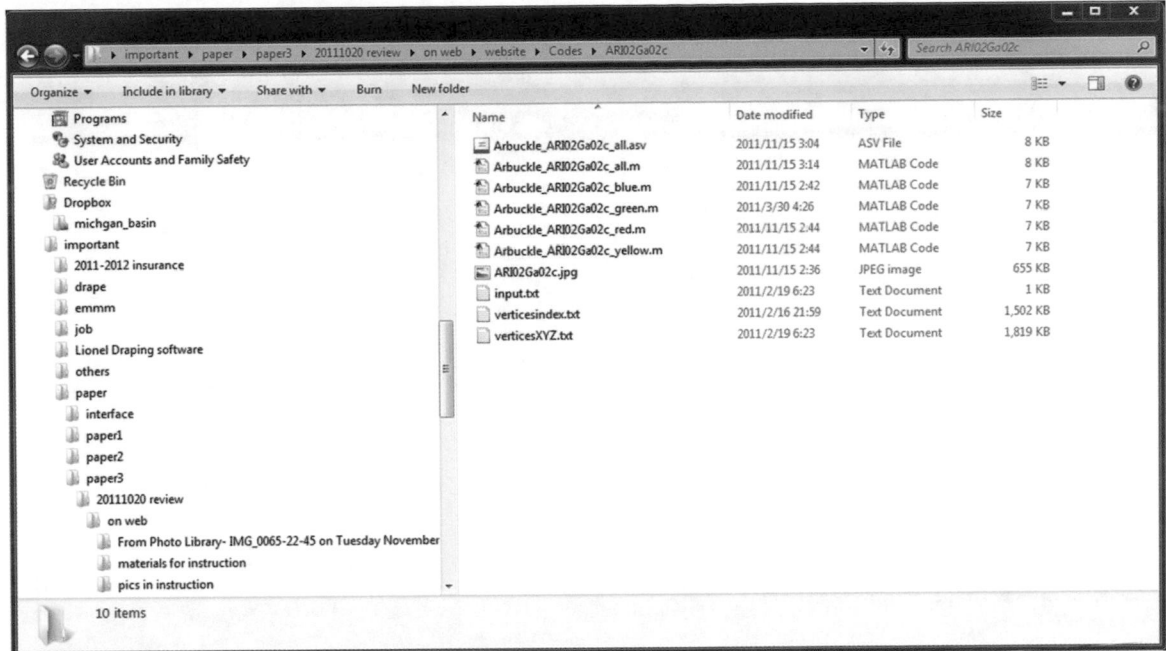

Figure AIII_1.

```
%%%%%%%%%%%%%%%%%%%%%%%%%%%%%%%%%%%%%%%%%%%%%%%%%%%%%%%%%%%%%%%%%%%%%%%%%%
% This program is divided into 2 sections: feature extraction and      %
% 2D to 3D transformation. Feature extraction is a program applied in  %
% photos edited on iPad. The concept is extracting features based on its %
% pixel value (RGB) and get its UV information (camera coordinates).Load %
% this result into the second program--Transformation. The concept is  %
% projecting 3D triangle mesh into 2D plane (UV coordinate)by collinearity %
% equations. Using image conversion to set 2D features and 2D triangles %
% into the same coordinate (obj UV coordinate). Then using kd-tree and %
% knn search to get nearest neighbor between 2D features and 2D triangles. %
% Then using the baricentric coordinate of those triangles to get 3D points %
% related to 2D features.                                              %
%                                                                      %
% create by miao wang, 03/28/2011                                      %
%%%%%%%%%%%%%%%%%%%%%%%%%%%%%%%%%%%%%%%%%%%%%%%%%%%%%%%%%%%%%%%%%%%%%%%%%%


clc;clear
a=imread('ARI02Ga02c.JPG');
[Num_R,Num_C,Num_color]=size(a);
edge=zeros(Num_R,Num_C,3);

k=0;m=0;t=0;l=0;
tic
for r=1:Num_R
    for c=1:Num_C

% extract red lines
        if (a(r,c,1)<=175+10)&&(a(r,c,1)>=175-10)&&(a(r,c,2)<=0+10)&&(a(r,c,2)>=0)
            edge(r,c,1)=175;
            m=m+1;
            line1(1,m)=r/Num_R;
            line1(2,m)=c/Num_C;

    end
        end
end
```

Figure AIII_2.

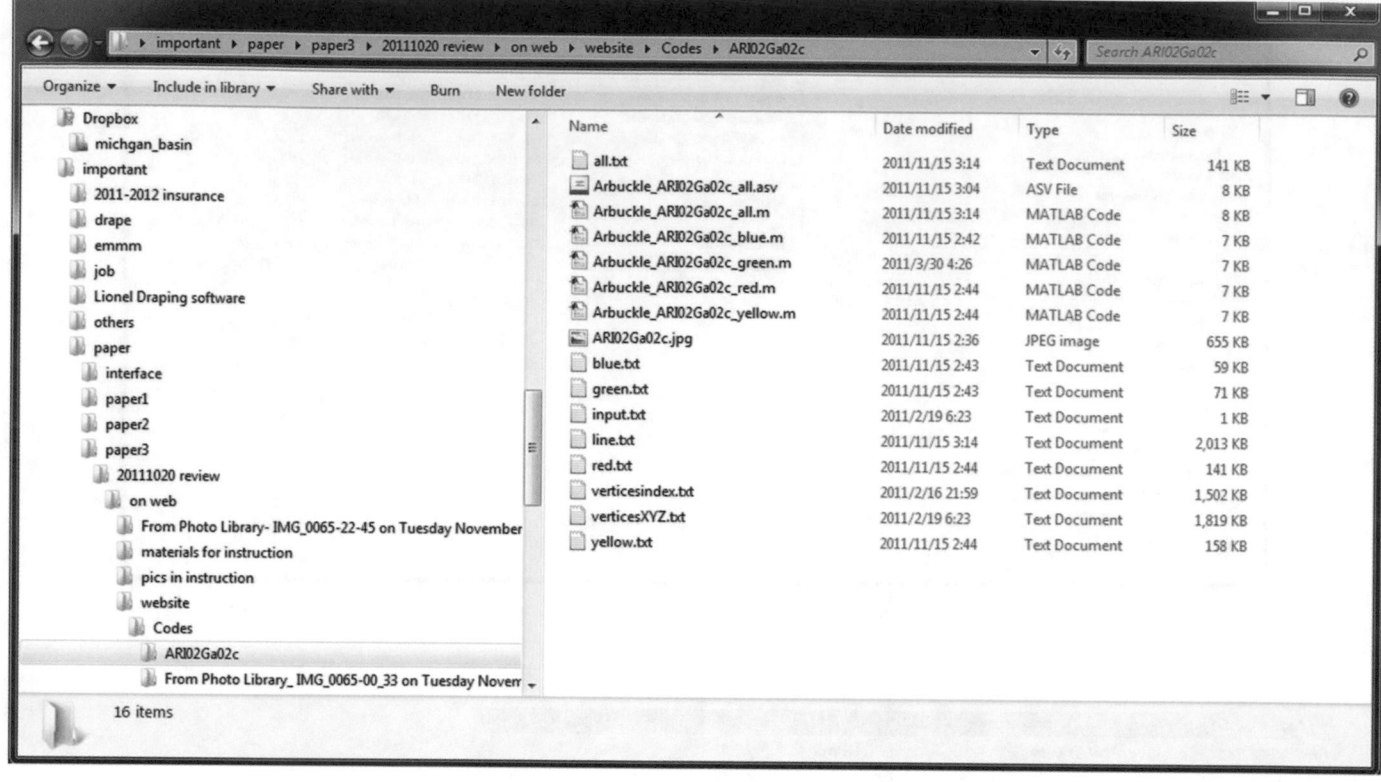

Figure AIII_3.

REFERENCES CITED

Aiken, C., Xu, X., Thurmond, J., Abdelsalam, M., Olariu, M., Olariu, C., and Thurmond, A., 2004, 3D laser scanning and virtual photorealistic outcrops: Acquisition, visualization and analysis: AAPG Short Course no. 3: Tulsa, Oklahoma, American Association of Petroleum Geologists, 100 p.

Alfarhan, M., 2010, Geosciences information system (GeoIS): A geospatial paradigm for real and virtual 3D worlds [Ph.D. thesis]: University of Texas at Dallas, 140 p.

Alfarhan, M., White, L., Dean Tuck, D., and Aiken, C., 2008, Laser rangefinders and ArcGIS combined with three-dimensional photorealistic modeling for mapping outcrops in the Slick Hills, Oklahoma: Geosphere, v. 4, no. 3, p. 576–587, doi:10.1130/GES00130.1.

Bellian, J.A., Kerans, C., and Jennette, D.C., 2005, Digital outcrop models: applications of terrestrial scanning Lidar technology in stratigraphic modeling: Journal of Sedimentary Research, v. 75, p. 166–176, doi:10.2110/jsr.2005.013.

Buckley, S.J., Howell, J.A., Enge, H.D., and Kurz, T.H., 2008, Terrestrial laser scanning in geology: Data acquisition, processing and accuracy considerations: Journal of the Geological Society, v. 165, p. 625–638, doi:10.1144/0016-76492007-100.

Clegg, P., Trinks, I., McCaffrey, K.J.W., Holdsworth, R.E., Jones, R.R., Hobbs, R., and Waggot, S., 2005, Toward the virtual outcrop: Geoscientist, v. 15, p. 8–9.

Dunn, F., and Parberry, I., 2002, 3DMath Primer for Graphics and Game Development, illustrated edition: Wordware Publishing Inc., 429 p.

Enge, H.D., Buckley, S.J., Rotevatn, A., and Howell, J.A., 2007, From outcrop to reservoir simulation model: Workflow and procedures: Geosphere, v. 3, p. 469–490, doi:10.1130/GES00099.1.

Fabuel-Perez, I., Hodgetts, D., and Redfern, J., 2009, A new approach for outcrop characterization and geostatistical analysis of a low-sinuosity fluvial-dominated succession using digital outcrop models: Upper Triassic Oukaimeden Sandstone Formation, central High Atlas, Morocco: AAPG Bulletin, v. 93, no. 6, p. 795–827, doi:10.1306/02230908102.

Fabuel-Perez, I., Hodgetts, D., and Redfern, J., 2010, Integration of digital outcrop models (DOMs) and high resolution sedimentology—workflow and implications for geological modelling: Oukaimeden Sandstone Formation, High Atlas (Morocco): Petroleum Geoscience, v. 16, p. 133–154, doi:10.1144/1354-079309-820.

Jones, R.R., McCaffrey, K.J.W., Wilson, R.W., and Holdsworth, R.E., 2004, Digital field data acquisition: towards increased quantification of uncertainty during geological mapping, *in* Curtis, A., and Wood, R., eds., Geological Prior Information. Gelogical Society Special Publication, v. 239, p. 43–56.

Labourdette, R., and Jones, R.R., 2007, Characterization of fluvial architectural elements using a three-dimensional outcrop data set: Escanilla braided system, South-Central Pyrenees, Spain: Geosphere, v. 3, no. 6, p. 422–434, doi:10.1130/GES00087.1.

McCaffrey, K.J.W., Jones, R.R., Holdsworth, R.E., Wilson, R.W., Clegg, P., Imber, J., Holliman, N., and Trinks, I., 2005, Unlocking the spatial dimension: Digital technologies and the future of geoscience fieldwork: Journal of the Geological Society, v. 162, p. 927–938, doi:10.1144/0016-764905-017.

Mikhail, E.M., Bethel, J.S., and McGlone, C.J., 2001, Introduction to Modern Photogrammetry: New York, John Wiley & Sons, 496 p.

Munnecke, A., Cramer, B.D., Boon, D.P., Kharwat, R., Aiken, C.L., and Schofield, D.I., 2012, The Digital Integrated Stratigraphy Project (DISP): Journal of Geosciences, Prague, Czech Republic (in press).

Olariu, M., Ferguson, J.F., and Aiken, C.L.V., 2008, Outcrop fracture characterization using terrestrial laser scanners: Deep-water Jackfork sandstone at Big Rock Quarry, Arkansas: Geosphere, v. 4, p. 247–259, doi:10.1130/GES00139.1.

Olariu, M.I., Aiken, C.L.V., Bhattacharya, J.P., and Xu, X., 2012, Interpretation of channelized architecture using three-dimensional photo real models, Pennsylvanian deep water deposits at Big Rock Quarry, Arkansas:

Marine and Petroleum Geology, v. 28, p. 1157–1170, doi:10.1016/j .marpetgeo.2010.12.007.

Oldow, J.S., Walker, J.D., Aiken, C.L.V., and Xu, X., 2006, Digital acquisition, analysis, and visualization in the earth sciences: Eos (Transactions, American Geophysical Union), v. 87, p. 351, doi:10.1029/2006EO350006.

Pringle, J., Gardiner, A., and Westerman, R., 2004, Virtual geological outcrops—fieldwork and analysis made less exhaustive?: Geology Today, v. 20, p. 67–71, doi:10.1111/j.1365-2451.2004.00450.x

Pringle, J.K., Brunt, R.L., Hodgson, D.M., and Flint, S.S., 2010, Capturing stratigraphic and sedimentological complexity from submarine channel complex outcrops to digital 3D models, Karoo Basin, South Africa: Petroleum Geoscience, v. 16, p. 307–330, doi:10.1144/1354 -079309-028.

Thurmond, J., Loseth, T., Rivenaes, J., Martinsen, O., Xu, X., and Aiken, C., 2005, Using outcrop data in the 21st Century–New methods and applications, with examples from the Ainsa Turbidite System, Ainsa, Spain, *in* Nilsen, T., et al., eds., Deep-water outcrops of the world atlas: Tulsa, Oklahoma: American Association of Petroleum Geologists Special Publication CD-ROM.

Trinks, I., Clegg, P., McCaffrey, K.J.W., Jones, R.R., Hobbs, R., Holdsworth, R.E., Holliman, N., Imber, J., Waggott, S., and Wilson, R., 2005, Mapping and analysing virtual outcrops: Visual Geosciences, v. 10, no. 1, p. 13–19, doi:10.1007/s10069-005-0026-9.

White, L.S., 2010, The Development Of Computer Algorithms for the Construction and Analysis of Photorealistic 3D Virtual Models of Geological Outcrops [M.S. thesis]: University of Texas at Dallas, 138 p.

Xu, X., 2000, Three-dimensional virtual geology: photorealistic outcrops, and their acquisition, visualization and analysis [Ph.D. thesis]: Bell & Howell Information and Learning, The University of Texas at Dallas, 189 p.

Xu, X., Aiken, C., Bhattacharya, J.P., Corbeanu, R.M., Nielsen, K.C., McMechan, G.A., and Abdelsalam, M.G., 2000, Creating virtual 3-D outcrop: Leading Edge, Society of Exploration Geophysicists, v. 19, no. 2, p. 197–202, doi:10.1190/1.1438576.

Manuscript Accepted by the Society 16 April 2012

The Geological Society of America
Special Paper 492
2012

Virtual fieldwork in geoscience teacher education: Issues, techniques, and models

Frank D. Granshaw*

Portland Community College, Department of Physical Science, P.O. Box 19000, Portland, Oregon 97280-0990, USA

Don Duggan-Haas*

Museum of the Earth at the Paleontological Research Institution, Ithaca, New York 14850, USA

ABSTRACT

Virtual field environments (VFEs) based on actual field sites are being used in professional development programs to familiarize teachers with field sites and give them the opportunity to practice investigative fieldwork, thus helping them make better use of limited field time. In other cases, the construction of VFEs provides a catalyst for actual fieldwork, and teacher workshop participants author VFEs that they can use with their own students. Virtual fieldwork development also improves technological skills relevant for the teaching of Earth system science. This article looks at what VFEs are, some of the practical and pedagogical issues involved in their design, and how they are used in teacher professional development to support and encourage field education.

INTRODUCTION

Investigative work in the field is one of the principal ways that geoscience knowledge is constructed and fieldwork is a "signature pedagogy" (Shulman, 2005) for professional preparation in Earth and environmental science. Despite its importance, field experience is often lacking at the K–12 level. This is partially due to limited budgets, liability, safety concerns, and accessibility issues for disabled students. But equally important is the limited field experience of K–12 teachers, particularly when it comes to inquiry-based field investigations. Furthermore, even if teachers do have field experience, they may have limited background in the pedagogy of fieldwork or have little opportunity to engage their own students in field activities.

With the advent of desktop virtual reality (VR), geospatial viewers, and digital panoramas, there is a growing interest in using VR for geoscience education. One type of VR based on actual field sites, the virtual field environment (VFE), is becoming increasingly useful to geoscience educators for a number of reasons. First, because much of geoscience is a "place-based" venture rooted in field investigation (Compton, 1985) there is a strong need for students to *experience* some of the places that they are studying and practice some of the data gathering and problem solving skills that are part of fieldwork (Butler, 2008; Hawley, 1997). Second, technologies are rapidly improving that facilitate the creation of virtual environments more quickly, less expensively, and in higher resolution than ever before. Third, when field time is available, having some type of orientation is critical

*Granshaw—fgransha@pcc.edu ; Duggan-Haas—dad55@cornell.edu.

Granshaw, F.D., and Duggan-Haas, D., 2012, Virtual fieldwork in geoscience teacher education: Issues, techniques, and models, *in* Whitmeyer, S.J., Bailey, J.E., De Paor, D.G., and Ornduff, T., eds., Google Earth and Virtual Visualizations in Geoscience Education and Research: Geological Society of America Special Paper 492, p. 285–303, doi:10.1130/2012.2492(20). For permission to copy, contact editing@geosociety.org. © 2012 The Geological Society of America. All rights reserved.

for helping students make effective use of that time. Fourth, if students are engaged in investigative fieldwork rather than traditional lecture style field trips, having a means of organizing, analyzing, and archiving field data once they have returned to the classroom is an important part of that experience.

Using a VFE for instruction addresses these issues by providing students virtual access to generally inaccessible sites, something that is particularly important if schools have little or no resources for field trips. If the virtual site is well designed, it gives students the ability to move around within and collect data from the site in ways that are similar to what they would do at the actual site. This is an aspect that lends itself well to inquiry-based educational activities. Furthermore, the use of a VFE for a site prior to visiting the actual site is useful for familiarizing students with a site, giving them an opportunity to practice some of the skills they will be using at that site, and helping them frame questions that they will address in the field. Finally, if the construction of a VFE is a major part of student fieldwork, it provides the students with a way of reviewing and archiving their experience.

Two facets of educational VFEs that have been largely unexplored are the cognitive aspects of their design and their construction as an educational field activity. The focus of this article is on examining those facets by discussing the following questions.

1. What do geoscience educators and students learn from constructing these environments?
2. How does the design of these environments impact what is learned?
3. Is it technically feasible for novices to construct useable virtual environments?
4. Can VFEs be constructed in such a way that allows novices to focus on the field science rather than the technical mechanics of VR design?

To address these questions, we look at several basic concepts and issues having to do with the design of VFEs and the activities that involve them. Since much of our experience has been in using and building VFEs in conjunction with teacher education, we examine these questions by looking at three models for incorporating virtual fieldwork into K–12 teacher education. While each model utilizes VFEs in unique ways, all three use them to support field education. VFE development is particularly valuable for helping teachers to become comfortable with fieldwork in the geosciences and to understand fieldwork's importance in teaching. Professional development focused upon VFE development and use also provides a venue for developing Technological and Pedagogical Content Knowledge (TPACK), the special set of skills and knowledge needed for teaching a particular content area (Mishra and Koehler, 2006; Thompson and Mishra, 2007).

TEACHER EDUCATION PROJECTS DISCUSSED IN THIS PAPER

The three projects presented in this article engaged teachers in VFE development in three quite different ways. In the first project, Regional and Local (ReaL) Earth Inquiry, teachers are presented with a variety of tools to build their own VFEs. This has been done through a series of professional development programs focused largely on actual fieldwork coupled with VFE design and construction. Each program begins with a face-to-face workshop within the educators' home region and continues with online working groups for the year following the workshop. Examples may be reviewed online at virtualfieldwork.org. The second project, Astrobiology at Yellowstone, involved a research team composed largely of middle and high school teachers investigating the hydrothermal ecology of a hot spring at Yellowstone National Park. In this project the team worked to produce a single VFE that was used to organize, present, and archive their field data. Project staff members are currently working with the teachers to create activities for their students that would use the VFE. The final project, Teachers on the Leading Edge or TOTLE, was a professional development program focused on familiarizing K–12 teachers with the geology and geologic hazards of the Pacific Northwest (TOTLE, http://orgs.up.edu/totle/, accessed 18 August 2011.). The program recently completed three years (2008–2010) of summer workshops for middle school Earth science teachers in Oregon and Washington as part of the EarthScope program's presence in the Pacific Northwest. EarthScope is a National Science Foundation–funded research program investigating the geophysical structure and evolution of the North American continent (EarthScope, 2011). As part of the project, a VFE was designed by a TOTLE staff member to prepare the program participants for field trips that were part of summer workshops focusing on seismic and volcanic processes and hazards in the Pacific Northwest. Though the program participants were not involved in the construction of the VFE, they were involved in guiding its development via a series of focus group sessions and educational experiments conducted throughout the three years of the project.

VIRTUAL FIELDWORK: CONCEPTS AND ISSUES

Before further discussion of how VFE construction was integrated into these programs, it is important to examine some fundamental questions about virtual fieldwork and VFEs, in particular what virtual fieldwork is, its benefits and limitations, and what makes for an effective VFE.

What is Virtual Fieldwork?

Virtual fieldwork is observation, data gathering, and problem solving using a computer-generated representation of an actual field site. It is a commonplace activity for planetary geologists (Fig. 1) and other researchers who work with remote sensing and geophysical data (Head et al., 2005; Xu and Aiken, 2000). It is also a pedagogical tool that allows instructors to bring the field into the classroom when it is not practical to take students into the field. Virtual fieldwork also allows for pre-trip debriefing and reflection.

We differentiate virtual fieldwork from using a virtual field trip (VFT) by looking at the difference between traditional field trips and investigative fieldwork. In a traditional "Cooks Tour" type field trip (Orion, 1993) the central activity is instructors explaining to students what they are seeing. The student part of these trips is usually limited to listening, taking notes, and asking questions. In investigative fieldwork, the focus is on students devising and addressing questions about what they are seeing. Since many VFTs tend to be an electronic version of the traditional field trip, they tend to be linear in structure with a single, predetermined path through the field site. Virtual fieldwork, on the other hand, is essentially a way of doing investigative fieldwork in a simulated field environment or VFE. Consequently, VFEs are built as digital models of a field site that allows users freedom of navigation as they collect data in ways that are similar to what they would do in the field. Hence VFEs tend to more readily support guided inquiry instruction.

Why Virtual Fieldwork?

Gathering data about geophysical phenomena in its native context is one of the principal means by which geoscience knowledge is constructed. Consequently, it is an essential part of the professional preparation of many geoscientists (Shulman, 2005). Yet despite its critical significance, scientific field experience is uncommon in K–12 education and often lacking in introductory and non-major courses on the college level. Chief among the reasons for this are field trip costs and budget limi-

tations, concerns about safety and liability, scheduling limitations, and the issue of accommodating students with disabilities (Cook et al., 2006; Smith, 2004; Fisher, 2001; Orion, 1993). Another important factor may be K–12 teachers' lack of experience with scientific fieldwork (Granshaw, 2011).

Engaging teachers in the development and use of VFEs is a rich opportunity for professional development across the spectrum of issues included in TPACK (Mishra and Koehler, 2006; Thompson and Mishra, 2007). In order to make a VFE—at least in the projects described here—the author must:

- closely study the environment with an eye toward engaging students in fieldwork;
- complete actual fieldwork at the site, and/or gather data remotely; and
- use educational and geospatial technologies in the assemblage and presentation of the VFE.

In the context of one of the teacher education programs discussed in this paper (ReaL Earth Inquiry) the driving question in VFE construction is "Why does this place look the way it does?" The educator participants are to create VFEs near their schools that facilitate answering that question from a variety of perspectives. Creation and use of these environments offers a way to bring the local environment to topics across the curriculum. VFEs serve as documentation of all of the above and may also document students doing fieldwork. The VFE also provides evidence of the teachers' use of technology, pedagogical approach, and Earth systems science content knowledge.

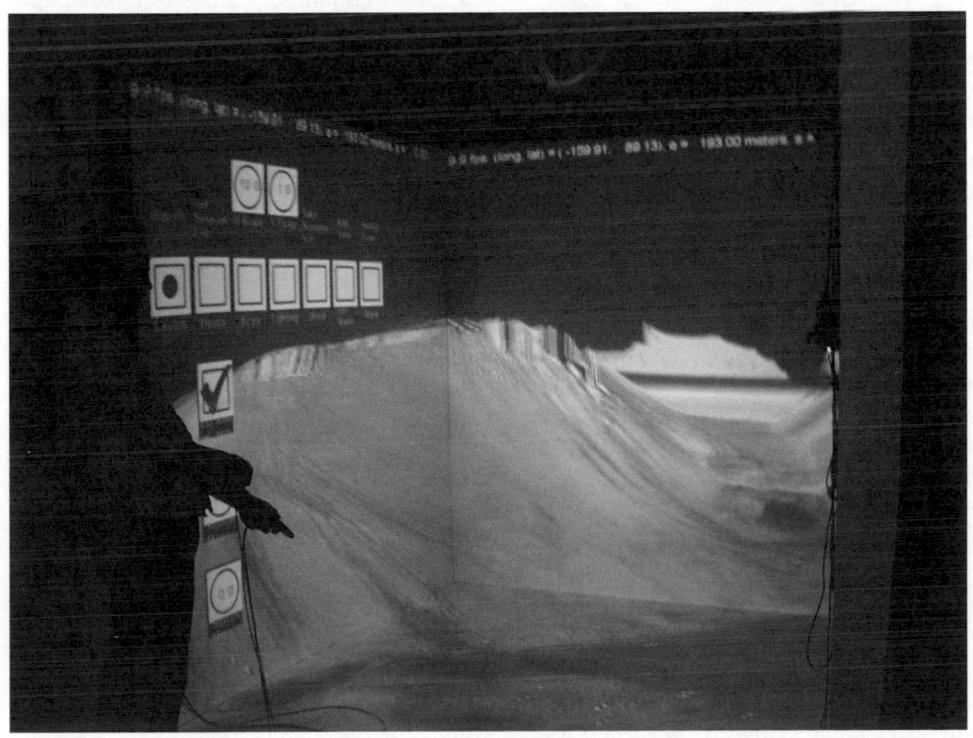

Figure 1. ADVISOR system in operation. Brown University, Providence, Rhode Island. ADVISOR is a data visualization system designed for viewing planetary geology data. This photograph shows a stereographic image of Victoria Crater on Mars appearing on the walls and floor of the system's visualization cave. The person on the left (visible as a silhouette) is holding a calculator like controller that they use to "fly" through and analyze the virtual terrain. The features of the controller are superimposed on the image on the left wall.

What are the Characteristics of an Effective VFE?

A VFE is an exploratory digital environment that possesses some of the visual/spatial characteristics of the actual sites on which it is based. An educational VFE is a virtual environment designed for students that enables them to practice key observation, measurement, and problem solving skills practiced by field geoscientists. Consequently it is an exploratory environment rather than a digital narrative. To build such an environment and associated activities requires addressing several key questions related to field-based education. Chief among these are the following:

- What constitutes a field study?
- What makes field study an effective means of geologic problem solving?
- What kinds of information are gathered in such a study?
- What principal cognitive processes are involved in geologic observation and problem solving?

Beginning with the first question, geologic field study is the process of observing and interpreting geologic phenomena in their natural context. So, while as a form of investigation it does not provide the degree of control over variables that laboratory modeling or analysis does, it does provide a more holistic and complete view of the phenomenon in question. Compton (1985) makes the claim that field studies have several unique advantages in solving geologic problems. The first of these is that Earth materials and structures are more easily identified when seen in context. He goes on to say that interpretations made in the field can be checked immediately against what they are intended to predict. He concludes his list by saying that studying actual associations of materials and structures often leads to the discovery of new kinds of features or relations. Finally, he states that geologic fieldwork is based on three kinds of information: (1) raw data derived from direct observation and measurement, (2) compositional and structural interpretations, and (3) age relations.

What is inherent in the last statement is that there are a number of cognitive skills that experienced geologists have learned that enable them to go from direct observation and measurement to interpretations of composition, structure, and age relationships. These skills involve three types of thinking: spatial, temporal, and systemic (Manduca and Mogk, 2006).

- Spatial thinking in geology means recognizing physical patterns, classifying objects on the basis of those patterns, making and using maps, and envisioning processes in three dimensions (Kastens and Ishikawa, 2006). For geologic novices, the problems that they encounter in looking at geologic features often revolve around filtering patterns from visual complexity, visualizing 3D structures being represented two dimensionally, visualizing 3D subsurface structures based on surface patterns, and thinking on several spatial scales simultaneously.
- Temporal thinking in geology means constructing chronological sequences from lithologic sequences and structures while thinking on different temporal scales simultaneously

(Dodick and Orion, 2006). In this case, the challenge for the novice is threefold: connecting geologic events and chronologies to visible geologic features, developing a sense of the large time spans involved in these chronologies, and linking these time spans to the much shorter time spans of directly observable processes.

- Finally, systemic thinking means seeing the dynamic and structural relationship between individual, directly observable materials, structures, and processes (Herbert, 2006). While such relationships are often apparent to trained geologists, geologic novices tend to be more immediate and isolated in their thinking.

The principal implication of these ideas for VFE design is that in addition to providing students with an authentic sense of the actual site, the environment should be built with scaffolding to help them "see" the site as a geoscientist would see it. In other words the environment and supporting activities should help students develop interpretive filters that enable them to make sense of what they are seeing.

One of the key aspects of a VFE that is useful in helping students develop spatial cognition is its scalability. Many VFEs are built in such a way that students can view a site from ground level via interactive panoramas or viewpoints. These panoramas allow them to look in various directions and even move between viewpoints within a virtual site. This part of the VFE simulates what students would see at the actual site and how they might navigate in it. This level provides a rather limited view of a site in that students do not see it in its broader context, thus limiting the questions they are able to ask or answer about the site. Consequently, many VFEs are designed to allow the user to move above and fly over a site. A common example of this is popular geospatial viewers such as Google Earth (http://www.google.com/earth/index.html), NASA World Wind (http://worldwind.arc.nasa.gov/java/; NASA Learning Technologies Project 2006), and Virtual Ocean (Marine Geoscience Data System, http://www.virtualocean.org/, accessed 17 August 2011). All three of these programs enable the user to view any place on the planet from varying distances. This allows students to see a location in its planetary, regional, and local context. The ability to fly over a landscape and view it from different directions helps them to understand the three-dimensional character of the landscape that contains a site. By integrating interactive panoramas such as GigaPan (http://www.gigapan.org/) or Street View (http://maps.google.com/help/maps/streetview/) into these viewers students are able to move from ground level to near Earth orbit, thus seeing a field site in its various contexts.

One additional advantage of VFE scalability is that it enables students to enlarge elements of a viewpoint allowing for closer inspection. This is done in a variety of ways. Panoramas like GigaPans rely on their extremely high resolution to allow users to zoom in on selected portions of a scene. The ability to look across scales offered by VFEs goes beyond or enhances what is possible with actual fieldwork alone. It allows the learner to see across scales not visible while simply standing in the field, and being able to zoom in and zoom out improves upon what can be done with paper maps. Another approach is to link features of interest in a scene to higher magnification views of those features. The advantage of structuring a VFE this way is that much higher magnifications are possible since increasingly higher magnification photomicrographs of a feature of interest can be linked to its location in a human-scale viewpoint. In either case, this scalability is reflective of a skill shared by the many field geoscientists: the ability to understand what is seen on the human scale at mul-

tiple scales (Fig. 2). An example of this is the field geologist who examines a rock hand sample through a magnifying glass while looking at the outcrop it came from and locating that outcrop on a geologic map.

An aspect of VFE design that is useful in helping students develop pattern recognition ability is image overlays and annotations. Classic geology texts such as *Geology Illustrated* (Shelton and Shelton, 1966) and *Guide to the Geology Olympic National Park* (Tabor, 1975) feature untouched

Figure 2. GigaPan of Balanced Rock Trail in Devil's Lake State Park, Wisconsin, embedded in Google Earth. Embedding GigaPan, or other high-resolution imagery, within Google Earth allows for exploration across multiple scales. (A) The moraine-dammed Devil's Lake from within Google Earth. The blue box in A shows the approximate area of the image in B, the near full view of the GigaPan image. The blue box in B shows the approximate area of the image shown in C, and C shows cross-bedding within the Baraboo Quartzite.

photographs next to sketches of the same scenes highlighting major structural features that included interpretative labeling. A more contemporary example of this is digital movies that allow users to overlay a translucent "field sketch" of the scene onto a photograph (Marshak, 2006). By alternating between the photograph and the sketch, students are shown major patterns a field geologist would regard as important. In two of the teacher education programs presented in this paper, the VFEs that were built included three overlays for each panorama, respectively designated "identify," "measure," and "enhance." The identification layer contained place names for features visible in each scene. The second layer, used for measurement, imposed distances, size, or elevations onto key features. The final layer highlighted patterns or identified materials within the scene (Fig. 3).

A significant challenge in VFE construction is designing an environment that helps students understand a field site as a part of a system. At its most fundamental, this involves understanding that a map is a model of a place and that it can be contextualized within a larger region. In progressively more abstract ways, it means placing the site within a regional, structural, and eventually dynamic context (tectonic and geomorphic causes and effects). To address this challenge, some VFEs include or are linked to geospatial browsers that are used to access ground views of field sites. Within the browsers, users can select what information they wish displayed on a map or aerial view. While accessing ground views from the browser is useful for seeing the location and geography of a site, the map overlay aspect of the browser helps users to see the materials, structure, and events shaping the landscape in which

the site is located (Fig. 4). An important question for educational researchers interested in VFE design is how effectively these features aid novices in thinking about what they see at a field site as part of a geophysical system. Linking of ground views to both maps and map overlays are common facets of geographic information systems and many geospatial browsers. However, since this software was originally designed for professional users, the question arises as to whether or not this design provides enough scaffolding for novice users to see the geographic, structural, and dynamic links between a ground location and the area it is part of. In other words, do users see the site as part of a system or just a picture pasted onto a map? A derivative question is whether or not the novices transfer information gathered from the map overlays to ground views of the field site, and if so how? If not, what design features are needed to help them make this transfer? For instance, how might scene enhancements be designed to reinforce information being given by map overlays?

In ReaL Earth Inquiry, participants use Google Earth to emulate the classic Eames' film *Powers of Ten* (1977). Using the program, they create a tour that begins at their school or a clearly recognized local landmark. By beginning with a familiar landmark, the Powers of Ten Google Earth tour helps students to place the local setting within the context of the state, nation, and world, and to see a smooth transition from the familiar place to the abstract map (Fig. 5). Additionally, ideas of scale and of exponential growth (in terms of the area shown) can be taught with this fairly easy to create and personalize resource. This assignment is completed before the face-to-face workshop both to give participants an introduction to Google Earth and to

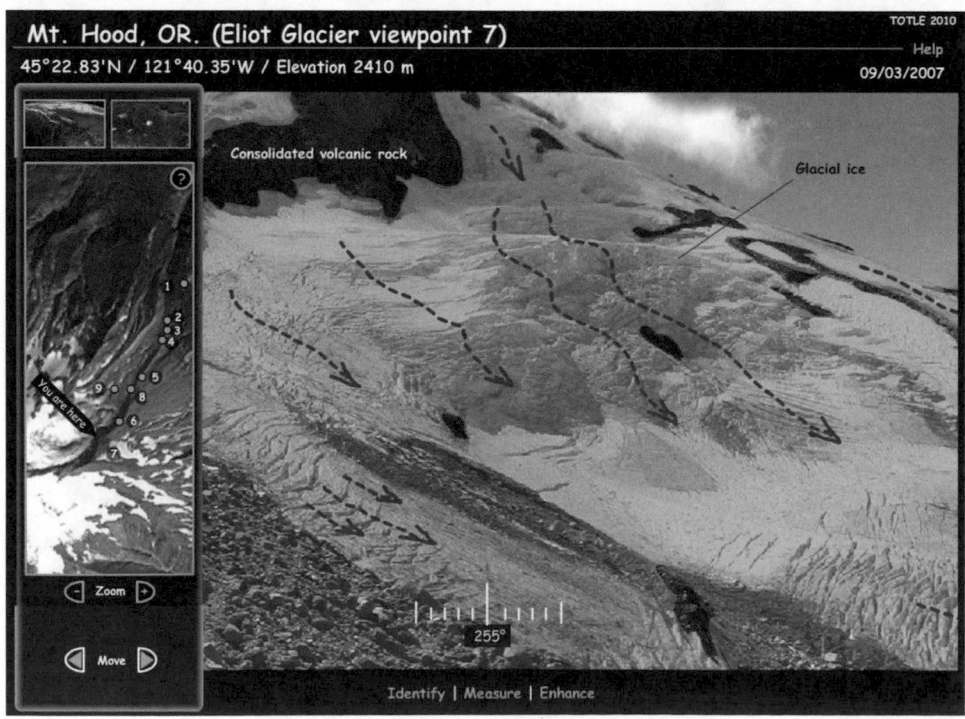

Figure 3. Enhanced overlay for a viewpoint overlooking Eliot Glacier, Mt. Hood, Oregon. The colored mask distinguishes glacial ice (blue) from consolidated (red) and unconsolidated (yellows) rock. Arrows on the glacial surface indicate the direction of ice flow.

create a useable instructional resource tailored to the needs of their students.

Another major challenge in VFE design is developing features that help students acquire temporal thinking skills. While this challenge could be met by building a VFE that allows students to move through time as easily as they move through the virtual space, constructing such an environment is difficult given

the time needed to see significant changes in an actual field location. Furthermore, many of the changes students are asked to visualize in geoscience courses take place over thousands if not millions of years. One approach to this problem involves a viewpoint from a VFE site for the Niawiakum River in Southwestern Washington (Granshaw, 2011). The centerpiece in this viewpoint is a riverbank showing a sequence of sediment underlying a tidal

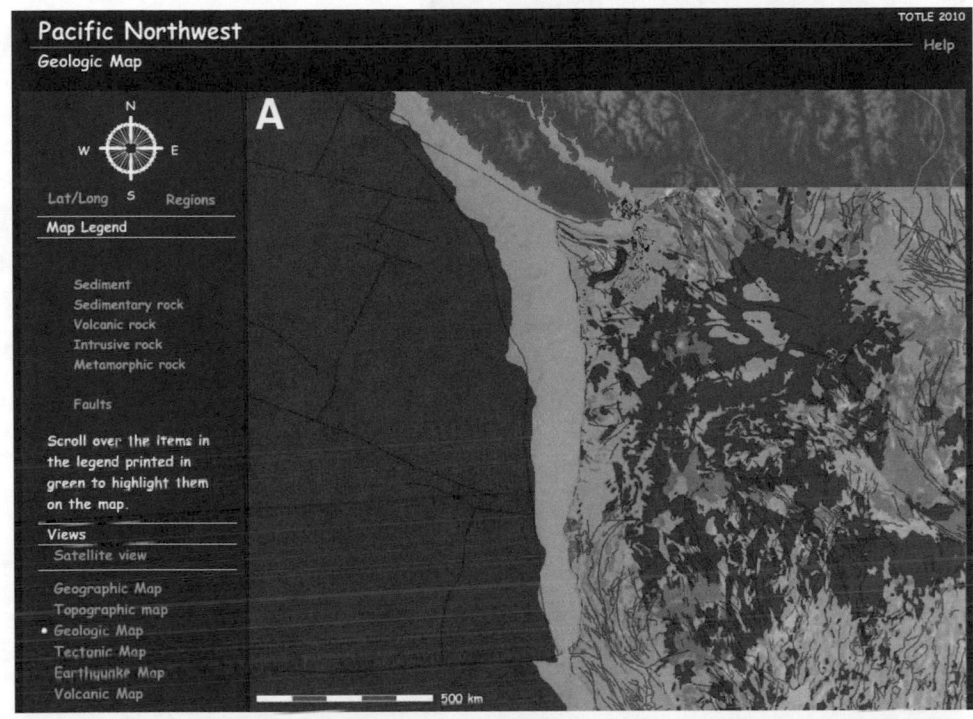

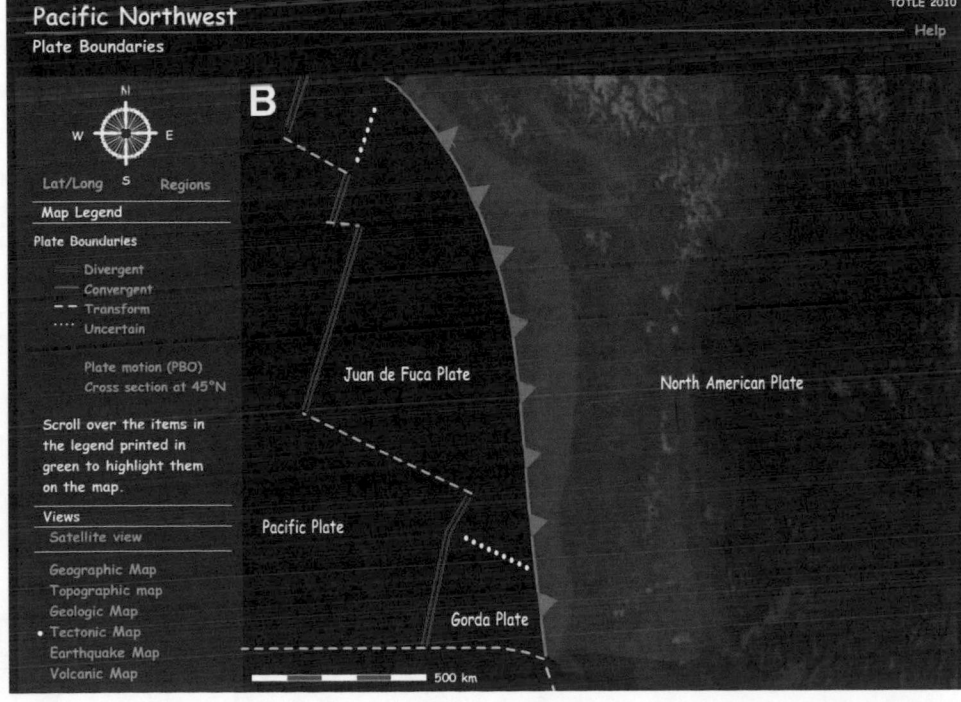

Figure 4. Scenes from the Map module of TOTLE in the Field. (A) Geologic map showing both lithology and structure (distribution of faults). (B) Tectonic map showing plate boundaries and movement. (*Continued.*)

Figure 4 (*continued*). (C) Volcanic map showing the distribution of active, dormant, and identifiable extinct volcanoes.

marsh. Contained within this sequence are layers of mud, peat, and sand indicative of tsunami (Fig. 6). What is significant about this scene is that many of the depositional environments associated with the sediment layers are visible in the contemporary environment. To help students understand and use this information, a guided inquiry exercise was designed that moved the students from description to interpretation of the sequence. The activity involved having students first describe the setting, the structure of the outcrop, and the physical characteristics of sediment in each layer.

INTEGRATING VIRTUAL FIELDWORK INTO TEACHER EDUCATION—THREE MODELS

The approaches to VFE development described here share many common attributes, but there are differences as well that relate to the pedagogical settings in which they arose. The use of VFEs in three different teacher professional development programs reflect attributes of how those programs differ in goals, duration, setting, and cost. A brief overview is provided in Table 1, and each program is briefly introduced prior to the discussion of VFE concepts and issues. In what follows, we offer a more detailed description of the three programs.

Model 1—ReaL Earth Inquiry

ReaL Earth Inquiry is a National Science Foundation–funded program (DRL 0733303) that provides professional development for Earth science educators and is building a range of curriculum support materials. The driving question

for the work is: *Why does this place look the way it does?* ReaL Earth Inquiry is intended to nurture place-based, inquiry-oriented teaching that ultimately enables learners to interpret novel landscapes.

ReaL Earth Inquiry uses a three-pronged approach to meet the project goals:

(1) A series of regional "Teacher-Friendly" guides to the Earth system science of the United States is under development. The United States is divided into seven regions with a guide for each. These are written in an accessible style, acknowledging the fact that Earth science has many teachers teaching out of the field of their primary training. The first two guides (to the northeastern and southeastern United States) were completed with earlier grants. For more information, see: http://teacherfriendlyguide.org/.

(2) Two cohorts of educators within each region participate in professional development programming that begins with a face-to-face workshop at a site that is interesting from an Earth systems science perspective. The program continues through online conferencing for the following 10 months during which time participants collaboratively complete a VFE of the workshop site and then each participant individually authors a VFE of a site local to their school or institution. These VFEs typically include guiding questions throughout the environment. The process of VFE creation is intended to both yield VFEs (the third prong of the project) while engaging teachers in the close study of their local environment with an eye toward bringing students out to do actual fieldwork. The workshop provides a mentored introduction to both fieldwork

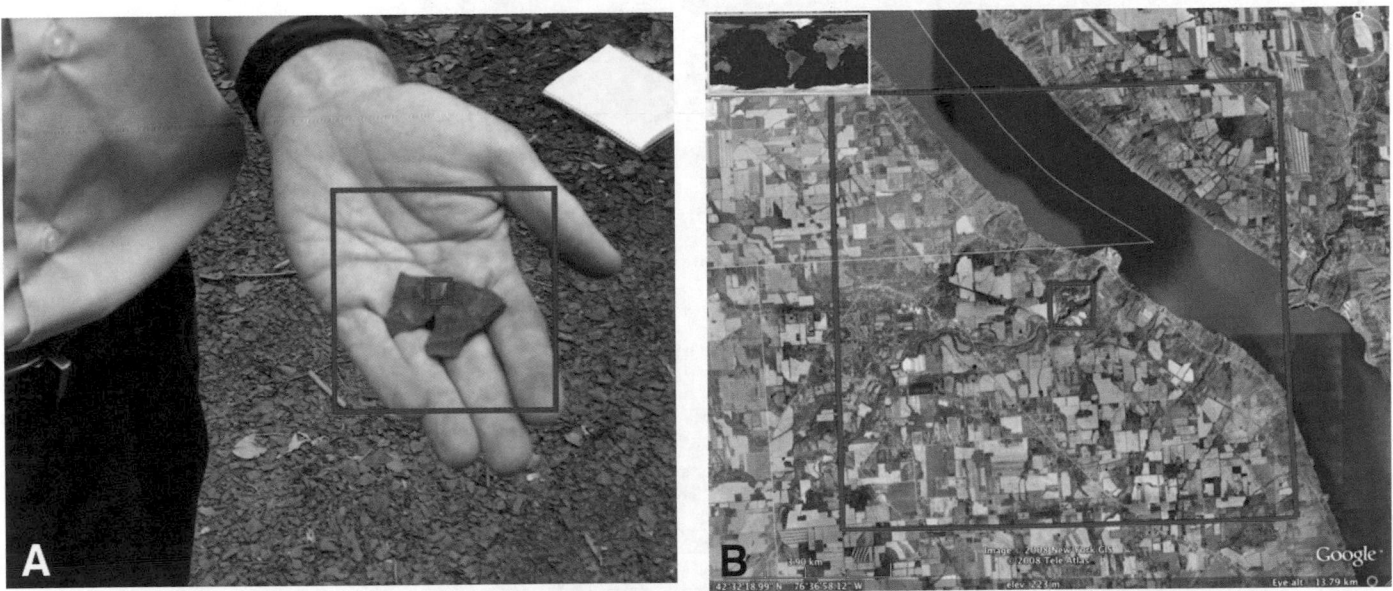

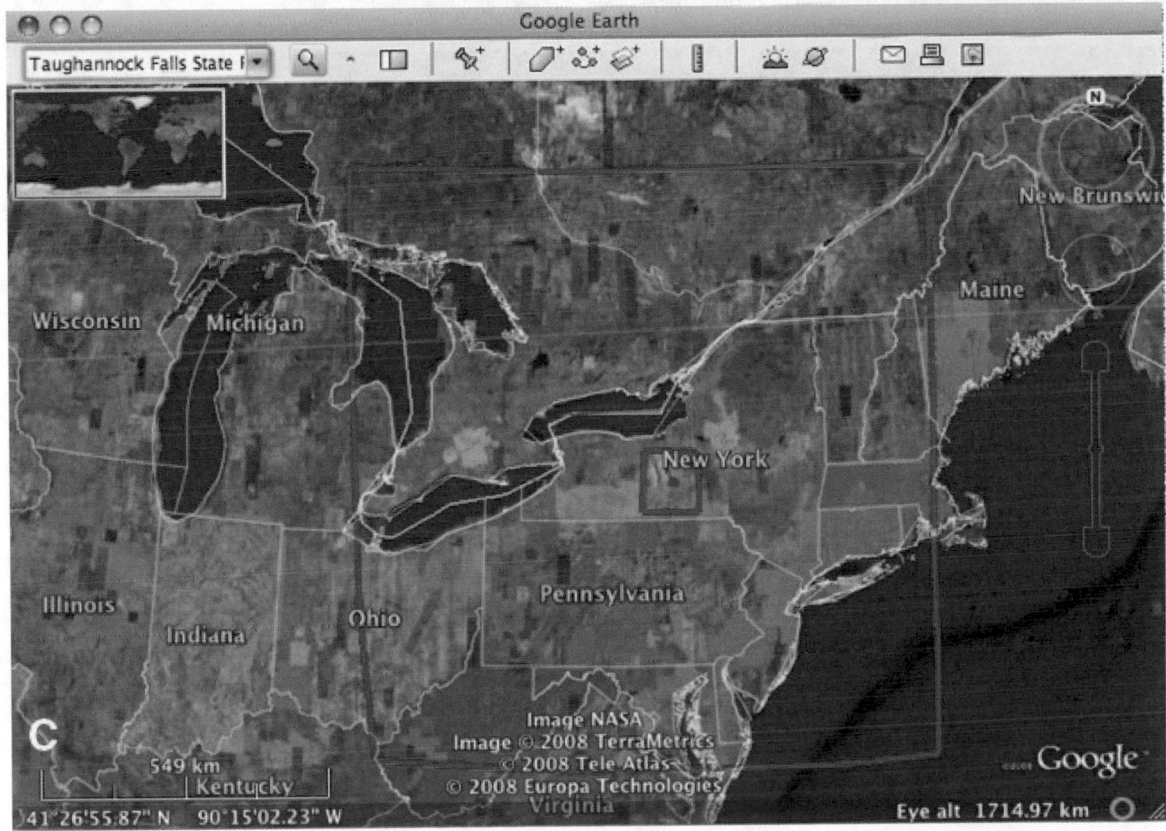

Figure 5. Three selections from the Taughannock Falls Powers of Ten Google Earth tour. The tour begins with 1 × 1 cm square and moves back stepwise progressively showing squares larger by powers of ten until the whole Earth is visible. A set of tutorial videos aids participants in creating their own tours centered on a local landmark. (A) 1 × 1 cm and 10 × 10 cm. (B) 1 × 1 km and 10 × 10 km. (C) 100 × 100 km and 1000 × 1000 km.

Figure 6. Scenes the Niawiakum branch of TOTLE in the field VFE This sequence is located in a bank of the Niawiakum River of southwestern Washington. (A) Close-up of tsunami-related sands sandwiched between a layer of peat and a layer of gray to reddish mud. (B) The same outcrop seen from several meters distant. The green plus sign on the bank marks where the sequence shown is located. In addition to showing the location of the sequence, this image also shows types of environments in which the peat and mud layers were deposited. In other words, the marsh corresponds to the peat layer, while the stream corresponds to the sediment in the upper mud layer. The sand is foreign to this environment, since the nearest source of sand is several kilometers to the west.

and the technologies to virtually replicate it. See http:// www.virtualfieldwork.org/.

(3) A database of VFEs is being assembled as a resource that allows for both quick and more in-depth comparisons of a wide array of field sites. See: http://virtualfieldwork .org/A_VFE_Database.html. The architecture of the database is under development, but many examples have

been uploaded and are accessible through a Google Map that is fed content through Google Spreadsheet Mapper. The database will ultimately be presented through two interfaces: the Spreadsheet Mapper and a color-coded tabular format that allows sorting. This presentation will allow for rapid comparison of places based on characteristics including but not limited to latitude, maximum

TABLE 1. COMPARISON OF THREE MODELS FOR VIRTUAL FIELD ENVIRONMENT (VFE) CONSTRUCTION IN TEACHER PROFESSIONAL DEVELOPMENT (PD)

Program characteristic	ReaL Earth Inquiry	Astrobiology at Yellowstone	TOTLE in the field
Primary goal(s) of PD	Foster place-based, inquiry-oriented, technologically rich teaching of Earth system science by engaging teachers in close study of local and regional environments and in the creation of virtual representation of those environments.	To engage teachers in creating a virtual reality environment in which they embed field data from multi-year geomicrobiology research. To engage teachers in creating student activities that use the environment.	Provide field experiences that familiarize teachers with geography and geology of the Northwest and with guided inquiry activities that will be used during the program and in participants' teaching
Program structure	2.5 day face-to-face workshop including actual fieldwork and classroom time, followed by 10 months of monthly online study groups.	A weeklong field experience for science teachers and majors that focuses on the biology, geology, and micro ecology of hot springs in the park.	A weeklong summer workshop that included two field trips. VFEs were used in field orientations that were usually an hour in length twice a week.
Audience for PD	Upper elementary through undergraduate instructors and PD providers.	Practicing middle school and high school teachers.	Primarily middle school Earth science teachers.
Locus of VFE development	Development is a shared activity among program staff and participants.	Participants collect data, take photographs and provide significant input into VFE design. Program staff assemble VFE.	Participants provide occasional input into VFE design and use. Program staff collect data, take photographs, and assemble the VFE.
Role of VFE in PD programming	VFE development is a program focus. Teachers collaborate on a VFE of the workshop site and then individually create VFEs sites local to their schools.	VFE is used for the assemblage and presentation of teacher data.	VFE use is to prepare teachers for work in the field. No VFE writing was done during the course as the goal was to use them for orientation, not to build them.

July temperature, minimum January temperature, bedrock type, minimum and maximum elevation, soil pH, and more. Using color codes based on common maps of these characteristics coupled with sortability and with the ability to select the fields you wish to include allows users to sort and identify in a glance sites that share common attributes but differ in specific ways. When populated, the sorting along with color-coding will also show associations between characteristics like latitude and temperature. This is intended to allow users to determine how changing particular attributes of the lithography or climate (for example) change the lay of the land. By sorting records and finding your local VFE within the context of the sort, you can see, for example, that the sites above and below your location share 8 of 10 attributes, and by then exploring the VFEs, you gain perspective on those differences and what they signify. This appears to be a promising tool for using the local to understand the global.

The remaining discussion of ReaL Earth Inquiry will focus on professional development and its outcomes. The outcomes are both in the form of curriculum materials and enhanced TPACK for the participating educators. Professional development that focuses on the development of VFEs is self-documenting in that the result represents a synthesis of each component of a teacher's TPACK-related inquiry-oriented teaching about their local environment. This yields a systems approach to the teaching of Earth systems. Just as the teaching of Earth system science attends to the connections among Earth science's different processes, topics, and big ideas, VFE-focused professional development draws attention to the interplay among teaching's fundamental processes, topics, and big ideas—the connections among geology, pedagogy, and technology. ReaL Earth Inquiry gives explicit attention to a small set of Earth system science big ideas and overarching questions (Duggan-Haas and Clark, 2009; Duggan-Haas, 2010, 2011; Ross, 2010) and regards deep understanding of these questions and ideas as the hallmark of Earth system science literacy.

The work is informed by an understanding that what you do is what you learn and that one cannot make a reasonable representation of something without studying it closely—an idea well captured by Samuel Scudder in his description of Louis Agassiz's teaching approach, written well over 100 years ago (1874). Agassiz is quoted as saying, "a pencil is one of the best of eyes." The creation of quality media that represents a place, event or thing, whether with a pencil or through the integration of computers, digital imagery and the array of available data still requires the media to be a great set of eyes.

To develop facility with the technological media in the service of Earth system science education, the project works with and assists teachers in a series of activities beginning just prior to the face-to-face workshop. For homework prior to the workshop, teachers create a Powers of Ten Google Earth tour (described above). At the workshop, two half-days are spent in the field with the remaining time in the classroom, focused on taking what was captured in the field (primarily in the form of photographs and notes) and beginning to transform it into a VFE. After the workshop, teachers participate in online study groups and work among those study groups to initially contribute pieces of the workshop site VFE. Study group focus then shifts to the development of VFEs that are local to participants' schools. Teachers receive a stipend for participation in the program that comes in two installments, one paid at the end of the workshop and the remaining paid upon submission of an acceptable local VFE and participation in at least four online study group sessions.

VFEs are never viewed as completed, but rather they gradually become more ready for more types of use. Just as the study of a field site can never completely explain all aspects of how the site came to be the way that it is, a virtual representation of a site can never capture all features of the site. The minimum expectations for a satisfactory VFE are shown in Table 2.

These criteria are indicators that the participant is studying the local environment in a way that supports engaging students in both virtual and actual fieldwork and grounds the study of at least some aspects of the curriculum in a locally relevant context.

While certain parameters of VFEs are prescribed, the specific form is left quite open. The author of a particular VFE is likely to be the heaviest user of that particular VFE. Furthermore, the participants vary tremendously in what they teach, their technological skill levels, and their level of content expertise. Following the practices of differentiated instruction (Tomlinson and McTighe, 2006; Tomlinson, 2001) and adaptive professional development (Trautmann and MaKinster, 2010), and recognizing that curriculum implementation is in the control of the teacher, program staff encourage participants to create VFEs that are, in the teachers' judgment, best suited to their classroom and that build their own knowledge and skills. Thus, there is no set format for the VFEs created in the project. Openendedness, however, can be frustrating and proved to be so for some participants.

This expressed frustration led to the creation of a template for VFE development that raises an array of supporting questions for the project's driving question: "Why does this place look the way it does?" Like the driving question, the supporting questions are phrased in such a way that they may be asked about any site. At the conceptual center of the VFE template is the graphic organizer shown in Figure 7. The template is available in PowerPoint, Apple Keynote, and Prezi formats, but documents built using the template are not intended to be used as presentations. Instead, they are activities for pairs or small groups of students to work through. The questions within the template are also available in a worksheet. Further, the template has a nonlinear structure—each box and arrow is hyperlinked to pages that include further questions and images for the relevant content and a thumbnail image of the graphic organizer that links back to this starting page in the PowerPoint and Keynote versions. The Prezi version has questions and images embedded within the graphic organizer, and users zoom in and out of different elements.

Teachers are using these materials to more effectively investigate the question, "Why does this place look the way it does?" with their students and are also investigating aspects of the answer themselves, thus enriching their own content knowledge. The ability to raise and investigate questions is regarded as important pedagogically as answering them, however, especially since the question is essentially bottomless—it can never be answered completely.

The simplest, "entry level" use of this template is for users to identify relevant photographs and substitute them into the file in the appropriate spots and delete slides and questions that are not directly related to the teacher's instructional goals. The imagery is more relevant when it is geographically contextualized, which participants are doing in at least two ways—by including hyperlinked maps within PowerPoint presentations and by exporting the slides as images that are then embedded in Google Earth Placemarks (Fig. 8). The creation of materials that are simply tailored to a specific location not only allows for the rapid creation of quality place-based resources, it also demonstrates that there are approaches to reading a landscape that are applicable anywhere.

Through participation in ReaL Earth Inquiry, educators are raising and investigating locally relevant Earth systems science,

TABLE 2. REAL EARTH INQUIRY REQUIREMENTS FOR A SATISFACTORY VIRTUAL FIELD ENVIRONMENT (VFE)

Teacher-authored:	Student-authored or collaboratively authored with students
✓ An abstract for the VFE concisely describes the relevant science and the technologies needed for using the VFE. ✓ An iconic image—a picture representative of what comes to mind when the site is remembered—is included. ✓ The VFE will take students at least a class period to complete. ✓ At least five of the included photographs were taken by the teacher, by colleagues in the school, or by the teacher's students.	
✓ The site is local enough to the school to assure that students will have either some familiarity with it prior to instruction or access to it during instruction. ✓ The VFE allows some student decision-making in what they explore. ✓ The VFE includes at least seven photos embedded in a Google Earth Tour, a PowerPoint document, a Prezi, or file of another software platform.	✓ The VFE includes evidence of students completing fieldwork.

Note: VFEs must meet all the requirements in one column of the table. The first four items are expectations for all VFEs completed for the program.

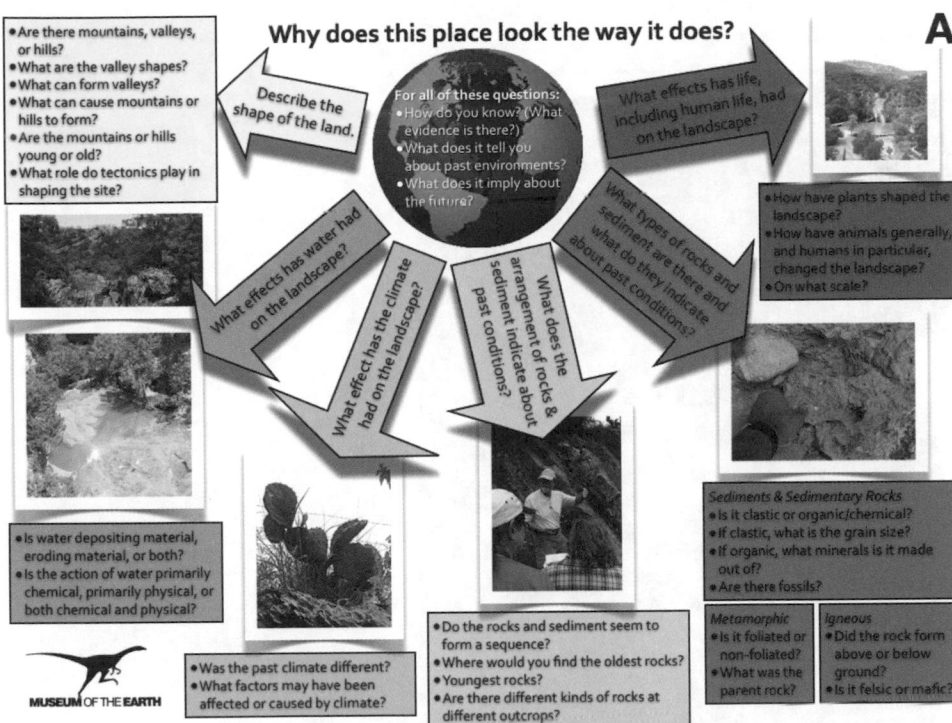

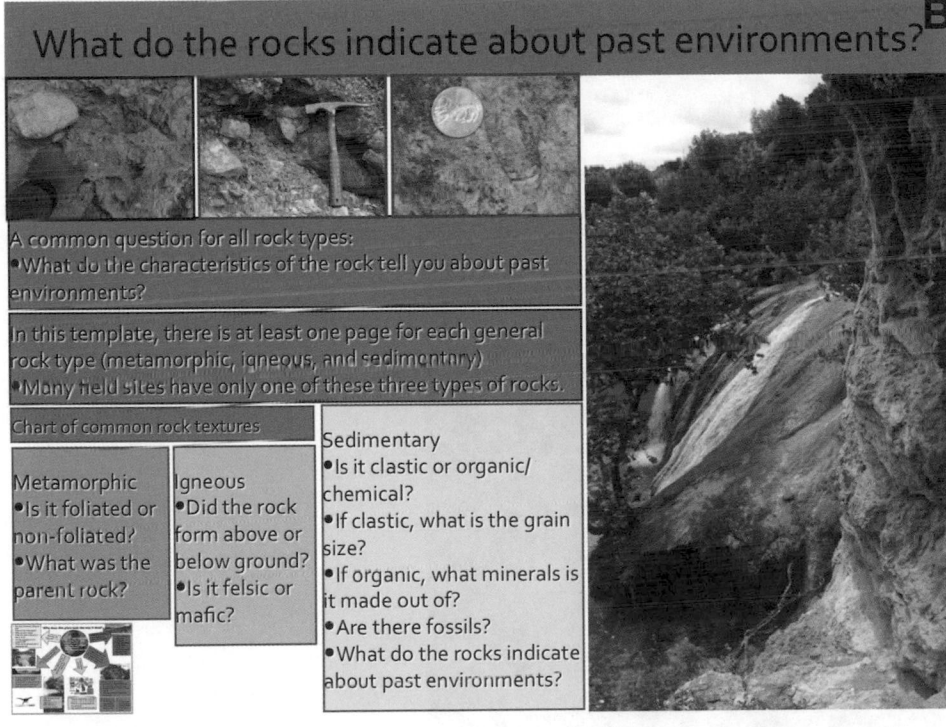

Figure 7. (A) The Graphic Organizer for the VFE PowerPoint Template. The template serves for the creation of "entry level" virtual fieldwork experiences. Boxes and arrows are linked to other slides with further questions, more pictures, and in some cases, further guidance. Questions in the template are phrased generically so that they may be asked of any site. Users simply substitute pictures from a field site of their choice and customize questions to address special features of the site. (B) Rock Slide #1. The arrow regarding rock types in A links to this slide. Each box outlined in blue (both images and text) link to other slides with more information and/or questions. The thumbnail of the graphic organizer links back to that slide.

producing materials that allow their students to do the same, and creating resources for others around the country to use in comparison to their own local sites. They are documenting this through the creation of VFEs that are themselves indicators of the educators' close study of local environments and of their technological and pedagogical aptitude.

Model 2—Astrobiology at Yellowstone

G410/510: "Mars Analogs in Yellowstone" is a summer course offered through the Geology Department at Portland State University. A major theme in this course is thermophilic bacterial communities as an analog for possible life on other planets (e.g.,

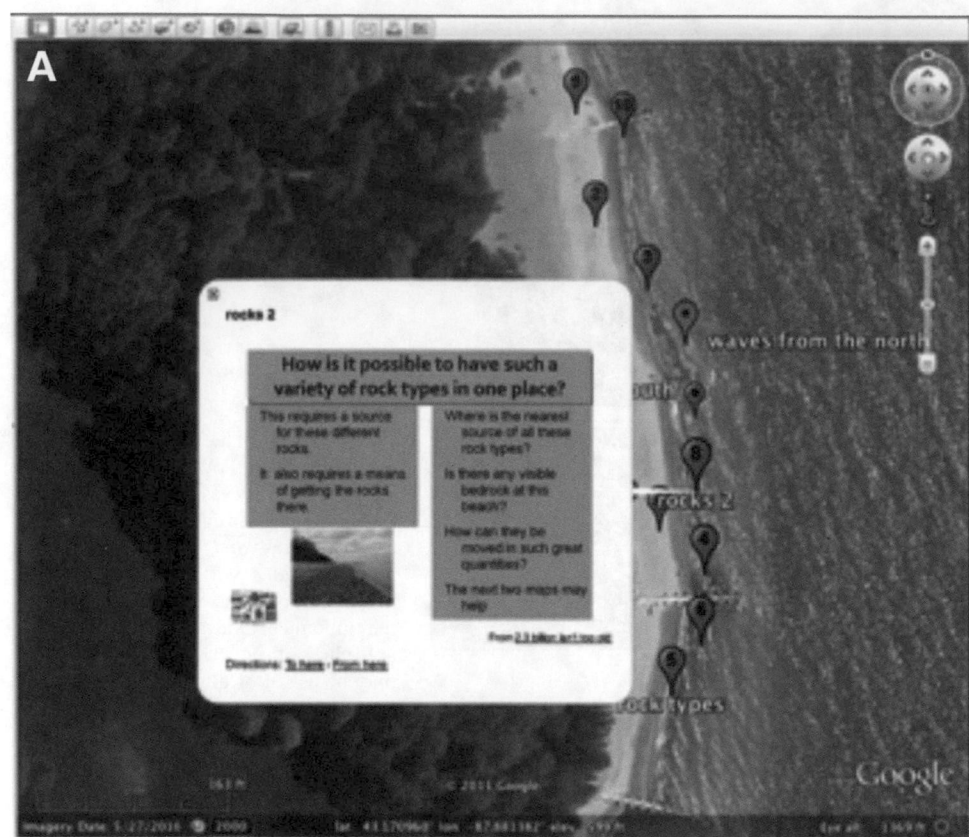

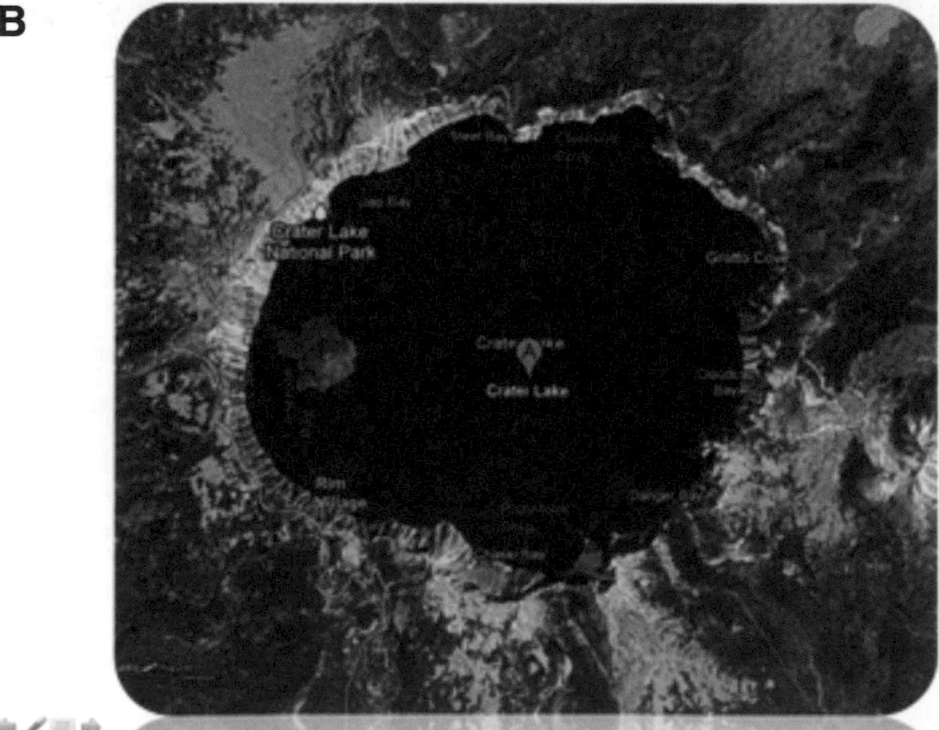

Figure 8. (A) A participant-created VFE that uses customized slides from the VFE Template as images embedded in Google Earth placemarks. (B) A participant-created PowerPoint VFE slide with a hyperlinked map. Clicking on the place names (in purple text) connects to relevant slides.

Mars). During the summers of 2009 and 2011, teachers who were course participants were given the task of constructing a VFE for Queen's Laundry, a hot spring in the park's Lower Geyser Basin. In this project, the participants were divided into small research teams, with each team collecting data from a specific location to address a research question of their choosing. The data they gathered included panoramic and macro-photography, survey measurements, and measurements of environmental conditions (e.g., pH, temperature, conductivity, and morphology of dominant organisms). These were incorporated into a single VFE by a consultant who helped guide the collection of the data and assembled the VFE. The result was a virtual environment consisting of several viewpoints accessible by an interactive map or a geospatial browser (Google Earth or Virtual Ocean). Each viewpoint is a 360° panorama for individual research locations that include informational and interpretational overlays based largely on the data provided by the research teams. In each panorama, users can access magnifications of features of interest within the scene. For example, in one viewpoint, users examined a bacterial colony in the spring outflow channel by accessing a series of photomicrographs of samples taken from that channel (Fig. 9).

Because this program occurred during two summers (2009 and 2011) the VFE construction process evolved significantly between the first and second years. During both years the course participants did much of the fieldwork necessary to construct it. In addition to gathering the data, they were engaged in determining the type of data gathered at each location and the research questions guiding that collection. Also during both years the consultant was responsible for assembling the collected data into a coherent virtual environment. The major way that this process evolved was the degree to which the participants were engaged in the assembly of the environment. During the first year (2009) the participants had little or no input on the design of the VFE beyond contributing photography and data. This occurred largely because the environment was assembled after rather than during the course. During the second year (2011), the participants worked with the consultant to create informational and interpretative overlays and locate points of interest within their viewpoint. To accomplish this task the construction of viewpoints was streamlined so that the virtual environment could be assembled as the participants were doing their fieldwork.

To date, much of our evaluation of the impact of VFE assembly on the course has been informal and focused on the mechanics of the environment assembly. The project has consisted of two pilot studies aimed at testing the feasibility of having students build VFEs. Based on conversations with the students and other faculty and the photography produced by the students, here are some preliminary conclusions:

1. It is technically viable for students with little or no field experience and limited photographic skills to collect photography and other data needed to construct a VFE. This conclusion was based on the quality of the photography and data collected by the students. Principal questions in this determination included "Did they produce a viable 360° panorama and set of nested images?" and "What kinds of non-photographic data did they collect, and how was it related to their viewpoint?"

2. The VFE construction was an activity that the participants found highly engaging. This was gauged by their responses during debriefing with individual teams, questions and comments they shared with the consultant in the field, and the types and detail of the data they collected.

3. The construction of the VFE provided a focus for fieldwork in that students could produce a tangible product that would provide a virtual reality archive of the site and the activities the participants led. In other words, the VFE that they helped produce is something that could be used by other students in the same class or classes taught subsequent seasons or years. This conclusion is based on discussions with the course instructors and the teacher participants who were enrolled in the course.

4. Whether and how the teachers have used the VFEs in their courses is currently an unanswered question. This is primarily because our focus has been on the logistics of fieldwork organization and construction. With the inclusion of a curriculum specialist in course staff, we are in the midst of working with the teachers from the 2011 workshop to develop activities they can use in their classrooms that involve the VFE.

TOTLE in the Field

The final example of teacher education involves the design and use of a VFE for TOTLE. This VFE, entitled "TOTLE in the Field," was constructed to represent the field sites visited by the participants during summer workshops. It was designed as a tool to familiarize the participants with the locations they would be visiting as well as preparing them for some of the field activities that they would be engaging in at those sites. It was also built as a curriculum resource enabling them to take elements of their summer field experience back to their own classrooms.

Unlike the previous two models, this VFE was designed and built by one of the program staff. However, since it was built concurrently with the workshops, it was designed with input from the program participants. In the first year of the program, participants were presented with a prototype of the VFE during focus groups conducted during the summer workshop. At this time, they were asked to provide feedback on the architecture of the VFE, present ideas on how they might use the VFE with their students, and state what features they wanted to see added to the VFE to use it in the ways they were proposing. This feedback was used to design a second prototype based on the sites for the following year's field trips. The first year of the program was in Oregon; the second and third years were in the state of Washington.

In year two of the program, the second prototype was used in orientations for both field trips. Here it was used to familiarize the participants with the field sites they would be visiting and to give them a vehicle for practicing guided inquiry activities they

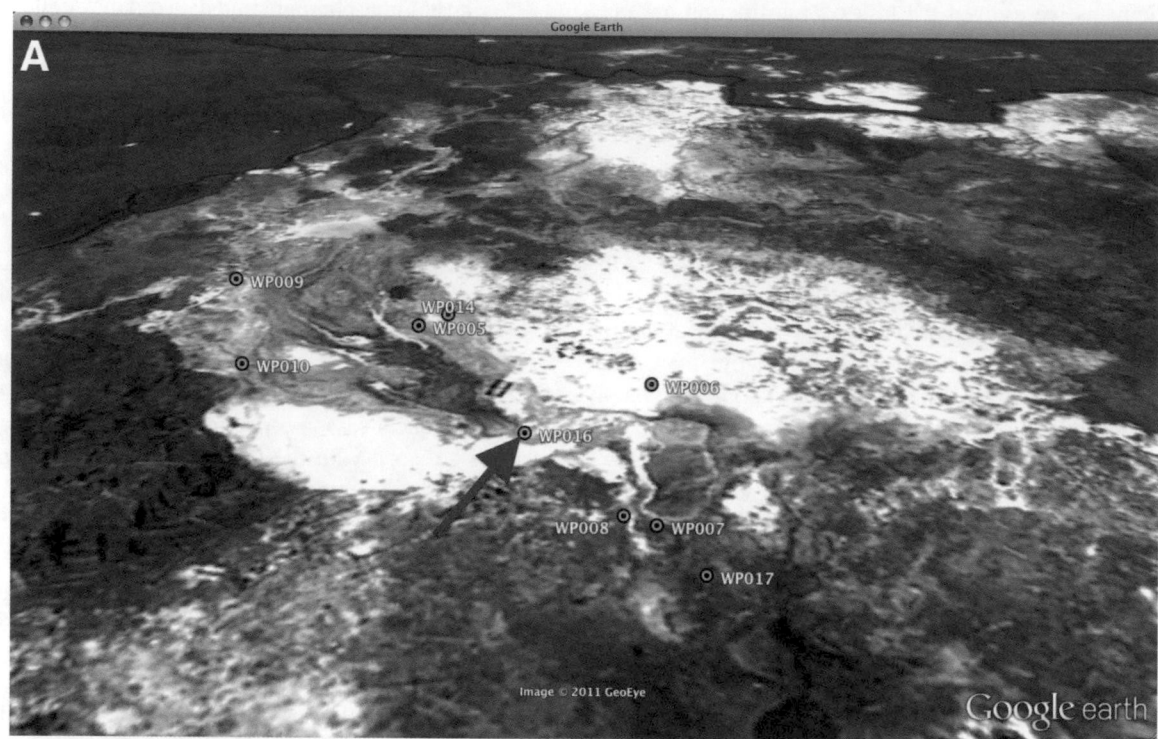

Figure 9. The Queens Laundry VFE. A student-constructed virtual environment based on a field survey of the Queens Laundry hot spring in Yellowstone National Parks Lower Geyser Basin. (B) Part of a panoramic ground view with photomicrographs of samples taken from the site marked by the white arrow. Clicking on a hotspot at the tip of the arrow accesses the micrographs. (A) Aerial view of Queen's Laundry showing the location of the scene and other scenes making up the VFE.

would be engaged in during their visit. During a follow-up survey, participants were asked to evaluate the effectiveness of the environment by ranking it on the basis of how well it helped them with the following.

A. Visualize the location of the sites and the route to it.

B. Visualize the geographic layout of the field sites.

C. Understand the link between the geology visible at the field sites and the regional geologic structures and processes.

D. Extract structural patterns from complex outcrops and landscapes.

E. Visualize the processes that shaped the geologic structures and landscapes visible at the field site.

Using a ranking scale of 0–4 (0 being not useful, 4 being highly useful) at least 74% of the participants (n = 35) ranked the VFE as 3 or higher for helping them with all five tasks. Of the five tasks, the participants gave the highest ranks for items A and B and the lowest ranks for C and E indicating that the environment was less successful at helping the teachers with the more abstract skills of process visualization and linking to local to regional geologic structure. These data and narrative responses from the survey were used to design the final version of the VFE and a set of guided inquiry activities to accompany them.

In year three, the participants were presented with the final VFE and an accompanying teacher's manual. This final version was a combination and revision of the two previous prototypes. The manual contained relevant background information about regional geology, volcanism, and seismic processes, in addition to a set of student activities accompanying many of the viewpoints and aerial views contained in the environment. As in year two, the environment was mainly used as a field trip orientation tool, though more time was spent on introducing the participants to how the environment could be used in their classrooms. Because the third year's participants received the final copy of the VFE, no formative evaluation was conducted with this group.

Though the involvement of teachers in developing the "TOTLE in the Field" VFE was much more limited than that of the teachers in the two professional development programs discussed earlier, this model did provide some important insights into VFE design as an instructional activity. While some of these insights were derived from the focus groups and field trip orientation experiments, others were derived from informal conversations with the workshop participants. In all three years of the program, several of the participants expressed strong interest in either building their own environments or having templates that their students could use for building an environment. The principal challenge in responding to this interest was in developing a strategy for building VFEs that does not require a high level of multi-media or programming experience from the participants or their students and is technically simple enough that they and their students can focus on the science of the fieldwork rather than being focused on computer-related production. Unlike the two workshop models described earlier, the design and use of the VFE was one among many activities that program staff and participants were involved in together. Consequently, there was little opportunity to pursue "user friendly" pedagogically rich construction strategies. Instead participant involvement was limited to providing input into the environment design at strategic points in its development. A principal question at this point is given the goals, structure, and schedule of the workshop, was the involvement of participants in the VFE design an educationally useful activity? If so, what did it provide to the participants? This question was addressed by a series of exit interviews conducted with TOTLE program staff. In these interviews, four of the seven key program staff were asked to comment on selected aspects of the VFE development process. During these interviews, the interviewer recapped the development procedure asking the interviewees what aspects of the procedure contributed to the participant's overall workshop experience and which detracted from it or complicated the administration of the workshop. They were also asked to suggest alternative development strategies. Though all four staff agreed that the development presented administrative challenges for the program, some stated that developing the environment concurrently with the program gave the participants a "sense of ownership" in the final product. The accuracy of this perception is challenged somewhat by several post-workshop surveys that show that in the first year of the program only 23% (n = 29) of the participants used the VFE in their classrooms. This did improve somewhat in the second year where 40% (n = 20) reported using the VFE at least once during the school year. Some of the reported reasons for not having using the VFE could be correlated with the fact that in first and second year of the program, participants received partially functional prototypes. Only in the final year did participants receive a complete version of the VFE. Another comment made by one of the interviewees is that the VFE development in its various forms provided the participants with insight into how educational research and development is conducted. While not an outcome directly related to the issue of VFE construction as a means of geoscience instruction, it does indicate that development is a means of engaging teachers in educational research rather than simply presenting them with its theoretic framework.

DISCUSSION AND SUMMARY

The three models discussed in this article illustrate how VFE design and construction is used to nurture technological pedagogical and content knowledge (TPACK) amongst geoscience educators. What is significant about all these models is that VFEs are used to reinforce or encourage fieldwork rather than replace it. Though each engages workshop participants in design and construction to a different degree, they share four principal characteristics.

(1) Each engages teachers in the design and/or construction of a form of virtual reality that simulates and reinforces their field experience.

(2) This activity provides an engaging focus for and means of recording their experience.

(3) Engaging the participants in the development of VFEs and supporting activities provides them with a means of transferring what they learn in the field back to their classrooms.

(4) Teacher development of VFEs and associated activities reinforces the significance of fieldwork in geoscience, and introduces the principal geocognitive skills involved in that work.

Despite the existence of virtual field trips in science education for nearly two decades, and similar ideas through slide shows long before that, the design and use of virtual reality for geoscience learning is still very much in its infancy. Unlike the latter, VFEs are much more complex and interactive, making them more adaptable to inquiry-style education. Because they are simulated environments rather than linear narratives they have a greater potential for providing teachers and students with a more authentic representation of a field site and how geoscience happens in a field situation. Furthermore, engaging teachers and students in designing and using these environments can be valuable for teaching cognitive skills fundamental to the geosciences and reinforcing the importance of fieldwork in the construction of geoscience knowledge. However, to realize this potential, additional research and/or development are needed in the following areas:

(1) Identifying the essential design features of a VFE that provide learners with an authentic representation of what they would do and see in the field.

(2) Identifying essential design features that provide learners with an extended or augmented view of a field location, which is an important step in teaching geocognitive skills.

(3) Reducing the "technological overhead" in VFE construction so that learner/developers focus on geoscience and scientific fieldwork rather multi-media development/computer programming. While this problem has lessened during the past decade due to technological advances, simplifying VFE construction remains a significant challenge to its usefulness as an educational field activity.

ACKNOWLEDGMENTS

• *ReaL Earth Inquiry* is supported by NSF DRK-12 Grant DRL 0733303 awarded to Robert M. Ross and Don Duggan-Haas of the Paleontological Research Institution. Richard Kissel and Don Duggan-Haas are the lead instructors for the ReaL Earth Inquiry Program. This work would not have been possible without Robert Ross and Richard Kissel, and their continued support and collaboration is deeply appreciated.

• *Astrobiology at Yellowstone*—The consultant activities of F. Granshaw were supported, in part, by grants awarded to S.L. Cady, professor of geology, Portland State University, lead instructor of the "Mars Analogs in Yellowstone" course taught during the summers of 2009 and 2011,

through the National Science Foundation Geoscience Education, grant NCE0808211 and NASA/Oregon Space Grant Consortium, grant NNX10AK68H.

• *TOTLE in the Field*—TOTLE was supported by NSF/EarthScope Geoscience Education grants #0745692, #0745681, #0745526, and #0745570 awarded to R. Butler of University of Portland, J. Whitman of Pacific University, B. Pratt-Sitaula of Central Washington University, and F. Granshaw of Portland Community College, respectively.

Any opinions, findings, and conclusions or recommendations are those of the authors and do not necessarily reflect the views of the National Science Foundation.

LINKS TO SOME VIRTUAL FIELDWORK RESOURCES

• Why does the Earth Look the Way it Does? The website of Virtualfieldwork.org http://www.virtualfieldwork.org/Welcome.html.

• Resources for virtual fieldwork (Granshaw, 2011). Links to virtual fieldwork projects and resources for constructing virtual field environments: http://www.artemis-science.com/VFW/.

REFERENCES CITED

Butler, R., 2008, Teaching Geoscience through Fieldwork, GEES Learning and Teaching Guide: Plymouth, UK, The Higher Education Academy Subject Centre for Geography, Earth, and Environmental Science, University of Plymouth, 56 p.

Compton, R.R., 1985, Geology in the Field: New York, Chichester, Brisbane, Toronto, Singapore, John Wiley & Sons, 398 p.

Cook, V.A., Phillips, D., and Holden, J., 2006, Geography fieldwork in a "risk society": Area, v. 38, no. 4, p. 413–420, doi:10.1111/j.1475-4762.2006.00707.x.

Dodick, J., and Orion, N., 2006, Building an understanding of geologic time: A cognitive synthesis of the "macro" and "micro" scales of time, *in* Manduca, C., and Mogk, D.W., eds., Earth and Mind: How Geologists Think and Learn about the Earth: Geological Society of America Special Paper 413, p. 77–93, doi:10.1130/2006.2413(06).

Duggan-Haas, D., 2010, Big ideas in earth system science: What are the most important ideas to understand about the earth?: American Paleontologist, v. 18, no. 1, p. 26–28.

Duggan-Haas, D., 2011, Bigger earth system science ideas and the next generation of science standards: Geological Society of America Abstracts with Programs, v. 43, no. 1, p. 142.

Duggan-Haas, D., and Clark, S.K., 2009, Forest for the trees: Earth systems science literacy initiatives and the need for a smaller integrated set of principles: Geological Society of America Abstracts with Programs, v. 41, no. 7, p. 712.

EarthScope, 2011, EarthScope: Exploring the Structure and Evolution of the North American Continent, http://www.earthscope.org/ (accessed 18 August 2011).

Fisher, J.A., 2001, The demise of fieldwork as an integral part of science education in United Kingdom schools: a victim of cultural change and political pressure?: Pedagogy, Culture & Society, v. 9, no. 1, p. 75–96, doi:10.1080/14681360100200104.

Granshaw, F.D., 2011, Designing and Using Virtual Field Environments to Enhance and Extend Field Experience in Professional Development Programs in Geology for K–12 Teachers [doctoral dissertation]: Portland State University, Department of Geology, 213 p.

Hawley, D., 1997, Being There: A Short Review of Field-Based Teaching and Learning, *in* Proceedings, UK Geoscience Fieldwork Symposium: UK

Geosciences Education Consortium, http://www.gees.ac.uk/essd/field.htm (accessed June 2012).

Head, J.W., van Dam, A., Fulcomer, S.G., Forsberg, A., Prabhat, Rosser, G., and Milkovich, S., 2005, Adviser: Immersive scientific visualization applied to Mars research and exploration: Photogrammetric Engineering and Remote Sensing, v. 71, no. 10, p. 1219–1225.

Herbert, B.E., 2006, Student understanding of complex Earth systems, *in* Manduca, C., and Mogk, D.W., eds., Earth and Mind: How Geologists Think and Learn about the Earth: Geological Society of America Special Paper 413, p. 95–104, doi:10.1130/2006.2413(07).

Kastens, K.A., and Ishikawa, T., 2006, Spatial thinking in the geosciences and cognitive sciences: A cross-disciplinary look at the intersection of the two fields, *in* Manduca, C., and Mogk, D.W., eds., Earth and Mind: How Geologists Think and Learn about the Earth: Geological Society of America Special Paper 413, p. 53–76, doi:10.1130/2006.2413(05).

Manduca, C., and Mogk, D.W., editors, 2006, Earth and Mind: How Geologists Think and Learn about the Earth: Geologic Society of America Special Paper 413, 188 p.

Marshak, S., 2006, Essentials of Geology—Student Website: W.W. Norton & Company, http://www.wwnorton.com/college/geo/egeo2/sitemap.asp (accessed 17 August 2011).

Mishra, P., and Koehler, M., 2006, Technological pedagogical content knowledge: A framework for teacher knowledge: Teachers College Record, v. 108, no. 6, p. 1017–1054, doi:10.1111/j.1467-9620.2006.00684.x.

Orion, N., 1993, A Model for the Development and Implementation of Field Trips as an Integral Part of the Science Curriculum: School Science and Mathematics, v. 93, no. 6, p. 325–331, doi:10.1111/j.1949-8594.1993.tb12254.x.

Ross, R.M., 2010, Big Ideas in earth system science: The foundation for earth system literacy: American Paleontologist, v. 18, no. 1, p. 24–26.

Scudder, S.H., 1874, In the Laboratory with Agassiz: Every Saturday, v. 16, p. 369–370.

Shelton, J.S., and Shelton, H., 1966, Geology Illustrated: San Francisco, London, W.H. Freeman, 283 p.

Shulman, L.S., 2005, Signature pedagogies in the professions: Daedalus, v. 134, no. 3, p. 52–59, doi:10.1162/0011526054622015.

Smith, D., 2004, Issues and trends in higher education biology fieldwork: Journal of Biological Education, v. 39, no. 1, p. 6–10, doi:10.1080/00219266.2004.9655946.

Tabor, R.W., 1975, Guide to the Geology of Olympic National Park: Seattle Washington/London, University of Washington Press, 144 p.

Thompson, A.D., and Mishra, P., 2007, Breaking News: TPCK Becomes TPACK!: Journal of Computing in Teacher Education, v. 24, no. 2, p. 38.

Tomlinson, C.A., 2001, How to differentiate instruction in mixed-ability classrooms, 2nd edition: Alexandria, Virginia, Association for Supervision and Curriculum Development, ISBN 978-0-87120-512-4.

Tomlinson, C.A., and McTighe, J., 2006, Integrating differentiated instruction & understanding by design: Connecting content and kids: Alexandria, Virginia, Association for Supervision and Curriculum Development, ISBN 978-1-4166-0284-2.

Trautmann, N., and MaKinster, J., 2010, Flexibly Adaptive Professional Development in Support of Teaching Science with Geospatial Technology: Journal of Science Teacher Education, v. 21, no. 3, p. 351–370, doi:10.1007/s10972-009-9181-4.

Xu, X., and Aiken, C., 2000, Digital field mapping and the virtual outcrop at UTD: Geological Society of America Abstracts with Programs, v. 32, no. 7, p. A514.

MANUSCRIPT ACCEPTED BY THE SOCIETY 16 APRIL 2012

The Geological Society of America
Special Paper 492
2012

Developing virtual field experiences for undergraduates with high-resolution panoramas (GigaPans) at multiple scales

Jennifer L. Piatek*
*Department of Physics and Earth Science, Central Connecticut State University, 1615 Stanley Street,
New Britain, Connecticut 06053, USA*

Candace L. Kairies Beatty
William L. Beatty
Department of Geoscience, Winona State University, Winona, Minnesota 55987, USA

Michael C. Wizevich
Alex Steullet
*Department of Physics and Earth Science, Central Connecticut State University, 1615 Stanley Street,
New Britain, Connecticut 06053, USA*

ABSTRACT

Field experiences are the cornerstone of a successful geoscience education, but these activities can be difficult (if not impossible) to include in many geoscience courses due to practical concerns. Virtual field exercises, presented through a series of high-resolution zoomable panoramas created with a GigaPan® robotic camera mount and associated software, allow students to gain experience interpreting outcrops and landscapes when physical travel to a site is not feasible. Exercises incorporating GigaPan panoramas have been developed for a number of undergraduate courses at different levels within the geoscience curriculum. Students in introductory-level courses are presented with exercises that explore local geology and illustrate basic concepts such as faulting and cross-bedding. Exercises for intermediate-level courses include analysis of geomorphic features in relation to bedrock type, the influence of landforms on historical events, and interpretation of shear stress orientations and magnitudes from small-scale structural features in outcrop. More advanced exercises, utilizing multiple-tier panoramas that range from outcrop to thin-section scales, have been developed from existing field research projects. These examples represent the initial effort to develop an extensive catalog of interactive self-paced exercises that will be incorporated into classes across the geoscience curriculum.

*piatekjel@ccsu.edu

Piatek, J.L., Kairies Beatty, C.L., Beatty, W.L., Wizevich, M.C., and Steullet, A., 2012, Developing virtual field experiences for undergraduates with high-resolution panoramas (GigaPans) at multiple scales, *in* Whitmeyer, S.J., Bailey, J.E., De Paor, D.G., and Ornduff, T., eds., Google Earth and Virtual Visualizations in Geoscience Education and Research: Geological Society of America Special Paper 492, p. 305–313, doi:10.1130/2012.2492(21). For permission to copy, contact editing@geosociety.org. © 2012 The Geological Society of America. All rights reserved.

INTRODUCTION

Geology is an inherently visual science. Most students' learning is promoted by visual and interactive aids, and educators are increasingly encouraged to use active and inquiry-based learning and visualization strategies in their classrooms, prompting students to make observations and collect and analyze data to solve problems, rather than take the traditional passive learning approach (Brunkhorst, 1996; National Research Council, 2000; Libarkin and Brick, 2002; McConnell et al., 2003; and Apedoe et al., 2006). In spite of this, many geoscience courses are presented in traditional lecture formats, with verbal presentations of material that students dutifully memorize and repeat on exams. Real-world applications of lecture material are difficult to present in classroom settings where the available visual resources are often limited to static photographs. Even though hand samples can provide additional experience, they are difficult to use in large classroom settings, especially when university collections are small or incomplete. Field observations, where geologic concepts and processes can be examined from large (landscape), intermediate (outcrop), small (hand sample), and potentially microscopic scales, allow students to explore diverse geologic settings to collect and analyze fundamentally different kinds of data. This experience enhances their understanding of geological processes and fosters the development of higher-order learning skills, including the ability to solve problems in varying situations (Kern and Carpenter, 1986; Hurst, 1998; Boyle et al., 2007; Elkins and Elkins, 2007; Kelso and Brown, 2009). Unfortunately, significant amounts of field work are often difficult to incorporate into classes, particularly large introductory courses that serve a wide student audience, for both financial and logistical reasons.

This project addresses some of these issues, common to many geoscience educators, by developing virtual multi-scaled field experiences that enable users to view outcrops or landscapes at scales from the panoramic wide scale down to the macroscopic hand sample and even to the microscopic thin-section level. Virtual field experiences are effective ways to incorporate active and inquiry-based learning and visualization strategies into both small and large classes. Field sites have been chosen to illustrate key concepts in introductory geoscience courses as well as further students' understanding of local and regional geologic history. Exercises emphasize observation and problem solving using online GigaPan panoramas. These explorable images actively engage the user and are well suited for use in open-ended inquiry-based exercises at all levels of geoscience education.

CREATING THE PANORAMAS

Virtual field experiences are built mainly around panoramas created using a GigaPan robotic camera mount (imager) and associated stitching software. Assembled panoramas (or other large image files) can be uploaded to the GigaPan website (http://www.gigapan.org or http://www.gigapan.com), where they can be viewed using a Flash-enabled browser. Most of the landscape,

outcrop, and hand sample panoramas discussed here were created using the GigaPan Epic model and a compact "point and shoot" digital camera (microscopic images are discussed later in this section). Other robotic camera mounts for use with larger cameras are also available (http://gigapan.com/cms/shop/store). The camera is mounted on the GigaPan imager, which utilizes a motor-driven lever (robotic arm) to press the shutter release and take photos automatically within a predefined grid. The upper left and lower right corners of the grid are selected by the camera operator, and the imager determines the appropriate number of rows and columns within the grid. The grid spacing is defined by the imager field of view, which the operator sets to allow for identifiable overlap between successive photographs. The imager automatically aims the camera and takes a photo at each grid location. The digital camera is typically set to a 3–4 times zoom to capture small-scale features. At this zoom setting, a typical outcrop or landscape pan (100–150° field of view) requires 200–400 photos and takes 30–60 min for setup and image acquisition. Higher zoom settings require more photos and therefore more time. The GigaPan imager is capable of acquiring 360° pans and can be used to create panoramic anaglyphs (red-cyan 3-D images) with two appropriately spaced pans and some additional image processing. Detailed information about camera settings and adjusting the robotic camera mount are specific to the camera and GigaPan models used: such instructions are included with GigaPan units and can also be downloaded from the website (http://gigapan.com/cms/manuals).

Once individual images are acquired, the final panorama is assembled using the GigaPan Stitch application. (The license for this software is included with the GigaPan imager.) The software utilizes common features within the overlap zone to align adjacent images and assemble them into a single panorama. Although it was designed to work with images captured by the GigaPan imager, the stitching software can also create panoramas from any set of overlapping images. This allows "micropanoramas" to be constructed from photomicrographs of thin sections acquired using microscope-mounted cameras. Like the GigaPan imager-generated panoramas, these micropanoramas are best made from images with ample overlap. Consistently sufficient overlap (about one-quarter to one-third of the image) of photomicrographs is facilitated using a point-counting stage. Completed pans can be viewed within the Stitch application as well as via the website (after uploading).

While interacting with the stitched pans, a viewer can move vertically and horizontally around the finished image as well as zoom in or out. The interactivity of the panoramas facilitates exercises that encourage inquiry and engage students to examine areas of interest outside those specifically queried by the assignment. The website also allows panorama owners to provide descriptive captions and geolocate pans in Google Earth. Pans can be viewed in the Google Earth application by following a link from the GigaPan website or from a saved placemark file. Registered users of the GigaPan site can create "snapshots" in panoramas—static views of pans at specific zoom levels with associated comments or questions. Subsequent viewers can click on the snapshots, be taken directly to the location and zoom of the snapshot, and add additional comments. Panorama owners have the ability to make pans private (accessible only to those who know the URL of the pan) and make snapshots visible only to the owner and the user who created the snapshot. This allows the owner to make a "game" of the image by challenging users to find specific locations within the pan by submitting snapshots that cannot be seen by future users, preserving the game aspect. Although these two options are not utilized in this particular work, they provide functionalities that are useful in the context of creating classroom exercises.

In addition to panoramas acquired with the robotic camera mount, large image files can be uploaded via the GigaPan Uploader (a free download from the GigaPan website). The viewer is able to explore these images in detail without needing to download the image file. This application is utilized to include maps or other image data sets within panorama-based exercises.

APPLICATIONS

Exercises utilizing GigaPan panoramas have been implemented in a number of courses, ranging from introductory to upper levels. Example exercises and associated images are described here. Links to the full GigaPan panoramas are also provided. Additional panoramas created by the authors can be found on the website by searching for appropriate tags (e.g., CCSU for Central Connecticut State University or WSU for Winona State University) or by clicking the usernames associated with these pans.

A variety of panoramas can be used at the introductory level to introduce basic concepts in the geosciences. For example, the panorama of Sheepeater Cliff at Yellowstone National Park (Fig. 1) is used to engage students in a short in-class discussion of the properties of lava and the formation of columnar joints. Students are asked to make observations about how this outcrop differs from other extrusive igneous features in the area and think about what might influence the cooling behavior of lava.

A panorama of an outcrop near Kodachrome Basin State Park in south-central Utah is used to illustrate concepts to introductory level students at multiple scales (Figs. 2A–2C). Physical geology students utilize snapshots to guide them to particular locations and are asked to observe, describe, and make simple interpretations of some of the features present, including faults, cross-beds, and talus slopes. Students are also asked to observe the outcrop in its entirety to interpret relative ages of stratigraphic features. Outcrop scale panoramas such as those in Figures 1 and 2A can also serve as examples for students learning to interpret field-based data sets such as stratigraphic sections and 3-D models derived from outcrop LiDAR measurements. The GigaPan panoramas do not replace the quantitative data derived from these methods, but serve as a complement to those trying to analyze those data.

Panoramas of multiple locations in Gettysburg National Military Park (e.g., Fig. 3), along with representative hand samples,

Figure 1. Panorama of Sheepeater Cliff, Yellowstone (http://gigapan.org/gigapans/59868/), used to engage introductory students in a short, in-class discussion of the properties of lava and the formation of columnar joints.

are used in an upper level geomorphology class as the basis of a "virtual field work" exercise, which allows students to perform more in-depth investigations of geologic problems. In this exercise, students investigate the role of geology and topography in the Battle of Gettysburg by exploring field sites keyed to different panoramas. Students visit each of the sites marked on a topographic map, explore the associated panorama, orient themselves using Google Earth, make field observations (including locations of outcrops, weathering patterns of rocks and topography of the area) and identify rocks in outcrop based on their identification of hand samples. After visiting the sites, students compile their data and create a geologic map of the battlefield. They then use the map to analyze how local geology and topography might have affected the strategies of the Union and Confederate armies, the fighting around the Pennsylvania town, and the heavy casualties suffered by both sides.

An outcrop in Harrison, Montana, (Fig. 4) is explored virtually by students in several courses. Introductory students are asked to perform tasks similar to those described above for the Kodachrome Basin State Park panorama. Pans of this nature are useful to help students develop their skills in rock identification, determination of grain size and sorting, and identification of sedimentary structures and textures. Students in an intermediate-level field methods course observe and explore the outcrop and use those observations as the basis for their field notes. They then create field sketches, drawn to scale, of both the entire outcrop and a smaller portion of the outcrop that was selected in a snapshot. By utilizing the GigaPan image, students are able to practice taking good field notes and making field sketches without traveling to the outcrop site (in this case, more than 1100 miles from campus). In subsequent weeks, students apply these skills firsthand at local outcrops. An exercise for students in an upper-level sedimentology course is currently being developed that will ask them to use information gathered from the pan along with additional regional context to reconstruct the depositional history of the Harrison Basin.

Panoramas of a gneiss from the Paleozoic Tatnic Hill Formation (part of the Willimantic Dome exposed in Coventry, Connecticut) illustrate the multiple-scale approach (Figs. 5A–5C). The outcrop panorama gives students an overview of the macroscale deformation, while smaller structures are visible in the hand-sample pan. Micro-scale structures can be described from the thin-section pan. While virtual exercises such as this are not replacements for field experiences for petrology or structural geology students, they give students additional exposure to real-world problems without incurring the time and financial costs or risks sometimes associated with fieldwork (in this case the out-

Figure 2. (A) Part of a panorama of an outcrop near Kodachrome Basin State Park (south-central Utah) that is used in an exercise for introductory geology students. Locations of insets (B and C) are outlined on the larger image. Full pan and snapshots at http://www.gigapan.org/gigapans/58281/. (*Continued.*)

Figure 2 (*continued*). (B) Students are guided to this area and asked to examine and explain the thin layers to the right of the hammer (noting the orientation changes) and to observe and speculate on the composition and formation process of the blob to the left of the hammer. (C) Snapshots in this area focus on the offset strata and the nature of the material at the base of the outcrop. Students are asked to sketch the offset layers and identify the structures and consider the stresses responsible. The rest of the exercise requires the student to describe the nature of the material at the base of the outcrop, discuss the source of the material and speculate on the future of this material.

Figure 3. The view from Little Round Top (http://gigapan.org/gigapans/33213/) is part of an exercise in which students use observations of the Gettysburg Battlefield to create a geologic map of the area and investigate the role that geology played in the battle.

Figure 4. Students of various skill levels observe this panorama of an outcrop in Harrison, Montana, (http://gigapan.org/gigapans/84749/) in different courses and use it as the basis of simple observations, detailed field notes, and analysis of depositional environments.

Figure 5. (A) Outcrop of deformed gneiss that forms part of the Willimantic Dome (http://www.gigapan.org/gigapans/50957/). Deformation structures can be observed from this scale (full pan) down to small, zooming within the pan and then linking to the associated hand sample (B) and thin section (C). (B) Pan of hand sample from Willimantic Dome (http://www.gigapan.org/gigapans/72884/). Utilizing a zoomable panorama allows viewers to focus on small scale details that might not be noticed using a static photograph. (*Continued.*)

Figure 5 (*continued*). (C) Thin section of Willimantic Dome hand sample (B), crossed polars (http://www.gigapan.org/gigapans/72751/). As with the hand sample pan, thin section scale pans allow for discussion of samples in a group setting.

crop is located alongside a busy highway that is difficult to access during field trips). In this exercise, the additional pans allow the students to investigate multiple scales of data. As with other pans, instructors may use the hand-sample and thin-section pans to augment group discussions.

Regional Field Trips

Virtual field trips can bring sites in faraway locations to students, but they're also a useful way to incorporate local geology into exploratory and inquiry-based exercises. A trial project to create a library of panoramas from regionally important outcrops that illustrate the geologic history of Connecticut (http://www.physics.ccsu.edu/piatek/ct_geo/) is currently being developed at Central Connecticut State University. This collection highlights local outcrops that are commonly visited during class field trips, and can be used to continue the discussion about outcrop- and small-scale features after returning to the classroom. In addition, students who are unable to attend field trips can use the panoramas to gain some of the same field experiences as their classmates. In addition to outcrop panoramas, the site links to images derived from LiDAR elevation data that were submitted to the GigaPan site using the Upload application. Expanding this collection will allow advanced students to explore the utility of GigaPan panoramas—those interested in geoscience education would identify new sites to image and create their own exercises, while students engaged in research projects can use the virtual tour to explore how to share their results with a general audience. Ultimately, the panoramas developed for the project will be made available to both teachers and the general public to facilitate a better understanding of regional geology.

CONCLUSIONS

Although we have focused on using GigaPan technology to develop the exercises discussed here, the concept of the virtual field experience is open to any technology that helps to recreate a field environment for students (e.g., GIS, LiDAR, Google Earth,

Fusion Tables, etc.). By utilizing these tools, geoscience educators can develop pedagogies that move away from traditional "chalk and talk" lectures toward those that encourage active exploration of and interaction with the natural world. When we cannot bring our students to the field, we can instead bring the field to our students.

The goal of the virtual field experience is to bring field work into the classroom, making it more accessible to more students. Virtual field exercises are not meant to replace traditional field experiences, although they are functional alternatives when field trips are not possible. They are best used as supplements that allow students to develop skills in observation, analysis, and interpretation of landscapes, outcrops, and thin sections. High resolution zoomable panoramas like those created with the GigaPan system allow viewers to explore features at multiple scales, and when combined with hand samples and thin sections (or their image equivalents) provide the best field experiences students can have without leaving the classroom.

ACKNOWLEDGMENTS

We wish to thank the Global Connection Project at Carnegie Mellon University and the Fine Foundation Outreach for Science project for including the authors in the beta program for the GigaPan Imager. GigaPan® is a registered trademark of Carnegie Mellon University licensed by GigaPan Systems.

REFERENCES CITED

Apedoe, X.S., Walker, S.E., and Reeves, T.C., 2006, Integrating inquiry-based learning into undergraduate geology: Journal of Geoscience Education, v. 54, p. 414–421.

Boyle, A., Maguire, S., Martin, A., Milsom, C., Nash, R., Rawlinson, S., Turner, A., Wurthmann, S., and Conchie, S., 2007, Fieldwork is good: The student perception and the affective domain: Journal of Geography in Higher Education, v. 31, p. 299–317, doi:10.1080/03098260601063628.

Brunkhorst, B.J., 1996, Assessing student learning in undergraduate geology courses by correlating assessment with what we want to teach: Journal of Geoscience Education, v. 44, p. 373–378.

Elkins, J.T., and Elkins, N.M.L., 2007, Teaching geology in the field: Significant geosciences concept gains in entirely field-based introductory geology courses: Journal of Geoscience Education, v. 55, p. 126–132.

Hurst, S.D., 1998, Use of "virtual" field trips in teaching introductory geology: Computers & Geosciences, v. 24, p. 653–658, doi:10.1016/S0098-3004(98)00043-0.

Kelso, P.R., and Brown, L.M., 2009, Integration of field experiences in a project-based geoscience curriculum, *in* Whitmeyer, S.J., Mogk, D.W., and Pyle, E.J., eds., Field Geology Education: Historical Perspectives and Modern approaches: Geological Society of America Special Paper 461, p. 57–64, doi:10.1130/2009.2461(06).

Kern, E.L., and Carpenter, J.R., 1986, Effect of field activities on student learning: Journal of Geological Education, v. 34, p. 180–183.

Libarkin, J.C., and Brick, C., 2002, Research methodologies in science education: visualization and the geosciences: Journal of Geoscience Education, v. 50, p. 449–455.

McConnell, D.A., Steer, D.N., and Owens, K.D., 2003, Assessment and active learning strategies for introductory geosciences courses: Journal of Geoscience Education, v. 51, p. 205–216.

National Research Council, Committee on the Development of an Addendum to the National Science Education Standards on Scientific Inquiry, 2000, Inquiry and the national science education standards: A guide for teaching and learning, Olson, S., and Loucks-Horsley, S., eds.: Washington, D.C., National Academy Press, http://books.nap.edu/openbook.php?record_id=9596&page=R1.

MANUSCRIPT ACCEPTED BY THE SOCIETY 16 APRIL 2012

The Geological Society of America
Special Paper 492
2012

Avatars and multi-student interactions in Google Earth–based virtual field experiences

Mladen M. Dordevic*
Steven C. Wild
Department of Physics, Old Dominion University, Norfolk, Virginia 23529, USA

ABSTRACT

We have developed object-oriented programming methods to enable avatar movement across the Google Earth surface in response to student actions. Students travel on their own, or in groups attached to a field vehicle avatar (a Jeep). Students communicate using text messages sent from their web pages to balloons that pop up from the avatars in Google Earth. Students can be located locally in a lab class or at great distances from one another, as in a distance education course.

Our programming methods help to create a more engaging virtual field trip in which the students take the lead and decide where to go rather than simply reading text and viewing graphics in a tour designed by their instructor. The user interactivity via avatars is controlled by JavaScript and PHP. Since the position of each avatar is known, it is possible to track their movements and offer text-message advice when students stray off-task or wander about aimlessly. Our methods will be included in new virtual field trips being developed for Iceland, Hawaii, and other locations.

INTRODUCTION

Google Earth (Google; see Appendix 1) comes in two forms: (*i*) a stand-alone application available for Windows, Macintosh, and Linux platforms, and (*ii*) a web browser plug-in compatible with a variety of JavaScript-enabled web browsers including Chrome, Firefox, and Safari (Google; see Appendix 1). The plug-in permits the programmer to incorporate one or more instances of Google Earth in a web page and to control each with familiar hypertext markup language (HTML) interface elements such as buttons, text fields, and sliders. This paper focuses on the web browser plug-in form of Google Earth because of its extensive JavaScript application programming interface (API) and the possibility for client-server-client communication.

The majority of work done in the area of Google Earth–based virtual field trips involves a single person using a computer to view images, read text, etc. (Simpson and De Paor, 2010). Activities include following prerecorded tours or clicking on placemarks and reading associated balloon content. Few tools are available for users (such as students, teachers, and administrators) to actually interact with the virtual surroundings other than panning and zooming the camera view. What we have done is to

*mdordevi@odu.edu

Dordevic, M.M., and Wild, S.C., 2012, Avatars and multi-student interactions in Google Earth–based virtual field experiences, *in* Whitmeyer, S.J., Bailey, J.E., De Paor, D.G., and Ornduff, T., eds., Google Earth and Virtual Visualizations in Geoscience Education and Research: Geological Society of America Special Paper 492, p. 315–321, doi:10.1130/2012.2492(22). For permission to copy, contact editing@geosociety.org. © 2012 The Geological Society of America. All rights reserved.

simultaneously bring multiple users together and allow them to interact and explore on the same virtual globe, thereby simulating a real field experience where each student would be able to communicate with colleagues and collaborate on collective tasks. Interaction in a virtual environment or Google Earth is not new (Roush, 2007) (Google; see Appendix 1). However, the hybridization of Google Earth API for a virtual-interactive geological environment is. To achieve the above goals, a client-side application alone is not sufficient. Being able to synchronize multiple client instances of Google Earth over the Internet requires server-side programming as well. The server has to log and process incoming traffic from clients. For this purpose, the PHP scripting language was chosen for its flexibility. First, it has the ability to generate HTML pages. PHP scripts can also be embedded into HTML pages. Finally, PHP scripts can manipulate MySQL-type databases. We might equally have chosen Python or Ruby-on-Rails for this task instead of PHP. The purpose of this paper is to demonstrate how to implement the programming tasks necessary to support user interaction on Google Earth. To this end, we will present the client-server-client communication code with a web-chat example. Passing of other data (e.g., avatar location) will be discussed along with server polling. Data logging is an extra benefit whose usefulness for educators and programmers will be explored. The combination of these parts makes creating virtual field trips possible.

WEB-CHAT EXAMPLE USING AJAX AND PHP

The backbone of the interactive Google Earth programming is the client-server-client communication. Once communication is established data such as chat messages, latitudes, longitudes, etc., may be exchanged among users. Communication between client and server is done via Ajax (Garrett; see Appendix 1). Ajax enables web pages to communicate with a server, send and receive data asynchronously without refreshing the page, and therefore avoid reloading Google Earth plug-in at every update. The code snippets that handle the Ajax interface (courtesy of http://icodesnip.com/search/ajax/1) are in the form of a function,

```
function Ajax_Send(GP,URL,PARAMETERS,
RESPONSEFUNCTION).
```

The function's parameters are as follows:
i. GP represents the type of request (POST or GET);
ii. URL is the address of the PHP script that will be executed;
iii. PARAMETERS is the string that contains variables and values stored in it (var1=value1&var2=value2&var3= value3); and
iv. RESPONSEFUNCTION is the function that will be evaluated upon server response with XMLHttpRequest. responseText as its argument.

To help the reader understand this AJAX-based communication, we will first explain the data structure and data handling on the server.

In the entry string, the tag <!@!> separated row variables, and "/n" denoted a new row. PHP retrieved variables values from a file containing this entry string with the explode() method (see the example in Table 1).

In our case, the single data packet from the client contains variables that tell the server the following:
i. what action is required;
ii. the integer value of the last received row from the table that contains chat messages and locations of the other avatars;
iii. the user name of the client making a chat request; and
iv. the user's group. Upon arrival of the data packet from the client, the server first decides what action to perform.

For example, if user1 (referred to here as the sender) starts a web-chat by sending a new message to user2 (the receiver), the data sent from the sender are:
i. user name of the message receiver (which can be a single other user or a whole group);
ii. the content of the message; and
iii. the name of the sender.

The code on the client side is a function defined as in Table 2.

The sender makes a connection with the server using the standard XMLHttpRequest protocol (Kesteren; see Appendix 1). The server then:
i. picks up the string with the variables mentioned above using the $_POST["variable name"] (in this case the string is called data);
ii. cleans it (a standard procedure of filtering client input so that corrupted, incorrectly formatted, or harmful data are not stored in the database); and
iii. stores the string into the database.

The next step is to ensure that the receiver is notified of the new message. All clients periodically query the server for updates by sending their user name, group, and a number that tells what database row they last read (lastReceived). The frequency of queries is set by the native JavaScript function: setInterval(). The time interval between successive queries needs to be experimentally determined and fine-tuned. In our case, the function called updateInfo() queries every 800 ms (see Table 3).

When the server receives an update request from a client, it passes the query to the database with values from the user making the update request:
i. the number (lastReceived) must be smaller than the current queue number in the database table;
ii. the receiver name must match the client user name making a request for an update; and
iii. the entry must have been posted with a maximum time interval (currently 50 minutes).

Entries matching the query are sent to the client who made the request in JSON format (json; see Appendix 1) by the procedure called echo(). The snippet of PHP code that the server uses to do this is shown in Table 4.

TABLE 1. PHP

```
/*example of data entry on the server in the data.txt file: */
1299265835.4478<!@!>mike<+@+>Hi<+@+>24.999<+@+>-39.999<+@+>25494105.783<+@+>0<+@+>0.0000018 /n


/*PHP script that checks if your chat box is up to date */
/*Takes the time (in UNIX format) from client when he did his last update and assigns it to $lastreceived */
        $lastreceived = $_POST['lastreceived'];
        /*Finds the physical location of the file with the conversation from the server and assigns it to a array variable $data
where every new line is listed as the nest element of array */
        $data = file("data.txt",FILE_IGNORE_NEW_LINES);
        /*Checks the number of lines in $data, if 0, there is no need to continue execution */
        if(count($data) == 0){
                exit();
        }
        /*Start a loop over the $data line by line */
        for($line = 0; $line < count($data); $line ++){
                /*Separates the current line ($data[$line]) into array wherever the string <!@!> is encountered, in this case we
will have $messageArr[0] = "1299265835.4478" and $messageArr[1] = "mike<+@+>Hi<+@+>24.999<+@+>-
39.999<+@+>25494105.783<+@+>0<+@+>0.0000018" */
                $messageArr = explode("<!@!>",$data[$line]);
                /*If there is new entry in data.txt, send it to the client */
                if($messageArr[0] > $lastreceived) echo $messageArr[1]."<+newline+>";
        }
        /*else, only send the time of the last entry from the data.txt */
        echo "<SRVTM>".$messageArr[0];
```

TABLE 2. JavaScript

```
function sendMessage(){
        /*Define the string that is being sent to the server */
        data = "action=send&to_who=milisav&content=" + document.getElementById('message').value +
        "&sender=" + Avatar.userName;
        /*Function invoking XMLHttpRequest */
        Ajax_Send("POST", "stuIndex_ch.php", data, sentOk);
}
```

TABLE 3. JavaScript

```
function updateInfo(){
        /*Define string that is being sent to the server */
        var data = "action=updateStat&lastReceived=" + lastReceived + "&iam="+Avatar.userName +
        "&myGroup=milisav";
        /*Function invoking XMLHttpRequest */
        Ajax_Send("POST", "stuIndex_ch.php", data,sentOk);
}
```

Two functions are involved in processing the received data on a receiver's side. The first function is sentOk(), which is the fourth argument (RESPONSEFUNCTION) of the Ajax_Send() operation in updateInfo(). See Table 5.

This function looks for the part of the sent data that contains the message (p.message) and passes it to a function called populatingTables(). See Table 6.

Figure 1 shows two students chatting about differences in rocks at Indian Gardens in the Grand Canyon, which is the result of the completion of the above function that arrived from the server (see Table 7).

PASSING OTHER DATA: AVATAR MOVEMENT

Web-chat information is not the only type of data that can be sent using the above approach. Avatar location and movement are other examples. There are two aspects of avatar movement. The first is controlling the movement of one's personal avatar using an approach adapted from the Google Earth API sample code called Monster Milktruck (see Appendix 1). Keyboard input controls the local avatar movement and a function responsible for sending of this avatar's position even while the user is stationary is nested in setInterval().

TABLE 4. PHP

```php
/*One of the switches in an if statement */
case "updateStat":
/*Check to see if all variables in received string are defined true: if so, than proceed; false: send error and exit */
        if(isset($_POST['lastReceived'],$_POST['iam'],$_POST['myGroup'])){
/*Assign received values to local variables after cleaning them to prevent corrupt data from being passed */
        $lastReceived=clean($_POST['lastReceived']);
        $iam=clean($_POST['iam']);
        $myGroup=clean($_POST['myGroup']);
/*Make the mySQL query */
        $res=mysql_query("SELECT * FROM chat WHERE $lastReceived< id_chat AND ('$myGroup'=to_who OR
'$iam'=to_who) AND (5000>( NOW()+0. -time_of_send+0.))");
/*List results of the query */
        while($row = mysql_fetch_array($res)) {
/*Fill the messageStr array with listed results */
        $messageStr[]= $row['sender'].":  ".$row['content'];
        $lastReceivedNew=$row['id_chat'];
        }
        $res=mysql_query("SELECT username FROM user INNER JOIN online ON online.id_user=user.id");
        while($row = mysql_fetch_array($res)) {
        $online[]= $row['username'];   }
/*If database query returned any result for new chat, echo it with the online user list */
        if(isset($messageStr)){
        $json=array('message'=>$messageStr,'lastReceived'=>$lastReceivedNew,'online'=>$online);
        echo json_encode($json);
        }
/*Just echo online user list */
        else{
        $json=array('online'=>$online);
        echo json_encode($json);
        }
        exit();
        }
        echo "Error: variables in updateStat not defined";
        exit();
```

TABLE 5. JavaScript

```javascript
function sentOk(res){
        /*Ensure the received data from server is in JSON format */
        if(res.isJSON()){
                /*Convert received JSON string to the Object p */
                var p=res.evalJSON([sanitize = true]);
        /*Check that fields are defined in Object p */
                if(typeof p.lastReceived != 'undefined' && typeof p.message != 'undefined'){
                        /*Updates the last received row from the chat table */
                        lastReceived = p.lastReceived;
        /*Run the function that dynamically populates the chat box with newly arrived messages */
                        populatingTables(p.message,'chatBox');
                }
        /*… (more function) */
        }
}
```

TABLE 6. JavaScript

```javascript
/*arguments:      cont, array that contains messages to be applied to the chat box "User : message"
                who, string id of the field where the message is applied
function populatingTables(cont,who){
/*Goes through all elements of the cont and passes them to the anonymous function as argument n
        cont.filter(
                function(n){
                /*Creates the div element
                var el = document.createElement("div");
                /*Apply the text to that div
                el.innerHTML = n;
                /*Finds the element who and applies the div element
                var a = document.getElementById(who);
                a.appendChild(el);
                a.scrollTop = a.scrollHeight;
                }
                )
        }
```

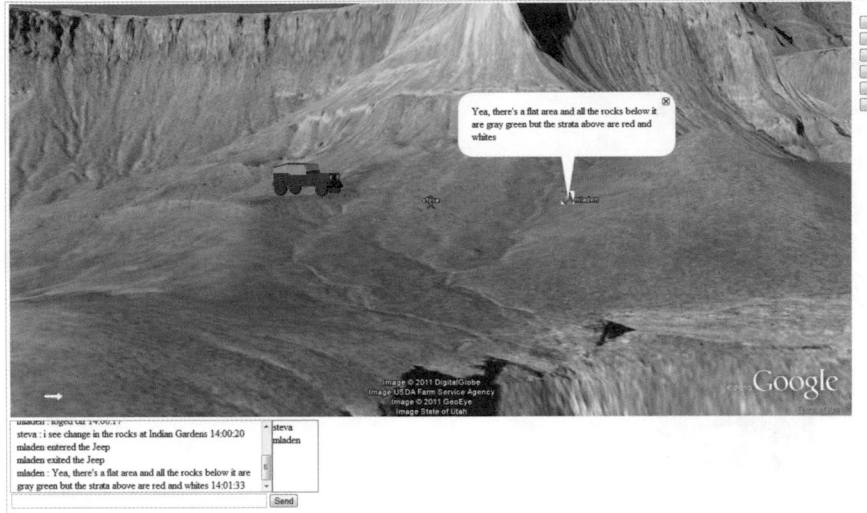

Figure 1. Two students discussing the geology near Indian Gardens in the Grand Canyon. Screenshot of "text balloon" after message is sent from one user to the server and then to the other user.

The second part of the avatar movement is updating positions of non-local avatars—that is, everybody but the local avatar. This is accomplished with the same approach for data handling used in the web-chat mentioned above. The same timing function that updates the avatar position is used to poll the server for updates to the non-local avatars' latitudes and longitudes. The server responds with the changes and the data are parsed and all non-local avatar positions are updated on the Google Earth terrain. Figure 2 contains an initial screen shot of two avatars and the field vehicle exploring in the Grand Canyon. An advantage of the virtual environment is that vehicles can travel anywhere using any means of transport—a set of horse icons could be substituted for the Jeep in this setting. Figure 3 shows updated positions of the avatars called "mladen" and "steva." This is the result of the "mladen" user:

   i. moving locally;

   ii. sending its position update to the server;

   iii. the server processing this update;

   iv. the second user "steva" getting an update from server; and

   v. "steva" locally parsing data and updating avatar locations.

The current iteration uses a MySQL database type instead of XML, which was used in the web-chat example.

DATA TRAFFIC

One challenge in client-server-client communications is polling, which relates to the timing of client-server or server-client communication. When the server receives more requests than in can fill, it stacks them and responds in order. During the response it hands out the most recent data set available. The issue that can arise at this moment is this: A local-user may miss an update of a non-local avatar by being ahead in the queue or miss a non-local position update because a non-local user had his position updated twice before the local user received the update. The first problem is not so major, as highly accurate positions of non-local avatars in some small time window are not vital to the virtual geology

TABLE 7. JavaScript

```
/*Google Earth update with the new message that pops above the sending avatar's head
function ballonMessage(message, user) {
        /*Try to remove any existing balloons first
        try{
                ge.setBalloon(null)
        }
        catch(err){};
    /*Create balloon and set it's content
        var balloon = ge.createHtmlStringBalloon('');
        /*Select the avatar of the associated balloon
        var placemark = Avatar.holder[Avatar.names.indexOf(user)].placemark;
        balloon.setFeature(placemark);
        balloon.setMaxWidth(300);
        balloon.setContentString(message);
        ge.setBalloon(balloon);
        /*Remove the balloon after 2 second
        setTimeout( function(try{ge.setBalloon(null)} catch(err){}},2000)
}
```

Figure 2. Screen shot of two students and the field vehicle. This is a reference shot taken on "steva's" screen to see the "mladen" avatar move in Figure 3. The region is again near Indian Springs in the Grand Canyon; same for Figure 3.

experience. The server is providing updates every 800 ms. The second problem is handled by linear interpolation to move the avatars from update to update. The interpolation helps minimize skipping a missed point by moving them on a linear path.

In future iterations we will try to implement new technologies for communication that have become available with the draft release of HTML 5 Web Socket (Hickson; see Appendix 1). While currently in the testing phase, Ajax does not create any problems in terms of polling, but for scalability it may require migration to more efficient solutions that would cut back unnecessary traffic and server load.

DATA-LOGGING

A major benefit gained from client-server-client communication is data-logging. The logging process occurs when users push their data to the server. The data recorded are latitude, longitude, and heading of every avatar with a time stamp and a log of their chat conversations with other students. Recording the users' actions has the potential to be useful to both code developers and educators.

Logging can help developers debug and optimize the code and therefore enhance the virtual field trip experience. Developers can also post-track users. Post-tracking means reviewing the user activity on a map either sequentially or as a scatter plot. When developing, it becomes important to know where and how much time the user spends on tasks. If a task is found to not be useful in the given educational benefit it could be removed or adapted to better suit the user.

Data-logging is also potentially valuable for teachers. The available feedbacks enable teachers to scaffold their student learning activities, assess learning outcomes, and evaluate students for credit if desired (Buckley et al., 2010; Horwitz et al., 2010; Sao Pedro et al., 2010; Jacobson and Reimann, 2010). Here again, post-tracking of users could be useful to help educators understand the ways in which students learn.

Figure 3. Second screen shot of user "steva's" screen. This shot is taken after user "mladen" has moved locally on his Google Earth. His movement is sent to the server and the server than sends the data to user "steva." The data are parsed locally and the "mladen" avatar is moved. The "mladen" avatar has moved many times to have noticeable change.

In future iterations of the application, we will be recording students' submitted work, such as locations of the sites they chose for data collection and their mapping efforts (identification of virtual specimens, drawing of contacts etc.). Also, as new tool sets become available for the client on different field trips, we will record their interactions within Google Earth.

CONCLUSIONS

Client-server-client communication enables virtual field trip developers to produce more interactive, efficient, and engaging learning experiences. Communication processes explored here include web-chat. This first process involves a client sending information to the server, the server processing the information, and the server returning a result. The second process involves sending a request for information update to the server, the server processing the request, and the server returning the information requested. Our program has been demonstrated at national meetings (Dordevic et al., 2011) and beta testing with structural geology students at Old Dominion University took place during the spring semester 2012. From early user reactions, we are confident that it will be a useful addition to the tools available to instructors in the geosciences. We anticipate that in future years, instructors will move away from virtual field trips in which the students are passive observers of content created by the instructor, toward interactive alternatives that constitute active virtual field work (Ross et al.; see Appendix 1).

ACKNOWLEDGMENTS

This research was support in part by NSF TUES 1022755, NSF GEO 1034643, and a 2010 Google Faculty Research Award. Any opinions, findings, and conclusions or recommendations expressed in this paper are those of the authors and do not necessarily reflect the views of the National Science Foundation or Google Inc.

We would like to thank Melissa Beebe for helping with the models, some of which were obtained from the Google 3D Warehouse. Janice Gobert helped greatly with the logging and scaffolding processes. The comments of two anonymous reviewers greatly improved the manuscript.

APPENDIX 1. WEB REFERENCES

Garrett, J.J., Ajax: A new approach to web applications, http://www.adaptivepath.com/ideas/ajax-new-approach-web-applications (accessed 22 November 2011).

Google, Google Earth API Demo Gallery, http://code.google.com/apis/earth/documentation/demogallery.html (accessed 22 November 2011).

Google, Google Earth plug-in, http://www.google.com/earth/explore/products/plugin.html (accessed 22 November 2011).

Google, What is Google Earth? http://earth.google.com/support/bin/answer.py?hl=en&answer=176145 (accessed 22 November 2011).

Hickson, I., HTML5, http://dev.w3.org/html5/spec/Overview.html (accessed 22 November 2011).

json, Introducing JSON, http://www.json.org/ (accessed 22 November 2011).

Monster Milktruck, http://en.wikipedia.org/wiki/Monster_Milktruck (accessed 22 November 2011).

Ross, R.M., Don Duggan-Haas, R.K., and Bessemer, C., Why does the earth look the way it does? http://www.virtualfieldwork.org/Welcome.html (accessed 22 November 2011).

van Kesteren, A., XMLHttpRequest, http://www.w3.org/TR/XMLHttpRequest/ (accessed 22 November 2011).

REFERENCES CITED

Buckley, B.C., Gobert, J., Horwitz, P., and O'Dwyer, L., 2010, Looking inside the black box: Assessments and decision-making in BioLogica: International Journal of Learning Technologies, v. 5, no. 2.

Dordevic, M.M., Wild, S.C., and De Paor, D.G., 2011, A geological mapping game in Google Earth: Geological Society of America Abstracts with Programs, v. 43, no. 5, p. 302.

Horwitz, P., Gobert, J.D., Buckley, B.C., and O'Dwyer, L.M., 2010, Learning genetics with dragons: From computer-based manipulatives to hypermodels, *in* Jacobson, M.J., and Reimann, P., eds., Designs for Learning Environments of the Future: International Perspectives from the Learning Sciences: Springer, p. 61–88.

Jacobson, M.J., and Reimann, P., 2010, Designs for learning environments of the future: International perspectives from the learning sciences: Springer, 268 p.

Roush, W., 2007, Second Earth: Technology Review, v. 110, no. 4, p. 39–48 (July/August).

Sao Pedro, M.A., Baker, R.S.J.d, Montalvo, O., Nakama, A., and Gobert, J.D., 2010, Using text replay tagging to produce detectors of systematic experimentation behavior pattern, *in* Baker, R.S.J.d., Merceron, A., and Pavlik, P.I., Jr., eds., Proceedings of the 3rd International Conference on Educational Data Mining, 11–13 June 2010, Pittsburgh, Pennsylvania, p. 181–190, http://educationaldatamining.org/EDM2010/?page_id=278.

Simpson, C., and De Paor, D.G., 2010, Restoring maps and memoirs to four-dimensional space using virtual globe technology: A case study from the Scottish Highlands, *in* Law, R.D., Butler, R.W.H., Holdsworth, R.E., Krabbendam, M., and Strachan, R.A., eds., Continental Tectonics and Mountain Building: The Legacy of Peach and Horne: Geological Society of London, Special Publication 335, p. 429–441, doi:10.1144/SP335.20.

Manuscript Accepted by the Society 16 April 2012

The Geological Society of America
Special Paper 492
2012

A geology-focused virtual field trip to Tenerife, Spain

Nicholas P. Lang*
Department of Geology, Mercyhurst University, 501 E. 38th Street, Erie, Pennsylvania 16546, USA

Kelley T. Lang
Center for Excellence in Teaching and Learning, Gannon University, 109 University Square, Erie, Pennsylvania 16541, USA

Brian M. Camodeca
Department of Information Technology, Mercyhurst University, 501 E. 38th Street, Erie, Pennsylvania 16546, USA

ABSTRACT

Virtual field trips provide a flexible, cost-effective way to teach or supplement course content. Here we describe a virtual field trip (VFT) to Tenerife, Spain, that emphasizes volcanism, but can be used to teach various geology-focused content areas. Specifically, using Google Earth as the platform, students can upload a Key-hole Markup Language (KML) file that will take them on a virtual transect across a volcanic terrain highlighting specific volcanic landforms and processes associated with the Caldera de las Cañadas and the Teide–Pico Viejo volcanic complex. The VFT is broken into 18 stops, each of which has embedded photographs and descriptions of each locality; YouTube videos are also included for some stops. Much as they would in a traditional "boots on the ground" field trip, students can go from stop to stop using their computers to make and tie together observations of diverse volcanic features at a variety of scales; students can complete this VFT either individually or in groups. Implementation of this VFT into two geology courses showed positive impacts on student learning, though testing on larger sample sizes should be conducted. The field trip and associated activities can be modified to suit a specific audience, class level, and/or learning objective. VFTs for Google Earth similar to the one we describe here can be constructed by: (1) directly authoring the KML code in a text editor, or (2) using graphical user interfaces in Google Earth.

**nlang@mercyhurst.edu

Lang, N.P., Lang, K.T., and Camodeca, B.M., 2012, A geology-focused virtual field trip to Tenerife, Spain, *in* Whitmeyer, S.J., Bailey, J.E., De Paor, D.G., and Ornduff, T., eds., Google Earth and Virtual Visualizations in Geoscience Education and Research: Geological Society of America Special Paper 492, p. 323–334, doi:10.1130/2012.2492(23). For permission to copy, contact editing@geosociety.org. © 2012 The Geological Society of America. All rights reserved.

INTRODUCTION

Educators and researchers have long known that students reap significant benefits from learning-by-doing, exploring independently, and interacting with the natural world. Field trips and laboratory experiments are two instructional approaches that have historically afforded students these types of learning opportunities in the STEM (science, technology, engineering, and mathematics) disciplines (Herrington and Kervin, 2007; Salend, 2009; Cauley et al., 2009; Lamb and Johnson, 2010; Marshall, 2009). However, decreasing budgets, heightened concerns about safety, increasing demands for content-specific learning, a rigorous testing environment, and mounting logistical challenges are among the many reasons educators have cited for the reduction or elimination of field trips from secondary schools, colleges, and universities across the country (Krepel and DuVall, 1981; Disinger, 1984; Stainfield et al., 2000). One way educators are attempting to provide interactive and experiential learning, while avoiding the difficulties associated with "boots on the ground" field trips, is through virtual field trips (VFTs) (Southworth and Klemm, 1985; Chance and LoBaugh, 1994; Bellan and Scheurman, 1998; Grunwald and Norton, 1999; Tuthill and Klemm, 2002; Mannel et al., 2007; Lamb and Johnson, 2010), which are defined as journeys taken without actually visiting the site (Woerner, 1999). Types of VFTs vary and have included travel brochures (Bellan and Scheurman, 1998), multimedia files (Tuthill and Klemm, 2002), and geographic information systems (GIS) (Mannel et al., 2007; Lamb and Johnson, 2010), to name a few.

VFTs providing the best educational experiences for students are those rich in graphic and interactive experiences (Gordin and Pea, 1995; Barraclough and Guymer, 1998; Renshaw and Taylor, 2000). This is especially necessary in the physical sciences such as geology where students typically rely on the ability to touch and pick up materials in order to help learn about them. Trips rich in graphic and interactive experiences positively impact the higher-order cognitive skills of students (Renshaw and Taylor, 2000) while also promoting the acquisition of factual knowledge (Kulik and Kulik, 1989; Ramasundaram et al., 2005). Due to its graphically rich and realistic nature, low cost, and ease of use, Google Earth (http://www.google.com/earth/index.html) provides perhaps one of the most efficient platforms for delivering VFT-type experiences. Here, we outline a Google Earth–based VFT to Tenerife, Spain, that highlights volcanic landforms and deposits around the island. The trip is intended to be used in a variety of geology courses at both the introductory and upper levels and can be used to teach lessons on volcanism, geologic mapping, and relative timing relations, to name a few. We describe two specific activities where we have used the VFT and discuss the gains in student learning through these activities. The described VFT is flexible in that it can be tailored to meet specific instructional needs and it provides a means for students to visit and interact with a locality they may otherwise not have the opportunity to visit and explore.

GEOLOGIC SETTING

The Canarian Archipelago is a chain of seven volcanic islands that are a Spanish autonomous community located in the Atlantic Ocean ~300 km west of North Africa (Fig. 1). They represent intraplate oceanic islands that are at various stages of their geologic evolution (Schmincke, 1979; Carracedo, 1999; Carracedo et al., 1998, 2001, 2002). The chain may have formed either over a mantle plume (Morgan, 1983) or from diapiric upwelling of the mantle along a 180–190 m.y. suture between oceanic and continental lithosphere (Schmincke, 1982). Regardless of their origin, the islands reside on very old, slow-moving crust near a passive continental margin (Schmincke and Sumita, 2010). As a result, island subsidence processes are not significant here so that the islands typically reside above sea level for time periods longer than other intraplate oceanic islands (e.g., the Hawaiian Islands) (Schmincke, 1979, 1982; Carracedo, 1999; Carracedo et al., 1998, 2002). Their presence on slow-moving crust and slow degradation allows for long magmatic histories, which, in turn, provides a favorable condition for the differentiation of large volumes of magma (Klügel et al., 1997, 1999; Nelson et al., 2005). This has then led to the formation of differentiated stratovolcanoes as well as phonolitic lavas and tephras and basanite lavas (Schmincke and Sumita, 2010). The location of the Canary Islands near continental crust, however, may have also played a role in these occurrences (Schmincke and Sumita, 2010, and references therein).

The largest island in the Canarian Archipelago is Tenerife (Fig. 1), which has an area of 2058 km² and is the destination of the VFT described here. Tenerife consists of parts of at least five major overlapping volcanic edifices, which, from oldest to youngest, are the central shield volcano (11.9–8.9 Ma), Teno shield volcano (6.1–5.2 Ma), Anaga shield volcano (4.9–3.9 Ma), Caldera de las Cañadas (~3.5–<1 Ma), and the stratovolcano Pico de Teide, which contains a smaller stratovolcano named Pico Viejo on its western flank (Guillou et al., 2004; Carracedo et al., 2007; Schmincke and Sumita, 2010); Pico de Teide and Pico Viejo are referred to as the Teide–Pico Viejo complex (Schmincke and Sumita, 2010). Based on the ages of the deposits from each of these volcanoes, the construction of Tenerife is interpreted to have occurred through an early shield building stage, which was responsible for forming the bulk of subaerial Tenerife (Ancochea et al., 1990, 1999), followed by more explosive volcanism as reflected by Caldera de las Cañadas and the Teide–Pico Veijo complex (Schmincke and Sumita, 2010). Between the shield building stage and the more explosive periods of volcanism was a 2–3 m.y. hiatus in volcanic activity that was characterized by mass wasting of the three shield volcanoes through both subaerial and submarine gravitational landslides (Carracedo et al., 2007; Schmincke and Sumita, 2010). These landslides likely removed >20,000 km³ of material from Tenerife during that time (Carracedo et al., 2007). Succeeding this hiatus was the formation of the Las Cañadas Volcano, which erupted mafic (alkalic) to phonolitic lavas and pyroclastics with minor basanite and phonotephrite lavas (Bryan et al., 1998). This volcano underwent several periods of collapse with the final, and largest, event occurring at ca. 0.18 Ma (Cantagrel et al., 1999). This collapse event formed Caldera de las Cañadas (or Cañadas Caldera) and left a large horseshoe-shaped scar that opens to the north and is prominent in aerial views of the island (Schmincke and Sumita, 2010) (see Fig. 1). Following the caldera-forming event was the growth of the Teide–Pico Viejo stratovolcano complex ~150,000 years ago (Ancochea et al., 1999), which has erupted dominantly phonolitic lavas and domes as well as basanite and intermediate flows (Carracedo et al., 2007). The Teide–Pico Viejo complex occurs at the junction of northwest- and northeast-trending rift zones, which have been active since the construction of the Las Cañadas Volcano. Eruptions along the rift zones are noted by numerous scoria cones that date to the Pleistocene (Schmincke and Sumita, 2010). Although smaller eruptions have occurred along the rift zones in the fifteenth, eighteenth, and twentieth centuries (Schmincke and Sumita, 2010), the most notable recent volcanic activity on Tenerife is an explosive and effusive phonolitic eruption from a source named Montana Blanca at the southeastern base of Teide

~2000 years ago (Ablay et al., 1995) and a phonolitic eruption from Pico Viejo in 1798 (Carracedo et al., 2007). Deposits from both of these events will be observed on this VFT.

The geologic description of Tenerife and the Canarian Archipelago presented here is only intended as a broad introduction to the complex geology represented by this island chain. We only mean to provide users of this VFT a proper reference frame for what they will be observing on the trip and conducting in post-trip exercises. For a complete and more detailed description of the geologic evolution of Tenerife and the rest of the Canarian Archipelago, please see Schmincke (1979), Ancochea et al., (1990, 1999) Carracedo (1999), Carracedo et al. (1998, 2001, 2002, 2007), Schmincke and Sumita (2010), and references therein.

THE VIRTUAL FIELD TRIP (VFT)

With the goal of using volcanic processes and landforms to teach various geology oriented lessons, we have constructed a VFT to the Cañadas Caldera area of Tenerife, Spain (Fig. 2). This

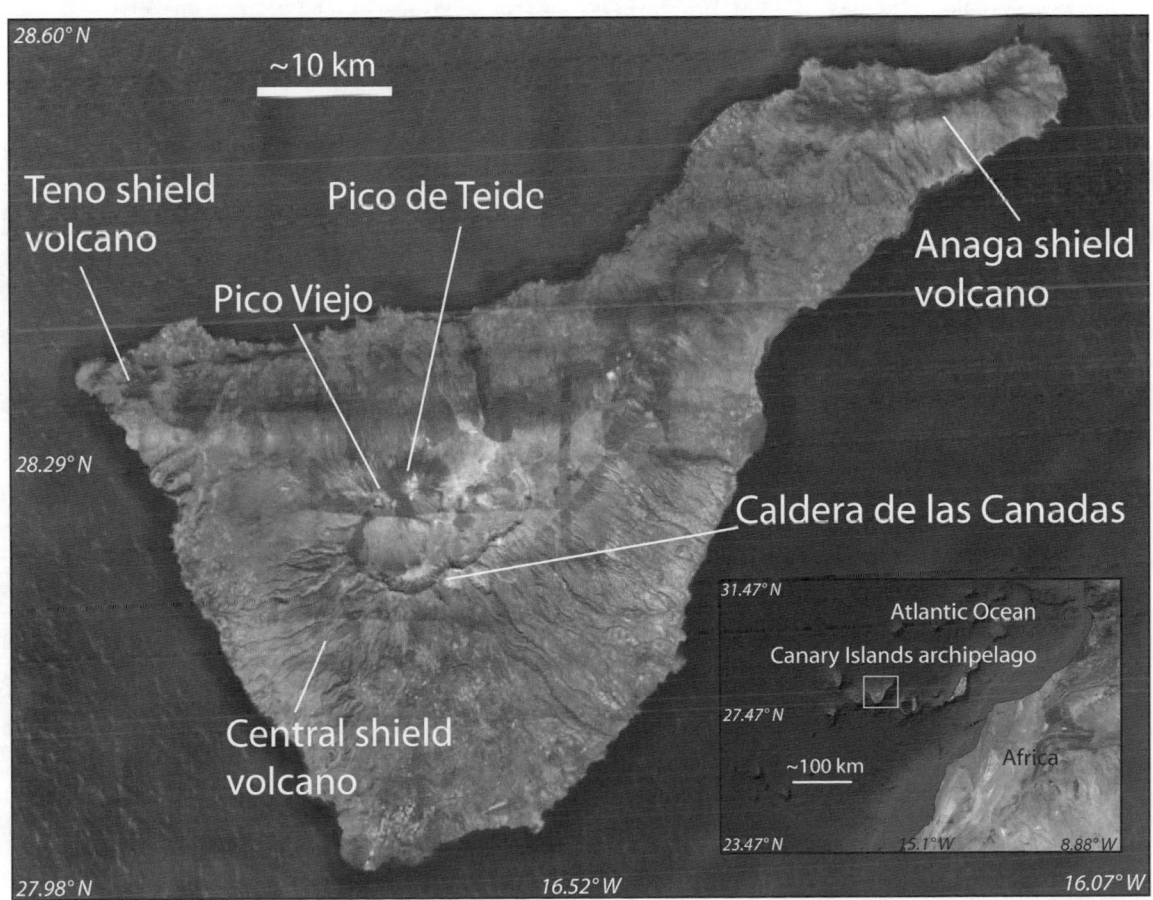

Figure 1. Satellite view of the island of Tenerife highlighting some of the key volcanic landforms comprising the island; the island is made up of four overlapping volcanic edifices with the youngest, and currently active, being Pico de Teide and Pico Viejo. Inset image in the lower right shows the location of Tenerife in context within the Canarian Archipelago and with Africa; white box shows the location of Tenerife within the archipelago. Image is from Google Earth.

area has an arid climate and has a paucity of vegetation, meaning that various volcanic units and structures are well exposed and easily observed in photographs and satellite imagery. Our field trip is contained within a Keyhole Markup Language (KML) file that can be downloaded from http://math.mercyhurst.edu/~nlang/vft/. Opening the KML file will produce an 18-stop field trip within Google Earth. Each stop is denoted by a volcano icon. Clicking once on an icon will produce a pop-up balloon that contains one or more photographs of deposits or features that are observable at that, or from that, locale. Clicking on one of the photographs will provide a larger version of the photo that will either fill the Google Earth screen or appear in a new browser window. A caption is associated with the photographs within the pop-up balloon that briefly summarizes what is observed in the photos; some balloons also contain YouTube videos to further illustrate what is observed at each stop. Double-clicking on an icon will fly the user into the stop so that their view is of the same perspective as that in the associated photograph(s). Several stops occur along highways TF-21 and TF-24, which pass through the middle of the caldera and its northern edge, respectively. Google has collected a continual stream of street-level view photographs along these roads and entering into street-level view in Google Earth, either by: (1) zooming into the road, or (2) clicking on the pegman located by the controls and dragging it to the road, will

reveal a 360°, high-resolution panoramic view along the highway. These photographs provide another viewing option of some of the features emphasized at each of the VFT stops and provide a perspective as if one were driving along the highway to each stop.

Figure 3 shows a simplified view of the geology in and around the Cañadas Caldera area in relation to the 18 stops on this trip. Specific geologic observations that can be made at each stop are outlined below. Also included below are observations that can be made while traveling between the stops along highways TF-21 and TF-24.

Stop 1 (28° 12′ 28″ N, 16° 40′ 46″ W)

Stop 1 is located along the western of edge of Cañadas Caldera and highlights pyroclastic deposits that comprise part of the outside base of the Cañadas Caldera wall. They are likely associated with the building of the Pleistocene-aged Las Cañadas volcano (Schmincke and Sumita, 2010) and appear to include both ground hugging and airfall deposits. The deposits are layered with the base of the deposits exhibiting large-scale cross-stratification. Both pumice and lithic fragments are contained within the deposits. These are some of the oldest, if not the oldest, deposits that will be addressed on this VFT.

The highway from Stop 1 to Stops 2 and 3 passes several spires of pyroclastic deposits similar to those observed at Stop 1.

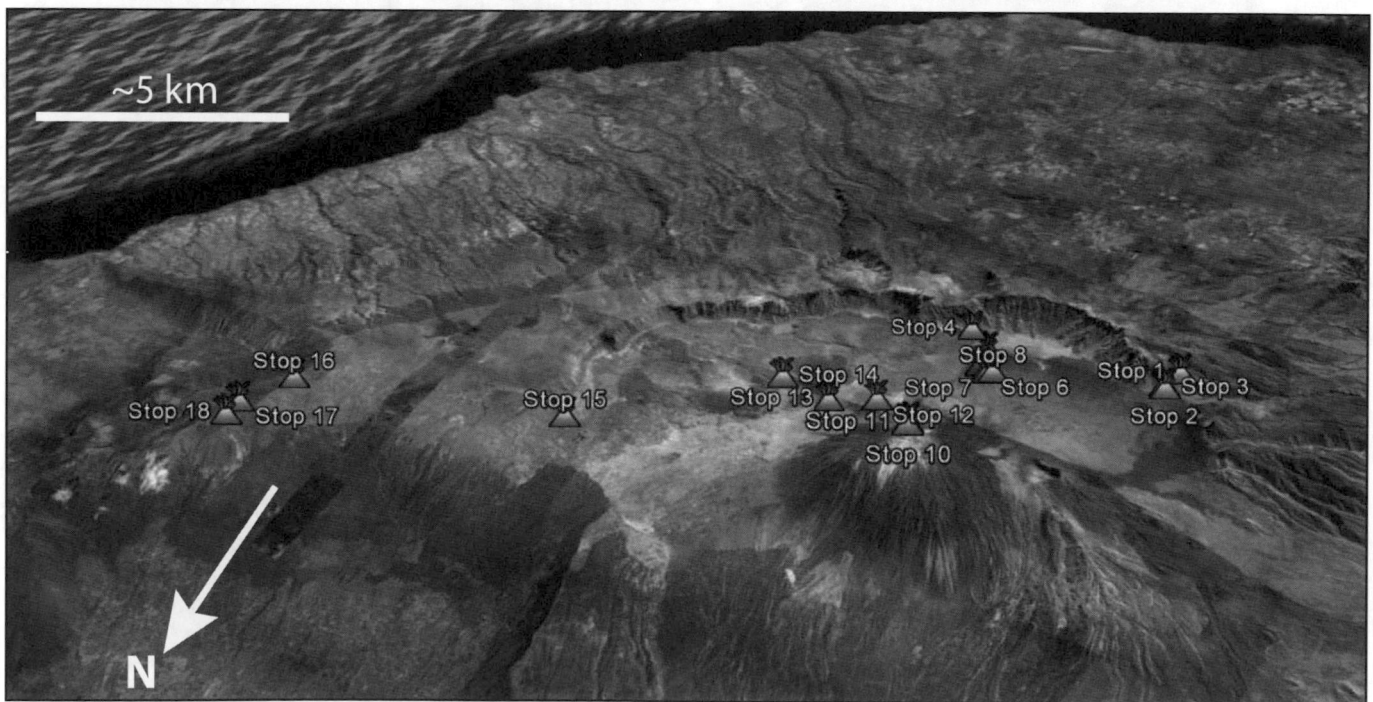

Figure 2. Oblique satellite view to the southeast of Tenerife showing how the virtual field trip is laid out in Google Earth. Each stop is denoted by a volcano icon and associated stop number. Clicking on an icon will produce a balloon pop-up with one or more photographs of landforms and/or processes that can be observed at or from that locality. Captions accompany each pop-up briefly explain what is observed in the photographs; links to YouTube videos are provided for some stops to help visualize processes discussed at that locality. Double clicking on a volcano icon will fly the user in to the stop so that they are viewing in a direction that the photographs were taken. The image is from Google Earth and is centered at 28.25° N, 16.62° W. The KML file for this tour can be accessed online at http://math.mercyhurst.edu/~nlang/vft/.

The highway then passes through a gap in the caldera wall and drops down onto the caldera floor, which has been covered by Pleistocene and Holocene basic, intermediate, and felsic (phonolitic) lavas and pyroclastic deposits. The relative ages of the lavas observed from the highway can be broadly distinguished based upon their colors—the older the flow, the lighter brown it generally appears. The youngest flow at this locality is the 1798 phonolite flow from Pico Viejo, which appears as the darkest of all the flows here. The Teide–Pico Viejo stratovolcano complex stands up from the caldera floor to the north. At the junction of TF-38 and TF-21 are Stops 2 and 3.

Stop 2 (28° 12′ 50″ N, 16° 40′ 40″ W)

Stop 2 highlights a columnar joint that is part of a set within an erosional remnant outcropping along TF-21; going into street-level view at this location will provide a better sense of how this joint fits into the overall outcrop. The columnar joint set appears to represent a mafic(?) east-west–trending dike that intruded into

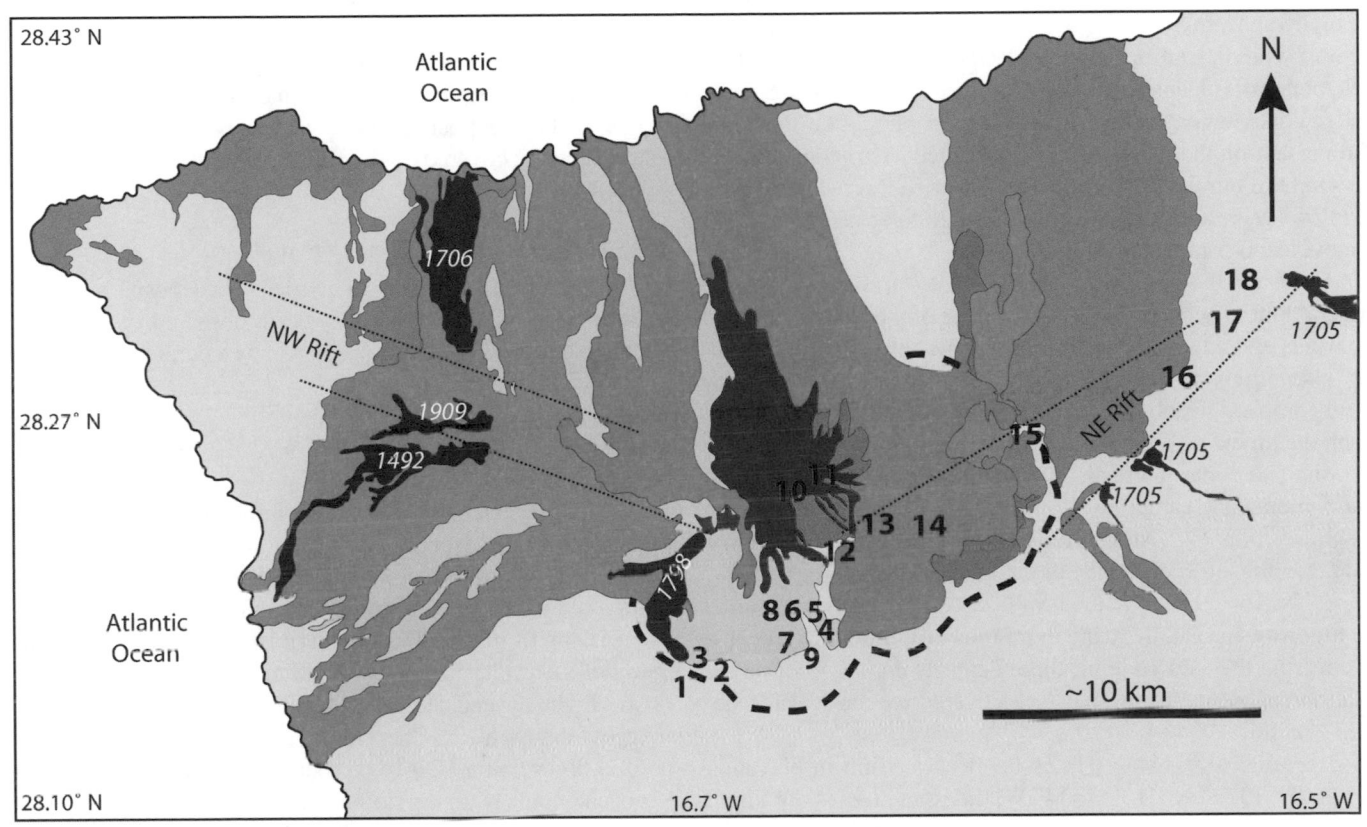

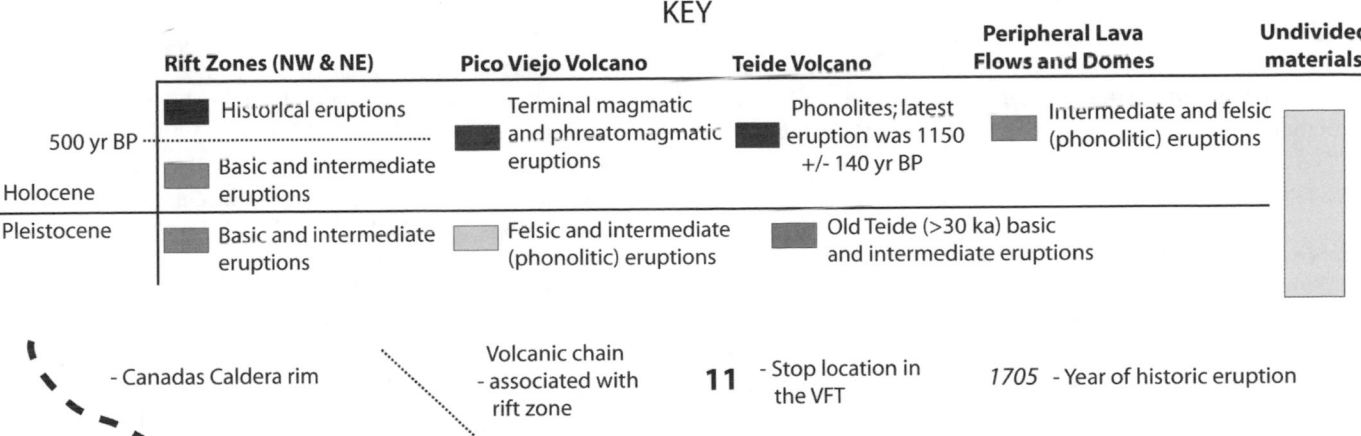

Figure 3. Geologic map of a portion of Tenerife showing the 18 stops associated with this virtual field trip. Map is modified from Carracedo et al. (2007).

pyroclastic deposits similar to those observed at Stop 1. Similar to the pyroclastic deposits, the dike is likely associated with activity at the Las Cañadas volcano. Columnar joints that are part of another intrusive suite associated with pyroclastic deposits can be observed at Stop 9.

Stop 3 (28° 12′ 51″ N, 16° 40′ 42″ W)

Turning 180° from Stop 2 reveals Stop 3, which is a panoramic view to the north-northeast toward the Teide–Pico Viejo complex. In the middle ground are the Pleistocene-aged phonolitic 'a'a flows erupted from Pico Viejo that can be observed from the highway from Stop 1. The dark-colored deposits to the left of the image are flows from the 1798 eruption of Pico Viejo and are the youngest volcanic deposits addressed on this VFT. Several cinder cones embayed by Holocene-aged phonolitic flows are also present on the flanks of the Teide–Pico Viejo complex near the southern termination of the northeast rift; these cinder cones and flows represent parasitic eruptions that have occurred at this stratovolcano complex.

Highway TF-21 between Stops 3 and 4 traverses along the caldera floor in between the southern edge of the caldera wall and the fronts and edges of several Pleistocene felsic and intermediate (phonolitic) 'a'a flows. To the left (north) is the Teide–Pico Viejo complex, which has sourced many of the flows observable along the highway. To the right (south) are numerous pyroclastic deposits and some interbedded lavas (and possible intrusions?) that comprise the caldera wall; many of the pyroclastics outcrop as spires similar to those seen along the highway between Stops 1 and 2. On the left (north) side of the highway beginning near 28° 12′ 42″ N, 16° 39′ 11″ W the lava flows do not come right up to the highway. Instead there are exposures of loose pumice deposits covering the caldera floor; these deposits do not occur on the lavas indicating that they must be covered by the Pleistocene-aged lava flows.

Eventually highway TF-21 begins to climb in elevation and at 28° 13′ 2″ N, 16° 37′ 34″ W, the road crosses onto light colored deposits that appear to represent pyroclastic deposits similar to those observed at Stop 1. The deposits observed along the highway here are part of a larger L-shaped outcropping of light-colored pyroclastic deposits; isolated outcrops of what appears to be these same pyroclastic materials occur immediately east of this "L" along the base of the caldera wall. Stops 4–9 will examine various exposures in and around this L-shaped outcrop.

Stop 4 (28° 13′ 7″ N, 16° 37′ 39″ W)

Stop 4 highlights a road cut exposure of a dike that has intruded along a fault zone and has subsequently been altered to an orange color. The fault itself is a south-dipping normal fault that juxtaposes pyroclastic deposits (hanging wall) against a volcanic megabreccia (footwall). Going into street-level view at this location will provide a longer segment of higher resolution photographs of the road cut and gives a better sense of the characteristics of the breccias, dike, and pyroclastics.

Stops 5, 6, and 7 (Centered at 28° 13′ 48″ N, 16° 38′ 17″ W)

Stops 5, 6, and 7 all highlight parts of a Pleistocene-aged phonolitic pahoehoe flow sourced from the Teide–Pico Viejo complex that has embayed the L-shaped outcropping of light-colored pyroclastics. Ropy and shelly pahoehoe textures can be observed at these three stops as can phenocrysts of zoned feldspar. This flow is barely observable within the aerial/satellite views in Google Earth and can only be observed if, literally, on top of the flow. A large reason for this is because the flow has been locally covered by younger Pleistocene 'a'a lavas such as the one immediately east of Stop 5 and the channelized flow immediately west of Stop 7. Also observable in one of the photographs for Stop 7 is a dark-colored band that is an intrusion (possibly mafic?) that has cut through pyroclastic deposits; these deposits are the same ones observed at Stop 8 and are locally covered by the pahoehoe flow observed at Stops 5, 6, and 7.

Stop 8 (28° 13′ 54″ N, 16° 38′ 22″ W)

An erosional remnant of pyroclastic deposits similar to (and possibly the same age as?) those observed at stops 1, 2, and 4 is seen at Stop 8. This appears to be a pumice-rich deposit with few lithics.

Stop 9 (28° 13′ 12″ N, 16° 37′ 57″ W)

Stop 9 highlights columnar joints that are part of a suite of intrusive dikes and sills that have cut through the weakly welded pyroclastic deposits observed at Stop 8. The orientation of the dikes observed at this stop broadly parallel the trend of the long portion of the "L" of these pyroclastic outcroppings.

Driving from Stop 4 toward Stop 10 along TF-21 will afford views of more Pleistocene phonolite 'a'a flows to the left (west) of the highway and Holocene phonolitic coulees to the right (east) of the highway. The Pleistocene flows are mostly channelized flows that appear to be sourced from Teide whereas the Holocene coulees appear to be sourced from unexposed vents located east of the highway (at 28° 14′ 57″ N, 16° 37′ 21″ W and 28° 13′ 52″ N, 16° 37′ 2″ W, respectively). Although it is difficult to see from the highway due to the thickness of the flows, a cinder cone has been mostly buried by a coulee at 28° 14′ 11″ N, 16° 37′ 3″ W; an overhead view of the coulee will reveal the buried cinder cone. At 28° 14′ 45″ N, 16° 37′ 44″ W the road passes by the southeast front of a channelized Holocene 'a'a flow sourced from Teide. More views of similar deposits will be observed at Stops 10 and 12.

Stop 10 (28° 16′ 12″ N, 16° 38′ 18″ W)

This stop is a view to the east-southeast of a channelized Holocene phonolite 'a'a flow. The flow is sourced from the summit cone on Teide and flowed over ~30,000-year-old Teide-sourced lavas (Carracedo et al., 2007); the front of this flow can be viewed at Stop 12 at the base of Teide. Visible in the distance at the bottom of Teide are ~2000-year-old flows sourced from Montana Blanca and that can be seen at Stops 13 and 14.

Stop 11 (28° 16′ 14″ N, 16° 38′ 19″ W)

Stop 11 highlights a piece of phonolite lava that comprises part of the 'a'a flow observed at Stop 10. Visible within the piece of lava are vesicles and phenocrysts of feldspar.

Stop 12 (28° 15′ 17″ N, 16° 37′ 19″ W)

Stop 12 visits the front of the phonolitic 'a'a flow observed at Stop 10. The rubbly front of the flow as well as the central channel and levees are readily observed at this location; entering into the street view at this stop will provide another view of the 'a'a flow. Other Holocene 'a'a flows are also visible on the flanks of Teide as dark gray elongate patches; these flows locally cover ~30,000-year-old Teide flows (Carracedo et al., 2007), which appear as light brown, sparsely vegetated patches on the flanks of Teide.

Highway TF-21 between Stops 12 and 13 climbs up through the toe of another Teide-sourced Holocene 'a'a flow. This flow has traveled directly on top of another Holocene flow (light brown color) that appears to have extended southeast of the highway. The highway then traverses across another Holocene flow before following along the base of Montaña Blanca, which has been the source of several phonolitic eruptions during the Holocene. Stop 13 is a view of one of these flows.

Stop 13 (28° 15′ 46″ N, 16° 36′ 46″ W)

Stop 13 visits ~2000-year-old phonolitic lavas that erupted from a northwest-trending fissure at the top of Montana Blanca. These deposits are part of a phonolitic flow and dome complex that make up a large part of the eastern portion of the Cañadas Caldera and are observable in the distance at Stop 10.

Highway TF-21 between Stops 13 and 14 provides additional views of the Holocene flows sourced from Montaña Blanca. Most of the flows the highway traverses along are likely associated with the ~2000-year-old eruption of Montaña Blanca. Stop 14, though, is located on top of a flow that pre-dates the ~2000-year-old eruption.

Stop 14 (28° 15′ 34″ N, 16° 35′ 52″ W)

Stop 14 is a view to the southeast toward the Cañadas Caldera wall. Within the caldera wall can be seen pyroclastic and lava deposits; the pyroclastic deposits appear as light colored bands across the image and the lavas are the darker bands. The pyroclastic deposits are seemingly truncated by a dome that is observed in the central part of the photograph and sticks up above the caldera wall. In the middle ground are similar phonolitic deposits to those observed at Stop 13.

Between Stops 14 and 15, highway TF-21 broadly parallels the northeast rift and traverses across more of the Holocene phonolite flows sourced from Montaña Blanca. With the exception of a small outcropping of ~30,000-year-old Teide-sourced flows in the southeast corner of the caldera, much of the eastern portion of the caldera is covered by Montaña Blanca–sourced flows. Viewable on both sides of the highway are cinder cones that have been embayed by the Montaña Blanca–sourced flows; these cones are likely affiliated with the northeast rift zone.

Stop 15 (28° 17′ 40″ N, 16° 33′ 35″ W)

Stop 15 highlights a cinder cone that has formed on top of the northeast rift zone. This is just one of numerous cinder cones that occur across the caldera floor and are concentrated within the northeast rift zone. This cinder cone is partially embayed by Teide-sourced Holocene basic and intermediate lavas.

The highway between Stop 15 and Stops 16, 17, and 18 traverses out of the eastern end of the caldera and onto a high volcanic plain littered with several cinder cones and tephra deposits. The cinder cones are likely Pleistocene in age and are affiliated with the northeast rift zone (Carracedo et al., 2007; Schmincke and Sumita, 2010). The cinder cones are also the likely source of the tephra deposits along the highway here (which has now become TF-24), though other vents in this area likely contributed to these deposits as well (see Carracedo et al., 2007; Schmincke and Sumita, 2010)

Stops 16, 17, and 18 (28° 19′ 2″ N, 16° 29′ 1″ W)

These stops are located to the northeast of the caldera and emphasize tephra deposits exposed in road cuts. Stop 16 shows a mafic lava flow that unconformably overlies pyroclastic tephra. Stop 17 highlights mafic pyroclastic deposits that consist of pieces of scoria and other lithic clasts; the road cut is undergoing differential erosion, which emphasizes the various grain sizes that make up the various layers comprising the deposit. Stop 18 visits a pyroclastic deposit that consists of alternating mafic and phonolitic tephra giving the outcrop a striking black and white appearance.

INCORPORATING THE VFT INTO GEOLOGY COURSEWORK

The overarching motivation for constructing this VFT to Tenerife, Spain, was to provide students a means to see and experience geology that they may not normally have an opportunity to observe and study. This is especially useful for students in places where opportunities to examine volcanic products and processes are limited. Subsequently, we have constructed this trip to be used in introductory (e.g., physical geology and natural disasters) as well as in upper-level (e.g., petrology) courses. In this section we discuss ideas of how to potentially incorporate this VFT into geology courses and describe sample assessments that may accompany those assignments. We then discuss the impact this VFT had on student learning through implementation of two different assignments in two different geology classes.

Ideas for VFT Activities and Possible Assessments

Perhaps one of the most simple and straightforward uses of this VFT is to have students learn to recognize various volcanic landforms and associated deposits. Specifically, after taking the trip, students will be able to recognize landforms such as stratovolcanoes and cinder cones and lava textures such as pahoehoe and 'a'a. An assignment could therefore be a class, lab, or homework task for physical geology, natural disasters, petrology, or volcanology where students, either on their own or in groups, take the tour or make observations and notes about each locality. Students could then be presented questions that make them tie together their observations and begin to make testable predictions as to what landforms they observed and how they formed. Depending upon the class level, these tie-together questions could be written at various levels of difficulty. Another step could be to have the class discuss their observations and predictions; this step could be used as an informal assessment as to how well the students comprehend the presented course content. A more formal assessment could be having the students take a quiz where they are presented with a picture of a landform or texture observed on the VFT, and they have to state what it is.

A more advanced use of this VFT could include quantifying physical properties of various examined volcanic deposits. In undergraduate or graduate courses in petrology or volcanology, students could use values for eruption temperature, effusion rates, and density as found in the literature to begin calculating emplacement times and viscosities for examined flows. Such exercises would require integrating Google Earth's ruler tool and topographic profile function to determine flow slopes, areas, and volumes. Similar to the previous assignment, this activity could be done as part of an in class, homework, or lab activity where students are provided detailed calculation instructions. Assessing student learning in such an activity would be straightforward if the assignment was written in a way that makes the students show their work (e.g., Johnson et al., 2011). An example of such an activity is also available at http://math .mercyhurst.edu/~nlang/vft/.

The previous two examples were volcanology-centered assignments. However, the tour could also be used to teach principles in other realms of geology such as geologic mapping and timing relations. To elaborate, the photographs observed at each stop and the street-level view photographs from along the highway essentially provide outcrop-scale views of what the students would traditionally see in the field, and the satellite images allow for the placement of the outcrop-scale view into a broader context. The tour could then give the students an introduction to the general geology of this region and the types of rocks they would be mapping. The resolution of the imagery in Google Earth and lack of vegetation in the tour area allows students to observe timing relations between units through cross-cutting and superposition relationships. These relations would then provide constraints for students to draw contacts on their maps and put flows into a stratigraphic sequence; based on the outcrop-scale views (i.e., the photographs), the students could also begin writing unit descriptions. Mapping could focus on just one particular part of the field trip area or it could extend across the entire region. Assessing student learning in this exercise could be accomplished by comparing their maps with published mapping of Tenerife (e.g., Carracedo et al., 2007). In fact, comparison of the geologic maps could be used as a vehicle for class discussion about geologic maps, including how data is presented on a map, and why several

geologic maps of the same area may all show variations. In addition, the mapping may be used in a volcanology-centered context where a morphologic map of a specific flow is constructed to illustrate how lava flows are emplaced. Such lessons may be useful for courses in structural geology, field methods, geomorphology, and stratigraphy, as well as petrology and volcanology.

Measures of Student Learning

We have outlined options for incorporating this VFT into geology courses, but how much does this VFT impact student learning? Are there benefits to incorporating this VFT into geology lesson plans over other means of instruction such as lectures? To address these questions we performed two studies where this VFT was incorporated into two lessons in two geology courses. Specifically, we used the VFT in an introductory physical geology course to teach a lesson about volcanic process and landforms and then in an upper-level stratigraphy course to teach about geologic mapping and correlating lithostratigraphic units. We present and discuss our results below.

Physical Geology

For our first study testing the effectiveness of this VFT, we implemented it into a three-hour laboratory section of an introductory geology class (Physical Geology) to teach a lesson on volcanic processes and landforms; the students had not had any previous introduction to igneous processes in this course prior to this study. Objectives for this assignment were to have the students be able to: (1) recognize different types of volcanic constructs (e.g., cinder cone and stratovolcano), (2) relate specific volcanic constructs with specific eruption styles, (3) recognize volcanic structures and textures (e.g., columnar jointing and 'a'a, and pahoehoe textures), and (4) define viscosity. The class size was 25 students, which we randomly divided into two groups; Group 1 consisted of 12 students and Group 2 had 13 students. Both groups completed a 12-question pre-test to establish their knowledge of volcanoes and volcanic processes. Group 1 then listened to a typical introductory-level lecture about volcanic processes and landforms while Group 2 used the VFT and an associated activity to guide their explorations—Group 2 students had the option of either working on their own or in small groups. Group 2 was given a 15-minute introduction to volcanic processes and how to use Google Earth and the VFT prior to their beginning the exercise; all students completed the assignment within two hours with a majority of the students finishing in a little over an hour. Both groups completed a post-test immediately following completion of their assigned instructional activity. We then calculated the mean difference between pre- and post-test scores for each group (Table 1). Both the lecture and the VFT demonstrated an increase in their scores, but, on average, Group 1 achieved a 3.4 point gain (with a standard deviation of 2.95) on the post-test compared to the pre-test, whereas Group 2 achieved a 4.2 point gain (with a standard deviation of 2.96). A two-sample t-test analysis on these point gains, however, demonstrated that

TABLE 1. INTRODUCTORY GEOLOGY PILOT STUDY RESULTS
WITH TWO SAMPLE T-TEST ANALYSIS RESULTS

Lesson type	Number of students	Mean difference	Standard deviation
Lecture (Group one)	12	3.4	2.95
Google Earth (Group two)	13	4.2	2.96
Two sample t-test analysis:		*t-statistic = 0.803*	*p-value = 0.5*

there was no statistically significant difference between the mean scores of the two groups where the t-statistic was 0.803 and the p-value was 0.5. This suggests that this VFT does not significantly enhance student learning over traditional lectures when teaching about volcanism to in introductory level course.

Stratigraphy

Our second study involved using this VFT to teach about geologic mapping and lithostratigraphic correlation to an upper-level stratigraphy class of 12 students. The students were to create a geologic map of the Cañadas Caldera area in order to achieve five learning objectives: (1) identify eruptive centers, (2) identify the contacts between the various exposed material units, (3) identify the types of material units represented in the caldera (e.g., 'a'a lava flow, pyroclastic deposit), (4) identify tectonic structures, and (5) write a geologic history. To measure the effectiveness of this VFT in helping to achieve these five objectives, we broke the study into two parts, each conducted over two 90-minute class periods. For part one (the first class period) the students were presented a 15-minute lecture on geologic mapping and were provided a static two-dimensional satellite image of the caldera; the students were to create their map using only the static image during the remaining class time. The students' maps and geologic histories were then scored on a scale of 0–4 for each of the five objectives; the closer a student came to meeting a specific objective, the higher they scored on the rubric. For part two of the study (the second class period), the same group of students used the VFT to create a geologic map of the same area as they did for part one; the students were then scored on the same 0–4 scale for each of the five objectives. Students were only given 90 minutes to complete this second part, but all students indicated they could have used at least another hour to finish their map—zooming in and out of Google Earth, examining all of the photographs, and watching the YouTube videos severely slowed the students compared to the first part of the assignment. Table 2 summarizes the results of the two parts of this study, but the mean rubric scores for the first mapping exercise was 9.9, and the mean rubric score for the VFT mapping exercise was 12.4—an increase in learning of 2.5 points. A paired t-test analysis on these mean scores

revealed that the 2.5 point increase on rubric scores was statistically significant between the two mapping approaches where the t-statistic was 1.98 and the p-value was 0.05. This suggests that using this VFT does significantly enhance students' abilities to construct a geologic map over using just a static image.

Discussion

Both studies indicated that student learning occurred when using this VFT. In the case of the Physical Geology course, both approaches (the lecture and the VFT) resulted in increased student learning. However, there was no statistical significance between the two modes of instruction. In the case of the Stratigraphy course, we did observe a statistically significant difference in the amount of student learning that occurred between activities. Because our sample size for both studies was small, it is difficult to draw wide overarching conclusions from these differences in scores. However, we postulate that this difference may be tied to the specific learning objectives of the assignments. To elaborate, the learning objectives for the Physical Geology assignments (study 1) were aimed at lower levels of the cognitive domain in Bloom's Taxonomy (Bloom, 1956)—specifically, the knowledge and comprehension levels. Students only needed to be able to recognize certain types of volcanoes and textures as well as define terms such as viscosity. In both the lecture format and the VFT, students were shown pictures of different volcano types and textures and were provided descriptions of those features; definitions for various terms were also directly provided for them in both learning environments. The only difference between the two environments was that Group 1 was presented the information from an instructor and Group 2 gained their information through a more active means; in both cases the students performed equally on the assessment instrument. Perhaps, then, students' abilities to acquire knowledge about volcanic landforms, structures, and textures at the factual level does not depend on the actual environment in which the knowledge is presented (i.e., lecture versus VFT).

This idea is seemingly consistent with our results from the second study where there was a significant gain in student rubric scores when students used Google Earth to construct their geologic maps. The biggest gains in this study occurred with objectives 1 and 5, which involved identifying eruptive centers (i.e., cinder cones and stratovolcanoes) and writing a geologic history based on their created map, respectively. In each case, there was an average increase in student scores of approximately one point from the first to second map. This increase in student scores between the two tasks is noteworthy and, based on informal conversations with students, we attribute it to image clarity

TABLE 2: UPPER LEVEL GEOLOGY EXERCISE RESULTS
WITH PAIRED SAMPLE T-TEST ANALYSIS RESULTS

Exercise	Number of students	Mean rubric score
Static image	12	9.88
Google Earth	12	12.42
Paired t-test analysis:	*t-statistic = 1.98*	*p-value = 0.05*

and the interactive nature of Google Earth. To elaborate, geologic mapping is a highly interactive undertaking that involves the continual formation, testing, and reworking of hypotheses (Wilhelms, 1990; Hansen, 2000) and calls on the upper cognitive domain levels of Bloom's Taxonomy. The more "field data" that is available to the mapper (e.g., a third dimension) the more accurate geologic map they can construct. With regard to the first part of this activity, the static image used for mapping severely hampered students' abilities to develop and test hypotheses; the students could not obtain a third dimension critical for differentiating between various volcano types nor could they zoom in close enough to examine contact relations. When they were able to use Google Earth and the VFT, the students were able to use the third dimension that Google Earth provides to see topography and to see specific contact relations. Students were also able to zoom into units and use the photographs provided at each stop as well as from the street-level view to observe the various units from a ground-based perspective. With these functions and perspectives, the students had the components critical for developing and testing the hypotheses necessary for creating their geologic map and writing their geologic history.

As mentioned earlier, our study size was small and any conclusions discussed above need to be additionally tested using larger sample sizes. That said, the results of our two studies indicate that student learning is positively impacted with this VFT. This is further supported by student surveys and informal interviews conducted after each study. Students from both studies indicated positive experiences with this VFT, and multiple students mentioned a preference to hands-on type learning experiences such as this VFT over traditional in class teaching approaches such as lecturing. In both cases students indicated that a lecture preceding the actual assignment was critical to their success; the students were able to take notes during the lecture and refer back to them during the assignment, which greatly aided their ability to complete the assignments. Students who did not take notes during the pre–field trip lecture struggled with the assignment compared to those who did take notes. Multiple students indicated that without a lecture they likely would have been lost in conducting assignments on this VFT. Some technical difficulties did arise during both assignments, which slowed down and frustrated some students. However, many of these difficulties arose from unfamiliarity with Google Earth, and most of the rest arose from navigating in Google Earth—in both studies, many students employed the VFT using their own laptops, and they drove through the tour using the touchpad on their computer. Using the touchpad added complications to zooming in and out of Google Earth, but once students were advised to drive and navigate using a mouse, most of the technical difficulties were eliminated. Even with technical difficulties, students in both studies were able to complete their assignments in reasonable time frames. For activities such as the one performed in study 1, two hours was enough time to successfully complete the assignment. For activities such as mapping, however, students may need upwards of three hours to complete the assignment.

CREATING VFTS FOR GOOGLE EARTH

There are two basic methods for creating Google Earth VFTs. The first is by manually writing the field trip in KML code and the second is by utilizing the Google Earth interface. In this section, we briefly outline how to create a Google Earth tour using both approaches. We only intend this section to be a quick, informal introduction to creating tours, as this is a topic worthy of its own study (see Treves and Bailey, this volume).

Writing Tours using KML Code

Building a tour directly from KML code requires knowing how to program in KML, but it also allows for the most control over the construction of the tour (e.g., bolding or italicizing specific text, having pictures appear in a stop's pop-up balloon, and eliminating the directional line at the bottom of a pop-up balloon). Programming in KML can be accomplished using a text editor program such as Notepad++ for Microsoft Windows, and we encourage those interested in creating a VFT straight from KML code to read Google's online tutorial on KML at http://code .google.com/apis/kml/documentation/, and/or *The KML Handbook: Geographic Visualization for the Web* (Wernecke, 2008). One does not need to be fluent in KML to do much programming, however. Any preexisting KML program can be opened and modified in a text editor program. For example, opening the VFT described in this paper in Notepad++ will reveal all the details of the program, such as each stop's latitude and longitude as well as the stop's name and associated text. These parameters can all be easily changed and saved within the text editor. The changes will then occur after the file is saved and then opened in Google Earth. Writing VFTs from scratch in KML is a time-consuming process, and the approach of modifying preexisting KML files does save a lot of time and may be a useful first step for those learning to program in KML. Modifying preexisting KML files also provides the flexibility of tailoring a VFT for a specific audience (e.g., Tuthill and Klemm, 2002). For example, text for certain stops can be rewritten for a specific audience level; geochemical data or thin sections of rocks could also be included as additional data tables or photographs or even as audio or movie files. Further, stops could be added to, or eliminated from, the tour based on the purpose of the trip. Users of our VFT should feel free to edit the KML file of the trip to suit the needs of their particular audience or assignment objective.

Creating Tours within Google Earth

Creating a VFT using the Google Earth interface is straightforward compared to writing one directly in KML; no ability to write KML code is required to construct VFTs through the Google Earth interface. To do so simply requires following these steps: (1) pick the field trip locations; (2) click on the "Record a Tour" button along the toolbar at the top of Google Earth (the button that looks like a video camera; pressing this button will bring up a window at the bottom of the Google Earth screen with two buttons—one with a red circle and the other with a microphone); (3) press the button with the red circle to begin recording; (4) navigate to each of the field trip locations; (5) press the circle with the red button again to stop recording (stopping the recording will bring up a new window that provides the opportunity to play back the recording); (6) check the recording to test that it suffices; and, if it suffices, (7) save the tour by pressing the "save" button on the right hand side of the window (this will bring up a new window that allows you to name the tour and add specific details describing the tour you just created). Audio narration can also be added when constructing this trip by clicking on the button that looks like a microphone and that is located next the button with red circle.

Each field trip site selected during step 1 can be marked with a placemarker to add additional information about that stop. To do this, click on the "Add Placemark" button on the toolbar at the top of the Google Earth screen (the button that appears as a yellow push pin) and drag the icon that appears on the screen to the location of interest. Clicking on the "Add Placemark" button will also bring up a window that allows you to name and add details regarding that specific stop. Photographs and videos can be linked to the placemark if they have an associated URL. If adding photographs, be sure that they are sized to appear on a regular computer screen; otherwise, they may appear oversized. Clicking on the yellow push pin button on this new window will allow you to choose a placemarker icon of your choice. Clicking "OK" at the bottom of the window will then save the marker to a temporary folder that appears in the left hand panel of the Google Earth window. After placing markers at each field stop, all of the stops can be saved into one file by right mouse clicking on the "Temporary Places" label and selecting "Save place as...." You can save the file as one of two file types: a KMZ file or a KML file. Saving as a KMZ file will produce a file that is smaller in size, but cannot be manually edited in a text editor program at a later time. Saving as a KML file allows the ability to be able to go back and manually edit the file at a later time using a text editor program. Specific placemarkers can also be edited on the fly by right mouse clicking on them either on the Google Earth map or in the left hand panel and selecting "Properties."

Following the instructions provided here will produce a "simplified" Google Earth tour. However, for more information on producing a tour using the Google Earth interface, please see: http://earth.google.com/outreach/tutorials.html.

SUMMARY

We have described here a VFT to Tenerife, Spain, that can be taken using Google Earth. The field trip is written as a KML file and is broken into 18 stops that emphasize volcanism associated with Cañadas Caldera and the Teide–Pico Viejo volcanic complex. The field trip is structured such that students can, on their own or in groups, virtually travel from stop to stop exploring specific volcanic landforms and processes. Testing in an introductory-level and upper-level geology course showed that student learning was positively impacted by this VFT, but our sample sizes were small and further testing needs to be done to more accurately interpret our results. The trip is flexible, though, in that it can be applied to the learning of several different geoscience-related topics at a variety of knowledge levels and provides students the opportunity to explore a region that they may otherwise not have the opportunity to visit. Similar VFTs can be constructed either from scratch by directly authoring the KML in a text editor or through the Google Earth interface, which requires no programming abilities.

ACKNOWLEDGMENTS

The idea for this paper stemmed from attendance by NPL at the "Cities on Volcanoes" conference in Tenerife, Spain, in 2010. Funding to the conference was provided by Mercyhurst University as a Faculty Travel Grant. Many thanks to two anonymous reviewers whose thoughtful and thorough comments enhanced and improved the quality of this manuscript. We also thank K. Evans for discussions on how to test the learning effectiveness of this VFT. We would also like to thank S. Morealli for loaning us her Physical Geology class to conduct our testing; we also want to thank the students in the winter 2012 Physical Geology and Stratigraphy courses at Mercyhurst University for being willing and agreeable participants in our studies. We finally want to thank J. Bailey, D. De Paor, T. Ornduff, and S. Whitmeyer for their work as editors on this volume.

REFERENCES CITED

Ablay, G.J., Ernst, G.G.J., Marti, J., and Sparks, R.S.J., 1995, The ~2 ka subplinian eruption of Montaña Blanca, Tenerife: Bulletin of Volcanology, v. 57, p. 337–355, doi:10.1007/BF00301292.

Ancochea, E., Fúster, J.M., Ibarrola, E., Cendrero, A., Coello, J., Hernan, F., Cantagrel, J.M., and Jamond, C., 1990, Volcanic evolution of Island of Tenerife (Canary Islands) in the light of new K-Ar data: Journal of Volcanology and Geothermal Research, v. 44, p. 231–249, doi:10.1016/0377-0273(90)90019-C.

Ancochea, E., Huertas, M.J., Cantagrel, J.M., Coello, J., Fúster, J.M., Arnaud, N., and Ibarrola, E., 1999, Evolution of the Cañadas edifice and its implications for the origin of the Cañadas Caldera (Tenerife Islands, Canary Islands): Journal of Volcanology and Geothermal Research, v. 88, p. 177–199, doi:10.1016/S0377-0273(98)00106-1.

Barraclough, A., and Guymer, I., 1998, Virtual reality—a role in environmental engineering education?: Water Science and Technology, v. 38, p. 303–310, doi:10.1016/S0273-1223(98)00668-4.

Bellan, J.M., and Scheurman, G., 1998, Actual and virtual reality: Making the most of field trips: Social Education, v. 62, p. 35–40.

Bloom, B.S., ed., 1956, Taxonomy of educational objectives: The classification of educational goals: Chicago, Susan Fauer Company, Inc., p. 201–207.

Bryan, S.E., Marti, J., and Cas, R.A.F., 1998, Stratigraphy of the Bandas del Sur Formation: An extracaldera record of Quaternary phonolitic explosive eruption from the Las Cañadas edifice, Tenerife (Canary Islands): Geological Magazine, v. 135, p. 605–636, doi:10.1017/S0016756897001258.

Cantagrel, J.M., Arnaud, N.O., Ancochea, E., Fúster, J.M., and Huertas, M.J., 1999, Repeated debris avalanches on Tenerife and genesis of Las Cañadas caldera wall (Canary Islands): Geology, v. 27, p. 739–742, doi:10.1130/0091-7613(1999)027<0739:RDAOTA>2.3.CO;2.

Carracedo, J.C., 1999, Growth, structure, instability, and collapse of Canarian volcanoes and comparisons with Hawaiian volcanoes: Journal of

Volcanology and Geothermal Research, v. 94, Special Issue, p. 1–9, doi:10.1016/S0377-0273(99)00095-5.

Carracedo, J.C., Day, S., Guillou, H., Rodríguez Badiola, E., Canas, J.A., and Pérez Torrado, F.J., 1998, Hotspot volcanism close to a passive continental margin: The Canary Islands: Geological Magazine, v. 135, p. 591–604, doi:10.1017/S0016756898001447.

Carracedo, J.C., Rodríguez Badiola, E., Guillou, H., De La Nuez, J., and Pérez Torrado, F.J., 2001, Geology and volcanology of La Palma and El Hierro (Canary Islands): Estudios Geologicos, v. 57, p. 175–273.

Carracedo, J.C., Pérez Torrado, F.J., Ancochea, E., Meco, J., Hernan, F., Cubas, C.R., Casillas, R., Rodríguez Badiola, E., and Ahijado, A., 2002, Cenozoic volcanism II: The Canary Islands, *in* Gibbons, W., and Moreno, T., eds., The Geology of Spain: London, Geological Society [London], 632 p.

Carracedo, J.C., Rodríguez Badiola, E., Guillou, H., Paterne, M., Scaillet, S., Pérez Torrado, F.J., Paris, R., Fra-Paleo, U., and Hansen, A., 2007, Eruptive and structural history of Teide Volcano and rift zones of Tenerife, Canary Islands: Geological Society of America Bulletin, v. 119, p. 1027–1051, doi:10.1130/B26087.1.

Cauley, F.G., Aiken, K.D., and Whitney, L.K., 2009, Technologies across our curriculum: A study of technology integration in the classroom: Journal of Education for Business, v. 85, no. 2, p. 114–118, doi:10.1080/08832320903258600.

Chance, E.W., and LoBaugh, L., 1994, Electronic field trips; using technology to enhance classroom instruction: Paper presented at the Annual Convention of the National Rural Education Association (86th), Salt Lake City, Utah.

Disinger, J.F., 1984, Field instruction in school settings: Environmental Education Digest, v. 1, p. 2–3.

Gordin, D.N., and Pea, R.D., 1995, Prospects for scientific visualization as an educational technology: Journal of the Learning Sciences, v. 4, p. 249–279, doi:10.1207/s15327809jls0403_1.

Grunwald, S., and Norton, L.D., 1999, An AGNPS-based runoff and sediment yield model for two small watersheds in Germany: American Society of Agricultural Engineers, Transactions of the ASAE, v. 42, p. 1723–1731.

Guillou, H., Pérez Torrado, F.J., Hansen Machin, A.R., Carracedo, J.C., and Gimeno, D., 2004, The Plio-Quaternary volcanic evolution of Gran Canaria based on new K-Ar ages and magnetostratigraphy: Journal of Volcanology and Geothermal Research, v. 135, p. 221–246, doi:10.1016/j.jvolgeores.2004.03.003.

Hansen, V.L., 2000, Geologic mapping of tectonic planets: Earth and Planetary Science Letters, v. 176, p. 527–542, doi:10.1016/S0012-821X(00)00017-0.

Herrington, J., and Kervin, L., 2007, Authentic learning supported by technology: ten suggestions and cases of integration in classrooms: Educational Media International, v. 44, no. 3, p. 219–236, doi:10.1080/09523980701491666.

Johnson, N.D., Lang, N.P., and Zophy, K.T., 2011, Overcoming assessment problems in Google Earth–based assignments: Journal of Geoscience Education, v. 59, p. 99–105, doi:10.5408/1.3604822.

Klügel, A., Hansteen, T.H., and Schmincke, H.U., 1997, Rates of magma ascent and depths of magma reservoirs beneath La Palma (Canary Islands): Terra Nova, v. 9, p. 117–121, doi:10.1046/j.1365-3121.1997.d01-15.x.

Klügel, A., Hoernle, K.A., Schmincke, H.U., and White, J.D.L., 1999, The chemically zoned 1949 eruption on La Palma (Canary Islands): Petrologic evolution and magma supply dynamics of a rift-zone eruption: Journal of Geophysical Research, v. 94, p. 267–282.

Krepel, W.J., and DuVall, C.R., 1981, Field trips: a guide for planning and conducting educational experiences: Analysis and Action Series, U.S. District of Columbia, p. 7–31.

Kulik, J.A., and Kulik, C.C., 1989, Effectiveness of computer-based instruction: School Library Media Quarterly, v. 17, p. 19–21.

Lamb, A., and Johnson, L., 2010, Virtual expeditions: Google Earth, GIS, and geovisualization technologies in teaching and learning: Teacher Librarian, v. 37, no. 3, p. 81–85.

Mannel, S., Winkelman, K., Phelps, S., and Fredenberg, M., 2007, Applications of a GIS program to tribal research: its benefits, challenges, and extensions to the community: Journal of Geoscience Education, v. 55, no. 6, p. 574–580.

Marshall, S.P., 2009, Re-imagining specialized STEM academies: Igniting and nurturing decidedly different minds, by design: Roeper Review, v. 32, no. 1, p. 48–60, doi:10.1080/02783190903386884.

Morgan, W.J., 1983, Hot spot tracks and the early rifting of the Atlantic: Tectonophysics, v. 94, p. 123–139, doi:10.1016/0040-1951(83)90013-6.

Nelson, B.K., Carracedo, J.C., Rodríguez Badiola, E., Hamilton, A., and Guetschow, H., 2005, Spatial and temporal isotopic gradients in the Western Canary Islands: American Geophysical Union Fall Meeting Abstracts, p. 1471.

Ramasundaram, V., Grunwald, S., Mangeot, A., Comerford, N.B., and Bliss, C.M., 2005, Development of an environmental virtual laboratory: Computers & Education, v. 45, p. 21–34, doi:10.1016/j.compedu.2004.03.002.

Renshaw, C.E., and Taylor, H.A., 2000, The educational effectiveness of computer-based instruction: Computers & Geosciences, v. 26, p. 677–682, doi:10.1016/S0098-3004(99)00103-X.

Salend, S.J., 2009, Technology-based classroom assessments: Teaching Exceptional Children, v. 41, no. 6, p. 48–58.

Schmincke, H.U., 1979, Age and crustal structure of the Canary Islands: Journal of Geophysics, v. 46, p. 217–224.

Schmincke, H.U., 1982, Volcanic and chemical evolution of the Canary Islands, *in* Von Rad, U., et al., eds., Geology of the northwest African continental margin: New York, Springer Verlag, p. 273–276.

Schmincke, H.-U., and Sumita, M., 2010, Geological Evolution of the Canary Islands: A Young Volcanic Archipelago Adjacent to the Old African Continent: Gorres-Verlag Publishing, Germany; 196 p.

Southworth, J.H., and Klemm, E.B., 1985, Increasing global understanding through telecommunications: NASSP Bulletin, v. 69, p. 39–44, doi:10.1177/019263658506948008.

Stainfield, J., Fisher, P., Ford, B., and Solem, M., 2000, International virtual field trips: a new direction?: Journal of Geography in Higher Education, v. 24, p. 255–262, doi:10.1080/713677387.

Treves, R., and Bailey, J.E., 2012, this volume, Best practices on how to design Google Earth tours for education, *in* Whitmeyer, S.J., Bailey, J.E., De Paor, D.G., and Ornduff, T., eds., Google Earth and Virtual Visualizations in Geoscience Education and Research: Geological Society of America Special Paper 492, doi:10.1130/2012.2492(28).

Tuthill, G., and Klemm, E.B., 2002, Virtual field trips: Alternatives to actual field trips: International Journal of Instructional Media, v. 29, p. 453–464.

Wernecke, J., 2008, The KML handbook: Geographic visualization for the web: Boston, Massachusetts, Addison Wesley Longman, 340 p.

Wilhelms, D.E., 1990, Geologic mapping, *in* Greeley, R., and Batson, R.M., eds., Planetary Mapping: New York, Cambridge University Press, p. 208–260.

Woerner, J.J., 1999, Virtual field trips in the earth science classroom, *in* Rubba, P.A., Rye, J.A., and Keig, P.F., eds., Proceedings of the Annual Conference of the Association for the Education of Teachers in Science, Austin, Texas, p. 1232–1249.

MANUSCRIPT ACCEPTED BY THE SOCIETY 16 APRIL 2012

The Geological Society of America
Special Paper 492
2012

Moving New York State Geological Association guidebooks into Google Earth

Otto H. Muller*

Geology Department, Alfred University, 1 Saxon Drive, Alfred, New York 14802, USA

ABSTRACT

The introductions and road logs from field trips offered by the New York State Geological Association (NYSGA) over the past 55 years are being transformed into kml files. These files are maintained as Google Fusion Tables, accessible to the public.

This paper begins by briefly summarizing the kinds of data being transformed, their strengths, and their limitations. It then details the procedures used to accomplish the transformation, from scanning the original document to uploading the data to Fusion Tables. By using a subset of available kml (Keyhole Markup Language) fields, and establishing a numbering convention for the placemarks, an efficient system has been developed where sufficient metadata is embedded within each placemark to permit mixing and matching of any of the placemarks. Using this system, additional field trip guides from GSA (Geological Society of America), AAPG (American Association of Petroleum Geologists), NEIGC (New England Inter-Collegiate Geological Conference), etc., might be transformed, increasing the size and value of the Fusion Tables database. The information provided will permit others to do this, producing kml files and Fusion Tables which will be consistent with those already done.

The paper discusses searching, merging, and adding photos to Fusion Tables and some of the ways in which Fusion Table data can be displayed on websites dynamically. The paper concludes by describing how the Fusion Tables from this project can export custom-made field trips, can be manipulated by other GIS applications, and can be used in a classroom setting to produce crude geologic maps.

INTRODUCTION

Much of the data that geologists deal with is found in outcrops and other exposures in the field. Field trips take scientists to locations the leaders select, and the guidebook descriptions often explain why the leaders believe the locations are significant. These guidebooks are usually the only repository of these descriptions. Although commonly found in the academic libraries of the regions visited by the field trips, guidebooks are less likely to exist in distant libraries.

Finding which stops (on which trips done in what year) contain descriptions of a particular fossil, or formation, or quarry, generally required a tedious search through the hard copies in the library.

*fmuller@alfred.edu

Muller, O.H., 2012, Moving New York State Geological Association guidebooks into Google Earth, *in* Whitmeyer, S.J., Bailey, J.E., De Paor, D.G., and Ornduff, T., eds., Google Earth and Virtual Visualizations in Geoscience Education and Research: Geological Society of America Special Paper 492, p. 335–345, doi:10.1130/2012.2492(24). For permission to copy, contact editing@geosociety.org. © 2012 The Geological Society of America. All rights reserved.

Seeing the spatial relationships between stops from different years has not been convenient. An individual field trip will often include a map showing the stop locations, so that the spatial relationships within a single trip are clear. Occasionally, guidebooks include index maps showing the stops for all of the trips in that guidebook. Some compendia exist, such as the New York State Geologic Highway Map (Rogers et al., 2000), which include the routes taken by many field trips; however, the individual stops are not shown.

This paper describes a project which endeavors to improve on this situation. Guidebooks from the New York State Geological Association, from 1956 to the present, are being transformed into a searchable database with spatial displays. Although less than halfway completed, the project has several thousand locations entered into its database. These provide a useful archive, and present a model which may be of use to others doing similar work with other field trips.

The New York State Geological Association (NYSGA)

Established in 1928, this organization facilitates interactions between professional, academic, student, and hobby geologists through annual field conferences, publications, etc. Relying largely on volunteer efforts, it obtains most of its funding through the sale of its guidebooks. One goal of this project is to encourage future sales.

Some time ago, the NYSGA decided to convert some (1957–1969) of its guidebooks to pdf format and to distribute them, for free, via the web at http://www.nysga.net/Guidebooks.html. All of these have been entered into the database described in this paper. More recent guidebooks are available for sale from the same website. Some of these (1980–1985) have already been entered into the database, and the rest will be included eventually.

Field Conferences and Guidebooks

NYSGA Field Conferences are held every year over a weekend, generally hosted by a college or university. Initially they were held in the spring, after the academic year was over for most participants. Starting in 1972, they have been in the fall. Trying to provide coverage for all of New York State, yet depending on institutions to volunteer as hosts, the organization moves these conferences around, but without a systematic rotation schedule.

An editor, usually from the host institution, seeks field trip leaders who will put together the trips, write up guidebook descriptions and road logs, and then lead the trips (sometimes twice, sometimes two different trips during the same conference). Efforts are made to include many different fields of geology and to visit the classic localities within a reasonable distance of the host institution, but also to include trips which address topics which are of current interest, either because of the geologic issues they address, or because of social or political concerns.

Participants are usually a mix of professionals, students, and hobbyists. Generally there is a great deal of educational activity going on, with no one feeling that the field trip leader has a

monopoly on what theories or ideas can be brought to bear at a particular outcrop. Knowing this, many field trip leaders include stops which have baffled them, hoping that the collective knowledge of the group may provide them with useful insights.

The guidebooks are the official record for these outings. Participants may have their own field notes, photos, measurements, samples, etc., but at present there is no way to access these. (See the discussion below on how these might be added to this project.) Guidebooks vary from less than 80 to 590 typescript pages or 397 two-column offset printed pages. They typically contain less than two dozen field trips. Sometimes the same trip is run during more than one field conference. This can be convenient, as participants can only attend two trips in any given year, and trips often have limits on how many can participate.

Field Trips and Road Logs

Most field trips begin with a short paper which explains the bigger picture addressed by the field trip. Often this paper begins with a paragraph or two of introduction, and if so, this is included in the description field of the path showing the route of the field trip.

The leader puts together a field trip visiting stops in a particular sequence. This sequence may result solely from geographic constraints, or it may be intended to introduce participants to features in a particular order. To preserve this sequence, the route taken by a field trip is preserved as a Google Earth path. Field trips which have only one stop, or which do not include routes from one stop to the next, will not include this path.

Most field trips include a road log, giving directions from one stop to the next, descriptions of the stops, and often descriptions of features which may be seen along the way. A trip is followed using the instructions provided in the road log. If the first descriptive information is far from the starting location, the path representing that trip may begin near the first feature described.

Road log instructions vary. Some include road and highway names, frequent road intersections, plenty of landmarks such as water towers, railroad crossings, etc. Others locate stops only by mileage, sometimes with distances of several miles between identifiable landmarks. Not surprisingly, these tend to be in areas with few roads or other anthropogenic features, often in the Adirondack Mountains. Working forward and backward from the landmarks may reveal errors of several tenths of a mile. In such cases the exact placemark locations are guessed at using topography, Street View images, etc., but substantial errors may occur. Some field trips, particularly those run after GPS (global positioning system) began to be used, include UTM coordinates or latitude and longitude.

The Nature of the Data

"Grey" literature. Papers in the guidebooks are reviewed, often by colleagues at the host institution, and particularly by the editor. Because the field trip leader is a volunteer, and rejection of the manuscript would probably mean eliminating that field trip, it is not likely that many rejections or serious revisions have occurred. Nonetheless, as the reputation of the host institution is

at stake, it seems unlikely that many egregious errors remain in these guidebooks.

Many of the field trip leaders are well-respected geologists with many peer-reviewed papers to their credit. They may use the guidebooks as a place to make field data available, as more formal publications often limit how much of this can be included. Some led trips as young scientists with limited experience, and their frequent guidebook contributions can reveal how their thinking evolved over their careers.

Sometimes speculative. Field trip leaders may propose explanations or theories which are new, and which may be tentative, seeking input from participants. Sometimes this is stated clearly in the descriptions; however it seems likely that this is not always the case.

Sometimes just wrong. On occasion, participants may inform a field trip leader that she or he has made errors, and these can be substantive. A bit like a review, and sometimes a brutal one, this may modify the leader's thinking, and it often affects how the participants think. There is currently no way to know about this, unless you were on the trip. The Fusion Tables produced by this project include a way for users to add comments. Hopefully this will be used to make such information available. As the Fusion Tables will not be closely monitored, users adding comments are advised to email me to bring them to my attention.

Contains the location of a feature of interest. Given these concerns, what may be safely gleaned from this data? At the time of the field trip, a location displayed something which the field trip leader thought was worth showing to the participants. What it showed, and why showing this was important, are the interpretations of the field trip leader. Like any interpretations, they would have been influenced by the education, experience, perspectives, etc., of that leader. As the earth sciences evolve, new models appear and interpretations change. But the features at a particular location generally do not. In this project, the location, the road log summary of what was to be observed there, the name of the field trip leader and the year of the field trip are kept together in each record. This should permit current and future users to find the feature, virtually or in reality mode, and to have some idea of the biases inherent in the descriptions presented.

Feature Locations

Google Earth provides a powerful interface to translate road log instructions into location data. Recent field trips, with good road logs, permit locations to be identified with confidence. Field trips from decades ago, however, even with excellent road logs, suffer from the changes which nature and society inflict on objects of geologic interest.

Roads change. Roads are re-routed, interstates disrupt older highway systems, and often a single stretch of road may have more than one name. Although Google Earth may not display the same name used in the road log, searching for the road log name will often identify the road correctly. Current U.S. Geological Survey quadrangle maps are easily imported into Google Earth, but often it is useful to refer to older editions. These are available

for New York and New England, as jpg downloads, from the University of New Hampshire libraries (http://docs.unh.edu/nhtopos/nhtopos.htm).

Outcrops disappear. Many of the stops on field trips are ephemeral. The working face of an active quarry is an obvious example. "Recent" land failure surfaces, flood deposits, etc., cannot remain recent. Other stops disappear because nature reclaims the exposures, covering them with soil, vegetation, or other detritus. People also modify road cuts. They remove them, intentionally obscure them with vegetation, and erect barriers in front of them. Quarrying operations may remove glacial or fluvial deposits, and usually restore disturbed areas when they finish. Builders fill in excavations; shopping malls fill in low-lying areas. Manicured lawns of suburbs sprawl over what were once natural areas, dotted with outcrops.

No ground truthing. Few efforts are made in this project to identify these changes. If Street View images indicate that long stretches of the field trip are obscured by sound barriers, there may be a note to this effect, but there has been no effort to ground truth the descriptions.

To the extent that they record features which no longer exist, some of the descriptions in this project may represent a unique, irreplaceable, data set. As these are identified, a virtual depiction of such "ghost" features might be constructed from photos, maps, sketches, etc., produced by people who visited the feature before its disappearance.

Feature Descriptions

The level of detail and amount of material included in road log descriptions vary considerably. The description field in this project contains the entire text from the road log (sometimes exceeding 2500 words) and most tabular data. They do not include figures, maps or references. If the text in the road log is essentially, "See text," or "Details will be discussed at the site," and if a reasonably concise description can be found in the text, this will be included. Efforts focus on including enough information to perform useful searches, produce geospatial representations, etc., while trying to avoid replacing the guidebooks.

Since Google Earth 4.3, a subset of HTML (hypertext markup language) has been permitted in the description field. Uses of HTML in this project include:

1. Most of the tabular data has been translated into simple HTML tables. These include stratigraphic columns, mining output, lists of fossils, minerals, etc.
2. The <i> </i> tags have been restricted to those fossil names which were italicized in the road log description. (Typically these have "sp." in front of them.) Other places where italics are used in the road logs, including species names of extant organisms, have <b> </b> tags, as do words initially in bold. This permits searching for "<i>" to find placemarks with descriptions containing fossil species names.
3. Editorial comments, added to explain routing choices, to point out the presence of sound barriers, etc., are colored

blue, and will usually indicate the year the information was added.

4. Warnings, admonitions and safety related comments, often in bold caps in the road logs, are in red.

Translating Field Trips

Overview

What follows describes the process of converting field trips to kml (Keyhole Markup Language) files. Although the NYSGA field trips are used here, the procedure will work with field trips from other sources, including those produced locally for in-house use. Suggested procedures for translating guidebook data into Google Earth paths and placemarks are included in the Appendix.

There are two different products, aimed at two target audiences:

A kmz file, to be loaded directly into Google Earth. For smaller collections of field trips, this approach has the advantage of being easily accessible to any Google Earth user without additional training or effort. The file for an individual year (Guidebook) contains folders for each field trip. These can be combined inside other folders. Individual stops can be moved from folder to folder, using only Google Earth, permitting users to create custom field trips. Some searching is possible, but only within the name field. Important metadata is moved with the stops so that the year, trip, leader, and stop number appear on the balloons. One variant is self-contained, incorporating the image and icon files into a kmz file (a zipped kml file), and requires no access to websites other than that for Google Earth. This is suitable for users who do not have a website where they can put image and icon files. The other variant retrieves image and icon files from a publicly accessible website where the user has stored them. Currently, this is required for the mobile version of Google Earth. The kmz files produced for this project are available at http://ottohmuller.com/nysga2ge/Files.html.

A Google Fusion Table. The kmz file can be imported into a spreadsheet or database, some additional data can be added, and the result converted into a csv (comma separated values) file which can be uploaded to Google Fusion Tables. There it can be searched, combined with other files, and subsets or supersets can be exported as static or dynamic files. This is the more powerful approach, and to be preferred for larger data sets, but the additional steps required may deter some users. The Google Fusion Table produced for this project is available at http://www.google.com/fusiontables/DataSource?dsrcid=1178844.

These two products both require the same initial kml file. To produce it you need to convert the road logs to text, reformatting as necessary, follow the road log directions to locate stops and views in Google Earth, and insert the appropriate text into the name and description fields for the Google Earth placemark you establish at that location. The resulting kml file is modified to add the metadata and remove unused fields. Styles are then added, with links to icon and image fields on the web if desired. If you need only the Fusion Table, the style steps can be omitted.

Guidebooks to Text Files

Much of the labor involved in this project involves converting images into text. The NYSGA supplied pdf files or hard copies of the guidebooks used in this project. Before 2004, the pdf files were scans of the hard-copy pages, but since then most of the guidebooks have been produced from electronic files. Some of these have been sent to me by the NYSGA, others by the relevant editors.

If no pdf of the guidebook exists, the relevant pages of the hard copy of the guidebook are scanned into images. These images, or those from the early pdf files, are analyzed using optical character recognition (OCR) software. The text files produced are then corrected for OCR errors, stripped of linefeeds, and formatted for Google Earth as described above.

The earlier guidebooks are photocopies of typewritten text, often from a variety of manual typewriters. OCR errors were plentiful; considerable efforts were made to find most of them, but many undoubtedly remain. As electric typewriters replaced manual ones, consistency improved and the error rate dropped. As with any translation, the user should exercise caution and consult the original guidebook if accuracy is essential.

Road Logs to Default KML Files

Road logs from guidebooks in any format are used to plot the paths taken by the field trips, and the locations of stops and views.

A "stop" is used here to refer to something the field trip leader called a Stop. They are usually numbered sequentially, with complicating factors when alternate stops are listed, two different stops are given the same number by mistake, etc. On many trips, the leaders include information about features which can be observed as the trip progresses from one stop to the next. Usually the trip does not stop at these features, and they are referred to here using the term "view." Mineral or fossil collecting trips tend to have few views, while trips studying glacial or geomorphological features may have far more views than stops.

The Appendix contains step-by-step instructions for entering the paths, stops, and views in Google Earth.

Using the Results

This project began as an effort to translate NYSGA field trips into Google Earth–compatible files. It soon became apparent that putting individual stops and views into a searchable database would make these translations more accessible. Google Fusion Tables provide an excellent interface for just such a database.

The Guidebooks and KMZ Files

Although the power and flexibility of Google Fusion Tables permits many uses which the original field trip leaders may not have envisioned, there are still good reasons to revisit the field trips as they were originally run:

1. The articles in the guidebooks which accompany the road logs are usually designed to provide context and background to a particular field trip. As these articles are

available in many libraries, and are available from the NYSGA either as a free download or for a nominal fee, they should be used in conjunction with the kmz files produced by this project to enhance the value of both.

2. The creative efforts of the field trip leaders include determining the sequence of stops, and sometimes this is important in appreciating the information being presented. It is not uncommon for leaders to show obvious examples of some feature in the earlier stops, and more subtle expressions of the same feature at later stops, counting on the earlier observations to inform the later ones. Nor is it uncommon for leaders to move through the stratigraphy or facies changes in a systematic way.

3. Many field trips try to show a "big picture" by means of examining many stops and views. If only some of these are visited, this "big picture" may be missed.

Retracing Original Field Trips

Users with access to guidebooks should be able to retrace field trips virtually. The quality of this experience will depend a great deal on the details of the field trip.

Virtual trips may be better. A trip exploring geomorphology, glacial geology, etc., may contain stops and views which are better seen from a distance, perhaps at an oblique angle, than from the highway. Virtual users can fly around these, seeing them from various perspectives, something which would otherwise require an aircraft. Vertical exaggeration can be adjusted to enhance minor variations in topography, something which is only possible on a virtual trip. Maps in the guidebook articles can be scanned, turned into Google Earth overlays, and draped over the topography, further enhancing the virtual experience.

Trips visiting ephemeral features, such as excavations or landslides, may be better experienced virtually, particularly if the historic views available from the Google Earth Time Slider contain remnants of the features. In addition to showing features which no longer exist, Time Slider images also reveal seasonal variations, as they are often taken at different times of the year. Outcrops may appear in images taken in March or April which are obscured by foliage or snow during much of the rest of the year. Streams may appear at various stages in different images, revealing fluctuations which can only be guessed at during a visit in person.

Locations normally closed to the public, where special permission was acquired for the NYSGA visit, may not be accessible except virtually. Dangerous traffic, rock falls, dogs, and other hazards are absent on virtual field trips. This can be of particular concern for larger groups, especially those containing participants who have difficulty heeding safety precautions.

Virtual trips may be equivalent. Stops on a petrology trip, where mineral or chemical analyses are discussed, may be just as informative when seen on Google Earth as when visited in person, with the obvious drawback that collecting samples is not possible.

Stops and views may have been obscured by sound barriers, vegetation, etc., since the NYSGA trip visited them. There may

be no advantage in visiting these in person. A quick check with Street View, assuming it is available for the locations of interest, can help.

Virtual trips may be inferior. Some stops visit outcrops which require careful examination. Structures, fossils, mineralization, etc., are often three-dimensional features at a scale of centimeters to meters. Even if fine photographs exist, from different angles and at different scales, they cannot compare with being there in person. Still, the field trip article often contains sketches or enhanced photos which can guide observations. These, and a quick check with Street View to see a recent photo of the outcrop, can help the user prepare for a real field trip.

NYSGA participants. Users who have participated in a NYSGA field trip may find value in retracing it. Samples, notes, and photos, carefully acquired and annotated during the trip, have a way of becoming ignored later when daily routines return. Running through the trip virtually can refresh and refocus the user's mind.

Those planning to lead field trips can study earlier ones to see how stops progressed, what accommodations were made for lunch or other rest stops, whether cars or buses were used, etc.

Using Google Fusion Tables

Google Fusion Tables provide an accessible, georeferenced, searchable database very well suited for this project. As described above, the table for this project contains only placemark data (stops, views, and paths). It is dynamic, changing as material is added or revised; hence it contains current versions of the data. The kmz files available on the website contain a little additional information—generally the introductions and other prefacing material from the guidebooks—but are not dynamic and may not be current. (Revision dates are listed on the website.)

Each Fusion Table is assigned a Data Source ID, called a "dsrcid," which remains with the table even if its name changes. Links to the table use this number. The NYSGA database has a dsrcid of 1178844 as seen in the link shown above.

As of this writing, that table is named "NYSGA 1956-1969 and 1980-1985" and as additional data is entered, its name will change to reflect its contents. Currently containing 2917 records, this Fusion Table can be exported and merged with other tables, and users can add comments to it. The csv file containing this data is 4.3 MB. Google currently has a quota of 250 MB for a user's Fusion Tables, so there is little question that field trip data from many additional sources can easily be accommodated. Anyone considering adding stops from other sources should contact me, and we can work out editor permissions or other collaborative arrangements.

With Fusion Tables, a user can construct a set of search queries which limit the data being displayed or output. For example, a user interested in stops containing "Marcellus" in the <description> field would start by clicking "options." The "Filter" tab opens and presents a drop-down menu on the left, from which the user selects "description." The user then selects "contains" from the next drop-down menu to the right, and types "Marcellus" into the available text entry box. One more click, on "Apply," and in a moment the 11 stops or views containing "Marcellus" are displayed. If the user wishes to further restrict data to field trips led by Selleck, clicking on "Add condition," then selecting "leader" from the first drop-down menu, "contains" from the second, and typing "Selleck" into the text entry box results in two records being selected.

Clicking "Visualize" on the Fusion Tables menu bar and selecting "Map" from the drop-down menu brings up a map view where these two placemarks are displayed. By now clicking on "Download KML link" a kml file is downloaded which contains the following:

```
<?xml version="1.0" encoding="UTF-8"?>
<kml xmlns="http://www.opengis.net/kml/2.2">
<NetworkLink>
<name><![CDATA[NYSGA 1956-1970, 1972, and 1980-
1985]]></name>
<Link>
<href>
https://www.google.com/fusiontables/exporttable?query=
select+col7+from+10moOMdycbyMvSweEV8i6anOL0cNPb
9e5SFgGcM+where+col1+contains+'Marcellus'+and+col3+
contains+'Selleck'&amp;o=kmllink&amp;g=col7</href>
</Link>
</NetworkLink>
</kml>
```

This link, and the "Get embeddable link" as well, contain the name, the encrypted ID (which in this case is 10moOMdy-cbyMvSweEV8i6anOL0cNPb9e5SFg-GcM) and search queries. Hence they are dynamic and will fetch new data as it is added to the database. With the 1970s and everything since 1985 yet to be added, it is very likely that Selleck will have led other trips to outcrops of the Marcellus Shale, and when these are translated and added to the database, those links will display the new Stops and Views.

Combining Data in Fusion Tables

"Merge" permits users to join together Fusion Tables which have the same unique values in a column. Users with additional information (links to photos, chemical analyses, etc.) about some set of stops can construct their own Fusion Table with this data, being sure that there is a column where each stop is given the ID as described above. It is then a simple matter to merge this Fusion Table with that of the NYSGA.

As an example, an Excel file containing URLs for three photos from NYSGA 1957, Trip A, was uploaded to Google Fusion Tables where it became the table, "ThreePics.xls." The data it contained were:

ID	Link
NYSGA1957-A-1.00	http://ottohmuller.com/nysga2ge/images/1957/A/Stop1.jpg
NYSGA1957-A-2.00	http://ottohmuller.com/nysga2ge/images/1957/A/Stop2.jpg
NYSGA1957-A-3.00	http://ottohmuller.com/nysga2ge/images/1957/A/Stop3.jpg

This was merged with the NYSGA Fusion Table to give a resulting table, "Only with pics" at:

http://www.google.com/fusiontables/
DataSource?dsrcid=1192568

To get access to the photos you need to insert the following code into the "Configure Info Window," "Custom" tab:

```
<div class='googft-info-window' style='font-family: sans-
serif; max-height: 600px; overflow: auto;'>
  <img src="http://ottohmuller.com/nysga2ge/images/
NYSGALogo.jpg" alt="NYSGA logo" width="89"
height="60" align="right" />
  <font size="+2">{name}</font><br></br><br></br>
    <div align="center">
      <a href="{Link}"><img src="{Link}" style="width:
250px"></a>
    </div>
    {description}
    <div align="right">
      <p>{leader}<br></br>NYSGA {year} Trip {trip} Stop
{stop}</p>
    </div>
</div>
```

This will result in a balloon as shown in Figure 1. Clicking on the thumbnail image will open the image as it exists on the web. Exporting as a kml file, a kml network link, or an embeddable link preserves this formatting.

Merges are dynamic, so if additional rows are added to the "ThreePics.xls" table, the merged table, "Only with pics" updates automatically to reflect this. The kml network links and embeddable links retrieving data from this Fusion Table will update automatically, also.

Fusion tables can also be extended vertically, using "Import more rows." It is necessary to have editor permissions to do this (see above).

Examples of Uses for Professional Geologists

Georeferencing photos from past trips. Participants from early trips may have photos or slides which were carefully annotated with the date, trip, stop number, etc., but are not geocoded. They can scan them, put them on a site on the web which is accessible to the public, and construct a spreadsheet or database with IDs in one column (constructed as described above, so they are of the form: NYSGA1980-A-3.01) and links to their web locations in other columns. This can be uploaded to Google Fusion Tables and then merged with the Fusion Table of this project. If the result is exported as a kml file, it will include the coordinates of the stops which can then be added to the metadata of the photos. In addition, the downloaded file lets the user click on the thumbnail to see the original photo as it resides on the web.

Finding poorly remembered stops. After participating in a number of NYSGA field trips, a user might have a recollection of something about a stop, but not enough to recall where it was, or which trip it was on. For example, the user may recall having seen some *Diplocraterion Yoyo* burrows. A search on Google Fusion Tables shows two stops with these terms in the description field on a trip in 1983.

Creating new field trips from selected stops and views. Users can select stops for their own, real, field trips by selecting stops and views from the NYSGA Fusion Table. (Once selected, of course, they should be visited to ensure that the features described in the road guide are still present and accessible.) Often, trips to the same outcrops, led by different leaders or run at different times, will emphasize different features or different interpretations of the same features. Being able to compare and combine parts of several descriptions, enhanced by referring to

the relevant guidebooks, may permit important improvements over any single field trip description.

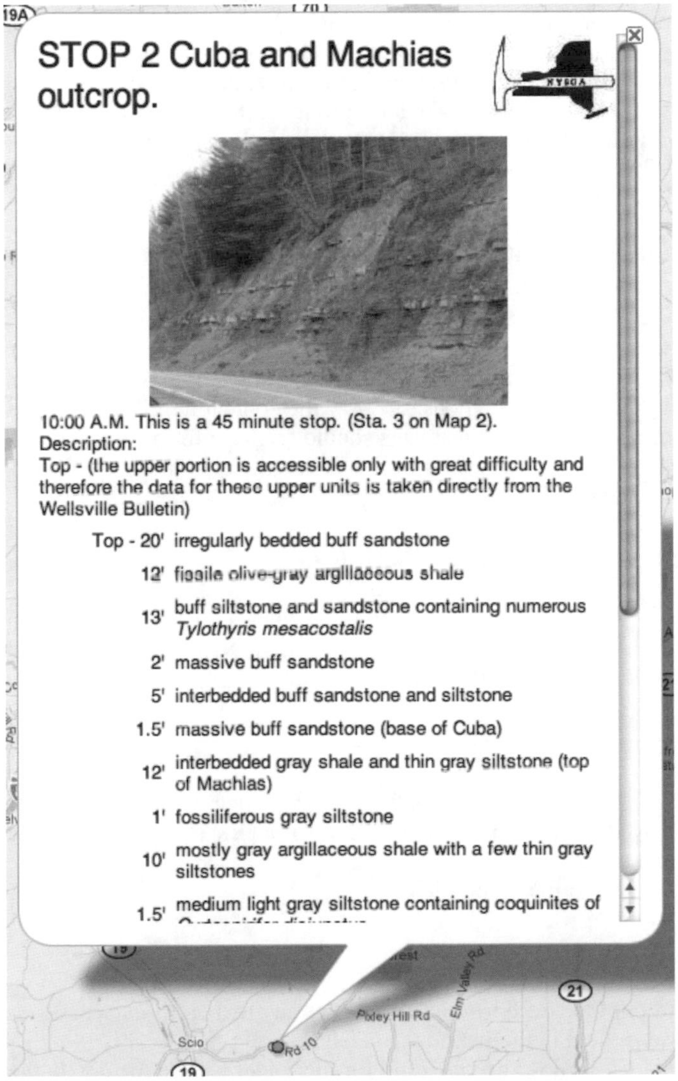

Using other GIS applications. The files produced by this project can be used in GIS (geographic information systems) applications other than those produced by Google. For example, Avenza makes a program called MAPublisher which can import kml files. Designed as an add-in for Adobe Illustrator, this program provides a number of ways to display and print high quality maps. One option is to produce geospatial pdf files. These geo-referenced files can have many layers produced from a variety of GIS applications (GRASS, ArcView, etc.). By putting NYSGA paths, stops, and views on top of layers showing bedrock or surficial geology, they can be seen in context. An example is included in the supplemental information for this paper[1]. Avenza provides a free app, "PDF Maps," to view such maps on an iPhone, iPad, or iPod, permitting the GPS capabilities of those devices to plot the user's location in real time.

Examples of Educational Uses

Emulation. The best use of this project would be for others to emulate it. Using the techniques described here, a high school earth science class could produce a local field trip and upload it to Google Fusion Tables. Finding additional outcrops, revisiting NYSGA stops and views, adding photos with additional Fusion Tables—all are worthwhile educational activities within the skill set of many high school students. Making appropriate observations and interpretations of the outcrops might be more of a stretch, but the commenting capability on the Fusion Table encourages suggestions for revisions from those with more training and experience. The NYSGA field trips list leaders and their affiliations, facilitating this kind of collaboration. School logos and abbreviations would replace those of the NYSGA, but otherwise it could be the same. Making it publicly accessible might be an incentive to students, and once a field trip is up on the web, later classes at that school could use it with an eye to improving it.

Geologic Map. A crude geologic map can be constructed from the Fusion Table in the following way: Select stops and views in which the <description> field contains the name of a particular geologic time period. Export the resulting kml file and, in Google Earth, set the color for its icon to one representing that time period. Repeat for the other time periods represented by the rocks of the area being mapped.

CONCLUSIONS

This paper gives detailed instructions for translating field trips, with their stops and views, into Google Earth paths and placemarks, and for deploying the results on Google Fusion

Figure 1. Display balloon in Google Fusion Tables, showing result of merging a table with a link to a photo, as well as part of a table containing stratigraphic information. Clicking on the thumbnail photo opens a web page with the photo at full size.

[1]GSA Data Repository item 2012304, supplementary information, is available at http://www.geosociety.org/pubs/ft2012.htm, or on request from editing@geosociety.org.

Tables. Although it uses NYSGA Guidebooks as the source of these trips, the techniques should be suitable for trips from many sources, even from fields outside of the earth sciences. As technology evolves, the capabilities of Google Earth and Fusion Tables will undoubtedly improve, but the bare-bones nature of the approach described should help to ensure a reasonably long useful life for the results obtained and an easy upgrade path for future development.

ACKNOWLEDGMENTS

Alan Benimoff, executive secretary of the NYSGA, Bill Kelly, state geologist of New York (retired) and Marion Weaver, earth science teacher (retired) have been enthusiastic supporters of this project since its inception. Their help and encouragement are gratefully acknowledged. This paper was enhanced by comments from an anonymous reviewer. The paper and the code in Fusion Table Info Windows were greatly improved by comments from reviewer Don Duggan-Haas.

APPENDIX

After using a variety of approaches, this procedure is recommended as the most efficient way of getting the road log data into Google Earth:

1. In Google Earth, open a folder and give it the name of the field trip. In this project, the name consists of the trip number, followed by a colon, followed by the name in caps.
2. Within that folder, create a path, giving it the same name of the field trip used for the folder. Navigate through the field trip, making a crude path, finding the roads and turns used to get from one stop to the next. Your path will be made of long straight line segments, connected by sharp angles. By focusing on the route, ignoring stops and views, you avoid a variety of annoying errors.
3. Refine the path, adding enough intermediate points to keep it within the roadway. This should allow you to measure distances along the route reasonably accurately.
4. Locate the stops and views sequentially. Google Earth will preserve this order, and it will be used later to number each stop and view. If you search, or otherwise interrupt the process, the next stop may not be where you want it, so check the sidebar frequently to ensure everything is going into the folder in the correct order.

    Locating stops and views may require some effort. The write-ups in the guidebooks sometimes contain useful maps and diagrams. Older 15′ USGS Quadrangle maps can be a great help for older field trips. Use the Google Earth Time Slider to see other images of the area you are working in. (Photos taken in early spring are usually the best.) Street View images are also a great help. Many stops are in parks and detailed maps of many parks are online.
5. Insert the name and description in the relevant boxes which Google Earth presents to you. (These are the results you ended up with after translating guidebooks

to text files.) If you modify the contents of the name box, be sure to click somewhere else afterwards, or your changes will not be saved. When done, click the OK button and locate the next stop or view.
6. Save the folder as a kml file. Google Earth defaults to saving as a kmz file, but you can choose the kml option, avoiding an additional step later.

    This default kml file will have a different style for each stop and view, as well as some fields of little direct use for this project. We will correct these next.

The Default KML File

A kml file is a text file. Tags are enclosed in angle brackets (< >) and fields are the text in between matching tags. Tag names are case sensitive and any quotation marks within them are required. Quotation marks within a kml file should be straight ones (" "), not the curly typographic ones (" ").

As with many other electronic documents, kml files separate content from appearance. Styles determine how a path or placemark looks and behaves in Google Earth, and include all of the fields with "style" in their name, plus one called <snippet>. The fields with content, produced by Google Earth in the default kml files, are <name>, <description>, <Point>, <LineString>, and <coordinates>. In addition we will add <ExtendedData> fields for "year," "trip," and "leader."

To modify this generic kml file, open it in a text editor which has GREP search and replace capabilities. BBEdit or the free TextWrangler work well on a Mac.

Modify Content Fields

In the examples below, a particular field trip, Trip A-10 from 1985, is used to demonstrate how the content fields are modified for this project. Adjustments should be made for other trips.

First we put an ID in the "Folder" tag. For this field trip, the tag will be:

```
<Folder "id= NYSGA 1985 A10">
```

As an example of how this would be adjusted for trips from a different organization, Trip 5 from an NEIGC conference in 2006 would have the tag:

```
<Folder "id= NEIGC 2006 5">
```

This folder begins with the <name> field (which will be displayed on the sidebar in Google Earth) and it should have that name in it already.

Insert the empty field:

```
<snippet maxLines="0"></snippet>
```

to tell Google Earth not to display any text from the <description> field in the list on the sidebar.

The <description> field for the folder is used to contain the prefacing material for the field trip, if it exists, plus the names and affiliations of the field trip leaders. The <description> field permits HTML code, and the <!CDATA[[...]]> construction keeps it more easily read by humans:

```
<description>
  <![CDATA[
    <center>
    DAVID J. DE SIMONE <br />
    Department of Geology, Rensselaer Polytechnic Institute<br />
    Troy, New York 12180-3590 <br /><br />
    ROBERT G. LA FLEUR <br />
    Department of Geology, Rensselaer Polytechnic Institute<br />
    Troy, New York 12180-3590 <br /><br />
    </center>
  ]]>
</description>
```

Next will be a series of placemarks, one for the path and one for each stop or view. Each placemark tag needs an ID added to it, and doing this manually, instead of with some search and replace scheme, seems to be most efficient. The numbering system for these IDs is key to retrieving data efficiently. An ID of 0.00 is used for the path. Whole numbers are used for stops; the decimal part is used for views. Views are numbered from the preceding stop. So an ID of 3.04 refers to the fourth view after stop 3:

```
<Placemark id="3.04">
```

For the path placemark the <name> field should already contain the same text as the <name> field for the folder. Copy the contents of the <description> field of the folder into the <description> field of the path. This insures that the prefacing material for each trip will be uploaded to Google Fusion Tables.

The rest of the placemarks will have the <name> and <description> fields filled with the data you entered when you created them in Google Earth.

Placemarks in the default kml file will have either a <LookAt> field or a <Camera> field. Within these fields are a number of fields specifying how Google Earth is to view the placemark, none of which are useful for this project. We replace these fields with our <ExtendedData> fields by selecting the folder for the trip, and then "Replace All," replacing:

```
<LookAt>(?s).+?</LookAt>
```

with:

```
<ExtendedData>
  <Data name="year">
    <value>1985</value>
  </Data>
  <Data name="trip">
    <value>A10</value>
  </Data>
  <Data name="leader">
    <value>D.J.DeSimone, R.W.LaFleur</value>
  </Data>
</ExtendedData>
```

(you should first substitute the appropriate values for "1985," "A10," and "D.J.DeSimone, R.W.LaFleur"...) Note that all leaders' names consist of initials, periods, and last names, with no spaces.

If the default kml file contains <Camera> fields, they should be replaced with the ExtendedData fields in the same way.

Any elements prefixed with "gx:" should be removed.

The placemark records are now complete. If the only output desired will use Google Fusion Tables, skip the next section.

Modify Style Fields in the Default KML File

Styles determine how the content in the kml file is displayed. A Google Fusion Table does not import styles, and it exports kml files with a generic icon and whatever information is set up for the table with "Configure info window."

Eleven styles are defined in this project. The first, with id="msn_Title" is to display the prefacing material for field trips and guidebooks. The rest, with id= "msn_Color" (where the word "Color" is replaced by any of ten color names) are used for placemarks. These are identical to each other except for the colors used for icons and labels. These colors, and their names, come from early HTML usage: Aqua, Lime, Violet, Yellow, Tomato, Magenta, HotPink, DarkSeaGreen, DodgerBlue, Orange. They were selected to stand out on Google Earth images and to be different from each other.

Each placemark style definition consists of a <StyleMap> field containing, in <StyleUrl> fields, the names of two <Style> fields. One of these <Style> fields, with id= "sn_Color" is what Google Earth normally displays. Here the icons have different colors and the scale of the <LabelStyle> is set to zero, so labels are not displayed. The other, highlight, style with id= "sh_Color" is what Google Earth displays when a mouse is over the icon for the placemark. When this happens, the icon turns red, gets a little larger, and the label is displayed.

Part of the style definition, the <BalloonStyle> field may contain a <text> field which can contain entities of the form $[name]. When displayed, the entity will be replaced by the value of the field called "name." This lets us display our extended data for each placemark.

Two images are used in this project: one for the icon and the other for the logo on the Balloons. If you put them inside a folder named "images," zip this together with the kml file, and then change the extension from "zip" to "kmz" you will produce a readable .kmz file from which the desktop version of Google Earth can retrieve these images. For an icon named "circle2.png" and a logo named "NYSGALogo.jpg" this would result in a <Style id = "sh_Aqua"> field such as:

```
<Style id="sh_Aqua">
  <IconStyle>
    <color>ff0000ff</color>
        <scale>0.5</scale>
    <Icon>
      <href>images/circle2.png</href>
    </Icon>
  </IconStyle>
  <BalloonStyle>
    <text>
      <![CDATA[<img src="images/NYSGALogo.jpg" alt="NYSGA logo"
width="89" height="60" align="right" /><font size="+2">$[name]</font><
br><br>$[description]<p align="right">$[leader]<br>NYSGA $[year]
Trip $[trip] Stop $[id]</p>]]>
    </text>
  </BalloonStyle>
  <LabelStyle>
    <color>ffffff00</color>
    <scale>1</scale>
  </LabelStyle>
  <LineStyle>
    <color>ffffff00</color>
    <width>2</width>
  </LineStyle>
</Style>
```

Alternatively, if you have write permissions on a publicly accessible website, you can upload the "images" file and link to it. Google Earth Mobile is not currently able to extract images from a kmz file, so this is the only approach which will work with the mobile version. In this case, using the Google icon called "placemark_circle.png," our style field becomes:

```
<Style id="sh_Aqua">
   <IconStyle>
      <color>ff0000ff</color>
      <scale>1.1</scale>
      <Icon>
         <href>http://maps.google.com/mapfiles/kml/shapes/
placemark_circle.png</href>
      </Icon>
   </IconStyle>
   <BalloonStyle>
      <text>
         <![CDATA[<img src="http://ottohmuller.com/nysga2ge/
images/NYSGALogo.jpg" alt="NYSGA logo" width="89" height="60"
align="right" /><font size="+2">$[name]</font><br><br>
<br>$[description]<p align="right">$[leader]<br>NYSGA $[year]
Trip $[trip] Stop $[id]</p>]]>
      </text>
   </BalloonStyle>
   <LabelStyle>
      <color>ffffff00</color>
      <scale>1</scale>
   </LabelStyle>
   <LineStyle>
      <color>ffffff00</color>
      <width>2</width>
   </LineStyle>
</Style>
```

Either one will result in the balloon shown in Figure A1.

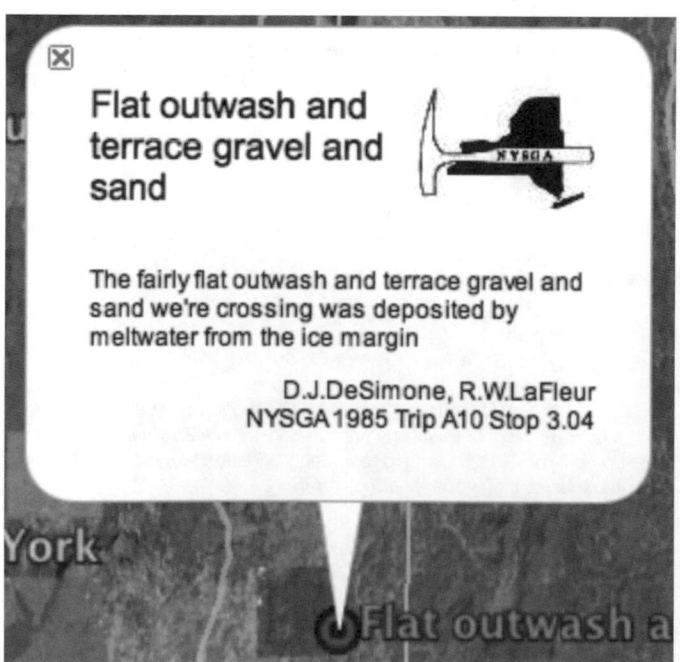

Figure A1. Basic display balloon as seen in Google Earth, showing name and logo at the top, then description, then leaders, year, trip and stop number at the bottom.

If a different organization were translating their field trips, they would use their own logos and replace "NYSGA" with their abbreviation.

Once you have adjusted the style files to your liking, throw away all of the <StyleMap> and <Style> fields in the generic kml file and paste in the ten sets of adjusted <StyleMap> and <Style> fields for either the stand alone or mobile version.

Next you need to change all the <styleUrl> fields. Again, select all of the Placemarks for a trip, and then replace:

```
<styleUrl>(?s).+?</styleUrl>
```

with:

```
<styleUrl>#msn_Color</styleUrl>
```

where "Color" is one of the ten mentioned above.

At this point, your kml file should be done, so save it. Zip it with your images folder if you are making a stand-alone version, zip it by itself if not, then change the extension to kmz. Test this in Google Earth.

Prepare KML Files for Upload

If you try to upload your kml file to Google Fusion Tables, many of the fields you have produced will not be imported. Furthermore, you can enhance the searching and merging capabilities in Google Fusion Tables by adding a few more fields. What you need is a program which will import your kml file, modify some of the fields, and then export csv data. Most spreadsheet or database programs should suffice. As I used Filemaker Pro 8.5, it is discussed here, however the steps and xslt (Extensible Stylesheet Language Transformations) files can be adjusted to accommodate other software. A clone (no records) of the database and the xslt file are included in the supplemental information (see footnote 1).

To begin, modify the kml file by deleting (or commenting out) the second line, "<kml xmlns= . . . >" and the final line, "</kml>" and then save the file with an .xml extension. Import it using the appropriate xslt file. Within your database/spreadsheet, create five new columns for ID, Longitude, Latitude, Geometry, and FusionStyle and populate these with calculations as follows:

ID: Concatenate the name of your organization (in the present case "NYSGA") with the contents of the fields for <year>, < trip>, and <stop>, separating these with hyphens. The result will be similar to "NYSGA1980-A-3.02" and as this will be used to merge tables it is important that it is formatted in just this way.

```
ID ="NYSGA" & year & "-" & trip & "-" & stop
```

Longitude and Latitude will be used to filter stops and views spatially. This data comes from the <coordinates> field, extracting text from the left and right of the comma and converting it into a number.

```
Longitude = GetAsNumber (Left (coords; Position (coords ;","; 1; 1)))
Latitude = GetAsNumber (Middle (coords; Position (coords ;","; 1;
1); Position(coords ;","; 1; 2) - Position (coords;".";1;1)))
```

The <geometry> field is used by Google Fusion Tables to plot routes as well as stops and views. Our csv upload needs to format this field the same way Google Earth would format it.

```
geometry = If (LineString="" ; "<Point><coordinates>"&coords&"
</coordinates></Point>"; "<LineString><tessellate>1</tessellate>
<coordinates>"&LineString&"</coordinates></LineString>")
```

The following code converts the ten colors used for styles into the five which Fusion Tables permits:

```
FusionStyle=Case(style="#msn_Aqua";"small_red";style=
"#msn_Lime";"small_blue";style=
"#msn_Violet";"small_yellow";style=
"#msn_Yellow";"small_green";style=
"#msn_Tomato";"small_purple";style=
"#msn_Magenta";"small_red";style=
"#msn_HotPink";"small_blue";style=
"#msn_DarkSeaGreen";"small_yellow";style=
"#msn_DodgerBlue";"small_green";style=
"#msn_Orange";"small_purple";"small_red")
```

Errors in your data are often much easier to see and correct in the database/spreadsheet format than in the kml format. Once you are satisfied, output your data as a csv file.

When Filemaker Pro exports data it converts returns into vertical tabs. That is, the non-printable character $0D (Carriage Return) is output as $0B (Vertical Tab). A Fusion Table ignores this and functions well, but includes these $0B characters when it exports a kml file. Then, perhaps because the contents of the <description> field are inside of the <![CDATA[...]]> con-struction, these kml files will fail to load in Google Earth. To avoid this, the output csv file needs to have all of its $0B characters found and replaced by $0D characters. (Other database or spreadsheet programs may have different glitches.)

Once this is done, upload the file to Google Fusion Tables. Although you will have the option of adjusting which columns are imported and their order, it is probably wise to export a small set of stops from the Fusion Table you wish to emulate, import it into your database or spreadsheet, and adjust the columns you produced and the export order for your database, to match.

In Fusion Tables, the "Import" dialog box lets you choose whether data can be exported. It is necessary to leave this enabled for the network links and embeddable links described above to function. It is also necessary to change the Fusion Table "Sharing" status from the default "Private" to either "Public" or "Unlisted."

REFERENCES CITED

Rogers, W.B., Rickard, L.V., Lauber, J.M., Landing, E., and Isachsen, Y.W., eds., 2000, Geology of New York: A Simplified Account, Second Edition. New York State Museum Educational Leaflet No. 28, 284 p, 4 pls. (Geologic Highway Map).

MANUSCRIPT ACCEPTED BY THE SOCIETY 16 APRIL 2012

The Geological Society of America
Special Paper 492
2012

Benedict Arnold's march to Quebec in 1775: An historical characterization using Google Earth

Bruce F. Rueger*
Emma N. Beck*
Department of Geology, Colby College, 5806 Mayflower Hill, Waterville, Maine 04901-8858, USA

ABSTRACT

Google Earth provides a dynamic mechanism with which the impact of geographic features on historical events is illustrated effectively. Using Benedict Arnold's ill-fated surprise attack on British forces in Quebec (Canada) during the American Revolution (1775) as an example, various features of Google Earth are used to support this relationship. Traveling up the Kennebec and Dead Rivers in Maine (USA), the expedition encountered topographic and/or geologic obstacles that ultimately led to failure.

Illustrated placemarks were created in Google Earth to document significant sites along the route. Placemarks were linked via the Tour feature producing a flyover, where the placemarks opened to provide a narrative along the route. At the Chopps near the mouth of the Kennebec River, a geologic map overlay was applied to document the geology and structural influences on this constriction of the river that causes navigational hazard. Along the portage route of the Great Carrying Place the trail was marked and an elevational profile created to illustrate the severity of the terrain. Anthropogenic changes along the route were documented by overlays of topographic maps produced both before and after the Dead River was dammed in 1950. Finally, an historical map from 1761 showing known landforms and their relative positions was overlain to document Arnold's understanding of the route in 1775.

This approach in teaching and interactive learning has proven very effective locally at all levels. Illustrations, overlays, and tours created provide improved understanding of the interaction between geography and historic events. Creation of Google Earth files such as these can be readily used and shared by schools, historical societies, and individuals to greatly enhance learning.

*bfrueger@colby.edu; enbeck@colby.edu

Rueger, B.F., and Beck, E.N, 2012, Benedict Arnold's march to Quebec in 1775: An historical characterization using Google Earth, *in* Whitmeyer, S.J., Bailey, J.E., De Paor, D.G., and Ornduff, T., eds., Google Earth and Virtual Visualizations in Geoscience Education and Research: Geological Society of America Special Paper 492, p. 347–354, doi:10.1130/2012.2492(25). For permission to copy, contact editing@geosociety.org. © 2012 The Geological Society of America. All rights reserved.

INTRODUCTION

In the fall of 1775, 34-year-old Colonel Benedict Arnold was given command of a group of 1,100 marginally trained American colonial militiamen (Smith, 1903; Desjardin, 2006). Authorized by George Washington, Arnold's mission was to lead these troops up the Kennebec River in Maine (USA) through sections of dense wilderness to the Dead River, cross the Height of Land, and march along the Chaudiere River in Quebec (Canada) to launch a surprise attack on the British forces holding the fortress of Quebec (Nelson, 2006). The expedition traveled over some very rugged terrain, and troops were subjected to severe hardships from both the topography and the weather (Desjardin, 2006). Use of features available in Google Earth can greatly enhance the understanding of historical events such as this. Overlays of topographic, geologic, and historical maps, illustrated placemarks, and tour options were constructed to demonstrate how history can be influenced by geography.

DEVELOPING THE GOOGLE EARTH FILES

To begin this historical characterization, sites of significance to Arnold's expedition through Maine were determined using historical accounts and road logs (Smith, 1903; Clark, 2003; Desjardin, 2006; Nelson, 2006; Lefkowitz, 2008). At these sites, pho-tographs were taken and longitude-latitude coordinates marked using GPS. From the GPS data and photographs, illustrated placemarks were created in Google Earth using the Add Place-mark feature (Fig. 1). Using the computer curser in Google Earth to tentatively locate these sites, each placemark was inserted and its position refined using GPS coordinates. In the dialog box for the placemark, text, photographs, and Internet links, if neces-sary, were added (Fig. 2). Once the text and photographs were inserted, the placemark was saved in Google Earth in My Places. When highlighted and clicked on, an illustrated balloon tied to its longitude-latitude coordinates appeared describing the attributes of the site (Fig. 3).

During the expedition, one of the first obstacles encountered occurred when the sailing ships bringing the troops to Maine had to navigate the Chopps. The Chopps represent a narrowing of the Kennebec River as it passes through an exposure of the Ordovician aged Cape Elizabeth Formation (Fig. 4). This resis-tant rock unit is composed of garnet schists and biotite schists at the southern end of Merrymeeting Bay that has very strong tidal influence on both incoming and outgoing tides (Hussey and Marvinney, 2002). The strong currents at the Chopps exist due to the fact that six rivers empty into Merrymeeting Bay and all exit through the Chopps. Immediately west of the Cape Eliza-beth Formation exposed at the Chopps lies a thrust fault that may be related to the uplift and exposure of this unit (Fig. 4). More

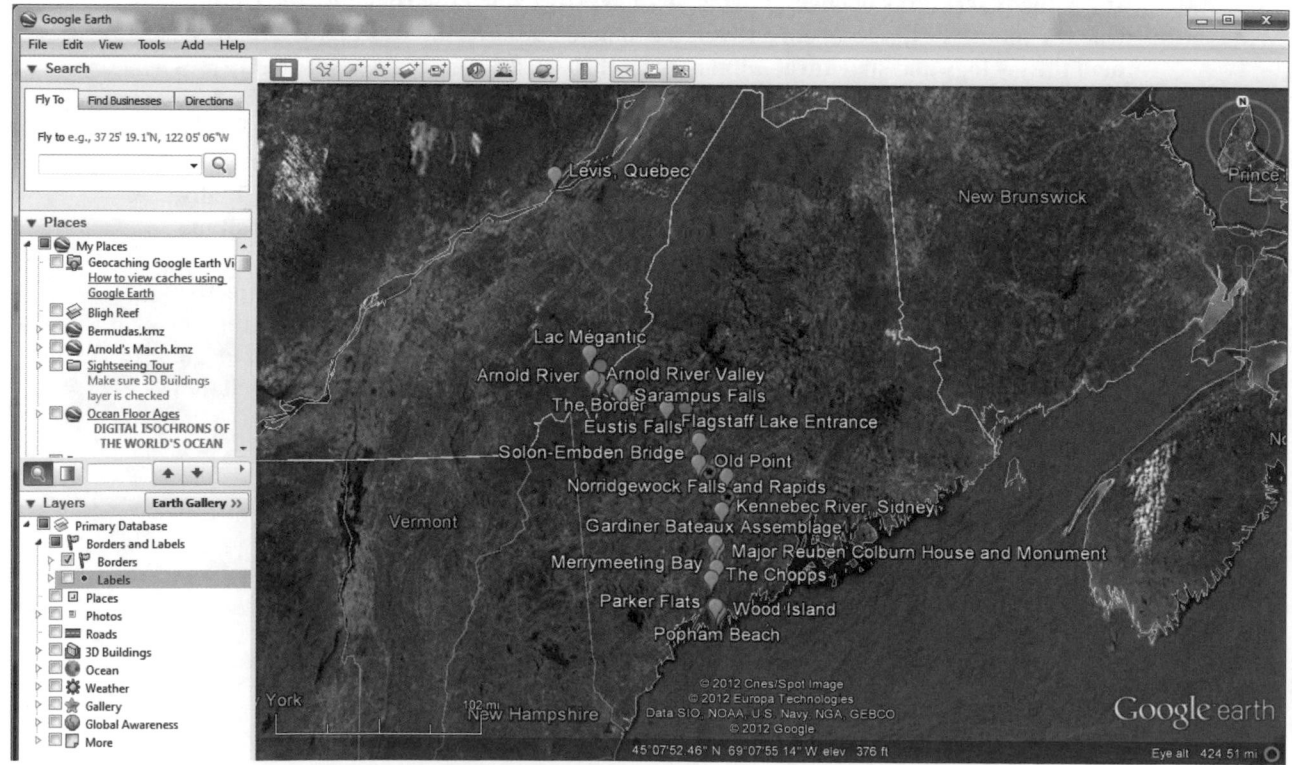

Figure 1. Computer screen capture of a map created using Google Earth to illustrate of the route taken by the Arnold expedition as it traveled through Maine to launch a surprise attack on the fortress of Quebec.

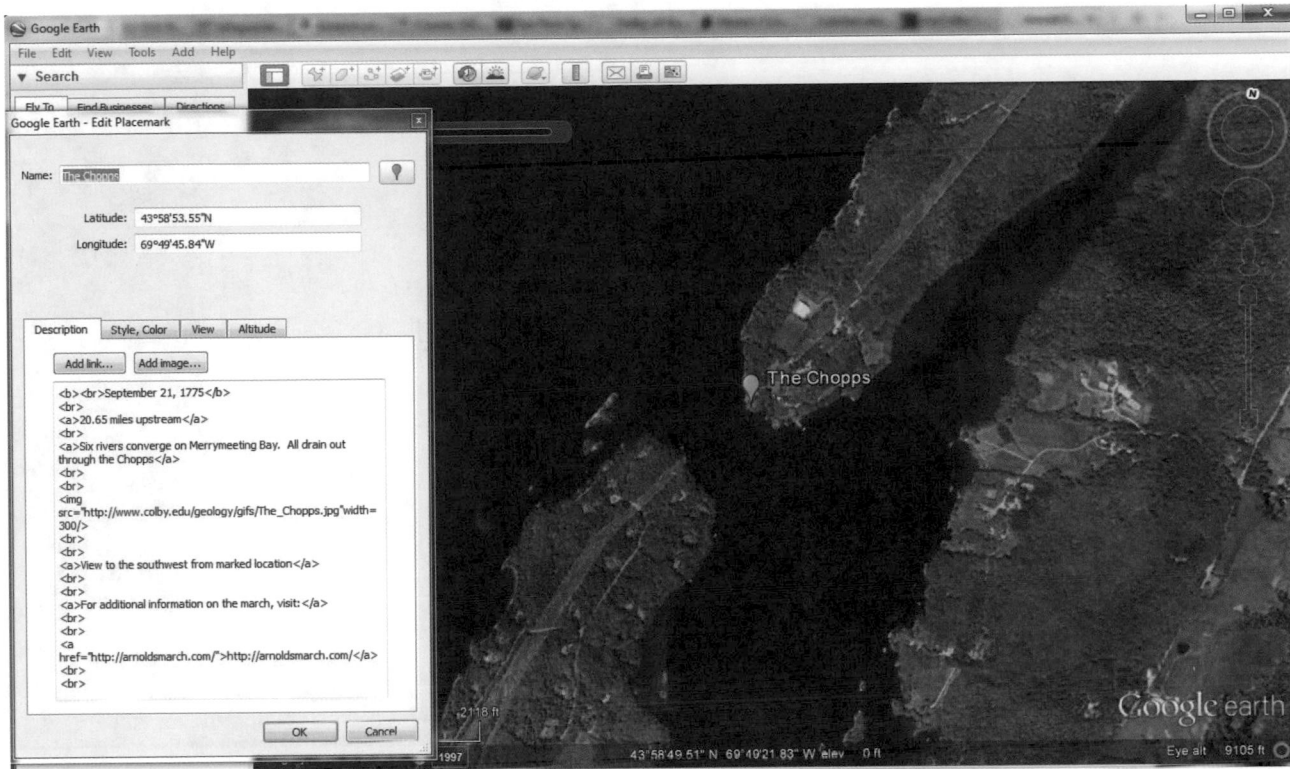

Figure 2. A computer screen capture illustrating the development of a placemark in Google Earth at the Chopps, a narrowing of the Kennebec River causing significant increases and decreases of water velocity as the tides ebb and flow as the discharge of six rivers enters the ocean causing navigational hazard.

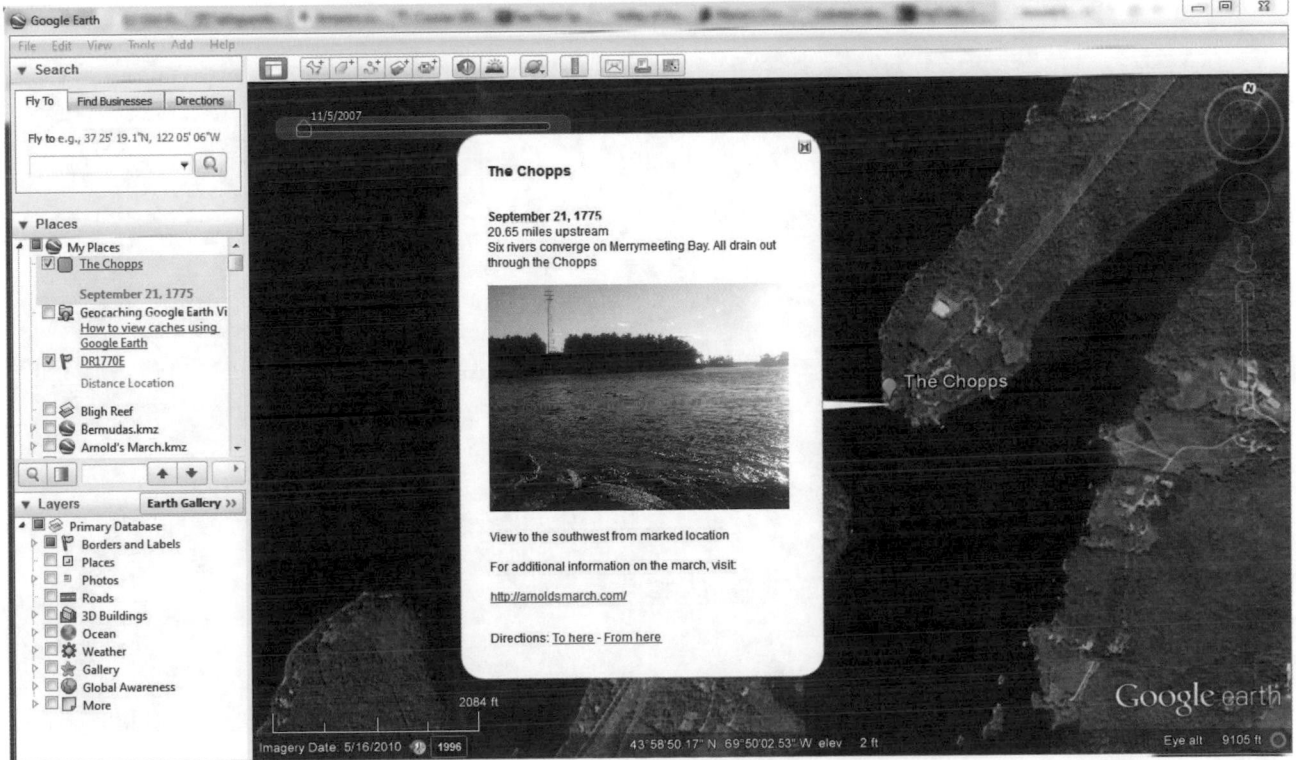

Figure 3. A computer screen capture illustrating the placemark bubble with illustrations and text for the Chopps.

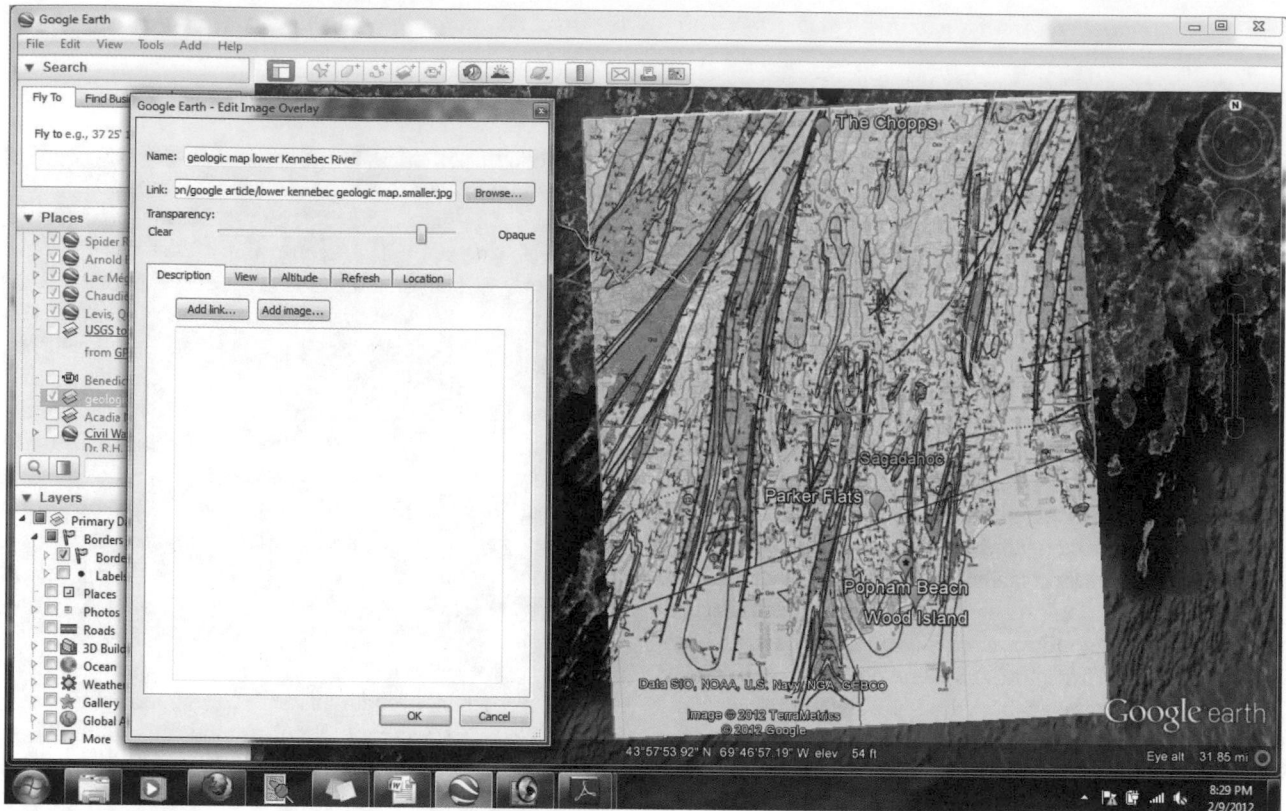

Figure 4. A computer screen capture illustrating the development of a portion of the Geologic Map of the Bath, Maine, 1:100 000 quadrangle featuring the constriction of the river at the Chopps by the Cape Elizabeth Formation and nearby thrust fault (from Hussey and Marvinney, 2002).

Figure 5. Computer screen capture illustrating the overlay of the Geologic Map of the Bath, Maine 1:100 000 quadrangle (from Hussey and Marvinney, 2002).

recently, the advance and retreat of the continental ice sheet that left coastal Maine ~14,000 years ago could have influenced the Kennebec drainage as well. These three attributes may have led to the restricted and extreme variation in flow of the Kennebec as it flows through the Chopps.

To illustrate this relationship, an overlay of the Bath, Maine, 1:100 000 scale Geologic Map (Hussey and Marvinney, 2002) was created by opening a .pdf version of the map (http://www.maine.gov/doc/nrimc/mgs/pubs/online/bedrock/bd-bath100.pdf) and taking a snapshot of the map area needed. This snapshot was opened in a photo-editing program and cropped to the desired size, and the portion to be used in making the overlay was saved as a .jpeg. Using the Add Image Overlay feature in Google Earth, the .jpeg file was imported. Opening that feature, the .jpeg of the map was selected using the browse option and imported into Google Earth. The location of the image was manipulated and refined using movement options around the illustration to make it fit the underlying landscape (Figs. 4 and 5).

Moving northward, the expedition arrived at the confluence of Great Carrying Place Stream and the Kennebec on 11 October 1775. Basing his decision on earlier reconnaissance of the region, Arnold decided to leave the Kennebec and begin an overland portage to the Dead River to avoid the possibility of slack water farther north. The portage over the Great Carrying Place consisted of four separate portages, crossed three ponds and was ~21 km in length (Fig. 6).

The Great Carrying Place portage represented the most challenging section of the journey (Clark, 2003; Desjardin, 2006). The first 1.2 km of the route included a 240 m rise in elevation (Clark, 2003). Over the entire portage, there was an elevational change of 360 m (Fig. 6). To move all gear, watercraft, and troops to the first pond on the route (Fig. 6) required seven to eight round trips from the Kennebec to the pond and took five days to complete. The total distance of the five trips to complete this part of the portage was 67 km. Additionally, the trail had to be widened over the entire route to accommodate the passage of the troops, boats, and gear.

Figure 6 was created using the Add Path feature in Google Earth. The menu option was selected and the desired path over the portage route was traced directly in Google Earth and saved. To produce the topographic profile for the portage route, the cursor was placed directly on the path, and by right-clicking, an option menu appears allowing the Show Elevation Profile feature to be selected, and the profile appears. On the profile, elevational change and distance traveled can be determined by moving the cursor along the trace.

West of the Great Carrying Place portage route, the expedition reached the Dead River (Fig. 7). The Dead River was once an easy flowing, meandering stream (Fig. 7), but that character

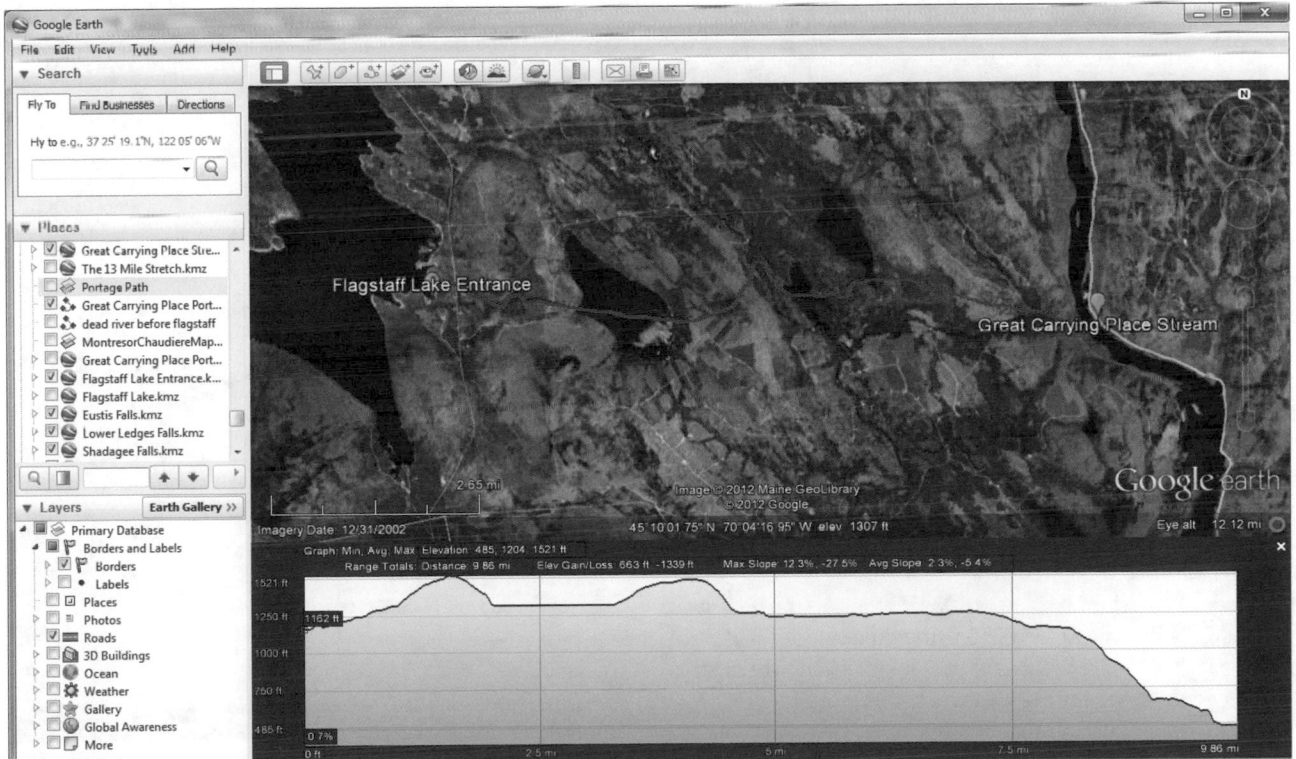

Figure 6. A computer screen capture of the Great Carrying Place portage route going from the Kennebec River in the east to the Dead River in the west. An elevational profile of the route can be seen at the bottom of the figure showing a dramatic rise in the eastern portion.

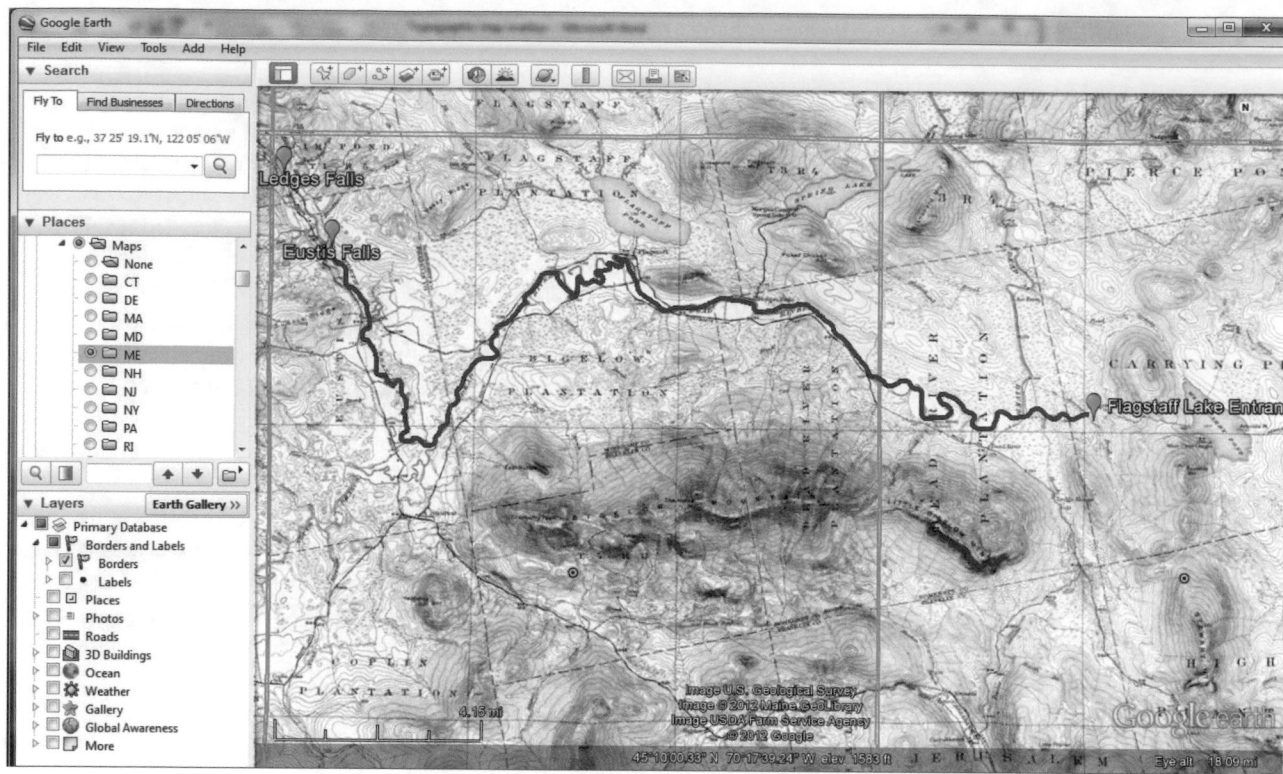

Figure 7. Computer screen capture with overlays of the Stratton, Maine (1932) and Dead River, Maine (1928) 15' topographic quadrangles showing of a portion of the Dead River valley prior to flooding by the construction of a dam in 1950. The heavy line documents the route taken by the Arnold expedition in 1775. Historic topographic maps were obtained as digital overlays from http://www.gelib.com/historic-topographic-maps.htm.

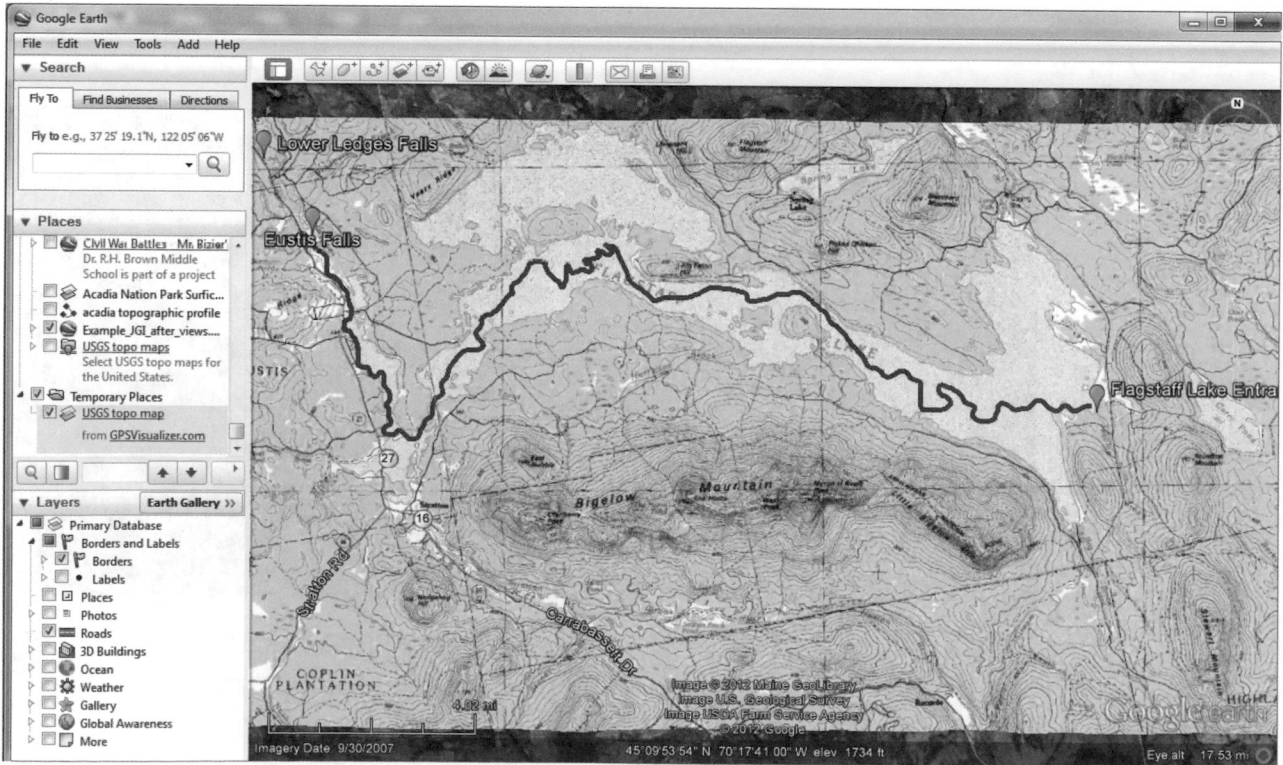

Figure 8. Computer screen capture with topographic overlay and the Arnold expedition route created after the damming of the Dead River in 1950 and consequent creation of Flagstaff Lake. The topographic overlays used here were created by inputting longitude and latitude coordinates on the GPSVisualizer website: http://www.gpsvisualizer.com/kml_overlay.

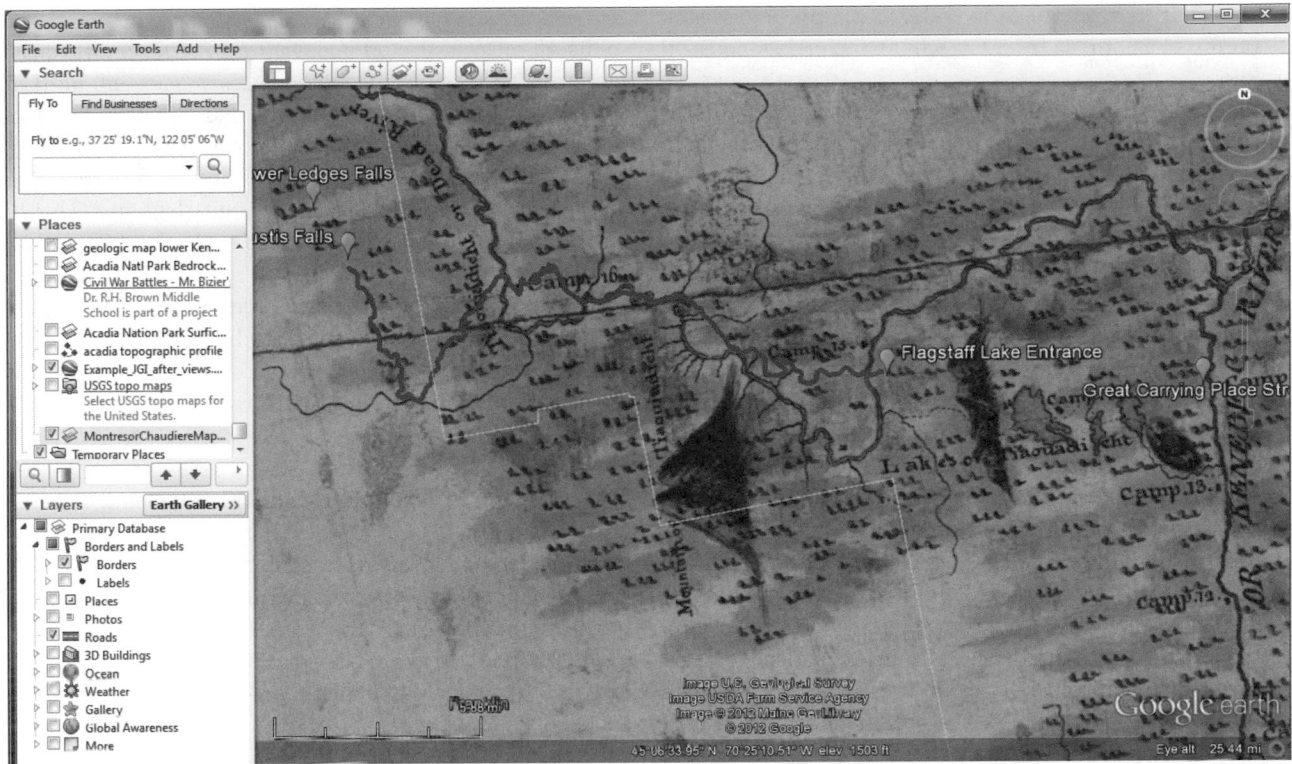

Figure 9. Computer screen capture with an overlay of a map created in 1761 by Lt. John Montressor (Royal Engineers). This map was utilized by Arnold in 1775 to navigate over the Great Carrying Place. Map obtained from http://upload.wikimedia.org/wikipedia/commons/1/1b/MontresorChaudiereMap1760.jpg with an overlay of the Dead River between Eustis Falls in the west and Flagstaff Lake Entrance in the east (solid line).

changed when the river was dammed in 1950 and Flagstaff Lake was created (Fig. 8).

The only readily available map for use by Arnold during the expedition was that produced by Lt. John Montressor of the Royal Engineers in 1761 (Roberts, 1938; Clark, 2003). A digital copy of this map, available from http://upload.wikimedia.org/wikipedia/commons/1/1b/MontresorChaudiereMap1760.jpg, was imported as an overlay to document the changes in mapping since 1761 (Fig. 9).

To create a flyover tour of these placemarks, the Add Tour menu function was selected in the menu bar. When the record bar appeared, the red record button was selected, allowing navigation to each of the placemarks. When a placemark was reached, it could be selected and the associated balloon would appear and the open balloon would be recorded. The balloon can be collapsed and then the tour can move on to the next placemark. This process is continued until all placemarks have been recorded. Stopping the recording and saving the tour results in the tour appearing as a feature in the sidebar. If you want to save the tour to another location, right click on the tour and select Save Place As. Then another location can be selected to store your tour.

OBSERVATIONS AND CONCLUSIONS

This Google Earth compilation has found widespread use on a local scale as it is of extreme interest to historical societies and scholars along the Kennebec River. The greatest utility has been in the classrooms of nearby schools as this type of local history is part of the Maine State Learning Standards. Google Earth files are easily transportable and lend themselves well to classroom and meeting hall venues possessing electricity and Internet connections.

This work has also found usefulness in the college classroom as a learning tool and avenue of communication. Once students understand how to illustrate features on Google Earth and import maps and other media, it is relatively easy for them to develop virtual field trips and presentations suitable for a variety of disciplines and to share information. For example, efforts locally have led to the development of a virtual field guide for the geology of a multi-use open space area that will become available on the Internet for educational and recreational purposes (Morgan, et al., 2011), and a virtual field guide to Civil War monuments in Maine is in preparation by Rueger.

REFERENCES CITED

Clark, S., 2003, Following their footsteps: A travel guide and history of the 1775 secret expedition to capture Quebec: Shapleigh, Maine, Clark Books, 123 p.

Desjardin, T.A., 2006, Through a Howling Wilderness: New York, New York, St. Martin's Press, 240 p.

Hussey, A.M., II, and Marvinney, R.G., 2002, Bath Quadrangle, Maine: Maine Geological Survey Geologic Map No. 02-152, 1:100 000, 1 sheet.

Lefkowitz, A., 2008, Benedict Arnold's Army: The 1775 American invasion of Canada during the Revolutionary War: El Dorado Hills, California, Savas Beatie, 384 p.

Morgan, B.J., Rueger, B.F., and Croft, M., 2011, Virtual geologic field guide to the Kennebec Highlands, Central Maine: Geological Society of America Abstracts with Programs, v. 43, no. 5, p. 476.

Nelson, J.L., 2006, Benedict Arnold's navy: Thomaston, Maine, International Marine/Ragged Mountain Press, 416 p.

Roberts, K., 1938, March to Quebec: Journals of the members of Arnold's expedition: Garden City, New York, Doubleday and Company, Inc., 720 p.

Smith, J.H., 1903 (reprinted 1989), Arnold's March from Cambridge to Quebec: A critical study, together with a reprint of Arnold's journal: Bowie, Maryland, Heritage Books, Inc., 498 p.

MANUSCRIPT ACCEPTED BY THE SOCIETY 16 APRIL 2012

The Geological Society of America
Special Paper 492
2012

Google Earth mashup of the geology in the Presidential Range, New Hampshire: Linking real and virtual field trips for an introductory geology class

J. Dykstra Eusden Jr.*
Department of Geology, Bates College, Lewiston, Maine 04240, USA

Mathieu Duvall*
Bates College Imaging and Computer Center, Bates College, Lewiston, Maine 04240, USA

Marita Bryant*
Department of Geology, Bates College, Lewiston, Maine 04240, USA

ABSTRACT

A Google Earth–based virtual field trip, part of an introductory geology class, has been developed to illustrate the geology of the Presidential Range, New Hampshire. During a class field trip to Mt. Washington, the highest peak in the Northeast, students record GPS locations of exposures and collect information in the form of field notes and digital images from outcrops. Students upload the GPS waypoints into Google Earth and their images into a class PicasaWeb album, and they also make video clips that are uploaded into a class YouTube account. In Google Earth, the students embed and geologically annotate their images and embed their video clips. The final product is a Google Earth .kmz file or what is termed a mashup. The mashup provides a permanent record of the excursion and, if made available on the Internet, allows any user the ability to easily view the geology at any time. Constructing the mashup from the real field trip initiated reflective, independent, student-motivated learning and group work using technology that the students regularly use and enjoy doing. The resulting mashups have been very good, with an appropriate level of geologic content for an introductory course. Grading, which normally is onerous, is actually enjoyable, entertaining, and easy.

*deusden@bates.edu; mduvall@bates.edu; mbryant@bates.edu

Eusden, J.D., Jr., Duvall, M., and Bryant, M., 2012, Google Earth mashup of the geology in the Presidential Range, New Hampshire: Linking real and virtual field trips for an introductory geology class, *in* Whitmeyer, S.J., Bailey, J.E., De Paor, D.G., and Ornduff, T., eds., Google Earth and Virtual Visualizations in Geoscience Education and Research: Geological Society of America Special Paper 492, p. 355–366, doi:10.1130/2012.2492(26). For permission to copy, contact editing@geosociety.org. © 2012 The Geological Society of America. All rights reserved.

INTRODUCTION

Google and its suite of cloud-based applications (Gmail, YouTube, Picasa web, and Google+, to name a few) are widely adopted by computer users everywhere. Many in academia rely on these tools for their scholarship and teaching. For example, for many of us in the geosciences, Google Earth is the tool when we need a quick satellite image of an area. We use it to create a base map for fieldwork, to check out mountain belts for structural features, or to fly up glaciers and coastlines for geomorphology. Although their motivations can be different, students use Google Earth frequently because they find this free service to be fun, easy, and useful. To teach students about a variety of geologic environments and to keep them engaged using tools they enjoy and are at ease using, we have developed a Google Earth–based real and virtual field trip in the Presidential Range of New Hampshire as part of our introductory geology class.

We start with a class field trip to a nearby mountain to collect outcrop information including field notes, GPS waypoints, and digital images from Paleozoic, Mesozoic, and Quaternary geologic settings. Later, in the lab, the students produce a Google Earth .kmz file or what is termed a mashup. Wiki defines a mashup as a "web page or application that uses and combines data, presentation or functionality from two or more sources to create new services." Mashups allow for easy and fast integration of a variety of data sources to produce an enriched end product that students in particular are good at and really enjoy creating.

WHY GOOGLE EARTH?

The benefits of using Google Earth to understand geospatial data and Earth system processes are abundantly clear to all geoscientists. Resources are available to assist geoscientists in their teaching using Google Earth. For example, the website "Teaching with Google Earth" (http://serc.carlcton.edu/NAGTWorkshops/teaching_methods/google_earth/index.html) offers a wide variety of teaching examples using the software. Virtual field trips using Google Earth are now becoming more common in the earth sciences and many online resources exist for those wanting to explore this technology. For example, "Resources for Virtual Fieldwork" (www.artemis-science.com/VFW/) and "The Trail Guide Project" (http://serc.carleton.edu/research_education/trail_guides/index.html), offer resources and examples of virtual field trips. The many papers in this volume further demonstrate the power of using Google Earth applications in geoscience education and research.

By incorporating YouTube videos into our Google Earth mashup we have broadened the functionality of virtual field trips to include animation. Most of the virtual field trips we have seen use static imagery and supporting text. Video clips allow the field trip designers the ability to illustrate geologic phenomena much more effectively by using analog models of processes coupled with simple oral explanations. The trilogy of Google applications—YouTube, PicasaWeb, and Google Earth—all accessed by one Google account, make linking and embedding an array of different types of media into a synergistic application relatively easy.

LINKING THE REAL TO THE VIRTUAL

Students' understanding of any geologic setting or process markedly improves when they see it in the field. Their notes, maps, and sketches allow them to put a field site within the broader context of the material covered in an introductory course. By creating a virtual trip or mashup that is linked to a real trip, students have to go through another important step that deepens their understanding of the geologic setting. They must synthesize, organize, and present the information in a way that others will understand and do so by using technology that their generation has great expertise in. The real value of the mashup exercise is that it is fun for the students to do, they are good at doing it, and their understanding of the geologic processes seen on a field excursion is genuinely enhanced.

GEOLOGY OF THE FIELD AREA

We take the entire class to Mt. Washington in the Presidential Range of northeastern New Hampshire, the highest peak in the northeastern United States at an elevation of 6288 ft (1900 m) (Fig. 1). The region is one of the most popular recreational areas in New England and has a great variety of Paleozoic, Mesozoic, and Quaternary geologic features to observe. During the fall when we run our course, the foliage is spectacular and the weather generally cool, but not snowy. All of these aspects make the trip seem more like a recreational hike to the students than a traditional lab in the urban setting of our campus.

Bedrock Geology

The Presidential Range of northern New Hampshire formed as part of the Acadian Orogeny, an ~50-million-year-long tectonic event resulting from the collision of the Gander and Avalon microcontinents with mainland Laurentia in the Late Silurian and Early Devonian (Hibbard et al., 2006; Bradley et al., 2000; Bradley and Tucker, 2002). During the Acadian Orogeny, marine turbidites deposited in a foreland basin were metamorphosed during a high-T–low-P, Buchan-type regional event, and by granitic intrusions, perhaps triggered by slab delamination (West et al., 2007; Groome et al., 2006; Solar and Brown, 1999; Solar et al., 1998).

The Presidential Range is located on the western flank of the Central Maine Terrane (Lyons et al., 1997), a basin stretching from Connecticut to New Brunswick. It has been mapped at a scale of ~1:3000 (Eusden, 2010; Eusden et al., 1996, 2000) which allows the different lithologic members of the Rangeley and Littleton Formations and the timing and spatial extent of the multiple pulses of Acadian deformation and metamorphism to be determined in great detail. Mesozoic rocks are limited to a few dozen scattered basalt dikes, while structures of that age are characterized by a ubiquitous joint system.

Surficial Geology

The surficial geology of the Presidential Range is dominated by an early phase of Late Wisconsinan alpine glaciation that was followed by continental glaciation. These events are characterized by many cirque basins, a blanket of till deposition, and limited striations. Lack of moraines in most valleys suggests the continental ice down-melted in place. Post-glacial weathering is common in the form of frost shattering and heaving, patterned ground, block fields, block terraces, and modern landslides and

rockfalls. Evidence of pre-Wisconsinan, post-Illinoian weathering exists above 5200 ft elevation in the form of massive block terraces resting above deeply deteriorated soil horizons. New evidence has also been found for post–late Wisconsinan cirque glaciation in Great Gulf in the form of eroded morainal complexes (Fowler, 2010).

THE FIELD TRIP

The all-day weekend field trip (normally on a Sunday to avoid conflict with athletic competitions) takes ~40 students, 2–3 instructors, and 2–4 teaching assistants in vans up the Mt. Washington Auto Road to the summit. There, we briefly visit the

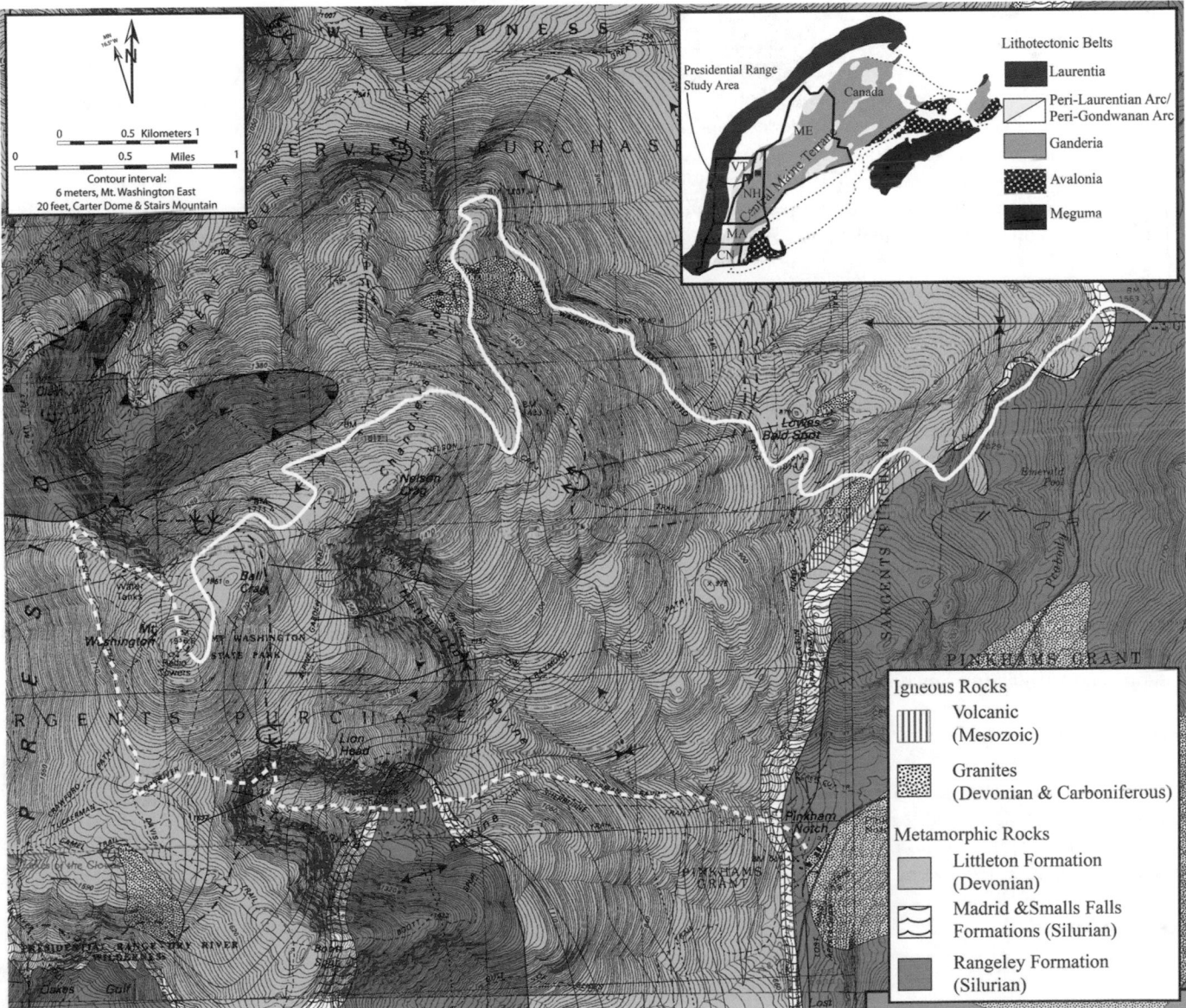

Figure 1. Generalized geologic map of the Presidential Range, New Hampshire (for details, see Eusden, 2010). The solid white line traces the Mt. Washington Auto Road and the dashed white line traces the hiking route for the field trip.

Mt. Washington Observatory, the state park, and other summit displays, discuss the geologic setting of the Presidentials, determine our route down the mountain, and complete a final safety and gear check. The group hikes down to Pinkham Notch, a distance of ~5 miles. We normally hike down the Great Gulf trail to the Gulfside Trail and on to the Westside Trail, across to Tuckerman's Junction and from there down the Tuckerman Ravine trail. (See Figure 1 for the road and trail route.)

On the hike down, the class sees a wide variety of geologic features including: D4 folds in the Devonian Littleton Formation; graded beds that show inverted bedding (Fig. 2A); coarse-grained, metamorphic andalusite in schists that are nicely aligned parallel to the D1 fold hinges; joints, fractures, and basalt dikes (Fig. 2B) in Great Gulf that a senior geology major has studied for thesis research on paleostresses from the Pangea rifting; an upside-down section of the Silurian Smalls Falls and Madrid Formations that controls the distribution of the endangered plant *Potentilla robinsiana*; a glacial cirque (Fig. 2C); glacial striations; post-glacial block fields and patterned ground; D1 isoclinal folds; a basalt dike at Tuckerman Ravine; modern landslide scars and deposits, and finally, a Mesozoic volcanic vent breccia.

Working in groups of 2 or 3, the students record a GPS waypoint at each stop, take their own digital images of outcrops and features, make field sketches, and take notes about the geology.

WORKING ON THE MASHUP

We dedicate one 3-hour, on-campus lab session to the work needed to create the mashup. During this time the student groups organize their data (upload images, downloaded waypoints, collate and condense field notes), research their geologic topics, and write scripts for videos.

Images

We create a single PicasaWeb account for the class into which each photographer places his or her pictures. Each photographer uses a separate photo album, and each album is made available to the entire class. Each student group is responsible for selecting and annotating its own images.

Waypoints and Tracks

Each group is responsible for collecting its own GPS waypoints. The groups download the GPS waypoints they collected on the trip into Google Earth on an overlay of the bedrock geologic map (Fig. 2D).

Research Materials

We provide the class with readings and geologic maps for the project. We require that they read a minimum of one paper or map on the bedrock history and another on the surficial geology (e.g., Eusden, 2010; Fowler, 2010).

Videos

Each group also makes a minimum of four 30-second videos, using their own or a borrowed camera, that describes in their words, and with props, aspects of the geology. Content for the videos comes from their field trip notes and the maps and papers they have read about the bedrock and surficial geology. The students choose four video topics from this list: (1) the overall Paleozoic geologic and tectonic setting of the Presidential Range; (2) a description of the Acadian orogeny folding in the Presidential Range; (3) a description of Acadian Paleozoic metamorphism, and of andalusite in particular, in the Presidential Range; (4) a history of Mesozoic basalt dikes and fractures in the Presidential Range; (5) a description of the Quaternary glacial and post-glacial history of the Presidentials; and (6) modern landslides in the Presidentials. Students are told to rehearse the videos before taking the final footage and to use simple animations, maps, blackboard drawings, and/or rock samples in the videos. Teaching assistants are available in the evenings to help groups with content and video production. When complete, the video clips are uploaded to the class YouTube website.

Mashing It Up

For each waypoint the students embed a digital image or a video clip, or both, in Google Earth that characterizes the geology at that waypoint. All the waypoints are accompanied by text, geared for a scientifically literate person of their ability, that describes the geology seen at that waypoint and/or of the image, and the images are annotated to point out salient geologic features. Drafts of the student's text and preliminary annotations are reviewed by the instructors and teaching assistants during the lab time block.

THE FINAL PRODUCT

Students ultimately produce a mashup as a .kmz file in Google Earth and upload this to our college-wide, web-based learning management system. The files are conveniently small in size, 10–20 KB, as all the image and video memory is in the Internet cloud. In the class following the assignment due date, each student group gives a short (5–10 minute) oral presentation and demonstration of their mashup. These make for a fun and engaging hour for both the class and instructors. Examples of

Figure 2. (A) An exposure of the Devonian Littleton Formation along the Tuckerman Ravine Trail showing a graded quartzite and andalusite schist couplet with an inverted topping direction to the lower right. (B) Mesozoic basalt dike intruding the country rock above the headwall on the Tuckerman Ravine trail. (C) View from the Gulfside trial looking east of Great Gulf, an early Wisconsinan glacial cirque and valley. (D) GPS waypoints from the field trip plotted on a geologic map-overlay in Google Earth (2× vertical exaggeration).

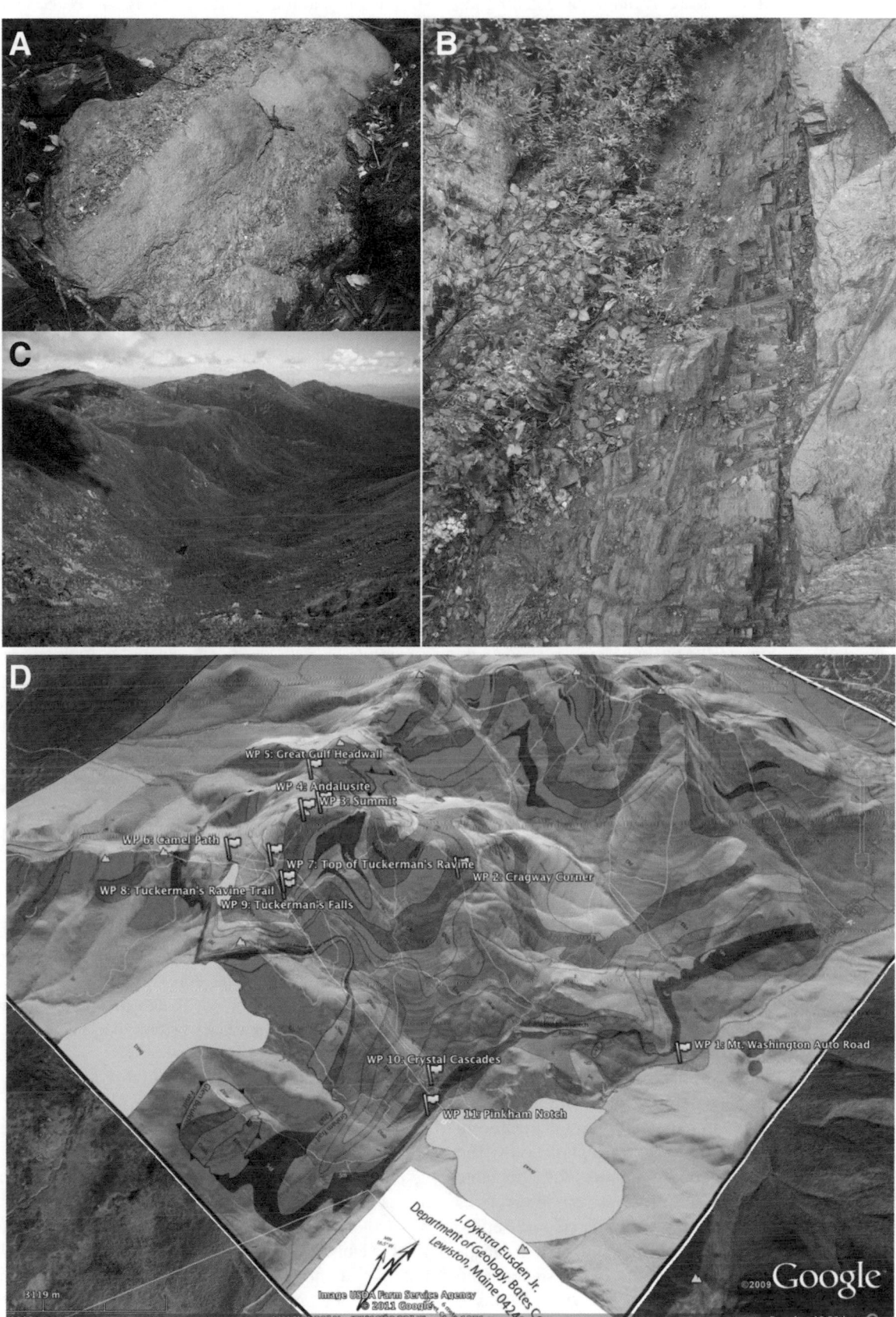

finished student mashup .kmz files are available by emailing the first author (deusden@bates.edu).

EVALUATION OF THE MASHUP

To evaluate the mashup we download the .kmz files and open them in Google Earth. As we view the virtual field trip mashup together, the instructors use a grading rubric for evaluation. The rubric includes the following categories:

5%—Do you have the correct number of stops and video clips in your mashup?

20%—What is the level of content in the video? (A = excellent use of the readings/maps and field-trip notes; B = good use of the readings/maps and field- trip notes; C = adequate use of the readings/maps and field-trip notes)

20%—How well choreographed and designed is the video? (A = excellent choreography, design, sound track, and use of props, B = good, C = adequate)

25%—What is the level of content in the stop descriptions and image captions? (A = excellent descriptions and captions; B = good; C = adequate)

25%—What is the quality of Google Earth markup in the mashup? (A = lots of bells and whistles, B = good stuff, C = minimum)

5%—Syntax and grammar

The resulting mashups have been very good, with an appropriate level of geologic content for an introductory course and showing a great deal of high quality, independent, student-motivated learning and group work. Grading, which normally is onerous, is actually enjoyable, entertaining, and easy. None of us instructors can ever remember laughing so much (at the videos) and having so much fun while grading.

SUCCESSES

Overall, the students are able to easily navigate through the technology, they learn a great deal about the local geology, and they successfully work together in small groups. The videos demonstrate the most independent creativity in that students clearly have practiced their skits, making for well-rehearsed and well-composed videos. They make use of simple props like cardboard and clay or running shoes and backpacks to show plate collisions, dorm hallway rugs to demonstrate folding, blackboard work to draw diagrams of P-T paths or basalt and joint formation, and even a PowerPoint lecture on some aspect of the geology that they have filmed (Fig. 3). Some do the filming in their dorm rooms and some outdoors, while others work in classrooms. The texts describing the images of geologic features are also quite good and range from detailed, descriptive outcrop observations to more general comments about the overall geologic setting (Fig. 4). The assigned background readings and geologic maps are clearly used effectively by every group. Most students are very comfortable in using the basic Google Earth editing tools to create placemarks, to determine font type, font size, and image

size, and to add the embed tags for the YouTube video clips and PicasaWeb images at each waypoint.

SHORTCOMINGS

Generally, the annotations of the images have been poor as many students are unable to annotate an image once it has been uploaded to PicasaWeb. As a result they do not label features on most images. For certain groups the audio quality of the videos has been poor and nearly inaudible; therefore, make sure that students speak into the microphone when recording their video clips. If students submit a .kmz file with extraneous placemarks that are not part of the field trip, Google Earth flies to seemingly random areas on the globe, making zooming in to the field area somewhat of a nuisance. Occasionally, students do not efficiently save their .kmz files using the "File>Save>Save Place As" command in Google Earth. As a result, many older versions of their mashups are included in their .kmz file, resulting in numerous redundant placemarks, or, more rarely, missing placemarks. Either way, this creates lots of stress within certain groups, and it makes it difficult for the instructors to know which placemark belongs in the final edition.

THE "HOW TO..." SECTION

One benefit of using the Google tools is that they are very accessible and very well documented. The tools are also extremely "friendly" for this sort of work. For example, both YouTube and PicasaWeb provide html code that the user can simply copy and paste into a Google Earth placemark balloon. Below we briefly outline the steps necessary to do this lab, but leave the heavy explanations to the web services themselves.

GPS Waypoints

In Google Earth, use the "Tools>GPS file" menu, and select the GPS device from which you are downloading (e.g., Garmin, Magellan, etc.). Google Earth will download your waypoints (and tracks if you want), and display them on the page. NOTE: If you have "other" locations stored in your GPS device, they will be included in this download. To avoid some of the locational pitfalls, clear your GPS memory prior to beginning your field work.

PicasaWeb

To upload material to PicasaWeb and YouTube requires a Google account. We created one for the entire class to use. If you do not have a Google account, create one first by going to http://picasaweb.google.com.

Upload your images to PicasaWeb then select the image you want to embed. Click on the photo, then click on the "Link to this Photo" text and you should see both the "Link" location as well as the "Embed image" address. This is a good time to select the size of the image you would like. The options are: thumbnail, small, medium, and large. Now, select and copy the embedded address and open up Google Earth (Fig. 5).

YouTube

If you do not have a YouTube account, create one by going to http://www.youtube.com. Upload your videos to your YouTube account. Select the video you want to embed in Google Earth, click on the "Share" button, and then the "Embed" button so that

the embed address appears. Select and copy the embed address and open Google Earth (Fig. 6).

Accessing Placemark Description "Balloons" in Google Earth

Right click or "ctrl" click the placemark where you want to embed the image or video or to add text. In the menu that pops up, select the "Get Info" (or "Properties" in some versions)

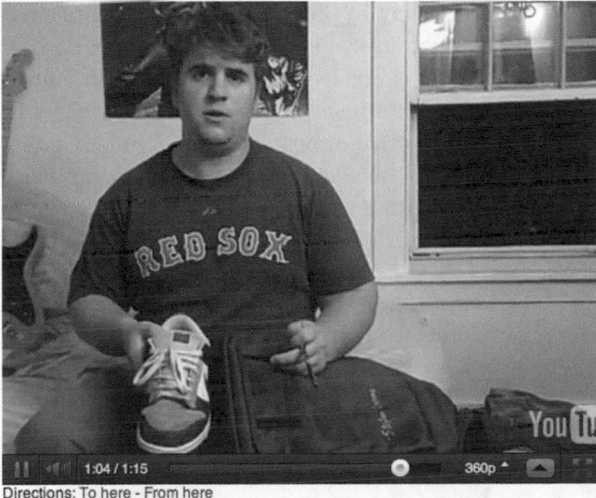

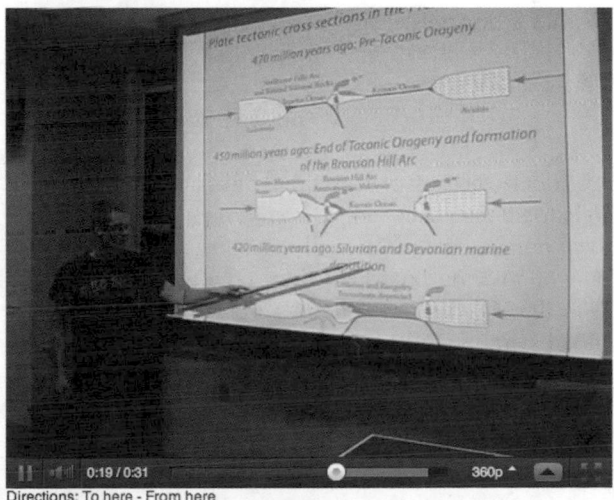

Figure 3. Examples of embedded YouTube videos. (A) Using simple props to explain the collision between Laurentia and Avalon and the deformation of the marine sediments in the Iapetus Ocean (green clay balls). (B) Using a conventional blackboard and simple props to explain the D1 phase of folding in the Presidential Range. (C) The classic sneaker and backpack subduction zone model of the Acadian Orogeny! (D) A student using a PowerPoint to illustrate the plate tectonic history of the Appalachians.

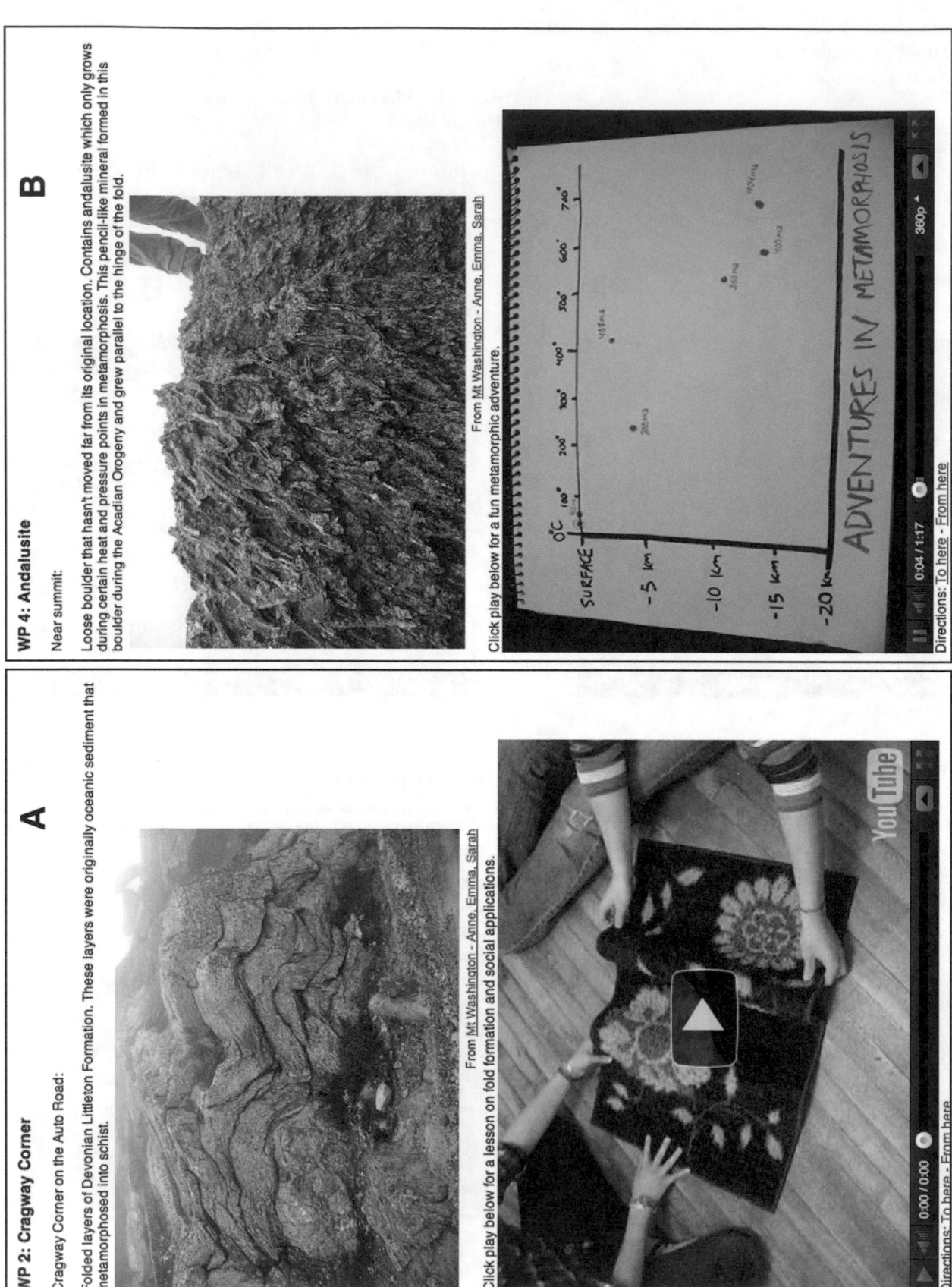

Figure 4. Examples of waypoints with both embedded images and videos. (A) D4 folding at Cragway Corner of the Auto Road with video below using a dorm carpet as an analog model. (B) Coarse-grained, aligned andalusite in Littleton Formation schists near the summit of Mt. Washington with video of the P-T path being completed in a PT diagram.

Figure 5. Image showing glacial striations near Tuckerman's Junction above the headwall as viewed in PicasaWeb. The embed address is highlighted by the red oval.

option. An "Edit Placemark" box appears where you may now change the name and style of the placemark. While still in the "Edit Placemark" box, select the "Description" field. Any text or image/video embed codes pasted in this field will show in the placemark's description Balloon (Fig. 7).

Formatting Text in Description Balloons

Google Earth placemark descriptions render text using HTML markup codes. It is not necessary that these be complex. For the purpose of this project, we outlined nine markup codes to allow for basic text formatting. These codes are the six text headings as a substitute for font size (<h1>really big text</h1> … <h6>really small text</h6>), the paragraph delimiter (<p>much text as a paragraph</p><p>followed by another paragraph</p>), a basic line break (<br />), and a tag that allows for the center alignment of content (<center>this content would be centered</center>).

SUMMARY

We will definitely incorporate more Google Earth mashup projects in the future as it has been largely successful for all involved. Student feedback on course evaluations has also been very positive about this experience. Linking the real and the

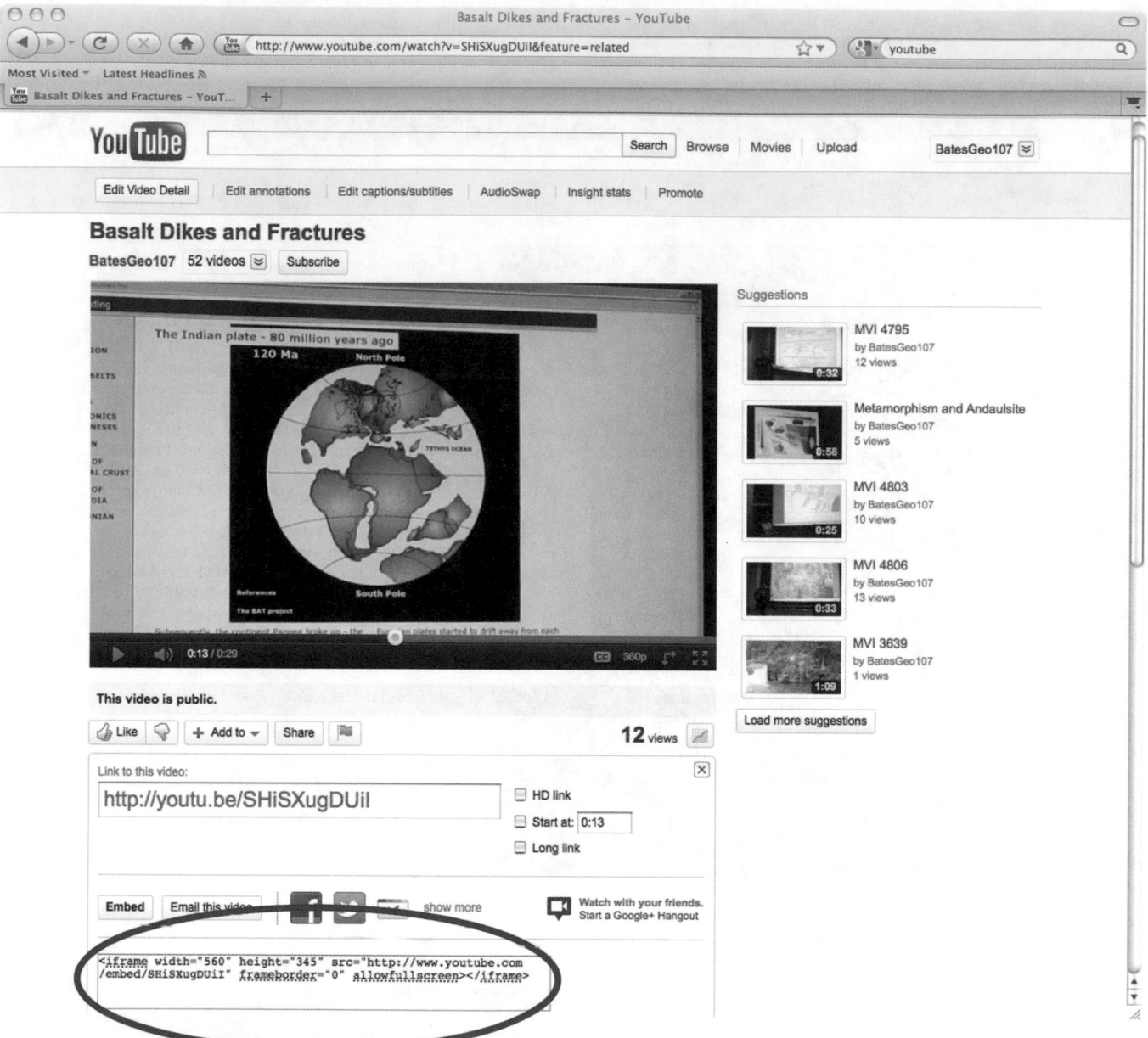

Figure 6. YouTube video of plate tectonics filmed from http://ansatte.uit.no/kku000/webgeology/. The embedded address is highlighted by the red oval.

virtual field trip in Google Earth makes students reflect upon and develop a deeper understanding of the geology. They essentially become virtual field trip leaders with all the responsibilities that go with it. The cooperation within the groups is probably some of the best we have seen, with an equitable mix of effort spent by each individual. Some students are better at writing text, others at designing videos, and others at Google Earth programming. The variety of Google products used—YouTube, PicasaWeb, and Google Earth—are very easy for the students to learn and operate.

ACKNOWLEDGMENTS

Thanks very much to the editors of this special paper on Google Earth—Steve Whitmeyer, John Bailey, Declan De Paor, and Tina Ornduff—great idea, folks! Many thanks to Dr. Peter J. Thompson for his thoughtful review of the manuscript. Thanks also to the anonymous reviewer; your comments were quite helpful to us. Most of all, thanks to the Bates College Geology 107 class, "Katahdin to Acadia: Field Geology in Maine," for learning with us how to make Google Earth mashups.

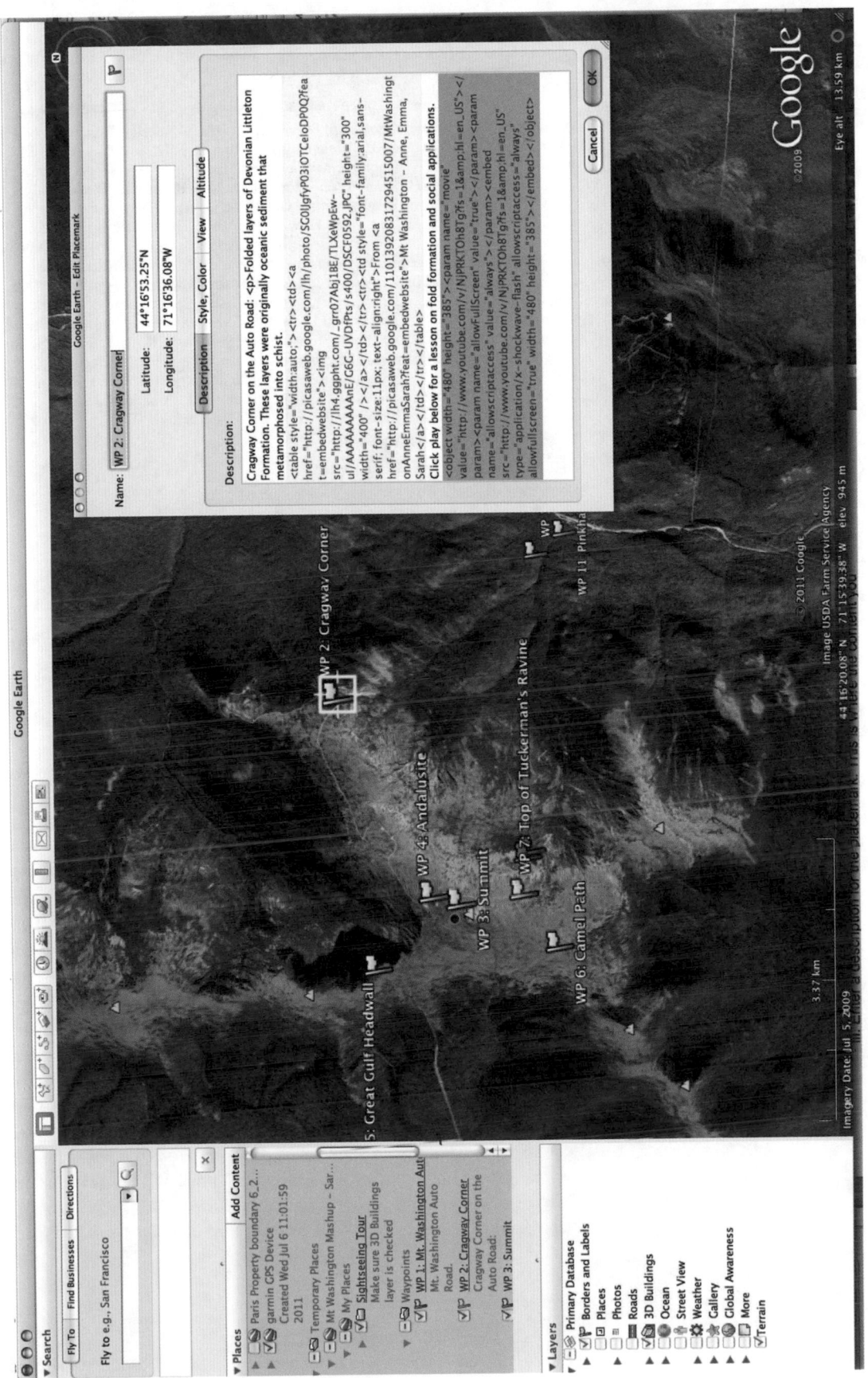

Figure 7. The "Edit Placemark" window for waypoint WP2: Cragway Corner showing the embed address for a PicasaWeb image highlighted in yellow and the embed address of a YouTube Video highlighted in green. Other text that is not highlighted appears as captions when the placemark is clicked or opened.

REFERENCES CITED

Bradley, D., and Tucker, R.D., 2002, Emsian Synorogenic Paleogeography of the Maine Appalachians: The Journal of Geology, v. 110, p. 483–492, doi:10.1086/340634.

Bradley, D.C., Tucker, R.D., Lux, D.R., Harris, A.G., and McGregor, C.C., 2000, Migration of the Acadian Orogen and Foreland Basin across the Northern Appalachians: U.S. Geological Survey, Professional Paper 1624, 49 p.

Eusden, J.D., 2010, The Presidential Range: Its Geologic History and Plate Tectonics: Lyme, New Hampshire, Durand Press, 62 p. and 1:20,000 scale bedrock map.

Eusden, J.D., Jr., Garesche, J.M., Johnson, A.H., Maconochie, J., Peters, S.P., O'Brien, J.B., and Widmann, B.L., 1996, Stratigraphy and ductile structure of the Presidential Range, New Hampshire: Tectonic implications for the Acadian orogeny: Geological Society of America Bulletin, v. 108, p. 417–436, 2 inserts, doi: 10.1130/0016-7606(1996)108<0417:SADSOT>2.3.CO;2.

Eusden, J.D., Jr., Guzofski, C.A., Robinson, A.C., and Tucker, R.D., 2000, Timing of the Acadian Orogeny in Northern New Hampshire: The Journal of Geology, v. 108, p. 219–232, doi:10.1086/314396.

Fowler, B.K., 2010, Surficial Geology of Mount Washington and The Presidential Range, New Hampshire: Lyme, New Hampshire, Durand Press, 1:24,000 surficial map.

Groome, W.G., Johnson, S.E., and Koons, P.O., 2006, The effects of porphyroblast growth on the effective viscosity of metapelitic rocks; implications for the strength of the middle crust: Journal of Metamorphic Geology, v. 24, p. 389–407, doi:10.1111/j.1525-1314.2006.00644.x.

Hibbard, J.P., van Staal, C.R., Rankin, D.W., and Williams, H., 2006, Lithotectonic Map of the Appalachian Orogen, Canada–United States of America: Geological Survey of Canada "A" Series, Report 2906A, 2 sheets.

Lyons, J.B., Bothner, W.A., Moench, R.H., and Thompson, J.B., 1997, Bedrock Geologic Map of New Hampshire: U.S. Department of Energy and the State of New Hampshire, New Hampshire Geological Survey, Eugene L. Boudette, State Geologist, Concord, New Hampshire.

Solar, G.S., and Brown, M., 1999, The classic high-T–low-P metamorphism of west-central Maine: Is it post tectonic or syntectonic? Evidence from porphyroblast-matrix relations: Canadian Mineralogist, v. 37, p. 311–333.

Solar, G.S., Pressley, R.A., Brown, M., and Tucker, R.D., 1998, Granite ascent in convergent orogenic belts: Testing a model: Geology, v. 26, p. 711–714, doi:10.1130/0091-7613(1998)026<0711:GAICOB>2.3.CO;2.

West, D.P., Jr., Tomascak, P.B., Coish, R.A., Yates, M.G., and Reilly, M.J., 2007, Petrogenesis of the Ultrapotassic Lincoln Syenite, Maine: Late Silurian–Early Devonian Melting of a Source Region Modified by Subduction Driven Metasomatism: American Journal of Science, v. 307, p. 265–310, doi:10.2475/01.2007.08.

MANUSCRIPT ACCEPTED BY THE SOCIETY 16 APRIL 2012

The Geological Society of America
Special Paper 492
2012

Google Venus

Declan G. De Paor*
Department of Physics, Old Dominion University, Norfolk, Virginia 23529, USA

Vicki L. Hansen
Department of Geological Sciences, University of Minnesota Duluth, Duluth, Minnesota 55812, USA

Mladen M. Dordevic
Department of Physics, Old Dominion University, Norfolk, Virginia 23529, USA

ABSTRACT

Google Earth includes digital elevation models and surface imagery for the Earth, Moon, and Mars, but not for Venus. To help geoscientists visualize Venusian geology, geophysics, and geodynamics, we have built a "Google Venus" virtual globe on a Google Earth foundation. We present here details of how this was done and offer regional samples to show the power of the virtual globe, combined with space mission imagery, and COLLADA models in displaying surface data and global, crust-to-core cross sections. We show how web data sources can be linked to Venusian locations in an engaging, interactive format. Our approach could be adapted to other planets and moons of the Solar System and to models of exoplanets.

INTRODUCTION

Anecdotal evidence from conversations with students and colleagues suggests that, of Earth's two planetary neighbors, Venus is the less well known and understood by geoscientists. This may reflect Venus's shroud of cloud, the relative paucity of space missions (especially landings), a smaller cohort of researchers, and a lack of earthly analogues for many of Venus's surficial and tectonic structures. Despite the fact that Venus is almost as large as Earth whereas Mars is closer to the size of Earth's core, nevertheless earth-like surface features such as river channels, sand dunes, etc., are ubiquitous on Mars but completely absent from Venus. This presents a challenge both to the educator and the presenter of research results. In fact, there is a huge repository of exciting data on Venus from National Aeronautics and Space Administration (NASA) and European Space Agency (ESA) orbital missions but it is accessed mainly by dedicated researchers.

To help rectify this situation, we have created a Google Venus virtual globe built on a Google Earth foundation with extensive use of COLLADA models (http://collada.org). Our model includes atmospheric imagery, emergent blocks, and cross sections of the planet's crust and mantle (cf. De Paor and Pinan-Llamas, 2006) that illustrate current interpretations of its internal structure. We showcase some of Venus's unique tectonic structures, notably the Ishtar region and the Artemis super-plume structure. Google Venus effectively conveys the benefits of virtual globe style touring and zooming in helping geoscientists to truly understand planetary structures on a range of scales.

---
*ddepaor@odu.edu

De Paor, D.G., Hansen, V.L., and Dordevic, M.M., 2012, Google Venus, *in* Whitmeyer, S.J., Bailey, J.E., De Paor, D.G., and Ornduff, T., eds., Google Earth and Virtual Visualizations in Geoscience Education and Research: Geological Society of America Special Paper 492, p. 367–382, doi:10.1130/2012.2492(27). For permission to copy, contact editing@geosociety.org. © 2012 The Geological Society of America. All rights reserved.

PLANET VENUS

Orbiting at 0.72 AU (astronomical units) from the Sun, Venus is 95% the diameter and 81.5% the mass of the Earth. Its slow retrograde rotation makes the Venusian day longer than its year (243 versus 225 Earth days). There are no seasons on Venus; its rotation axis is less than 3° off vertical, and its orbit is even less eccentric than Earth's; consequently, solar insolation is everywhere virtually constant year-round.

Venus differs from Earth in that it lacks a magnetic field, and its upper atmosphere super-rotates at an estimated 300 km/h. Atmospheric composition (96% CO_2), surface pressure (~95 bars), and surface temperature (up to 475 °C) vary radically from equivalent values on Earth. The thick insulating atmosphere results in negligible diurnal temperature variations and an enhanced global greenhouse effect, making terrestrial-style surface processes impossible. Venus thus lacks the weathering, erosion, and transportation that lead to sedimentary deposits on Earth (Phillips et al., 2001).

Venus also lacks any evidence of plate tectonics or of regions analogous to Earth's continents (Solomon et al., 1991; Phillips and Hansen, 1994, 1998). Clearly, the planet cools through mechanisms other than seafloor spreading and subduction. Geomorphological and geochemical arguments, along with data from Soviet landers, are consistent with dry basaltic crust (Grimm and Hess, 1997). The crust is stronger than Earth's basaltic rocks despite temperatures that would promote plastic deformation in wet conditions (Mackwell et al., 1998). A wide variety of volcanic landforms occur (Crumpler et al., 1997; López, 2011) but there are no arcuate volcanic arcs comparable to the products of plate-tectonics on Earth. Long, low-viscosity, basaltic lava flows are common (Bridges, 1995, 1997; Stofan et al., 2000). Volcanic shields, 1–20 km in diameter, occur in shield fields and as shield terrains distributed across millions of square kilometers.

Although Venus is ultra-dry today, water may have played a role in the past. Isotopic data are consistent with extensive water reservoirs more than 1 b.y. ago that have since been lost to space (Donahue and Russell, 1997; Donahue et al., 1997; Hunten, 2002).

Venus and Earth are described as sister planets because they share planetary scale attributes such as size and composition despite very different surface processes. In the absence of weathering, surface features record past processes on Venus that hold clues to its evolution, and might also provide clues to early Earth history. Clearly this planet merits study with all the visualization tools available today, including virtual globes.

GOOGLE VENUS

Version 6.2 of Google Earth includes three virtual globes with digital elevation models (DEMs) and surface imagery (also known as primary databases) for the Earth, Mars, and Moon. In the desktop application, alternate globes (as well as the Google Sky option) are accessed via the "View >> explore…" menu item or the "Switch between Earth, Sky, and other planets" icon on the

TABLE 1. JAVASCRIPT CODE OPTIONS FOR SELECTING AMONG PLANETARY DATABASES IN THE API

```
google.earth.createInstance('map3d', initCB, failureCB,
{ database: 'http://khmdb.google.com/?db=earth' });

google.earth.createInstance('map3d', initCB, failureCB,
{ database: 'http://khmdb.google.com/?db=mars' });

google.earth.createInstance('map3d', initCB, failureCB,
{ database: 'http://khmdb.google.com/?db=moon' });
```

Note: One of these options is added to the Javascript initialization function.

toolbar. When an instance of Google Earth is embedded in a web page using the Google Earth plugin and its JavaScript application programming interface (API), the planet selection is made by the person designing the web page by specifying the primary database in the JavaScript initialization function and it cannot be changed thereafter (Table 1)

In personal communications, Google engineers contrast Venus with the Moon and Mars in terms of missions and especially landings, with their human-interest stories and toy-like, youth-attracting rovers, and they convey little likelihood of a Google Venus option being built into future versions of Google Earth, therefore if we are to explore Venus as a virtual globe, we must use Google Earth with the atmosphere and 3-D terrain switched off, hide the primary database, and superimpose images and models to represent features of the Venusian atmosphere, surface, and interior (Fig. 1).

THE VENUSIAN ATMOSPHERE

Viewed through an earthbound telescope, or even from an orbiting spacecraft, Venus is permanently shrouded in cloud. Its dense atmosphere is opaque to visible light between 30 and 60 km altitude. Sulfuric acid droplets produced in the cloud tops reflect 70% of incoming visible light, making Venus the third brightest object in the sky after the Sun and Moon. Curved beige cloud bands, accentuated in ultraviolet photography, are the only discernible features outside of the polar regions.

Atmospheric Super-Rotation

We present the atmosphere using a ground overlay image elevated 45 km and covering 90° to −90° latitude and 180° to −180° longitude (Table 2; to hide the underlying Earth terrain, we cover the ground with a black image). In this image, the effect of atmospheric super-rotation (Durand-Manterola, 2010) is seen in the curvature of cloud bands indicating differential rotation at the equator (a day on Venus is 243 Earth days long, but the atmosphere rotates in only 4 Earth days). The Keyhole Markup Language (KML) file linking to the atmospheric image (venus_atmos.kml) is available in the GSA Data Repository[1].

---

[1]GSA Data Repository item 2012305, KML files, is available at http://www.geosociety.org/pubs/ft2012.htm, or on request from editing@geosociety.org.

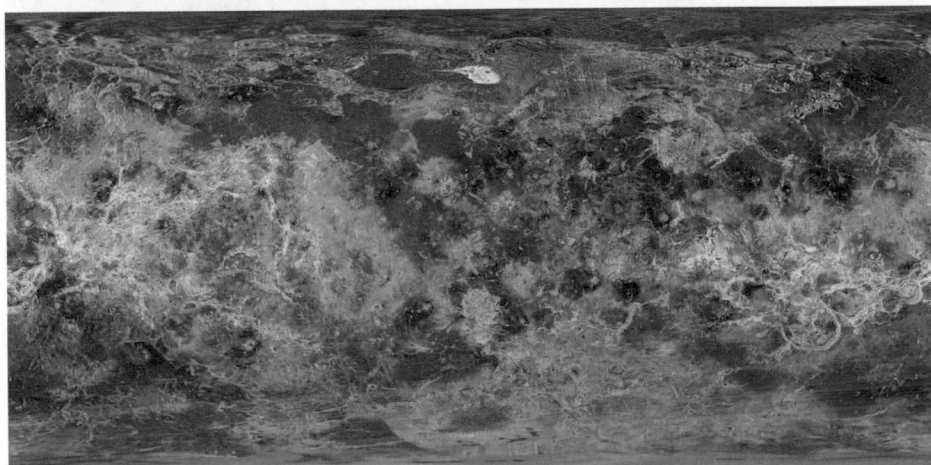

Figure 1. Google Venus created by draping visible light cloud tops and Magellan radar altimetry over the Google Earth surface imagery. Mercator, Miller, or Plate Carrée (Simple Cylindrical) projections work best.

TABLE 2. CODE SNIPPET FROM
http://www.digitalplanet.org/gv/venus_atmos.kml.

```
GroundOverlay>
        <name>atmosphere</name>
        <Icon>
                <href>images/venus_atmos.jpg</href>
        </Icon>
        <altitude>25000</altitude>
        <altitudeMode>absolute</altitudeMode>
        <LatLonBox>
                <north>90</north>
                <south>-90</south>
                <east>180</east>
                <west>-180</west>
        </LatLonBox>
    </GroundOverlay>
    <GroundOverlay>
        <Icon>
                <href>images/black.tiff</href>
        </Icon>
        <altitude>0</altitude>
        <altitudeMode>clampedToGround</altitudeMode>
        <LatLonBox>
                <north>90</north>
                <south>-90</south>
                <east>180</east>
                <west>-180</west>
        </LatLonBox>
    </GroundOverlay>
```

Note: The black ground overlay conceals Earth's surface imagery. The image file "venus_atmos.jpg" is wrapped around the entire planet at an elevation of 25 km (25,000 m).

Atmospheric Polar Vortices

NASA and ESA orbital missions have provided many images of atmospheric vortices at Venus's poles (e.g., http://www.windows2universe.org/venus/ venus_polar_atmosphere. html). However, it is not possible to drape these images as ground overlays because ground overlays cannot enclose the poles. This problem could be solved in the future were Google Earth programmers to adopt the Universal Transverse Mercator (UTM) projection system which switches to stereographic projection in the polar regions.[2] Meanwhile, we must use COLLADA models (Tables 3A and 3B) to display polar vortex images as illustrated in Figure 2. Double vortices are discussed in Lebonnois et al. (2006) and can be explored in the KML file "venus_vortices.kml" in the GSA Data Repository (see footnote 1).

[2]There is an option to output UTM coordinates in the Google Earth preferences panel, but the green ground overlay placement handles that appear with the "add image overlay" command still converge to a singularity at the poles.

370 *De Paor et al.*

TABLE 3A: NORTH POLAR VORTICES ARE DISPLAYED USING THE nPole.dae COLLADA MODEL

```
<Placemark>
        <name>Double vortex at North Pole</name>
        <visibility>1</visibility>
        <description><![CDATA[
            More...
            <br> <br>
            http://www.youtube.com/watch?v=WtxqkmIvikU
]]></description>
        <Model>
            <altitudeMode>absolute</altitudeMode>
            <Location>
                <longitude>0</longitude>
                <latitude>90</latitude>
                <altitude>-230000</altitude>
            </Location>
            <Scale>
                <x>0.5</x>
                <y>0.5</y>
                <z>0.15</z>
            </Scale>
            <Link>
                <href>models/nPole.dae</href>
            </Link>
        </Model>
    </Placemark>
```

Note: The scale element was adjusted iteratively to fit the image used.

TABLE 3B. SOUTH POLAR VORTICES ARE DISPLAYED USING THE sPole.dae COLLADA MODEL

```
<Placemark>
    <name>Double vortex at South Pole</name>
    <visibility>1</visibility>
    <description><![CDATA[
        More...
        <br> <br>
        http://www.esa.int/esaMI/Venus_Express/
                        SEMYGQEFWOE_0.html
        <br><br>
        http://www.atm.ox.ac.uk/main/Science/

posters2006/2006cw2.pdf
]]></description>
    <Model>
        <altitudeMode>absolute</altitudeMode>
        <Location>
            <longitude>-90</longitude>
            <latitude>-90</latitude>
            <altitude>-1600000</altitude>
        </Location>
        <Link>
            <href>models/sPole.dae</href>
        </Link>
    </Model>
</Placemark>
```

Figure 2. The dipole vortex at Venus's north pole is added as a COLLADA model because ground overlays cannot be draped across the poles. Despite the option of displaying UTM coordinates (white), Google Earth does not project surface imagery using UTM.

Solar Weather

Venus does not have a magnetic field to protect its atmosphere from the solar wind; consequently, solar weather has an impact close to the ground. Recently, hot flow anomalies have been detected on Venus (http://www.nasa.gov/mission_pages/sunearth/news/venus-explosions.html). We display these using a COLLADA model (file "venus_hfa.kml" in the GSA Data Repository) that extends to twice the diameter of the planet (Fig. 3 and Table 4).

This is just one example of a wide range of atmospheric data that can be displayed on Google Venus with COLLADA and KML.

SURFACE IMAGERY

Magellan Altimetry as Primary Database

To serve as a primary database, we used a global altimetry image from NASA's Magellan mission (Fig. 4) which is available in simple cylindrical (Plate Carrée) projection, the appropriate projection for Google Earth. Because it covers the entire globe, we call this the tablecloth (file "tablecloth.kml" in the GSA Data Repository).

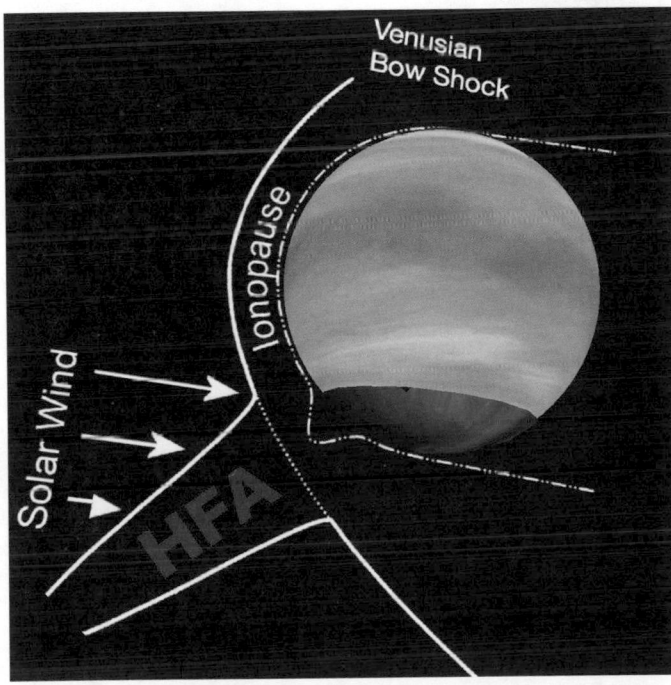

Figure 3. Solar weather on Venus is intense at low atmospheric altitudes owing to the lack of magnetic field protection. Recently discovered hot flow anomalies (HFAs) are illustrated using a COLLADA model that extends twice the diameter of the planet. Source: http://www.nasa.gov/mission_pages/sunearth/news/venus-explosions.html. The colored image at the south pole is a COLLADA model of the southern dipolar vortex.

Note that, owing to file size, we did not simply drape an image as a ground overlay as in the case of the atmosphere. Instead, we linked to the root of an image pyramid which sharpens the focus of the surface as the viewer zooms in, by loading progressively more detailed image tiles, as explained in the next section.

Surface Imagery Management

A fundamental feature of Google Earth is the efficient management of surface imagery across a range of scales. When the user zooms in from the elevation of the Space Shuttle to that of a helicopter, progressively higher resolution images are automatically draped on the surface DEM. If the most detailed ground imagery were loaded regardless of viewer elevation, there would be terabytes of data to cache and the application would grind to a halt even on a supercomputer. When creating Google Venus, we had to emulate Google Earth's surface image management system. We did so by downloading the highest resolution surface images available from NASA and making image tiles (also known as image pyramids or super-overlays) using a freely available application called MapTiler (http://www.maptiler.org). For example, the source of the Magellan tablecloth covering the Earth's surface imagery is a 5 MB file. MapTiler converted this into a pyramid of tiles each less than 256 KB. Successive levels in the pyramid structure cover different surface areas of the virtual globe. At level zero, a single 256 KB tile covers 90° to −90° latitude and all 360 degrees of longitude, which is displayed only when the view is zoomed all the way out.

The file and folder structure for an image pyramid is shown in Figure 5A, and sample code is listed in Table 5. Each KML document in the pyramid contains an image tile and links to four nested documents. The Region element checks the level-of-detail (Lod) to ascertain whether the current tile occupies a minimum of 128 pixels of the main viewer window space and if it does, that tile is loaded. Setting maxLodPixels to "−1" denotes no cut-off

TABLE 4. A COLLADA MODEL EXTENDING TWICE THE DIAMETER OF THE PLANET IS USED TO DISPLAY HOT FLOW ANOMALIES CAUSED BY SOLAR WIND

```
<Document>
    <name>venus_hfa</name>
    <LookAt>...</LookAt>
    <description>...</description>
    <visibility>1</visibility>
    <Placemark>
        <name>Hot Flow Anomaly</name>
        <Model>
            <altitudeMode>absolute</altitudeMode>
            <Location>
                <latitude>90</latitude>
                <longitude>0</longitude>
                <altitude>0</altitude>
            </Location>
            <Link>
                <href>models/hfa.dae</href>
            </Link>
        </Model>
    </Placemark>
</Document>
```

Figure 4. Google Venus's primary database was created by draping Magellan altimetry over the Google Earth surface imagery. White graticule shows longitude convention for Venus (see text). Latitude is measured identically on both planets. Venus has an equator and prime meridian but no tropics or arctic/antarctic circles.

level-of-detail, so the image remains visible as one zooms in to ground level. Consequently, in an inclined view, the tiles near the viewer will be relatively high resolution draped over the larger, lower resolution tiles visible in the distance. The code snippet in Table 5 tiles one-eighth of the globe's surface in the southeast. Figure 5B illustrates the structure of a 4-level image pyramid. Here, image tile files were purposely removed, causing Google Earth to substitute a red X for each missing image.

Venusian Lat-Lon Grid

Planets have a natural origin of latitude—their equator—but not of longitude. Earth's prime meridian, passing through Greenwich, England, was selected arbitrarily. On Venus, longitude is measured from 0° to 359°, increasing eastward from the prime meridian, which is arbitrarily set to pass through the peak of Ariadne Crater. Because Venus's axial tilt is less than three degrees off vertical, there are no useful Venusian analogs of the Tropics of Cancer and Capricorn nor of the Arctic and Antarctic Circles, which reflect Earth's 23.5° tilt. (Strictly speaking, Venusian tropics border the equator at 2.7° north and south, and the circles encircle the poles at the same angle. However, there are no seasonal climatic implications analogous to Earth). To avoid the confusion that might result from

using the "Grid" feature built into Google Earth, we created a graticule appropriate to Venus ("venus_latlon.kml," see footnote 1; it is turned on in Fig. 4):

Note that Google Earth's W180° to E180° longitude convention must be used when creating content in KML code. Correct longitude conversion can be checked by turning on and comparing the custom Venus grid and the built-in Google Earth grid.

Earth's Continental Outlines as Geographic Reference

Because viewers unfamiliar with Venus's surface may find it difficult to orient themselves in a global context, we created a layer that projects the outlines of Earth's continents onto the Google Venus surface ("venus_conts.kml," see footnote 1; Fig. 6). This method was previously shown to help students with spatial awareness (Brooks et al., 2010)—their verbal reports suggest that the outlines helped develop their sense of place and scale.

Google Earth Tools and Layers

The built-in Google Earth ruler tool can be used for measurements on Google Venus bearing in mind that the ruler overestimates surface distances and areas because Venus is slightly smaller than Earth. The mean radius of Venus is 6052 km

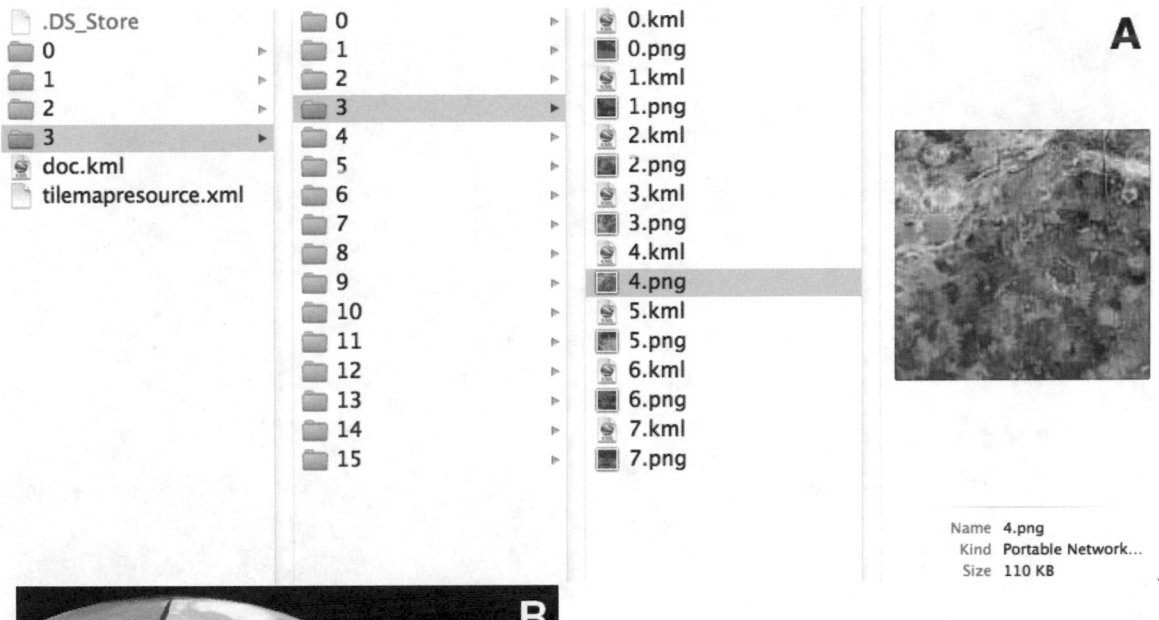

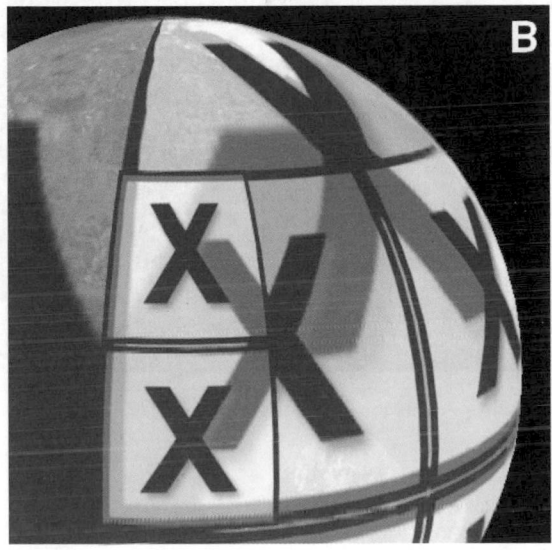

Figure 5. (A) Image tiles or pyramids are essential for file size management. This figure shows the nested folder and file system created by MapTiler for the Magellan Tablecloth image pyramid (B) The structure of an image pyramid revealed by red X symbols replacing missing image files.

TABLE 5. SAMPLE NETWORK LINK FROM AN IMAGE PYRAMID

```
<NetworkLink>
    <name>1/0/0.png</name>
    <Region>
        <Lod>
            <minLodPixels>128</minLodPixels>
            <maxLodPixels>-1</maxLodPixels>
        </Lod>
        <LatLonAltBox>
            <north>0</north>
            <south>-90</south>
            <east>-90</east>
            <west>-180</west>
        </LatLonAltBox>
    </Region>
    <Link>
        <href>../../1/0/0.kml</href>
        <viewRefreshMode>onRegion</viewRefreshMode>
    </Link>
</NetworkLink>
```

Note: The Region element determines whether the current image (1/0/0.png) will be displayed based in the level of detail (Lod). If this image occupies more than 128 pixels, a more detailed image links to ../../1/0/0.kml is tested recursively.

compared to Earth's 6378 km, therefore great circle distances between equivalent points on Venus are ~95% of the lengths of those arcs on Earth, and surface areas are ~90%.

The "Search" box and "Layers" in the sidebar of the Google Earth desktop application are of no use on Venus and should be kept collapsed, leaving only the "Places" sidebar panel expanded. Also, driving instructions in placemark balloons can be suppressed by creating custom balloon styles (Table 6).

MISSIONS TO VENUS

It may come as a surprise that since the 1960s, more than 20 successful missions to Venus have been launched by the former Soviet Union (USSR), the United States (NASA), and the European Space Agency (ESA). For details, see http://planetary.org/explore/topics/venus/missions.html. Of these, only the Soviet Venera and Vega missions placed landing craft

Figure 6. Outlines of Earth's continents (white) help students locate features on Venus.

on the surface. The Japanese Aerospace Exploration Agency (JAXA) mission will reattempt to enter a Venusian orbit in 2012 after a 2010 failure (http://www.jaxa.jp/projects/sat/planet_c/index_e.html) and there are plans for a joint ESA-JAXA mission called BepiColombo that will fly by Venus en route to Mercury shortly thereafter (http://sci.esa.int/science-e/www/object/index.cfm?fobjectid=48871).

TABLE 6. CREATING A CUSTOM BALLOON STYLE SUPPRESSES DEFAULT DRIVING INSTRUCTIONS.

```
<Placemark>
        <name>Any Placemark</name>
        <Style>
                <BalloonStyle>
                        <text>
                                [Balloon text here]
                                [No driving instructions!]
                        </text>
                </BalloonStyle>
        </Style>
        <Point>
                <coordinates>0,0,0</coordinates>
        </Point>
</Placemark>
```

Venera Landing Sites

We have included four Venera landing sites in Google Venus, each with a photo overlay taken from the surface (Fig. 7). They are grouped in a radiobutton-style folder and can be visited as a miniature virtual field trip ("venera_landing.kml"; see footnote 1).

Sample code for the Venera 9 photo overlay is shown in Table 7. Fortunately, the camera angle and view parameters do not require coding by hand—they can be generated automatically by adding a photo in the desktop application and adjusting the settings in the New Photo Overlay box. When a photo overlay is saved to "MyPlaces" or "Temporary Places," the automatically generated code can be accessed by selecting the photo overlay in the "Places" sidebar; right-clicking and copying; switching to a text editor or code editor; and pasting into a KML document.

Fly-bys, Probes, and Orbital Missions

Since the 1960s, there have been several fly-by missions to Venus, some of which dropped probes into the atmosphere

Figure 7. Surface image of Venus from the Soviet Venera mission shows a dry, fine-grained, probably basaltic crust with no sign of the varied granitic rock types that characterize Earth's continents. This placemark is at the Venera 14 landing site. ©Google; image source: http://solarsystem .nasa.gov/multimedia/gallery/venera14TOP.jpg.

TABLE 7. CODE SNIPPET FOR THE
PHOTO OVERLAY AT THE VENERA 9 LANDING SITE.

```
<PhotoOverlay>
    <name>Venera 9</name>
    <description><![CDATA[
        Photo from Venera 9 landing site, 31.01N,
291.64E.
        <br/>
        <img src="images/venera_9.jpg" width="400">
]]></description>
    <Camera>
        <longitude>-68.36000000099992</longitude>
        <latitude>30.99999999999999</latitude>
        <altitude>194.8600000003038</altitude>
        <heading>2.099355619496519e-05</heading>
        <tilt>87.7595142797512</tilt>
        <roll>-8.844200208292416e-10</roll>
    </Camera>
    <Style>...</Style>
    <Icon>
        <href>images/venera_9.jpg</href>
    </Icon>
    <ViewVolume>
        <leftFov>-45.615</leftFov>
        <rightFov>45.615</rightFov>
        <bottomFov>-12.615</bottomFov>
        <topFov>12.615</topFov>
        <near>68.201</near>
    </ViewVolume>
    <Point>
        <coordinates>-68.36,31,194.86</coordinates>
    </Point>
</PhotoOverlay>
```

(http://nssdc.gsfc.nasa.gov/planetary/chronology_venus.html). These collected useful information on extreme atmospheric temperatures, crushing pressures, and the lack of a magnetic field. However, the most important missions were those that placed spacecraft in orbit and mapped the surface remotely most notably the 1989 Magellan mission (Ford et al., 1993; Bindschadler, 1995).

Although opaque to visible light, the Venusian atmosphere is transparent in microwave or radio frequencies. Synthetic aperture radar (SAR) imagery collected as part of the Magellan mission provides spectacular views of the surface, covering 98% of the planet with resolutions up to 100 m/pixel. SAR data taken together with Magellan altimetry reveal basic surface features stereoscopically. Magellan altimetry represents the major global data set that we use as a "base map" for Google Venus, with SAR images projected onto the surface in specific areas. The altimetry data have low-resolution SAR data imbedded within the color-coded output. These data allow recognition of highland and lowland regions, and global-scale zones of localized tectonism and volcanism.

SAR Image Pyramids

Because of the very large file sizes involved, SAR images must be tiled before being draped over the surface in order to avoid slowing the performance of Google Venus. For example,

De Paor et al.

a 60 MB image of Bonnevie crater was tiled to 7 levels-of-detail to create a deeply zoomable viewing experience (Fig. 8A). The image pyramid may be loaded using the root file "bonnevie.kml" (see footnote 1).

Care must be taken, however, when more than one image pyramid is viewed, as tiles may interfere. In Figure 8B, for example, some tiles are from the gray SAR image pyramid but others are from the colored Magellan tablecloth image pyra-mid. As one flies over this region, tiles may compete for promi-nence, as illustrated by the zigzag boundary and the gap in the gray tiles. The solution would be to carefully control the KML <drawOrder> used in different image pyramids. However, this process is not automated and would be tedious to implement by hand, given dozens of nested KML documents. A simpler solu-tion is to turn off all but one image pyramid at a time. We have created the image pyramids from SARs files (posted to the GSA

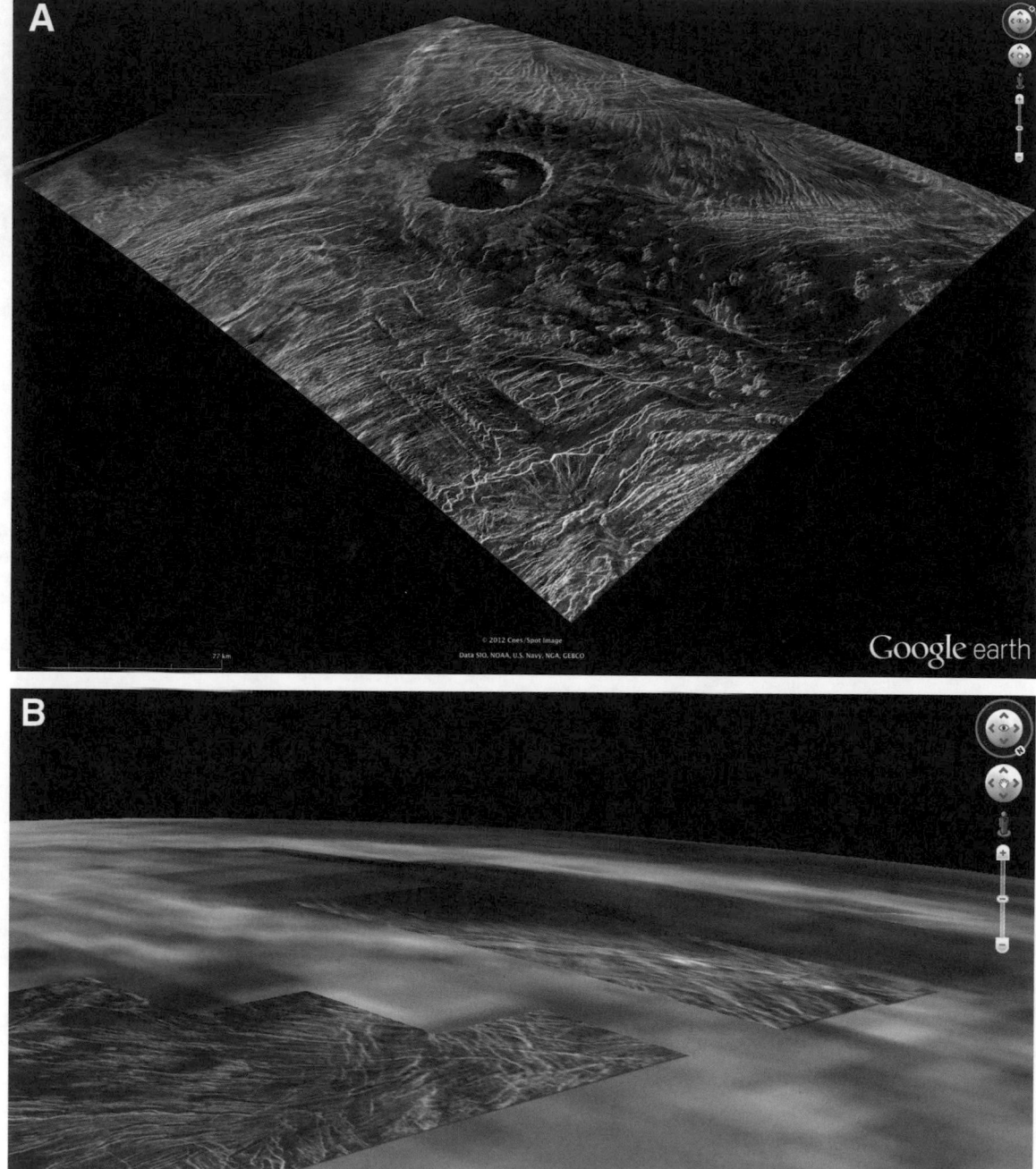

Figure 8. (A) A 60 MB PNG file of Venus's Bonnevie Crater was tiled to 7 levels to create a surface image that is zoom-able. (B) Superposition of tiled images sometimes causes competition for drawOrder. See text for discussion.

Data Repository; see footnote 1). The first three are named for use in student exercises, while the remainder are named for their SARs database references. (See listing in Data Repository.) It is important to load these sequentially, not simultaneously; otherwise, Google Venus responsiveness will grind to a halt. If performance deteriorates gradually, it may help to clear the caches.

EXPLORING THE SURFACE

Global topography, defining the shape of the surface, provides fantastic clues to how a planet evolved and to dynamic interior processes. Topography on Venus is keyed to mean planetary radius, or MPR (6051.9 km), since sea-level is meaningless on a dry planet. The range of topography is similar to Earth, with the highest point (Maxwell Montes) ~11 km above MPR and the lowest (Diana-Dali Chasma) ~7 km below MPR. However the distribution of Venus's topography is quite different from Earth's. Venus has a unimodal distribution of topography with most of the planet close to MPR, unlike Earth's bimodal topographic division into low oceans and high continents.

Magellan altimetry and SAR data reveal basic surface features and permit the definition of three types of terrain based on altitude—lowlands, mesolands, and highlands—as well as distinctive features such as craters and uniquely Venusian structures discussed below.

Lowlands

Just as most of the Earth's surface (72%) is oceanic, most of Venus's surface (80%) comprises lowlands which include relatively smooth plains called planitiae, deformation belts that can extend hundreds of kilometers, and narrow channels or "canali" ranging from tens or hundreds of kilometers long up to 6900 km long in the case of Baltis, according to Banerdt et al. (1997). Lowland regions also host thousands of inliers of ribbon tessera terrain (see below).

Lowland surfaces are deformed by so-called wrinkle ridges and linear fractures. Long, sinuous ridges mark the crest of wrinkles, or folds. Linear fractures are the surface expression of dikes at depth where magma traveled possibly thousands of kilometers (Ernst et al., 2001; Grosfils and Head, 1994a, 1994b; McKenzie et al., 1992a). Wrinkle ridges and linear fractures define patterns traceable across tens of millions of square kilometers, and mapping out these patterns can provide critical clues to Venus's volcanic and tectonic processes. Coronae (see next section) occur rarely as isolated features in the lowlands.

Mesolands

The mesolands together with the highlands account for the remaining 20% of the surface. These middle altitudes host most of the planet's ~500 coronae—quasi-circular features generally ranging from 60 to 800 km in diameter and typically forming chains spatially associated with troughs called chasmata (Stofan et al., 1997). Although coronae are widely accepted as representative of diapirism, formation by impact has also been proposed (e.g., Vita-Finzi et al., 2005).

Highlands

Highland features include volcanic rises and crustal plateaus. Volcanic rises are large domed regions of radial volcanic flows ranging from 1500 to 2500 km in diameter and 1–3 km in elevation. They are thought to be thermally supported expressions of deep mantle plumes under the thick lithosphere (e.g., Phillips and Hansen, 1994; Smrekar et al., 1997; Nimmo and McKenzie, 1998). Coronae form clusters associated with some volcanic rises (e.g., Themis Regio, Bell Regio, and Eastern, Western, and Central Eistla Regio). The KML file, "venus_rises.kml" (see footnote 1) features numerous volcanic rises.

Crustal plateaus, similar in planform to rises, are steep-sided, rising 0.5–4 km above their surroundings. Widely accepted as relatively old features, they are likely to be isostatically supported by thick crust, or low-density mantle residuum (e.g., Phillips and Hansen, 1994; Simons et al., 1997; Hansen, 2006). They are marked by ribbon-tessera terrains consisting of short folds with long orthogonal "ribbon" structures and graben that imply layer extension perpendicular to fold crests. Lava that floods low points indicates accompanying volcanic activity (Phillips and Hansen, 1994; Ivanov and Head, 1996; Hansen and López, 2010). Ribbon tessera terrains can be viewed at two Google Venus locations: Ovda Region Tessera Terrain and in the Bell Regio Area. Several plateaus are described in the "venus_plateaus.kml" file (see footnote 1).

Impact Craters

Venus has some 1,000 craters randomly distributed in Lowlands, mesolands, and highlands. Craters range in diameter from 1 to 270 km (Schaber et al., 1992; Phillips et al., 1992; Herrick et al., 1997) and show evidence of progressive degradation (Izenberg et al., 1994; Basilevsky and Head, 2002) which allows for delineation of relative age relations (Phillips and Izenberg, 1995; Hansen and Young, 2007; Herrick and Rumpf, 2011). Venus lacks small craters due to screening by its thick atmosphere which unfortunately hampers crater density dating (see Hartmann, 2005). Low crater density implies a mean model surface age of 0.5–1.0 b.y. (McKinnon et al., 1997).

The early interpretation that most of Venus's craters were pristine—not faulted or volcanically embayed—led to the widely accepted idea that Venus was catastrophically resurfaced at ~500 Ma (Schaber et al., 1992; Bullock et al., 1993; Strom et al., 1994). Catastrophic resurfacing was defined as a near-global event that flooded at least 80% of the planet with lava to a depth of 1 km or more within a 10–100 million year period, wiping out essentially all of Venus's early history. This resurfacing event was assigned a major role in Venus's climate evolution (e.g., Bullock

De Paor et al.

and Grinspoon, 2001; Taylor and Grinspoon, 2009). However, a growing body of data is difficult to reconcile with catastrophic resurfacing hypotheses, and it is likely that Venus's surface preserves a rich record of past geological processes (see Guest and Stofan, 1999; Hansen and Young, 2007; Hansen and López, 2010; Herrick and Rumpf, 2011; Bjonnes et al., 2011).

Unique Features—Ishtar and Artemis

Structures such as impact craters, volcanic cones, and wrinkle ridges are common on planets and moons of the Solar System. However, Venus has features unlike anything seen elsewhere. Prominent among these are the mutually antipodal Ishtar Terra and Artemis structures (Hansen et al., 1997).

Ishtar

Ishtar Terra forms a huge continent-like region represented by the 4-km-high plateau of Lakshmi Planum, surrounded by expanses of tessera terrain and mountain belts (or "montes"), including Venus's highest point, Maxwell Montes. The region has been proposed to form above a huge downwelling mantle convection cell which resulted in low-density mantle melt resid-

uum uplifting the thickened crust like an iceberg (Hansen and Phillips, 1995).

Artemis

Artemis, as historically defined, includes an interior topographic high surrounded by Artemis Chasma, a large narrow circular trough, and an outer rise that transitions to the surrounding lowland (Fig. 9). However, recent geologic mapping indicates that these are only the center of a more expansive feature, including a 5000 km diameter topographic trough and suites of radial dikes and concentric wrinkle ridges that extend for 14,000 km (Hansen and Olive, 2010). Artemis's global footprint thus affects more than one fourth of the planet's surface–perhaps the largest tectonic feature in the Solar System.

Hypotheses for the formation of Artemis Chasma and surrounding regions have included subduction, bolide impact, and a mantle plume (e.g., Griffiths and Campbell, 1991; McKenzie et al., 1992b; Brown and Grimm, 1995, 1996; Hamilton, 2005, 2007; Schubert and Sandwell, 1995; Smrekar and Stofan, 1997; Hansen, 2002). Detailed geologic mapping of Artemis Chasma favor a plume hypothesis (Bannister and Hansen, 2010), however, the recognition of the Artemis super-structure suggests an

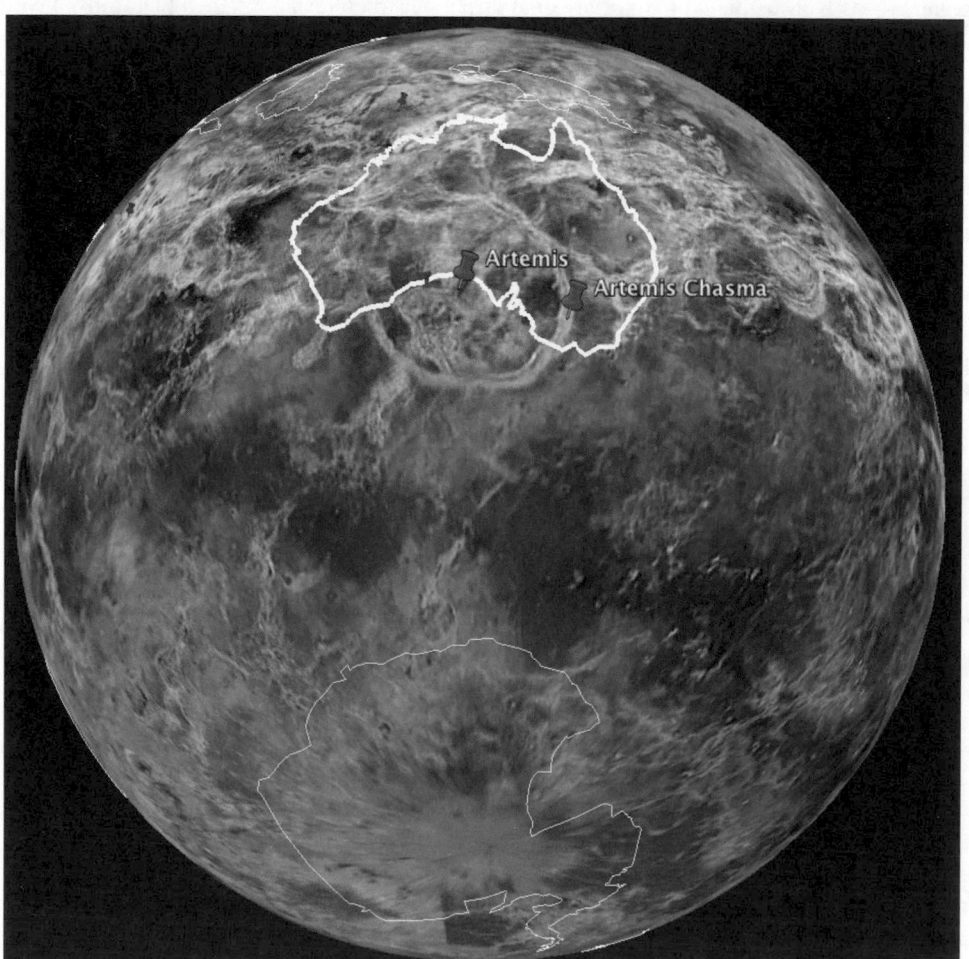

Figure 9. The Artemis Structure and Artemis Chasma. Outline of Australia conveys the scale. ©Google; image source: http://solarsystem.nasa.gov//multimedia/display.cfm?IM_ID=9603.

additional super-plume component (Hansen and Olive, 2010; Courtillot et al., 2003) affecting ~25% of the surface of Venus with radial dikes and concentric wrinkle ridges. The size and radial nature of the dike swarm and the concentric nature of the wrinkle ridge pattern provide evidence of a huge axisymmetric global stress field (Watters, 1992; Mége and Reidel, 2001). One way to consider the size of the Artemis super-plume signature is by way of comparison. When scaled to the host planet's size, the Artemis radial fracture suite is proportionally similar to the Tharsis radial dike system on Mars which includes the famous Olympus Mons volcano—familiar to many as the largest volcano in our Solar System (Wilson and Head, 2002). This comparison might provide clues to possible global-scale mantle flow patterns in terrestrial planets lacking plate tectonics.

The areal extent of radial dikes and wrinkle ridges record the effect of the Artemis super-plume across the surface, however, it is also likely that the super-plume affected flow patterns within the mantle. Global-scale upwelling associated with the super-plume would flow outward in a radial pattern from Artemis Chasma, with flow along the base of the lithosphere. Ultimately, return flow might converge in a global-scale downwelling. Just as the upwelling would be centered on Artemis, one might expect a global downwelling to be antipodal. Ishtar Terra is located in an antipodal position to Artemis, and could represent a surface expression of a global-scale downwelling. These processes can perhaps be best viewed in global-scale COLLADA models, which allow one to view both surface and interior features simultaneously.

VENUS'S INTERIOR

Despite surficial differences, models for Solar System evolution suggest that Venus and Earth share similar rocky mantles and metallic cores (Wetherill, 1990). Given a lack of critical data, particularly seismic, models of Venus's deep structure are based on analogy with Earth. Lack of data affects our understanding of internal structure which has implications for thermal evolution. Mechanisms of heat transfer (conduction, convection, and advection) depend in part on rheology. Terrestrial planets have a stiff lithosphere that conducts, and a deeper mantle that convects heat. Venus, unlike Earth, exhibits strongly correlated long-wavelength gravity and topography fields (Phillips and Lambeck, 1980) implying either a ~300-km-thick strong lithosphere or mantle convection beneath an ~150-km-thick lithosphere. To the first order, lithosphere thickness is inversely proportional to heat flow and thus is relevant to present-day heat loss which remains unknown. A reasonable working hypothesis assumes Venusian effective mantle viscosity is similar to Earth, with strong temperature-dependence. Large viscosity contrasts are likely across thermal boundary layers, notably across the lithosphere and core-mantle boundaries. Magellan gravity and topography data do not require a low viscosity asthenosphere (e.g., Simons et al., 1997), likely contributing to a lack of plate tectonics on Venus.

VISUALIZING THE INTERIOR

Concepts that many Venus scientists accept are illustrated in speculative crust-to-core cross-sections (e.g., Schubert et al., 1997; Smrckar et al., 1997) as COLLADA models in the Google Venus environment. Lowlands lie above mantle downwellings; volcanic rises are thermally supported and underlain by deep-mantle plumes; crustal plateaus garner isostatic support due to either thick crust or low-density mantle melt residuum. Chasmata chains (e.g., Diana-Dali, Parga, and Hecate Chasmata) represent large cylindrical mantle upwellings; coronae can be spawned within the shallow part of this upwelling environment, or within some plume heads (e.g., Bell or Themis Regiones). COLLADA models allow us to view the mantle, gaining a global 3-D understanding of the planet (Fig. 10A).

Schematic cross sections were created in a drawing program and saved as portable network graphic (PNG) image files. These were imported into SketchUp (http://www.sketchup.com) and used as surface textures on spherical-slice models. The models were then exported as Digital Asset Exchange (.dae) files. For the desktop application version, the models may be loaded from the "venus_mantle.kml" file (see footnote 1).

Loading this file prompts the viewer to switch from Google Earth to Google Mars. The radius of Mars is thought to be a good approximation to Venus's core; therefore we draped the Google Mars surface with an orange PNG file and superimposed COLLADA models of Google Venus's mantle and crust.

The three cross sections allow one to postulate spatial—and by extension, temporal—relations between surface geology and mantle. Two longitudinal sections cut from the north pole to the south pole: the 135E longitude section cuts through Artemis, and the 45E longitude section cuts the western edge of the surface expression of the Artemis super-plume, as marked by the concentric wrinkle ridge suite. The third section parallels the equator. Sections are easily identified when the lat-lon grid is turned on (it appears on the surface of the core, Fig. 10B).

The 135E Longitude Section

This section cuts from the surface to the core, slicing through the Artemis plume, represented by the faded pink thermal plume. The Diana-Dali Chasma upwelling affects this region just north of Artemis. Thetis Regio with thickened crust or low-density upper mantle occurs within this cross section. North of Thetis and south of Artemis, mantle downwellings draw the surface downward forming lowland regions of Niobe Planitia and Zhibek Planitia. South of Artemis, Zhibek and Imapinua Planitiae are separated by a small corona-chasmata chain that might signal a local mantle upwelling.

The 45E Longitude Section

Just north of the center of this section, a mantle plume that supports Eastern Eistla Regio rises from the core. A similar plume

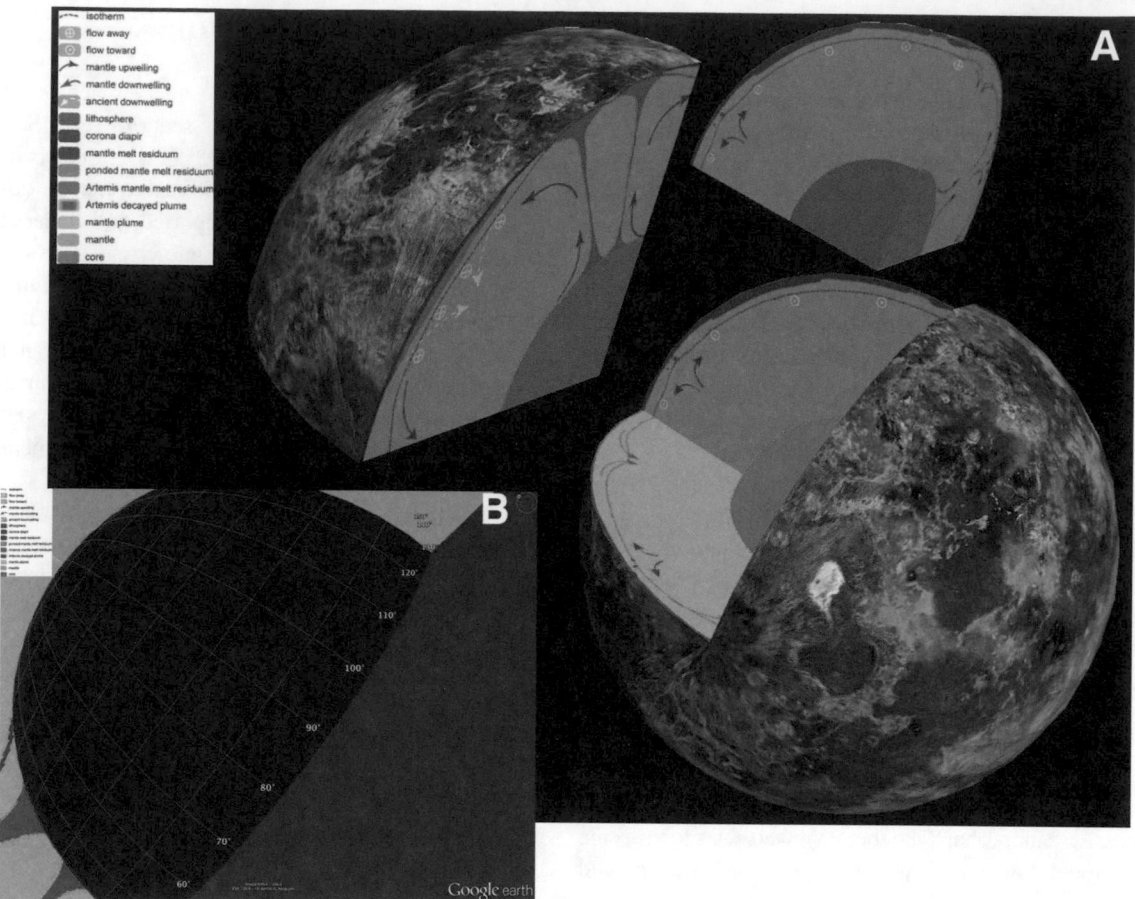

Figure 10. (A) COLLADA models of the interior of Google Venus. (B) Turning on the Venusian lat-lon grid on the core identifies the mantle cross sections in the COLLADA models of the interior of Google Venus.

supports the corona-dominated volcanic rise Bell Regio to the north. Bell Regio is centered slightly east of Eastern Eistla, and therefore, depending on thickness of a plume root (unknown), a root for Bell Regio might not lie within the plane of the section. South of Eastern Eistla, mantle return flow results in formation of broad lowlands extending south to Fauna Planitia.

The Equatorial Section

The equatorial section cuts through much of Aphrodite Terra, the largest highland region on Venus. Near 45E, remnants of thickened crust and/or mantle melt residuum associated with the formation of ancient ribbon tessera terrain are likely. Farther east, such a unit is clearly present, providing isostatic support for the huge mass of Ovda Regio (90E), Venus's largest crustal plateau. Within this section the strongest flow is likely outward to the north and south from the generally WNW-trending Diana-Dali Chasma.

Venus's Core

Little is known of Venus's core and little is likely to be discovered from missions in the near future. Its existence is inferred from the planet's bulk density (5.2 g/cc) which is similar to Earth's (5.5 g/cc). Since the observed basaltic crust and presumably peridotitic mantle are less dense, an ~3000 km metallic core is needed to yield this bulk value. We have no way of knowing whether there is an inner core, but the lack of a magnetic field implies that either the core is entirely solid or if a solid inner core exists, it does not rotate fast enough to generate an outer core dynamo as on Earth. Our plain orange ground overlay thus aptly represents the extent of our knowledge of Venus's core.

CONCLUSIONS

Google Venus provides a new and exciting means to explore Earth's sister planet, with lessons for both Venus and Earth. Virtual exploration allows students and researchers to appreciate a wide range of spatial relationships, on the surface, and in three dimensions at a truly global scale and in a way not previously possible. We hope that Google Venus will continue to grow and evolve as a community resource.

Venus's surface, long veiled by thick, heretofore impenetrable clouds, can now be viewed by anyone with an Internet connection by loading the "venus.kml" file (see footnote 1).

As space missions deliver more detailed surface images and data for the planets and moons of the Solar System and for exoplanets, especially "super earths" orbiting other stars, there will be new opportunities to create virtual globes for numerous bodies. Following the methods described here, readers could create their own Google Europa, Google Titan, etc.

ACKNOWLEDGMENTS

Hansen acknowledges support from the McKnight Foundation and NASA grants NNX06AB90G and NNX11AQ86G. De Paor and Dordevic were supported by NSF TUES 1022755, NSF GEO 1034643, and a Google Faculty Research Award. Any opinions, findings, conclusions or recommendations expressed in this paper are those of the authors and do not necessarily reflect the views of NASA, the National Science Foundation, the McKnight Foundation, or Google Inc. Comments of reviewers Eric Pyle and Jen Piatek greatly improved the manuscript.

REFERENCES CITED

Banerdt, W.B., McGill, G.E., and Zuber, M.T., 1997, Plains tectonics on Venus, *in* Bouger, S.W., Hunten, D.M., and Phillips, R.J., eds., Venus II—Geology, Geophysics, Atmosphere, and Solar Wind Environment: University of Arizona Press, p. 901–930.

Bannister, R.A., and Hansen, V.L., 2010, Geologic map of the Artemis quadrangle (V 48), Venus: U.S. Geological Survey Scientific Investigations Map SIM-3099, scale 1:5,000,000.

Basilevsky, A.T., and Head, J.W., 2002, Venus: Timing and rates of geological activity: Geology, v. 30, p. 1015–1018, doi:10.1130/0091-7613 (2002)030<1015:VTAROG>2.0.CO;2.

Bindschadler, D.L., 1995, Magellan—a new view of Venus geology and geophysics: Reviews of Geophysics, v. 33, p. 459–467, doi:10.1029/95RG00281.

Bjonnes, E.E., Hansen, V.L., James, B., and Swenson, J.B., 2011, Equilibrium resurfacing of Venus is possible: Results from new Monte Carlo modeling and implications for Venus surface histories: Icarus, v. 217, p. 451–461, doi:10.1016/j.icarus.2011.03.033.

Bridges, N.T., 1995, Submarine analogs to Venusian pancake domes: Geophysical Research Letters, v. 22, no. 20, p. 2781, doi:10.1029/95GL02662.

Bridges, N.T., 1997, Ambient effects on basalt and rhyolite lavas under Venusian, subaerial, and subaqueous conditions: Journal of Geophysical Research, v. 102, p. 9243–9255, doi:10.1029/97JE00390.

Brooks, W.D., Dordevic, M.M., and De Paor, D.G., 2010, Exploring planets and moons using Google Earth to convey a sense of place, scale, and time: Geological Society of America Abstracts with Programs, v. 42, no. 5, p. 421.

Brown, C.D., and Grimm, R.E., 1995, Tectonics of Artemis Chasma: a Venusian "plate" boundary: Icarus, v. 117, p. 219–249, doi:10.1006/icar.1995.1155.

Brown, C.D., and Grimm, R.E., 1996, Lithospheric rheology and flexure at Artemis Chasma, Venus: Journal of Geophysical Research, v. 101, p. 12697–12708, doi:10.1029/96JE00834.

Bullock, M.A., and Grinspoon, G.H., 2001, The recent evolution of climate on Venus: Icarus, v. 150, p. 19–37, doi:10.1006/icar.2000.6570.

Bullock, M.A., Grinspoon, D.H., and Head, J.W., 1993, Venus resurfacing rates—Constraints provided by 3-D Monte-Carlo simulations: Geophysical Research Letters, v. 20, no. 19, p. 2147–2150.

Courtillot, V., Davaille, A., Besse, J., and Stock, J., 2003, Three distinct types of hotspots in the Earth's mantle: Earth and Planetary Science Letters, v. 205, p. 295–308, doi:10.1016/S0012-821X(02)01048-8.

Crumpler, L.S., Aubele, J.C., Senske, D.A., Keddie, S.T., Magee, K.P., and Head, J.W., 1997, Volcanoes and centers of volcanism on Venus, *in* Bouger, S.W., Hunten, D.M., and Phillips, R.J., eds., Venus II Geology, Geophysics, Atmosphere, and Solar Wind Environment: Tucson, Arizona, The University of Arizona Press, p. 697–756.

De Paor, D.G., and Pinan-Llamas, A., 2006, Application of novel presentation techniques to a structural and metamorphic map of the Pampean Orogenic

Belt, Northwest Argentina: Geological Society of America Abstracts with Programs, v. 38, no. 7, p. 326, paper no. 131-12.

Donahue, T.M., and Russell, C.T., 1997, The Venus atmosphere and ionosphere and their interaction with the solar wind: An overview, *in* Bouger, S.W., Hunten, D.M., and Phillips, R.J., eds., Venus II—Geology, Geophysics, Atmosphere, and Solar Wind Environment: Tucson, University of Arizona Press, p. 3–31.

Donahue, T.M., Grinspoon, D.H., Hartle, R.E., and Hodges, R.R., 1997, Ion/neutral escape of hydrogen and deuterium: Evolution of water, *in* Bouger, S.W., Hunten, D.M., and Phillips, R.J., eds., Venus II—Geology, Geophysics, Atmosphere, and Solar Wind Environment: Tucson, University of Arizona Press, p. 385–414.

Durand-Manterola, H.J., 2010, Superrotation on Venus: Driven by Waves Generated by Dissipation of the Transterminator Flow: Cornell University, arXiv, http://arxiv.org/abs/1005.3488.

Ernst, R.E., Grosfils, E.B., and Mege, D., 2001, Giant dyke swarms on Earth, Venus and Mars: Annual Review of Earth and Planetary Sciences, v. 29, p. 489–534, doi:10.1146/annurev.earth.29.1.489.

Ford, J.P., Plaut, J.J., Weitz, C.M., Farr, T.G., Senske, D.A., Stofan, E.R., Michaels, G., and Parker, T.J., 1993, Guide to Magellan Image Interpretation: National Aeronautics and Space Administration Jet Propulsion Laboratory Publication.

Griffiths, R.W., and Campbell, I.H., 1991, Interaction of mantle plume heads with the Earth's surface and onset of small-scale convection: Journal of Geophysical Research, v. 96, p. 18,295–18,310, doi:10.1029/91JB01897.

Grimm, R.E., and Hess, P.C., 1997, The crust of Venus, *in* Bouger, S.W., Hunten, D.M., and Phillips, R.J., eds., Venus II—Geology, Geophysics, Atmosphere, and Solar Wind Environment: University of Arizona Press, p. 1205–1244.

Grosfils, E.B., and Head, J.W., 1994a, The global distribution of giant radiating dike swarms on Venus: Implications for the global stress state: Geophysical Research Letters, v. 21, p. 701–704, doi:10.1029/94GL00592.

Grosfils, E.B., and Head, J.W., 1994b, Emplacement of a radiating dike swarm in western Vinmara-Planitia, Venus—Interpretation of the regional stress-field orientation and subsurface magmatic configuration: Earth, Moon, and Planets, v. 66, no. 2, p. 153–171, doi:10.1007/BF00644129.

Guest, J.E., and Stofan, E.R., 1999, A new view of the stratigraphic history of Venus: Icarus, v. 139, p. 55–66.

Hamilton, W.B., 2005, Plumeless Venus has ancient impact-accretionary surface, *in* Foulger, G.R., et al., eds., Plates, Plumes, and Paradigms: Geological Society of America Special Paper 388, p. 781–814, doi:10.1130/0 -8137-2388-4.781.

Hamilton, W.B., 2007, An alternative Venus, *in* Foulger, G.R., and Jurdy, D.M., eds., Plates, Plumes, and Planetary Processes: Geological Society of America Special Paper 430, p. 879–911, doi:10.1130/2007.2430(41).

Hansen, V.L., 2002, Artemis: Surface expression of a deep mantle plume on Venus: Geological Society of America Bulletin, v. 114, no. 7, p. 839–848, doi:10.1130/0016-7606(2002)114<0839:ASEOAD>2.0.CO;2.

Hansen, V.L., 2006, Geologic constraints on crustal plateau surface histories, Venus: The lava pond and bolide impact hypotheses: Journal of Geophysical Research, v. 111, no. E11010, p. doi:10.1029/2006JE002714.

Hansen, V.L., and López, I., 2010, Venus records a rich early history: Geology, v. 38, no. 4, p. 311–314, doi:10.1130/G30587.1.

Hansen, V.L., and Olive, A., 2010, Artemis, Venus: The largest tectonomagmatic feature in the solar system?: Geology, v. 38, no. 5, p. 467–470, doi:10.1130/G30643.1.

Hansen, V.L., and Phillips, R.J., 1995, Formation of Ishtar Terra, Venus: Surface and gravity constraints: Geology, v. 23, no. 4, p. 292–296, doi:10.1130/0091-7613(1995)023<0292:FOITVS>2.3.CO;2.

Hansen, V.L., and Young, D.A., 2007, Venus's evolution: A synthesis, *in* Cloos, M., Carlson, W.D., Gilbert, M.C., Liou, J.G., and Sorensen, S.S., eds., Convergent Margin Terranes and Associated Regions: A Tribute to W.G. Ernst: Geological Society of America Special Paper 419, p. 255–273, doi:10.1130/2006.2419(13).

Hansen, V.L., Willis, J.J., and Banerdt, W.B., 1997, Tectonic overview and synthesis, *in* Bouger, S.W., Hunten, D.M., and Phillips, R.J., eds., Venus II—Geology, Geophysics, Atmosphere, and Solar Wind Environment: University of Arizona Press, p. 797–844.

Hartmann, W.K., 2005, Martian cratering 8: Isochron refinement and the chronology of Mars. Icarus, v. 174, no. 2, p. 294–320.

Herrick, R.R. and Rumpf, M.E., 2011, Postimpact modification by volcanic or tectonic processes as the rule, not the exception, for Venus craters: Journal of Geophysical Research, v. 116, E02004, doi:10.1029/2010JE003722.

Herrick, R.R., Sharpton, V.L., Malin, M.C., Lyons, S.N., and Feely, K., 1997, Morphology and morphometry of impact craters, *in* Bouger, S.W., Hunten, D.M., and Phillips, R.J., eds., Venus II—Geology, Geophysics, Atmosphere, and Solar Wind Environment: University of Arizona Press, p. 1015–1046.

Hunten, D.M., 2002, Exospheres and Planetary Escape, *in* Mendillo, M., Nagy, A., and Waite, J.H., eds., Atmospheres in the Solar System: Comparative Aeronomy: American Geophysical Union Geophysical Monograph, v. 130, p. 191–202.

Ivanov, M.A., and Head, J.W., 1996, Tessera terrain on Venus: A survey of the global distribution, characteristics, and relation to surrounding units from Magellan data: Journal of Geophysical Research, v. 101, p. 14861–14908, doi:10.1029/96JE01245.

Izenberg, N.R., Arvidson, R.E., and Phillips, R.J., 1994, Impact crater degradation on Venusian plains: Geophysical Research Letters, v. 21, p. 289–292.

Lebonnois, S., Luz, D., Wilson, C.F., Hueso, R., Drossart, P., Piccioni, G., Sanchez-Lavega, A., Titov, D., Baines, K.H., and Taylor, F.W., and theVIRTIS/Venus Express Team, 2006, Venus atmospheric dynamics from VIRTIS on Venus Express—preliminary results: European Planetary Science Congress, Berlin, Germany, p. 24, http://articles.adsabs.harvard.edu/full/2006epsc.conf..424L.

López, I., 2011, Embayed intermediate volcanoes on Venus: Implications for the evolution of the volcanic plains: Icarus, v. 213, no. 1, p. 73–85, doi:10.1016/j.icarus.2011.02.022.

Mackwell, S.J., Zimmerman, M.E., and Kohlstedt, D.L., 1998, High-temperature deformation of dry diabase with application to tectonics on Venus: Journal of Geophysical Research, v. 103, p. 975–984, doi:10.1029/97JB02671.

McKenzie, D., McKenzie, J.M., and Saunders, R.S., 1992a, Dike emplacement on Venus and on Earth: Journal of Geophysical Research, v. 97, p. 15977–15990, doi:10.1029/92JE01559.

McKenzie, D., Ford, P.G., Johnson, C., Parsons, B., Sandwell, D., Saunders, S., and Solomon, S.C., 1992b, Features on Venus generated by plate boundary processes: Journal of Geophysical Research, v. 97, no. E8, p. 13533–13544, doi:10.1029/92JE01350.

McKinnon, W.B., Zahnle, K.J., Ivanov, B.A., and Melosh, H.J., 1997, Cratering on Venus: Models and observations, *in* Bouger, S.W., Hunten, D.M., and Phillips, R.J., eds., Venus II—Geology, Geophysics, Atmosphere, and Solar Wind Environment: University of Arizona Press, p. 969–1014.

Mège, D., and Reidel, S.P., 2001, A method for estimating 2D wrinkle ridge strain from fault displacement scaling applied to the Yakima folds: Geophysical Research Letters, v. 28, no. 18, p. 3545–3548, doi:10.1029/2001GL012934.

Nimmo, F., and McKenzie, D., 1998, Volcanism and tectonics on Venus: Annual Review of Earth and Planetary Sciences, v. 26, p. 23–51, doi:10.1146/annurev.earth.26.1.23.

Phillips, R.J., and Hansen, V.L., 1994, Tectonic and magmatic evolution of Venus: Annual Review of Earth and Planetary Sciences, v. 22, p. 597–656, doi:10.1146/annurev.ea.22.050194.003121.

Phillips, R.J., and Hansen, V.L., 1998, Geological evolution of Venus: Rises, plains, plumes and plateaus: Science, v. 279, p. 1492–1497, doi:10.1126/science.279.5356.1492.

Phillips, R.J., and Izenberg, N.R., 1995, Ejecta correlations with spatial crater density and Venus resurfacing history: Geophysical Research Letters, v. 22, no. 12, p. 1517–1520.

Phillips, R.J., and Lambeck, K., 1980, Gravity fields of the terrestrial planets: long-wavelength anomalies and tectonics: Reviews of Geophysics and Space Physics, v. 18, p. 27–76, doi:10.1029/RG018i001p00027.

Phillips, R.J., Raubertas, R.F., Arvidson, R.E., Sarkar, I.C., Herrick, R.R., Izenberrg, N., and Grimm, R.E., 1992, Impact craters and Venus resurfacing history: Journal of Geophysical Research, v. 97, p. 15923–15948.

Phillips, R.J., Bullock, M.A., and Hauck, S.A., II, 2001, Climate and interior coupled evolution on Venus: Geophysical Research Letters, v. 28, p. 1779–1782, doi:10.1029/2000GL011821.

Schaber, G.G., Strom, R.G., Moore, H.J., Soderblom, L.A., Kirk, R.L., Chadwick, D.J., Dawson, D.D., Gaddis, L.R., Boyce, J.M., and Russell, J., 1992, Geology and distribution of impact craters on Venus: What are they telling us?: Journal of Geophysical Research, v. 97, p. 13257–13302.

Schubert, G., and Sandwell, D.T., 1995, A global survey of possible subduction sites on Venus: Icarus, v. 117, p. 173–196, doi:10.1006/icar.1995.1150.

Schubert, G.S., Solomatov, V.S., Tackely, P.J., and Turcotte, D.L., 1997, Mantle convection and thermal evolution of Venus, *in* Bouger, S.W., Hunten, D.M., and Phillips, R.J., eds., Venus II—Geology, Geophysics, Atmosphere, and Solar Wind Environment: University of Arizona Press, p. 1245–1288.

Simons, M., Solomon, S.C., and Hager, B.H., 1997, Localization of gravity and topography: constraints on the tectonics and mantle dynamics of Venus: Geophysical Journal International, v. 131, no. 1, p. 24–44, doi:10.1111/j.1365-246X.1997.tb00593.x.

Smrekar, S.E., and Stofan, E.R., 1997, Corona formation and heat loss on Venus by coupled upwelling and delamination: Science, v. 277, no. 5330, p. 1289–1294, doi:10.1126/science.277.5330.1289.

Smrekar, S.E., Kiefer, W.S., and Stofan, E.R., 1997, Large volcanic rises on Venus, *in* Bouger, S.W., Hunten, D.M., and Phillips, R.J., eds., Venus II—Geology, Geophysics, Atmosphere, and Solar Wind Environment: University of Arizona Press, p. 845–879.

Solomon, S.C., Head, J.W., Kaula, W.M., McKenzie, D., Parsons, B., Phillips, R.J., Schubert, G., and Talwani, M., 1991, Venus tectonics: Initial analysis from Magellan: Science, v. 252, p. 297–312, doi:10.1126/science.252.5003.297.

Stofan, E.R., Hamilton, V.E., Janes, D.M., and Smrekar, S.E., 1997, Coronae on Venus: Morphology and origin, *in* Bouger, S.W., Hunten, D.M., and Phillips, R.J., eds., Venus II—Geology, Geophysics, Atmosphere, and Solar Wind Environment: University of Arizona Press, p. 931–968.

Stofan, E.R., Anderson, S.W., Crown, D.A., and Plaut, J.J., 2000, Emplacement and composition of steep-sided domes on Venus: Journal of Geophysical Research, v. 105, no. E11, p. 26757–26771, doi:10.1029/1999JE001206.

Strom, R.G., Schaber, G.G., and Dawson, D.D., 1994, The global resurfacing of Venus: Journal of Geophysical Research, v. 99, p. 10899–10926.

Taylor, F., and Grinspoon, D., 2009, Climate evolution of Venus: Journal of Geophysical Research, v. 114, doi:10.1029/2008JE003316.

Vita-Finzi, C., Howarth, R., Tapper, S., and Robinson, C., 2005, Venusian craters, size distributions and the origin of coronae, *in* Foulger, G.R., et al., eds., Plates, Plumes, and Paradigms: Boulder, Colorado, Geological Society of America Special Paper 388, p. 815–823, doi:10.1130/0-8137-2388-4.815.

Watters, T.R., 1992, System of tectonic features common to Earth, Mars and Venus: Geology, v. 20, p. 609–612, doi:10.1130/0091-7613(1992)020<0609:SOTFCT>2.3.CO;2.

Wetherill, G.W., 1990, Formation of the Earth: Annual Review of Earth and Planetary Sciences, v. 18, p. 205–256, doi:10.1146/annurev.ea.18.050190.001225.

Wilson, L., and Head, J.W., 2002, Tharsis-radial graben systems as the surface manifestation of plume-related dike intrusion complexes: Models and implications: Journal of Geophysical Research, v. 107, no. E8, p. 5057–5081, doi:10.1029/2001JE001593.

Manuscript Accepted by the Society 16 April 2012

The Geological Society of America
Special Paper 492
2012

Best practices on how to design Google Earth tours for education

Richard Treves*

School of Geography and the Environment, University of Southampton, Southampton SO17 1BJ, UK

John E. Bailey

Scenarios Network for Alaska & Arctic Planning, University of Alaska Fairbanks, Fairbanks, Alaska 99709, USA

ABSTRACT

Google Earth tours (GETs) are recorded flights around Google Earth. They are highly engaging to watch and have great potential for communicating spatially in a teaching environment. They also benefit from being easy for an educator to produce but they can be ineffective if they are designed poorly. With this in mind, in this paper we cover three main topics: (1) we consider how best to produce GETs, (2) we deconstruct them as a communication media and finally (3), we consider the larger educational context in which they are used. By reviewing literature relevant to these areas we produce 19 best practices for using GETs in education. The amount of evidence we can show in support of our best practices varies. Those that were generated by comparing GETs to the well-researched area of educational animations are highly reliable because they are based on empirical evidence. Those associated with the virtual flights between locations within a GET are more open to interpretation as they have been less well studied. We conclude that further work should be focused on investigating virtual flight within a GET.

INTRODUCTION

A Google Earth tour (GET) is a recorded camera flight around Google Earth (GE), which can be saved, replayed, and edited. Similar "tour" functionality has previously existed in geographic information system (GIS) software (e.g., Shephard, 2003) but GETs are much easier to produce when compared to this functionality. To record a tour, the user simply clicks a record button, navigates around in GE's three-dimensional (3-D) browser window as desired and then clicks "stop" when the tour is complete. An audio narrative can be added and author-created features (e.g., place marks) stored in the Places column can be turned on and off by the tour. The low level of skill required to author a GET is one of the primary reasons for advocating the use of GETs for education as it puts it within reach of the average educator.

A GET within GE can be usefully used to *visualize* spatial data in a narrative format. It is important to differentiate this from teaching involving GIS which is concerned mostly with *analyzing* spatial data. So a sensible use of a GET may well be to take the results from a GIS analysis and output these in a GET for presentation to interested parties.

The literature concerning both GETs proper and other techniques of producing "tours" within GE, such as recording a video of the screen, has generally dealt with them in a positive, non-critical, descriptive manner (Bomar, 2009; Walden, 2011; Green

*R.W.Treves@soton.ac.uk

Treves, R., and Bailey, J.E., 2012, Best practices on how to design Google Earth tours for education, *in* Whitmeyer, S.J., Bailey, J.E., De Paor, D.G., and Ornduff, T., eds., Google Earth and Virtual Visualizations in Geoscience Education and Research: Geological Society of America Special Paper 492, p. 383–394, doi:10.1130/2012.2492(28). For permission to copy, contact editing@geosociety.org. © 2012 The Geological Society of America. All rights reserved.

and Mouatt, 2008; Stott et al., 2009). The exception is Priestnall and Cowton (2009) who do address some issues of use. This non-critical approach does little to help educators understand the true potential of GETs and what educational contexts they are best suited to. A similar criticism has been leveled at the literature concerning the educational use of virtual worlds by Dalgarno and Lee (2010) as we discuss later. In this paper, we aim to examine literature related to GETs and advocate a set of best practices that educators can then apply to using GETs in the classroom.

We stated above that ease of use for *authors* is an important characteristic of GETs in education. Ease of use for *users* is the next most important characteristic. The user controls a GET through a set of buttons on screen similar to those on a video-cassette recorder (VCR). This operation is similar to playing a video clip and is therefore a very user-friendly experience. Figure 1 illustrates a GET using a series of screen shots.

GE has been used in other educational contexts (Heavner et al., 2011; Turk et al., 2011), and in these situations users can

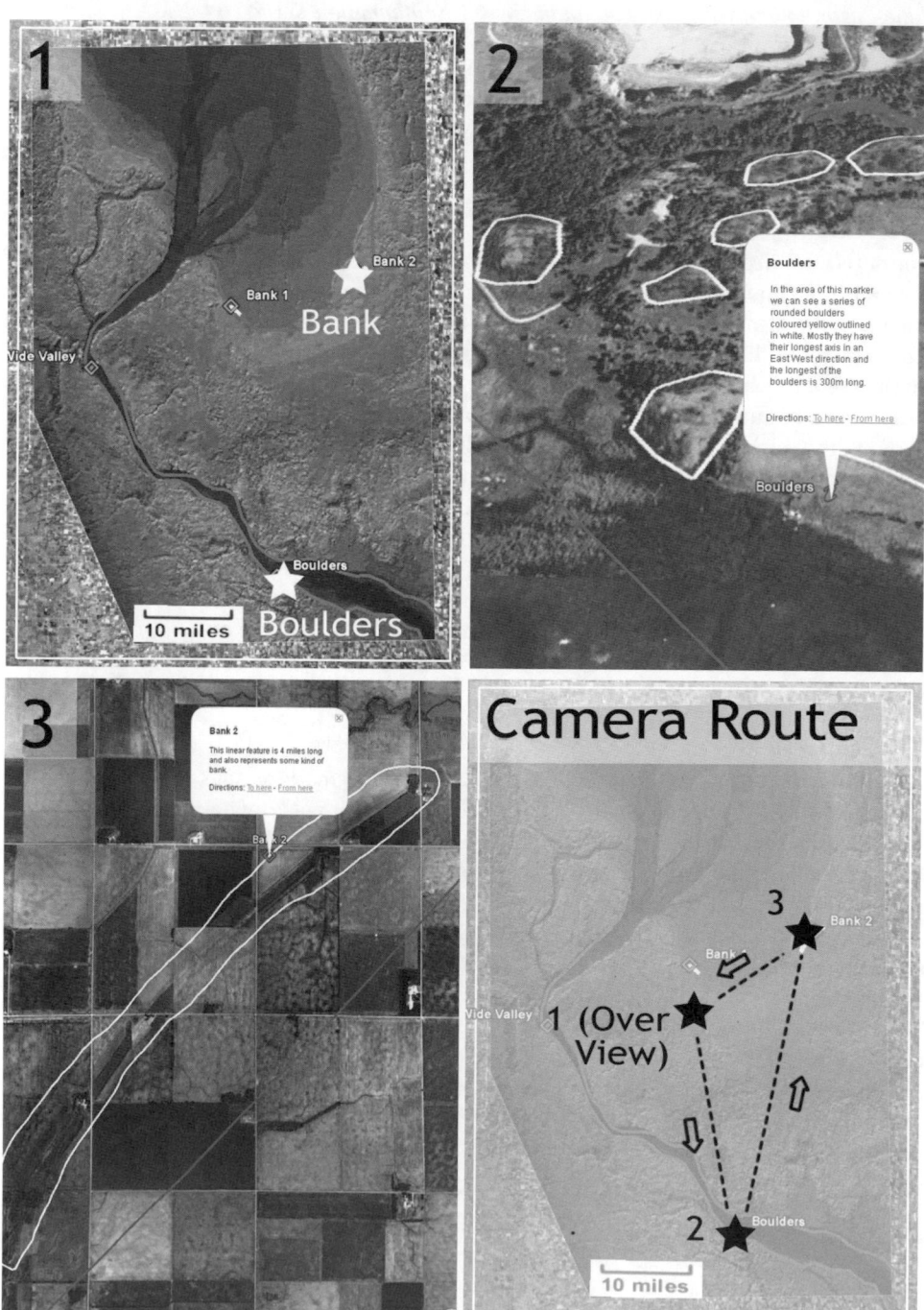

Figure 1. A Google Earth tour explaining paleo-glacial landscape is illustrated by a series of screen shots. In educational use it would be supported by an audio narration. It starts at a high altitude view showing topography (panel 1) with two locations to be visited marked ("Bank," "Boulders"). This initial view is followed by the camera zooming in to visit the Boulders location (panel 2) showing rounded boulders and a feature description. The camera then flies in a looped path out to high altitude and back down to investigate a bank feature (panel 3). After audio description of this feature the camera flies back to panel 1 and the geographical evidence (topography, boulders, bank) is interpreted.

navigate around the globe using five degrees of freedom (altitude, latitude, longitude, camera bearing, camera pitch angle). Within a GET, these five degrees of freedom are reduced to one: play speed (which encompasses play, fast forward, rewind). This is an advantage of GETs because when given multiple degrees of freedom in a 3-D environment such as GE there a number of potential problems for students:

- Getting "lost" in virtual space (Hanson et al., 1997).
- Encountering "desert fog" where all that is shown is a blank white or colored screen with no visual clues to the student's location because they have moved too close to a virtual surface such as a wall (Jul and Furnas, 1998).
- Navigating past key visual information, e.g., flying quickly past the crater of the volcano, failing to notice it and concluding it is a mountain rather than a volcano.
- Flying inside a model, such as a house, that was constructed to be viewed from the outside. Not only is the visualization on the inside confusing, the flight through the wall can be disconcerting.

A well-designed GET using an appropriate narrative and visual annotations either completely solves or at least mitigates these issues. The central purpose of this paper is primarily to outline how to design an effective GET for education using best practices that avoid the above issues, allowing the advantages of 3-D visualization to be leveraged effectively.

The approach taken in this paper is directed toward producing GETs with tangible, educational benefit rather than attention-grabbing aesthetic value. GETs can use advanced Keyhole Markup Language (KML) to create "complex" visualizations such as 3-D blocks rising out of the ground to reveal geological cross sections (De Paor And Whitmeyer, 2011; Fig. 2). However, in this discussion we eschew these complex GETs and discuss only "simple" GETs because these are within the grasp of the average educator to produce. We define a simple GET as one which does not suffer from significant view-impeding problems from objects within the tour and which uses functionality that can be produced by recording the tour with GE alone without the editing of any KML code. Important features that require editing of KML code and therefore are beyond our definition include the ability to insert pauses in the GET (so the user is forced to press "play" to make the tour continue) or functionality such as animated updates, by which objects can be moved or resized during a tour.

In addition to limiting discussion to simple tours, the technical and practical details about how to produce a tour are not part of this paper, for three reasons: first, a number of tutorials are freely available on the web (e.g., http://bit.ly/s82Vem); second, by keeping discussion of technical details to a minimum, the principles outlined in this paper will remain valid regardless of GE interface changes in the future; in addition, they will also apply to GET-like functionality appearing in other virtual globes such as ArcGIS Explorer.

In the rest of this paper we begin by reviewing relevant literature to GETs in education. We then move on to consider the process of producing a GET leading to the first of a set of best practices. Next we explain two self-defined terms useful to our discussion: "GET slides" and "GET flights," and use these to deconstruct GETs as a communication media proposing further best practices. This is followed by a deconstruction of how GET characteristics best benefit the student and then a section exploring the applications of GETs: in what educational situations are

Figure 2. A screenshot of a three-dimensional block that is animated to rise out of the ground in Google Earth. Courtesy DigitalPlanet.org (http://www.digitalplanet.org/API/SOS/index.html).

they particularly well-suited? In the conclusion we review the reliability of our best practices and suggest new avenues suitable for further research.

BACKGROUND LITERATURE

We have searched for discussion of how to best produce GETs or 3-D tours similar to GETs in the literature. We found few examples, so we broadened our search beyond "tour" literature to other non-tour topic areas which we consider relevant. We have grouped the resulting literature into five groups below: pan and zooming, navigation in virtual globes and worlds, tours, cartographic practices, and educational reviews.

Pan and Zooming

There is a significant set of literature on the effective use of "large information spaces" where navigation around two-dimensional (2-D) space such as a map or a graphic such as a family tree is achieved by panning and zooming. The applications used vary widely and are often custom built for the experiment concerned. Furnas and Bederson (1995) discuss a graphical system of representing zoom and pan tracks. They also performed informal tests concerned with flights from point to point; they found that zooming out and then back in with some panning was more popular than routes using pure panning. Ahmed and Eades (2005) discuss automatic camera generation in a zoom and pan environment. They state that it is important to keep landmarks in view at all times but do not offer any evidence for this hypothesis. Igarashi and Hinckley (2000) created automatic zooming so that when scrolling through a large document at speed the user was automatically zoomed out. Conversely, as the scrolling slowed they were zoomed back in again. Informal testing showed that users preferred it to normal scrolling. Hornbaek et al. (2002) investigated pan and zoom searches on maps with and without overview maps. Although users preferred having an overview available, the results from their testing showed that in some cases it slowed down their ability to perform searches.

Van Wijk and Nuij (2003) discussed how best to calculate a zoom and pan path from a close up of map location "A" toward a close up of map location "B." This was done via a looped path, which zooms out and then zooms in again. Unfortunately they make assumptions about the cognitive efficiency of the flight, such as the prime importance of making the path smooth and continuous, without any reference to user tests. They did perform an informal test to find user preferences for speed and shape of the loop between two targets. Their results found there was a wide range of preferences in both variables, but it is noticeable that they did not test users' performance in any way.

Navigation in Virtual Globes and Worlds

There is a distinction between the terms *virtual globe* and *virtual world*. The former are either attempts to mirror the "real-world" (e.g., Google Earth) or present a specialized representation of data (e.g., Software MacKiev's 3-D Weather Globe and Atlas). A virtual world does not follow real-life constraints and creates an imaginary landscape (e.g., Second Life). The common feature between the two is that the user must navigate through a 3-D environment (Bailey, 2010; Warburton, 2009).

A review of Human Computer Interaction literature reveals that researchers in this field are focused on two topics relevant to our discussion. The first is wayfinding tasks in 3-D environments, which is analogous to the task of finding your way to a given location in an unknown city in real space. To be able to follow such a route, users have to have far more freedom of movement in these environments than they do with a GET so the literature has limited relevance to our consideration of GETs. However, Darken and Sibert (1993) is relevant, as they investigated how good users were at search and navigation of a 2-D virtual world when provided with a radial grid or an overview map of the world. Broadly, they discovered that both overview maps and grids aided search and location tasks. A grid is a type of landmark that we discuss later.

The second topic area is reducing the degrees of freedom available in a 3-D environment. Hanson et al. (1997) discuss constraining the six degrees of freedom (three location and three camera degrees of freedom) commonly available in a virtual world in various manners. For example, they produce an environment where movement is constrained to two dimensions. An example application is inspecting the features of a tower. User movement is constrained to the surface of a cylinder surrounding the tower at a set distance and the user's view is forced to point at the tower wall. These constraints mean that users avoid navigating through the walls of the tower, which can be very confusing. It also means they cannot turn their camera to look away from the tower and lose sense of where the tower has gone. The informal tests results showed that constraining degrees of freedom proved effective for learning. Drucker and Zeltzer (1994) implement a similar idea, where they designed an interface which uses "intelligent" cameras that align themselves with useful views as users move in 3-D space in order to avoid lost-in-space issues.

Tours

There is a lack of literature on tours in education similar to GETs whether in GE or other 3-D software or in real, rather than virtual, space. Chittaro et al. (2003) briefly discusses tours in a virtual world but their discussion is centered on using avatar tour guides and also discusses problems of tours indoors where avoiding furniture by the best path is important. Object avoidance is not generally an issue in GETs. Priestnall and Cowton (2009) discuss a GET they produced (in video format) showing landscape drawings. Their design-related discussion is limited to noting that during flights a key landmark should be kept in view to help the user fix their movement and the finding that fast, low flights are disorienting.

Wu et al. (2009) discuss wayfinding within an urban virtual world populated with 3-D buildings. They tested their users' ability to navigate virtual streets on foot aided by three different tools: an overview map, text instructions that appeared on screen while the user was moving, and an initial "tour" similar to a GET in which users flew from point "A" to point "B" giving them the benefit of an aerial view. Users then attempted to navigate back from "B" to "A" on foot through streets. The results show that text instructions were the least helpful aid to navigation, and the flying tour was also less useful than an overview map. We think this unfairly measures the efficacy of the GET-like tour because in the tour condition users had to remember spatial information from what they had seen (flying from "A" to "B") as they attempted to return from "B" to "A" on foot. In the overview map condition they could access the spatial information at all times from the overview map.

Goldin et al. (1981) set up an experiment in real space where they sent users on a tour of an urban area and tested their memories of it afterwards. The major conditions were comparing a real bus tour to a film taken from the bus. They also supplemented these two conditions with extra tools; some users used a landmark overview map and others had audio narration that informed users of angles of turns and distances from base while on the real or film tour. Despite the lack of richness of the film, during subsequent testing, users recalled similar amounts to those on the real bus journey. Goldin et al. (1981) explain these finding in two ways. First, on the bus tour, maintaining a clear forward view was difficult. Second, the cameraman taking the film focused the tour view on the correct objects thus minimizing distractions.

Cartographic Practices

Users' understanding of GETs is strongly affected by the effectiveness of symbols and the cartography used within GETs. Of particular interest is the change of symbolization when zooming in or out of a map and how this impacts students' comprehension (Zhang, 2005). We also concur with Harrower (2003) that generally symbolization involved in a map animation should be simpler than on a static map because of the extra cognitive load in tracking the movement of an animation. However, because the same practices apply equally to the educational use of any type of dynamic map system and are not specific to GETs, we do not cover such topics here and refer the reader to Brewer (2005).

Educational Reviews

There are two papers that are relevant to our discussion from the general educational literature: the first discusses virtual worlds and the second, guided learning.

Virtual worlds can be defined as 3-D environments in which user has multiple degrees of freedom to move and the ability to look in whatever direction they choose. They differ from GE by using avatars and representing imaginary landscapes whereas GE usually does not involve the use of avatars and mirrors our own planet's landscape. Dalgarno and Lee (2010) review and comment on the literature relating to education using virtual worlds. They criticize the discussion as being overly descriptive and lacking proper critical appraisal of the strengths of these applications as educational tools.

Mayer (2004) reviews the literature comparing "discovery learning," where students are left with great freedom to effectively teach themselves, to "guided learning," where students are engaged in learning activities but also receive guidance and feedback by an educator. He uses empirical evidence from a number of studies spanning three decades to show that discovery learning is inferior to guided learning. In the background to the paper he gives an overview of the widely accepted theory of constructivism and he also accepts this theory to the extent that to learn effectively students should be engaged in activities rather than just passive listening or watching.

PROCESS OF AUTHORING GETs

In this section, we examine the context of GETs in education before getting into the detail of GET characteristics in later sections. We begin by discussing the process of testing, iterating, and maintaining focus on the educational value of a GET. This process is applicable to any technology-enhanced learning intervention but is particularly important in producing GETs because they are a complex interaction of layers, camera angles, and flight speed. Our discussion, both of this and later sections, is summarized as a collection of best practices (Table 1). GE is a highly usable tool with which educators can produce GETs. Unfortunately, it is still possible for educators to create a confusing tour with poor usability. To understand why this happens, it is helpful to consider usability more generally. Gould and Lewis (1985) outline three principles of usability that should apply to any computer interface design development process:

- *Test the design*—The design should produce prototypes and these should be tested by real users who use them to solve realistic problems.
- *Iterate*—Results from testing should be used to redesign the system in an iterative manner.
- *Focus on users*—Involve the end users of the system in discussion with the designers from the beginning and ensure the designers react to their input.

Although Gould and Lewis (1985) were discussing interfaces and a GET can only loosely be described as an interface, we think their discussion is relevant when designing GETs because GETs have the potential to be complex from the user perspective because of the multiple layers, scale, camera angles, camera movement, types of symbology, and labels available. With this in mind, we address Gould and Lewis's (1985) three points in turn below.

For educators, testing the design of a new lesson, or part of a lesson, is often done through the use of student feedback questionnaires. The change is introduced, student feedback collected, and the instructors then reflect on the impact of the change. This

TABLE 1. BEST PRACTICES FOR GOOGLE EARTH TOURS (GETs) IN EDUCATION

No.	Subject area		Best practice description	Evidence
	Major	**Minor**		
1	Producing process	Iteration and testing	GETs should be generated in an iterative process that incorporates user testing.	Scholarly
2		Student focus	Production of GETs should be student-focused; in particular, educational value should be promoted above flashy presentation.	Scholarly
3	GET slides	Narration, annotations and labels	Within a GET slide, narration should be delivered by audio narration alone.	Empirical
4			Audio narration should relate to what is on screen at the time.	Empirical
5			Labels and annotations should be close in space and time to the map elements they are describing.	Empirical
6			Labels and annotations should be used often but without impacting visual clarity.	Empirical
7		Chart junk	In formal educational contexts, graphical and audio elements not directly connected with the educational message should be removed.	Empirical
8		Personalization	Within a GET narration, less formal language should be used.	Empirical
9		Using animations	Within a GET, rates of change of variables such as population growth are better depicted on a graph than on a map as a changing color or symbol.	Scholarly
10			The visual complexity of GETs should be made simpler than comparative static maps where possible.	Scholarly
11	GET virtual flights	Speed of flight	Speed in flights should be slower with rising complexity on screen and should be in the range of 0.5 to 8 scales per second.	Scholarly
12		Camera angles	The camera in a GET should be kept orientated northward and vertically downward, unless the subject of the GET requires viewing from an angle.	Empirical
13		Looped paths and overviews	In GET flights between two low-altitude slides, GET flight paths should always pass through a high loop which encompasses both locations in the same view.	Empirical
14			If possible, a GET should start and end at high altitude capturing the important points, lines or areas the GET describes in one view.	Empirical
15		Acceleration and deceleration	GET flights between waypoints should accelerate at the start and decelerate at the end.	Scholarly
16	Combined GET slides and virtual flights	Grids for navigation and scale	In a landscape, which has few strong natural landmarks such as roads or coastlines, a static, north-south–oriented grid should be added unless it creates a cluttered view.	Scholarly
17			The grid should have a line separation of a suitable round figure to give a sense of scale. If a grid is not applicable, scale can be included as a line on the ground representing some round figure.	Scholarly
18	Embedding GETs in earth science teaching	Topics that are effective when presented as GETs	The use of a GET should be particularly considered when illustrating three-dimensional topography, data over a range of scales/locations, and/or introducing a GE map collection.	Logical
19		Creating activities for GETs	GETs should be used to support activity-based teaching.	Empirical

has value but when the change involves introducing a GET we would advocate doing extra testing in addition to the usual feedback and reflection. Our choice would be to set up a test and observe a single student trying out the GET prior to use in the classroom. The methodology outlined by Nielsen (1994) on testing websites is applicable.

In terms of iteration, the single student observation we suggest has the advantage that the results can be used to edit the GET, improving it before it is used in a real classroom situation. In our own practice using this technique we concur with Nielsen's (1994) findings that asking students what they thought of an educational program after they have used it reveals less than observing a student using it directly. These two points lead to best practice 1:

BP1] GETs should be generated in an iterative process that incorporates user testing.

Focusing on users' needs is of central importance and we would advocate it as being the most important of the three principles listed above. When introducing a GET, educators need to consider, "What exactly do I expect my students to get from this GET and how is that better than using a familiar teaching medium such as PowerPoint?" A common problem with GETs based on the authors' own experiences is that it is tempting when producing one to focus on flashy and attention-grabbing effects, thereby losing focus on the educational objectives. For example, rapid and extreme camera movements between two camera locations are possible within a GET. These grab student's attention but a more simple curved flight up and down between the two locations is simpler to follow and perceptually gives students important knowledge about the relative positions of the two sites (Priestnall and Cowton, 2009). While an attention-grabbing GET may be appropriate for an informal learning situation like a museum, we think it is inappropriate for a formal education situation. Faced with "flashy" tours in every lecture, students will soon tire of the effect and any attention benefit disappears. This point leads to best practice 2:

BP2] Production of GETs should be student-focused; in particular, educational value should be promoted above flashy presentation.

SLIDES AND GET FLIGHTS

In order to discuss how to author effective GETs for education we use a framework that splits a GET into two parts, "GET slides" and "GET flights." In this section, we will first define what these terms mean then discuss best practices for producing these two different parts of a GET separately. We then move on to deal with issues that apply to both parts.

Definitions

In presentational software such as PowerPoint, the content is split into slides with transitions between them such as fading or a turning-page animation. In our following discussion a "GET slide" denotes a map-based animation taking place in GE where the camera is static but other layers or elements such as labels may turn on/off. A "GET flight" is a transition between slides where the camera pans and/or zooms to a different position but where other changes, such as layer visibility, do not change. These two concepts are illustrated graphically in Figure 3 and the three panels of Figure 1 also serve as an example of three GET slides.

The definitions of slides and flights are helpful because GET slides are very similar to other animations in education that have been extensively researched. Also, the usability issues of following movement in space are very different from those of an animation where the camera is fixed.

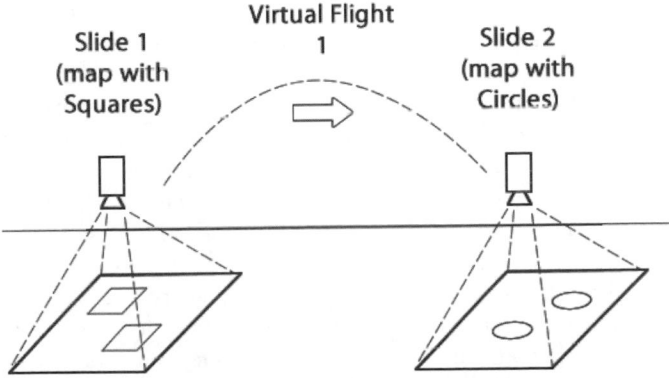

Figure 3. An illustration showing the relationship of Google Earth tour (GET) slides with GET flights. In an imaginary GET, the Google Earth camera first shows a map animation on slide 1 involving squares, then it follows a GET flight to slide 2 where it follows a map animation involving circles.

Best Practices for GET Slides

The use of animations in learning was investigated through a series of empirical experiments by Mayer and Moreno (2003). They produced a series of best practices that can be applied to the GET toolset. Many of these principles have been discussed elsewhere either in whole, or part, but we prefer to use Mayer and Moreno's (2003) framework because of the empirical evidence they have to back up their discussion. We consider their principles in turn below, grouping them together for ease of discussion.

Mayer and Moreno (2003) define *multimedia, modality*, and *redundancy principles*, which state that in educational animations, audio narration outperforms on-screen text narration. Just audio narration also outperforms on-screen text plus the same content delivered as audio narration. In this case there is no significant difference between an educational animation and a GET, so this leads to best practice 3:

BP3] Within a GET slide, narration should be delivered by audio narration alone.

Mayer and Moreno (2003) also define *spatial and temporal contiguity principles*. These state that linked concepts are most clearly associated in users' minds when watching an educational animation if they are placed close together in space and/or time. Practical examples include map annotations such as a red circle marking certain areas of a complex map being discussed by the audio narrative, labels being placed close to the cities they relate to, and when elements on screen are changing, describing the change happening in real time rather than in retrospect. There is obviously a line to be drawn between what constitutes a label and what constitutes on-screen narration, which would contradict BP1. Best practices 4 and 5 are thus:

BP4] Audio narration should relate to what is on screen at the time.

BP5] Labels and annotations should be close in space and time to the map elements they are describing.

Combing previous discussions, the findings of Drucker and Zeltzer (1994), Mayer and Moreno (2003), Hanson et al. (1997), and Goldin et al. (1981), suggest that annotations and labels have great value in guiding students to what is important within 3-D space, and in our case GET slides. This is particularly true when the map view presented within a slide is visually complex, e.g., a geological map with many irregular polygons (Fig. 4).

However, there is a balance to be struck between the value of annotations and having so many of them that the view becomes confused. Treves and Martin (2008) discuss how geographical material can be presented in simple animations including discussion of what constitutes too much labeling and annotation on animated maps. In a related discussion, Harrower (2003) discusses the advantages and disadvantages of flashing or moving

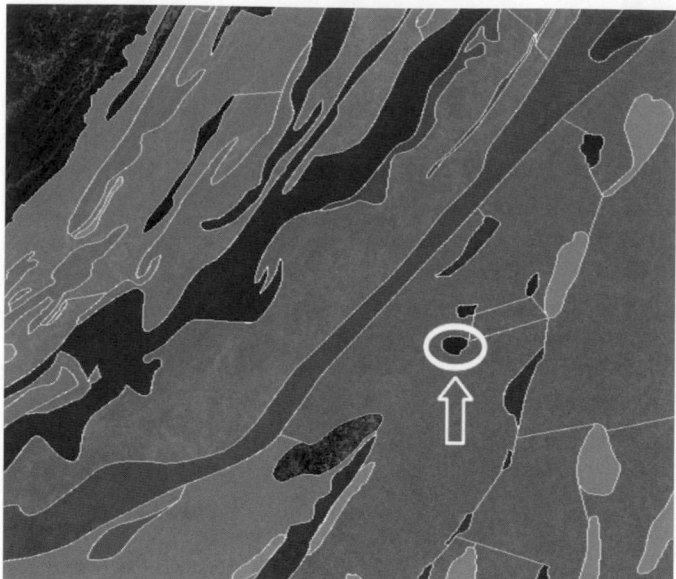

Figure 4. It would be very difficult to pinpoint the polygon shown without the use of an annotation. Image Courtesy DigitalPlanet.org (http://www.digitalplanet.org/API/SOB/index.html).

annotations to direct the users view effectively. This leads us to best practice 6:

BP6] Labels and annotations to guide the eye of the student should be used often but without impacting visual clarity.

Mayer and Moreno (2003) also found clear evidence that graphical and audio elements not directly connected with the educational message made it less effective. They called this their *coherence principle* and examples would be decorative graphics and background music. This is clearly linked to our previous discussion about being student focused and that "flashy" GETs are attention grabbing but not necessarily effective.

BP7] In formal educational contexts, graphical and audio elements not directly connected with the educational message should be removed.

The final principle from Mayer and Moreno (2003) that we will consider is the *personalization principle*. This says that language within an animation should be informal, e.g., employ the use of "you" and "I." The authors reason that this produces a more effective educational message, as students find it more personal and friendly and so it is more likely to engage them.

BP8] Within a GET narration, less formal language should be used.

Harrower (2003) discusses animations in graphs based on his experiences as a map producer, user, and from some formal tests. Several points he makes deserve discussion here. He proposes that graphs are better at showing spatial patterns rather than depicting

rates of change. This idea is exemplified by the Gapminder software (http://www.gapminder.org), which shows changes in world population by country on a graph with links to a map.

BP9] Within a GET, rates of change of variables such as population growth are better depicted on a graph than on a map as a changing color or symbol.

Harrower (2003) also discusses techniques to simplify maps, which is relevant to slides in a GET. We concur with his view that simplification is necessary. Static maps can be more complex because users have time to view them. In a GET, a user has limited time to make sense of the slide before it changes. The author suggests that visual complexity can be reduced by data smoothing, data filtering, and data class collation. His paper pre-dates the recent rapid increase in map mashups where data is often plotted over visually complex base maps. This is probably why he does not discuss the importance of simplifying the background map, which is another important way of reducing complexity on a map in a GET slide.

BP10] The visual complexity of GETs should be made simpler than comparative static maps where possible.

Best Practices for GET Flights

Unlike GET slides, there have been relatively few studies that relate to GET flights. We believe GET flights are the key to the power of GETs for education as they can be watched by a student who can track change in location and scale with little effort because the GET flight taps into the student's visual system. By comparison, a traditional textbook must use graphical structure such as overview maps to communicate the same information and as discussed. Results from Hornbaek et al. (2002) indicate that the mental work associated with processing such graphical structures is significant.

Treves and Engelbrecht (2011) investigated the usability of GET flights, including camera angles, camera paths, and camera speed. They did this by placing two place markers in a GE landscape with few clear landmarks. It should be remembered that the satellite data used to make up a GE base map is much more complex visually than other maps such as the road base map used in Google Maps. Users were flown from a close-up view of one point to a close-up view of the other. The points were then turned off, the view zoomed out, and users had to identify where the points were. Based on the results from these experiments, several best practices, which we discuss below, can be suggested.

When the speed of the GET flight was increased, students became less accurate when identifying the point locations. The impact of changing the speed varied depending on the other factors in the test so the authors do not identify an optimum speed for GETs. However, the middle of the three speed values used by Treves and Engelbrecht (2011) did not appear to cause users performance problems in the test while being twice as fast as the slower speed. Speed in Google Earth is relative, but one way to quantify it is in terms of zoom "scales." The middle speed in the Treves and Engelbrecht (2011) study was 0.5 scales per second, i.e., when zooming out it took two seconds for a 10-pixel-length line to reduce on-screen to five pixels long. This compares with another study (Guo et al., 2000) where the virtual world was much simpler visually than the types of view GE generates, and a zoom speed of eight scales per second was found to be optimum.

BP11] Speed in GET flights should be slower with rising visual complexity. They should be in the range 0.5–8 scales per second.

It is possible in a GET to orient the camera at any compass bearing and angle to the ground. A simple flight between two points in GE mimics the behavior in web mapping systems such as Google Maps where the camera is kept vertically downward and oriented northward throughout. A more complex movement could have the camera angle and orientation changing in addition to the zoom and pan changes. Within the Treves and Engelbrecht (2011) test, there was a condition where users were flown from point to point with the camera angle varying through the flight. When compared to the simple GET flight, the complex flight made it more difficult for users to remember the points. As before, confusion was increased when combined with other variables, e.g., changing the speed.

However, there are some examples of use where an angled, non-vertical camera view is important, e.g., when illustrating the view down a U-shaped valley. These two considerations lead to:

BP12] The camera in a GET should be kept orientated northward and vertically downward, unless the subject of the GET requires viewing from an angle.

This best practice also makes it possible to enhance the users' locational sense as they are more likely to recognize a coastline (for example) if it is presented in a north upward orientation.

The final condition of the Treves and Engelbrecht (2011) test was that the camera route between "A" and "B," two low-elevation points, was varied. In the "high case" it flew via high point where both "A" and "B" were clearly visible, in the "low case," it also flew to a higher point (than the markers), but from this point neither "A" nor "B" were visible. This condition represents the potential problem of low-altitude flights (Priestnall and Cowton, 2009). Relative to the changing speed and camera angles, the change of altitude had the strongest effect on students' ability to locate the points, which leads to the best practice:

BP13] In GET flights between two low-altitude slides, GET flight paths should always pass through a high loop which encompasses both locations in the same view.

Given this result, the positive reaction of users when given access to overview maps (Hornbaek et al., 2002), and Priestnall and Cowton's (2009) suggestion of keeping key landmarks in view, another best practice is suggested:

BP14] If possible, a GET should start and end at high altitude capturing the important points, lines, or areas the GET describes in one view.

A feature not well covered by Treves and Engelbrecht (2011) is that of GET flight acceleration. In a GET it is possible either to fly at a constant speed or with acceleration from the start and deceleration at the end. Mackinlay et al. (1990) describe the advantages of a flight speed that decelerates as it approaches its goal. Zhang (2005) goes further and suggests that there should be:

1. An initial acceleration.
2. A deceleration as the camera reaches its final destination.
3. Deceleration as the mid-point is approached.
4. Acceleration as the mid-point it is passed.

The applicability of parts 3 and 4 depends on if the GET flight follows the high loop mentioned in best practice 13. These stages are illustrated graphically in Figure 5.

Application of this approach represents an efficient use of time within a GET and also helps meet the user's need to track their movement. In the middle of the flight the visual information the student is getting from the flight is low (they just need to track the movement of the GE camera), whereas at the start and end of the GET flight the information needs rise as the student attempts to make sense of where they are and what other layers are being shown on screen.

BP15] GET flights between waypoints should accelerate at the start and decelerate at the end.

Here a "waypoint" indicates either a slide or a mid-point of a high loop as noted in BP13.

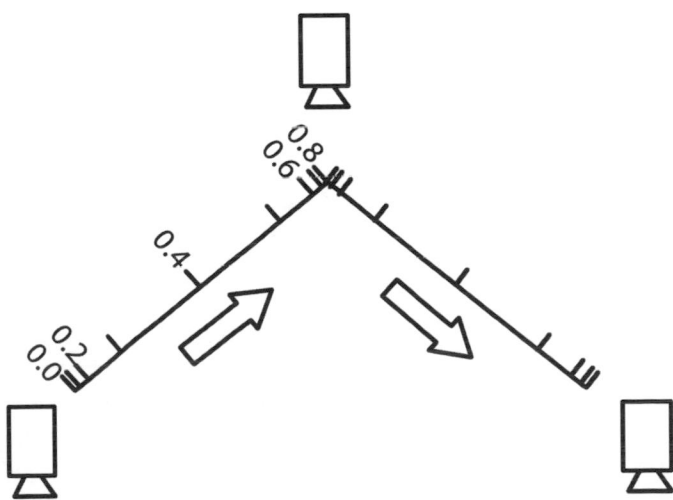

Figure 5. The suggested accelerations and decelerations of a Google Earth tour flight shown as tick marks illustrating camera position every 0.1 seconds.

Best Practices Applicable to both GET Slides and GET Flights

The use of landmarks for navigation in virtual worlds has been comprehensively reviewed by Vinson (1999). A GET differs from his stated problem in that the degrees of freedom are restricted to one dimension but we believe the discussion remains relevant as users in a virtual world or in a GET are trying to make sense of the landscape they view.

One of Vinson's (1999) main recommendations is that a grid acts as a system of landmarks to assist navigation which is in agreement with Darken and Sibert (1993) who tested the effect. In a GET, a grid could be made up of polygon squares fixed to the ground over the study area. GE currently allows a latitude/longitude grid to be turned on via its menu system but we suspect this feature would not prove effective as a landmark system because it is dynamic: as you zoom out, label positions and the scale of the grid change. Darken and Sibert (1993) described how a map grid helps reduce spatial distortions, specifically clustering and axes distortions that impact users. Clustering is where users, without a strong set of alternative landmarks, remember locations based on groupings. An example would be a bus stop: if you tend to remember a location such as a café by its relation to a particular bus stop then you are likely to distort the café to bus stop distance to be smaller than it actually is. An axis distortion is the tendency to associate points with line landmarks such as a coastline and then to distort the line landmark as being more vertical or horizontal than it actually is. In distorting the line, the points associated with that line also get distorted to an incorrect position.

Laying a grid over a region will reduce these spatial distortions. However, this presupposes that the map has few strong landmarks such as roads. We hypothesize, that if a grid is overlain on such a landscape it will be less effective because the natural landmarks will compete with the grid for attention. It is our intention to explore this, along with testing our ideas about dynamic grids, as a part of future studies. It should also be noted that a grid might clutter the view creating problems of visual overload instead of reducing the effect of distortions. These points can be summarized as a best practice:

BP16] In a landscape which has few strong natural landmarks such as roads or coastlines, a static, north–south oriented grid should be added unless it creates a cluttered view.

An additional benefit of a grid is that it can help users gauge the scale of a camera view which shows a scale invariant landscape providing the square size of the grid is a round figure e.g., 500 km. The problem occurs with certain landscapes viewed in GE, for example, along the Antarctic coast; with no roads or houses, there is little visual information to help gauge absolute distances. A grid would provide this visual information, but if a grid is not appropriate, a line drawn on the surface of GE, measuring a given length, is also sufficient.

BP17] The grid should have a line separation of a suitable round figure to give a sense of scale. If a grid is not applicable, scale can be included as a line on the ground representing some round figure.

GE version 6 has a scale bar that can be made visible at the bottom of the 3-D viewer. Like the GE grid it is dynamic and (at the time of writing) it cannot be forced to show round figures. Trying to relate the length of an object with a scale bar showing ticks at non–round figure intervals (e.g., 648 m), involves a great deal of unnecessary mental effort. For this reason we discourage its use. Scale could be shown as a screen overlay (a graphic that is fixed to the screen rather than the virtual ground in GE), however, a line fixed to the virtual ground in GE is very easy to implement in GE and keeps the correct scale as the camera zooms in and out (which would be problematic to implement for a screen overlay).

Embedding GETs in Earth Science Teaching

In our consideration of best practices so far, our focus has been on how to produce effective GETs by deconstructing them. However, a GET does not make an earth science lesson or lecture on its own and we now turn to consider a more holistic view of the place of GETs in education beyond our previous discussion of the process of producing a GET.

Examples of Topics that are Effective When Presented as GETs

Learning benefits depend on what earth science topic is being taught. We believe that GETs are particularly suited to some types of teaching. We identify these areas and give specific examples below. However, it should be noted that our argument rests on logical argument and is not backed up by empirical research.

Using the 3-D topography of Google Earth. GE is particularly good at showing 3-D topography because its terrain model is detailed and the refresh rate of the screen is rapid, giving the illusion of actually moving above the ground. Examples where this could be useful are illustrating U-shaped glacial valleys and showing how geology outcrops interact with topography (Tewksbury, 2008).

Using GETs to introduce a GE map collection. An excellent resource for education or educational outreach is a GE map collection by which we mean a set of related map layers. Such a collection can be used to illustrate all manner of earth science topics. For example, tracking tropical hurricanes would contain satellite imagery on one layer and a track of the eye of the hurricane on another. Common problems with such a collection are that it is best viewed across a number of scales, and by definition it is multi-layered and requires the understanding of map symbolization (e.g., lightning strikes, wind speed symbols). When first opening such a project in GE, students are likely to be put off by the overwhelming amount of data and affordances that are available to them. We believe that authoring a GET to guide a student through the project can be an effective solution (e.g., Storm Tracking with Google Earth, http://youtu.be/2C8IIzFY-oM).

Multiple scales and locations. When students have to comprehend earth science evidence about multiple locations and across multiple scales, GETs can be an excellent aid. In Treves and Engelbrecht (2011), a GET was used to present geographical evidence from a number of locations which was then interpreted on a larger scale (Fig. 1 is derived from this GET). We believe that the GET flights between the three slides (in Fig. 1) are highly effective at explaining the changes in location and scale for students.

BP18] The use of a GET should be particularly considered when illustrating 3-D topography, data over a range of scales/locations, and/or introducing a GE map collection.

Creating Activities for GETs

A key finding from an earlier investigation into GETs in education found that students reacted to them in a passive manner (Treves, 2009). Even when explicitly encouraged to use the VCR pause/play controls in watching a GET, most participants didn't use them and allowed the GET to just play through unhindered. As discussed in the literature section (Mayer, 2004), to achieve optimum learning, students need both an effective communication medium, such as GETs, but also need to make sense of the information in a wider context by taking part in learning activities.

An example of how this can be achieved is a self-assessment question (SAQ). SAQs are a well-used format of learning where students read text and then are asked to problem solve, applying the knowledge they have just gained. Cook et al. (2006) reported empirical evidence that SAQs aided students learning. Our suggestion is that short GETs are accompanied by GE-based activities. For example, having been taught the difference between a V- and a U-shaped valley, students can be tasked with identifying examples of both landforms in a delimited area. However, if an activity within GE cannot be authored easily, a normal text SAQ can still be used in conjunction with a GET.

BP 19] GETs should be used to support activity-based teaching.

CONCLUSIONS

In this paper we set out to examine how best to author GETs to produce effective learning for students. Our approach was to focus on student usability rather than technical details. We achieved this by deconstructing what a GET offers in terms of educational communication and identifying its important characteristics. These are then cross-referenced with relevant literature to produce the best practice points listed in Table 1. These best practices both promote the strengths of GETs but also represent a critique of them as an educational tool.

The importance of discussing a technology critically in terms of its usability is strengthened by Dalgarno and Lee's (2010) discussion about virtual worlds in education. They argue that the educational literature on the topic could be improved because it has been largely descriptive with little deconstruction and critique of the value of the technology from a user's point of view. We concur with that view, and observe the same pattern in the literature concerning GETs as outlined in the introduction of this paper. We believe our table of best practices (Table 1) meets their call for critical evaluation of 3-D learning technologies.

We introduced the idea that allowing multiple freedoms of movement navigation in virtual worlds can create serious user problems, such as getting lost. Ahmed and Eades (2005), Drucker and Zeltzer (1994), Goldin et al. (1981), Hanson et al. (1997), and Igarashi and Hinckley (2000) have all identified the same issue across a range of software using 3-D information space. These five sets of authors have suggested independent solutions to the issue. Our contention is that GETs solve the "too many degrees of freedom" problem for students and, crucially for educational use, do so in a way that is easy to implement by educators without high-level information technology skills. This enables educators to leverage the power of 3-D visualizations into their teaching without having to devote an unacceptable amount of time to software-specific training.

Leading on from Gould and Lewis's (1985) usability discussion we also stated that just producing and using a GET in a teaching situation is not sufficient to produce a good educational experience for students. The GET must be produced and embedded into learning in a student-focused manner that demands testing and iteration (see best practices 1 and 2; Table 1).

Following this processed-based view of a GET in education, we moved to deconstruct a GET as a communication media. We defined a GET slide and GET flight, and by splitting the sections of a GET into these two different categories we were able to relate GET slides to the wealth of evidence-based literature on animations in learning. Our best practices 3–10 (Table 1) come out of this discussion, and we believe them to be highly reliable because they are based on evidence-based research. In contrast, the best practices associated with GET flights (BPs 11–15, Table 1) are less reliable. They refer mostly to one empirical user experiment (Treves and Engelbrecht, 2011) that directly addresses GETs and some other literature only loosely associated with GETs. Further investigative work in this area is required; for example, the best practice related to tour speed is vague and highly dependent on the complexity of the map view concerned. There are also a huge variety of routes and patterns of acceleration/deceleration that can be used in a GET flight between points, which also needs further investigation.

As well as considering the process of producing GETs and deconstructing them, it is also useful to consider the context of their use in education. There is no direct literature on embedding GETs in earth science education but there is indirectly relevant work on the benefits of guided learning. Our best practices (BP 18–19; Table 1) in this subject area rely on logic and inference from related studies. In the case of active learning, user experimentation is planned (by the authors) but other investigations into the value of GETs are encouraged. A particular experiment of interest would be to compare the effectiveness of student learning where students produce presentations supported by GETs to when they present the same material using a PowerPoint presentation.

REFERENCES CITED

Ahmed, A., and Eades, P., 2005, Automatic camera path generation for graph navigation in 3D, *in* Hong, S-H., ed., Proceedings, Asia Pacific Symposium on Information Visualization: Sydney, Australia, Conferences in Research and Practice in Information Technology, v. 45, p. 27–32.

Bailey, J.E., 2010, Entry for "Virtual Globe," *in* Warf, B., ed., Encyclopedia of Geography: Sage Publications, p. 3528.

Bomar, S., 2009, The Genocide Project: Knowledge Quest, v. 37, no. 4, p. 11.

Brewer, C.A., 2005, Designing Better Maps: A Guide for GIS Users: Redlands, California, ESRI Press, 203 p.

Chittaro, L., Ranon, R., and Leronutti, L., 2003, Guiding visitors of Web3D worlds through automatically generated tours, *in* Proceedings of the Eighth International Conference on 3D Web Technology: 9–12 March, Saint Malo, France, doi:10.1145/636593.636598.

Cook, D.A., Thompson, W.G., Thomas, K.G., Thomas, M.R., and Pankratz, V.S., 2006, Impact of self-assessment questions and learning styles in Web-based learning: A randomized, controlled, crossover trial: Academic Medicine, v. 81, no. 3, p. 231–238, doi:10.1097/00001888-200603000-00005.

Dalgarno, B., and Lee, M.J.W., 2010, What are the learning affordances of 3-D virtual environments?: British Journal of Educational Technology, v. 41, no. 10–32, doi: 10.1111/j.1467-8535.2009.01038.x.

Darken, R.P., and Sibert, J.L., 1993, A toolset for navigation in virtual environments, *in* Hudson, S.E., Pausch, R., Zanden, B.V., and Foley, J.D., eds., Proceedings of the 6th Annual Association of Computing Machinery (ACM) Symposium on User Interface Software and Technology: 3–5 November, Atlanta, Georgia, p. 157–165.

De Paor, D.G., and Whitmeyer, S.J., 2011, Geological and geophysical modeling on virtual globes using KML, COLLADA, and Javascript: Computers & Geosciences, v. 37, no. 1, p. 100–110, doi:10.1016/j.cageo.2010.05.003.

Drucker, S.M., and Zeltzer, D., 1994, Intelligent camera control in a virtual environment, *in* Proceedings of Graphics Interface: 18–20 May, Banff, Alberta, Canada, p. 190–199.

Furnas, G.W., and Bederson, B.B., 1995, Space-scale diagrams: Understanding multiscale interfaces, *in* Proceedings of Human Factors in Computing Systems: 7–11 May, Denver, Colorado, Association of Computing Machinery (ACM) Press, p. 234–241.

Goldin, S., Thorndyke, P., and Berman, J., 1981, Simulating navigation for spatial knowledge acquisition: Santa Monica, California, Rand Corporation Technical Report R-1675-ARMY.

Gould, J.D., and Lewis, C., 1985, Designing for usability: Key principles and what designers think: Communications of the Association of Computing Machinery (ACM), v. 28, no. 3, p. 300–311, doi:10.1145/3166.3170.

Green, D., and Mouatt, J., 2008, The Digital Globe: Using Google Earth for Virtual Fieldtrips of Coastal Environments, *in* Milson A., and Alibrandi M., eds., Digital Geography: Geospatial Technologies in the Social Studies Classroom, p. 147–163.

Guo, H., Zhang, V., and Jing Wu, 2000, The Effect of zooming speed in a zoomable user interface: Report from SHORE 2000, Student Human Computer Interaction Online Research Experiments (http://otal.umd.edu/SHORE2000/zoom/).

Hanson, A.J., Wernert, E.A., and Hughes, S.B., 1997, Constrained navigation environments, *in* Proceedings of Scientific Visualization: 9–13 June, Dagstuhl, Germany, p. 95–104.

Harrower, M., 2003, Tips for designing effective animated maps: Cartographic Perspectives, v. 44, p. 63–65.

Heavner, M.J., Fatland, D.R., Hood, E., and Connor, C., 2011, SEAMONSTER: A demonstration sensor web operating in virtual globes: Computers & Geosciences, v. 37, no. 1, p. 93–99, doi:10.1016/j.cageo.2010.05.011.

Hornbaek, K., Bederson, B.B., and Plaisant, C., 2002, Navigation patterns and usability of zoomable user interfaces with and without an overview, *in* Proceedings of the 6th annual ACM Symposium on User Interface Software and Technology: 2–4 November, Marina del Rey, California, USA, p. 362–389. doi:10.1145/586081.586086.

Igarashi, T., and Hinckley, K., 2000, Speed-dependent automatic zooming for browsing large documents, *in* Ackerman, Mark S., and Edwards, K., eds., Proceedings of the 13th Annual Association of Computing Machin-

ery (ACM) Symposium on User Interface Software and Technology: 6–8 November, San Diego, California, USA, p. 139–148.

Jul, S., and Furnas, G.W., 1998, Critical zones in desert fog: aids to multiscale navigation, *in* Proceedings of the 11th Annual Association of Computing Machinery (ACM) Symposium on User Interface Software and Technology: 1–4 November, San Francisco, California, USA, p. 97–106.

Mackinlay, J.D., Card, S.K., and Robertson, G.G., 1990, Rapid controlled movement through a virtual 3D workspace: Computer Graphics, v. 24, no. 4, p. 171–176, doi:10.1145/97880.97898.

Mayer, R.E., 2004, Should there be a three-strikes rule against pure discovery learning? The case for guided methods of instruction: The American Psychologist, v. 59, no. 1, p. 14–19, doi:10.1037/0003-066X.59.1.14.

Mayer, R., and Moreno, R., 2003, Nine ways to reduce cognitive load in multimedia learning: Educational Psychologist, v. 38, no. 1, p. 43–52, doi:10.1207/S15326985EP3801_6.

Nielsen, J., 1994, Guerrilla HCI: Using discount usability engineering to penetrate the intimidation barrier, *in* Bias, R.G., and Mayhew, D.J., eds., Cost-Justifying Usability: Boston, Massachusetts, Academic Press, p. 245–272.

Priestnall, G., and Cowton, J., 2009, Putting landscape drawings in their place: Virtual tours in an exhibition context: 5th IEEE International Conference on E-Science Workshops, 9–11 December, p. 148–151, doi:10.1109/ESCIW.2009.5407975

Shephard, N., 2003, Animation in ArcScene: ArcUser Magazine, Winter, p. 8–10.

Stott, T.A., Nuttall, A.-M., and Mccloskey, J., 2009, Design, development and student evaluation of a Virtual Alps Field Guide: Planet, v. 22, p. 64–71.

Tewksbury, B., 2008, Introducing geologic map interpretation and cross section construction using Google Earth in a structural geology course: Joint Meeting of The Geological Society of America, Abstract 308-1, 5–9 October, Houston, Texas.

Treves, R., 2009, Tours in Virtual Globes: Eos (Transactions, American Geophysical Union), v. 90, no. 52, Fall Meeting Supplement, Abstract IN22A-04, 14–18 December, San Francisco, California.

Treves, R., and Engelbrecht, P., 2011, User tests on Google Earth tour comprehension: Southhampton, UK, University of Southampton, 10 p.

Treves, R., and Martin, D., 2008, Simple geography-related multimedia, *in* Rees, P., MacKay, L., Martin, D., and Durham, H., eds., E-Learning for Geographers: Online Materials, Resources, and Repositories: Hershey, Pennsylvania, Information Science Reference, p. 204–221.

Turk, J.F., Hawkins, J., Richardson, K., and Surratt, M., 2011, A tropical cyclone application for virtual globes: Computers & Geosciences, v. 37, no. 1, p. 13–24, doi:10.1016/j.cageo.2010.05.001.

van Wijk, J.J., and Nuij, W.A.A., 2003, Smooth and efficient zooming and panning, *in* Proceedings of IEEE Symposium on Information Visualization: 19–21 October, Seattle, Washington, doi:10.1109/INFVIS.2003.1249004.

Vinson, N.G., 1999, Design guidelines for landmarks to support navigation in virtual environments, *in* Proceedings of Human Factors in Computing Systems: 15–20 May, Pittsburgh, Pennsylvania, Association of Computing Machinery (ACM) Press, p. 278–285. doi:10.1145/302979.303062.

Walden, D., 2011, The Bushtucker database, *in* Jones, D.R., and Webb, A., eds., Eriss Research Summary 2009–2010, Supervising Scientist Report 202: Darwin, Northern Territory, Australia, p. 133–135.

Warburton, S., 2009, Second Life in higher education: Assessing the potential for and the barriers to deploying virtual worlds in learning and teaching: British Journal of Educational Technology, v. 40, no. 3, p. 414–426, doi:10.1111/j.1467-8535.2009.00952.x.

Wu, A., Wei Zhang, and Xiaolong Zhang, 2009, Evaluation of wayfinding aids in virtual environment: International Journal of Human-Computer Interaction, v. 25, no. 1, p. 1–21, doi:10.1080/10447310802537582.

Zhang, X., 2005, Space-scale animation: Enhancing cross-scale understanding of multiscale structures in multiple views, *in* Proceedings of the 3rd International Conference on Coordinated & Multiple Views in Exploratory Visualization: 5 July, London, England, p. 109–120, doi:10.1109/CMV.2005.16.

MANUSCRIPT ACCEPTED BY THE SOCIETY 16 APRIL 2012

The Geological Society of America
Special Paper 492
2012

Building an education game with the Google Earth application programming interface to enhance geographic literacy

Tsan-Kuang Lee*

Education Technology Services, The Pennsylvania State University, University Park, Pennsylvania 16802, USA

Laura Guertin*

Earth Science, Penn State Brandywine, Media, Pennsylvania 19063, USA

ABSTRACT

As part of a course objective to improve the geographic literacy of students in higher education, *Penn State's Amazing Race*, a modified version of Google Earth's application programming interface (API) demo game, *Geo Whiz*, engages students in learning physical geography within a Google Earth browser plug-in. Students navigate around the Earth to identify on the globe the locations of various countries, major cities, United States national parks, and locations with features of geological significance. To better achieve the learning goal, several game elements were incorporated into the game interface: a timer to encourage concentration, a ranking board with scores of all players to motivate students to improve and to assess learning results, and a replay function for instructors to review students' performance and specific difficulties. Google Earth API is used to control the Earth movements and map display, while custom JavaScript code adds the function of a timer, recording/playback, and score keeping. Google Earth's browser plug-in does not provide a layer that contains political boundaries without state and country labels, so one additional feature added to *The Amazing Race* is Central Intelligence Agency (CIA)–published boundary data without the names of world countries and U.S. states converted to a zipped Keyhole Markup Language (KMZ) file. The framework of this game can be easily exported for application to other disciplines for various student levels and ages.

*Lee—geology@tklee.com; Guertin—uxg3@psu.edu.

Lee, Tsan-Kuang, and Guertin, L., 2012, Building an education game with the Google Earth application programming interface to enhance geographic literacy, *in* Whitmeyer, S.J., Bailey, J.E., De Paor, D.G., and Ornduff, T., eds., Google Earth and Virtual Visualizations in Geoscience Education and Research: Geological Society of America Special Paper 492, p. 395–401, doi:10.1130/2012.2492(29). For permission to copy, contact editing@geosociety.org. © 2012 The Geological Society of America. All rights reserved.

INTRODUCTION

A geographic literacy study of 500 young American adults between the ages of 18 and 24 found that this group has a limited understanding of the world beyond United States' borders (National Geographic–Roper Public Affairs, 2006). Sixty-three percent of the respondents could not find Iraq on a map, twenty percent believe Sudan is located in Asia, and fifty percent could not locate New York on a United States map. This lack of geographic context provides a challenge for instructors teaching students about global disasters, hazards, and natural resources. In order to address this lack of geographic literacy of place, we customized the Google Earth application *Geo Whiz* as an educational game for students to improve their geographic content knowledge in understanding the locations of global cities, states, and countries.

Educational Gaming

Royle (2008) defines educational games as games designed with specific curriculum objectives, often used to support the practice of factual information. Gaming is not new to higher education students or classrooms. Jones (2003), in a Pew Internet and American Life Project report, states that seventy percent of college students reported playing video, computer, or online games at least once in a while, and sixty-five percent of college students reported being regular or occasional game players. Games have found their way into the classroom beyond technology fields, from disciplines such as English to communications. Entire video game rooms are appearing in some campus libraries (Wieder, 2011). In 2011, Michigan State University posted a job advertisement for an assistant professor of games. Gaming is part of the life of students and culture on campus, which makes the introduction of a game appropriate to the challenge of increasing student geographic literacy.

Not all educators believe that educational gaming benefits students. Most reviews in the past twenty years have focused on what players learn from video games rather than how video games can be designed to facilitate learning (Dondlinger, 2007), and limited data have been cited beyond anecdotal statements (Gredler, 2004). Shaffer et al. (2004) concur that most educational games have not been grounded in learning theory or directly supported by any body of research, and Wolfe and Crookall (1998) report a notable lack of rigorously evaluated educational games. Despite the overall lack of design experiments and assessments of educational games in the literature, there are some exceptions within the disciplines of medicine and business (c.f. Washbush and Gosen, 2001; Kashibuchi and Sakamoto, 2001).

Interestingly, Denis and Jouvelot (2005) would not consider *Penn State's Amazing Race* an educational game but classify the program under their definition of "edutainment," a "skill-and-drill" exercise where users only practice repetitive skills and simple comprehension. Prensky (2006) prefers to call edutainment games "drill-and-kill" because they are based only on trial and error and provide little motivation to learn—unlike our *Amazing Race* game, which includes motivational features such as a timer, score, and performance rankings. Lee et al. (2004) document that even skill-and-drill or drill-and-kill games can demonstrate learning gains in students. For example, Squire et al. (2003) have documented secondary school students developing a deeper understanding and appreciation of geographic features through educational gaming.

Google Earth Application Programming Interface (API)

Google Earth is an interactive tool for geologists, educators, and hobbyists to explore the Earth, Moon, Mars, and space. The core engine of Google Earth powers data retrieval, image/model rendering, user interface, and other functions such as user-layer data submission. Google opens the access of these core functions to developers via API (application programming interface, http://code.google.com/apis/earth) so any developer can build his or her own application on top of Google Earth by using these functions in his or her program. To facilitate data communication with the Google server, each developer needs to obtain a free API key from Google (http://code.google.com/apis/maps/signup.html) and include that key in the network communication.

In addition to a standard API library, Google Earth uses an open data format of KML/KMZ so different users and applications can exchange data with each other in addition to the Google Earth/Map database. For example, one can author a KML (Keyhole Markup Language) file on the Google Earth desktop version to contain a personal annotation of the state parks in central Pennsylvania and have the Google Earth web version read the annotation automatically to present the tour to the world. Such external data is usually read into Google Earth as a "layer," which can be turned on or off.

Some custom Google Earth examples include a virtual tour annotated with multimedia (text, pictures, audio, video, etc.) by worldwide users, a flight simulation to fly over landscapes, a three-dimensional (3-D) tour to the moon, a Mars 3-D exploration, a historical tour of an event, and a geological tour to observe land and ocean changes in time. There are a great variety of customized Google Earth files, and Google provides some showcases on http://www.google.com/earth/explore/showcase and http://code.google.com/apis/earth/documentation/demogallery.html.

Geo Whiz

Geo Whiz is highlighted on the Google Earth API examples page (http://code.google.com/apis/earth/documentation/examples.html). When a user clicks to open *Geo Whiz*, the user is taken to a webpage (http://earth-api-samples.googlecode.com/svn/trunk/demos/geowhiz/index.html) with an embedded Google Earth window. One rule appears in a box across the screen: "Move the earth such that the reticle is on top of the geographic location requested. That's all there is to it!" The landmasses in the Google Earth window do not have any borders or labels. Navigation and zoom tools are not present. A national or international location appears in a dialog box for the user to navigate to. The game presents fifteen locations for the user to find. If the reticle is placed over the correct location, the location name will appear in green coloring in a results column on the right side of the page. Failure to locate a city in the allotted time causes the city name to appear in red under the results column. Every time the game is played, it is the same locations that appear in the same order.

The structure of the *Geo Whiz* game is ideal for engaging students in learning and demonstrating their knowledge of global locations. The game offers an alternative to the traditional map quizzes that have been a standard for the assessment of location knowledge. *Geo Whiz* also shares the four defining traits of games, as defined by McGonigal (2011): a goal, rules, a feedback system, and voluntary participation (knowingly and willingly accepting the previous three traits). As the existing *Geo Whiz* game contains many limitations that, in our case, did not match the teaching and learning objectives of our introductory-level earth science/geoscience/geography courses, modifications to the game were necessary.

CUSTOMIZING THE RACE

For the game to meet our pedagogical goals, we identified three major functions to implement on top of the existing *Geo Whiz*: the user interface, navigation recording, and score keeping.

The User Interface

Displaying Boundaries but Not Labels

Google Earth provides a layer that contains both country/state boundaries and their label names. For our purpose, we want to show only the boundaries but not the labels, as we want the students to find the locations without assistance. Since this layer is part of Google Earth's built-in function, we cannot manipulate the data partially; we can only turn the entire layer on or off. Therefore, we needed to provide our own boundary information. There are many public data libraries freely available for Google Earth. We used the Central Intelligence Agency (CIA) World Factbook for country boundaries (http://www.gelib.com/world-borders.htm), and a Google released data file for U.S. state boundaries (http://code.google.com/apis/kml/documentation/us_states.kml). The data files we obtained allow us to edit out the labels with a text editor. See Figure 1 for a screenshot of our modified game interface.

Since our application is designed to run in any browser, all the data are stored remotely on a server, and the large amount of data points in the two data files affect the loading and processing speed. In our application, very fine boundaries details are not necessary. We use the Ramer–Douglas–Peucker algorithm (http://en.wikipedia.org/wiki/Ramer%E2%80%93Douglas%E2%80%93Peucker_algorithm) to omit data points that are beyond the precision we need. The last step is to merge the two data files

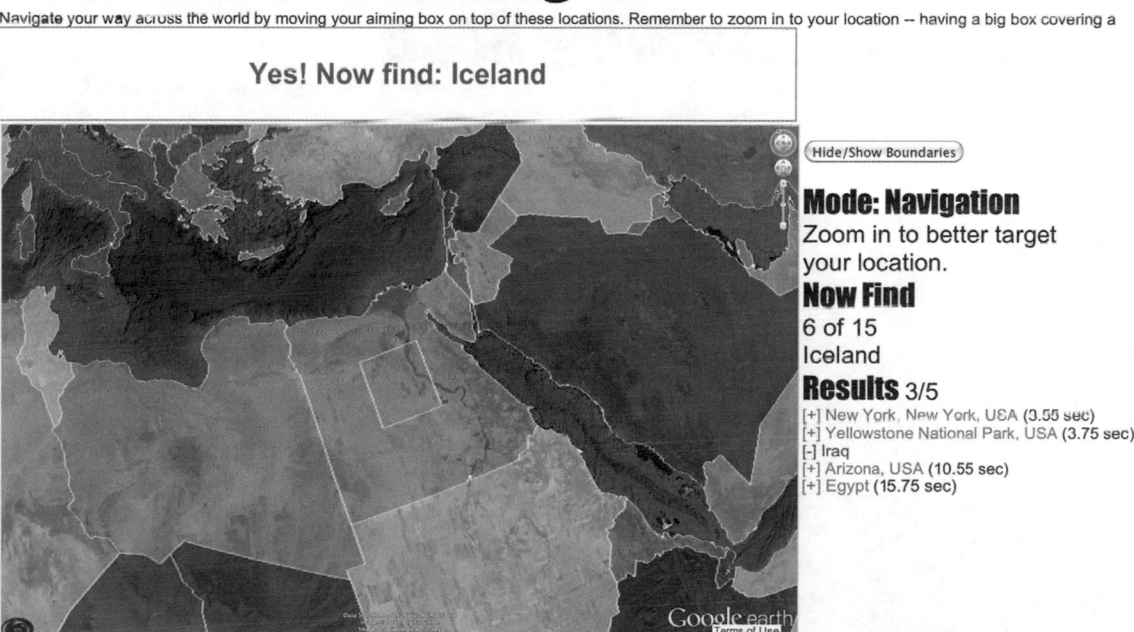

Figure 1. The interface of *Penn State's Amazing Race*, showing our additions of the standard Google Earth navigation controls and the state/country boundaries. The button above the Mode: Navigation text allows users to activate/deactivate the boundary layer. The reticle is in Egypt in this screenshot. The score and time spent finding each location are recorded in the right column.

into one KML file, and to compress it into a KMZ file for more efficient data transfer.

Game Timer and Scoreboard

We include a timer in the game to keep track of how much time students spend on each question. The timer starts after the user presses the start button, and ends when the user successfully navigates to the target or has used up the allotted time. Originally, we gave students three minutes to navigate to each location, but later found that one minute is enough for those who are familiar with navigation in Google Earth and the location of the target destination. Depending on the situation, the timer can be made visible to intensify the game atmosphere, or hidden from view to allow students to comfortably explore on their own. A scoreboard is shown at the right of the Google Earth window, so that students know where they are in the whole adventure, as well as how many right answers they have given (Fig. 1).

Aiming Reticle

We show an aiming reticle in the center of the Google Earth window to enable students to target specific locations (see Fig. 1). We use a rectangle to represent the aiming reticle as opposed to one single point because the targets are always regions (a country, a state, a park, etc.). Students navigate the background map to align the target region with the reticle.

The size of the reticle requires some consideration. Since the user can adjust the height of the viewpoint, having a reticle of a fixed size is not very meaningful because then students can easily move the viewpoint far away from the Earth and have everything included in the reticle and claim to hit the target. However, having the size of the reticle vary according to the height, either linearly or exponentially, is impractical as well, since we do not want the reticle to be too small to see on the screen when moving away from the Earth to a whole Earth view, neither do we want the reticle to be too big to cover too many regions when moving closer to the regions.

One possible way to solve this challenge is to apply another exponential function to smoothen the reticle resizing. We chose a simpler solution by using a fixed-size reticle and telling users whether the current height is within the acceptable range, similar to the concept of the "shooting range," i.e., the longest distance the ammunition can reach. Users then have to move closer to sea level in order to have a meaningful aiming.

Navigation Recording

Google API does not provide any native function for programmers to capture user screens. In order to replay how the students navigated during every view change, termed a *fired event*, we record the timestamp, the coordinates of the current camera position (viewpoint) and the viewed target. In the replay, we fire those coordinate changes according to the recorded timestamps (Fig. 2).

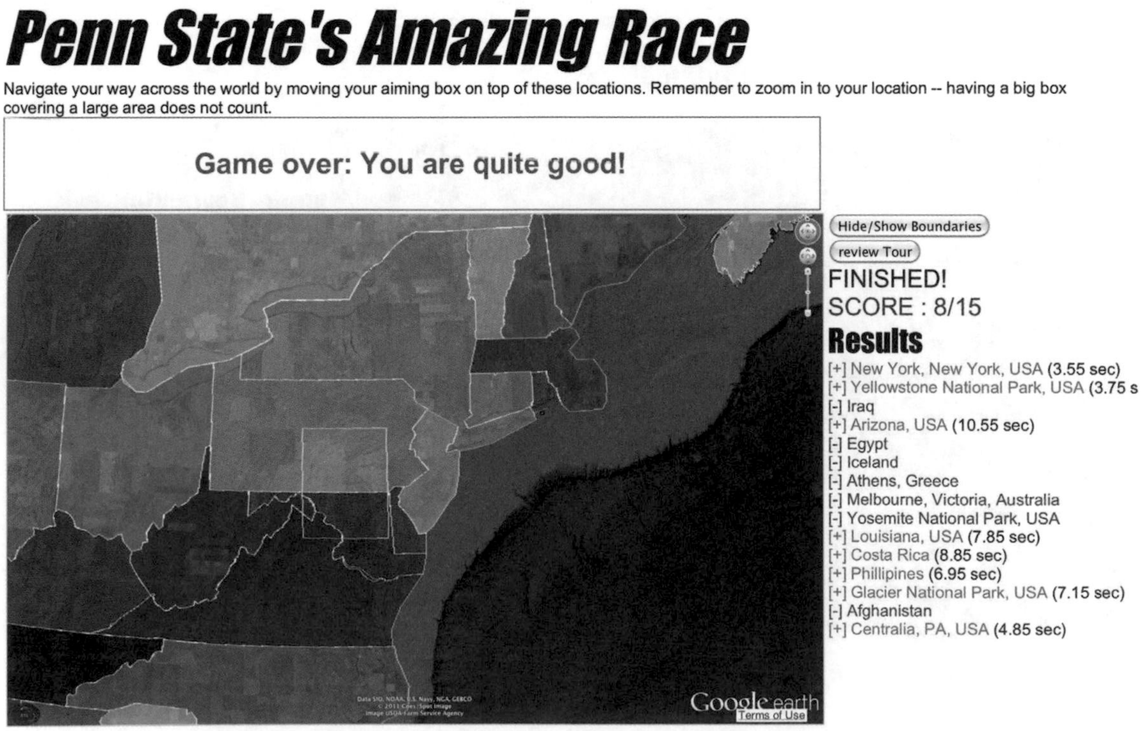

Figure 2. At the end of the game, a player can view a tour to watch their navigation of the reticle.

Score Keeping

Successful Navigation and Data Submission

To determine whether the students have successfully navigated to the target, we examine the current height first to determine whether we are within the "shooting range," and then check to see whether the target coordinates fall inside our reticle. At the end of the game, the program connects to our central server that supports authentication, and sends in the time spent on each question, whether the student was successful in their navigation, and the navigation steps they took. Since the navigation data size can be large, we compress the data before sending it out to the game score-keeping server, to minimize the waiting time for network communication.

USE OF THE RACE

Field Test of Design

Initial testing of *Penn State's Amazing Race* was done in an upper-division Earth science classroom with each student on his own computer so that we could make visual note of student interactions with the game. From the technical side, students did not experience many difficulties navigating in the Google Earth interface. This could be due to the fact that most of the students had experience with Google Earth in prior courses. Some students struggled with aiming the reticle over the correct location and zooming to the necessary level for the game to register a correct answer, but students who were in the right geographic region were able to eventually have their navigation scored as correct.

The focus of student frustration we witnessed was primarily centered on student unfamiliarity with geography. For example, during one of the tests, a student voiced out loud that the game was "messed up," as he was navigating over Costa Rica and the game was not registering his response. When the game ended, he looked on a map and realized he was actually navigating over Puerto Rico. His movements of the computer mouse also reflected this inaccurate navigation in the game playback. We witnessed students just sitting and staring at the computer screen when they had no idea where a location may be on the globe. Students who had several consecutive locations where they did not know where to navigate were especially frustrated but tended to sit in silence instead of acknowledging out loud their lack of geographic knowledge. The students reacted positively to the scoreboard after completing the game (Fig. 3). Students enjoyed comparing their scores to others and owned-up to how well and how poorly they performed on the game.

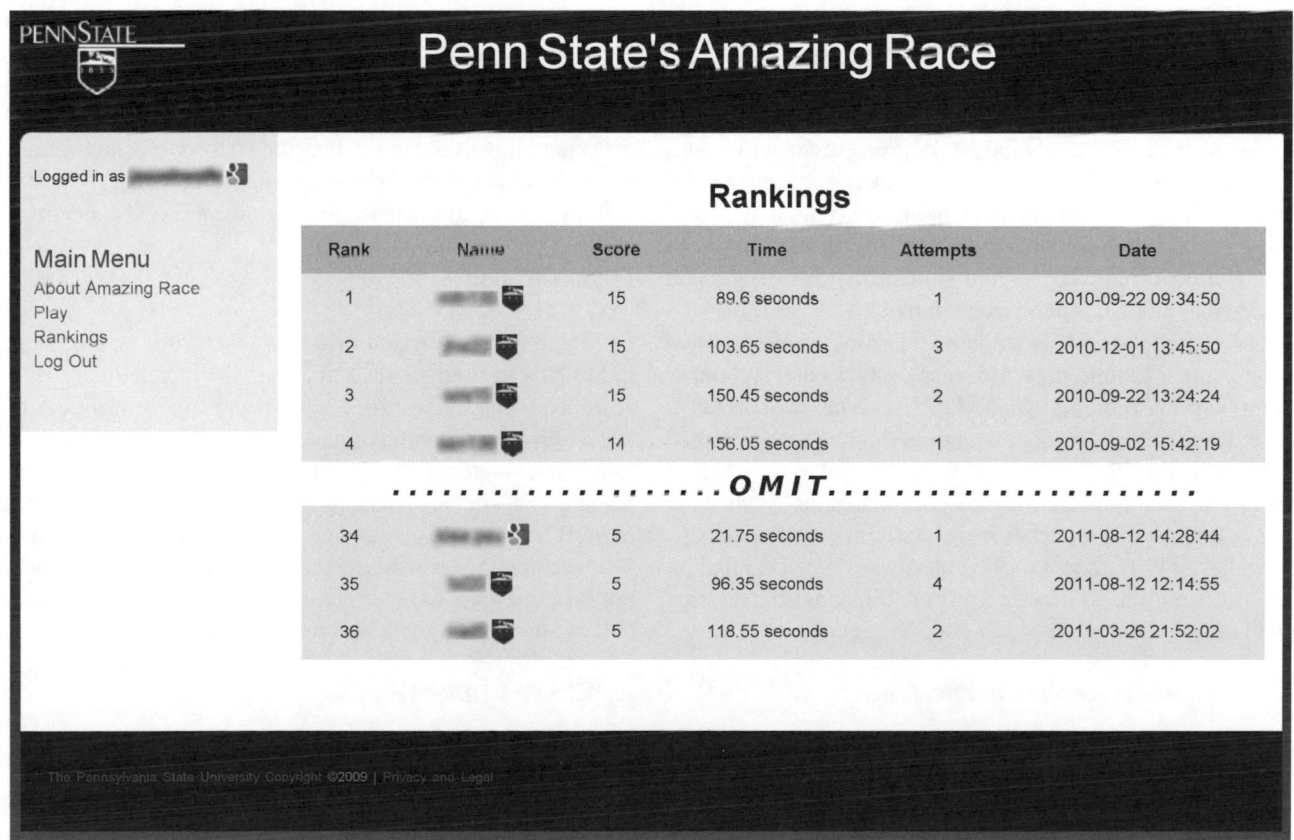

Figure 3. A results page shows the ranking, the scores, number of attempts, and the time of each game for each player.

Surface versus Deep Learning

Although we achieved our goal of students identifying a place, we agree with Dunn (2011) that students who correctly answer a simple location question using a traditional outline map do not necessarily know much more about the location from a spatial thinking perspective. The *Amazing Race* is only assessing surface learning (c.f. Chin and Brown, 2000)—student location of places—not student knowledge of the important attributes of these places that would be typical of deep learning.

The navigation of the Google Earth interface is a skill that students learn by repetition, another characteristic of reproducing memorized facts seen with surface learning. Going back to Prensky's (2006) drill-and-kill idea, familiarity with the game interface will help students get to the target destinations in a shorter time. However, the geographic knowledge of this game cannot be obtained via trial and error. Students need to learn where places are located—the overarching goal of the game—in order to get a higher score. Again, we are not claiming that merely knowing where locations exist on the planet demonstrates deep geographic knowledge. Along with Rhem (1995), we have observed classroom situations where the lack of basic prior knowledge impedes the connection and integration of new concepts for further learning.

Recommendations

Penn State's Amazing Race has several educational applications beyond being a game for recreational use. The *Amazing Race* can serve as a pre- and post-assessment in a course, where an instructor can track and compare the speed and accuracy of a student finding locations before a semester begins and after instruction of location-based content. A pre-assessment is valuable for the instructor to see where student misconceptions lie in terms of their perceived understanding of geography. For a student, completing the *Amazing Race* at the beginning of the semester will also reinforce to that student that his geographic knowledge may not be as good as he thinks.

Instructors may fear that students will look up the answers while the game is taking place. Because of the timing and tracking features of our modified version of *Geo Whiz*, we can tell if there is a delay before a student begins navigating to a prompted location, indicating that a student was engaged in a different activity before searching for the location. Interestingly, when the students completed the test trials in a classroom with global maps on the wall, not one student looked up toward a map to find an answer to a location. We feel comfortable having the students complete the game either inside or outside of class.

Suggested Future Interface Modifications

Region-Based Target Determination

In our prototype, the targets are specified by names, such as New York City, Yellowstone National Park, or Philippines. We rely on Google Earth to look up the coordinates of the targets for us. However, Google Earth only provides one single point

coordinates for each target, and that is usually the location of the target's label on Google Maps. This works most of the time when the label location of the target region is intuitive to the user. In few cases where there the target region is spread out (containing many islands, or having a long and thin outline), students may need to move the reticle around to get the right answer. We can improve this by providing a "region" data set for each question, i.e., a set of coordinates to describe boundaries, and then determine whether the center of the reticle falls into any of these boundaries.

Real-Time Race

Our current design keeps track of each individual user separately, where we know the performance of an individual player at the end of the game. To increase the competitiveness, we can show other players' current progress in real-time. For example, on a player's screen, we can show a synchronous percentage of completion of other players so each individual student can know whether they are ahead or behind. The concept can be pushed further to provide a virtual competition with "ghost players," where the other players do not need to play at the same time, as long as their completion percentage is recorded with timestamps.

Other Applications

The framework of the *Amazing Race* can easily adapt to other applications that help familiarize players with locations and maps. Instead of using Google's world map, various customized local maps can be used for different uses, such as a campus orientation for new students, a tour of a hometown for elementary school students, and travels along the Oregon Trail for a history class. The code is available in the Appendix[1] for anyone wishing to customize their own game from our original modifications to *Geo Whiz*.

Demonstration

Penn State's version of the *Geo Whiz* game records students' scores to a campus server and allows the instructors to view the scoreboard and the recorded movement of the students. A public stand-alone demo, and its source code, can be found at this URL (uniform resource locator; in the unlikely event of URL change, readers can contact the authors for updates): http://blog.tklee.org/2011/11/google-earth-game-amazing-race.html. This version keeps scores and records user movement during the game but does not save the score to an external server. A videocast is also available at http://www.youtube.com/watch?v=rJqFlzAV-8o.

ACKNOWLEDGMENTS

The authors would like to thank collaborators Jason Wolfe, Christopher Stubbs, and Christopher Millet for discussions and

---

[1]GSA Data Repository item 2012306, Appendix, is available at http://www.geosociety.org/pubs/ft2012.htm, or on request from editing@geosociety.org.

suggestions during development of *Penn State's Amazing Race*. This project was part of co-author Guertin's time as a Teaching and Learning with Technology Fellow with Education Technology Services at The Pennsylvania State University.

REFERENCES CITED

Chin, C., and Brown, D.E., 2000, Learning in science: A comparison of deep and surface approaches: Journal of Research in Science Teaching, v. 37, p. 109–138, doi:10.1002/(SICI)1098-2736(200002)37:2 <109::AID-TEA3>3.0.CO;2-7.

Denis, G., and Jouvelot, P., 2005, Motivation-driven educational game design: Applying best practices to music education: Valencia, Spain, Paper presented at the 2005 Association for Computing Machinery Special Interest Group on Computer Human Interaction (ACM SIGCHI) International Conference on Advances in Computer Entertainment Technology.

Dondlinger, M.J., 2007, Educational video game design: A review of the literature: Journal of Applied Educational Technology, v. 4, p. 21–31.

Dunn, J.M., 2011, Location knowledge: Assessment, spatial thinking, and new national geography standards: The Journal of Geography, v. 110, p. 81–89, doi:10.1080/00221341.2010.511243.

Gredler, M.E., 2004, Games and simulations and their relationships to learning, *in* Jonassen, D.H., ed., Handbook of Research on Educational Communications and Technology: Mahwah, New Jersey, Lawrence Erlbaum, p. 571–581.

Jones, S., 2003, Let the games begin: Gaming technology and college students: Washington, D.C., Pew Internet & American Life Project, http://www.pewinternet.org/Reports/2003/Let-the-games-begin-Gaming-technology-and-college-students.aspx (accessed 8 August 2011).

Kashibuchi, M., and Sakamoto, A., 2001, The educational effectiveness of a simulation/game in sex education: Simulation & Gaming: An Interdisciplinary Journal, v. 32, p. 331–343.

Lee, J., Luchini, K., and Michael, B., Norris, C., and Soloway, E., 2004, More than just fun and games: Assessing the value of educational video games in the classroom. Vienna, Austria, Paper presented at the CHI2004 (Conference on Human Factors in Computing Systems, 24–29 April 2004), Extended Abstracts.

McGonigal, J., 2011, Reality is Broken: Why Games Make Us Better and How They Can Change the World: New York, The Penguin Press, 400 p.

National Geographic-Roper Public Affairs, 2006, Final Report—2006 Geographic Literacy Study: New York, Growth from Knowledge National Opinion Poll (GfK NOP), http://www.nationalgeographic.com/roper2006/pdf/FINALReport2006GeogLitsurvey.pdf (accessed 8 August 2011).

Prensky, M., 2006, Don't Bother Me, Mom—I'm Learning: Saint Paul, Minnesota, Paragon House, 350 p.

Rhem, J., 1995, Deep/surface approaches to learning: An introduction: The National Teaching & Learning Forum, v. 5, p. 1–3.

Royle, K., 2008, Game-based learning: A different perspective: Innovate, v. 4, http://www.webcitation.org/5cbDTFiih (accessed 9 November 2011).

Shaffer, D., Squire, K., Halverson, R., and Gee, J., 2004, Video games and the future of learning: Madison, Wisconsin, USA, University of Wisconsin, http://www.academiccolab.org/resources/gappspaper1.pdf (accessed 11 November 2011).

Squire, K., Jenkins, H., Holland, W., Miller, H., O'Driscoll, A., Tan, K.P., and members of the Games-to-Teach Research Team, 2003, Design principles of next-generation digital gaming for education: Educational Technology, v. 33, p. 17–23.

Washbush, J., and Gosen, J., 2001, An exploration of game-derived learning in total enterprise simulations: Simulation & Gaming: An Interdisciplinary Journal, v. 32, p. 281–296.

Wieder, B., 2011, 2011, Video-game rooms become the newest library space invaders: The Chronicle of Higher Education, April, p. 24, http://chronicle.com/article/Video-Game-Rooms-Become-the/127229/ (accessed 8 August 2011).

Wolfe, J., and Crookall, D., 1998, Developing a scientific knowledge of simulation/gaming: Simulation & Gaming: International Journal (Toronto, Ontario, Canada), v. 29, p. 7–19.

MANUSCRIPT ACCEPTED BY THE SOCIETY 16 APRIL 2012

The Geological Society of America
Special Paper 492
2012

Developing a scope and sequence for using Google Earth in the middle school earth science classroom

Heather Almquist*

College of Arts and Sciences, University of Montana, Missoula, Montana 59812, USA

Lisa Blank

Department of Curriculum and Instruction, University of Montana, Missoula, Montana 59812, USA

Jennifer Estrada

College of Arts and Sciences, University of Montana, Missoula, Montana 59812, USA

ABSTRACT

Google Earth–based learning activities have become increasingly popular among K–12 educators in recent years. However, most of these activities are short-term, singular events within a more traditional curriculum and involve students' passive observation of pre-developed Google Earth "tours." The Cyber-Enabled Earth Exploration (CE3) project is developing an immersive, Google Earth–based, middle school curriculum on plate tectonics. The curriculum is designed to scaffold students' technical skills, analytical abilities, and scientific content knowledge, as well as increase students' understandings about the nature of science. Through the curriculum, students progress from making simple observations in the Google Earth environment to using Google Earth as a transformative data analysis tool. Here, we describe how the curriculum was developed, its essential elements, and how it was received by students in grades five through nine. Results including teachers' reflections and student notebook entries reveal what skills and concepts were most challenging for students at each grade level and provide a preliminary roadmap for designing a scope and sequence for using Google Earth to teach core concepts related to volcanoes, earthquakes, and plate tectonics in the middle school classroom.

*heather.almquist@umontana.edu

Almquist, H., Blank, L. and Estrada, J., 2012, Developing a scope and sequence for using Google Earth in the middle school earth science classroom, *in* Whitmeyer, S.J., Bailey, J.E., De Paor, D.G., and Ornduff, T., eds., Google Earth and Virtual Visualizations in Geoscience Education and Research: Geological Society of America Special Paper 492, p. 403–412, doi:10.1130/2012.2492(30). For permission to copy, contact editing@geosociety.org. © 2012 The Geological Society of America. All rights reserved.

INTRODUCTION

Use of technology in learning activities is an increasingly popular approach that has been shown to improve student motivation and increase achievement (Cradler et al., 2002; Andrews, 2006). This approach not only engages students, but also helps better prepare them for the twenty-first–century workforce. In recent decades, new remote sensing technologies have resulted in the rapid growth of observational data sets. In fact, it is estimated that 80% of all data available today can be geospatially referenced (National Research Council, 2006). These advances have dramatically increased the need for knowledge workers who can frame problems in spatial terms, manage spatially referenced data, create and use spatial representations, and conduct appropriate spatial analyses to infer Earth system processes of the past, present, and future. These skills support and are supported by other critical skills of twenty-first–century knowledge workers, including critical thinking, computer and communication skills, and a global frame of reference (Halpern, 1999; National Research Council, 2008; Wagner, 2008).

Educators are well aware that the vast majority of today's students are "digital natives." They show great skill in surfing the web to collect information, using web-based communication tools, video games, and other applications. Computer applications that give students a measure of autonomy, flexibility, and creative expression increase their engagement and proficiency, leading them to become more self-sufficient, more willing to share knowledge, and more enthusiastic toward learning (Fedisson and Braidic, 2007).

For content areas with a spatial domain, such as earth, life, and social sciences, fairly sophisticated geospatial technologies may be employed. Anything that can be referenced to a specific geographic position becomes a candidate for investigation. Thus, the use of geographic information systems (GIS) or virtual globes expands the scope of potential science investigations leading to authentic learning contexts at local, regional, and global scales. Geospatial technologies can provide hands-on learning experiences that require students to make observations, reflect on their findings, ask questions from a variety of perspectives, and formulate their own hypotheses. Such a learning process is now understood to improve problem solving skills and learning outcomes (Bransford and Donovan, 2005), perhaps because these technologies allow students to formulate conceptual, rather than definitional understandings of central science concepts (Audet and Abegg, 1996).

In the classroom setting, however, many teachers face significant barriers in providing such learning opportunities to students. In a recent professional development program for Montana middle school teachers (Almquist et al., 2007, 2008; Stanley and Almquist, 2008), we trained teachers in the application of geospatial technologies, including Google Earth, global positioning systems (GPS), and GIS, to conduct technology-embedded, inquiry-based learning activities with their students. Among other things, the project revealed that few Montana teachers are well versed in the use of computer technologies. Some might even be considered technology-phobic. Even after many hours of intensive training in the use of ArcGIS software, most teachers felt far more comfortable using the more intuitive Google Earth virtual globe to display and manipulate georeferenced data. This finding was significant because in a classroom environment, few teachers are willing to delve into content areas or use technologies that they are uncomfortable with, even if they sense that their students may be up to the task. This tendency is all for the best perhaps, for in the case of GIS, although it can be used by teachers with little prior experience, research shows that students perform better and respond better to the technology experience if taught by a teacher with a high level of GIS competency (Purcell et al., 2006).

Teachers often also lack appropriate inquiry-based, technology-embedded, curricular materials, and few have the time or expertise to develop such materials on their own. Although some recent materials have earned praise for college-level instruction (e.g., Marshak, 2007), those aimed at lower grades primarily include individual lessons (e.g., http://gelessons.com/lessons/) that do not support the systematic study of a topic. Other excellent, open-access Google Earth files exist (e.g., http://earthquake.usgs.gov/learn/kml.php), but do not include accompanying lesson plans. Moreover, teachers have little class time in which to incorporate new materials beyond the required curriculum, unless the new materials directly address educational standards and can replace outdated ones. The future may be brighter, however, as an increasing number of pre-service teacher education programs are addressing these barriers. But for in-service science teachers today, particularly in underserved schools or regions of the country where time demands on teachers are many and professional development opportunities are few, immediate and practical interventions are needed. These teachers require new resources that reflect new and effective modes of teaching, boost teacher competence and confidence in the content area, make use of relatively intuitive technologies, and fit within the required curriculum with little or no additional time demands. The pay-offs for improving science instruction in this manner reach far beyond science achievement scores. Studies show that improvements can be expected in students' writing, mathematics, and higher-order thinking skills as well (Amaral et al., 2002).

PROJECT GOALS

Plate tectonics is a fundamental topic in the middle school earth science curriculum. Most leading textbooks (e.g., Allison et al., 2006; Cooney et al., 2006; Feather et al., 1995; Feather et al., 2008; McDougall Littell, 2006; Morrison et al., 1997; Todd, 2001) cover plate tectonics, earthquakes, and volcanoes, in that order, using a consistently didactic approach with explanation of the underlying phenomena presented first, followed by descriptions of observable geologic outcomes.

A goal of the Cyber-Enabled Earth Exploration (CE3) project is to create a curriculum that allows students to learn about plate tectonics in a more active manner, as a scientist would. We envision students using their own observations, measurements, calculations, and reasoning abilities to embark on a journey of discovery that allows them to construct their own understanding of the underlying causal mechanisms and outcomes of plate tectonics.

In contrast to most curricula addressing this content area, the CE3 curriculum begins with volcanoes and ends with plate tectonics. This is a deliberate attempt to help students experience the process of discovery that the scientific and lay communities have undertaken through the ages. That is, people's first understandings of natural phenomena are based upon what they observe and experience at Earth's surface. These experiences foster curiosity about the essential nature of these events—how and why they happen. Over time, through concerted and cumulative observations, advances in technology, and new ways of thinking, we begin to build an understanding of processes occurring beyond our immediate experience.

In addition, through their own process of knowledge construction, students gain a deeper understanding of the nature of science—that science is an ongoing, dynamic pursuit. Google Earth provides an intuitive, stimulating, and immersive environment in which such knowledge construction can occur. However, there has been little if any systematic study of how well students at different grade levels can acquire the technical skills, analytical abilities, and scientific understandings needed to master a content area. In this project, we assess these aspects of student learning within the context of a pilot curriculum on plate tectonics in order to help identify a scope and sequence for using Google Earth in the middle school earth science classroom.

THE CURRICULUM DEVELOPMENT PROCESS

In this project, University of Montana curriculum and instruction specialists, geoscientists, and Montana K–12 teachers are working in concert to develop, test, and refine a curriculum covering volcanoes, earthquakes, and plate tectonics. The stages of development have included (1) initial conceptualization by a University of Montana–based curriculum development team, (2) review of the concept and curriculum outline by an external, three-member, middle school teacher focus group, (3) full-scale development of a pilot curriculum by the development team, (4) review of the curriculum by a three-member university panel including a geoscientist, geographer, and curriculum design expert, (5) review of the curriculum by the Montana classroom teachers who would be implementing it in their own classrooms, (6) subsequent additional revision of the curriculum, (7) classroom testing by those nine teachers, and (8) analysis of teacher feedback and student outcomes from those implementations. Student outcomes included pre- and post-content assessments developed specifically for the CE3 curriculum, as well as detailed student "field notebooks" covering each exercise. In addition, one 7th grade teacher tested the Google Earth–based curriculum against traditional methods for groups of students of differing abilities. Results of this analysis elucidate how Google Earth can be used in differential instruction and will be reported on elsewhere. These collective results will form the basis for further revision and larger scale classroom testing of the curriculum.

The nine Montana teachers who conducted the initial classroom implementation were recruited from a pool of teachers who had participated in the co–principal investigators' previous professional development projects involving geospatial technologies. Although they were not considered to be a group of master teachers, they all had a working knowledge of Google Earth, and otherwise represented a broad range of teacher and classroom characteristics commonly encountered in Montana schools. The teachers included four males and five females. Most were experienced teachers, with seven having taught earth science for more than 15 years and only two in their first or second year of teaching. Three teachers taught grade 6, one taught six sections of grade 7, two taught grade 8, and three taught grade 9. A range of classroom settings were represented, with class sizes ranging from 6 to ~30 students. Teachers also employed a range of instructional strategies depending on their own teaching styles and preferences and the computer facilities available to them. An in-depth look at the effects of these confounding factors will be reported elsewhere. The teachers used the curriculum with over 320 students. All classes were reported by teachers to be of mixed ability.

This initial analysis of the CE3 curriculum was designed to help define a hierarchy of technical skills, analytical abilities, and core science concepts required to teach about volcanoes, earthquakes, and plate tectonics. These outcomes are being used to revise the curriculum by grade band. These revisions should in turn be subjected to additional classroom testing focusing on fidelity of implementation with significantly higher numbers of student subjects.

DESCRIPTION OF THE CURRICULUM

The curriculum comprises three sequential modules containing three investigations each (http://www.spatialsci.com/ce3; Table 1). Each investigation is presented in a context of what type of scientist studies the particular topic and why. This context is meant to motivate students and help them relate to science as something real, meaningful, and current. Investigation support materials include a teacher's guide, student "field notebook," challenge activities (assessments), grading rubrics, and Google Earth (.kmz [zipped Keyhole Markup Language]) files through which the content is delivered. The .kmz files contain a series of placemarks and overlays. The placemarks have "balloons" that include text with a pop-up glossary, pronunciation guides for foreign words, graphics, videos, animations, and GigaPans (Schott, 2009), creating a rich learning environment. Animations and videos include text and audio to accommodate different learning styles. After exploring the information provided through these various formats, students are asked to record specific observations in their field notebooks. Students are expected to collect information from all of the materials presented, including

TABLE 1. OUTLINE OF CYBER-ENABLED EARTH EXPLORATION (CE3) MODULES AND INVESTIGATIONS

Module 1: Introduction to Volcanoes

Investigation 1: Volcano Hazards and Benefits	This investigation motivates students to learn about volcanoes by exploring several historic eruptions in Google Earth and comparing the particular hazards posed by each and their effects on local, regional, and in some cases, global populations.
Investigation 2: How Volcanoes Work	Here we use animations to explain the three basic morphological/compositional types of volcanoes and ask students to use this information to classify a series of volcanoes they visit and explore "in the field" using Google Earth.
Investigation 3: Predicting Eruptions	Students explore the variety of methods scientists use to monitor volcanoes. They also explore the abundance and distribution of Earth's active volcanoes, which will link to subsequent modules.

Module 2: Introduction to Earthquakes

Investigation 1: Earthquake Hazards	As with volcanoes, this module begins by visiting sites of important historical events and seeing how people were affected. Students learn how earthquake magnitude and intensity are described, and explore how geologic substrate affects earthquake intensity.
Investigation 2: Geologic Faults	Students begin to explore what happens below the Earth's surface during an earthquake by visiting and interpreting fault exposures. An animation is used to describe the three basic types of faults, and students test their understandings "in the field" by classifying a mystery fault in Google Earth. Students explore Nevada's Basin and Range to see how faulting can affect the Earth on a landscape scale.
Investigation 3: Seismic Waves and Earth Structure	Students further explore what happens below ground during an earthquake and how energy travels through the Earth to create the shaking experienced during earthquakes. An animation illustrates the elastic properties of rock and behavior of different types of seismic waves travelling through various substrates. Students use this understanding to "map" Earth's internal structure based on seismic data. Finally, they take a Google Earth tour of seismic stations around the world.

Module 3: Plate Tectonics

Investigation 1: Continental Drift	Students use Google Earth to explore geographic, fossil, and rock evidence supporting the continental drift hypothesis. They create their own artifacts in the Google Earth environment to compare and analyze these data.
Investigation 2: Sea-Floor Spreading	Students create a topographic profile and trace the extent of mid-ocean ridges in Google Earth. They learn about geomagnetism, "magnetic stripes," the sea-floor spreading hypothesis, and review maps of sea-floor age that support this hypothesis.
Investigation 3: Plate Boundaries	Students explore the geographic distribution of volcanoes and earthquakes to infer plate boundaries, and use their understanding of Earth's structure to visualize lithosphere plates. Animations illustrate subsurface processes at the various types of plate boundaries. Students use this information to classify a series of mystery plate boundaries that they observe in Google Earth.

making their own observations and measurements in the Google Earth environment. They then respond to a series of questions using a Claims, Evidence, and Reasoning format (Berland and McNeill, 2010; McNeill, 2011; McNeill and Martin, 2011), which allows a more thorough examination of students' ability to perform the required tasks, analyze their data, and understand the science concepts.

The sequencing of topics within the curriculum allows for concepts of increasing complexity to be introduced. The scientific content related in each investigation builds upon the previous one and lays the groundwork for the next. Throughout the curriculum, students are encouraged to continually refer back to material previously covered and recorded in field notebooks, allowing them to develop understanding as a scientist would.

Google Earth includes an ever-increasing array of tools and features that can be employed to enhance observation and analysis. These include tools for navigation, measuring, drawing, and overlaying data and images, among others. The CE3

curriculum is designed to gradually build these technical skills. It also attempts to scaffold students' analytical abilities, including measuring and calculating, spatial analysis and visualization, and scientific reasoning. The CE3 curriculum utilizes 17 discrete technical skills (Table 2) and 39 tests of analytical abilities (Table 3), while introducing 22 specific science concepts (Table 4).

RESULTS OF THE CLASSROOM IMPLEMENTATION

Two data sources are used to assess students' response to specific elements within the curriculum. These include teachers' daily recorded reflections as they implemented the curriculum, and student responses to specific questions in their field notebooks. Two 6th grade teachers, the 7th grade teacher, two 8th grade teachers, and two 9th grade teachers submitted detailed reflections on their classroom experiences. They reported on what portions of the curriculum they were teaching, how they were teaching it (e.g., as a class, with students working in pairs,

TABLE 2. SCAFFOLDING OF TECHNICAL SKILLS USING
GOOGLE EARTH IN THE CYBER-ENABLED EARTH EXPLORATION (CE3) CURRICULUM

Category	Skill	Module Volcanoes investigation			Earthquakes investigation			Plates investigation		
		1	2	3	1	2	3	1	2	3
Navigation and display	Pan, zoom, tilt, rotate	6	–*	–	–	–	–	–	–	–
	Fly to/search	6	–	–	–	–	–	–	–	–
	Turn layers on/off	6	–	–	–	–	–	–	–	–
	Set elevation units	6	–	–	–	–	–	–	–	–
	Show terrain	6	–	–	–	–	–	–	–	–
	Set elevation exaggeration	6	–	–	–	–	–	–	–	–
	Set latitude and longitude units	6	–	–	–	–	–	–	–	–
	Zoom to see country labels	6	–	–	–	–	–	–	–	–
	Play embedded videos/animations	6	–	–	–	–	–	–	–	–
	Navigate among folders/placemarks	8	–	–	–	–	–	–	–	–
Observing and recording	Take careful notes	7	–	–	–	–	–	–	–	–
	Make meaningful diagrams/sketches		6	–	–	–	–	–	–	–
	Explore GigaPans		6		6	6				
	Identify craters and lava flows from satellite imagery		6							
	Explore external web links			6	6			6		6
	Explore historical imagery				6					
	Organize content in folders								9	

Note: Filled boxes show where in the curriculum the skill or concept is introduced or reinforced. Numbers refer to lowest grade level at which each skill or concept appeared not to be problematic for students.
*Dashes denote continued use of skill.

or individually), and where students were having difficulty. They also outlined extra supports that they used to help students grasp key concepts. Using those reports for reference, the students' responses to questions posed in their "field notebooks" were examined. Each question requires students to employ a specific technical skill, analytical ability, and/or science concept. Further, each question requires students to justify their answer, allowing a more thorough examination of their thinking process and depth of understanding. We examined a total of 136 student notebooks, although not all of them covered all investigations and questions. The total number of completed questions ranged from 9 to 19 for 6th grade students, 19–86 for 7th grade students, 6–17 for 8th grade students and 6–14 for 9th grade students. Marked gaps in notebook entries were generally explained by the teachers as deliberate omissions because of either activity difficulty or time constraints. Nevertheless, the notebook data combined with teacher reflections provide an informative look at the performance of the curriculum with respect to students' cognitive development.

Teacher Reflections

All of the teachers reported that the CE3 curriculum took longer than anticipated to implement than the developers had anticipated and all made useful suggestions regarding corrections or alterations that could be made in the revision. Several teachers from smaller, more rural schools found that access to computers and/or inadequate bandwidth presented problems for full and consistent implementation of the curriculum, and made adjustments accordingly. Teachers' reflections on specific technical,

analytical, and cognitive issues encountered by students using CE3 are summarized below in relation to grade level.

Technical Skills

Navigation and display. One of the 6th grade teachers reported that her students had trouble moving among folders, placemarks, and overlays in Google Earth and were confused about what to click on. All 7th and 8th grade teachers reported that students had little trouble using Google Earth. Some of the 7th graders discovered the flight simulator, which was not used in the curriculum, and became distracted "flying" everywhere. The 7th graders also became a little confused about which files they were working on because there were so many. However, teachers reported that the students enjoyed using the Google Earth technology and "learned a lot."

Observing and recording. In the "Plate Boundaries" investigation, one 6th grade teacher had to coach students to record their observations on the animations they viewed so that they could successfully identify the mystery sites. She reported that the students ultimately did a good job identifying the mystery plate boundaries. One 8th grade teacher mentioned students having trouble placing the paths they had created in the "Continental Drift" investigation into the appropriate folder. Another stated that she appreciated having students conduct Internet research on the various types of fossils in that investigation.

Analytical Abilities

Measuring and calculating. 6th grade students struggled with many of the mathematical operations required in CE3 and additional supports were needed. For example, the concept of

TABLE 3. SCAFFOLDING OF ANALYTICAL ABILITIES IN THE
CYBER-ENABLED EARTH EXPLORATION (CE3) CURRICULUM

Category	Ability	Volcanoes investigation			Earthquakes investigation			Plates investigation		
		1	2	3	1	2	3	1	2	3
Measuring and calculating	Set ruler units	6								
	Measure lines/paths	6			6					
	Measure three-dimensional feature width/height		9							
	Calculate slope		9							
	Use circle tool				6					
	Determine area/radius of a circle				8					
	Calculate rate				7–9					
	Create a topographic profile					6–9		7		
	Create a graph of x y data						7			
	Interpolate a value from a line graph						6			
Spatial analysis and visualization	Determine cardinal/relative direction	7								
	Estimate population density	7								
	Visualize volcano internal structure		7							
	Interpret InSAR-derived images			9						
	Interpret infrared images			6						
	Interpret a lateral offset from a photo				6					
	Compare changes over time				6					
	Interpret a Shakemap				6					
	Interpret a simple geologic map				6					
	Visualize a fault penetrating the earth					7				
	Match stratigraphic layers by eye					7				
	Interpret a topographic profile					6–8		7		
	Match coastlines by eye						7			
	Create a path to associate data						9			
	Trace a feature across the globe							7		
	Relate bandwidth to spreading rate on a sea-floor age map								7	
	Visualize negative rates as movement in opposite directions									7
Scientific reasoning	Use information on ash deposition at specific locations to determine wind direction during a volcanic eruption	6								
	Use various lines of evidence to understand the nature of pyroclastic flows	6								
	Use physical evidence to discern relative temperature	6								
	Understand the relative threat posed by lahars to valley vs. hillside settlements	6								
	Use characteristics to classify volcanoes		7							
	Predict volcano eruptive behavior based on volcano type		7							
	Assess the risk a volcano poses to a population		7							
	Classify earthquake intensity from physical evidence using the Mercalli scale				N*					
	Classify fault type in a GigaPan photo					7				
	Use fossil age to determine when continents were connected							6		
	Use characteristics of fossil organisms to determine how dispersal occurred							6		
	Correlate temporal and spatial events									9
	Classify plate boundaries based on physical evidence							6–8		

Note: Filled boxes show where in the curriculum the skill or concept is introduced or reinforced. Numbers refer to lowest grade level at which each skill or concept appeared not to be problematic for students. InSAR—interferometric synthetic aperture radar.
*N signifies that no grade levels appeared competent with the element.

slope was new for them and they needed it to be explained and modeled. In the "Sea-Floor Spreading" investigation, students in one 6th grade class were unfamiliar with how to set up graphs with appropriate scales to construct the trans-Atlantic topographic profile. Students in the other 6th grade class also had trouble. Although they had "done fine" constructing the transect graph of the Basin and Range topography in the "Earthquake" module, they struggled understanding the reverse scale of the vertical axis in the "Sea-Floor Spreading" exercise until the teacher drew the ocean surface at the top of the graph. Once the students realized the relationship between water depth and the vertical axis, they were able to map the transect across the ocean floor and determine that the scar-like feature that they could see in Google Earth was a mid-ocean ridge.

For the "Earthquake Hazards" investigation, rather than use the mathematical formula for the area of the circle, one 6th grade

TABLE 4. SCAFFOLDING OF SCIENCE CONCEPTS IN THE CYBER-ENABLED EARTH EXPLORATION (CE3) CURRICULUM

Science concept	Volcanoes investigation			Earthquakes investigation			Plates investigation		
	1	2	3	1	2	3	1	2	3
Relationship between atmospheric warming and the risk of lahars	7								
How gases reaching the stratosphere can be carried around the globe	8	8							
How atmospheric gases affect global temperatures	8								
How and why gas and vapor pressure change as magma rises		7							
How seismographs are used to predict eruptions		6							
How electronic distance measurement is used to monitor ground movement prior to eruptions		7							
Volcanic gases in the atmosphere move and dissipate over time		8							
Earthquakes can cause landslides				6					
Earthquake intensity dissipates away from the source				6	6				
Earthquake intensity is related to soil type				6		6			
Tsunami increase in amplitude in shallow water				6					
Relationship between seismic wave velocity and material density					7	9			
How compression and extension forces work					8				
Minimum age					8		8		
Substrate affects seismic wave amplitude						9			
Increasing lag time between P and S waves with distance						9			
Geologic forces have split continents apart over time							6		
New sea floor is created at the mid-oceanic ridges								7	
Reversals of Earth's magnetic field have occurred at specific points in time								7	
The age of the sea floor increases from the mid-ocean ridge to the coastlines								7	
Composition of tectonic plates									6
Radioactive decay, convection currents, and plate movement									6

Note: Filled boxes show where in the curriculum the skill or concept is introduced or reinforced. Numbers refer to lowest grade level at which each skill or concept appeared not to be problematic for students.

class used the circle tool, expanding the circle around the epicenter until the reported damage area was covered, making note of the circle radius, and then creating a circle with the same radius around their town. The other 6th grade teacher simply explained the concept first until she felt the students understood it.

The concept of area was also new to many 7th grade students, and they had trouble thinking about area at the landscape scale. They also struggled with creating topographic profiles, and the teacher improvised by replacing these graphing-by-hand exercises with using the elevation profile feature available in Google Earth. (This tool had been deliberately avoided by the curriculum developers so that students would be forced to construct their own profiles and gain a stronger sense of what the profiles represented.)

Some 8th grade students had difficulty measuring the width of the volcanoes and calculating their slopes. They also struggled with calculating rates in the exercise on tsunami. As with the 7th graders, they needed help setting up the graph to map the topography of the sea floor.

A 9th grade teacher reported that he liked the graphical nature of the "Sea-Floor Spreading" and "Plate Boundaries" investigations, as well as the incorporation of math skills. No issues with measuring and calculating were reported for 9th grade students.

Spatial analysis and visualization. Some 6th grade students had trouble with the cardinal directions in the analysis of the Mount Vesuvius eruption ("Volcanic Hazards and Benefits" investigation). The teacher improvised by drawing a compass rose over Mount Vesuvius to help students visualize the directional relationships among the volcano, Pompeii, and Hercu-

laneum. In the analysis of Mt. Fuji ("How Volcanoes Work" investigation), it was intended for students to observe the towns, buildings, and roads around the base of the volcano and make a descriptive assessment about the population density. However, some 6th graders did not seem able to do this. Instead, they wanted population facts to be given. In tracing the mid-oceanic ridge ("Sea-Floor Spreading" investigation), some 6th grade students had trouble seeing the feature outside of the Atlantic and Indian Ocean basins, while others had no problem tracing the feature around the globe.

The 7th grade students didn't tilt, maneuver, and spin the Google Earth images as much as the teacher had expected they would. For example, on the Basin and Range overlay ("Geologic Faults" investigation) students didn't manipulate the view, which made analysis more challenging for them. Students enjoyed creating their own paths and folders, for the "Continental Drift" investigation, although in some cases the paths were not supported with relational data.

Some of the 8th graders had trouble creating paths in the "Continental Drift" investigation. Some students created far too many paths or began answering the questions prior to making the paths as they had preexisting knowledge about Pangaea. That being said, the teacher felt that the paths were an excellent visual aid for students. In the "Plate Boundaries" investigation, the 8th grade students had trouble figuring out the significance of the break in the graph of GPS monument movement, not realizing that it coincided with a nearby earthquake. The 9th grade teachers reported no obvious problems with spatially related tasks.

Scientific reasoning. Some 6th grade students had trouble classifying the mystery volcanoes ("How Volcanoes Work" investigation) even after completing the first example as a class. The teacher needed to go over the important characteristics and how they are used to differentiate among volcano types. In the rate of plate movements exercise ("Plate Boundaries" investigation), 6th grade students had a hard time understanding that negative rates mean movement in the opposite direction as most of these students had had little previous experience with negative numbers. The 7th grade teacher reported that many students seemed to stop at one- or two-word answers to the questions posed in the notebooks, so the teacher's challenge was getting students to elaborate on their thinking. The teacher felt this was not an artifact of the curriculum as much as a "perennial developmental challenge." All of the 8th grade students were able to classify the mystery plate boundaries with little difficulty.

Science Concepts

6th grade students had trouble with many of the science concepts presented, including faults, sea-floor features, reversing magnetic fields, seismic waves, and classification of plate boundaries. These topics were either covered as a class or skipped entirely. 7th grade students struggled most with the concept of geologic faults and their relationship to earthquakes. Students conceptualized faults more as plate boundaries than as cracks in Earth's crust. None of the 8th or 9th grade teachers reported problems with students being able to grasp the science concepts. One 9th grade teacher did report that the CE3 curriculum represented a "new kind of thinking" for her students.

Student Field Notebooks

Questions posed in the student field notebooks were graded on a nine-point scale. Up to three points could be awarded for a correct and complete claim, and up to six points could be awarded for providing appropriate evidence and reasoning. Average scores on each question ranged from 3.0 to 8.4. Many students struggled with expressing appropriate justifications for their claims in part because this style of questioning was very foreign to them (McNeill and Martin, 2011). Therefore, scores of 5.0 or higher were considered evidence of the ability of a student to master the specific skill or concept. However, these scores were reviewed in light of how the curriculum was covered by each teacher. For example, 6th grade students who had answered the questions as a class because the teacher felt that the concepts were too difficult for them were not considered to be reliable indicators of 6th grade student ability. Tables 2, 3, and 4 outline the order in which various technical skills, analytical abilities, and science concepts were introduced in the CE3 curriculum, respectively, and the grade level at which students appeared to be competent to use and understand them. For some skills, the grade levels at which students demonstrated competency varied widely and in those cases a range of grades is presented.

DISCUSSION

The teacher reflections and results from students' field notebooks, outlined above, reveal how well equipped students at each grade level were to employ each skill or ability and to understand each concept presented. Overall, technical skills using Google Earth posed the fewest problems for students while science concepts posed the greatest challenges (Fig. 1).

Specifically, most of the technical skills required for this earth science curriculum are well within the reach of 6th graders. Only a few items, including folder management and complete and accurate data recording require more advanced (7th or 8th grade) students or special attention from the instructor for younger students. However, various analytical abilities are more challenging for students. Creating a graph from x-y data is too advanced for most 6th graders. 7th grade students are generally unfamiliar with using the equation for the area of a circle, and measuring volcano height, width, and slope was problematic for many students, including 8th graders. Students' ability to conduct some operations varied considerably. For example, many students in 8th grade struggled with calculating the rate at which a tsunami would travel, while many 7th graders had no trouble solving the algebraic problem. Students also varied widely in their ability to create a topographic profile of the Basin and Range topography, but in a subsequent exercise when asked to draw a topographic profile of the Atlantic sea floor, most students at the 7th grade level or higher were able to successfully complete the activity.

Of the 17 discrete spatial analysis and visualization skills required in the CE3 curriculum, the most difficult appear to be interpreting InSAR (interferometric synthetic aperture radar)-derived images and creating paths to associate data in Google Earth. These activities are being revisited in the revision of the curriculum as they may not be adequately explained or modeled. The National Research Council's 2006 report, "Learning to Think Spatially," calls for development of systems to support spatial thinking in K–12 education to ensure that all students have the opportunity to develop these crucial skills and understandings. These support systems should provide an interactive

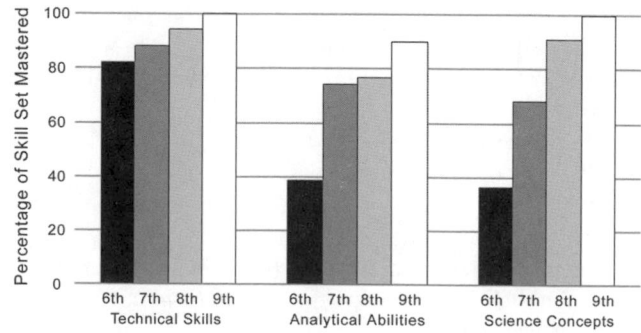

Figure 1. Percentage of each skill set students in each grade level were able to master. Analytical abilities that varied widely among grades are not included.

environment that allows orderly and sequential introduction of basic concepts and skills and guidance in the inquiry process.

In the earth science content domain, a specific set of geospatial skills and practices are used to understand spatial and temporal patterns at local, regional, and global scales. These skills include, for example, setting theoretical relations between data sets (e.g., union, intersection), data interpolation and extrapolation, creating abstractions (e.g., contouring), describing spatial distribution in terms of density or dispersion, pattern analysis (e.g., clustering), structural analysis (e.g., nearest neighbors), and process modeling (expressing spatial patterns as a function of distance or time). These geospatial skills are used in addition to spatial visualization skills germane to the geosciences, including (1) spatial relations (ability to mentally rotate an object about its center), (2) spatial manipulation (ability to mentally manipulate an image into another arrangement), and (3) visual penetrative ability (ability to mentally imagine what is inside of a solid object) (Titus and Horsman, 2009; Almquist et al., 2011). In many instances, people apply these skills and practices when thinking through a problem without even realizing it, but how and when students develop these abilities in largely unknown. Numerous studies do suggest that although people vary in their innate spatial thinking abilities, practicing spatial operations can help build spatial competencies (Titus and Horsman, 2009; Kastens et al., 2009; Hegarty et al., 2006; Piburn et al., 2002; Sorby, 2001; Orion et al., 1997; Kali and Orion, 1996; Lord, 1987; Lord, 1985), especially when combined with timely and informative feedback (Ishikawa and Montello, 2006).

Attention should also be paid to students' scientific reasoning abilities. It is unclear why the CE3 activity involving classifying earthquake intensity from physical evidence using the modified Mercalli intensity scale proves so difficult for students. Clearly, restructuring of the activity or additional supports are needed. Interpreting topographic profiles also proved difficult for many students at multiple grade levels. A more in-depth analysis of how different teachers approached these activities might provide useful clues regarding best practices. Another difficult scientific reasoning exercise was relating a jump in the rate of movement of a GPS monument with a nearby earthquake that occurred on the same day. This activity required that students (1) understand that the break in the line on the graph signified an abrupt movement of the GPS monument, (2) hypothesize that an earthquake could cause this type of movement, and (3) examine the map of earthquakes near the monument to identify one of significant magnitude occurring at the same time. This higher-order thinking was displayed only by 9th grade students, but might have been exhibited by younger students with appropriate prompting. Students' ability to classify plate boundaries based on physical characteristics observed in Google Earth also varied widely. While one 6th grade teacher reported students having "no problem," the 7th graders found this task difficult.

Science concepts presented were sequenced according to the content covered. However, students' ability to understand these concepts did not consistently follow this progression. The most difficult concepts included the nature and behavior of seismic waves, which only 9th graders seemed to grasp, followed by the behavior and effects of volcanic gases in the atmosphere, the concept of minimum age, and compressional and extensional forces, which were approachable only by 8th and 9th graders. Most 7th graders could comprehend the basic concepts related to sea-floor spreading, how electronic distance measurement (EDM) works, and how gas and vapor pressure inside the magma chamber changes as magma rises, and could link together the concepts of atmospheric warming, glacial retreat, and reduced risk of lahars. Because the content is organized to build student understanding cumulatively, disconnects in students' ability to grasp specific topics has potential to disrupt this progression, leading to only superficial understanding of the causes and implications of plate tectonics.

Restructuring the curriculum to better align and scaffold specific analytical abilities and science concepts will likely involve creating separate, sequential units for various grade levels, as well as thoroughly revising or eliminating some activities altogether. These changes will also be beneficial because although teachers were generally very positive about their experiences using the CE3 curriculum, all of them felt that it took too long to complete. This resulted in portions of the curriculum being rushed through or skipped completely, and a certain amount of fatigue for students. Even so, most of the pilot teachers elected to implement the curriculum again in the following year even before the next round of revisions could be made.

CONCLUSIONS AND RECOMMENDATIONS

Google Earth represents an intuitive, powerful, and highly adaptable technology that excites students and teachers alike. However, its effective use to convey critical science content requires considerable, iterative curriculum design. The fact that Google Earth is easy for even young students to use does not automatically translate into their being able to analyze and understand the implications of their observations.

Classroom testing is a useful approach for identifying specific disconnects among technology, analysis, and conceptualization, as this method reveals the reality of how teachers and students interact with the technology within a specific science context. Insights gained through these results are now being utilized to revise the curriculum and again examine classroom outcomes. Over time, this iterative design cycle will result in a comprehensive scope and sequence of technical skills, analytical abilities, and science concepts for the study of plate tectonics. While some of this scope and sequence will be domain specific, other portions (i.e., technical skills) may be more broadly applied. Thus, the Cyber-Enabled Earth Exploration project represents a significant advance toward a universal scope and sequence of technical skills using Google Earth, and a domain-specific scope and sequence for Google Earth–supported analytical abilities and science concepts concerning volcanoes, earthquakes, and plate tectonics in the middle school classroom.

ACKNOWLEDGMENTS

We sincerely thank the dedicated classroom teachers who made this project possible. This material is based upon work supported by the National Science Foundation, Grant No. DRL-0918683. Any opinions, findings, and conclusions or recommendations expressed in this material are those of the authors and do not necessarily reflect the views of the National Science Foundation.

REFERENCES CITED

Allison, M.A., DeGaetano, A.T., and Pasachoff, J.M., 2006, Earth Science: Austin, Texas, Holt, Rinehart and Winston, 962 p.

Almquist, H., Stanley, G., and Hendrix, M., 2007, Paleontology exploration project: Inquiry-based learning project focusing on technology and field-based paleontology and sedimentology for middle-school teachers and students: Geological Society of America Abstracts with Programs, v. 39, no. 6, p. 251.

Almquist, H., Blank, L., Stanley, G., and Hendrix, M., 2008, Paleo Exploration Project: Teachers, GIS, and Fossils in Montana: San Diego, California, Environmental System Research Institute (ESRI) 2008 International User Conference, http://proceedings.esri.com/library/userconf/educ08/educ/abstracts/a1506.html

Almquist, H., Stanley, G., Hendrix, M., Blank, L., Hanfling, S., and Crews, J., 2011, An integrated field-based approach to building teachers' geosciences skills: Journal of Geoscience Education, v. 59, p. 31–40, doi:10.5408/1.3543926.

Amaral, O.M., Garrison, L., and Klentschy, M.P., 2002, Helping English learners increase achievement through inquiry-based instruction: Bilingual Research Journal, v. 26, p. 213–239.

Andrews, G., 2006, Laptops + challenging curriculum = student success: Techniques: Connecting Education & Careers, v. 81, no. 5, p. 42–44.

Audet, R.H., and Abegg, G.L., 1996, Geographic information systems: Educational implications for problem solving: Journal of Research in Science Teaching, v. 33, no. 1, p. 21–45, doi:10.1002/(SICI)1098-2736(199601)33:1<21::AID-TEA2>3.0.CO;2-R.

Berland, L.K., and McNeill, K.L., 2010, A learning progression for scientific argumentation: Understanding student work and designing supportive instructional contexts: Science Education, v. 94, no. 5, p. 765–793, doi:10.1002/sce.20402.

Bransford, J.D., and Donovan, M.S., 2005, Scientific inquiry and how people learn, *in* Donovan, M.S., and Bransford, J.D., eds., How students learn: Science in the classroom: Washington D.C., National Academies Press, p. 27–52.

Cooney, T., Cummins, J., Flood, J., Foots, B.K., Goldston, M.J., Key, S.G., et al., 2006, Scott Foresman Science: Earth Science: New York, Pearson/Scott Foresman, 360 p.

Cradler, J., McNabb, M., Freeman, M., and Burchett, R., 2002, How does technology influence student learning?: Learning and Leading With Technology, v. 29, no. 8, p. 46–52.

Feather, R.M., Jr., Snyder, S.L., and Hesser, D.T., 1995, Merrill Earth Science: New York, Glencoe/McGraw-Hill, 744 p.

Feather, R.M., Jr., Snyder, S.L., and Zike, D., 2008, Earth Science: New York, Glencoe/McGraw-Hill, 854 p.

Fedisson, M., and Braidic, S., 2007, PowerPoint presentations increase achievement and student attitudes toward technology: International Journal of Information and Communication Technology Education, v. 3, no. 4, p. 64–75, doi:10.4018/jicte.2007100106.

Halpern, D.F., 1999, Teaching for critical thinking: Helping college students develop the skills and dispositions of a critical thinker: New Directions for Teaching and Learning, v. 80, p. 69–74, doi:10.1002/tl.8005.

Hegarty, M., Montello, D.R., Richardson, A.E., Ishikawa, T., and Lovelace, K., 2006, Spatial abilities at different scales: Individual differences in aptitude-test performance and spatial-layout learning: Intelligence, v. 34, p. 151–176, doi:10.1016/j.intell.2005.09.005.

Ishikawa, T., and Montello, D.R., 2006, Spatial knowledge acquisition from direct experience in the environment: Individual differences in the development of metric knowledge and the integration of separately learned places: Cognitive Psychology, v. 52, p. 93–129, doi:10.1016/j.cogpsych.2005.08.003.

Kali, Y., and Orion, N., 1996, Spatial abilities of high-school students in the perception of geologic structures: Journal of Research in Science Teaching, v. 33, p. 369–391, doi:10.1002/(SICI)1098-2736(199604)33:4<369::AID-TEA2>3.0.CO;2-Q.

Kastens, K.A., Agrawal, S., and Liben, L.S., 2009, How students and field geologists reason in integrating spatial observations from outcrops to visualize a 3-D geological structure: International Journal of Science Education, v. 31, p. 365–393, doi:10.1080/09500690802595797.

Lord, T.R., 1985, Enhancing the visuo-spatial aptitude of students: Journal of Research in Science Teaching, v. 22, p. 395–405, doi:10.1002/tea.3660220503.

Lord, T.R., 1987, A look at spatial abilities in undergraduate women science majors: Journal of Research in Science Teaching, v. 24, p. 757–767, doi:10.1002/tea.3660240808.

Marshak, S., 2007, Earth: Portrait of a Planet, 3rd edition: New York, W.W. Norton & Co Inc., 832 p.

McDougall Littell, 2006, Earth Science: USA, McDougal Littell, 784 p.

McNeill, K.L., 2011, Elementary students' views of explanation, argumentation and evidence and abilities to construct arguments over the school year: Journal of Research in Science Teaching, v. 48, no. 7, p. 793–823, doi:10.1002/tea.20430.

McNeill, K.L., and Martin, D.M., 2011, Claims, evidence and reasoning: Demystifying data during a unit on simple machines: Science and Children, v. 48, no. 8, p. 52–56.

Morrison, E.S., Moore, A., Armour, N., Hammond, A., Haysom, J., Nicoll, E., and Smyth, M., 1997, Science Plus Level Red: New York, Holt Reinhart and Winston, 559 p.

National Research Council, 2006, Learning to Think Spatially: Washington, D.C., National Academies Press, 313 p.

National Research Council, 2008, Research on Future Skill Demands: Washington, D.C., National Academies Press, 112 p.

Orion, N., Ben-Chaim, D., and Kali, Y., 1997, Relationship between earth-science education and spatial visualization: Journal of Geoscience Education, v. 45, p. 129–132.

Piburn, M.D., Reynolds, S.J., Leedy, D.E., McAuliffe, C.M., Birk, J.P., and Johnson, J.K., 2002, The hidden Earth: Visualization of geologic features and their substrate geometry, http://reynolds.asu.edu/pubs/NARST_final.pdf (accessed May 2012)

Purcell, A.D., Ponomarenko, A.L., and Brown, S.C., 2006, A fifth grader's guide to the world: Science and Children, v. 43, no. 8, p. 24–27.

Schott, R.C., 2009, GigaGeology: Virtual field trips in a Web2.0 world: Geological Society of America Abstracts with Programs, v. 41, no. 7, p. 165.

Sorby, S.A., 2001, A course on spatial visualization and its impact on the retention of female engineering students: Journal of Women and Minorities in Science and Engineering, v. 7, p. 153–172.

Stanley, G., and Almquist, H., 2008, Spatial analysis of fossil sites in the northern plains: A unique model for teacher education: GSA Today, v. 18, no. 2, p. 24–25, doi:10.1130/1052-5173(2008)18[24:SAOFSI]2.0.CO;2.

Titus, S., and Horsman, E., 2009, Characterizing and improving spatial visualization skills: Journal of Geoscience Education, v. 57, p. 242–254, doi:10.5408/1.3559671.

Todd, R.W., ed., 2001, Holt Science & Technology Earth Science: Austin, Texas, Holt, Rinehart and Winston, 784 p.

Wagner, T., 2008, The global achievement gap: Why even our best schools don't teach the new survival skills our children need and what we can do about it: New York, Basic Books, 344 p.

MANUSCRIPT ACCEPTED BY THE SOCIETY 16 APRIL 2012

The Geological Society of America
Special Paper 492
2012

Google Earth geo-education resources: A transnational approach from Ireland, Iceland, Finland, and Norway

Ronán Hennessy*

Earth and Ocean Sciences, National University of Ireland Galway, University Road, Galway, Ireland

Thorvardur Arnason

University of Iceland, Hornafjordur Regional Research Centre, Iceland

Ilkka Ratinen

Department of Teacher Education, University of Jyväskylä, Finland

Lena Rubensdotter

Geological Survey of Norway, Trondheim, Norway

ABSTRACT

The Northern Environmental Education Development (NEED) project was a transnational geo-education resource development initiative cooperated by four European countries: Ireland, Finland, Norway, and Iceland. NEED supported the development of novel geological information resources, and the establishment of geo-education networks in the four project regions. An analysis of the national geography and science curriculum in each region was conducted. This analysis identified which geographic learning skills were common to each curriculum. The development of innovative geographic information system (GIS) resources to help support the development of geographic skills was considered a potentially valuable outcome of the NEED project. Google Earth was chosen as the technology with which to develop and serve the GIS information and learning resources. The resultant Google Earth learning resources provide free access to geoscientific information about the project regions for educators and students. While the development of these learning resources represents an important step in providing locally relevant geo-educational resources, further research is necessary to determine the pedagogical effectiveness of the learning content.

*ronan.hennessy@nuigalway.ie

Hennessy, R., Arnason, T., Ratinen, I., and Rubensdotter, L., 2012, Google Earth geo-education resources: A transnational approach from Ireland, Iceland, Finland, and Norway, *in* Whitmeyer, S.J., Bailey, J.E., De Paor, D.G., and Ornduff, T., eds., Google Earth and Virtual Visualizations in Geoscience Education and Research: Geological Society of America Special Paper 492, p. 413–418, doi:10.1130/2012.2492(31). For permission to copy, contact editing@geosociety.org. © 2012 The Geological Society of America. All rights reserved.

INTRODUCTION

The role of graphical illustrations and visualizations for the analysis and communication of information has long been a fundamental tool in geoscience. Agricola (1556) presented a series of woodcuts on the layout and workings of mines in central medieval Europe, and Rudwick (1976) explored the significant role of visual language in the fields of geology and mineralogy during the eighteenth and nineteenth centuries. The emergence of geospatial technologies such as satellite imagery, light detection and ranging (LIDAR), geographic information systems (GIS), and Google Earth in more recent times have provided geoscientists and educators with unprecedented access to vast archives of geological and environmental information. The availability of geoscientific information through GIS and Internet technologies represents a paradigm shift in how geoscientific information is distributed, investigated, and communicated. For the school student or novice geoscientist, fundamental geographic skills such as map and scale interpretation, a comprehension of two-dimensional (2-D) and three-dimensional (3-D) geometries, and an appreciation of temporal data, are essential to understanding geoscientific information. The development of these geographic skills is common to many geographic and scientific school curricula (Northern Environmental Education Development [NEED], 2011).

Freely available, interactive geospatial visualization technologies such as Google Earth, NASA (National Aeronautics and Space Administration) World Wind, and ArcGIS Online, present new opportunities to develop tools and resources to support teaching and learning. Tversky et al. (2002) noted that while visual representations are not always more successful than textual information in facilitating comprehension and learning, the use of interactivity can help overcome difficulties in perception and understanding. Dutrow (2007) noted that the preferred style of learning of over 60% of students is of a visual nature, while Dale (1969) claimed that people extract and retain more information from visual presentations than from written or spoken prose. The development of Google Earth visualizations in the transnational NEED project represents one example of how Google Earth can be used to deliver geoscientific information to support geographic and scientific curricula in four regions of northwest Europe.

SCIENCE VISUALIZATION IN EDUCATION

Visualization is fundamental to communicating processes, concepts, and phenomena in many science and engineering disciplines. The role of visualizations in science education is documented in Gilbert (2007). Reynolds et al. (2007) discussed the significance of visualizations in aspects of geological research, communication, and teaching. Lisle (2006) suggested that tools such as Google Earth demonstrated potential for aiding in the teaching of map interpretation and the comprehension of 3-D physiographical features. Further discussions on the application of Google Earth for geoscientific student instruction are provided in Allen (2008) and Patterson (2007).

Digital visualizations and animations allow geoscientists to observe phenomena at a variety of scales, from microscopic to planetary, and at varying timescales. Dutrow (2007) proposed that visual literacy is among a set of cognitive skills necessary to correctly interpret geological visualizations. Avgerinou and Ericson (1997) provide several definitions for visual literacy, of which Horton's (1983) concise definition is considered suitably applicable to the geosciences: "visual literacy is the ability to understand (read) and use (write) images and to think and learn in terms of images, i.e., to think visually." Piburn et al. (2002) further explore the importance of developing visual and spatial skills for geoscience learning and emphasize the potential of visualizations to help students develop these skills.

Mayer and Moreno (1998) examined the cognitive constraints on learning with multimedia and proposed that learners engage in three main cognitive processes. The first cognitive progress, selecting, is applied to incoming verbal information to yield a text base and is applied to incoming visual information to yield an image base. The second cognitive process, organizing, is applied to the word base to create a verbally based model of the topic being learned and is applied to the image base to create a visually based model of the topic being learned. The third process, integrating, occurs when the learner builds connections between corresponding events in the verbally based and visually based models. As a multimedia learning tool, Google Earth first enables users to select, or visualize, the information displayed in the viewer. Users can then organize the information, by deciphering what is being visualized (e.g., mountains, rivers, coastlines), before proceeding to integrate the information by analyzing spatial relationships, field-of-view perspectives, and using these inferences to extract or predict meaningful information from the visualization.

Rapp (2007) remarked on the challenges of developing effective educational visualization experiences, and on the necessity for technologists, scientists, and cognitive scientists to collaborate in the development of effective visualizations for teaching and learning. Tversky et al. (2002, p. 249) recommended that effective graphics for teaching and learning follow the "Congruence Principle," wherein "the structure and content of the external representation should correspond to the desired structure and content of the internal (mental) representation." Tversky et al. (2002, p. 255) advised that the "Apprehension Principle" be considered equivocally when developing visualizations for learning, wherein "the structure and content of an external representation should be readily and accurately perceived and comprehended." Dale (1969) suggested that audio-visual learning content fosters effective learning for students with low verbal learning abilities; encourages active participation in the learning process; and contributes to the efficiency, depth, and variety of learning. Considering the application of audio-visual technologies and learning content to geoscientific learning, it can be argued that Google Earth satisfies many of these attributes.

NEED—NORTHERN ENVIRONMENTAL EDUCATION DEVELOPMENT

The Northern Environmental Education Development project (NEED) operated as a transnational geo-education initiative from 2008 to 2010 and involved four European regions: Ireland, Iceland, Finland, and Norway. The project was established to deliver novel resources for enhancing the communication and understanding of geo-knowledge in schools and communities in four peripheral European regions. The Burren region, on Ireland's west coast, is a designated UNESCO (United Nations Educational, Scientific and Cultural Organization)-supported Global Geopark. In Iceland, Vatnajökull National Park covers more than 13% of the entire country and is the largest national park in Europe. In north-central Norway, the Junkerdal and Saltfjellet-Svartisen national parks are located in the county of Nordland and are home to the Svartisen glacier, the largest ice-sheet in northern Scandinavia. In Finland, Koli National Park is located in the North Karelia region of eastern Finland. A project website (www.geoneed.org) was developed to provide information about the geo-education networks in each region and to make available the educational resources developed during the project term.

GEOGRAPHIC SKILLS IN NATIONAL CURRICULA

Collaboration between each national project team and their local schools was an integral element of NEED. Each partner conducted an analysis of their national core curriculum (primary and secondary levels) to identify how NEED could deliver curriculum-relevant education resources. The analysis focused on the aspects of the curriculum that were relevant to the five NEED themes (geology, landscape, natural hazards, geological materials in society, and climate change) (NEED, 2011). Many subjects and learning topics were found to be common to each curriculum. Significantly, the role of maps, globes, and models in developing geographic skills was identified as a fundamental component in

each of the four national geography and science syllabus. Four main sets of geographical skills were identified to be common to each curriculum (described below). In consideration of how information and communication technology (ICT) could be used to support teaching and learning in each curriculum, Google Earth was deemed a potentially valuable learning support tool for assisting the development geographic skills (Table 1).

Skill 1: Observing, Investigating, and Describing Processes and Phenomena

Google Earth allows students to observe and visually investigate 3-D terrain, surface features, and land-cover signatures at a range of scales, throughout the Earth, (as well as the Moon and Mars). As a learning resource, Google Earth provides students with free access to data pertaining to a range of subject matter covered in primary and secondary science and geography syllabi.

Skill 2: Retrieving, Analyzing, and Communicating Information

Google Earth is a free geospatial visualization and ICT tool that allows students to conduct their own investigations. The facility for students to direct their own investigation and analysis supports constructivist learning by allowing students to interact with their specific subject, particularly with regard to geospatial characteristics, relationships, and constraints of the subject matter. The ability for students to easily capture and save a map view, placemark a location, or record a tour demonstrates the suitability of Google Earth for recording, analyzing, retrieving, and communicating information.

Skill 3: Development of an Awareness of Place

As a geospatial map and location-information service, Google Earth provides access to a vast data archive of photographic

TABLE 1. SELECTED ELEMENTS FROM THE NATIONAL CURRICULA OF IRELAND, ICELAND, FINLAND, AND NORWAY, WHERE GOOGLE EARTH IS CONSIDERED TO BE OF VALUE IN SUPPORTING LEARNING OBJECTIVES

Curriculum	Age (yr)	Learning objective
Ireland		
Geography, science (P)	4–12	Compare and use maps, globes, airphotos, satellite and remotely sensed images.
Geography (S)	13–18	Map/airphoto interpretation, use of information and communication technology, investigative skills: accurate observation.
Iceland		
Mathematics (P, S)	6–15	Map reading/interpretation, use public data sources to find information about natural phenomena.
Finland		
Geography, biology, physics (P, S)	7–15	Develop environmental literacy, interpret maps, use pictures as sources of geographic information.
Cross-curricular themes (S)	16–18	Learn to measure, assess, and analyse changes in natural, and cultural environments.
Norway		
Geography (P)	6–13	Read and use digital maps and identify local and global geographical characteristics.
Geography (S)	13–16	Use pictures, maps, and models to understand scale and theme.

Note: P—Primary curriculum; S—Secondary curriculum.

imagery, satellite imagery, and 3-D topography, ranging from planetary scale views to the large scale, "Street View" mode. This facility can help students to develop an appreciation of their respective location within their own region, their own country, and within the world.

Skill 4: Acquiring a Sense of Environmental Responsibility; Considering the Human Relationship with Natural Resources and Ecosystems

The ability to visually investigate land-cover features in Google Earth renders the application a valuable learning tool for gaining an insight into, and an understanding of, human impacts and naturally occurring events on the natural environment.

GOOGLE EARTH FOR EDUCATION

Google Earth was chosen as the ICT through which geographic visualizations could be developed and delivered for NEED, primarily because of its free availability, but also because each project partner was aware that Google Earth was used by teachers in schools in their region. The Google Earth visualizations and learning-tasks developed for NEED are available for download on the project website (www.geoneed.org). Visualizations and learning activities were generated for the Burren region, with separate students and teacher handbooks, and a basic Google Earth user manual available for download. In Iceland, visualizations focused on recently developed glacial landforms in the south and west quadrants of Vatnajökull National Park (e.g., moraines, eskers) and national climatic and elevation maps. In Finland, Google Earth–based learning assignments were designed to support spatial thinking skills. Ratinen and Keinonen (2011) investigated the use of the Google Earth visualizations for problem-based learning processes in teacher education in Finland. The satellite imagery for the Nordland region of Norway was not of sufficient resolution (>15 m) to encourage the development of Google Earth–based visualizations; albeit a limited number of visualizations were developed to satisfy the terms of the project.

DISCUSSION

The Google Earth educational resource development component of NEED was a successful outcome of the project, in consideration of the provision of freely accessible geo-information that was relevant to the target (project) regions. The development of the NEED project website helped to serve the Google Earth visualizations alongside other educational resources developed throughout the project and secures open access to the resources into the future.

Observations made of students and teachers ("users") using or viewing the Google Earth educational resources tended to fall into three main categories: (1) users were stimulated and encouraged to engage with the geo-information presented to them in Google Earth after realizing that they could interact with the information (fade in/out maps; rotate views of maps and models; zoom to other places and features in their local area; save placemarks in interesting locations); (2) the ability to carry out simple, yet effective scientific observations and measurements (measuring distances, changing sun illumination) allowed students to actively participate in the learning process, and this encouraged students to direct their own investigations rather than relying on passive instruction; (3) teachers considered the availability of information and maps pertaining to their local area to be of particular value, mainly because geological, archaeological, or satellite data sets are difficult to access, due to issues in sourcing digital data from providers, data licensing limitations, requirements of understanding or purchasing advanced GIS software, or simply due to a lack of suitable teaching and learning information presented in an easy-to-use ICT application such as Google Earth. The Google Earth learning resources developed in NEED aimed to address these issues, wherein data providers such as the national geological surveys permitted open use of their data to be freely downloadable in easy-to-use Keyhole Markup Language (KML) format for viewing in the interactive Google Earth free geo-browser application.

For a transnational geo-education project such as NEED, where project partners use different languages, the development of a shared web portal to serve the educational resources yielded a challenge. In order to satisfy the terms of the project, the NEED project website was developed and served in English. While three of the project partners do not use English as their primary language, English is widely understood in these regions, and it was therefore the optimum common language. The idea of providing all the information on the project website in each of the four languages was considered, however this was deemed too great a task, and it was felt that it would separate the final resources into region specific sections. Instead, much of the content is served on the same page as is the case for the Google Earth resources. While three of the project partners subsequently provided the NEED project information pertaining to their region on their own website (e.g., www.burrenconnect.ie/geopark, www.need.is, www.uef.fi/need), the Norwegian partners considered the main project website to be sufficient to serve information on Nordland. For future transnational projects such as NEED, overcoming the language "issue" in the final project deliverable is likely to remain a challenge. Apart from the delivery of content on the website, differences in languages had no negative impact on the project.

Prior to commencing the development of the Google Earth learning content in the Burren region, initial school visits were conducted, wherein Google Earth demonstrations were shown to classes. Participating teachers who had previously used Google Earth in the classroom remarked that Google Earth provided a "quick, exciting, and rewarding experience" when displaying geographical information in the classroom. It is herein suggested that this user-friendly and rewarding experience is one of the successes of Google Earth, and this has arguably led to

Google Earth being hailed as facilitating the "democraticization" of GIS technologies (Butler, 2006). Following a demonstration of "my local geology" visualizations in three primary school classes (ages 9–11) in the Burren region in 2009, a class survey revealed that the majority of students (70%, 80%, and 88% in the respective classes) had prior experience in using Google Earth either at home or at school. The students remarked that it was the interactive "gaming-type" environment of Google Earth that made it appealing as an exploration and learning technology. The interactive capability of Google Earth was a major factor in initially choosing Google Earth for the NEED project. Tversky et al. (2002) propose that the introduction of interactivity into learning activities facilitates learning, and helps to overcome difficulties of perception and comprehension. The visualizations developed in NEED provide a mechanism for user interaction and self-guided tuition, and it is argued that this supports a constructivist approach to student learning.

While the development of the Google Earth visualizations was a core specification of NEED, subsequent analysis of how widely the learning resources have been adopted and used by educators or communities remains outstanding. Informal feedback gathered from teachers and educators in each project region indicated that the Google Earth content developed in NEED was a welcome teaching resource, however this feedback is not qualitative. To determine how successfully the NEED GIS learning resources integrated into the formal curriculum, it will be necessary to conduct a survey of schools wherein the Google Earth resources have been used. Such a survey is planned to be conducted in the Burren region in a select number of schools that are known to have used the Google Earth learning content. This research has the potential to reveal what elements of the learning resources were found to be of value in the classroom or home, and why other content was not utilized. Ratinen and Keinonen (2011) carried out an assessment of student teacher's geographical thinking skills in Finland. The results of the study revealed that while student-teachers may have inadequate ICT skills, the use of Google Earth improved student-teachers' geographical thinking skills, even though difficulties in interpreting maps and analyzing geological data remained.

The default imagery in Google Earth for each of the four regions did not provide a complete, high-resolution satellite imagery mosaic. Ireland (1–15 m), Finland (2.5 m), and Iceland (2.5 m) enjoyed sufficient high-resolution coverage, while imagery for the Nordland region of Norway was poor, with the entire region limited to 15 m resolution. While this resolution may be adequate for viewing medium and landscape-scale features, the imagery was of insufficient quality for displaying detailed surface features or cultural landmarks. The absence of high-resolution imagery in Nordland resulted in Google Earth not being successfully adopted as an educational tool by the Norwegian partners. For future multi-region projects (such as NEED) that endeavor to adopt Google Earth as a visualization tool, it is important to ensure that each region is represented by a relatively equal level, and sufficient coverage of, high-resolution imagery.

In retrospect, the generation and dissemination of a NEED project-specific KML template for implementing KML features and styles would have greatly enhanced the final deliverables of the educational visualizations. A project-specific template would have assisted the visualization developers, and reduced the time spent developing and testing KML files. A template could have established a common theme and format of displaying screen overlays, description HTML (hypertext markup language) in placemarks, and the use of regions and network links. In addition, a template could have helped to ensure best-practice in implementing the syntax of KML elements for the NEED project educational visualizations.

The NEED project demonstrated the successful adoption of Google Earth over the past decade for geographic visualization in the four partner regions of the NEED project. The regions are rural areas on Europe's northern periphery, and it is probable that (in the presence of access to high-speed broadband Internet connectivity) Google Earth is widely used in most rural and peripheral communities of Europe. This is encouraging for future projects and initiatives (such as the NEED project) that wish to employ Google Earth for educational, environmental conservation, or tourism-related programs. The use of project-themed templates, as well as ensuring that each partner enjoys adequate imagery resolution is recommended prior to embarking on a Google Earth–based program. In consideration of the development of effective educational content to develop geographic skills and comprehension, further research is necessary to determine the pedagogical effectiveness of the Google Earth visualizations for teaching and learning. A survey of the adoption of Google Earth is a valuable and rewarding educational tool, and if used properly, it has the potential to play an important role in the future development of educational curricula.

ACKNOWLEDGMENTS

NEED (Northern Environmental Education Development) was funded under the INTERREG IIIB Northern Periphery Programme 2007–2013 (www.northernperiphery.eu). The administrative partners of NEED were Burren Connect, Clare County Council, Ireland; Hornafjordur Regional Research Centre, University of Iceland; University of Eastern Finland; and Nordland National Park, Norway.

REFERENCES CITED

Agricola, G., 1556, De Re Metallica: http://books.google.com/books?id=Wm eddCtAobUC&dq=de+re+metallica&ie=ISO-8859-1&source=gbs_gdata (accessed 28 November 2011).

Allen, T.R., 2008, Digital terrain visualization and virtual globes for teaching geomorphology: The Journal of Geography, v. 106, p. 253–266, doi:10.1080/00221340701863766.

Avgerinou, M., and Ericson, J., 1997, A review of the concept of visual literacy: British Journal of Educational Technology, v. 28, p. 280–291, doi:10.1111/1467-8535.00035.

Butler, D., 2006, Virtual globes: The web-wide world: Nature, v. 439, p. 776–778, doi:10.1038/439776a.

Dale, E., 1969, Audio-Visual Methods in Teaching, 3rd Edition: New York, Holt, Rinehart & Winston, 108 p.

Dutrow, B.L., 2007, Visual communication: Do you see what I see?: Elements, v. 3, no. 2, p. 119–126, doi:10.2113/gselements.3.2.119.

Gilbert, J.K., 2007, Visualization in Science Education: Dordrecht, The Netherlands, Springer, 346 p.

Horton, J., 1983, Visual literacy and visual thinking, *in* Burbank, L., and Pett, D., eds., Contributions to the Study of Visual Literacy: Texas, International Virtual Learning Academy, p. 92–106.

Lisle, R.J., 2006, Google Earth: A new geological resource: Geology Today, v. 22, p. 29–32, doi:10.1111/j.1365-2451.2006.00546.x.

Mayer, R.E., and Moreno, R., 1998, A cognitive theory of multimedia learning: Implications for design principles: Paper presented to the annual meeting of the World Conference on Educational Multimedia, Hypermedia and Telecommunications, Montreal, Canada, www.unm.edu/~moreno/PDFS/chi.pdf (accessed 28 November 2011).

Northern Environmental Education Development (NEED), 2011, Curriculum analyses: National core curricula analysis of the NEED project target areas: www.uef.fi/need/curriculum-analyses (accessed 28 November 2011).

Patterson, T.C., 2007, Google Earth as a (not just) geography education tool: The Journal of Geography, v. 106, p. 145–152, doi:10.1080/00221340701678032.

Piburn, M.D., Reynolds, S.J., Leedy, D.E., McAuliffe, C.M., Birk, J.P., and Johnson, J.K., 2002, The Hidden Earth: Visualization of geologic features and their subsurface geometry: Proceedings of the annual meeting of the National Association for Research in Science Teaching, New Orleans, Louisiana, http://reynolds.asu.edu/pubs/NARST_final.pdf (accessed 28 November 2011).

Rapp, D.N., 2007, Mental models: Theoretical issues for visualizations in science education, *in* Gilbert, J.K., ed., Visualization in Science Education: Dordecht, The Netherlands, Springer, p. 43–60.

Ratinen, I., and Keinonen, T., 2011, Student-teachers' use of Google Earth in problem-based geology learning: International Journal of Environmental and Geographical Education, v. 20, p. 345–358, doi:10.1080/10382046.2011.619811.

Reynolds, S.J., Johnson, J.K., Piburn, M.D., Leedy, D.E., Coyan, J.A., and Busch, M.M., 2007, Visualization in undergraduate geology courses, *in* Gilbert, J.K., ed., Visualization in Science Education: Dordecht, The Netherlands, Springer, p. 253–266.

Rudwick, M., 1976, The emergence of a visual language for geological science 1760–1840: History of Science, v. 14, p. 149–195.

Tversky, B., Morrison, J., and Betrancourt, M., 2002, Animation: Can it facilitate?: International Journal of Human-Computer Studies, v. 57, p. 247–262, doi:10.1006/ijhc.2002.1017.

MANUSCRIPT ACCEPTED BY THE SOCIETY 16 APRIL 2012

The Geological Society of America
Special Paper 492
2012

Using Google Earth to teach geomorphology

Holly A.S. Dolliver*

*University of Wisconsin–River Falls, Department of Plant and Earth Science, 410 S. Third Street, 307 Agricultural Science,
River Falls, Wisconsin 54022, USA*

ABSTRACT

Google Earth is a free, easy-to-use geobrowser that has become a popular tool for observing planet Earth. Several features within Google Earth can enhance teaching of geomorphology concepts. The ability to tilt a scene and view a landscape three-dimensionally, along with the capability to make measurements and construct an elevation profile, can greatly facilitate the identification and characterization of land-forms and geomorphic mapping. Historical imagery allows users to access and analyze imagery dating back to the 1940s in some locations. This time-series of imagery is useful for studying natural and anthropogenic geomorphic processes and change. In addition to the geospatial data provided by Google Earth, supplementary data such as U.S. Geological Survey topographic or geologic maps, can be imported and easily georeferenced to provide an opportunity for more comprehensive analysis. Although not as powerful as commercial geographic information system (GIS) software, Google Earth is a dynamic venue for students to explore and analyze the geomorphology of the entire planet.

INTRODUCTION

Geomorphologists study the nature of landscapes and the complex processes acting to create and modify the land surface over time. In addition to natural physical, chemical, and biological processes, geomorphologists must also understand the impact of humans on Earth's surface. Climate change, urban development, deforestation, and agriculture have all been shown to accelerate geomorphic change (Williams et al., 1990; Wendland, 1996; Hooke, 2000; Knox, 2001; Jones et al., 2009). Understanding the rate and magnitude of surface processes and geomorphic change is essential knowledge for sustainable land and management decision making.

One of the most challenging aspects of geomorphology is that geomorphic processes occur over a broad range of spatial and temporal scales (Brunsden and Thornes, 1979; Phillips, 2009). Geomorphologists have an increasingly wide array of data sets and tools for both qualitative and quantitative study of landforms and geomorphic processes.

(1) *Imagery:* The use of ground and aerial photographs (Fig. 1A) by geomorphologists to investigate landscape change over time is well documented (Trimble and Cooke, 1991; Bierman et al., 2005; Trimble, 2008; Cerney, 2010). In many places, imagery is available from as early as the 1930s, providing a long temporal record of landscape features and change (Trimble and Cooke, 1991). A challenge in using aerial photography is that spatial and temporal coverage, as well as resolution and quality, are highly variable depending on the objectives of the organization collecting the imagery (Morgan et

*holly.dolliver@uwrf.edu

Dolliver, H.A.S., 2012, Using Google Earth to teach geomorphology, *in* Whitmeyer, S.J., Bailey, J.E., De Paor, D.G., and Ornduff, T., eds., Google Earth and Virtual Visualizations in Geoscience Education and Research: Geological Society of America Special Paper 492, p. 419–429, doi:10.1130/2012.2492(32). For permission to copy, contact editing@geosociety.org. © 2012 The Geological Society of America. All rights reserved.

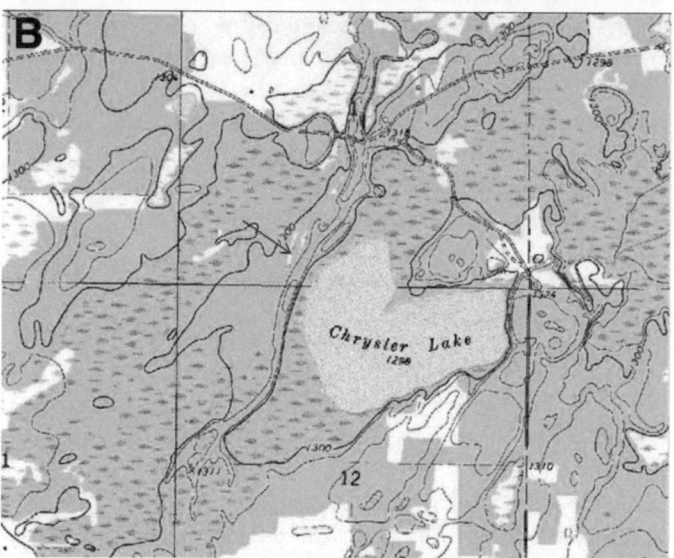

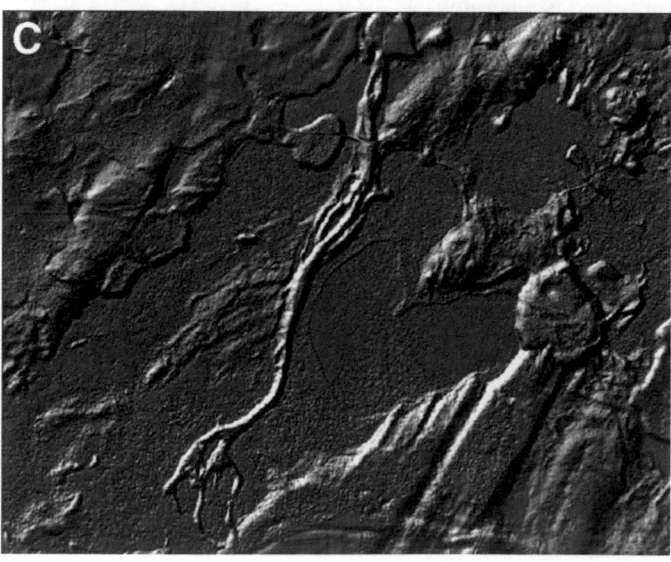

al., 2010). Although satellite imagery such as Landsat can provide imagery of the entire planet, resolution generally limits its use to regional analysis (scales greater than 1:100,000) (Smith and Pain, 2009). Using aerial photographs to study landscape morphology is largely limited to stereoscopic viewing (using overlapping "stereo pairs" of aerial photographs). Accurately quantifying topographic relief from stereo photography is difficult and time-consuming.

(2) *Topographic maps*: The elevation data on topographic maps (Fig. 1B) make them ideal for studying and mapping landscape morphology and have been extensively used by geologists for over a century. The United States Geological Survey (USGS) began producing topographic maps in the 1880s, although coverage of the United States at 1:24,000 (7.5 min quadrangle) wasn't complete until 1992 (Moore, 2000). As with any physical map, the scale and contour interval sets constraints on the amount of information included in the map. For the standard 7.5 min quadrangle, areal extent of a feature must generally be greater than 0.05 km^2 to be delineated on the map, limiting their use for identifying small landforms. Topographic maps also contain land use and cultural information that can be useful in geomorphic investigations of human impacts (Trimble and Cooke, 1991), although inaccuracies can result from generalization during map production (Vitek et al., 1996). Despite their utility, topographic maps are largely ineffective for studying geomorphic change. Only in rare cases does a series of topographic map editions exist for the same area across three or more dates. Of the more than 50,000 7.5 min quadrangles covering the United States, less than 1,500 are revised annually (Moore, 2000). Furthermore, the standard revision of a single 7.5 min quadrangle rarely includes updating contour information (Moore, 2000).

(3) *Light detection and ranging (LiDAR)*: LiDAR (Fig. 1C) has recently emerged as a powerful technology for capturing high-resolution topographic information (Haugerud et al., 2003; Smith and Pain, 2009; Church, 2010; Höfle and Rutzinger, 2011). Once acquired, data are generally used to construct a so-called bare earth digital elevation model (DEM) that can be used for quantitative analysis and modeling of surface terrain, processes, and dynamics. LiDAR data are very precise with vertical accuracies usually less than 0.25 m (Smith and Pain, 2009). Comparatively, a DEM derived from a 7.5 min USGS

Figure 1. Aerial photograph (A), U.S. Geological Survey 7.5 min quadrangle (B), and LiDAR (light detection and ranging) hillshade (C) of an esker (indicated by arrow) in Crow Wing County, Minnesota. Images courtesy of Betsy Oehlke, Natural Resources Conservation Service.

quadrangle has a vertical accuracy tolerance up to 15 m (USGS, 2011). LiDAR has recently been used for geomorphic mapping in a variety of settings (Woolard and Colby, 2002; Haugerud et al., 2003; Smith et al., 2006; Jones et al., 2007; Van Den Eeckhaut et al., 2007; Notebaert et al., 2009). LiDAR can also be used to precisely quantify geomorphic change (Thoma et al., 2005; Kessler et al., 2012). Although LiDAR has the potential to revolutionize geomorphic analysis, limited availability of LiDAR data has restricted its widespread use (Höfle and Rutzinger, 2011). Coverage is largely limited to local and regional-levels, although progress is being made toward the development of a national LiDAR data set in the United States (Stoker et al., 2008).

(4) *Geographic information systems (GIS)*: GIS is a powerful tool for compiling, organizing, manipulating, and analyzing geospatial data. GIS has widespread use in geomorphology as a tool for measuring and mapping landforms, monitoring of geomorphic processes and geomorphic change, and landscape modeling (Vitek et al., 1996; Wilson and Gallant, 2000). GIS is also extensively used in map production (Gustavsson et al., 2006). GIS provides the flexibility to easily customize maps for a specific application and efficiently edit maps as new data is acquired. One of the great benefits of GIS is the ability to process and analyze large and complex data sets (maps, imagery, surveys, etc.).With modern computing power, a GIS can efficiently and quickly perform geostatistical and spatial analyses on regional to global scales, which is impractical or impossible to do with traditional paper maps and imagery (Church, 2010). However, GIS software is expensive (a license can exceed $5,000) and is time-consuming to learn. Initiatives such as "The National Map" by the USGS, are improving availability and access to digital geospatial data (Sugarbaker and Carswell, 2011); however, it can be challenging to obtain data from other parts of the world.

Despite the availability of new tools and technologies, the traditional approach to teaching geomorphology relies predominantly on paper copies of topographic maps and aerial photographs for landform identification and interpretation of geomorphic processes (Easterbrook and Kovanen, 1998; Zaprowski et al., 2002; Lillquist, 2006). The availability of maps and images in sufficient quantities for classroom use determines the locations and geomorphic processes that can be examined. This largely constrains activities to locations within the United States. While integration of GIS into geomorphology curricula would allow for more sophisticated quantitative analysis, it may not be practical. Even if the software is provided by an institution, it is challenging or potentially not possible to allow students to access the software remotely from their personal computers. Furthermore, GIS software is time-consuming to learn, making it difficult to balance with demands of teaching the science of geomorphology, especially at the undergraduate level.

USING GOOGLE EARTH TO TEACH GEOMORPHOLOGY

Since its initial release in 2005, Google Earth has revolutionized access to, and visualization of, spatial data and has quickly become an important tool for scientific research and teaching (Butler, 2006; Patterson, 2007; Goodchild, 2008; SERC, 2009a). Google Earth is a free, virtual globe or "geobrowser" that accesses satellite and aerial imagery along with DEM (i.e., topographic) data over the Internet to represent the Earth in three dimensions. Google Earth seamlessly integrates geospatial data into a dynamic environment that allows users to easily navigate to any location on the planet to make observations and perform basic quantitative analysis. Google Earth has the capability to enhance teaching of a wide array of geomorphology concepts.

(1) *Landform visualization, identification, and characterization:* One of the most challenging skills for geology students to learn is how to visualize three-dimensional reality from two-dimensional maps (Vitek et al., 1996; Rapp et al., 2007). In some cases, the inability to do this can be a barrier to learning the science. Visualization is one of the strongest attributes of Google Earth (Goodchild, 2008). In addition to an overhead view, a user can tilt a scene in Google Earth to obtain a three-dimensional view of the landscape with imagery draped over a DEM, which greatly facilitates landform visualization and identification (Fig. 2). The degree of elevation exaggeration can be set to optimize visualization in both low and high relief terrains.

In places where DEM data provided by Google Earth do not have sufficient resolution for landform identification,

Figure 2. Three-dimensional view of a glacial cirque in Glacier National Park, Montana.

supplemental data, such as USGS 7.5 min topographic maps, can be imported into Google Earth as image overlays. Once imported, overlays can be easily georeferenced using suitable control points (roads, intersections, fixed structures, etc.). There are also resources available for accessing and integrating publically available geospatial data such as USGS quadrangles, relief maps, and aerial photographs that are georeferenced and ready for direct viewing in Google Earth (Davis and Turner, 2009; Google Earth Library, 2011). In addition, when the base scene is viewed three-dimensionally, the overlay is draped over the DEM. In the case of USGS topographic maps, this creates a three-dimensional model of the map (Fig. 3A). By control-

ling the transparency of the overlay, users can partially view the underlying base scene. With topographic maps, students can see the contour line patterns and their relationship to the three-dimensional topography. Although there is some terrain distortion, this could provide an alternative to expensive raised-relief models for teaching and improving topographic map reading and interpretation skills.

Along with visual interpretation, Google Earth provides a measurement tool for basic quantitative analysis of landforms and features (Fig. 4). This is useful for both identification and interpretation of geomorphic processes and dynamics. More recently, Google Earth added the capability to create an elevation profile of a particular landform or feature from a path (a series of two or more points) (Fig. 3B). Moving a cursor along the elevation profile graph simultaneously pinpoints the location and corresponding statistics (distance, slope, etc.) on the map to aid in interpretation. Elevation profiles may be especially useful when landform classification is based on more subtle differences in landform morphology (Fig. 5).

(2) *Geomorphic mapping:* Geomorphic maps are an important tool for landscape interpretation. Although maps are largely produced using commercial GIS and graphics software, Google Earth allows for presentation and/or production of geologic maps (Conroy et al., 2008; Whitmeyer et al., 2010). Geomorphic maps can be imported as an image overlay and georeferenced as previously described or created directly in Google Earth using placemarks (points), paths (lines), and/or polygons (Fig. 6A). To facilitate mapping directly within Google Earth, data can be collected with a global positioning system (GPS)

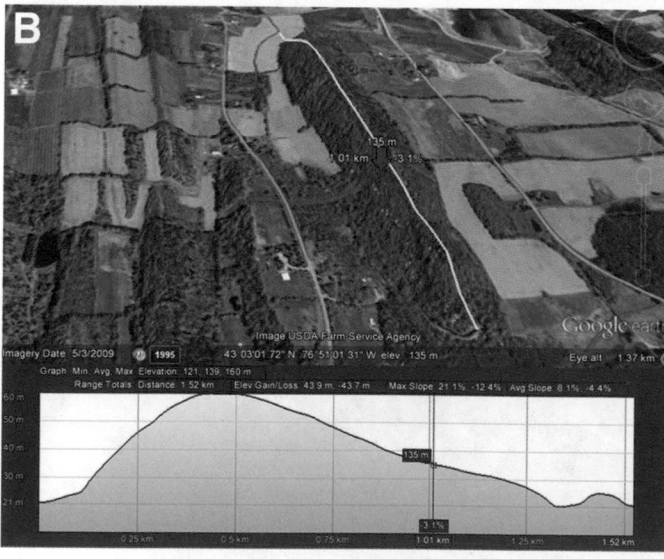

Figure 3. Three-dimensional view of a portion of the Savannah, New York, 15 min quadrangle showing a drumlin (indicated by red arrow) (A) and elevation profile of the drumlin (B). The morphology of the drumlin indicates that ice flow was from north to south.

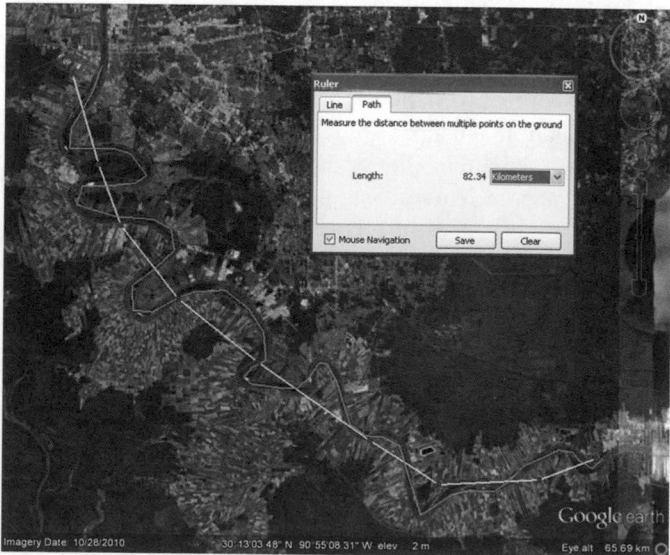

Figure 4. Image of the Mississippi River in Louisiana. The blue line delineates the center of the channel, while the yellow line delineates the valley length. Using the measurement tool for both, sinuosity of the river can be calculated (1.6 for this portion of the river).

and imported into Google Earth. Balloons (pop-up information window linked to a placemark, polygon, or path) can be used to provide additional information, such as sample characteristics or metadata (Fig. 6B). Although it is cumbersome to create maps in Google Earth, it provides an opportunity to expose students to the process of digitization and map production and resembles the technique used to produce maps using GIS (McCaffrey et al., 2005; de Donatis et al., 2008).

(3) *Understanding natural geomorphic processes and change:* By default, Google Earth displays the most recent high-quality image, which is generally not more than three years old. Historical imagery within Google Earth allows users to quickly access imagery dating back to the early 1990s, although some locations have

imagery available from the 1940s. Google Earth historical imagery can be used to study topics such as glacier advance/retreat, coastal erosion, steam dynamics, sand dune migration, and mass movements. For many locations, more than six historical images are available making Google Earth a valuable tool for studying geomorphic processes and change over short time scales ($<10^2$ yr) (Fig. 7). It is common in many places to find yearly or even sub-yearly imagery available for the past five years. This temporal frequency can even allow for studying of seasonal dynamics in some instances (Fig. 8). Supplemental historic maps, such as topographic maps, can be used for studying change over longer time scales or to improve understanding of the frequency or magnitude of geomorphic change (Fig. 9). Google Earth is also

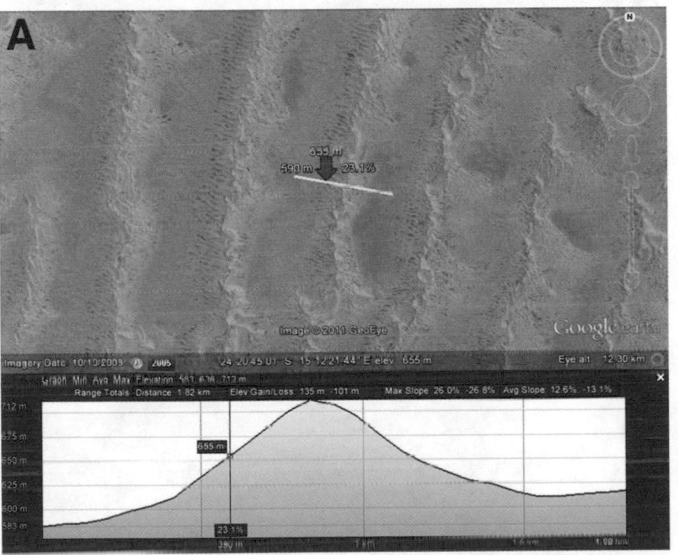

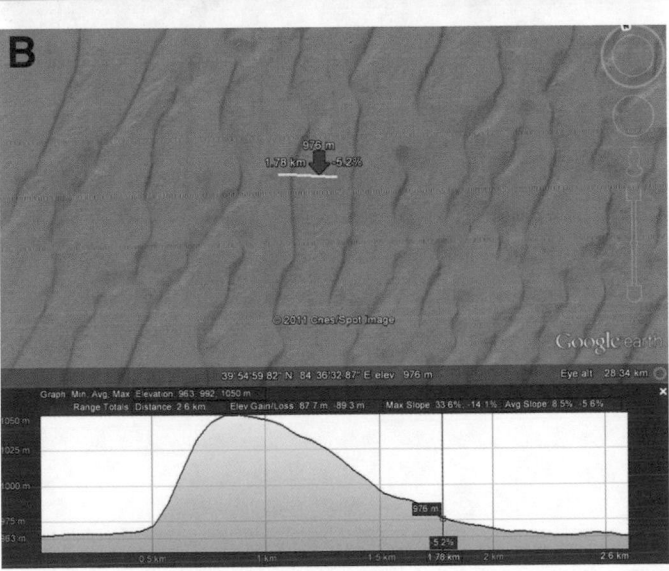

Figure 5. Symmetrical elevation profile of a linear dune in Namibia (A) and an assymetrical profile of a barchanoid ridge dune in China (B).

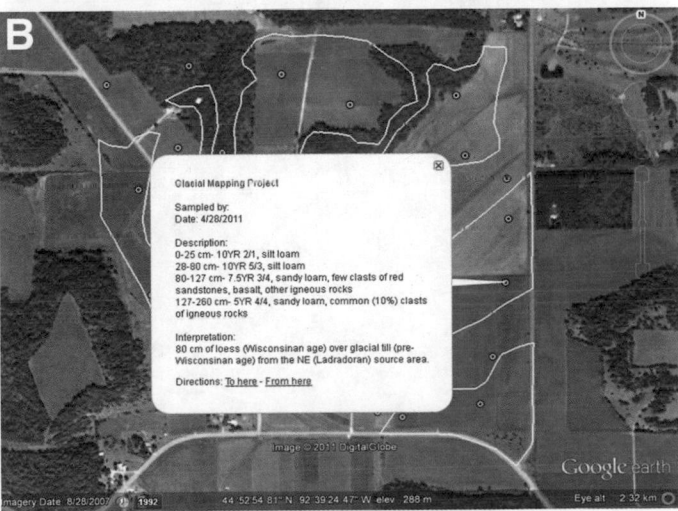

Figure 6. Map units created using polygons for a glacial geology mapping project (A). Circular markers (points) indicate sampling locations. A balloon linked to each marker contains a description and interpretation of the sediment at each location (B).

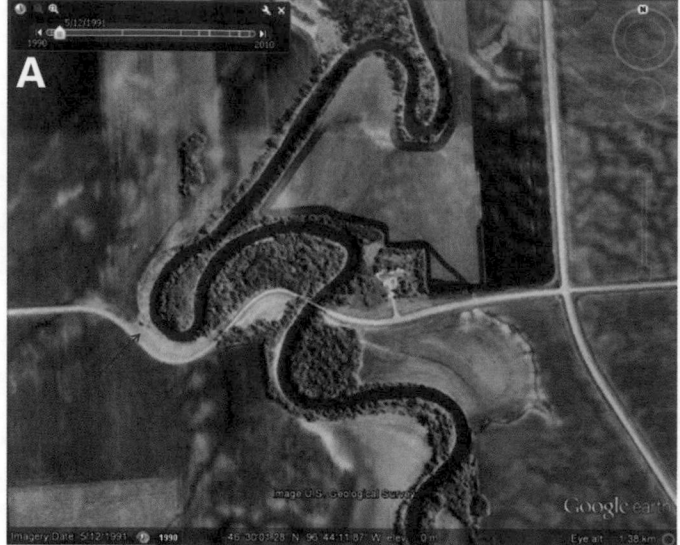

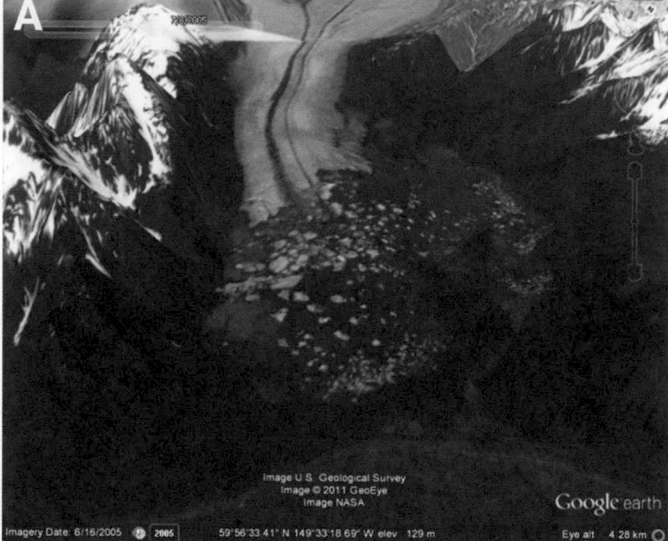

Figure 8. Image of a glacier in Alaska in June (A) and April (B).

Figure 7. Image of the Red River (political boundary between Minnesota and North Dakota) in 1991 (A), 2003 (B), and 2009 (C) showing the formation of an oxbow (indicated by arrow).

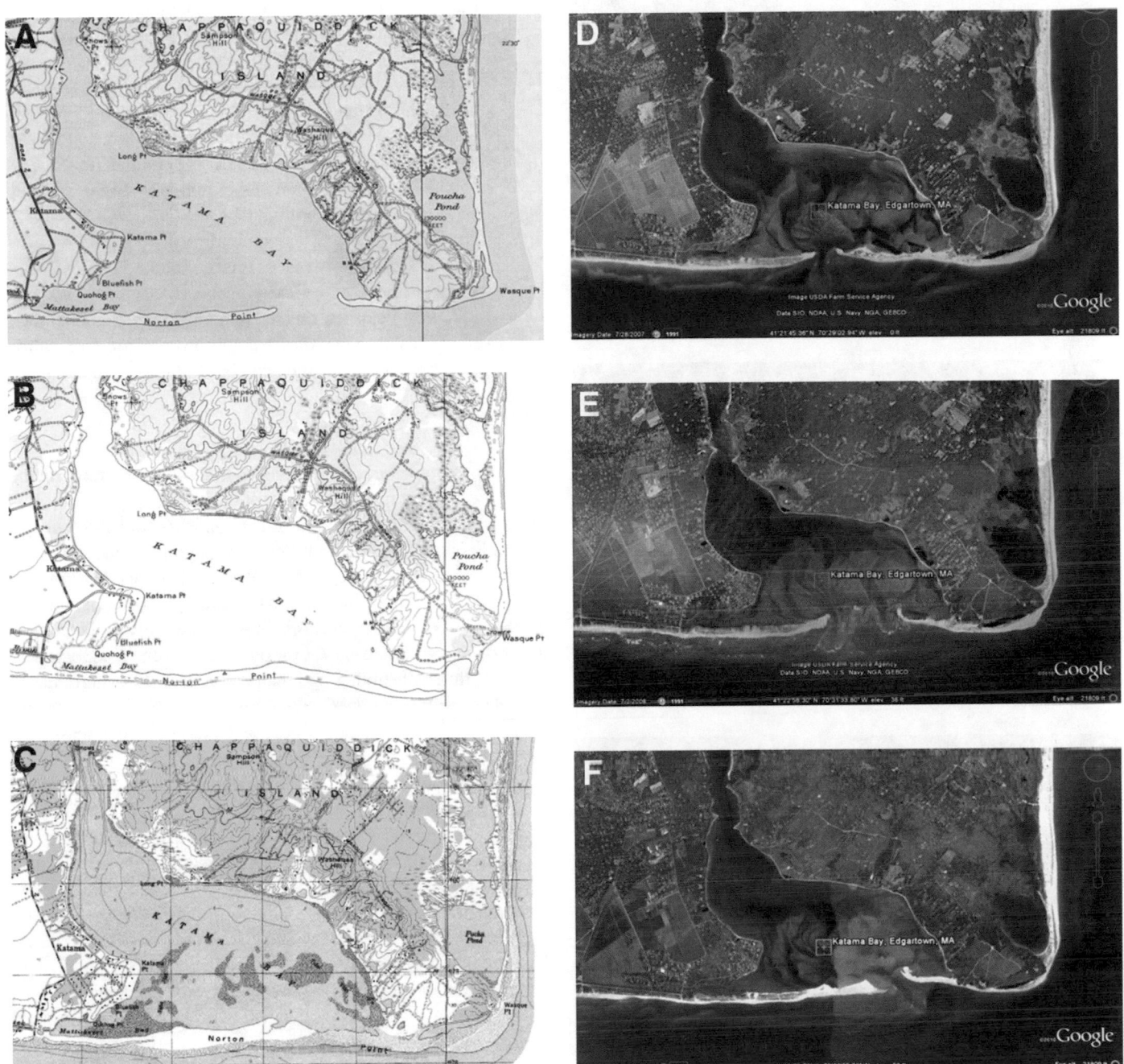

Figure 9. Southwest portion of the Edgartown 7.5 min U.S. Geological Survey (USGS) quadrangle from the 1944 edition (A), 1951 edition (B), and 1972/1979 edition (C), which leads to the conclusion that longshore transport of sand has progressively led to the deposition of sand across Norton Point (south end of Katama Bay) until the south shore of Martha's Vineyard and Chappaquiddick were connected in the 1970s. However, Google Earth images from 2007 (D), 2008 (E), and 2010 (F) show catastrophic erosion of Norton Point during a powerful storm in 2007. It is unclear from the decadal temporal scale of the USGS topographic maps whether similar extreme erosion events occurred in the past. Collectively using the USGS topographic maps and Google Earth offers a much clearer perspective on the dynamics of coastal change.

Figure 10. Images of a portion of Miami Beach, Florida in 1995 (A). A series of breakwater structures were installed in 2000 (B). By 2010 (C), breakwater structures had significantly impacted longshore transport of sand (from north to south) resulting in a 50% increase in beach width north of the structures and a 60% reduction in beach width south of the structures.

ideal for incorporating current events, since new imagery is continually being added.

(4) *Humans and geomorphology:* Human interactions with Earth's land surface are complex. Humans are both geomorphic agents and affected by geomorphic change (Hooke, 2000; Farris et al., 2007). These concepts are especially relevant to geomorphologists involved in land use planning and management. Although topographic maps include some cultural information, it is limited and generally not sufficient for analysis of human impacts. As with studying natural processes and geomorphic change, historical imagery within Google Earth can provide a time-series of images documenting the impact of humans on geomorphic processes (Fig. 10) and how humans are directly impacted by extreme geomorphic events (Fig. 11). In some cases, it is ideal to have background knowledge for interpretation (structure installation date, hurricane timing, etc.), which can be challenging to obtain.

DISCUSSION

The examples included in the previous section illustrate the breadth of concepts that can be taught using Google Earth. It is important to acknowledge that for the most part, these concepts can be taught using traditional paper topographic maps and aerial photographs. However, there are several advantages to using Google Earth. One of the major advantages is the ease and efficiency with which geomorphic analysis can be conducted. For instance, constructing an elevation profile from a topographic map is a time consuming and tedious process. In Google Earth, this takes less than a minute, allowing time for students to explore concepts in greater depth. Because Google Earth is easy and quick to use, activities can even be integrated in short lecture sessions rather than traditional lab periods.

Google Earth also provides tools for easily importing geospatial data beyond what is provided, allowing for more comprehensive landscape analysis. Once georeferenced, all data are presented at the same scale, helping overcome the difficulty that students have managing and appropriately interpreting data at different scales (i.e., topographic map at 1:24,000 and aerial photograph at 1:15,840) (Vitek et al., 1996). In addition, Google Earth has an embedded browser. Links to websites, photographs, and video can easily be integrated into a placemark and viewed directly within Google Earth. This allows an instructor to integrate dynamic content to customize and tailor the learning experience.

With Google Earth, students also have the ability to explore the entire planet. Traditional map activities often feature "classical" locations (i.e., Glacier National Park). While this is partly because processes and landforms are well-expressed, it is also because the maps are easily available in large quantities for teaching purposes. Google Earth makes the entire planet accessible, providing an opportunity for students to explore areas of both local and global interest, thus fostering a richer, more diverse learning experience.

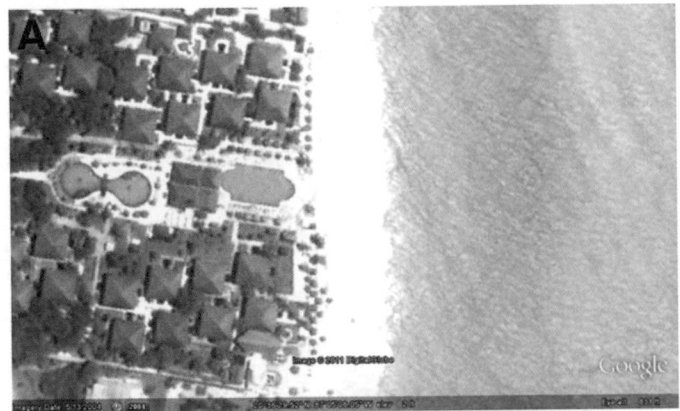

Google Earth provides students with virtual, on-demand access to geospatial data, greatly facilitating student learning outside of the classroom. Paper topographic maps and aerial photographs generally constrain learning to a physical classroom. Maps and imagery can be expensive, limiting the number of copies that can be purchased. Although materials are often made available for viewing outside of classroom hours, it is often not practical to allow materials to be removed from the classroom (difficult to manage, potential damage, etc.). Since Google Earth is free, using it can reduce the cost of establishing and/or maintaining expensive map and imagery collections.

Google Earth can also be used for communication of scientific information. Within a balloon, a user can include text, links to websites, videos, images, etc. In addition to simple balloons (plain text), more elaborate balloons can be created using the templates provided or fully customized with an HTML (hypertext markup language) editor. Recorded tours (fly-through and narration) can also be created to showcase a project within Google Earth. Collectively, these tools could provide an alternative to a traditional term paper or project (Fig. 12).

Despite the benefits of using Google Earth to teach geomorphology, there is no debate that commercial GIS software allows

Figure 12. Portion of a student term project showing a balloon containing scientific information and imagery related to the geomorphology of the location indicated by the placemark.

Figure 11. Images of a resort property south of Cancun near Playa del Carmen, Mexico in 2004 (A). Hurricane Wilma devastated the region in 2005 causing extensive shoreline erosion, which is evident on the 2007 image (B). By 2009 (C), the resort pool was shortened nearly 5 m to protect the structure. In late 2009, the government undertook a large beach restoration and nourishment project, replenishing nearly 50 m of sand as seen on the 2010 image (D).

for more sophisticated quantitative analysis (Patterson, 2007; Goodchild, 2008). However, the popularity of Google Earth is unmatched. Since 2005 when Google Earth was first introduced, it has been downloaded more than 1 billion times—a remarkable testament to its popularity (Google Inc., 2011). Google Earth provides an opportunity to communicate scientific information and data to the general public.

FINAL REMARKS AND CONCLUSIONS

Earth is a dynamic planet. Understanding the complex array of landforms and natural geomorphic processes, coupled with the impact of humans on Earth's landscapes, is challenging. As technology evolves and data sets improve, geomorphic investigations are becoming increasingly more complex and quantitative (Murray et al., 2009; Church, 2010). While it is important to expose students to the traditional maps and imagery, it is critical that modern data and tools, such as Google Earth, be integrated into geomorphology curricula (Shroder et al., 2002; Murray et al., 2009).

Textbook companies have begun to produce workbooks and manuals of Google Earth activities and locations (e.g., Marshak and Wilkerson, 2007; Kluge, 2008). These products are primarily marketed to introductory courses and generally address basic rather than specialized concepts that are needed in more advanced courses like geomorphology. A few online resources are available (e.g., SERC, 2009b), but databases of good teaching locations are limited at this point. This leaves the instructor with the arduous task of scanning the planet for suitable teaching locations. A significant amount of time was invested in finding and developing the examples featured in this paper. Enhancing existing resources (e.g., SERC, 2009b) would greatly facilitate the integration of Google Earth into the curriculum.

NOTE

This manuscript was developed using Google Earth version 6.1. Historical imagery was introduced in version 5.0 and elevation profiling was introduced in version 5.2.

ACKNOWLEDGMENTS

My sincere appreciation and thanks to my geomorphology students for their willingness and patience to try new activities and for providing candid feedback and thoughtful insights. I would also like to thank Brett and the staff of 2300 for their tremendous support and encouragement. This manuscript benefited greatly from the comments and suggestions made by peer reviewers.

REFERENCES CITED

Bierman, P.R., Howe, J., Stanley-Mann, E., Peabody, M., Hilke, J., and Massey, C.A., 2005, Old images record landscape change through time: GSA Today, v. 15, no. 4, p. 4–10.

Brunsden, D., and Thornes, J.B., 1979, Landscape sensitivity and change: Transactions of the Institute of British Geographers, v. 4, no. 4, p. 463–484, doi:10.2307/622210.

Butler, D., 2006, The web-wide world: Nature, v. 439, p. 776–778, doi:10.1038/439776a.

Cerney, D.L., 2010, The use of repeat photography in contemporary geomorphic studies: An evolving approach to understanding landscape change: Geography Compass, v. 4, p. 1339–1357, doi:10.1111/j.1749-8198.2010.00376.x.

Church, M., 2010, The trajectory of geomorphology: Progress in Physical Geography, v. 34, no. 3, p. 265–286, doi:10.1177/0309133310363992.

Conroy, G.C., Anemone, R.L., Van Regenmorter, J., and Addison, A., 2008, Google Earth, GIS, and the Great Divide: A new and simple method for sharing paleontological data: Journal of Human Evolution, v. 55, p. 751–755, doi:10.1016/j.jhevol.2008.03.001.

Davis, T.G., and Turner, R., 2009, USGS quadrangles in Google Earth: The American Surveyor, October, 3 p.

de Donatis, M., Susini, S., and Delmonaco, G., 2008, Digital geologic mapping methods: From field to 3D model: International Journal of Geology, v. 2, p. 47–52.

Easterbrook, D.J., and Kovanen, D.J., 1998, Interpretation of landforms from topographic maps and air photographs: A laboratory manual: Englewood Cliffs, New Jersey, Prentice Hall, 193 p.

Farris, G.S., Smith, G.J., Crane, M.P., Demas, C.R., Robbins, L.L., and Lavoie, D.L., eds., 2007, Science and the storms—the USGS response to the hurricanes of 2005: U.S. Geological Survey Circular 1306, p. 113–118.

Goodchild, M.F., 2008, The use cases of digital earth: International Journal of Digital Earth, v. 1, p. 31–42, doi:10.1080/17538940701782528.

Google Earth Library, 2011, USGS topographic maps, http://www.gelib.com/usgs-topographic-maps-2.htm (accessed July 2011).

Google Inc., 2011, Google Earth downloaded more than one billion times: http://googleblog.blogspot.com/2011/10/google-earth-downloaded-more-than-one.html (accessed December 2011).

Gustavsson, M., Kolstrup, E., and Seijmonsbergen, A.C., 2006, A new symbol-and-GIS based detailed geomorphological mapping system: Renewal of a scientific discipline for understanding landscape development: Geomorphology, v. 77, p. 90–111, doi:10.1016/j.geomorph.2006.01.026.

Haugerud, R.A., Harding, D.J., Johnson, S.Y., Harless, J.L., and Weaver, C.S., 2003, High-resolution Lidar topography of the Puget Lowland, Washington—A bonanza for earth science: GSA Today, v. 13, no. 6, p. 4–10.

Höfle, B., and Rutzinger, M., 2011, Topographic airborne LiDAR in geomorphology: A technological perspective: Zeitschrift fur Geomorphologie, v. 55, p. 1–29, doi:10.1127/0372-8854/2011/0055S2-0043.

Hooke, R., 2000, On the history of humans as geomorphic agents: Geology, v. 28, no. 9, p. 843–846, doi:10.1130/0091-7613(2000)28<843:OTHOHA>2.0.CO;2.

Jones, A.F., Brewer, P.A., Johnstone, E., and Macklin, M.G., 2007, High-resolution interpretative geomorphological mapping of river valley environments using airborne LiDAR data: Earth Surface Processes and Landforms, v. 32, p. 1574–1592, doi:10.1002/esp.1505.

Jones, B.M., Arp, C.D., Jorgenson, M.T., Hinkel, K.M., Schmutz, J.A., and Flint, P.L., 2009, Increase in the rate and uniformity of coastline erosion in Arctic Alaska: Geophysical Research Letters, v. 36, L03503, doi:10.1029/2008GL036205.

Kessler, A.C., Gupta, S.C., Dolliver, H.A.S., and Thoma, D.P., 2012, Lidar quantification of bank erosion in Blue Earth County, Minnesota: Journal of Environmental Quality, v. 41, p. 1–11.

Kluge, S., 2008, Encounter Earth: Interactive Geoscience Explorations: Englewood Cliffs, New Jersey, Prentice Hall, 72 p.

Knox, J.C., 2001, Agricultural influence on landscape sensitivity in the Upper Mississippi River Valley: Catena, v. 42, p. 193–224, doi:10.1016/S0341-8162(00)00138-7.

Lillquist, K., 2006, Teaching with catastrophe: Topographic map interpretation and the physical geography of the 1949 Mann Gulch, Montana: Journal of Geoscience Education, v. 54, no. 5, p. 561–571.

Marshak, S., and Wilkerson, M.S., 2007, Geotours Workbook: New York, W.W. Norton & Company, 160 p.

McCaffrey, K.J.W., Jones, R.R., Holdsworth, R.E., Wilson, R.W., Clegg, P., Imber, J., Holliman, N., and Trinks, I., 2005, Unlocking the spatial dimension: Digital technologies and the future of geoscience fieldwork: Journal of the Geological Society, v. 162, p. 927–938, doi:10.1144/0016-764905-017.

Moore, L., 2000, The U.S. Geological Survey's revision program for 7.5-minute topographic maps, http://thor-f5.er.usgs.gov/topomaps/revision_overview.pdf (accessed July 2011).

Morgan, J.L., Gergel, S.E., and Coops, N.C., 2010, Aerial photography: A rapidly evolving tool for ecological management: Bioscience, v. 60, p. 47–59, doi:10.1525/bio.2010.60.1.9.

Murray, A.B., Lazaru, E., Ashton, A., Baas, A., Coco, G., Coulthard, T., Fonstad, M., Haff, P., McNamara, D., Paola, C., Pelletier, J., and Reinhardt, L., 2009, Geomorphology, complexity, and the emerging science of the Earth's surface: Geomorphology, v. 103, p. 496–505, doi:10.1016/j.geomorph.2008.08.013.

Notebaert, B., Verstaeten, G., Govers, G., and Poesen, J., 2009, Qualitative and quantitative application of LiDAR imagery in fluvial geomorphology: Earth Surface Processes and Landforms, v. 34, p. 217–231, doi:10.1002/esp.1705.

Patterson, T.C., 2007, Google Earth as a (not just) geography education tool: The Journal of Geography, v. 106, p. 145–152, doi:10.1080/00221340701678032.

Phillips, J.D., 2009, Changes, perturbation, and responses in geomorphic systems: Progress in Physical Geology, v. 33, p. 17–30, doi:10.1177/0309133309103889.

Rapp, D.N., Culpepper, S.A., Kirkby, K., and Morin, P., 2007, Fostering students' comprehension of topographic maps: Journal of Geoscience Education, v. 55, p. 5–16.

Science Education Resource Center (SERC), 2009a, Why teach with Google Earth?, http://serc.carleton.edu/sp/library/google_earth/why.html (accessed July 2011).

Science Education Resource Center (SERC), 2009b, Examples of Google Earth activities, http://serc.carleton.edu/NAGTWorkshops/teaching_methods/google_earth/examples.html (accessed July 2011).

Smith, M.J., and Pain, C.F., 2009, Applications of remote sensing in geomorphology: Progress in Physical Geography, v. 33, p. 568–582, doi:10.1177/0309133309346648.

Smith, M.J., Rose, J., and Booth, S., 2006, Geomorphological mapping of glacial landforms from remotely sensed data: An evaluation of the principal data sources and an assessment of their quality: Geomorphology, v. 76, p. 148–165, doi:10.1016/j.geomorph.2005.11.001.

Shroder, J.F., Jr., Bishop, M.P., Olsenholler, J., and Craiger, J.P., 2002, Geomorphology and the World Wide Web: Geomorphology, v. 47, p. 343–363, doi:10.1016/S0169-555X(02)00097-1.

Stoker, J., Harding, D., and Parrish, J., 2008, The need for a national lidar dataset: Photogrammetric Engineering and Remote Sensing, v. 74, no. 9, p. 1066–1068.

Sugarbaker, L.J., and Carswell, W.J., Jr., 2011, The National Map: U.S. Geological Survey Fact Sheet 2011-3042, 4 p.

Thoma, D.P., Gupta, S.C., Bauer, M.E., and Kirchoff, C.E., 2005, Airborne laser scanning for riverbank erosion assessment: Remote Sensing of Environment, v. 95, p. 493–501, doi:10.1016/j.rse.2005.01.012.

Trimble, S.W., 2008, The use of historical data and artifacts in geomorphology: Progress in Physical Geography, v. 32, p. 3–29, doi:10.1177/0309133308089495.

Trimble, S.W., and Cooke, R.U., 1991, Historical sources for geomorphological research in the United States: The Professional Geographer, v. 43, p. 212–228, doi:10.1111/j.0033-0124.1991.00212.x.

U.S. Geological Survey (USGS), 2011, Digital elevation model, http://eros.usgs.gov/#/Guides/dem (accessed December 2011).

Van Den Eeckhaut, M., Poesen, J., Verstraeten, G., Vanacker, V., Nyssen, J., Moeyerson, J., van Beek, L.P.H., and Vandekerckhove, L., 2007, Use of LIDAR-derived images for mapping old landslides under forest: Earth Surface Processes and Landforms, v. 32, p. 754–769, doi:10.1002/esp.1417.

Vitek, J.D., Giardino, J.R., and Fitzgerald, J.W., 1996, Mapping geomorphology: A journey from paper maps, through computer mapping to GIS and virtual reality: Geomorphology, v. 16, p. 233–249, doi:10.1016/S0169-555X(96)80003-1.

Wendland, W.M., 1996, Climate changes: Impacts on geomorphic processes: Engineering Geology, v. 45, p. 347–358, doi:10.1016/S0013-7952(96)00021-X.

Whitmeyer, S.J., Nicoletti, J., and De Paor, D.G., 2010, The digital revolution in geologic mapping: GSA Today, v. 20, p. 4–10, doi:10.1130/GSATG70A.1.

Williams, S.J., Dodd, K., and Gohn, K.K., 1990, Coasts in Crisis: U.S. Geological Survey Circular 1075, 32 p.

Wilson, J.P., and Gallant, J.C., eds., 2000, Terrain Analysis: Principles and Applications: John Wiley and Sons, 479 p.

Woolard, J.W., and Colby, J.D., 2002, Spatial characterization, resolution, and volumetric change of coastal dunes using airborne LiDAR: Cape Hatteras, North Carolina: Geomorphology, v. 48, p. 269–287, doi:10.1016/S0169-555X(02)00185-X.

Zaprowski, B.J., Evenson, E.J., and Epstein, J.B., 2002, Stream piracy in the Black Hills: A geomorphology exercise: Journal of Geoscience Education, v. 50, no. 4, p. 380–388.

MANUSCRIPT ACCEPTED BY THE SOCIETY 16 APRIL 2012

The Geological Society of America
Special Paper 492
2012

Development of a web-based hydrologic education tool using Google Earth resources

Emad Habib*
Department of Civil Engineering, University of Louisiana at Lafayette, P.O. Box 42991, Lafayette, Louisiana 70504, USA

Yuxin Ma
Douglas Williams
Center for Innovative Learning and Assessment Technologies (CILAT), University of Louisiana at Lafayette, P.O. Box 42051, Lafayette, Louisiana 70504, USA

ABSTRACT

This study reports on the development of a web-based hydrologic educational system (HydroViz) that supports students' learning in hydrology or related earth science subjects. HydroViz (http://hydroviz.cilat.org/hydro/) is designed as a virtual hydrologic observatory and is based on the integration of field data, remote sensing observations, and computer simulations of hydrologic variables and processes. HydroViz can run on a typical personal computer with Internet access and does not require any specific software package, which makes it easy to utilize. HydroViz employs the free Google Earth browser-based plug-in and its JavaScript application programming interface (API) to enable presentation of geospatial data layers in Google Earth and embed them in web pages that have the same look and feel of Google Earth. The decision to use Google Earth within the HydroViz project was driven by the great deal of geospatial data and visual capabilities it provides for hydrologic educational applications. Besides being freely accessible to a wide user community, Google Earth offers the ability to place and visualize hydrologic technical information on a three-dimensional model of Earth, which facilitates students' interactive and visually supported learning. Within a HydroViz setting, students can use Google Earth navigation capabilities to explore the watershed, either on their own or by using the embedded inquiry-based investigations and the supporting layers of hydrologic information. Cascading style sheets (CSS) and hypertext markup language (HTML) describe the look and formatting of each HydroViz web page. With the aid of Google Earth API, it was also possible to create customized buttons and panels for students to interact with and display

*habib@louisiana.edu

Habib, E., Ma, Y., and Williams, D., 2012, Development of a web-based hydrologic education tool using Google Earth resources, *in* Whitmeyer, S.J., Bailey, J.E., De Paor, D.G., and Ornduff, T., eds., Google Earth and Virtual Visualizations in Geoscience Education and Research: Geological Society of America Special Paper 492, p. 431–439, doi:10.1130/2012.2492(33). For permission to copy, contact editing@geosociety.org. © 2012 The Geological Society of America. All rights reserved.

the data. HydroViz is populated with several educational modules that range from basic activities (e.g., exploring watershed characteristics) to advanced analysis of field data and simulations. Each module is self-contained where instructions and technical questions are embedded within the same screens that show the watershed and its visual displays. HydroViz has been implemented in several classrooms, and evaluation data showed its potential value as a tool for supporting learning.

INTRODUCTION

Hydrology is the science that deals with the occurrence, distribution, and circulation of water and its interaction with the different physical, chemical, and biological processes of the earth system. Over the last several decades, hydrology has evolved to become a multidisciplinary science that includes various physical, chemical, and biological processes and extends across a wide spectrum of scales. As such, hydrologists nowadays deal with complex problems that are embedded within natural ecosystems with a multitude of inter-connected processes (e.g., rainfall-runoff-infiltration dynamics, ground-surface water interactions, ocean-atmospheric teleconnections, land-surface-atmosphere interactions, and human alterations to the hydrologic and geomorphic setting). Recently, the hydrologic research community has strived to formulate a science vision and research agenda for achieving real advances in the theory and practices of hydrologic sciences (Gupta, 2001; Hooper and Foufoula-Georgiou, 2008; CUAHSI, 2010). Key elements of this agenda include advances in new observational settings (e.g., critical zone and water, sustainability and climate observatories), instrumentation, hydrologic information systems, and modeling methods. Such advances need parallel efforts and reform in hydrologic education, especially at the undergraduate level. Numerous national and international reports (Nash et al., 1990; National Research Council, 2000) have highlighted the need for improving existing undergraduate hydrology curriculum, especially in two areas: observation and modeling. The lack of field and data experience has profound consequences on the hydrology education process, especially at the undergraduate level. Deficiencies in field components will likely lead to adverse effects on the skills of graduating students. For example, students will lack appreciation of spatial and temporal variability of hydrologic processes and will lack the ability to develop self-learning skills and intuitive understanding. Despite their numerous benefits, data and field resources can be unavailable to educational institutions because of high equipment costs, inaccessibility to measurement sites, and class time constraints. Limitations on how much students learn from data can be compensated for through the use of simulation models (Beven, 2001). An important supplement to the educational power of real field data is the use of computer simulation models. The use of such models has multiple benefits from a hydrologic educational point of view. Computer simulation models provide a user-friendly, interactive environment for hypothesis. For example, students and instructors can use simulation models to test scenarios of land-change effects and flooding impacts with different storm cases. The use of these models also compensates for inherent limitations of field data which are related to sensors limited spatial sampling density and coverage. However, it is well recognized that all hydrologic models, regardless of their degree of complexity, are subject to certain limitations caused by model assumptions and simplifications and lack of parameter identification, among others. Therefore, it is critical that such models be used in combination with field data and other sources of observations so that students can appreciate the challenges associated with representing nature by such models. Given the complementary benefits of both field data and modeling techniques, it is natural that educators should focus on combining these resources for improving hydrology and water resources education.

Hydrology is usually taught in a single course in most engineering programs where instructors do not have enough time to teach modeling software or expose students to field instruments. Therefore, innovative educational approaches need to be developed to address these challenges. Interactive visualization and geospatial navigation techniques that are based on the principles of active engagement and discovery can enhance learning by exploiting the visual senses of students and thus engaging their interests. Google Earth technologies provide a practical and efficient avenue for addressing the required integration of different sources of information (in situ data, observations, model simulations, and other analysis methods) into flexible and adaptable educational modules that are necessary for reforming the education of hydrology as a multi-faceted discipline. This paper describes the development of a Google Earth–based active learning environment with a real-world problem context. The development of this learning platform leverages emerging research and monitoring by the hydrologic research and professional community with recent advances in the fields of scientific visualization and web-based geospatial technologies (Cunningham, 2004; Zia, 2004).

The main purpose of the educational Google Earth–based system developed herein (called HydroViz) is to capitalize on research advances in hydrologic data and simulations and enhance students' learning of basic and applied hydrologic concepts. To allow for wide dissemination, HydroViz was built as a browser-based, web-accessible system. It leverages the power of free geospatial and visualization resources (Google Earth embedded in a web browser) to provide authentic and hands-on inquiry-based students' activities. Following the concept of using real-world applications for facilitating students' learning, HydroViz was built for a local watershed case study providing a "virtual" hydrologic educational observatory. Using HydroViz, students can download and analyze field data from various instruments, and can also examine simulations of rainfall-runoff processes generated using a physically distributed hydrologic model on the same watershed. To facilitate usage of HydroViz in classrooms, a total of 13 modules were embedded into HydroViz. The modules ranged from basic activities (e.g., exploring watershed characteristics) to advanced analysis of field data and model simulations. Each module is self-contained where instructions and technical questions are embedded within the same screens that show the watershed and its visual displays. This paper describes the development of HydroViz, its overall structure and technical components, and the educational modules embedded into it. The paper also presents some results on the evaluation of HydroViz after it was implemented and assessed in various classrooms. The paper closes with a summary and conclusions section.

CONCEPTUAL DESIGN AND FUNCTIONALITY OF HYDROVIZ

Figure 1 shows a schematic representation of the HydroViz virtual observatory and its different components. The concep-tual design of the system is based on several functional criteria and requirements. First, the "Observatory" should integrate observational and modeling aspects of the hydrologic watershed. The purpose is to instill in the students the concept that data and models complement each other and that model predictions are not perfect and should be checked against independent data. The "observational" component of the Observatory should mimic students' learning process during a real visit to the watershed. The Observatory should give students an opportunity to "virtually" download and analyze data. The "simulation" component should be built on a distributed process-based hydrologic model to allow for physical representation of important rainfall-runoff processes. The idea is to exploit the capabilities of these models and their informational richness without overwhelming the students with the vast amount of generated data and output. The design of the Observatory is also based on separating procedural aspects of the modeling exercise (i.e., learning the model software) from the processes and applications that a student can learn about by using the model (e.g., variability and interactions of rainfall-runoff processes). Finally, the Observatory combines visual and quantitative information. The students should be able to visualize the watershed and its processes, but can also extract quantitative information to perform further analysis outside the system.

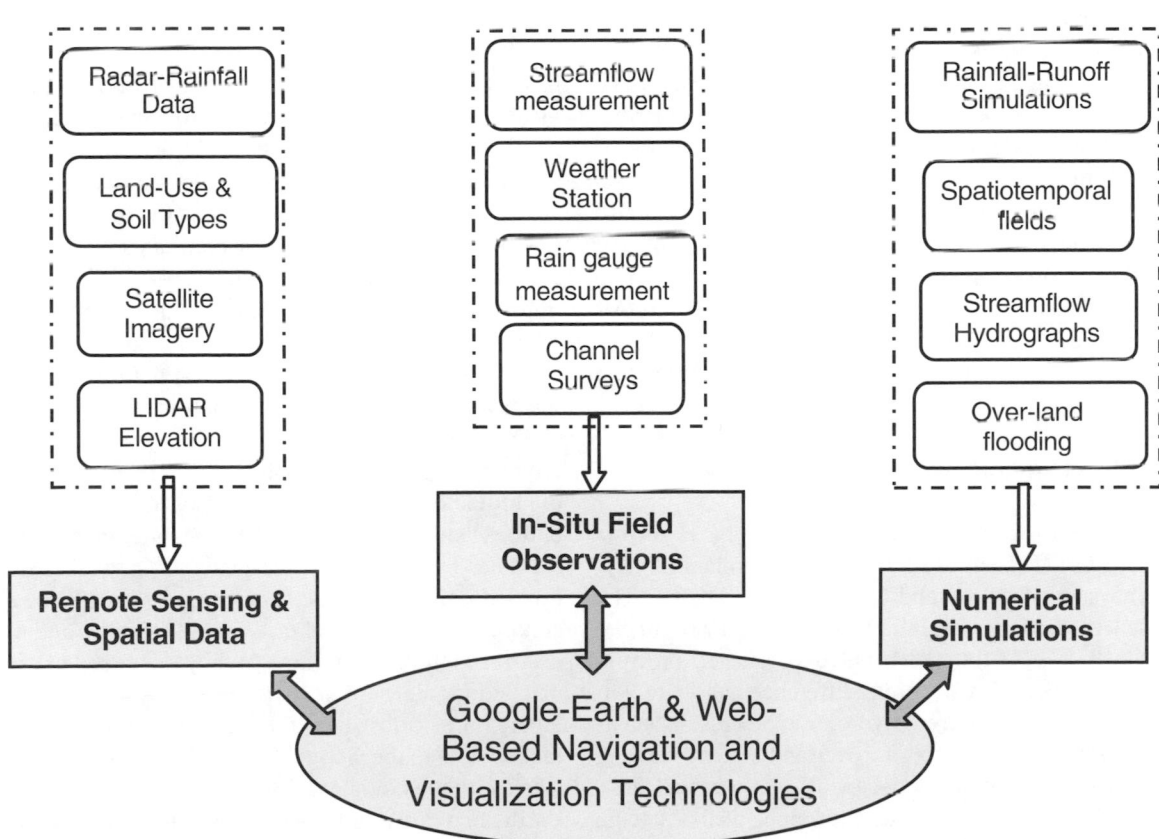

Figure 1. Overall structure of the HydroViz hydrologic observatory.

EXPERIMENTAL SITE, DATA, AND COMPUTE SIMULATIONS

HydroViz Field Site

The virtual-reality hydrologic system is developed using data and simulations that are based on the 35-km^2 Isaac-Verot (IV) experimental watershed (Fig. 2). The watershed, located in the vicinity of the campus of the University of Louisiana at Lafayette, is a sub-drainage area of the Vermilion river basin which drains into the Gulf of Mexico. The watershed is frequently subject to frontal systems, air-mass thunderstorms, as well as tropical cyclones with annual rainfall of ~140–155 cm and monthly accumulations as high as 17 cm. The main soil type in the watershed is silt loam with low to medium drainage capacity. Land-use in the watershed is composed of urban areas, cropland, pasture, and some forested areas.

In Situ and Remote Sensing Data

The Department of Civil Engineering at the University of Louisiana at Lafayette has deployed a dense experimental network of rainfall and runoff monitoring sites over the watershed (Fig. 2). A total of 13 tipping-bucket rain gauge sites are distributed over the watershed and every site has a dual-gauge setup for improved data continuity and quality (Habib et al., 2009). Streamflow measurements are collected at the outlet of the watershed, as well as at four interior locations using bi-directional acoustic velocity meters, from which discharge estimates can be obtained. Rainfall drop size distribution data (DSD) are collected at one of the experimental sites located close to the boundary of the IV watershed using a rain disdrometer. In addition to in situ data, the virtual system also includes rainfall remote-sensing data from weather radars and satellites. Other remote-sensing data that are integrated into the hydrologic virtual system include high-resolution LiDAR (light detection and ranging) topographic data, satellite-based imagery of land use and land cover, and soil type information assembled over the watershed site.

Hydrologic Model Simulations

The computer hydrologic simulations used by HydroViz are based on the Gridded Surface Subsurface Hydrologic Analysis (GSSHA) system (Downer and Ogden, 2004). The GSSHA model is used to develop a rainfall-runoff model for the IV watershed. GSSHA is a fully distributed–parameter, process-based hydrologic model. It uses finite difference and finite volume methods to simulate different hydrologic processes such as rainfall distribution and interception, overland water retention, infiltration, evapotranspiration, two-dimensional overland flow, one-dimensional channel routing, and different methods (e.g., Green and Ampt method, and Richards' equation) for modeling the soil moisture profile in the unsaturated zone. The watershed topographic and hydrologic properties are represented using a 25×25 m^2 Cartesian grid. Overland hydraulic properties (e.g., roughness parameters), soil hydraulic parameters (e.g., saturated hydraulic conductivity, soil suction head, and effective porosity), and evapotranspiration parameters (e.g., vegetation transmission coefficients and root depths) were initially assigned based on spatial variations in the combined classifications of soil type and land use maps. The parameters were further adjusted through model calibration.

The model output provides a suite of simulations results on various hydrologic variables and processes such as: (1) streamflow time series at the watershed outlet and at interior locations within the watershed, (2) spatial maps of overland runoff discharge and water depth, (3) spatial maps of infiltration rates, cumulative infiltrated water, and soil moisture. The spatial maps are provided at every time step (5 min interval) within the simulation period. A typical simulation period may consist of one single rainfall-runoff event (few hours) or a series of consecutive events over the span of several days.

HYDROVIZ SOFTWARE DESIGN

HydroViz is a web-based tool that employs the free Google Earth plug-in and its JavaScript application programming interface (API) to enable presentation of geospatial data layers in Google Earth and embed them in web pages that have the same look and feel of Google Earth (see Fig. 3 and Table 1).

The decision to use Google Earth within the HydroViz project was driven by the great deal of geospatial data and visual capabilities it provides for hydrologic educational applications. Besides being freely accessible to a wide user community, Google Earth offers the ability to place and visualize hydrologic technical information on a three-dimensional (3-D) model of Earth, which facilitates students' interactive and visually supported learning. Within a HydroViz setting, students can use Google Earth navigation capabilities to explore the watershed, either on their own or by using the embedded inquiry-based investigations and the supporting layers of hydrologic information. Cascading style sheets (CSS) and hypertext markup language (HTML) describe the look and formatting of each HydroViz web page. With the aid of Google Earth API, it was also possible to create customized buttons and panels for students to interact with and display the data. The HydroViz interface (see an example in Fig. 4) is divided into three areas: (1) Google Earth, (2) the educational content and educational tasks, and (3) layers and tools. As the user reads the content (area 2) they may need to turn layers on or conduct measurements (area 3). As the user turns various layers on or off or uses the measuring tools, JavaScript code makes calls to the Google Earth API which, in turn, communicates with the Google Earth plug-in (area 1). Layers in Google Earth are defined in Keyhole Markup Language (KML), a tag-based file format that defines the content to be displayed in Google Earth.

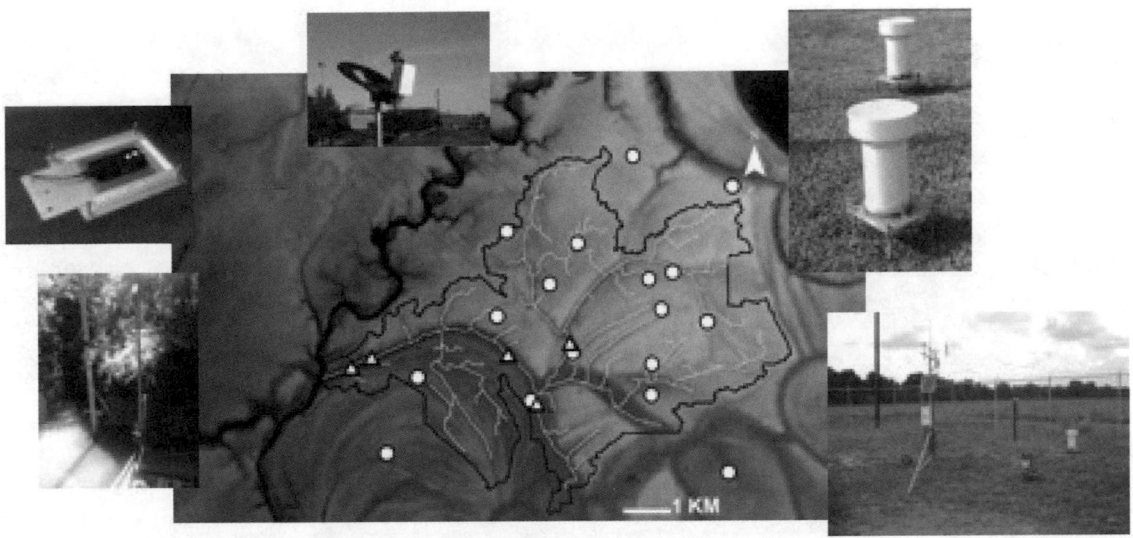

Figure 2. Map of the experimental watershed used in HydroViz showing the locations and some pictures of the in-situ hydrologic and meteorological instruments.

HYDROVIZ COURSE MODULES

To facilitate usage of HydroViz in classrooms, a total of 13 learning modules were embedded into HydroViz. The modules ranged from basic activities (e.g., exploring watershed characteristics) to advanced analysis of field data and simulations. Each module is self-contained where instruc tions and technical questions are embedded within the same screens that show the watershed and its visual displays. The following is a brief list of these modules; the reader is referred to http://hydroviz.cilat.org/hydro/ for full details on each module:

- Module 1: Getting Acquainted with the Watershed
- Module 2: Exploring Land-Use Land-Cover (LULC) Coverage Map
- Module 3: Exploring Soil Coverage
- Module 4: Exploring Land Elevation
- Module 5: Exploring Field Equipment
- Module 6: Working with a Real Rainfall Storm
- Module 7: Analysis of Rainfall Measurements during Storm Matthew
- Module 8: Measuring Rainfall Using Remote Sensing Techniques
- Module 9: Analysis of Streamflow Observations due to Storm Matthew
- Module 10: Runoff Analysis using Curve Number
- Module 11: TR55 Graphical Method
- Module 12: Setting up a Hydrologic Model
- Module 13: Hydrologic Model Simulations

Figure 3. Client-server architecture of the proposed HydroViz system. HTTP—hypertext transfer protocol; API—application programming interface; KML—Keyhole Markup Language; JSON—JavaScript object notation; CSV—comma-separated values.

TABLE 1. GOOGLE EARTH TECHNOLOGIES AND THEIR USE IN HYDROVIZ

Technology	Description	Use in HydroViz
Google Earth plug-in	A free plug-in that provides a 3-D digital globe that can be embedded in web pages.	Renders the hydrologic geospatial data on a 3-D model of Earth.
Google Earth JavaScript API	This API enables the creation of sophisticated 3-D map applications though simple JavaScript function calls.	Dynamically load hydrologic geospatial data as layers (KML files), create markers with hydrologic data, load time-series data, turn layers on/off, measure area and distance.

Note: 3-D—three-dimensional; API—application programming interface; KML—Keyhole Markup Language.

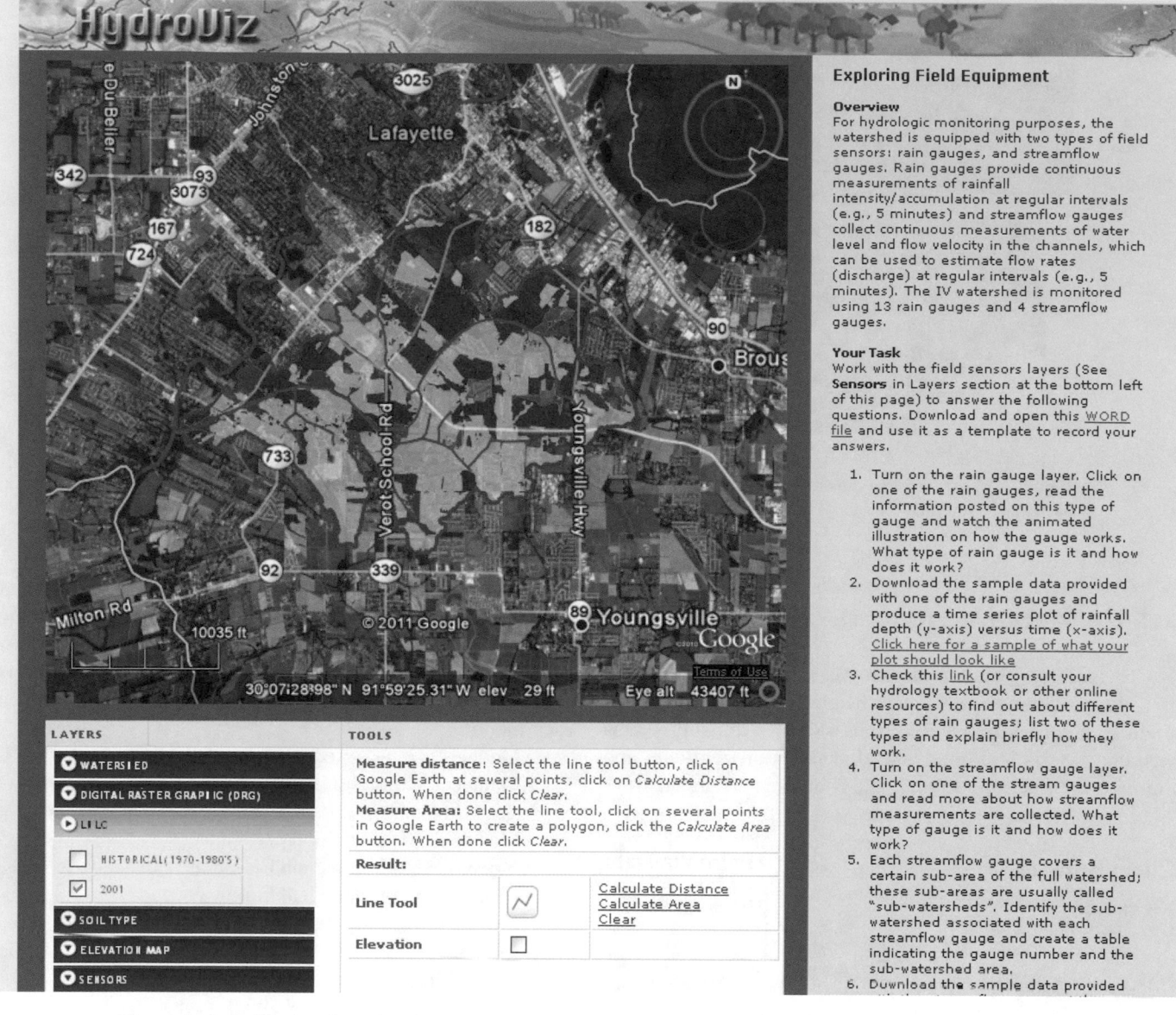

Figure 4. HydroViz interface showing the Google Earth plug-in (upper-left part), educational content and tasks (right-side part), and layers and tools (bottom part).

These modules were designed using an actual rainfall event, Tropical Storm Matthew, which swept across south Louisiana for several days (7–10 October 2004). The modules can be introduced to the students at different stages within a single course, or a subset of the 13 modules can be used across different classes. Each module starts with an introduction to the technical subject followed by a set of activities that the students need to complete. The activities are interactive and inquiry-based and include investigative tasks as well as quantitative and qualitative analyses. For most activities, a set of questions are embedded within the module and students are asked to record their activities and answers to such questions and type them in file to be submitted to the instructor. While HydroViz is primarily designed to be used in junior- or senior-level courses, it can also be used in freshmen introductory-level courses to expose new civil engineering and earth science students to basic watershed concepts, hydrologic variables, rainfall-runoff processes, spatial geographic data, and field measurements and sensors.

HYDROVZ IMPLEMENTATION AND EVALUATION

An evaluation study was conducted where the HydroViz system was implemented in several hydrology-related courses in three different universities (University of Louisiana at Lafayette, University of Texas at San Antonio, and Tennessee Tech University). A total of 182 students and six instructors in three universities participated in the evaluation study over the course of two semesters (spring 2010 and fall 2010). The courses included two senior-level engineering hydrology classes, one senior-level water resources class, and four freshmen-level introduction to civil engineering classes. An improvement-focused evaluation model (Posavac and Carey, 2003) informed the evaluation of HydroViz. This evaluation was both formative and summative in nature. The summative aspect focused on project effectiveness. The formative aspect examined the buy-in of the program and investigated program components in order to improve the project. The evaluation relied on three main sources of data: (1) students' answers to homework assignments embedded within the each of the 13 learning modules in HydroViz, (2) online surveys which include 17 statements, each on a five-point Likert scale instrument with five possible answers (strongly agree, agree, neither agree nor disagree, disagree, and strongly disagree), and (3) informal interviews with students and instructors. In the following sections we discuss the evaluation results which were focused on examining four main aspects of the HydroViz system.

How Effective is HydroViz in Facilitating Students' Learning?

To answer this question, students' HydroViz homework assignments were graded and descriptive statistics were calculated for all the sections. In addition, descriptive statistics were calculated for six survey items that ask about students' perceptions of how well HydroViz contributed to their learning. The results indicated that HydroViz was effective in facilitating students' learning and understanding of hydrologic concepts and increasing related skills as indicated by students' homework assignments. Figure 5 shows that an average of 66%–85% of the students understood the key concepts and increased their skills. It seems to be more effective for students in the senior-level

engineering hydrology or water resource engineering courses. An average of 69%–91% of them showed competency for these concepts. Concepts in the latter modules seem to be more difficult to grasp than those in the other modules. The percentages of students who received full or almost full scores for the last four modules were lower than the previous ones.

What Are Students' Perceptions of Various Features of HydroViz?

Descriptive statistics were calculated for seven survey items that were designed for this. Students' comments and informal interview notes were also analyzed to identify ideas and patterns. The results indicate that students had positive perceptions of various features of HydroViz (Fig. 6). Of all the participants, 79%–93% strongly agreed or agreed that they liked various features and characteristics of HydroViz. Students in the senior-level courses were slightly more positive. In these courses, the percentage of students who strongly agreed or agreed that they liked various features of the tool ranged from 83% to 97%, yet the percentage of students in the freshmen course who strongly agree or agree that they like various features of the tool range from 75% to 90%. Students commented that they like the tool because it is hands-on and it presents the technical subject within a real-world context.

What Are Students' Perceptions of HydroViz as a Part of the Curriculum?

Descriptive statistics were calculated for two survey items related to this question. Related students' comments and informal interview notes were also analyzed to identify ideas and patterns. Overall, students have positive perceptions of HydroViz as a part of the curriculum. About 85% of the students in the senior-level courses strongly agreed or agreed with

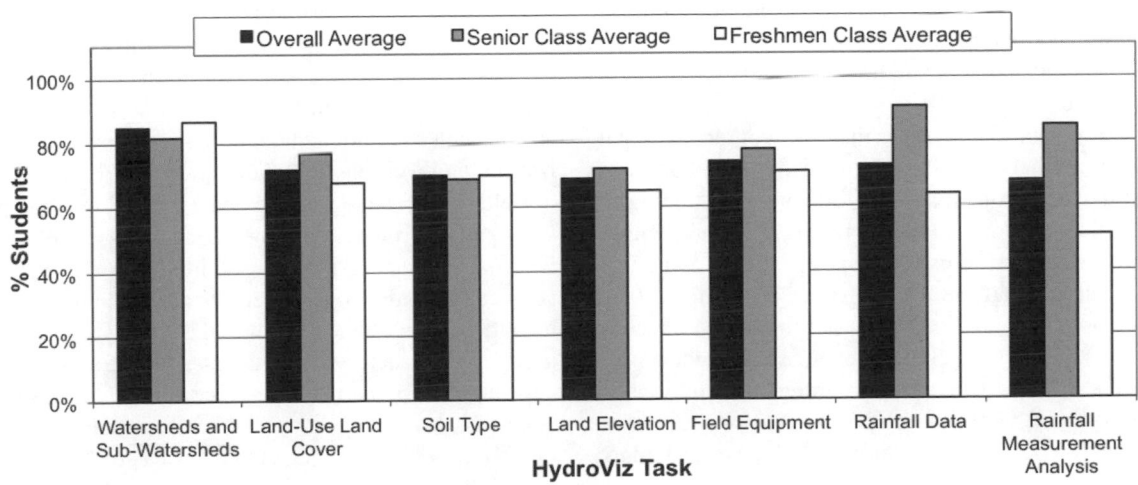

Figure 5. Percentages of students who received full or almost full scores on different tasks in HydroViz.

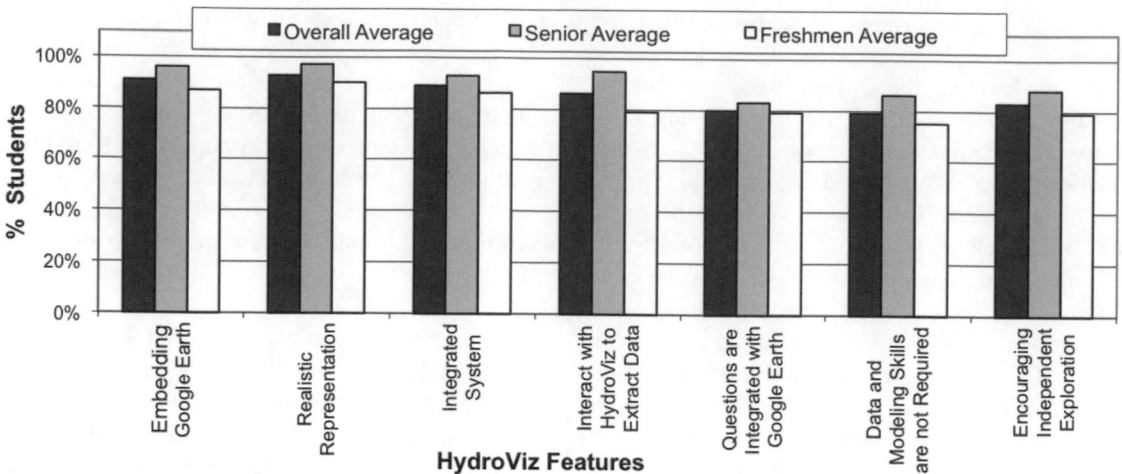

Figure 6. Percentages of students who strongly agree or agree that they like various features of HydroViz.

the following statement: "I find that HydroViz improves on current teaching tools/methods typically used in hydrologic engineering courses." Almost all of them (96%) strongly agreed or agreed that HydroViz fits well with the curriculum. The freshmen in the civil engineering introduction course found the tool less relevant. About 63% of them strongly agreed or agreed that HydroViz fits well with the curriculum.

What Can be Done to Improve HydroViz?

To answer this question, students' comments and informal interview notes were also analyzed to identify ideas and patterns. Students identified some issues that can be addressed to improve HydroViz. Many students commented on the heavy workload. In addition, for some classes the timing of the project was not optimal because it was the end of the semester and students had a commitment to assignments and finals for other classes. They wished that they had more time for the project. They said that they could enjoy the project more if they had more time. Students suggested that the instructor should introduce HydroViz early on in the semester and present it in smaller chunks to the students as they go through the semester.

Some freshmen did not think that the content was particular relevant to their specific engineering field. Some of them thought that the project was too long and challenging for them. A simplified version of HydroViz might be more appropriate for this group of students. One suggestion was that the tool include a demo that shows the overall functionality of HydroViz with some examples of the activities and expected results/outcomes. One of the challenges for some students was the required use of Excel in the assignment. Instructions on how to graph in Excel would be helpful to these students. The tasks for the last few sets of questions seemed to be particularly challenging. More guidance or instructions might be needed to improve students' learning.

CONCLUSIONS AND FUTURE WORK

This study reported on the development of an educational online system called HydroViz (http://hydroviz.cilat.org/). HydroViz is a web-based, student-centered, highly visual educational tool designed to support active learning in the field of engineering hydrology and other related earth science disciplines. The HydroViz system is built using Google Earth resources which provide valuable geospatial data and visual capabilities for hydrologic educational applications. Besides being freely accessible to a wide user community, Google Earth offers the ability to place and visualize hydrologic technical information on a 3-D model of Earth, which facilitates students' interactive and visually supported learning. HydroViz employs the free Google Earth plug-in and its JavaScript API to enable presentation of geospatial data layers in Google Earth and embed them in web pages that have the same look and feel of Google Earth. CSS and HTML were used to describe the look and formatting of the HydroViz web pages. With the aid of Google Earth API, it was also possible to create customized buttons and panels for students to interact with and display the data. The development of HydroViz is informed by recent advances in hydrologic data, numerical simulations and visualization, and web-based technologies. It is based on integration of field data, remote sensing observations, and computer simulations of hydrologic variables and processes. While HydroViz is primarily designed to be used in junior/senior/graduate level courses on hydrology, water resources engineering, or other related subjects, selected modules in HydroViz can also be used in introductory-level courses that focus on introducing new engineering and earth science students to basic hydrologic concepts. A total of 13 educational modules were embedded into HydroViz which range from basic activities (e.g., exploring watershed characteristics) to advanced analysis of field data and simulations. Each module is self-contained where instructions and technical questions are embedded within the same screens that show the watershed and its visual displays.

To evaluate its effectiveness in achieving desirable educational goals, the HydroViz system was introduced in several hydrology-related courses at three different institutions. Positive feedback was received from students and instructors showing a highly promising potential of HydroViz for enhancing students learning and their enthusiasm about hydrology. The web-based and visual and spatial navigation tools deployed in HydroViz through the Google Earth technologies were identified as the most attractive and effective features that appealed to both students and instructors. The evaluation results indicated some aspects of HydroViz that need further improvements such as providing more guidance to students and instructors in the form of video tutorials, templates for data entry, and example solutions that can be all embedded within the online system. The positive results of this pilot study indicate that further expansion of the HydroViz technical approach and educational contents can be valuable for enhancing students' learning and their enthusiasm for studying hydrology and earth sciences.

The design of HydroViz enables its adaptation by other users where it can be tailored toward other hydrologic basins and watersheds. In fact, after completing the current study, an independent institution contacted the authors and decided to adapt HydroViz design and functionality to develop a completely new application for a local water watershed on their campus. Besides using the design features of the current HydroViz system, this new adaptation experiment will further expand the technological basis of HydroViz to include more students' interactive activities for field-based investigations. Finally, the authors, in collaboration with other educators and researchers from two more universities, are currently in the process of developing an expanded version of HydroViz. The new system will cover three large-scale ecosystems (Coastal Louisiana, South Florida and the Everglades, and the Great Salt Lake basin in Utah), each presenting some unique hydrologic concepts and physical settings and will serve as case-based student-centered learning environments. The new HydroViz system will continue to use Google Earth resources and techniques to enhance HydroViz data and simulation capabilities and facilitate users' access and potential adaptation.

ACKNOWLEDGMENTS

Support for this work was provided by the National Science Foundation's Course, Curriculum, and Laboratory Improvement (CCLI) program under Award No. DUE-0737073, and by the National Science Foundation's Transforming Undergraduate Education in Science, Technology, Engineering and Mathematics (TUES) program under Award No. DUE-1122898. The authors acknowledge the participation of students and instructors at the University of Louisiana at Lafayette in evaluation of HydroViz. Instructors and students from the University of Texas at San Antonio (Dr. Hatim Sharif) and Tennessee Technological University (Dr. Faisal Hossain) are acknowledged for participating in independent evaluation of HydroViz in their classrooms.

REFERENCES CITED

Beven, K.J., 2001, Rainfall-Runoff Modelling: The Primer: Chichester, UK, Wiley, 372 p.

CUAHSI, 2010, Water in a Dynamic Planet: A Five-year Strategic Plan for Water Science, http://dx.doi.org/10.4211/sciplan.200711 (accessed May 2012).

Cunningham, S., 2004, Visualization in Science Education, *in* Invention and Impact: Building Excellence in Undergraduate Science, Technology, Engineering, and Mathematics (STEM) Education: The American Association for the Advancement of Science (AAAS) Press, p. 127–158.

Downer, C.W., and Ogden, F.L., 2004, GSSHA: Model to Simulate Diverse Stream Flow Producing Processes: Journal of Hydrologic Engineering, v. 9, no. 3, p. 161–174, doi:10.1061/(ASCE)1084-0699(2004)9:3(161).

Gupta, V.K., 2001, Hydrology (summary of the Water, Earth, and Biota initiative as a 2000 highlight for Geosciences): Geotimes, v. 46, no. 7, p. 25–26.

Habib, E., Larson, B., and Graschel, J., 2009, Validation of NEXRAD multisensor precipitation estimates using an experimental dense rain gauge network in south Louisiana: Journal of Hydrology (Amsterdam), v. 373, p. 463–478, doi:10.1016/j.jhydrol.2009.05.010.

Hooper, R., and Foufoula-Georgiou, E., 2008, Advancing the Theory and Practice of Hydrologic Science: Eos (Transactions, American Geophysical Union), v. 89, no. 39, doi:10.1029/2008EO390005.

Nash, J.E., Eagleson, P.S., Phillip, J.R., and van der Molen, W.H., 1990, The education of hydrologists (Report of the IAHS-UNESCO Panel): Hydrological Sciences Journal, v. 35, no. 6, p. 597–607, doi:10.1080/02626669009492466.

National Research Council (NRC), 2000, Inquiry and the National Science Education Standards: A Guide for Teaching and Learning: Washington, D.C., National Academies Press, 224 p.

Posavac, E.J., and Carey, R.G., 2003, Program Evaluation: Methods and Case Studies (6th ed.): Upper Saddle River, New Jersey, Prentice Hall, 344 p.

Zia, L.L., 2004, Web-enabled Learning Environments, *in* Invention and Impact: Building Excellence in Undergraduate Science, Technology, Engineering and Mathematics (STEM): Education: The American Association for the Advancement of Science (AAAS) Press, p. 161–182.

MANUSCRIPT ACCEPTED BY THE SOCIETY 16 APRIL 2012

The Geological Society of America
Special Paper 492
2012

Oceanography and Google Earth: Observing ocean processes with time animations and student-built ocean drifters

Alfred Hochstaedter*

Earth Science Department, Monterey Peninsula College, 980 Fremont Street, Monterey, California 93940, USA

Deidre Sullivan*

*Marine Advanced Technology Education (MATE) Center, Monterey Peninsula College, 980 Fremont Street,
Monterey, California 93940, USA*

ABSTRACT

Google Earth provides an easily accessible platform for students to view animations of oceanographic processes created by merging satellite, buoy, and student-built ocean-drifter data. The power of Google Earth is that many oceanographic properties can be displayed simultaneously over time such as atmospheric pressure, winds, surface currents, and sea-surface temperature. Lessons created from these activities address many of the principal outcomes of an introductory oceanography course, including the ability to analyze the interrelationship of ocean processes and understand how modern oceanography relies on technology to observe and measure the state of the oceans. In the classroom, the time-animation effort is paired with a drifter project where students build and release Global Positioning System–equipped drifters and watch their movement via satellite over the Internet. Because students see, touch, and feel the drifter in the classroom as they build and decorate it, they develop an inherent interest in its fate. Following the movement of the drifter fosters student interest in related oceanographic processes, many of which can be animated in Google Earth for the same time period, providing easy comparisons. The satellite data used in these animations are accessed through the National Oceanic and Atmospheric Administration (NOAA) Southwest Fisheries Science Center's ERDDAP (Environmental Research Division's Data Access Program) data server and displayed in Google Earth using keyhole markup language (KML) scripts. Satellite products are downloaded as .png image files and displayed as a series of image overlays to create time animations. Python scripts automate the process of generating KML scripts.

*ahochstaedter@mpc.edu; dsullivan@mpc.edu.

Hochstaedter, A., and Sullivan, D., 2012, Oceanography and Google Earth: Observing ocean processes with time animations and student-built ocean drifters, *in* Whitmeyer, S.J., Bailey, J.E., De Paor, D.G., and Ornduff, T., eds., Google Earth and Virtual Visualizations in Geoscience Education and Research: Geological Society of America Special Paper 492, p. 441–451, doi:10.1130/2012.2492(34). For permission to copy, contact editing@geosociety.org. © 2012 The Geological Society of America. All rights reserved.

INTRODUCTION

Oceanography Today

Studying earth systems and how they are interconnected has become a central theme in science starting in the 1970s with the Gaia hypothesis (Lovelock, 1979). The earth system is so interconnected that being able to fully understand and predict any part of it depends on integrating measurements and observations from many disciplines over a wide range of spatial and temporal scales (Rayner, 2010). Twenty-first century oceanography, as with many fields in science, has migrated from that of discrete sampling (such as observations made from a ship at a particular time and place) to continuous sampling made from floating buoys, autonomous underwater vehicles, and cabled underwater observatories augmented with remotely sensed data from satellites (Spinrad, 2006; Lubchenco, 2010). Ocean observing systems such as the U.S. Integrated Ocean Observing System (IOOS) and the international Global Ocean Observing System (GOOS) are coming of age to understand ocean processes and address issues of global concern just as accurate weather forecasts are dependent upon globally integrated measurements and observations of the atmosphere and oceans delivered on a consistent and timely basis.

The realization that it is the interconnection of earth system processes that is the dominant theme, and that these processes constantly undergo changes spatially and on seasonal, yearly, and decadal timescales, has prompted a shift in the way the oceans are studied, described, taught, and ultimately perceived.

Monitoring programs, such as IOOS and GOOS, have produced enormous quantities of data, much of which is available for public download over the Internet. Like never before, students, as well as the general public, are able to access and evaluate the data used to support or argue against scientific ideas as well as policy decisions, such as the placement of marine protected areas, climate change policy, oil-spill mitigation and fishing regulations. In addition, students can use these data to test hypotheses or make predictions during their own inquiry-based learning experiences. There are so many sources of large quantities of data that new organizations have been developed to coordinate and encourage the dissemination of these new sources of information. IOOS has been referred to as a "social network of organizations and people" that fosters the efficient and effective connection of ocean data and information to a large network of interested parties (U.S. IOOS Office, 2010).

The changes under way in the ocean sciences are representative of changes occurring in nearly every branch of science and technology. The ability to visualize complex interactions is a requirement across the science, technology, engineering, and mathematics (STEM) fields from earth sciences to materials sciences to engineering. In all of these fields, workplace skills require new employees to be able to find relevant information and data, acquire it many times via download, manipulate it, and visualize it in useful and flexible ways. The ability to investigate ocean processes using real-time data is becoming an essential skill in today's workplace.

Challenges that impair these efforts include the difficulty of downloading data from a wide variety of Internet portals and the lack of visceral experience that students have with important oceanographic properties such as atmospheric pressure, nutrient concentrations, or changes in sea-surface temperature. Although much data are readily available, the methods to download and the formats of the downloaded data are extremely variable. For many students, just downloading the data and seeing it in the right format becomes the primary task rather than learning something from the data itself.

The Opportunity—Student-Built Drifters and Visualizations using Google Earth Time Animations

To address these challenges the Marine Advanced Technology Education (MATE) Center has developed a two-pronged approach designed for introductory oceanography courses. Many of these students lack a tangible understanding of the scientific process, or previous experiences with ocean currents or changes in ocean properties over time or space. For many students, oceanography will be their only collegiate-level science course.

First, in an effort to give students direct exposure to ocean technology that enables much of modern oceanography, we have the students participate in the construction of relatively simple, ocean drifters equipped with Global Positioning System (GPS) transponders (i.e., two-way communications that can send and receive geographic position information). Once deployed, students can watch the movement of the drifter in real-time over the Internet. The notion of students watching the motion of an object that they helped build, decorate, and launch that is now in the ocean, helps spur interest in additional ocean characteristics that are changing on the same timescales that the drifter is observing.

The second part to our approach is downloading oceanographic data—sea-surface temperature, chlorophyll, sea-surface height, winds, and atmospheric pressure collected from satellites and oceanic buoys—and then displaying them in Google Earth for the same temporal and geographic extent as the drifter in the water. Google Earth lends itself to these tasks because of its familiarity to students, ease of use, and ability to create animations of changes over time.

The MATE ocean drifter project enables students to be involved in the collection of oceanographic data that is also of interest to local scientists. Through collaborations with research institutions, government agencies, and non-profit organizations, students can feel a part of an ongoing scientific study and instructors can reap the benefits of collaboration with science colleagues. Such collaborations enable the instructors, teachers, and educators to more fully understand the science behind the scientific study. These drifters have been deployed by colleges and universities to help support studies addressing: harmful algal blooms; the transport of fish, clam, lobster, and crab larvae; the movement of oil during the Deep Horizon spill; the movement of trash in rivers during flood stage; currents for tidal power; the dispersion of power plant effluent; and surface circulation and circulation model validation. These efforts often inspire instructors to be more enthusiastic and share their passion for science with the students.

The Impact

In this paper, we describe our efforts in building drifters, equipping them with GPS transponders, following their movement over the Internet, and creating Keyhole Markup Language (KML) scripts to display associated oceanographic data in Google Earth in animated formats. The result of these efforts is an effective method to teach the science of complex ocean systems. Students see data rather than generalized textbook diagrams and gain an appreciation of the complexity and variability of ocean processes from the experience. These efforts emphasize the reliance of modern ocean science on technology and open the doors of opportunity for collaboration between ocean scientists and educators.

STUDENT-BUILT DRIFTERS

Construction of the ocean drifters is based on designs developed by a consortium including the National Oceanic and Atmospheric Administration's (NOAA) Northeast Fisheries Science Center, Southern Maine Community College, and the Gulf of Maine Lobster Foundation (http://www.nefsc.noaa.gov/epd/ocean/MainPage/lob/driftdesign.html). Recent efforts follow this consortium's "Eddie" design slightly modified by the MATE

Center, and consisting of a central 2 × 4 wooden post through which fiberglass sail spars are inserted. Vinyl sails are attached to the spars and toggle floats attached to the top of the 2 × 4. The GPS transponder is attached to the end of the 2 × 4 and rests above water level (Fig. 1). Please visit http://coseenow.net/mate/ for more information on the MATE Drifter Project.

The drifter design is sufficiently simple and flexible to accommodate both small and large class sizes. A small class could make measurements, cut materials to size, and assemble the drifters. In larger classes with multiple lab sections, the drifter could be assembled by students in each lab section after being disassembled between sections by the instructor. In this way all students can experience some aspect of drifter construction and develop a tactile connection to an instrument that will soon be in the ocean recording ocean currents. In practice, the drifter is sufficiently bulky that its final assembly often occurs after transportation to the dock for deployment.

The transponder and associated satellite tracking system are purchased through Comtech Mobile Datacom Corporation, which provides a web-based tracking system. Real-time drifter locations can be viewed through the Comtech Web site. All location data can also be downloaded for further manipulation and loading into Google Earth using KML scripts.

TIME ANIMATIONS

Ocean processes are inherently complex. For example, Figure 2 shows how sea-surface temperatures (SSTs) vary from one location to another in the Monterey Bay region. Figure 2 also conveys the complexity and interrelationships SST has with

Figure 1. Cape Fear Community College students deploy a drifter during a cruise (left). A drifter deployed by Monterey Peninsula water drifts in the water shortly after deployment (right). The GPS transponder rests ~20 cm above the water level; the "sails" extend roughly a meter below the sea-surface level.

Figure 2. The distribution of sea-surface temperatures off the coast of central California is an example of interconnected ocean processes (the southern end of Monterey Bay is toward the top of the image). Sea-surface temperatures in this region depend on the vigor and location of upwelling, low-level winds, the shape of the coastline, and shallow bathymetry. In turn, sea-surface temperatures also influence the distribution and amount of biological productivity in the area. Image from the Monterey Bay Aquarium Research Institute: http://www.mbari .org/canon/ProcessStudies.htm.

wind, landforms, and bathymetry. What the image does not show, however, is the change in SSTs over time or any primary data. The image is excellent at conveying a generalized process that produces a commonly observed pattern of SSTs along the coast of California. To convey the temporal variability, one possibility could be to plot SSTs from a particular location against time. Real data could be used, but the significance of the geographic variability would be lost. Google Earth time animations go far beyond static images and allow students to view primary data, as well as simultaneously interpret the temporal and spatial variability of the data and relate it to other ocean properties such as wind, atmospheric pressure, and currents. To appreciate the power of time animations as compared to the static nature of Figure 2, the reader is encouraged to download the KML files from the GSA Data Repository[1] and interact with them. Screen grabs of these time animations in Figures 3–6 show progressively more complex relationships between the kinds of ocean properties emphasized here. None of the screen shots conveys the utility of the animations as well as viewing the animations themselves.

A comparison between Pacific and Atlantic drifter tracks over the same amount of time demonstrates the advantages of

employing Google Earth time animations (Fig. 3). The screen shots in Figure 3 immediately convey that the Gulf Stream moves much faster than the California Current. Based on in-class response, these animations effectively impress upon students the differences between western and eastern boundary currents.

Adding SSTs to the animations reveals some of the relationships between SSTs and surface currents. The Gulf Stream shoots along a boundary between warm and cool water in the north Atlantic. It follows a meandering path, generating a series of eddies, as it heads northward and eastward feeding the North Atlantic Current. A marine instrumentation class at Cape Fear Community College released a drifter into the Gulf Stream during fall 2009 (Fig. 4). After following the Gulf Stream, the drifter first flowed counterclockwise around a cold-water eddy (9/30) and then clockwise around a warm-water eddy (10/5). Later, the drifter moved counterclockwise around a persistent cold-water eddy (11/9 and 11/27).

The time animations demonstrate that a balance exists between the forces associated with the large-scale surface currents and those associated with the spatial distribution of sea-surface temperatures. Currents flowing around eddies follow the meandering isotherms prominent in the north Atlantic. Furthermore, watching the animations demonstrates that these eddies are relatively stable features, remaining in the same general vicinity for time scales of weeks or months.

In the northeastern Pacific along the west coast of California, the ocean dynamics are much different. A drifter released by College of the Redwoods in spring 2010 moved slowly in a southward direction (Fig. 5). Plotting wind data derived from National Data Buoy Center (NDBC) buoys along with the drifter data shows that the ocean currents, and hence the drifter motion, are influenced by the winds. During late spring and early summer, northwesterly winds dominate the California coast. These winds are created by clockwise rotation around the North Pacific High, an atmospheric high pressure system that establishes itself in the northeast Pacific during the summer months. During spring, low pressure systems occasionally approach the coast. As these low pressure systems approach the coast, the coastal winds reverse themselves and winds from the south dominate. The reversal in wind direction results from the counterclockwise rotation around atmospheric low pressure systems.

The movement of the College of the Redwoods drifter shown in Figure 5 illustrates this relationship. In this image, drifter locations are shown along with sea-level atmospheric pressure isobars and arrows representing the direction and magnitude of winds derived from the NDBC buoys. The buoy data shows that winds typically blow parallel to the isobars, clockwise around high pressure systems and counterclockwise around low pressure systems. During this time period, the most common atmospheric state is a clockwise rotating high pressure system centered offshore in the northeastern Pacific. Consequently, the most common movement of the drifter is in a southward direction as strong winds blow from the north along the coast. As low pressure systems occasionally approach the coast, the motion of the drifter

[1]GSA Data Repository item 2012307, KMZ files, is available at http://www .geosociety.org/pubs/ft2012.htm, or on request from editing@geosociety.org.

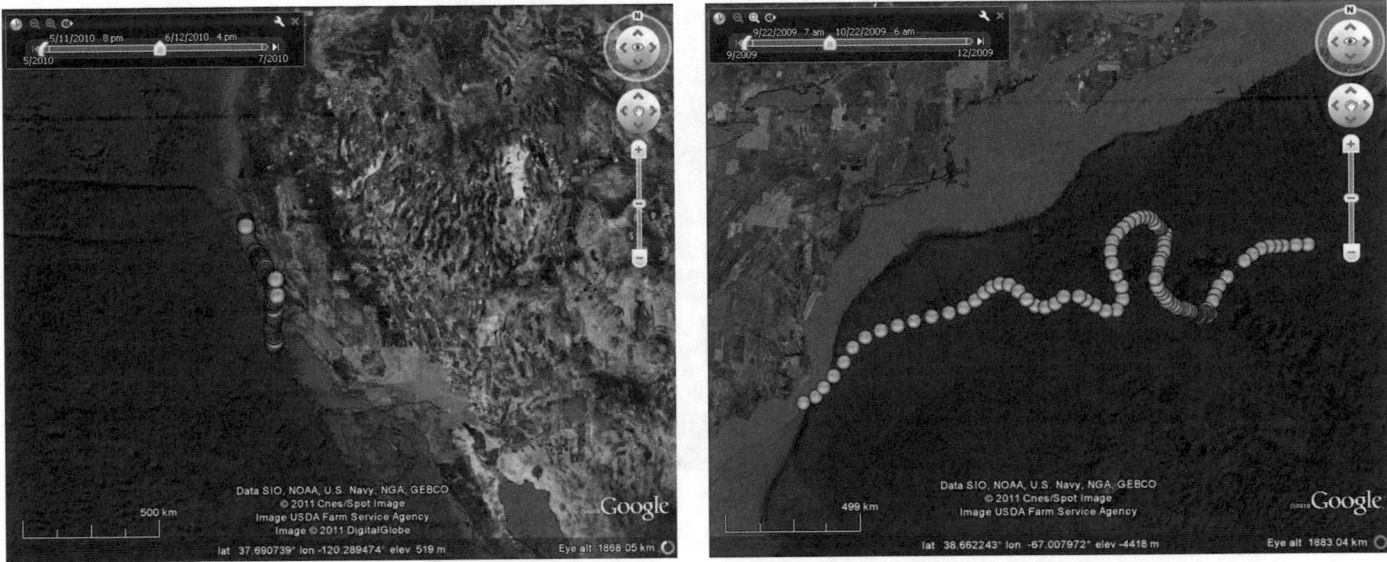

Figure 3. Google Earth screen shots show drifter tracks from the Pacific (left) and the Atlantic (right) at the same scale. Both images represent roughly one month. The icons in the Pacific example show locations every four hours, whereas those in the Atlantic show locations every six hours. Both images are shown at the same scale; the bar scale in the lower left represents ~500 km in each image.

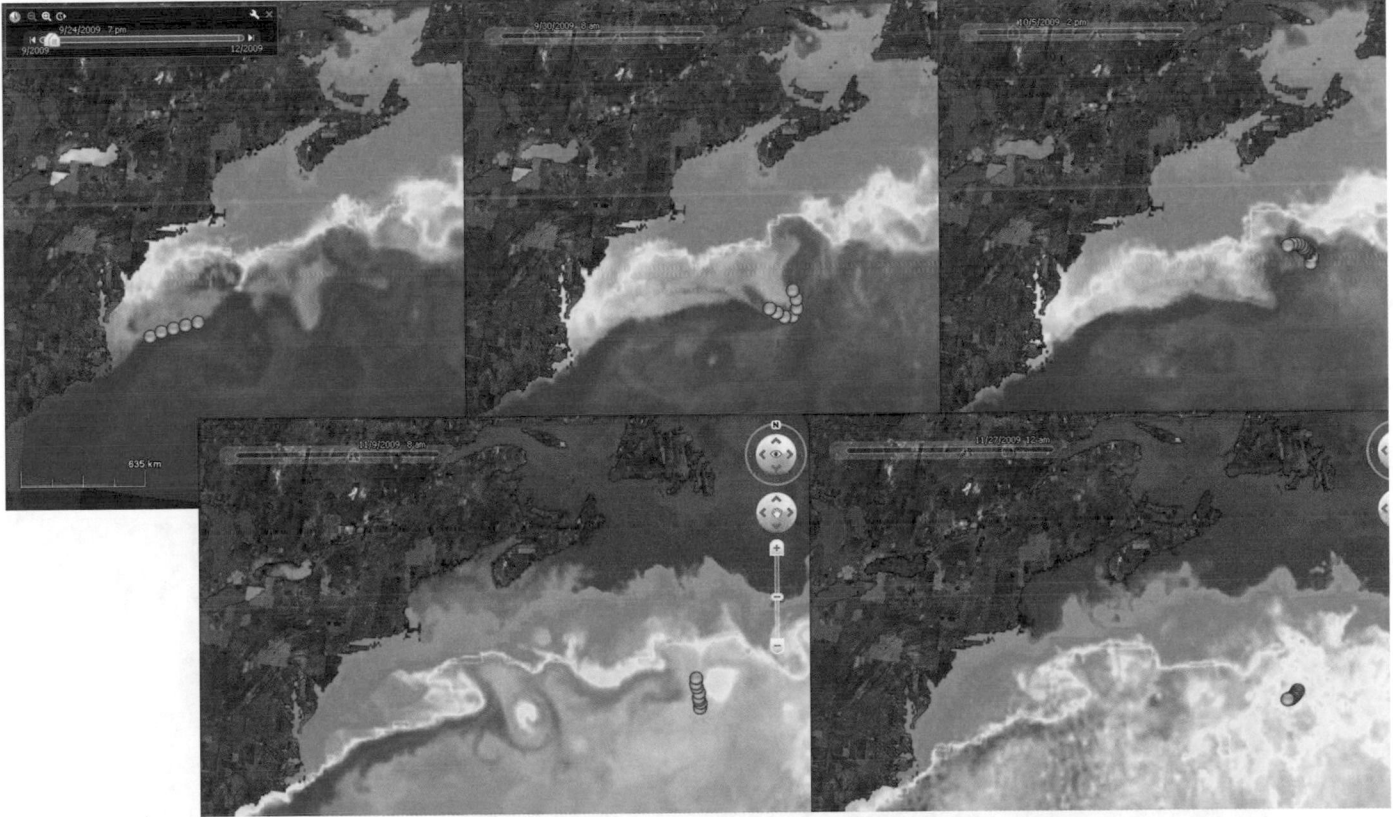

Figure 4. Screen shots from a Google Earth time animation show drifter movement superimposed on sea-surface temperatures (SSTs). SST data are blended from a variety of satellite sources for each day at 1-km resolution by the Jet Propulsion Laboratory (http://ourocean.jpl.nasa.gov/ SST/) and downloaded through ERDDAP. Drifter released by Cape Fear Community College.

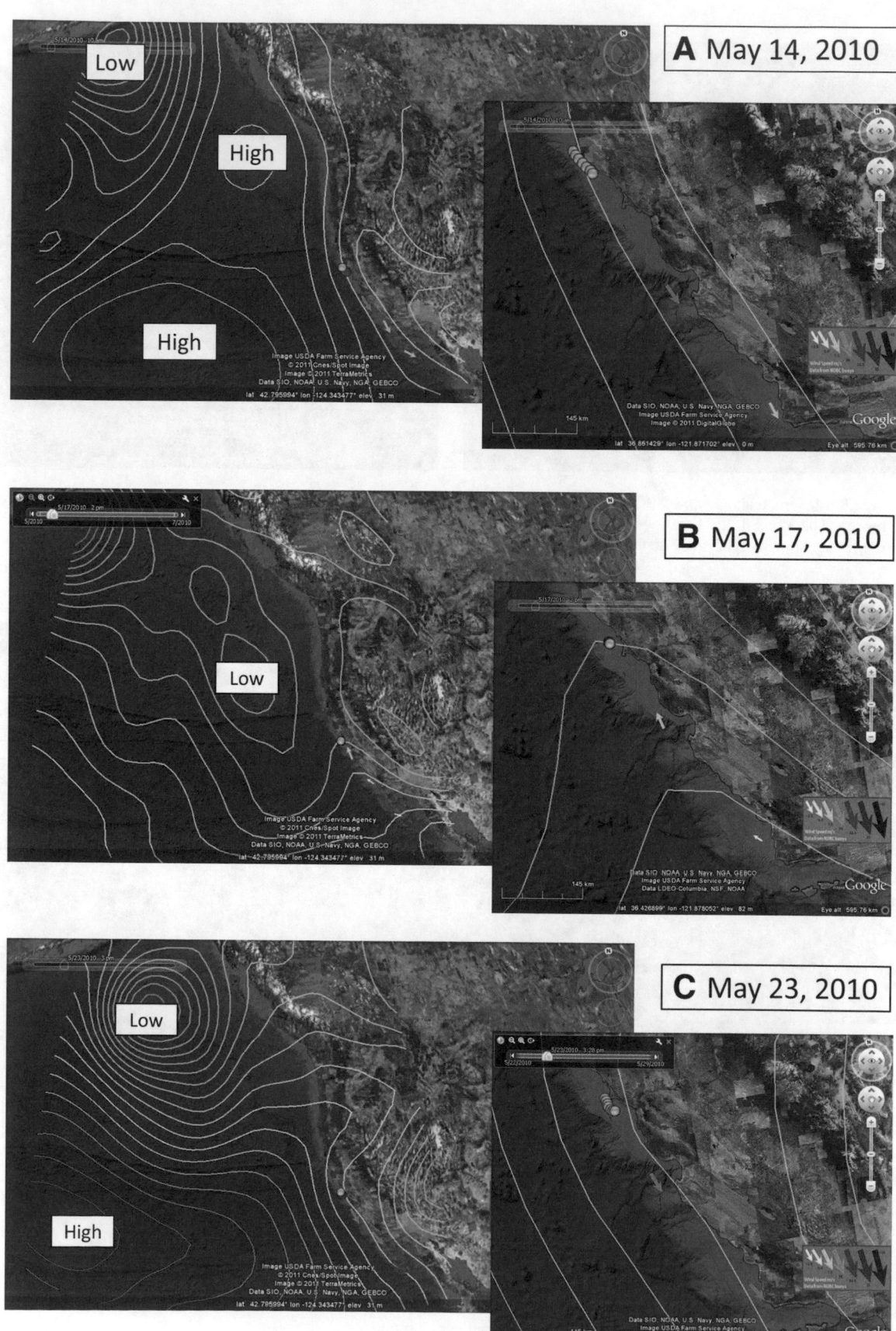

Figure 5. A series of Google Earth screen shots show the relationship between atmospheric pressure (colored isobars), wind direction and magnitude (colored arrows), and drifter motion (circular icons, with most recent location on top). In A and C, high-pressure systems are most prominent off of the California coast. Air rotating clockwise around the high-pressure systems caused wind to blow from northwest to southeast during these time periods. As a result, the drifter moved with ocean currents in a southeastward direction. In B, a low-pressure system entered the region off of the California coast. Counterclockwise flow around the low-pressure system caused a wind relaxation event in which winds blew weakly from south to north. As a result, the drifter stopped moving in a southeastward direction and remained relatively stationary during this time period.

slows dramatically or even reverses, moving slowly toward the north. As the low pressure system exits and high pressure once again dominates the region offshore California, the drifter continues its southward movement.

Spring conditions in the northeast Pacific offshore of California are characterized by coastal upwelling. This process is shown schematically in Figure 2, and with satellite- and buoy-derived data in Figure 6. Three different parameters are shown in Figure 6: isobars of atmospheric pressure at sea-level, wind direction and magnitude at three NDBC buoys, and SSTs. Whereas the screen shot shown in Figure 6 is a snapshot in time, the time animation available in the GSA Data Repository [see footnote 1] shows how all of these parameters continuously change with time. During the North American spring and summer, clockwise rotation around the high pressure system in the northeastern Pacific results in winds from the northwest along the

coast of North America. At any particular point in time, at locations where the isobars are closer together, the pressure gradient force is greater, resulting in stronger winds (compare the wind arrows off the coast of California to those off of the Columbia River in Fig. 6). Surface waters respond to the northwest winds as well as the Coriolis effect. The resulting Ekman transport causes near-surface water to flow in an offshore direction. Upwelling is the replacement of this warmer near-surface water by colder, and more nutrient-rich, deep water. Upwelling causes the colder water to appear at the surface near the coast, as seen in Figure 6.

DOWNLOADING OCEANOGRAPHIC DATA FOR THE TIME ANIMATIONS

Most of the oceanographic data used in the animations shown here is downloaded from the ERDDAP (Environmental Research Division's Data Access Program) data server maintained by NOAA's Southwest Fisheries Science Center (http://coastwatch.pfeg.noaa.gov/erddap/index.html). ERDDAP allows the user to specify the format of the downloaded data, as well as both the temporal and geographic extent of the downloaded data. The ERDDAP server re-serves data from a large variety of sources, including a variety of satellite products and the NDBC buoys. Currently hundreds of different data sets are available.

A key aspect of the ERDDAP portal is that data is retrievable in a variety of formats, making it easy to load into Google Earth. These formats include transparent .png image files which can be "draped" over the topography or sea surface as image overlays, and ASCII (American Standard Code for Information Interchange) grids, which can be manipulated further and displayed as

Figure 6. Sea-surface temperatures (SSTs), sea-level atmospheric isobars, and wind arrows show conditions favorable for upwelling along the California coast in June 2010. Upwelling indicated by the dark blue and purple along the central coast. In response to northwest winds created by an offshore atmospheric high, Ekman transport moves near-surface water away from the coast. This near-surface water is replaced by colder water from deeper levels. Thus, the cool SSTs along the coast are a result of the upwelling. Drifter locations removed for clarity. This is the process shown schematically in Figure 2.

paths (polylines) or icons. Most importantly, all of the formatting variables controlling the appearance of the downloaded image (color scales, geographic boundaries, time information, etc.) are contained directly within the URL (uniform resource locator) used to download the data. This URL coding within the ERD-DAP system enables the automation of data downloading. With a predictable, systematic system of communicating the formatting variables through the URL, simple scripts can be developed using Python, MATLAB, or similar languages, to download a series of data images, save them on a hard disk or server, and write KML scripts to load them into Google Earth in the correct chronological order, thus creating a time animation.

An easy way to download a chronologic series of data from the ERDDAP site and save the individual files on a local hard disk is to use the cURL utility, a command-line tool for transferring data using the URL syntax (http://curl.haxx.se/). cURL is open-source software and is free to download. More information on the specific cURL commands to use, as well as alternative techniques, are given on ERDDAP's documentation page (http://coastwatch.pfeg.noaa.gov/erddap/griddap/documentation.html).

CREATING KML SCRIPTS TO CONTROL THE TIME ANIMATIONS

In this section we briefly describe portions of the KML scripts that are used to create these animations. These examples are intended to illustrate the technique and inspire new animation ideas for those creating animations for beginning students or those providing instruction in KML itself to more advanced Earth and ocean science students. A full description of KML is beyond the scope of this paper. For a more detailed description of KML and descriptions of its use, see Wernecke (2009). Full KML scripts for the animations described in this paper are available in the GSA Data Repository (see footnote 1).

KML Basics

In the KML scripts, drifter locations are shown by point data, or icons. The buoy-derived wind data are also point data shown by icons. In the KML vernacular, icons, or point data, are loaded using the <Placemark> element. The yellow pushpin commonly seen in many Google Earth images is a familiar example of a <Placemark>. All of the other data, including SSTs, sea surface heights (SSHs), high-frequency radar current data, and chlorophyll data, are shown using .png image files downloaded from the ERDDAP web portal. The .png image files are loaded using the <GroundOverlay> element. A <GroundOverlay> is an image draped over the basic satellite and seafloor imagery native to Google Earth. The .png images are downloaded from the ERD-DAP web portal and saved to a local hard disk. The KML script then references these locally stored .png files.

The <Placemark> and <GroundOverlay> elements are the most important building blocks of the KML scripts that create the animations. Both <Placemark> and <GroundOverlay> are derived from the <Feature> element, and thus contain similar child elements. This means that, like any <Feature> element, both <Placemark> and <GroundOverlay> can contain child elements that define things like their <name>, <Style>, <description>, and, most importantly for animations, <TimeSpan>.

<Timespan> is a critical element in creating animations because it defines the dates and times for which the <Placemark> (i.e., icon) and <GroundOverlay> (i.e., .png images) should be visible to the viewer. The <Timespan> element has two child elements: <begin> and <end>, each of which contain a "dateTime" value. The values contained by the <begin> and <end> elements document when to begin and end showing a <Placemark> or <GroundOverlay>.

The <TimeSpan> element has the following format:

```
<TimeSpan>
        <begin>dateTime</begin>
        <end>dateTime</end>
</TimeSpan>
```

The *dateTime* value within the <begin> and <end> elements specifies a date ± time in UTC (Coordinated Universal Time). The *dateTime* has the following format:

```
yyyy-mm-ddThh:mm:ss
```

where:
- yyyy is a four-digit value that specifies the year (e.g., 2011).
- mm is a two-digit value that specifies the month between 01 and 12 (e.g., 08 for August).
- dd is a two-digit value that specifies the day of the month between 01 and 31 (e.g., 03 for the third day of the month).
- T is the separator between the date and the time.
- hh is a two-digit value for hours between 00 and 24.
- mm is a two-digit value for minutes between 00 and 60.
- ss is a two-digit value for seconds between 00 and 60.

An SST Animation Example

An example of KML script for displaying the .png files follows (Table 1). Each <GroundOverlay> element contains several child elements. In the order shown in the KML script, these include <name> which defines the name of the <GroundOverlay> as seen in the list view on the left-hand side of the Google Earth screen, <TimeSpan> which defines the date and time the <GroundOverlay> is visible as described above, <Icon> which contains the child element <href> which tells Google Earth which .png file and the location of the file to associate with this particular <GroundOverlay>, and <LatLonBox> which contains child elements that define location and spatial extent over which Google Earth should display the .png image file.

The <GroundOverlay> element is repeated for each .png file that needs to be displayed to create the animation. For the SST data used in these examples, a single image is produced for

TABLE 1.

```
<?xml version="1.0" encoding="UTF-8"?>
<kml xmlns="http://www.opengis.net/kml/2.2" xmlns:gx="http://www.google.com/kml/ext/2.2"
xmlns:kml="http://www.opengis.net/kml/2.2" xmlns:atom="http://www.w3.org/2005/Atom">
<Document>
 <name>EastPac SST Data and Legend</name>

<Folder>
 <name>SST Data</name>
 <GroundOverlay>
         <name>2011-07-05 Data</name>
         <TimeSpan>
             <begin>2011-07-05T00:00:00</begin>
             <end>2011-07-06T00:00:00</end>
         </TimeSpan>
         <Icon>
                 <href>C:/Folder/EastPacSST-2011-07-05.png </href>
         </Icon>
         <LatLonBox>
                 <north>48.005 </north>
                 <south>23.005 </south>
                 <east>-115.995 </east>
                 <west>-139.995 </west>
         </LatLonBox>
 </GroundOverlay>

 <GroundOverlay>
         <name>2011-07-06 Data</name>
         <TimeSpan>
             <begin>2011-07-06T00:00:00</begin>
             <end>2011-07-07T00:00:00</end>
         </TimeSpan>
         <Icon>
                 <href>C:/Folder/EastPacSST-2011-07-06.png </href>
         </Icon>
         <LatLonBox>
                 <north>48.005 </north>
                 <south>23.005 </south>
                 <east>-115.995 </east>
                 <west>-139.995 </west>
         </LatLonBox>
 </GroundOverlay>
```

Note relationship between the values in <name>, <begin>, <end>, and the .png file in <ref>.

End of the first <GroundOverlay> and beginning of the second <GroundOverlay>.

Note the repetitive pattern.

(...following lines omitted)

each day. Therefore, the <TimeSpan> element describes a 24-h window for display and a different <GroundOverlay> element is inserted into the KML script for each day of the animation. Each <GroundOverlay> element contains a <name> element. The value contained within each of these <name> elements is a date—the date that the SST data was collected. Thus, this <name> element helps the viewer identify when the data shown in each .png image was collected.

A Drifter Track Example

An example of a KML script for displaying drifter locations as point data within a <Placemark> element follows (Table 2). Each <Placemark> element contains several child elements. As with the <GroundOverlay> element described above, these child elements include the <name> and <TimeSpan> elements which behave similarly for <Placemark>. In addition, the <Placemark> element also contains two additional child elements. The <styleUrl> element contains a reference to a description of the style for the icon that is defined elsewhere in the KML script. This style contains a reference to a .png file to be used as an icon, the color and size of the icon, and other attributes defining what the icon should look like to the Google Earth viewer. The <Point> element contains a <coordinates> element that defines the location to display the icon with coordinates in the order of longitude, latitude, and altitude.

The <Placemark> element is repeated for each drifter location that is part of the animation. For the drifter data used in these examples, the <TimeSpan> element is manipulated so that the drifter track always has a "tail" of a few previous drifter locations. This "tail" helps the viewer's eye visualize the motion of the drifter. This "tail" is accomplished by giving each drifter location an <end> time that is *after* the <begin> time of the next drifter location. For example, in the KML script shown in the box, the <end> time of the first drifter location is 2009-09-23T00:00:00Z, which is 18 h *after* the <begin> time of the second drifter

TABLE 2.

(...previous lines omitted)

```
<Placemark>
<name>2009-09-22 00:00</name>
        <TimeSpan>
<begin>2009-09-22T00:00:00Z</begin>
<end>2009-09-23T00:00:00Z</end>
</TimeSpan>
        <styleUrl>#BeigeButton</styleUrl>
        <Point>
                <coordinates>-77.442241,33.691238,0</coordinates>
        </Point>
</Placemark>

<Placemark>
        <name>2009-09-22 06:00</name>
        <TimeSpan>
<begin>2009-09-22T06:00:00Z</begin>
<end>2009-09-23T06:00:00Z</end>
</TimeSpan>
        <styleUrl>#BeigeButton</styleUrl>
        <Point>
                <coordinates>-78.010182,33.894346,0</coordinates>
        </Point>
</Placemark>

<Placemark>
        <name>2009-09-22 12:00</name>
        <TimeSpan>
<begin>2009-09-22T12:00:00Z</begin>
<end>2009-09-23T12:00:00Z</end>
</TimeSpan>
        <styleUrl>#BeigeButton</styleUrl>
        <Point>
                <coordinates>-78.010182,33.894346,0</coordinates>
        </Point>
</Placemark>
```

> The overlap of times in the <begin> and <end> elements of successive <Placemark> elements creates the "tail" of the drifter tracks that helps convey its motion.

> Note the repetitive pattern of successive <Placemark> elements.

(...following lines omitted)

location: 2009-09-22T06:00:00Z. In this manner, the KML script always displays a "tail" of at least four icons.

CONCLUSIONS

A characteristic of effective teaching is relating new material to experiences familiar to the student, (e.g., contextual learning [Center for Occupational Research and Development, 2011]). Teaching topics in physical oceanography can be challenging because students often lack familiarity with oceanic and atmospheric variables like SSTs, high- and low-pressure systems, biological productivity and how these variables change over time and from one location to another. The first part of the two-pronged approach described here addresses contextual learning by involving students in the building, decorating, and launching of ocean drifters. This experience provides students a tactile experience with an instrument that is placed in the ocean and monitored. Following the drifter via satellite over the Internet is representative of the methods employed in twenty-first century

oceanography which largely rely on continuous sampling to discover patterns of variability over a variety of temporal and spatial scales. This tactile experience also enables students to make a personal connection with the instrument and increases their level of interest in its fate as it flows along with the ocean currents. As a result, students begin to ask questions about what causes ocean currents and the variations observed in SSTs, wind, SSH, biologic productivity, and other ocean properties.

The second part of this approach involves students using Google Earth to interact with time animations of oceanographic data obtained for the same time periods as the drifter data they have been monitoring real-time over the Internet. If watching the drifter prompted questions about the dynamics or nature of ocean processes, the animations provide insight into some of the answers. Viewing animations of oceanographic data, rather than static and generalized textbook illustrations, can show how many ocean properties—like the distribution of atmospheric high- and low-pressure systems, or the distribution of SSTs in a coastal upwelling system—can undergo significant change over a number of days, but that clear patterns emerge over a longer time frame. Furthermore, the natural variation in these processes can reveal interconnections in a more instinctive and profound manner than static drawings. The realization that ocean processes are interconnected is an important learning outcome in many introductory oceanography courses.

It is difficult to describe the utility of time animations using only images of screen grabs on the printed page. We urge readers to download the animations from the repository and interact with them for themselves. Only in this way will readers experience the utility of seeing many oceanographic variables changing together. Google Earth has been integral to our efforts because of its ease of use, broad distribution, and convenience of sharing animations using KML files.

ACKNOWLEDGMENTS

This project was funded in part by the National Science Foundation's Advanced Technology Education Program (DUE #0703197), the Division of Ocean Science's Centers for Ocean Science Education Excellence (OCE #0731046), and the Monterey Peninsula College Foundation. We would like to acknowledge the help and guidance of a number of individuals who have made this project possible: James Manning at the NOAA Northeast Fisheries Science Center; Bob Simons, Lynn Dewitt, Cara Wilson and Dave Foley in the Environmental Research Division at the NOAA Southwest Fisheries Science Center; and Sage Lichtenwalner with COSEE NOW (Centers for Ocean Sciences Education Excellence Networked Ocean World) at Rutgers University. Leslie Rosenfeld reviewed the manuscript and helped us clarify the presentation of the physics of ocean circulation. Two anonymous reviewers improved the quality of the manuscript.

REFERENCES CITED

Center for Occupational Research and Development (CORD), http://www.cord.org/contextual-learning-definition/ (accessed 30 August 2011).

Lovelock, J., 1979, Gaia: A New Look at Life on Earth: USA, Oxford University Press, 176 p.

Lubchenco, J., 2010, Ocean observations: Essential for good stewardship: Marine Technology Society Journal, v. 44, no. 6, p. 6–9, doi:10.4031/MTSJ.44.6.23.

Rayner, R., 2010, The U.S. Integrated Ocean Observing System in a global context: Marine Technology Society Journal, v. 44, no. 6, p. 26–31, doi:10.4031/MTSJ.44.6.1.

Spinrad, R.W., 2006, The evolution and revolution of ocean science and technology: Marine Technology Society Journal, v. 40, no. 2, p. 134–135, doi:10.4031/002533206787353312.

Wernecke, J., 2009, The KML Handbook Geographic Visualization for the Web: Boston, Massachusetts, Pearson Education, Inc., 339 p.

U.S. IOOS Office, 2010, http://www.ioos.gov/library/us_ioos_blueprint_ver1.pdf (accessed 30 August 2011).

MANUSCRIPT ACCEPTED BY THE SOCIETY 16 APRIL 2012

The Geological Society of America
Special Paper 492
2012

Testing the effects of prior coursework and gender on geoscience learning with Google Earth

Janice Gobert*

*Social Sciences and Policy Studies, Worcester Polytechnic Institute, 100 Institute Road,
Worcester, Massachusetts 01609-2280, USA*

Steven C. Wild*

Old Dominion University, Physics Department, 4600 Elkhorn Ave., Norfolk, Virginia 23529, USA

Lisa Rossi*

*Social Sciences and Policy Studies, Worcester Polytechnic Institute, 100 Institute Road,
Worcester, Massachusetts 01609-2280, USA*

ABSTRACT

Two sets of learning activities in Google Earth were developed for use by geoscience majors and non-science majors. The first activity aimed to foster undergraduate students' understanding of the geography and basic geology of Iceland. We tested the efficacy of this activity for learning with 300 undergraduates from a university in the southeastern part of the United States. In terms of post- versus pre-test scores we found: (1) overall learning gains when collapsing over type of prior knowledge and gender, (2) no differences in learning gains when comparing those with prior coursework in geology or geography to other students without such prior coursework, and (3) no differences in learning gains when comparing males and females. In terms of items completed during the lab exercise, again we found no differences by prior coursework (prior geology, prior geography, or none), and no differences by gender. Lastly, moderate positive correlations were found between students' pre-test and post-test scores, as well as between students' embedded lab scores and post-test scores.

For the second activity, we developed a laboratory activity about the classic Tonga region of the west Pacific in order to support undergraduate students' understanding of: (1) the physical geography of the Tonga Subduction System, (2) the dynamic geological processes involved in plate movement, subduction, magmatic arc evolution, and trench rollback, and (3) geological processes resulting from subduction, including volcanism, and earthquake formation. Using the program called Sketch-Up, we created 3-D COLLADA (three-dimensional COLLAborative Design Activity) models

*jgobert@wpi.edu; swild@odu.edu; lrossi@wpi.edu.

Gobert, J., Wild, S.C., and Rossi, L., 2012, Testing the effects of prior coursework and gender on geoscience learning with Google Earth, *in* Whitmeyer, S.J., Bailey, J.E., De Paor, D.G., and Ornduff, T., eds., Google Earth and Virtual Visualizations in Geoscience Education and Research: Geological Society of America Special Paper 492, p. 453–468, doi:10.1130/2012.2492(35). For permission to copy, contact editing@geosociety.org. © 2012 The Geological Society of America. All rights reserved.

that are viewable as four-dimensional animations in the Google Earth API (application programming interface; a web-based version of Google Earth) to help demonstrate several geophysical processes. These animations potentially have a wide range of learning application from basic to more abstract ideas. Specifically, the learning objects created involve the Pacific Plate subducting underneath the Australian Plate in the Tonga Region. These are designed to help show subduction, active and dormant volcanoes, back-arc spreading, trench rollback, and migration of the tear point that marks the northern termination of the subduction system. We tested the efficacy of this activity with 127 undergraduates from a university in the southeastern part of the United States. In terms of post- versus pre-test scores we found: (1) overall learning gains when collapsing over type of prior knowledge and gender, (2) no differences in learning gains when comparing those with prior coursework in geology or geography to other students without this prior coursework; and (3) no differences in learning gains when comparing males and females. For the lab activity itself, we found no differences by prior coursework (geology and/or geography versus none), but found a gender difference favoring males; however this learning did not show up as statistically significant at post-test (as previously mentioned). Lastly, moderate positive correlations were found between students' pre-test and lab scores.

Data is discussed with respect to Google Earth's utility to convey basic geoscience principles to non-geology undergraduates and its potential impact on public understanding. This is important and aligned with many current educational reform efforts (the American Association for the Advancement of Science, National Science Education Standards), which call for broader scientific literacy.

INTRODUCTION

Learning in this Domain: Why is it Difficult?

The domain chosen for this study is plate tectonics, the lead paradigm for understanding the origin and evolution of Earth's surface features including continents, oceans, and island arcs. This is a difficult topic to learn both because of the hidden mechanical processes, which are outside our direct experience, and because it involves several different types of knowledge including spatial, causal, and dynamic knowledge (Gobert and Clement, 1999; Gobert, 2000). Specifically, conceptual understanding in this domain requires understanding the spatial arrangement of the various material components of the earth (i.e., spatial/static information) as well as understanding the movements within these layers and their dynamic causes (i.e., primordial core) and radioactive sources of heat (i.e., mantle) that must escape Earth's deep interior (KamLAND Collaboration, 2011), convection of solid material through the mantle (Wilson, 1973), plate movements, divergence and convergence at plate boundaries, and the interaction of surface plates with deep mantle plumes (Morgan, 1972). In addition to acquiring two types of knowledge (spatial/static and kinematic/dynamic), several concepts need to be integrated into a complex causal chain to build a rich, 4-D mental model of the system (Gobert and Clement, 1999; Gobert, 2000). From these mental models, predictions and inferences can be made about the system's behavior: in the case of plate tectonics, explaining or depicting

locations of earthquakes and volcanoes, sea-floor spreading, mountain building, and island-arc evolution.

Among the most difficult concepts that we present to students are (1) plate-plume interaction as in Iceland (Ito and Lin, 1995), and (2) trench rollback as in the Tonga region (Isacks et al., 1968). Iceland stands high above sea level because it marks the intersection of the Mid-Atlantic Spreading Ridge and a deep mantle plume emanating from the core-mantle boundary. Students thus have to visualize two processes with very different length and time scales. Time scale is particularly difficult to understand, even for graduate students of geology (Jacobi et al., 1996). Tonga is the type locality for the process of trench rollback whereby the line along which the plate bends into a subduction zone migrates in the opposite direction to the material of the plate (Uyeda and Kanamori, 1979; Rosenbaum and Lister, 2004; Moores and Twiss, 1995). At Tonga, for example, the rocks of the Pacific plate move westward while the trench marking the initiation of subduction migrates east.

Relevant Work on Learning in Geoscience

Google Earth, a fairly new program (version 1 was released in 2005), constructs pictures of Earth by downloading satellite data from a remote terabyte server (Lisle, 2006) and rendering them on a virtual globe in real time. The program is interactive so that the location and size of the region viewed is under full control of the learner/user; the user can zoom in, pan, and tilt the terrain from any desired viewpoint, and the surface imagery communicates information in a format that is more intuitive and realistic than paper maps and cross-sections (Whitmeyer et al., 2010). This last feature makes them useful for learning and reasoning for experts in the domain, as well as for students and lay people, e.g., non-science majors (this is addressed more fully later in the paper).

It is argued that Google Earth is a tool that can help build scientific literacy on a broad scale because it and other geo-technologies are ways to give citizens basic knowledge of geography (Sanchez, 2009) and geoscience (Thompson et al., 2006). Secondly, in addition to basic content knowledge, some researchers claim that working in Google Earth can hone one's data analysis and interpretation skills, which, many argue, are becoming increasingly important in scientific and industrial fields. As an extension to this latter point, it has been further argued that the ability to use images and spatial technologies is necessary in order to participate in modern society (Bednarz et al., 2006) since information and data tends to be displayed in spatially oriented formats.

To date, there have been a fairly large number of studies that address learning in geoscience, but most of these have been conducted with a pre-college population (Libarkin and Anderson, 2005), and studies on college students or other adults only emerged within the last decade or so (cf. DeLaughter et al., 1998; Trend, 2000; Libarkin, 2001; Libarkin et al., 2005; Dahl et al., 2005). Amongst the research on this topic with an adult population, the research that is most closely related to the present research is the research on learning with visualizations in geoscience (cf. Hall-Wallace and McAuliffe, 2002; Whitmeyer et al., 2009; Thompson et al., 2006).

With respect to training students in geoscience specifically, recent reform efforts emphasize the need to utilize technology in teaching and learning (Stout et al., 1994; National Research Council, 1996), which has translated into greater demand for technology-based teaching methods (Cruz and Zellers, 2006). In parallel, there also have been demands for greater instructor accountability for students' learning at all levels, as decreasing enrollment trends continue in the STEM (science, technology, engineering, and mathematics) disciplines (McConnell et al., 2006). Although learning with Google Earth has been touted as having great potential for improving students' knowledge about geological phenomena, spatial skills, problem-solving, etc., and the fact that, intuitively it appears to have many affordances for geoscience learning (Cruz and Zellers, 2006; Whitmeyer et al., 2009), there are relatively few studies that either characterize the learning processes that students engage in while learning with Google Earth, or that address the efficacy of learning with Google Earth.

Characterizing Learning Processes with Google Earth

It has been noted that Google Earth (referred to herein as GE) offers a benefit over more traditional GIS (geographic information systems) in that GE can be implemented into classrooms at any level because it has relatively few tools and thus less over-head for the teacher in learning it (Patterson, 2007; Bodzin et al., 2012). In terms of the utility of GE for college professors and high school teachers, GE only requires a basic knowledge of scripting languages in order to construct materials (Whitmeyer et al., 2010). For example, GE has been used in high school classrooms for virtual exploration of geologic features to support students' understandings of geological processes (Fermann, 2006; Stahley, 2006). Similarly, Sanchez (2009) describes implementations in which a teacher developed a geological map that encompasses layers of data about earthquakes and volcanoes. Here, it was suggested that these implementations help students to identify different aspects of oceanic crust formation and understand the mid-ocean ridge system. Lastly, Patterson (2007), who has used GE for middle school instruction, suggested that GE's interactive exploration capacity helps students understand the spatial context of their location and engage in spatially oriented learning in an entertaining and meaningful manner.

Studies that have Addressed the Efficacy of Google Earth

One study compared GE to traditional textbook materials for undergraduates' learning of landforms (Cruz and Zellers, 2006). Findings revealed that students in the GE condition gained deeper understanding of the content compared to those in the traditional textbook condition. Furthermore, those students who had previous exposure to GE performed better than those who did not. Similarly, Martin and Treves (2008) showed that GE is effective to help students and the general public (i.e., non-majors) visualize both scientific data and science content in 3-D. Martin and Treves (2008) stressed the importance of promoting active learning and dissuaded the development of "flashy" 3-D animations, since students, who by definition lack expertise, do not know what is salient in order to engage in knowledge acquisition from information sources (Gobert, 2005a). Bodzin and Cirucci (2009) similarly noted that resources such as GE, when used in conjunction with appropriately designed instructional materials, show much potential in promoting students' spatial thinking.

In two innovative studies in which students constructed their own materials using GE, learning gains were obtained. First, Whitmeyer et al. (2009) had undergraduates use handheld computers to collect lithologic and structural data and then analyze it in order to construct geologic maps of their field areas. This approach, according to the authors, familiarizes students with GE tools, and in turn, can be useful in improving students' interpretations of field geology. Similar results have been found in which students constructed their own representations of geoscientific phenomena (Gobert and Clement, 1999; Gobert, 2000; Gobert and Pallant, 2004; Gobert, 2005b). In another study, Thompson et al. (2006) showed students how to create their own content in GE. Here, not only did students learn important design elements and skills, but students also reported that these skills were amongst the most important that they learned in their geoscience program.

These last studies described address an important issue underlying learning with visualizations; that is, that deep learning with visualizations typically requires accompanying materials, scaffolds, etc., in order to support and guide students in their learning processes. This is critical since students often do not know what is salient within these rich visual information sources (Lowe, 1993) because they present all information simultaneously (see Larkin and Simon [1987] and Gobert [2005a] for more on this topic).

RATIONALE

In our project, we address the learning gains for two different units developed in GE. In particular, in each study we address the efficacy of those with prior coursework in geology and geography, compared to non-majors with no prior coursework in these domains. Secondly, although it was not part of the original design of the research, we compare the learning gains of both males and females, since many studies have shown that females lag behind males in their learning of geoscience concepts due to their inherent spatial nature and females' oft-reported diminished spatial skills (Kahle et al., 1993; Dabbs et al., 1998; Burkham et al., 1997; Britner, 2008).

STUDY 1

Purpose

In the first study, we developed a GE activity to support students' understanding of the geography and basic geology of Iceland. We tested the efficacy of this activity in terms of post- versus pre-test scores for: (1) overall learning gains as measured by pre- and post-tests, (2) differences in learning gains when comparing those with prior coursework in geology or geography to other students without such prior coursework, and (3) differences in learning gains when comparing males and females. Lastly, we also compared students' learning on the lab activity itself (i.e., the pedagogical activities that were completed as part of the lab).

Method

Participants

A total of 225 undergraduate students from a southeastern university participated in this study as part of their coursework[1]; age data for the participants was not collected. All students were part of the same large lecture; there were nine sections of the lab that corresponded to the lecture from which our data was drawn.

Materials

Pre-Test/Post-Test

The pre-test and post-test consisted of the same set of 10 questions on basic geological and geographical knowledge of

TABLE 1. AN EXAMPLE OF TWO QUESTIONS FROM THE PRE- AND POST-TEST, ICELAND ACTIVITY
Q8 What is the principle rock type seen in Iceland (i) limestone (ii) basalt (iii) granite (iv) marble
Q9 Which best describes the geological origins of Iceland? (i) Iceland sits on top of both a deep mantle plume and a divergent plate boundary. (ii) Iceland is a fragment of continental crust, like Britain and Ireland, that detached from the European margin during North Atlantic spreading. (iii) Iceland is a volcanic island arc forming above a subduction zone. (iv) Iceland is a huge floating mass of ice drifting very slowly away from Greenland.

TABLE 2. AN EXAMPLE OF A QUESTION ASKED IN THE ICELAND LAB ACTIVITY
15.1 Wait for the images to load, then drag the time slider in order to reveal the deep mantle plume under Iceland (Fig. 15).
⊗ Compare the height (thickness) of the plume and the thickness of the lithosphere:
....
⊗ Estimate how deep the plume extends:
....

Iceland as well as one question asking about prior experience studying this topic (Table 1). Of these questions, nine were multiple choice, and one asked participants to locate Iceland on a provided map. The pre-test served to determine a baseline of prior knowledge that a participant had coming into the activity, while the post-test determined what knowledge was gained as a result of participating in the lab activity (Table 2). All items were developed by experts in the area of geoscience as part of three ongoing projects (NSF-CLLI #0837040, De Paor and Whitmeyer, 2008; NSF-GEO #1034643, De Paor and Whitmeyer, 2010; NSF-DUE #1022755, De Paor et al., 2010). Some examples of items on the pre- and post-test are given below; the full set of items is included in Appendix A[2].

Lab Activities for Iceland

The Iceland lab activity consisted of a series of tasks that were designed to develop students' understanding of the geography and geology of Iceland. Tasks included: locating Iceland in Google Earth, specifying its relationship geographically with respect to the Arctic Circle, using the time slider to observe the

---

[1]Data were collected, coded, and stored in compliance with the requirements as outlined by Federal Policy for the Protection of Human Subjects.

[2]GSA Data Repository item 2012308, Appendices A and B, is available at http://www.geosociety.org/pubs/ft2012.htm, or on request from editing@geosociety.org.

horizon, asking students what they would expect to see here at the Winter Solstice, observing the landscape, geological features (e.g., rock types), and other characteristics of Iceland's urban and rural landscapes by driving around in a virtual car, observing the formation of the Mid-Atlantic Ridge by using a time slider, noting how the Mid-Atlantic Ridge is displaced across the Gibbs Fracture Zone, and observing the deep mantle plume under Iceland.

Procedure

Students initially were given consent forms, with verbal explanation, and a tracking identification number was assigned. Identification numbers were based on the course lab number, the beginning five digits, and then some digits after that given by the lab teaching assistant. Students were informed to not use their university identification numbers. Labs and pre- or post-tests with university identification numbers were not used and were removed from the study. Once the consent forms were completed and tracking identification numbers administered, each student was given a pre-test. If students finished their pre-test early, they were asked to wait quietly while others finished.

The students worked in groups of two to four depending on the lab section, with lab sections having different numbers of students. Students were encouraged, sometimes with help from instructors, to take turns working on the computers. Instructors were only allowed to help if students had technical problems but not with lab material itself. The students were informed that the lab itself would not be graded as part of their lab score, which may have had an effect on the way students answered questions or participated during lab. As students completed the labs, they were collected and the students were asked to wait for their fellow classmates to finish.

The last part of the lab consisted of the post-test. The post-test is the same as the pre-test. Each student was given a post-test and upon completion was allowed to leave the lab. No collaboration was allowed during the pre- or post-tests. The pre-test, lab activity, and post-test were all completed in one, two-hour lab period.

Data scoring

Pre- and Post-Test Scoring

The pre- and post-tests consisted mainly of multiple choice questions and were scored on a partial- or full-credit basis. A participant could earn a maximum of two points on each question for choosing the correct answer, one point for choosing a partially correct answer, and zero points for choosing an incorrect answer. Some questions had more than one possible answer worth one point, as shown in Table 3. Answer "iii" is worth two points, answers "i," "iv," or "v" are each worth one point, and answer "ii" is worth zero points.

Lab Items Scoring

The lab activity consisted of seven open response or "yes/no" questions, which were scored on a partial credit basis out of a possible one, two, or three, depending on the question. The scoring scheme for a three-point question is shown below in Table 4. Each correctly circled answer earned one point, and the open response portion was scored as zero, 0.5, or one point based on accuracy and detail.

Results

Data were analyzed to address overall learning gains from the Iceland lab, as measured by pre- and post-tests, as well as to test whether there were any learning gain differences due to prior

TABLE 3. QUESTION 2 AND ITS CODING SCHEME FOR THE ICELAND ACTIVITY

Q2 Where is Iceland relative to the Arctic Circle?	Scoring Q2	
(i) Iceland lies entirely south of the Arctic Circle.	i.	1
(ii) Iceland lies entirely north of the Arctic Circle.	ii.	0
(iii) The Arctic Circle touches the northern coast or offshore islands.	iii.	2
(iv) The Arctic Circle touches the southern coast or offshore islands.	iv.	1
(v) The Arctic Circle goes through the center of Iceland.	v.	1

TABLE 4. QUESTION 4.4 AND ITS CODING SCHEME FOR THE ICELAND ACTIVITY

4.4 Visit various parts of Iceland and record your first impressions of the country here:	
Outiside of Reykjavik, is Iceland heavily populated/developed?	⊗ [Yes / No]
Do you see a lot of large-scale agricultural or industrial plants?	⊗ [Yes / No]
How would you describe the terrain?	

undeveloped or under-developed or poor land or barren or isolated or partly farmed or grassland or tundra or equivalent

TABLE 5. AVERAGE SCORES ON PRE-TEST BY TOTAL, GENDER,
AND PRIOR COURSEWORK FOR THE ICELAND ACTIVITY

	Overall	Female	Male	Geology and/or geography	No geology or geography
Mean pre-test score	6.69	6.32	7.05	7.44	6.57
Standard deviation	2.95	2.52	3.44	3.37	2.75

coursework in geology and/or geography. Gender differences were also analyzed, although this was not part of the original design of the study. Lastly, data were analyzed with respect to learning during the lab activity itself. Each analysis is presented and described in turn. The unit of analysis here was data from each individual student.

Were there Differences by Prior Coursework or Gender before the Iceland Learning Activity with Google Earth?

First, we addressed if there were any differences on the pre-test both by prior coursework and by gender. A univariate analysis of variance was computed for each of these analyses. First, the difference between the total scores on the pre-test was not statistically different when comparing those with prior coursework to those with no prior coursework (F = 1.838, p = 0.162). Secondly, the difference on the total pre-test score was not statistically different when comparing males and females (F = 1.890; p = 0.154). See Table 5 for means and standard deviations for each of these analyses.[3]

Did the Iceland Activity Yield Differences in Overall Pre-Post Comparisons?

Next we addressed if there were differences in overall post-test scores compared to pre-test scores collapsing over both prior coursework and gender categories. A paired t-test was computed for this analysis. The difference in overall pre-test score and overall post-test score was statistically significant, [t(224) = 13.33, p = 0.000; $\bar{X}$ pre = 6.69, SD = 2.95; $\bar{X}$ post = 9.68, SD = 3.58] ($\bar{X}$—mean; SD—standard deviation); this result demonstrates that on average, students had higher scores on the post-test than on the pre-test. See Table 6 and Figure 1.

Did Type of Prior Coursework Influence Learning in the Iceland Activity?

In order to address whether there were differences between the pre- and post-test scores when comparing those with prior coursework in geology or geography to those with no relevant prior coursework, a univariate analysis of variance was conducted with the total post-test as the dependent variable and type of prior coursework as the independent variable. Pre-test

was used as a covariate. The difference in post-test score by level of prior coursework was not statistically significant (F = 2.107; p = 0.124). Thus, both students with prior geology or geography coursework and those without this prior coursework learned approximately the same amount of content knowledge from the Google Earth Iceland activity, as measured by the post-test gains. The means and standard deviations can be seen in Table 7.

Were Post-Test Differences by Gender Yielded for the Iceland Activity?

In order to address whether the overall pattern observed was different when comparing males and females, a univariate analysis of variance was conducted with the total post-test as the dependent variable and gender as the independent variable. Pre-test was used as a covariate. The difference in post-test minus pre-test scores by gender was not significant (F = 0.436; p = 0.647). Thus, both males and females learned approximately the same amount of content knowledge, as measured by the post-test,

TABLE 6. OVERALL SCORES FOR
PRE- AND POST-TEST FOR THE ICELAND ACTIVITY

	Pre-test	Post-test	t (df)	P
Overall score	6.69	9.68	13.329 (224)	0.000
Standard deviation	2.95	3.58		

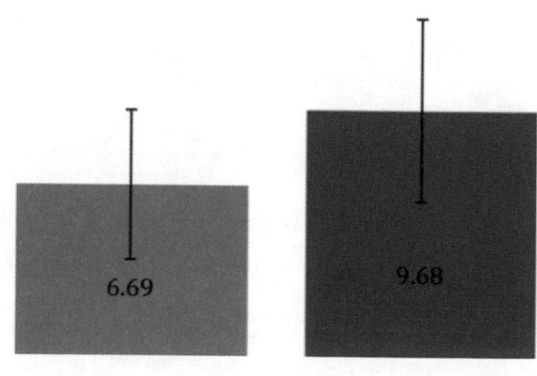

Overall Score

■ Pre-test ■ Post-test

Figure 1. Overall scores on both pre-test and post-test for the Iceland activity.

---

[3]Although the means appear different when both comparing males and females, and when comparing those with prior relevant coursework to those with no relevant prior coursework, the standard deviations associated with these means indicate that the dispersion of scores was large in both cases, thus no statistically reliable differences were found for either comparison.

TABLE 7. AVERAGE SCORES ON PRE-TEST AND POST-TEST
FOR THE ICELAND ACTIVITY
BY TYPE OF PRIOR COURSEWORK

	Geology and/or geography	No geology or geography
Mean pre-test score	7.44	6.57
Standard deviation	3.37	2.75
Mean post-test score	10.78	9.57
Standard deviation	3.22	3.58

TABLE 8. AVERAGE SCORES ON PRE-TEST AND
POST-TEST FOR THE ICELAND ACTIVITY BY GENDER

	Female	Male
Mean pre-test score	6.32	7.05
Standard deviation	2.52	3.44
Mean post-test score	9.37	10.21
Standard deviation	3.30	3.84

TABLE 9. AVERAGE SCORES ON THE ICELAND LAB ACTIVITY
BY GENDER AND PRIOR COURSEWORK

	Female	Male	Geology and/or geography	No geology or geography
Mean lab score	4.36	4.69	4.51	4.55
Standard deviation	1.62	1.87	1.52	1.77

TABLE 10. PEARSON CORRELATION VALUES
BETWEEN PRE-TEST, POST-TEST, AND LAB SCORES
FOR THE ICELAND ACTIVITY

	Pre-test	Post-test	Lab
Pre-test	1	0.483*	0.101
Post-test	0.483*	1	0.291*
Lab	0.101	0.291*	1

*Statistically significant at the 0.01 level of alpha.

holding the effects of the pre-test score constant. The means and standard deviations can be seen in Table 8.

Were There Any Differences on the Lab Scores for the Iceland Activity When Comparing by Prior Coursework or by Gender?

Next we addressed the differences on the lab activity scores both by prior coursework and by gender; in other words, whether there was a difference on students' performance in the lab activity by prior coursework in geology and/or geography, or by gender. A univariate analysis of variance was computed for each of these analyses. There was no statistically significant difference found between the total scores on the lab activity when comparing those with prior coursework to those with no prior coursework (F = 0.069, p = 0.934). Additionally, the difference on the total score for the lab activity was not statistically significant when comparing males and females (F = 1.109, p = 0.332). This result demonstrates that on average, males and females scored similarly on the lab activity (see Table 9).

Is There a Relationship between the Pre-Test Scores, the Lab Scores, and the Post-Test Scores for the Iceland Activity?

In order to establish whether there was a relationship between these learning measures, a Pearson correlation analysis was conducted between all the measures, namely, the pre-test, the lab scores, and the post-test. A statistically significant correlation was found between the pre-test and post-test (r = 0.483, p = 0.000, [two-tailed]), indicating a moderate, positive relationship between the pre-test and the post-test. Another statistically significant correlation was found between the lab scores and the post-test (r = 0.291, p = 0.000, [two-tailed]). The Pearson correlation values can be seen in Table 10. No statistically significant correlation was found between the pre-test score and the lab scores (r = 0.101, p = 0.132, [two-tailed]). The scatterplots for pre-test and post-test, and pre-test and lab scores can be seen in Figures 2 and 3, respectively.

STUDY 2

Purpose

In the second study, we developed a laboratory activity focused on the classic Tonga region of the west Pacific in order to support undergraduate non-geology majors' understanding of: (1) the geographic layout of the Tonga Subduction System, (2) the dynamic geological processes involved in plate movement, subduction, magmatic arc evolution, and trench rollback, and (3) geological processes related to subduction, including volcanism and earthquake formation.

Method

Participants

A total of 138 undergraduate students from a southeastern university participated in this study[4]; age data for the participants was not collected. All students were part of the same large lecture; there were nine sections of the lab that corresponded to the lecture, from which our data was drawn.

Materials

Pre-Test/Post-Test

The pre-test and post-test consisted of the same set of 11 questions on basic geological and geographical knowledge of the American-Samoa/Tonga region as well as one question asking about prior experience studying this topic. Of these questions, nine were multiple choice, one asked for the order of four listed events, and one asked participants to locate American-Samoa/Tonga on a map that was provided to them. All items were developed by experts in the area of geoscience as part

[4]Data were collected, coded, and stored in compliance with the requirements as outlined by Federal Policy for the Protection of Human Subjects.

Gobert et al.

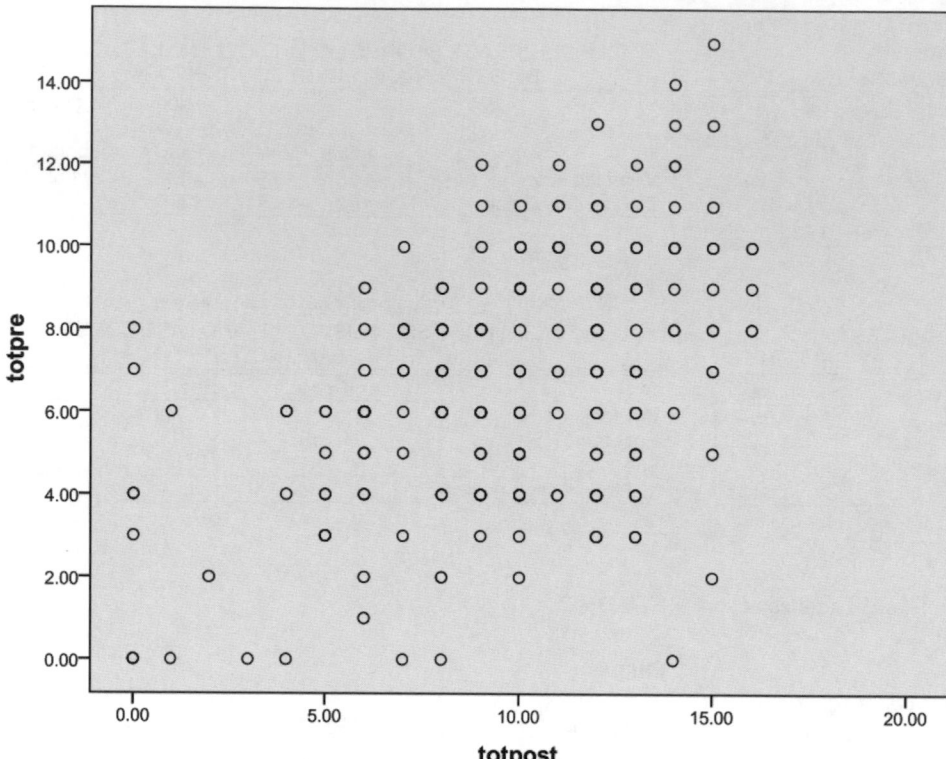

Figure 2. Scatterplot of correlation between total pre-test scores and total post-test scores for the Iceland activity.

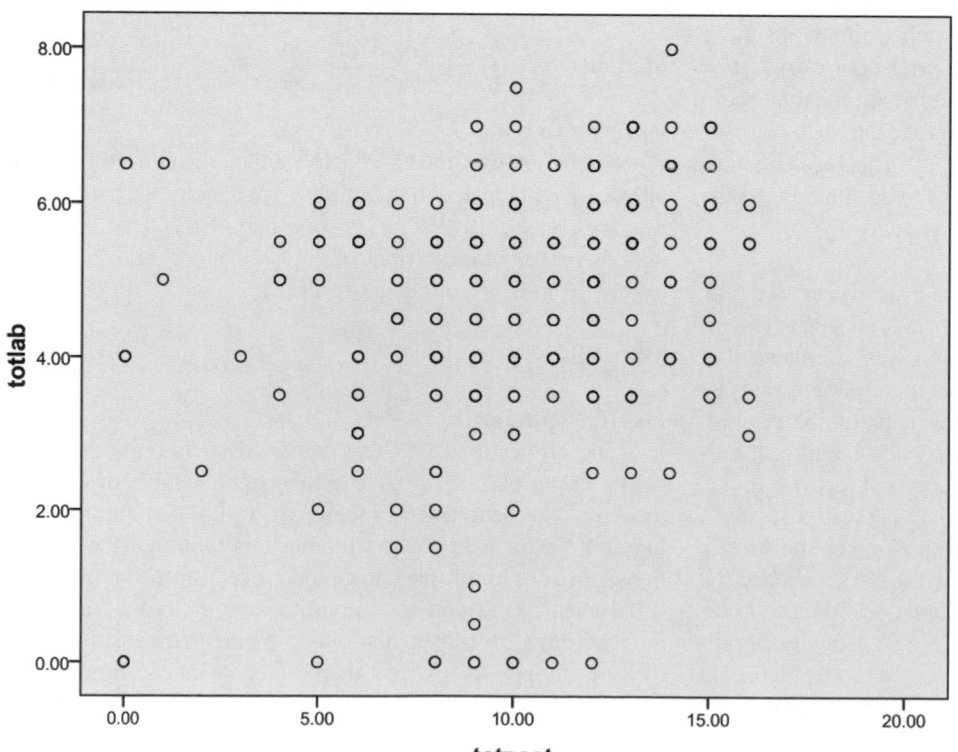

Figure 3. Scatterplot of correlation between total lab scores and total post-test scores for the Iceland activity.

TABLE 11. A SAMPLE OF A QUESTION FROM THE PRE- AND POST-TEST FOR THE TONGA ACTIVITY

Q5 Which of the following pictures shows the earthquake pattern for the American-Samoa/Tonga region. Where **A** represents the Australian Plate and **B** is the Pacific Plate. Plate B moves under Plate A.

With ● being deep earthquakes

▢ Are medium depth earthquakes

And **X** representing shallow earthquakes

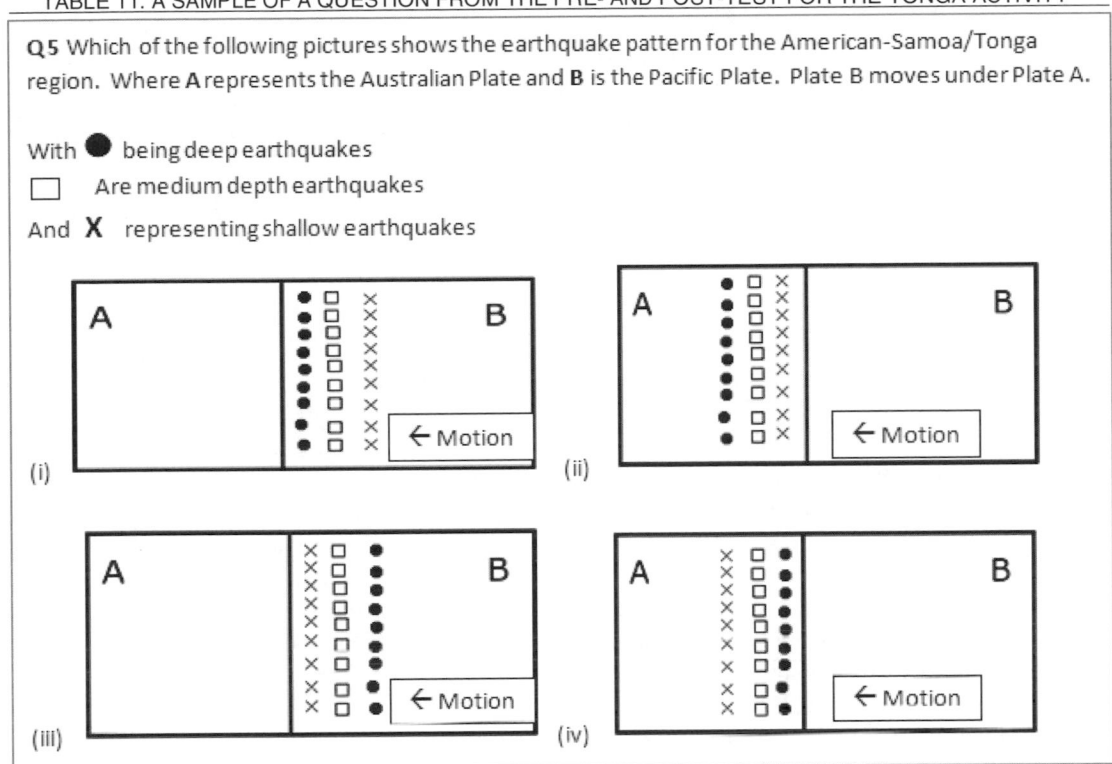

of three ongoing projects (NSF-CLL1 #0837040, De Paor and Whitmeyer, 2008; NSF-GEO #1034643, De Paor and Whitmeyer, 2010; NSF-DUE #1022755, De Paor et al., 2010). The pre-test served to determine a baseline of prior knowledge for each participant, while the post-test determined what knowledge gains were made after participating in the lab activity. See Tables 11 and 12 for sample items; all items are shown in Appendix B (see footnote 2).

Lab Activities for Tonga

The Tonga lab activity consisted of a series of tasks that were designed to develop students' understanding of the geology of the Tonga region in the western Pacific Ocean. Tasks included: locating the Tonga Region with respect to the Tropic of Capricorn, viewing and manipulating virtual block diagrams to observe animations of subduction, island arc formation, and

trench migration, answering questions about the relative location of volcanoes and earthquakes, and answering questions about plate movement, trench formation, and plate movement and trench rollback.

Procedure

The process of gathering student performance was done in the following manner. Students initially were given consent forms, with verbal explanation, and a tracking identification number was assigned. Identification numbers were based on the course lab number, the beginning five digits, and then some digits after that given by the lab teaching assistant. Students were informed to not use their university identification numbers. Labs and pre- or post-tests with university identification numbers were not used and were removed from the study. Once the consent

TABLE 12. AN EXAMPLE OF TWO QUESTIONS ASKED
IN THE PRE- AND POST-TEST FOR THE TONGA ACTIVITY

Q2 Name two features present on the surface during subduction.
  1.
  2.

Q3 When subduction occurs, do the volcanoes form on the down-going plate (on the east side of the trench in this case) or the over-riding plate (on the west side of the trench in this case)?

TABLE 13. QUESTION 6 AND ITS CODING SCHEME FOR THE TONGA ACTIVITY

Q6 The Tonga Trench's motion relative to the Pacific Plate is		Scoring Q6
(i) Moves forward with the Pacific Plate.	i.	1
(ii) Stationary (trench does not move).	ii.	1
(iii) Moves against plate motion.	iii.	2
(iv) There is no such thing as the Tonga Trench.	iv.	0

TABLE 14. QUESTION 1 AND ITS CODING SCHEME FOR THE TONGA ACTIVITY

Q1 What is your previous experience of the geology or geography of American Samoa/Tonga		Scoring Q1
(i) I have no significant previous study experience.	i.	0
(ii) I did a class project about the geology or geography of American Samoa/Tonga.	ii.	1
(iii) I participated in a real field trip or a holiday visit.	iii.	2
(iv) I am native to or lived in the American Samoa/Tonga region for an extended period.	iv.	3

forms were completed and tracking identification numbers administered, each student was given a pre-test. If students finished their pre-test early, they were asked to wait quietly while others finished.

The lab only had 10 MacBooks available for use, thus, students worked in groups of two to four depending on the lab section, with all lab sections having different numbers of students. Students were encouraged, sometimes with help from instructors, to take turns working on the computers. Instructors were only allowed to help if students had technical problems but not with lab material itself. The students were informed that the lab itself would not be graded as part of their lab score, which may have had an effect on the way students answered questions or participated during lab. As students completed the labs, they were collected and the students were asked to wait for their fellow classmates to finish.

The last part of the lab consisted of the post-test. The post-test is the same as the pre-test. Each student was given a post-test and upon completion was allowed to leave the lab. No collaboration was allowed during the pre- or post-tests. The pre-test, lab activity, and post-test were all completed in one, two-hour lab period.

Data Scoring

Pre- and Post-Tests

The pre- and post-tests consisted mainly of multiple choice questions and were scored on a partial- or full-credit basis. A participant could earn a maximum of two points on each question for choosing the correct answer, one point for choosing a partially correct answer, and zero points for choosing an incorrect answer. Some questions had more than one possible answer worth one point, as shown in Table 13 below. Answer "iii" is worth two points, either answers "i" or "ii" are worth one point, and answer "iv" is worth zero points.

The four choices for the question asking about the participant's prior knowledge on the subject were coded for categorization purposes as zero, one, two, or three. This scheme is illustrated in Table 14.

Lab Items Scoring

The lab activity consisted of 13 open response or matching questions, which were scored on a partial- or full-credit basis out of a possible one, two, or three, depending on the question. These questions ranged on topics covered in the activity including plate movement, subduction processes, and trench formation (see Tables 15 and 16).

Results

Data were analyzed to address overall learning gains from the Tonga lab, as measured by pre- and post-tests, as well as to test whether there were any learning gain differences due to prior coursework in geology and/or geography. Gender differences were also analyzed although this was not part of the original design of the study. The unit of analysis here was data from each individual student. Each analysis is presented and described in turn.

Were there Differences by Prior Coursework or Gender before the Tonga Learning Activity with Google Earth?

We first addressed if there were any differences on the pre-test by prior coursework and by gender; in other words, whether there was a difference on students' knowledge going into the pre-test either by prior coursework in geology and/or geography, or by gender. A univariate analysis of variance was computed for each of these analyses. The difference between the total scores on the pre-test was not statistically significant when comparing those with prior coursework to those with no prior coursework ($F = 3.052$, $p = 0.051$). Secondly, there was no statistically significant difference on the total pre-test score when comparing the males and females ($F = 1.831$; $p = 0.179$; see footnote 3). See Table 17 for means and standard deviations for each of these analyses.

Were Differences in Overall Pre-Post Comparisons Found for the Tonga Activity?

Next we addressed if there were differences in overall post-test scores compared to pre-test scores collapsing over prior

TABLE 15. QUESTION 4 AND ITS CODING SCHEME
FOR THE TONGA ACTIVITY

Q4 On which side of the trench do we expect to see earthquakes on, the down-going plate (on the east side of the trench in this case) or the over-riding plate (on the west side of the trench in this case)?

over-riding plate or west side (1 pt)

TABLE 16. QUESTION 5 AND ITS CODING SCHEME (1 POINT FOR EACH PROPER MATCH) FOR THE TONGA ACTIVITY

Q5 Match the depth of the earthquakes based on their distance from the trench. Put one distance for each depth.

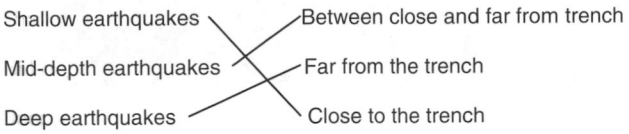

coursework and gender. A paired t-test was computed. The difference in overall pre-test score and overall post-test score was statistically significant, [t(136) = 6.591, p = 0.000; $\overline{X}$pre = 0.82, SD − 0.35; $\overline{X}$ post = 1.09, SD = 0.38]; this result demonstrates that on average, students had higher scores on the post-test than on the pre-test (see Table 18 and Figure 4).

Were Post-Test Differences by Type of Prior Coursework Found for the Tonga Activity?

We addressed whether there were differences between the pre- and post-test scores when comparing those with prior coursework in geology or geography to those with no relevant prior coursework. To do this, we conducted a univariate analysis of variance with the total post-test as the dependent variable and type of prior coursework as the independent variable. Pre-

test was used as a covariate. The difference in post-test score by type of prior coursework was not statistically significant (F = 0.692; p = 0.502). Thus, both students with prior coursework and those without prior coursework learned approximately the same amount of content knowledge from the Tonga lab, as measured by the post-test, when holding the pre-test scores constant. The means and standard deviations can be seen in Table 19.

Were Post-Test Differences by Gender Found for the Tonga Activity?

In order to address whether there were differences when comparing males and females on their post-test scores for the Tonga activity, a univariate analysis of variance was conducted with the total post-test as the dependent variable and gender as the independent variable. Pre-test was used as a covariate. The difference in post-test score by gender was not significant (F = 0.545; p = 0.462). Thus, both males and females learned approximately the same amount of content knowledge, as measured by the post-test. The means and standard deviations can be seen in Table 20.

WereThere any Differences on the Lab Scores for the Tonga Activity When Comparing Groups by Prior Coursework or by Gender?

Next, we addressed the differences on the lab activity scores both by prior coursework and by gender; in other words, whether

Overall Score

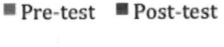

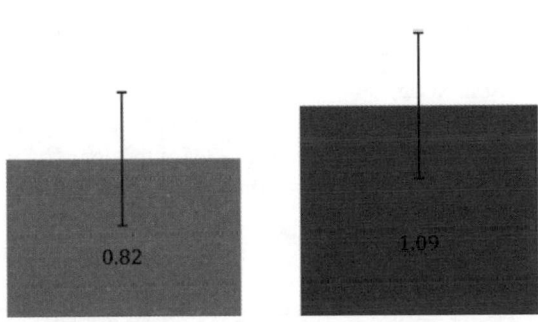

Figure 4. Overall scores on both pre-test and post-test for the Tonga activity.

TABLE 17. AVERAGE SCORES ON PRE-TEST BY TOTAL, GENDER, AND PRIOR COURSEWORK FOR THE TONGA ACTIVITY

	Female	Male	Geology and/or geography	No geology or geography
Mean pre-test score	0.88	0.80	0.86	0.85
Standard deviation	0.36	0.31	0.25	0.36
Mean post-test score	1.13	1.07	1.03	1.12
Standard deviation	0.37	0.39	0.39	0.35

TABLE 18. OVERALL SCORES FOR PRE- AND POST-TEST FOR THE TONGA ACTIVITY

	Pre-test	Post-test	t (df)	p
Overall score	0.82	1.09	6.591 (136)	0.000*
Standard deviation	0.35	0.38		
*Significant at the p < 0.001 level.				

TABLE 19. AVERAGE SCORES ON PRE-TEST AND POST-TEST FOR THE TONGA ACTIVITY BY TYPE OF PRIOR COURSEWORK

	Geology and/or geography	Non-geology and/or geography
Mean pre-test score	0.86	0.85
Standard deviation	0.25	0.36
Mean post-test score	1.03	1.12
Standard deviation	0.39	0.35

there was a difference on students' performance in the lab activity by prior coursework in geology and/or geography, or by gender. A univariate analysis of variance was computed for each of these analyses. The difference between the total scores on the lab activity was not statistically significant when comparing those with prior coursework to those with no prior coursework ($F = 0.738$, $p = 0.480$); however, the difference on the total scores on the lab activity was statistically significant when comparing males and females ($F = 8.463$, $p = 0.004$). This result demonstrates that on average, males outperformed females on the lab activity (see Table 21 and Fig. 5).

Correlations between Pre-Test, Lab, and Post-Test Scores

In order to establish whether there was a relationship between each of the three scores, a Pearson correlation analysis was conducted using the pre-test, post-test, and lab scores. A statistically significant correlation was found between the pre-test and lab scores ($r = 0.384$, $p = 0.000$, [two-tailed]), indicating a moderate, positive relationship between pre-test and lab scores. Pearson correlation values can be seen in Table 22. These results are depicted in Figure 6.

DISCUSSION

Summary of Goals and Approach

In this research and development effort, we report on two studies that examined the efficacy for learning with Google Earth lab activities. This involved examining students' prior knowledge, their knowledge acquired during the lab activity, and their post-test learning gains, thereby examining both the processes (answers to the lab exercises) and products of learning (post-test compared to pre-test); an approach that is important since it has the potential to inform instruction in the geosciences (Libarkin and Anderson, 2005).

Our goal in these studies was to compare learning during the lab activity, as well as the resulting learning gains by comparing pre- and post-test scores for those with prior geology and/or

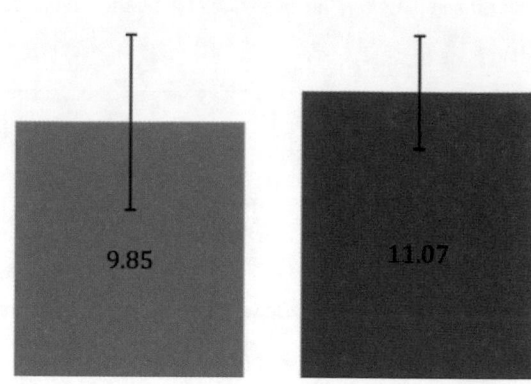

Figure 5. Total lab score by gender for the Tonga activity.

geography coursework to those with no such prior coursework. This research question is important in terms of addressing the efficacy of Google Earth as a learning tool for both majors and non-majors since Google Earth is potentially important to scientific literacy on a broad scale (AAAS, 1993; National Research Council, 1996). Our findings suggest that Google Earth can be an effective learning tool for non-majors, and thus, it also has potential efficacy for scientific literacy on a broad scale.

We also compared learning gains of males and females, although it was not part of the original research design. That is, since there have been a plethora of studies that have reported gender differences in science (Halpern and LaMay, 2000; Linn and Petersen, 1985; Maccoby and Jacklin, 1974; McGee, 1979), and in geoscience in particular (Kali and Orion, 1996; Downs and Liben, 1991; Piburn et al., 2005; Schofield and Kirby, 1994), it was a research question that we could address in the present studies. Similar to the issues around the type of prior knowledge students have coming into our studies, addressing whether there are differential learning gains yielded by males versus females allows us to address the efficacy of Google Earth as a teaching tool for both genders. If gender differences were to be borne out, we as a community of educators would need to begin to think about how to scaffold different learners to accommodate these differences.

TABLE 20. TONGA ACTIVITY RESULTS OF AVERAGE SCORES ON PRE-TEST AND POST-TEST BY GENDER

	Female	Male
Mean pre-test score	0.88	0.80
Standard deviation	0.36	0.31
Mean post-test score	1.13	1.07
Standard deviation	0.37	0.39

TABLE 21. AVERAGE SCORES ON THE TONGA LAB ACTIVITY BY GENDER

	Female	Male	p
Lab score	9.85	11.07	0.004*
Standard deviation	3.40	2.20	
*Significant at the $p < 0.005$ level.			

TABLE 22. PEARSON CORRELATION VALUES BETWEEN PRE-TEST, POST-TEST, AND LAB SCORES FOR THE TONGA ACTIVITY

	Pre-test	Post-test	Lab
Pre-test	1	0.123	0.384*
Post-test	0.123	1	0.153
Lab	0.384*	0.153	1
*Significant at the 0.01 level of alpha.			

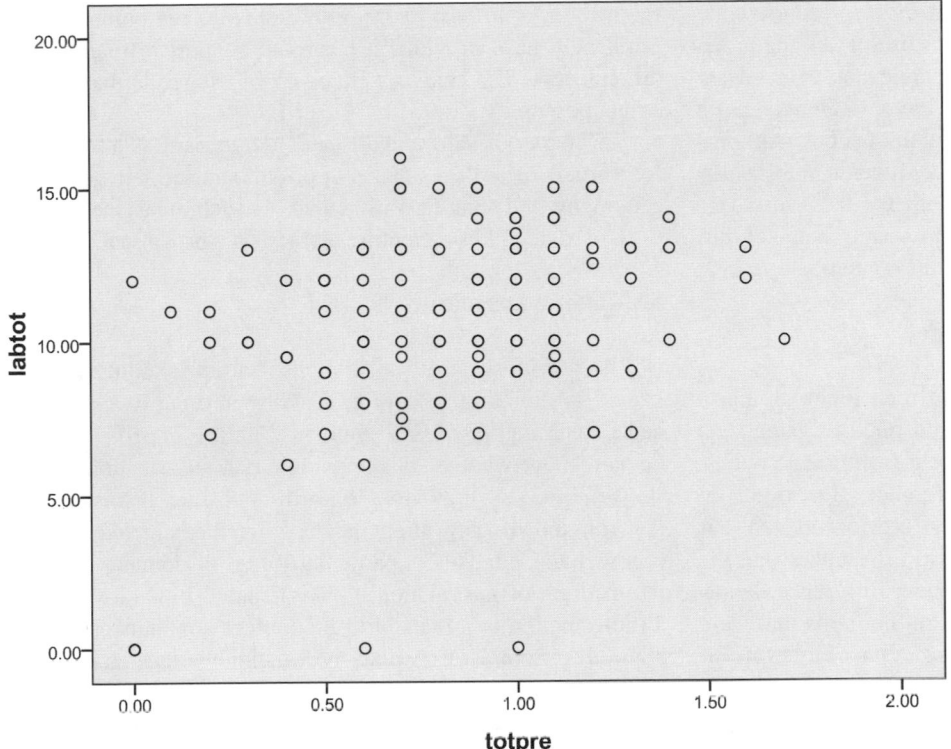

Figure 6. Scatterplot of correlation between total lab scores and total pre-test scores for the Tonga activity.

Overview of Findings Regarding Prior Coursework

In the first study, we used an activity developed in Google Earth by Declan De Paor and his group (De Paor and Whitmeyer, 2008, NSF-CLLI #0837040). The goal of this activity was to deepen students' understanding of the geography and geology of Iceland. The concepts and knowledge targeted here were: specifying Iceland's relationship geographically with respect to the Arctic Circle, using the time slider to observe the horizon, asking students what they would expect to see here at the Winter Solstice, observing geological features (e.g., rock types), observing the formation of the Mid-Atlantic Ridge by using a time slider, noting how the Mid-Atlantic Ridge is displaced across the Gibbs Fracture Zone, and observing the plate-plume interaction under Iceland. In the second study, we used an activity, also developed by Declan De Paor and his group (De Paor and Whitmeyer, 2008), which was more difficult than the first activity in terms of the geoscience content it targeted. Specifically, the activity consisted of: locating the Tonga region with respect to the Tropic of Capricorn; viewing and manipulating virtual block diagrams to observe animations of subduction; island arc formation and trench migration; and answering questions about the relative location of volcanoes, earthquakes, plate movement, trench formation, plate movement, and trench rollback.

Our results for the two studies were highly similar and thus will be summarized together, except for one measure for which significant differences exist.

Our data for both studies showed that there were overall gains in learning when comparing all students' scores—that is, collapsing over type of prior coursework and gender, all students, on average, had higher post-test scores compared to their pre-test scores for both the Iceland activity as well as the Tonga activity. Since there were no group differences on the pre-test by either type of prior coursework or by gender for either the Iceland or the Tonga activities, we can attribute our post-test gains as being due to the Google Earth labs for Iceland and Tonga, respectively.

When analyzing post-test gains by type of prior coursework (geology and/or geography versus no prior coursework in either of these areas), we found that there were no significant differences, on average, for those with prior coursework in geology or geography when compared to those with no prior coursework in these areas for either the Iceland lab or the Tonga lab. This suggests that the two Google Earth labs were effective as learning activities, regardless of the type of prior coursework.

This finding is important because prior studies that have used traditional methods of geoscience instruction often do not yield large learning gains (Libarkin and Anderson, 2005; Hall-Wallace and McAuliffe, 2002). Furthermore, our findings are commensurate with prior research that showed that rich dynamic visualizations such as GIS and Google Earth are successful at remediating students' misconceptions about 3-D geoscience phenomena such as ocean ridges and tsunamis (Hall-Wallace and McAuliffe, 2002); ocean ridges were targeted in our activities in both the Iceland and Tonga activity. Findings from our studies suggest that Google Earth appears to provide a means of deep learning for students that *does not* hinge on prior coursework, thus our data suggest that Google Earth is a useful tool for undergraduate education, regardless of prior relevant coursework. Thus, in terms of promoting scientific literacy, Google Earth may also be very effective with the general public, but additional research would need to be conducted since students in this study were self-selected by virtue of signing up for the geoscience course from which this subject population was drawn, and thus may have been favorably predisposed to this content, etc.

Overview of Findings Regarding Gender

When analyzing our post-test gains for each Google Earth activity by gender, we see that males and females gained, on average, about the same amount of content from the activities. Furthermore, because there were no differences due to gender on the pre-tests scores for either activity (see footnote 3), our data suggests that this effect is not due to differences that the students had before the activity. One gender difference, favoring males, was found for the Tonga lab on the items that were answered as students worked through the lab activity. Specifically, on the Tonga lab, the more difficult of the two labs, males outperformed females in terms of the number of correct items they answered. However, since there were no significant differences by gender yielded on the post-test for the Tonga activity, the differences favoring males on the lab items were not robust enough to be reflected in the males' post-test understanding.

These findings are important since many studies have shown that males tend to outperform females on spatially oriented tasks (cf. Halpern and LaMay, 2000; Linn and Petersen, 1985; Maccoby and Jacklin, 1974; McGee, 1979). In geoscience in particular, few studies regarding gender effects have been conducted (Kali and Orion, 1996; Downs and Liben, 1991; Piburn et al., 2002; Schofield and Kirby, 1994) although researchers have noted a need to address the relationship of spatial skills to specific sciences, rather than as science in the aggregate (Lau and Roeser, 2002). In terms of such studies, Dabbs et al. (1998) found that basic spatial skills contribute to geographic knowledge and that men tended to excel at mental rotation, whereas women tended to excel at object location. Black (2005) found a relationship between specific types of spatial skills, namely mental rotation and earth science misconceptions (Black, 2005). Black hypothesized that mental rotation is required to visualize the position of objects from varying vantage points, and further that this is the type of mental rotation needed for understanding both seasonal change and phases of the moon, two areas in which significant misconceptions have been found. In terms of the present study, we found only one gender difference of the several measures taken, and as previously stated, this difference favoring males was not robust enough to be maintained, as evidenced by the lack of differences due to gender on the post-test. Thus, from our data, it appears that Google Earth does not offer a differential bias for one gender over another. Furthermore, since Google Earth has features that permit students to manipulate the tilt of Earth in order to view it from different vantage points, Google Earth may have provided a means to support learners on this difficult task. The study by Black (2005) suggests that this is a strong possibility.

All told, our data suggest that Google Earth is a useful tool for learners regardless of level of prior coursework in geology or geography, and regardless of gender. As such, it has the potential to be used to address scientific literacy on a broad scale.

Scaffolding Learning

As previously stated, all complex learning should be accompanied by orienting tasks or scaffolding in order to support students' learning processes. Students, unlike experts, typically do not know what is salient within rich information sources (Lowe, 1993) such as Google Earth, and thus, if unscaffolded (i.e., unguided) they might not acquire the targeted information as intended. This is particularly true in domains in which the medium of information is visual-spatial in nature in which all information is presented to the learner simultaneously. This is in direct contrast to textual information sources in which the knowledge acquisition processes are guided by the structure of the text (Larkin and Simon, 1987; Gobert, 2005a). In prior work, Bodzin and Cirucci (2009) noted that resources such as Google Earth, when used in conjunction with appropriately designed instructional materials, show much potential in promoting students' spatial thinking. Our data, which yielded learning gains, also support this.

In the present study, a great deal of effort was taken to ensure that the lab exercises both oriented and scaffolded the students in order to deepen their learning. It is doubtful that learning gains would have been found for both those with and without prior coursework if the learning activities had not been well designed, although the activity with and without its scaffolding and orienting was not tested as part of this study. Thus, for those using Google Earth as a pedagogical tool at any level of education (K–graduate school), it is important that care is taken to guide students' knowledge acquisition processes in order to deepen their learning. Scaffolding is particularly important when novices are learning with visual information sources (Lowe, 1993). The materials developed by De Paor, Steve Whitmeyer, and their colleagues (NSF-CCLI #0837040, De Paor and Whitmeyer, 2008; NSF-GEO #1034643, De Paor, and Whitmeyer, 2010; and NSF-DUE #1022755, De Paor et al., 2010) that were used in the present research provide a good example for how such scaffolding is accomplished.

ACKNOWLEDGMENTS

This work was funded by the National Science Foundation by NSF-CCLI #0837040 awarded to Declan De Paor and Steven Whitmeyer, by NSF-GEO #1034643 awarded to Declan De Paor and Steve Whitmeyer, and by NSF-DUE #1022755 awarded to Declan De Paor, Steve Whitmeyer, and John Bailey.

Any findings or opinions expressed are those of the authors and do not necessarily reflect the views of the funding agency.

REFERENCES CITED

American Association for the Advancement of Science, 1993, Benchmarks for Science Literacy: New York, Oxford University Press, 448 p.

Bednarz, S.W., Acheson, G., and Bednarz, R.S., 2006, Maps and map learning in social studies: Social Education, v. 70, no. 7, p. 398–404.

Black, A.A., 2005, Spatial ability and earth science conceptual understanding: Journal of Geoscience Education, v. 53, no. 4, p. 402–414.

Bodzin, A., and Cirucci, L., 2009, Integrating geospatial technologies to examine urban land use change: A design partnership: Journal of Geography, v. 108, p. 186–197.

Bodzin, A., Anastasio, D., and Kulo, V., 2012, Designing Google Earth activities for learning earth and environmental science, *in* MaKinster, J.G., Trautmann, N.M., and Barnett, M., eds., Teaching Science and Investigating Environmental Issues with Geospatial Technology: Designing Effective Professional Development for Teachers: Dordrecht, Netherlands, Springer (in press).

Britner, S.L., 2008, Motivation in high school science students: A comparison of gender differences in life, physical, and earth science classes: Journal of Research in Science Teaching, v. 45, no. 8, p. 955–970, doi:10.1002/tea.20249.

Burkam, D.T., Lee, V.E., and Smerdon, B.A., 1997, Gender and science learning early in high school: Subject matter and laboratory experiences: American Educational Research Journal, v. 34, p. 297–331.

Cruz, D., and Zellers, S.D., 2006, Effectiveness of Google Earth in the study of geologic landforms: Geological Society of America Abstracts with Programs, v. 38, no. 7, p. 498.

Dabbs, J.M., Jr., Chang, E., Strong, R.A., and Milun, R., 1998, Spatial ability, navigation strategy, and geographic knowledge among men and women: Evolution and Human Behavior, v. 19, no. 2, p. 89–98, doi:10.1016/S1090-5138(97)00107-4.

Dahl, J., Anderson, S.W., and Libarkin, J.C., 2005, Digging into Earth Science: Alternative conceptions held by K-12 teachers: Journal of Science Education, v. 12, p. 65–68.

DeLaughter, J.E., Stein, S., Stein, C.A., and Bain, K.R., 1998, Preconceptions abound among students in an introductory earth science course: Eos (Transactions, American Geophysical Union), v. 79, p. 429, doi:10.1029/98EO00325.

De Paor, D., and Whitmeyer, S., 2008, Collaborative research: Enhancing the geoscience curriculum using GeoBrowsers-based learning objects: Proposal NSF-CCLI #0837040 funded by the National Science Foundation.

De Paor, D., and Whitmeyer, S., 2010, Collaborative research: Virtual 4-D field education in Google Earth: Proposal NSF-GEO #1034643 funded by the National Science Foundation.

De Paor, D., Whitmeyer, S., and Bailey, J., 2010, Collaborative research: Scaffolding undergraduate geoscience inquiry using new loggable Google Earth explorations: Proposal NSF-DUE #1022755 funded by the National Science Foundation.

Downs, R.M., and Liben, L.S., 1991, The development of expertise in geography: A cognitive-developmental approach to geographic education: Annals of the Association of American Geographers: Association of American Geographers, v. 81, p. 304–327, doi:10.1111/j.1467-8306.1991.tb01692.x.

Fermann, E.J., 2006, Google Earth–based lessons and lab activities for earth science classes: Poster presented at the 2006 Geological Society of America annual meeting, in Philadelphia, Pennsylvania.

Gobert, J., 2000, A typology of models for plate tectonics: Inferential power and barriers to understanding: International Journal of Science Education, v. 22, no. 9, p. 937–977, doi:10.1080/095006900416857.

Gobert, J., 2005a, Leveraging technology and cognitive theory on visualization to promote students' science learning and literacy, *in* Gilbert, J., ed., Visualization in Science Education: Dordrecht, The Netherlands, Springer-Verlag Publishers, p. 73–90.

Gobert, J., 2005b, The effects of different learning tasks on conceptual understanding in science: Teasing out representational modality of diagramming versus explaining: Journal of Geoscience Education, v. 53, no. 4, p. 444–455.

Gobert, J.D., and Clement, J.J., 1999, Effects of student-generated diagrams versus student-generated summaries on conceptual understanding of causal and dynamic knowledge in plate tectonics: Journal of

Research in Science Teaching, v. 36, p. 39–53, doi:10.1002/(SICI)1098-2736(199901)36:1<39::AID-TEA4>3.0.CO;2-I.

Gobert, J.D., and Pallant, A., 2004, Fostering students' epistemologies of models via authentic model-based tasks: Journal of Science Education and Technology, v. 13, no. 1, p. 7–22, doi:10.1023/B:JOST.0000019635.70068.6f.

Hall-Wallace, M.K., and McAuliffe, C.M., 2002, Design, implementation, and evaluation of GIS-based learning materials in an introductory geoscience course: Journal of Geoscience Education, v. 50, no. 1, p. 5–14.

Halpern, D.F., and LaMay, M.L., 2000, The smarter sex: A critical review of sex differences in intelligence: Educational Psychology Review, v. 12, p. 229–246, doi:10.1023/A:1009027516424.

Isacks, B., Oliver, J., and Sykes, L.R., 1968, Seismology and the new global tectonics: Journal of Geophysical Research, v. 73, no. 18, p. 5855–5899, doi:10.1029/JB073i018p05855.

Ito, G., and Lin, J., 1995, Oceanic spreading center–hotspot interactions: Constraints from along-isochron bathymetric and gravity anomalies: Geology, v. 23, no. 7, p. 657–660, doi:10.1130/0091-7613(1995)023<0657:OSCHIC>2.3.CO;2.

Jacobi, D., Bergeron, A., and Malvesy, T., 1996, The popularization of plate tectonics: Presenting the concepts of dynamics and time: Public Understanding of Science (Bristol, England), v. 5, p. 75–100, doi:10.1088/0963-6625/5/2/001.

Kahle, J.B., Parker, L.H., Rennie, L.J., and Riley, D., 1993, Gender differences in science education: Building a model: Educational Psychologist, v. 28, p. 379–404, doi:10.1207/s15326985ep2804_6.

Kali, Y., and Orion, N., 1996, Relationship between earth science education and spatial visualization: Journal of Research in Science Teaching, v. 33, p. 369–391, doi:10.1002/(SICI)1098-2736(199604)33:4<369::AID-TEA2>3.0.CO;2-Q.

KamLAND Collaboration, 2011, Partial radiogenic heat model for Earth revealed by geoneutrino measurements: Nature Geoscience, v. 4, p. 647–651, doi:10.1038/ngeo1205.

Larkin, J., and Simon, H., 1987, Why a diagram is (sometimes) worth ten thousand words: Cognitive Science, v. 11, p. 65–100, doi:10.1111/j.1551-6708.1987.tb00863.x.

Lau, S., and Roeser, R.W., 2002, Cognitive abilities and motivational processes in high school students' situational engagement and achievement in science: Educational Assessment, v. 8, p. 139–162, doi:10.1207/S15326977EA0802_04.

Libarkin, J.C., 2001, Development of an assessment of student conception of the nature of science: Journal of Geoscience Education, v. 49, no. 5, p. 435–442.

Libarkin, J.C., and Anderson, S.W., 2005, Assessment of learning in entry-level geoscience courses: Results from the geoscience concept inventory: Journal of Geoscience Education, v. 53, no. 4, p. 394–401.

Libarkin, J.C., Anderson, S.W., Dahl, J.S., Beilfuss, M., Boone, W., and Kurdziel, J.P., 2005, Qualitative analysis of college students' ideas about the Earth: Interviews and open-ended questionnaires: Journal of Geoscience Education, v. 53, p. 17–26.

Linn, M.C., and Petersen, A.C., 1985, Emergence and characterization of sex differences in spatial ability: A meta-analysis: Child Development, v. 56, p. 1479–1498, doi:10.2307/1130467.

Lisle, R.J., 2006, Google Earth: A new geological resource: Geology Today, v. 22, no. 1, p. 29–32, doi:10.1111/j.1365-2451.2006.00546.x.

Lowe, R., 1993, Constructing a mental representation from an abstract technical diagram: Learning and Instruction, v. 3, p. 157–179, doi:10.1016/0959-4752(93)90002-H.

Maccoby, E.E., and Jacklin, C.N., 1974, The Psychology of Sex Differences: Stanford, California, Stanford University Press, 634 p.

Martin, D.J., and Treves, R., 2008, Visualizing geographic data in Google Earth for education and outreach: American Geophysical Union, Fall Meeting 2008, abstract #IN41B-1145.

McConnell, D.A., Steer, D.N., Owens, K.D., Knott, J.R., Van Horn, S., Borowski, W., Dick, J., Foos, A., Malone, M., McGrew, H., Greer, L., and Heaney, P.J., 2006, Using concept tests to assess and improve student conceptual understanding in introductory geoscience courses: Journal of Geoscience Education, v. 54, no. 1, p. 61–68.

McGee, M.G., 1979, Human spatial abilities: Psychometric studies and environmental, genetic, hormonal and neurological influences: Psychological Bulletin, v. 86, p. 889–918, doi:10.1037/0033-2909.86.5.889.

Moores, E.M., and Twiss, R.J., 1995, Tectonics: Dayton, Ohio, W.H. Freeman, 415 p.

Morgan, W.J., 1972, Deep mantle convection plumes and plate tectonics: American Association of Petroleum Geologists Bulletin, v. 56, doi:10.1306/819A3E50-16C5-11D7-8645000102C1865D.

National Research Council (U.S.), 1996, National Science Education Standards: Washington, D.C., National Academy Press, 272 p.

Patterson, T.C., 2007, Google Earth as a (not just) geography educational tool: The Journal of Geography, v. 106, no. 4, p. 145–152, doi:10.1080/00221340701678032.

Piburn, M., Reynolds, S., McAuliffe, C., Leedy, D., Birk, J., and Johnson, J., 2005, The role of visualization in learning from computer-based images: International Journal of Science Education, v. 27, no. 5, p. 513–527.

Rosenbaum, G., and Lister, G.S., 2004, Neogene and Quaternary rollback evolution of the Tyrrhenian Sea, the Apennines and the Sicilian Maghrebides: Tectonics, v. 23, TC1013, doi:1010.1029/2003TC001518.

Sanchez, E., 2009, Innovative teaching/learning with geotechnologies in secondary education: Education and Technology for a Better World, v. 302, p. 65–74, doi:10.1007/978-3-642-03115-1_7.

Schofield, N.J., and Kirby, J.R., 1994, Position location on topographical maps: Effects of task factors, training, and strategies: Cognition and Instruction, v. 12, no. 1, p. 35–60, doi:10.1207/s1532690xci1201_2.

Stahley, T., 2006, Earth from Above: Science Teacher (Normal, Illinois), v. 73, no. 7, p. 44–48.

Stout, D., Bierly, E.W., and Snow, J.T., 1994, Scrutiny of undergraduate geoscience education: Is the viability of the geosciences in jeopardy?: American Geophysical Union Chapman Conference Proceedings, p. 55.

Thompson, K., Keith, J., Swan, R.H., and Hamblin, W.K., 2006, Linking geoscience visualization tools: Google Earth, oblique aerial panoramas, and illustrations and mapping software: Geological Society of America Abstracts with Programs, v. 38, no. 7, p. 325.

Trend, R., 2000, Conceptions of geological time among primary teacher trainees, with reference to their engagement with geosciences, history and science: International Journal of Science Education, v. 22, p. 539–555, doi:10.1080/095006900289778.

Uyeda, S., and Kanamori, H., 1979, Back-arc opening and the mode of subduction: Journal of Geophysical Research, v. 84, B3, p. 1049–1061, doi:10.1029/JB084iB03p01049.

Whitmeyer, S., Feely, M., De Paor, D., and Hennessy, R., Whitmeyer, Shelley, Nicoletti, J., Santangelo, B., Daniels, J., and Rivera, M., 2009, Visualization techniques in field geology education: A case study from western Ireland, *in* Whitmeyer, S.J., Mogk, D.W., and Pyle, E.J., eds., Field Geology Education: Historical Perspectives and Modern Approaches: Geological Society of America Special Paper 461, p. 105–115, doi:10.1130/2009.2461(10).

Whitmeyer, S.J., Nicoletti, J., and De Paor, D.G., 2010, The digital revolution in geologic mapping: GSA Today, v. 20, no. 4, p. 4–10, doi:10.1130/GSATG70A.1.

Wilson, J.T., 1973, Mantle plumes and plate tectonics: Tectonophysics, v. 19, no. 2, p. 149–164, doi:10.1016/0040-1951(73)90037-1.

MANUSCRIPT ACCEPTED BY THE SOCIETY 16 APRIL 2012